Fruit and Vegetables

Fruit and Vegetables

Harvesting, Handling and Storage

A. K. Thompson

Editorial Offices:
Blackwell Publishing Ltd, 9600 Garsington Road, Oxford OX4 2DQ, UK
Tel: +44 (0)1865 776868
Iowa State Press, a Blackwell Publishing Company, 2121 State Avenue, Ames, Iowa 50014-8300, USA
Tel: +1 515 292 0140
Blackwell Publishing Asia, 550 Swanston Street, Carlton, Victoria 3053, Australia
Tel: +61 (0)3 8359 1011

First published 2003 by Blackwell Publishing Ltd

Library of Congress Cataloging-in-Publication Data
Thompson, A. K. (A. Keith)
Fruit and vegetables : harvesting, handling, and storage / A.K. Thompson.
p. cm.
Includes bibliographical references and index.
ISBN 1-4051-0619-0 (Hardback : alk. paper)
1. Fruit--Postharvest technology. 2. Vegetables--Postharvest technology. I. Title.
SB360.T45 2003
634'.046--dc21
2003009287

ISBN 1-4051-0619-0

A catalogue record for this title is available from the British Library

Typeset in Minion and produced by
Gray Publishing, Tunbridge Wells, Kent
Printed and bound in the UK using acid-free paper
by The Bath Press, Bath, Avon

For further information on Blackwell Publishing, visit our website:
www.blackwellpublishing.com

Contents

(Colour plate section falls between pages 206 and 207)

To

Elara, Maya, Ciaran, Caitlin and Cameron
to whom I owe much more than they will ever know

Preface

The technology involved in getting fresh produce from the field to the consumer has been the subject of detailed research for over a century. It is enormously complicated because many of the crops are highly perishable and variable. This variability militates against simple solutions. The fresh produce trade would prefer not to be involved with this variation and complexity: they would prefer to be able to look up their particular crop on a chart, which will say it should be harvested, packaged and stored in a certain way. Information in this form is readily available but will rarely give the best results in terms of preserving the quality of the crop. The objective of this book is to provide a range of options from which the produce technologist can select. Additionally it puts into context our current state of knowledge on postharvest technology and thus identifies areas where research is needed.

The work is based on a selective review of the literature and my experiences since I was first formally involved in postharvest technology in 1967. Since that time postharvest technology has taken me all over the world doing short consultancies and long-term assignments, of up to three years, meeting particular challenges in research, training and development of the fruit and vegetable industry. Although much of my time has been spent as an academic and government or United Nations adviser, I have always worked closely with the horticultural industry. The information in this book and the way that it is presented are therefore largely what is required by the industry. Also, there is increasing pressure for universities to provide graduates who are more relevant to the needs of industry and most students of postharvest technology will eventually work in the industry or in some way be associated with it; so the book will also serve their needs.

For the produce technologists in Europe and North America, the range of fruit and vegetables with which they come into contact is constantly increasing. One of the reasons for this is that retailers are competing for customers and therefore they need constantly to find an edge to attract new customers. Fresh fruit and vegetables are a major factor in showing that one retailer is different; the fresh produce section is usually the first section inside a supermarket. This book therefore covers the whole range of produce from the major sellers to those that are of minor importance in industrial countries and to those that may become important in the future. The parts on the latter group of produce (often referred to by names such as 'exotic' or 'queer gear' by the trade in the UK) will also give some ideas to those in the trade of what crops might be developed for the future.

During the Second World War, Winston Churchill concluded a long and rambling oration with the words, 'I am sorry to have made such a long speech, but I did not have time to write a shorter one'. During her time as British Prime Minister, Margaret Thatcher always insisted that briefing notes from officials should be no longer than half a page. There is an enormous literature on postharvest science and technology of fruits and vegetables. Scientists have written much of this for other scientists not only to contribute to the scientific literature, but also to gain recognition or even promotion. To extract from this literature information that is useful to the industry in a concise form is a prohibitive task. There are high losses and variable quality in the fruits and vegetables offered to the consumer. One solution to this problem is to provide those concerned with the technology of marketing these crops with easily accessible information. This, in part, means information that is brief, easily understood and directly to the point. In this book I have tried to achieve this. I have searched relevant reviews and original research papers in order to extract relevant data and present it in a form that should be easily accessible to all those working in the industry.

The book is an update of one I wrote with Brian Clarke, which was published by Blackwells in 1996, but it is more focused on technology. The final chapter is based on the collected memoirs of Professor C.W. Wardlaw, published in 1938, when he and his colleagues did so much research on the postharvest technology of fruit and vegetables and the work of Dr J.M. Lutz and Dr R.E. Hardenburg published in the United States Department of Agriculture Bulletin 66.

Acknowledgements

To Mr Allen Hilton, Dr Wei Yuqing, Dr Dick Sharples, Professor Don Tindall, Dr Sulafa Musa, Dr Bob Booth, Dr Andy Medlicott, Dr Robin Tillet, Dr James Ssemwanga, Mr David Bishop, Mr Devon Zagory, Mr. Tim Bach, Silsoe Research Institute, FAO Rome, WIBDECO St. Lucia and Positive Ventilation Limited for use of photographs and other illustrative material. To Dr Graham Seymour, Dr John Stow, Mr John Love, Dr Nick Smith, Mr Derek Plilchar, Mr Gary Bradbury and Mr Graham Clampin for technical help and advice.

I wish to express my deep appreciation to Dr Chris Bishop who proof-read the book due to difficulties of communication while I was working in a village in the Central Lowlands of Eritrea.

1
Preharvest factors on postharvest life

Introduction

The quality of a crop at harvest can have a major effect on its postharvest life. There are numerous factors involved and these factors frequently interact, giving complex interrelationships. In tree crops, fruit produced on the same tree and harvested at the same time may behave differently during marketing or when stored. The issues that influence produce quality include obvious things, such as harvest maturity and cultivar or variety, but also the climate and soil in which it was grown, chemicals which have been applied to the crop and its water status. Many of these factors can also interact with time such as when fertilizers or irrigation is applied or the weather conditions near to the time of harvest.

An equation was proposed (David Johnson, personal communication 1994) to predict the probability of low temperature breakdown in apples in storage where variance accounted for 56%. This equation was based on preharvest factors such as temperature, rainfall and nutrient level in the leaves and fruit of the trees as follows:

$$8.2 + 4.5\, T_{max}\,[J] - 2.9\, T_{max}\,[A - S] + 0.11\ \text{rain}\ [A + S] - 16.4\ \text{leaf N} - 3.9\ \text{fruit P}$$

where:

T_{max} [J] = mean daily maximum temperature in June
T_{max} [A – S] = difference in mean daily maximum temperature in August and September
rain [A + S] = total rainfall in August and September
leaf N = level of nitrogen in the leaves
fruit P = level of phosphorus in the fruit.

Nutrients

The soil type and its fertility affect the chemical composition of a crop. Excess or deficiency of certain elements from the crop can affect its quality and its postharvest life. Many storage disorders of apples are associated with an imbalance of chemicals within the fruit at harvest (Table 1.1).

The relation between the mineral composition of fruits and their quality and behaviour during storage is not always predictable (Table 1.2), but in some cases the mineral content of fruits can be used to predict storage quality. For good storage quality of Cox's Orange Pippin apples it was found that they required the following composition (on a dry matter basis) for storage until December at 3.5°C or 4.5% calcium with

Table 1.1 Storage disorders and other storage characteristics of Cox's Orange Pippin apples in relation to their mineral content (source: Rowe 1980)

	Composition in mg per 100 g				
Disorder	N	P	Ca	Mg	K/Ca
Bitter pit			<4.5	>5	>30
Breakdown		<11	<5		>30
Lenticel blotch pit			<3.1		
Loss of firmness	>80	<11	<5		
Loss of texture		<12			

Table 1.2 Summary of the most consistent significant correlations between mineral composition (fruits and leaves) and storage attributes in a three-year survey (1967, 1968 and 1969) of Cox's Orange Pippin commercial orchards (source: Sharples 1980)

	Positive correlation years	Negative correlation years
Fruit firmness	Fruit P (68, 69)	–
Gloeosporium rot susceptibility	Fruit K/Ca:Mg Ca (67, 68, 69)	Fruit Ca (67, 68, 69)
Bitter pit	Fruit K/Ca: Mg Ca (67, 68, 69)	(67, 68, 69)
Senescent breakdown	–	Fruit Ca (67, 68, 69)
Core flush	Leaf K (67, 69) (August)	Leaf N (68, 69) (July)
Low-temperature breakdown	Fruitlet	Ca P (67, 68, 69) (July)

minimum storage in 2% oxygen and <1% carbon dioxide at 4°C until March (Sharples 1980):

- 50–70% nitrogen
- 11% minimum phosphorus
- 130–160% potassium
- 5% magnesium
- 5% calcium.

The physiological disorder that results in the production of colourless fruit in strawberries is called 'albinism'. The fruit, which were suffering from this physiological disorder, were also found to be softer. The potassium:calcium and nitrogen:calcium ratios were found to be greater in fruit suffering from albinism than in red fruit (Lieten and Marcelle 1993). Albinism was associated with the cultivar Elsanta and some American cultivars and the recommendation for control was either to grow only resistant cultivars or decrease the application of nitrogen and potassium fertilizers (Lieten and Marcelle 1993).

The application of fertilizers to crops has been shown to influence their postharvest respiration rate. This has been reported for a variety of fertilizers on several crops including potassium on tomatoes, nitrogen on oranges and organic fertilizers on mangoes. An example of this is that an imbalance of fertilizers can result in the physiological disorder of watermelon called blossom end rot (Cirulli and Ciccarese 1981).

Nitrogen

Generally, crops that contain high levels of nitrogen typically have poorer keeping qualities than those with lower levels. Application of nitrogen fertilizer to pome fruits and stone fruits has been shown to increase their susceptibility to physiological disorders and decrease fruit colour (Shear and Faust 1980). Link (1980) showed that high rates of application of nitrogen fertilizer to apple trees could adversely affect the flavour of the fruit. High nitrogen increased the susceptibility of Braeburn apple fruit to flesh and core browning during storage (Rabus and Streif 2000). In fertilizer trials on avocados, Kohne *et al.* (1992) showed that the application of nitrogen could reduce the percentage of 'clean' fruit, but where it was combined with magnesium and potassium there was no effect. Bunches of Italia grapes from vines treated with 35% nitrogen as urea and 65% as $Ca(NO_3)_2$ through fertigation had less water loss and less decay after 56 days of storage at 2–4°C and 90–95% r.h. than bunches from treatments that had higher levels of nitrogen (Choudhury *et al.* 1999). *Alternaria alternata*, *Cladosporium herbarum*, *Penicillium* spp., *Rhizopus* spp. and *Aspergillus niger* caused storage decay in those trials.

High nitrogen content in bulbs was associated with short keeping quality of shallots in Thailand (Ruaysoongnern and Midmore 1994). Pertot and Perin (1999) showed that excessive nitrogen fertilization significantly increases the incidence of rot in kiwifruit in cold storage, both in the year of application and in the following year. In contrast, Ystaas (1980) showed that the application of nitrogen fertilizer to pear trees did not affect the soluble solids content, firmness, ground colour or keeping quality of the fruit. In a field experiment in the Netherlands there were variable results to field application of nitrogen fertilizer. However, during storage at 12°C and 90% r.h., 10 days after the first harvest, nitrogen had no effect on the yellowing of small Brussels sprouts, but the application of 31 kg N $hectare^{-1}$ as calcium nitrate resulted in increased yellowing of large sprouts. At the second harvest, no effect of nitrogen was observed (Everaarts 2000).

Phosphorus

There is little information in the literature on the effects of phosphate fertilizers on crop storage. Singh *et al.* (1998) found that the application of 100 kg hectare^{-1} of phosphorus minimized the weight loss, sprouting and rotting in onions compared with lesser applications during 160 days of storage.

Phosphorus nutrition can alter the postharvest physiology of cucumber fruits by affecting membrane lipid chemistry, membrane integrity and respiratory metabolism. Cucumbers were grown in a greenhouse under low and high phosphorus fertilizer regimes by Knowles *et al.* (2001). Tissue phosphorus concentration of the low-phosphorus fruits was 45% of that of fruits from high-phosphorus plants. The respiration rate of low-phosphorus fruits was 21% higher than that of high-phosphorus fruits during 16 days of storage at 23°C and the low-phosphorus fruits began the climacteric rise about 40 hours after harvest, reached a maximum at 72 hours and declined to preclimacteric levels by 90 hours. The difference in respiration rate between low- and high-phosphorus fruits was as high as 57% during the climacteric. The respiratory climacteric was different to the low-phosphorus fruits and was not associated with an increase in fruit ethylene concentration or ripening.

Potassium

The application of potassium fertilizer to watermelons was shown to decrease the respiration rate of the fruit after harvest (Cirulli and Ciccarese 1981). In tomato fruits, dry matter and soluble solids content increased as the potassium rate increased, but there were no significant differences in titratable acidity at different potassium rates (Chiesa *et al.* 1998). Spraying Shamouti orange trees with 9% Bonus 13-2-44, a potassium fertilizer from Haifa Chemicals Ltd, increased leaf potassium concentration in the fruit and reduced the incidence of the physiological fruit storage disorder called superficial rind pitting (Tamim *et al.* 2000). Hofman and Smith (1993) found that the application of potassium to citrus trees could affect the shape of their fruits and increase their acidity, although this effect was not observed when potassium was applied to banana plants. High potassium generally increased acidity in strawberries, but this effect varied between cultivars (Lacroix and Carmentran 2001).

Calcium

The physiological disorder of stored apples called 'bitterpit' (see Figure 12.5, in the colour plates) is principally associated with calcium deficiency during the period of fruit growth and may be detectable at harvest or sometimes only after protracted periods of storage (Atkinson *et al.* 1980). The incidence and severity of bitterpit is influenced also by the dynamic balance of minerals in different parts of the fruit as well as the storage temperature and levels of oxygen and carbon dioxide in the store atmosphere (Sharples and Johnson 1987). Also, low calcium levels in fruit increased the susceptibility of Braeburn apples to flesh and core browning (Rabus and Streif 2000).

Dipping certain fruit and vegetables in calcium after harvest has been shown to have beneficial effects (Wills and Tirmazi 1979, 1981, 1982; Yuen *et al.* 1993) (see Chapter 6). There is some evidence in the literature that preharvest sprays can also be beneficial. The treatment of tomatoes with a foliar spray and a postharvest dip in calcium was the most effective at increasing cell wall calcium contents, which is associated with fruit texture. Niitaka pear fruits from trees that had been supplied with liquid calcium fertilizer were firmer after storage than fruit from untreated trees. Fruit weight loss was also reduced following liquid calcium fertilizer treatment, but there was no effect on soluble solids contents (Moon *et al.* 2000). Gypsum, applied to sapodilla trees at up to 4 kg per tree once every week for the 6 weeks prior to harvest, improved the appearance of fruit, pulp colour, taste, firmness, aroma and texture after storage in ambient conditions in India (Lakshmana and Reddy 1995). High calcium fertilizer levels reduced the acidity of strawberries and played a part in loss of visual fruit quality after harvest (Lacroix and Carmentran 2001).

Organic production

The market for organically produced food is increasing. There is conflicting information on the effects of organic production of fruit and vegetables on their postharvest characteristics. Organic production has been shown to result in crops having higher levels of postharvest diseases. Massignan *et al.* (1999) grew Italia grapes both conventionally and organically and after storage at 0°C and 90–95% r.h. for 30 days they found that organic grapes were more prone to storage decay than those grown conventionally. In another case there was evidence that organic production reduced disease level. In samples from organically cultivated Bintje and Ukama potato tubers, the gangrene disease (*Phoma foveata*) levels were lower compared with conventionally cultivated ones. However, there was no such difference in King Edward and Ulama tested 4 months later. The dry rot (*Fusarium solani* var. *coeruleum*) levels were generally lower in organically cultivated potatoes compared with tubers from the conventional system (Povolny 1995).

Producing cops organically can have other effects. Although harvested on the same day, conventionally produced kiwifruits were generally more mature, as indicated by soluble solids concentrations, but their average firmness did not differ significantly from those produced organically. Despite the differences in maturity, whole fruit softening during storage at 0°C did not differ significantly with production system. However, organic fruits nearly always developed less soft patches on the fruit surface than conventionally produced fruits (Benge *et al.* 2000). The effect of organic compost fertilization on the storage of Baba lettuce was evaluated by Santos *et al.* (2001). The organic compost was applied at 0, 22.8, 45.6, 68.4 and 91.2 tonnes per hectare on a dry matter basis, with and without mineral fertilizer. During storage at 4°C, lettuce grown in increasing rates of organic compost had reduced levels of fresh weight loss by up to 7%. The chlorophyll content decreased during storage when plants were grown with 45.6 and 91.2 tonnes per hectare of organic compost with mineral fertilizer. The fertilization with organic compost and mineral fertilizer altogether resulted in plants with early senescence during cold storage.

In a survey in Japan of fruit quality of Philippine bananas from non-chemical production, the problems highlighted all related to management practices and none to the effects of organic production on postharvest aspects (Alvindia *et al.* 2000). However, in Britain Nyanjage *et al.* (2000) found that imported organically grown Robusta bananas ripened faster at 22–25°C than non-organically grown bananas as measured by peel colour change, but ripe fruit had similar total soluble solids levels from both production systems. The peel of non-organic fruits had higher nitrogen and lower phosphorus contents than organic fruits. Differences in mineral content between the pulp of organic and non-organic fruits were much lower than those between the pulp and the peel.

Rootstock

For various reasons, fruit trees are grafted on to rootstocks and the rootstock can have a profound effect on the performance of the crop, including its postharvest life. Considerable work has been done, particularly at Horticultural Research International at East Malling in the UK, on the use of different rootstocks to control tree vigour and cropping. Tomala *et al.* (1999) found that the rootstock had a considerable effect on maturation and storage of Jonagold apples. Fruits from trees on the rootstock B146 had the lowest respiration rates and ethylene production after 2 and 4 months of storage at 0°C but not after 6 months. Fruits from trees on P60 and 62-396 started their climacteric rise in respiration rate 5–7 days earlier than fruits from PB-4. Fruits were yellower at harvest from trees on P60, 62-396 and M.26; fruit colour was weak on PB-4 and fruits from these trees coloured most slowly during storage.

Rootstocks also affect other fruit crops. In some work in South Africa on avocados (Kohne *et al.* 1992), it was shown that the cultivar Fuerte grown on seedling rootstocks showed a large variation in both yield and quality of fruit. There was also some indication that rootstocks, which gave a low yield generally, produced a higher proportion of low-quality fruit. Kohne *et al.* (1992) also showed similar results for the avocado cultivar Hass on different clonal rootstocks (Table 1.3). Rootstock studies conducted in Australia on Hass avocado by Willingham *et al.* (2001) found that the rootstock had a significant impact on postharvest anthracnose disease susceptibility. Differences in anthracnose susceptibility were related to significant differences in concentrations of antifungal dienes in leaves, and mineral nutrients in leaves and fruits, of trees grafted to different rootstocks.

Fruits of Ruby Red grapefruit, which had been budded on *Citrus amblycarpa* or rough lemon (*C. jambhiri*) rootstocks, were stored at 4 or 12°C for 6

Table 1.3 Effect of clonal rootstock on the yield and quality of Hass avocados (source: Kohne *et al.* 1992)

Rootstock	Yield in kg per tree	% fruit internally clean
Thomas	92.7	96
Duke 7	62.1	100
G 755	12.1	100
D 9	7.4	64
Barr Duke	3.1	70

weeks by Reynaldo (1999). Losses due to decay and chilling injury were generally lower in fruits from trees budded on rough lemon than on *C. amblycarpa* rootstock and there was an indication that rootstocks affected the metabolic activity of fruits during subsequent storage at 4°C.

Light

Fruits on the parts of trees that are constantly exposed to the sun may be of different quality and have different postharvest characteristics than those on the shady side of the tree or those shaded by leaves. Citrus and mango fruits produced in full sun generally had a thinner skin, a lower average weight, a lower juice content, a lower level of acidity but a higher total and soluble solids content (Sites and Reitz 1949, 1950a, b).

Woolf *et al.* (2000) showed that during ripening of avocados at 20°C, fruit that had been exposed to the sun showed a delay of 2–5 days in their ethylene peak compared with shade fruits. The side of the fruit that had been exposed to the sun was generally firmer than the none exposed side, and the average firmness was higher than that of shade fruits. After inoculation with *Colletotrichum gloeosporioides* the appearance of lesions on sun fruits occurred 2–3 days after shade fruits.

There is also some evidence that citrus fruits grown in the shade may be less susceptible to chilling injury when subsequent kept in cold storage. Specific disorders such as water core in apples and chilling injury in avocado can also be related to fruit exposure to sunlight (Ferguson *et al.* 1999).

Day length

Day length is related to the number of hours of light in each 24-hour cycle, which varies little near the equator but varies between summer and winter in increasing amounts further from equator. Certain crop species and varieties have evolved or been bred for certain day lengths. If this requirement is not met by using an unsuitable variety then the crop may still be immature at harvest. An example of this is the onion, where cultivars that have been bred to grow in temperate countries, where the day length is long and becomes progressively shorter during the maturation phase, will not mature correctly when grown in the tropics, where day length is shorter and less variable during the maturation period. In such cases the onion bulbs have very poor storage characteristics (Thompson 1985).

Temperature

The temperature in which a crop is grown can affect its quality and postharvest life. An example of this is pineapple grown in Australia. Where the night time temperature fell below 21°C, internal browning of the fruit could be detected postharvest (Smith and Glennie 1987). The recommended storage temperature for Valencia oranges grown in California is 3–9°C with a storage life of up to 8 weeks. The same cultivar grown in Florida can be successfully stored at 0°C for up to 12 weeks. Oranges grown in the tropics tend to have a higher sugar and total solids content than those grown in the sub-tropics. However, tropical grown oranges tend to be less orange in colour and peel less easily. These two factors seem to be related more to the lower diurnal temperature variation that occurs in the tropics rather than to the actual temperature difference between the tropics and subtropics.

The apple cultivar Cox's Orange Pippin grown in the UK can suffer from chilling injury when stored below 3°C whereas those grown in New Zealand can be successfully stored at 0°C. However, this may be a clonal effect since there are considerable differences in many quality factors, including taste and colour, between clones of Cox's Orange Pippin grown in the UK and those grown in New Zealand (John Love, personal communication 1994). In Braeburn apples, the growing conditions were shown to influence scald, browning disorder and internal cavities during storage. Hence following a cool growing season it was recommended that they should be stored in air at 0°C to avoid the risks of those disorders, but they may be stored in controlled atmospheres after warm seasons because this retains texture and acidity better (Lau 1990). Ferguson *et al.* (1999) found that in both apples and avocados, exposure of fruits to high temperatures on the tree could influence the response of those fruits to low and high postharvest temperatures. Specific disorders such as water core in apples and chilling injury in avocado can also be related to fruit exposure to high temperatures, and disorders such as scald in apples may be related to the frequency of low-temperature exposure over the season. Oosthuyse (1998) found that cool, humid or wet conditions on the date of harvest strongly favour the postharvest development of lenticel damage in mangoes. Conversely, dry, hot conditions discouraged the postharvest development of lenticel damage.

Water relations

Generally crops that have higher moisture content have poorer storage characteristics. For example, hybrid onion cultivars that tend to give high yield of bulbs with low dry matter content but only a short storage life (Thompson *et al.* 1972; Thompson 1985). If bananas are allowed to mature fully before harvest and harvesting occurs shortly after rainfall or irrigation, the fruit can easily split during handling operations, allowing microorganism infection and postharvest rotting (Thompson and Burden 1995). If oranges are too turgid at harvest the oil glands in the skin can be ruptured, releasing phenolic compounds and causing oleocellosis (Wardlaw 1938). Some growers harvest crops in late morning or early afternoon. In the case of leaf vegetables such as lettuce they may be too turgid in the early morning and the leaves are soft and more susceptible to bruising (John Love, personal communication). Also, too much rain or irrigation can result in the leaves becoming brittle with the same effect. Irrigating crops can have other effects on their postharvest life. In carrots, heavy irrigation during the first 90 days after drilling resulted in up to 20% growth splitting, while minimal irrigation for the first 120 days followed by heavy irrigation resulted in virtually split-free carrots with a better skin colour and finish and only a small reduction in yield (McGarry 1993). Shibairo *et al.* (1998) grew carrots with different irrigation levels and found that preharvest water stress lowered membrane integrity of carrot roots, which may enhance moisture loss during storage. The effects of water stress, applied for 45 or 30 days before flowering on Haden mangoes, which were stored at 13°C for 21 days after harvest, were studied by Vega-Pina *et al.* (2000). They found that the 45-day fruits exhibited a higher incidence and severity of internal darkening, were firmer, contained a higher content of titratable acidity and had redder skins than 30-day fruits.

In a study of the storage of onions grown in Tajikistan by Pirov (2001) under various irrigation regimes, it was found that if onions are to be used fairly quickly, then maximum yields can be achieved by keeping the soil at 80–90% of field capacity. However, when they were stored for 7 months at 0–1°C and 75–80% r.h. the best irrigation regime was 70% of field capacity throughout the growing season.

Tree age

Not much information could be found on the effects of tree age on the postharvest characteristics of fruit, but fruit from young Braeburn apple trees were more susceptible to flesh and core browning than those from older trees (Rabus and Streif 2000).

Flowering time

In the tropics, the flowering time of fruit trees can affect the postharvest life of fruits. Mayne *et al.* (1993) showed that jelly-seed, a physiological disorder of mangoes, is associated with flowering time in Tommy Atkins. They showed that delaying flowering by removing all the inflorescences from the tree greatly reduced jelly-seed in fruit, which developed from the subsequent flowering. These fruit were larger than those produced from trees where the inflorescences had not been removed but the number of fruit per tree was reduced.

Harvest time

Late-harvested Braeburn apples were more susceptible to flesh and core browning (Rabus and Streif 2000). Harvey *et al.* (1997) found that *Cucurbita maxima* cultivar Delica harvested at 7 kg force, which occurred between 240 and 300 growing degree days (base temperature 8°C) from flowering, required a postharvest ripening period to enhance sweetness and texture and to optimize sensory quality that was not necessary for fruits of later harvests. Ahmed *et al.* (2001) found very strong evidence that for Robusta bananas the fruit had much better organoleptic properties the more mature they were when harvested. Medlicott *et al.* (1987a) showed that early maturing mangoes tended to have better quality and postharvest characteristics than those that matured later. See also Chapter 2 for more details.

Preharvest infection

Crop hygiene can be important in reducing field infections and infestations that may be carried into storage or the marketing chain. This usually involves removal of rotting material from the field, especially fruit windfalls or tree prunings. It can also involve efficient weed control of species that might be alternative hosts for disease-causing organisms.

Frequently, crops are infected with microorganisms or infested with invertebrate pests during production. They may well be on or in the crop at harvest and taken into storage or through the marketing chain. Almost all postharvest pests originate from field infestations, and if the storage conditions are conducive they can multiply on or in the crop. Field infestation of yam tubers with parasitic nematode were shown to increase when the tubers were stored in tropical ambient conditions resulting in areas of necrotic tissues. However, when the tubers were stored at 13°C there was no increase in nematode population in the tubers and no increase in necrosis (Thompson *et al.* 1973a). The potato tuber moth (*Phthorimaea operculella*) may infest tubers during growth if they are exposed above the soil. They may also attack tubers postharvest, and it is therefore important to protect the stored tubers to prevent access to them by the moths. Mealy bugs on pineapples occur in the marketing chain from field infestations (Figure 1.1). Their presence may affect their acceptance on the market or the damage they cause may allow infection by microorganisms that can cause the fruit to rot.

Figure 1.1 Pineapple infested with mealy bug in a field in Sri Lanka.

Aspergillus niger infection in onions occurs during production but will only develop on the bulbs during storage where the conditions are conducive. Infection with bacteria such as *Erwinia carotovorum* can occur in the field on vegetables, especially where they have been damaged and cause postharvest soft rots (Thompson *et al.* 1972).

Chemical treatments

The control of pests and diseases is commonly achieved by spraying chemicals directly on to the crop, although this is becoming less prevalent with increasing use of techniques such as integrated pest management and integrated crop management. The control of field infection can have considerable effect on the postharvest life of the crop. An example of this is anthracnose disease that is caused by field infection by the fungus *Colletotrichium gloeosporioides* [*Glomerella cingulata*], which if not controlled it can cause rapid postharvest losses (Thompson 1987). The fruits look perfectly healthy at harvest and the disease symptoms develop postharvest. The time between infection and the symptoms of the disease developing may be lengthy, e.g. anthracnose (*Colletotrichium musae*) in bananas can take over five months (Simmonds 1941). Generally if a crop has suffered an infection during development its storage or marketable life may be adversely affected. Bananas may ripen prematurely or abnormally after harvest because of leaf infections by fungi during growth, which cause stress and therefore shorten their storage life. This can be manifest on the crop before harvesting or it may only be observed as a 'physiological disorder' postharvest. Fungicide applications in the field to control Sigatoka leaf spot (*Micosphaerella musicola*) were shown to reduce premature ripening (Thompson and Burden 1995).

Chemicals may also be applied to certain crops in the field to prevent them sprouting during storage and thus to extend their storage period. An example of this is the application of maleic hydrazide to onions. Because it is necessary for the chemical to be translocated to the apex of the growing point towards the centre of the bulb, it has to be applied to the leaves of the growing crop.

Growth-regulating chemicals have been applied to trees to increase fruit quality and yield. One such chemical, which has been the subject of considerable debate in the news media, is daminozide (*N*-dimethylaminosuccinamic acid), also called Alar, B9 or B995. When applied to Cox's Orange Pippin apples at 2500 μg $litre^{-1}$ in late June and mid August, they developed more red colour in the skin and were firmer than unsprayed fruits (Sharples 1967). Sprayed fruit were less susceptible to Gloeosporium rots but had more core flush during storage. There was some indication that sprayed fruits were slower to mature since daminozide tended to retard the climacteric rise in respiration. In a comparison between preharvest and postharvest application of daminozide to Cox's Orange Pippin apples, immersion of fruits in a solution containing 4.25 g $litre^{-1}$ for 5 minutes delayed the rise in ethylene production at 15°C by about 2 days, whereas orchard application of 0.85 g $litre^{-1}$ caused delays of about 3 days (Knee and Looney 1990). Both modes of application depressed the maximal rate of ethylene production attained by ripe apples by about 30%. Daminozide-treated fruit were also shown to be less sensitive to ethylene in the storage atmosphere than untreated fruit, but this response varied between cultivars (Knee and Looney 1990). Daminozole has been withdrawn from the market in several countries (John Love, personal communication).

2
Assessment of crop maturity

Introduction

The principles that underlie the stage of maturity at which a fruit or vegetable should be harvested are crucial to both its quality and its subsequent storage and marketable life. Maturity may be defined in terms of either their physiological maturity or horticultural maturity and is based on the measurement of various qualitative and quantitative factors. There are certain guiding principles to be followed when selecting fruit or vegetables to be harvested. Harvest maturity should be at a maturity that:

- will allow them to be at their peak condition when they reach the consumer
- allows them to develop an acceptable flavour or appearance
- allows them to have an adequate shelf-life
- gives a size acceptable to the market.
- is not toxic.

The methods used to assess the maturity of produce may be based on the subjective estimate of people carrying out the operation. To achieve this, sight, touch, smell, morphological changes and resonance may be used. These methods may be made more objective and perhaps more consistent by the use of aids such as colour charts (see Figure 12.107 in the colour plates). Chemical and physical analyses are also used. These depend on sampling procedures and can therefore be used only on crops where a small representative sample can be taken. Computation is also used by calculating such factors as time after flowering as a guide to when to harvest fruit. Many of the methods, which use a qualitative attribute of the crop, may also be used to determine its postharvest quality. Almost all the measurements described here can also have that function.

Field methods

Skin Colour

Skin colour is used for fruit where skin colour changes occur as the fruit ripens or matures, but in some fruits there are no perceptible colour changes during maturation. Colour changes may occur only on particular cultivars and not on others. Also, with some tree fruit the colour of the skin may be partly dependent on the position of the fruit on the tree or the weather conditions during production, which may confound its use as a maturity measurement. Instrumental methods of measuring the colour of fruit have been used for many years, but these tend to have been used in mainly in laboratories and only on harvested fruit (Medlicott *et al.* 1992). Commercial on-line colour sorters have been used for many crops (Figure 2.1).

Shape

The shape of fruit can change during maturation and this can be used as a characteristic to determine harvest maturity. In bananas the individual fruit become more rounded in cross-section and less angular as they develop on the plant. Mangoes also change shape during maturation on the tree: on very immature fruits

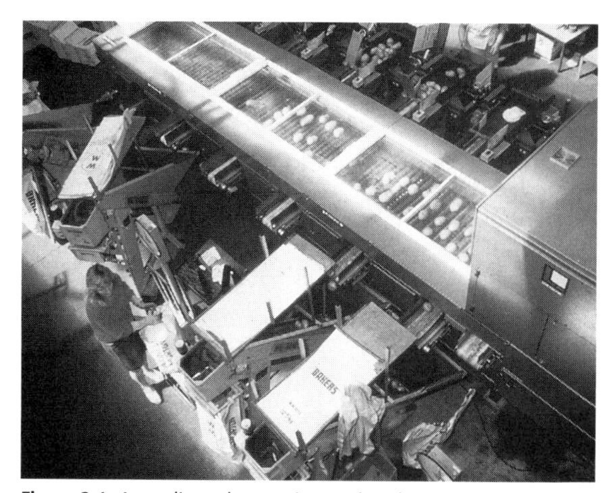

Figure 2.1 An on-line colour sorting machine being used on potatoes.

the shoulders slope away from the fruit stalk, on more mature fruit the shoulders become level with the point on attachment and on even more mature fruit the shoulders may be raised above the point of attachment (see Figure 12.74). Using this method of determining mango fruit maturity, Thompson (1971) showed that the percentage of fruit still unripe after storage at 7°C for 28 days was 68% for fruit with sloping shoulders, 57% for fruit with level shoulders and 41% for fruit with raised shoulders.

Size

The changes in size of a crop as it is growing are frequently used to determine when it should be harvested. In fruits this may simply be related to the market requirement and the fruit may not be physiologically mature, e.g. example in capsicums and aubergine. Partially mature cobs of *Zea mays saccharata* are marketed as sweetcorn while even more immature, and thus smaller, cobs are marketed as babycorn. In some crops fibres develop as they mature and it is important that they are harvested before this occurs. In crops such as green beans, okra and asparagus this relationship may be related to its size. In bananas the width of individual fingers can be used for determining their harvest maturity. Usually a predetermined finger from the bunch is used and its maximum width is measured with callipers, hence it is referred to as the calliper grade. The length of the same finger may also be measured for the same purpose. Both of these measurements are often used as quality criteria during marketing of fruit. Fruit size can also be used for determining the harvest maturity of litchi. In South Africa

Table 2.1 Effects of harvest maturity, as measured by fruit diameter, on weight, price and income from the fruit where 100 is a comparative base (source: Blumenfeld 1993)

	Fruit diameter (mm)		
	60	65	70
Weight	100	120	140
Price	100	115	130
Income	100	138	182

it was shown that the size of the fruit at harvest could have a major effect on its profitability during marketing (Table 2.1). However, the longer the fruits were left to mature on the tree, the higher were the postharvest losses, but even if 70 mm diameter fruit were harvested and had postharvest losses, it may still be economic (Blumenfeld 1993a). In longan fruit, size and weight were consistently shown to have a high correlation with eating quality (Onnap *et al.* 1993).

Several devices have been developed to aid size grading, including hand-held templates (Figure 2.2) and large-size grading machines used in packhouses.

Aroma

Most fruits synthesize volatile chemicals as they ripen. These may give the fruit its characteristic odour and can be used to determine whether a fruit is ripe

Figure 2.2 Templates used for size grading limes in Colombia.

or not. These odours may only be detectable to human senses when a fruit is completely ripe and therefore have limited use in commercial situations. This applies to several types of fruit, but in practice they are used in association with other changes. Equipment fitted with aroma sensors has been developed for postharvest measurement of fruit ripeness.

Computation (Ribbon tagging)

The time between flowering and fruit being ready for harvesting may be fairly constant. For many fruit crops grown in temperate climates, such as apples, the annual optimum harvest date may vary little from year to year, even though the weather conditions may differ considerably. In tropical fruit, flowering may occur at various times of the year, but the time between flowering and maturity may vary very little. With most fruit it is difficult to utilize this consistency in practice. In mangoes, for example, if flowers or young fruit are marked or tagged so as to identify their flowering or fruit-set time, they almost invariably shed that fruit before it is fully developed. In bananas it is different. At anthesis a plastic cover is placed over the bunch to protect the fruit as it is developing. In order to identify exactly when anthesis occurred, a coloured plastic ribbon is attached to the bunch (see Figure 12.18). The same colour is used for one week and changed to another colour the following week and so on. This means at the harvest time the age of is bunch is precisely known. Jayatilake *et al.* (1993) showed that the Ambul variety of banana grown in Sri Lanka reached physiological maturity 8–9 weeks after the flowers had opened. Fruit growth and development continued until the thirteenth week but changes in other physical and chemical parameters were minimal after 11 weeks. In Ecuador the maximum time from anthesis to harvest is usually 12 weeks and in the Windward Islands it is 13 weeks.

In apples the time of petal fall may be recorded. This gives an approximate guide to when fruit should be harvested. Harvest maturity for rambutans may be judged on the time after full flowering. In Thailand this is 90–120 days, in Indonesia 90–100 days and in Malaysia 100–130 days (Kosiyachinda 1968). In New Zealand, optimum harvest maturity of kiwifruit is some 23 weeks after flowering (Pratt and Reid 1974).

Leaf changes

This is a characteristic that is used in both fruit and vegetables to determine when they should be harvested. In many root crops the condition of the leaves can indicate the condition of the crop below ground. If potatoes are to be stored then the optimum harvest time is after the leaves and stems have died down. If they are harvested earlier the skins are less resistant to harvesting and handling damage and are more prone to storage diseases. Bulb onions that are to be stored should be allowed to mature fully before harvest, which is judged to when the leaves bend just above the top of the bulb and fall over. When the leaf dies in whose axis a fruit is borne in melons, then that fruit is judged to be ready for harvesting.

Abscission

As part of the natural development of fruit, an abscission layer is formed in the pedicel. This can be judged by gently pulling the fruit. However, fruit harvested at this maturity will be well advanced and have only a short marketable life.

Firmness

Fruit may change in texture during maturation and especially during ripening, when they may rapidly become softer. Excessive loss of moisture may also affect the texture of crops. These textural changes may be detected by touch, and the harvester may simply be able to squeeze the fruit gently and judge whether to harvest it. A non-destructive firmness test was investigated at Cranfield University, which simulated the practice of customers who may test a fruit's ripeness by feeling it. A narrow metal cylindrical probe was pressed on to the skin of the fruit (approximately 1 newton was sufficient) and the amount of the depression of the skin was measured very accurately on an Instron Universal Tester (Curd 1988; Allsop 1991). This was found to correlate well with maturation and ripening of the fruit and also caused no detectable damage. Similar studies had previously been carried out by Mehlschau *et al.* (1981), who used steel balls, one each on opposite sides of the fruit, to apply a fixed force. They then measured the deformation that was caused to the surface of the fruit. Perry (1977) described a device which applied low pressure air to opposite sides of fruit and then measured the surface deformation.

Firmness, or what is usually called 'solidity,' can be used for assessing harvest maturity in many leafy vegetables. The harvester who slightly presses vegetables such as cabbages and hearting lettuce with the thumb and fingers can do this by hand. Harvest maturity is assessed on the basis of how much the vegetable yields to this pressure. Normally the back of the hand is used for testing the firmness of lettuce in order to avoid damage (John Love, ,personal communication).

Postharvest methods

Firmness

In some cases, a representative sample of fruit may be taken from the orchard and tested in a device which will give a numerical value for texture; when that value reaches a predetermined critical level, then all the fruit in that orchard are harvested. These so-called 'pressure testers' were first developed for apples (Magness and Taylor 1925) and are currently available in various forms (see Figure 12.85). Hand-held pressure testers could give variable results because the basis on which they are used to measure the firmness of the crop is affected by the angle at which the force is applied. An experienced operator may be able to achieve consistent and reliable results, but greater reproducibility can be achieved if the gauge for measuring firmness is held in a stand so that the angle of force applied to the crop is always constant. The speed with which the probe presses against the fruit can also affect the measurement of firmness, so instruments have been developed which can control it (Figure 2.3). The performance of a firmness penetrometer developed by DeLong *et al.* (2000) was evaluated over two growing seasons with post-storage apples against the Effegi, Magness–Taylor and electronic pressure tester. Highly significant instrument–operator interactions indicated that the influence of operators on instrument performance was not consistent, but overall the newly developed penetrometer performed as well as or better than the other instruments tested. In a comparison between a penetrometer (puncture) test and a flat plate compression test, Sirisomboon *et al.* (2000) found that the penetrometer was superior for analysing the texture of Japanese pears.

In crops such as peas, a shear cell is used to measure texture and is called a 'tenderometer' (Knight

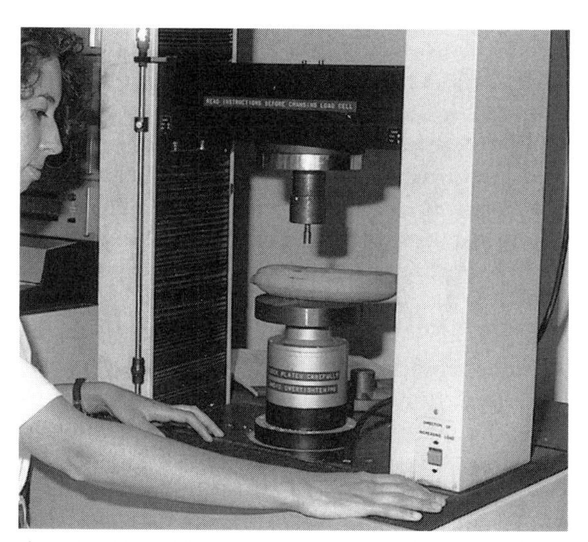

Figure 2.3 Testing the firmness of a banana with a pressure tester. Source: Mr A.J. Hilton.

1991). Pressure testers used for fruits and tenderometers are destructive tests which assume the sample taken is representative of the crop.

Juice

The juice content of many fruit increases as they mature on the tree. By taking representative samples of the fruit, extracting the juice in a standard and specified way and then relating the juice volume to the original mass of the fruit, it is possible to specify its maturity. In some countries legislation exists which specifies the minimum juice content before fruit can be harvested (Table 2.2).

Oil

This is probably only applicable to avocados, where the oil level increases as the fruit matures on the tree. Also, it is only applicable to those grown in the subtropics. This is because it is based on a sampling technique

Table 2.2 The minimum juice content levels for citrus fruits harvested in the USA

Type of citrus fruit	Minimum juice content (%)
Navel oranges	30
Other oranges	35
Grapefruit	35
Lemons	25
Mandarins	33
Clementines	40

where it is assumed that the sample of fruit on which the oil analysis has been taken is representative of the whole field. In the subtropics there are distinct seasons and flowering of avocados occurs after a cold season and the trees tend to flower and thus set fruit over a short period of time. Trees of the same variety in one orchard will have fruit that therefore mature at about the same time and so a representative sample can be taken. In the tropics the flowering period, even on the same tree, is over a much more protracted period and so there is a wide range of fruit maturities. It is rarely possible, therefore, to obtain a representative sample.

Sugars

In climacteric fruit, carbohydrates are accumulated during maturation in the form of starch. As the fruit ripens starch is broken down to sugars. In non-climacteric fruits it is sugars not starch that are accumulated during maturation. In both cases it follows that measurement of sugars in the fruit can provide an indication of the stage of ripeness or maturity of that fruit. In practice the soluble solids, also called °Brix, is measured in the juice of samples of fruit because it is much easier to measure. Usually sugars are the soluble solids that are in the largest quantity in fruit, so measuring the soluble material in samples of the juice can give a reliable measure of its sugar content. This is done either with a suitable Brix hydrometer or in a refractometer (see Figure 12.85). This factor is used in certain parts of the world to specify maturity of, for example, kiwifruit, honeydew melons, peaches and longan.

Starch

In apple and pears, carbohydrates are accumulated during maturation in the form of starch. The measurement of starch content in the developing fruit can provide a reliable method for assessing its harvest maturity, but it does not work for all cultivars. The method involves taking a representative sample of fruit from the orchard as the harvest approaches. These fruit are cut into two and the cut surface dipped in a solution containing 4% potassium iodide and 1% iodine. The cut surface will be stained a blue–black colour in the places where starch is present. It is possible, often with the use of Perspex templates marked with concentric rings, to determine the percentage starch (see Figure 12.2). Starch is converted to sugar as harvest time is approached. In practice in England, samples would be taken from pears from mid August, when the whole fruit surface should contain starch and harvesting should be carried out when samples show about 65–70% of the cut surface which has turned blue–black (Cockburn and Sharples 1979). Studies using this technique on apples gave inconsistent results in England, but it was very effective on several cultivars in Turkey.

Acidity

The acidity of many types of fruit changes during maturation and ripening. In many fruit acidity progressively reduces as the fruit matures on the tree. Taking samples of these fruit, extracting the juice and titrating it against a standard alkaline solution gives a measure that can be related to optimum time of harvest. It is important to measure acidity by titration and not by measuring the pH of the fruit because of the considerable buffering capacity in fruit juices. Normally acidity is not taken as a measurement of fruit maturity by itself. It is usually related to soluble solids, giving what is termed the °Brix:acid ratio.

Specific gravity

The specific gravity of solids or liquids is the relative gravity or weight compared with pure distilled water at 62°F (16.7°C), which is reckoned to be unity. By comparing the weights of equal bulks of other bodies with the weight of water their specific gravity is obtained. In practice, the fruit or vegetable is weighed in air and then in pure water and its weight in air is divided by the loss in weight in water, thus giving its specific gravity. As fruit mature their specific gravity increases. This parameter is rarely used in practice to determine when to harvest a crop, but it could be where it is possible to develop a suitable sampling technique. It is used, however, to grade crops into different maturities postharvest. To do this the fruit or vegetable is placed in a tank of water and those which float will be less mature than those which sink. To give greater flexibility to the test and make it more precise, a salt or sugar solution can be used in place of water in the tank. This changes the density of the liquid, resulting in fruits or vegetables that would have sunk in water floating in the salt or sugar solution. Lizada (1993) showed that a 1% sodium chloride solution was suitable for grading Carabao mangoes in the Philippines.

Diffuse light transmission

A strong light can be shone on a fruit, some of which may actually diffuse through it and can be collected and measured on the other side of the fruit (Figure 2.4). These light transmission properties of fruit can be used to measure their ripeness. The wavelength of peak transmittance of light through fruits was shown to have good correlation with their maturity (Birth and Norris 1958). This was tested on tomatoes and a portable, easy-to-use instrument was developed which was non-destructive and could be used on several types of fruit. Czbaffey (1984) and Czbaffey (1985) measured diffuse light transmission in the range 380–730 nm through cherries and apricots and found that the transmission of diffuse light was shown to be affected by the level of chlorophyll in the fruit that reduces during maturation. In apricots the correlation coefficients were in the range 0.84–0.93. The transmission of near-infrared light has been used as a non-destructive method for measuring the soluble solids content of cantaloupe melons (Dull *et al.* 1989).

Delayed light emission

Colour sorting equipment is commercially available for use in packhouses. Delayed light emission has also been used on fruits such as satsuma oranges (Chuma *et al.* 1977), bananas (Chuma *et al.* 1980), tomatoes (Forbus *et al.* 1985) and papaya (Forbus *et al.* 1987) objectively to grade fruit into different maturity groups postharvest. This is based on the chlorophyll content of the fruit, which is reduced during maturation. The fruit is exposed a bright light inside a dark enclosed compartment and then the light is switched off so the fruit is in the dark. A sensor measures the amount of light emitted from the fruit, which is proportional to its chlorophyll content and thus its maturity.

Figure 2.4 Experimental equipment used for testing changes in diffuse light in fruit. Source: Mr A.J. Hilton.

Body transmittance spectroscopy

Body transmittance spectroscopy was used for optical grading of papaya fruits into ripe and not ripe groups (Birth *et al.* 1984). With this method it was possible to grade fruits that were indistinguishable by visual examination.

Photo electric

A photo electric machine was used for colour sorting citrus fruit (Jahn and Gaffney 1972). They used an ESM Model G and were able to separate oranges and limes into different classes on the basis of their chlorophyll levels. Watada (1989) has expressed transmittance data on scattering samples as

$$\text{optical density} + \log_{10} (E_0/E)$$

where:

E_0 = the incident energy
E = the transmission energy.

Diffuse reflectance

Diffuse reflectance measures the reflected light just below the surface of the crop. It was shown to be effective with persimmon with a diffuse reflectance of 680 nm suitable for automatic grading lines (Chuma *et al.* 1980a). In lettuce there was a shift in diffuse reflectance from 640–660 nm to 700–750 nm during maturation (Brach *et al.* 1982).

Colour difference meters

Colour difference meters (Figure 2.5) can be used to measure the chlorophyll content of fruits and vegetables (Medlicott *et al.* 1992; Meir *et al.* 1992). Chlorophyll content can vary during maturation and senescence of crops and therefore it might be possible to adapt this method in objectively measuring crop maturity. A colour difference meter (Minolta Chroma CR-200) was successfully used to measure the surface colour of peaches and relate this to fruit pigments, soluble solids content and firmness (Kim *et al.* 1993). The range of dominant wavelengths varied between peach cultivars but they were all within the range 565–780 nm (Kim *et al.* 1993).

Figure 2.5 A colour difference meter being used to measure the colour of melons. Source: Mr A.J. Hilton.

Figure 2.6 Equipment which puts vibrational energy into fruit and measures its response being tested on apples. Source: Mr A.J. Hilton.

Heat units

Day degrees or heat units are used to compute harvest dates for vining peas and predict maturity dates for cauliflower and broccoli (John Love, personal communication).

Acoustic and vibration tests

The sound of a fruit as it is tapped sharply with the knuckle of the finger can change during maturation and ripening. Consumers sometimes use this method of testing when purchasing fruit. Fruits such as melons may be tapped in the field to judge whether they are ready to be harvested. This method may also be used postharvest to determine their maturity, for example in pineapples. The principle of the method has been applied in equipment that puts vibration energy into a fruit, and measures the response of the fruit to this input (Figure 2.6). Although much of this work is in the experimental stage, good correlations have been found using the second resonant frequency to determine apple, tomato, mango and avocado maturity (Darko 1984; Punchoo 1988). The following formula was used based on the work of Cooke (1970):

$$f_2^2 m^{2/3}$$

where

f_2 = the second resonant frequency
m = the mass of the fruit.

Commercial audio speakers have been used as vibration exciters with the response of the fruit (Mutsu apples and Nijitsu-seiki pears) measured on a computer-controlled strain gauge. During storage for 20 days it was found that the resonance frequency shifted from 110–80 Hz, suggesting that the method was measuring some quality change (Ikeda 1986). Velocity of propagation of mechanical pulses through whole apple fruits can be used to provide a non-destructive method of monitoring changes in the elastic properties of the tissue (Garrett and Furrey 1974). These elastic properties may be related to the degree of fruit maturity and other characteristics such as fruit firmness and toughness (Garrett and Furrey 1974). Self *et al.* (1993) have shown that there is a relationship between ultrasonic velocity and the maturity and ripeness of bananas, the ripeness of avocados and perhaps the internal browning in pineapples. Mechanical resonance techniques have been applied to apples and tomatoes (de Baerdemaeker 1989). The second resonant frequency in apples was shown to reduce with ripeness, with 600–750 Hz for non-attractive unripe fruits to 200–500 Hz for non-attractive, over-ripe fruit. Attractive, ripe fruit were in the frequency range 400–600 Hz. Similar results were shown for the stiffness factor. The stiffness factor in tomatoes was also shown to decrease during storage. Equipment developed by de Baerdemaeker at the Katholieke Universiteit in Leuven in Belgium uses a small microphone mounted in a piece of plastic pipe containing shaped polystyrene on which the fruit is placed about a centimetre from the microphone. The fruit is tapped lightly with a small object (a pencil will do) on the side of the fruit opposite to the microphone and the first resonant frequency, picked up by the microphone, is measured. The measurement was found to change with the changes in fruit ripeness.

Measurements of acoustic responses of apples were tested against human auditory sensing and found to correlate well in the first and second resonant frequencies (Chen *et al.* 1992). In Japan this technique, based on acoustic impulse responses, has been developed and installed in commercial packing facilities (Kouno *et al.* 1993c). It is being used for watermelons to detect both the ripeness and any hollowness in the fruit. Nine machines have been installed and they are as accurate in grading melons as skilled inspectors using 'the traditional slapping method.' Kouno *et al.* (1993b) described their ripeness and hollowness detector, which was called the MWA-9002.

A sensor system for automatic non-destructive sorting of firm (tree-ripe) avocado fruits used vibrational excitement of one side of the fruit, while measuring the transmitted vibration energy on the other. Special signal processing hardware and software were developed for computing several alternative firmness indexes, which were highly correlated with value obtained by the standard destructive piercing force test method. Optimal classification algorithms were developed whereby fruit could be classified into two or three firmness grades (Peleg *et al.* 1990).

While success has been shown with these methods, it is not always clear exactly what characteristic or group of characteristics of the fruit are being measured. Good correlation was shown between weight loss of fruit and changes in resonant frequency in apples (Ghafir and Thompson 1994). Terwongworkule (1995) showed that stiffness coefficient was related to weight loss in stored apples in that where the fruit were stored at high humidity the stiffness coefficient remained constant and when they were stored at low humidity it reduced in proportion to weight loss. It appears, therefore, that stiffness coefficient in apples is related to their moisture content or some related factor and would not be of use in measuring its maturity or ripeness.

Electrical properties

Studies have been carried out passing electrical currents through fruit. Some correlations have been shown between different characteristics of the fruit, some of which are related to fruit ripening, and the way in which the current passes through the fruit (Nelson 1983; Kagy 1989). Koto (1987) found a difference in electrical properties between fresh fruit and those that were spoiled or physically damaged. He showed that capacitance of deteriorated cells increased while resistance would decrease, and therefore these measurements could be used to determine the freshness or age of the fruit. McLendon and Brown (1971) also found that the dielectric properties (resistivity and conductivity) of peaches changed with fruit ripeness. At 500 Hz the dielectric constant of green peaches was 550 whereas for ripe peaches it was 150. At 5000 Hz the figures were 300 and 100, respectively. These figures appear to be high because the dielectric constant of water under these conditions is 80 and grain 4 (B.C. Stenning, personal communication). In watermelons, specific electrical resistance appeared to decrease with increasing sugar content (Nagai 1975). In work on honeydew melons, Kagy (1989) found no significant relationship between capacitance, loss tangent and resistance and weight loss, firmness and sugar content. The measurement of electrical impedance of Granny Smith apples after impact tests showed a good correlation with bruise levels, with no significant changes in the impedance properties of the fruit after the initial damage had occurred (Cox *et al.* 1993). These electrical properties are exploited widely in the measurement of moisture content of low-moisture products (Dull 1986), but there are no publications to indicate that the method has been sufficiently developed for use in determining maturity of fruits and vegetables.

Electromagnetic

Nuclear magnetic resonance (NMR) spectroscopy has been developed in human pathology to provide real-time images of the inside of the body. It can be used for detecting protons and the variation and binding state on water and oil. Such equipment can be used to provide similar micro-images of the internal structures of fruits (Williams *et al.* 1992). NMR data have also been shown to correlate well with the sugar content of bananas and apples (Cho and Krutz 1989) and the oil content in avocados (McCarthy *et al.* 1989). The technique of using a surface coil NMR probe to obtain the oil/water resonance peak ratio of the signal from a region of an intact avocado fruit produced the best result and has desirable features for high-speed sorting (Chen *et al.* 1993). NMR was also shown to be effective in detecting the physiological disorder water core in apples (Wang *et al.* 1988). Magnetic resonance imaging has been used to obtain images of bruises on apples, peaches, pears and onions, pits in olives and prunes and insect damage in pears (Chen *et al.* 1989a). Shewfelt and Prussia (1993) pointed out that its cost and speed of operation limit the great potential of magnetic resonance imaging. Williamson (1993) acquired three-dimensional data sets from using NMR microscopy with a high-field spectrometer (7.2 T; 300 MHz), and with surface rendering techniques was able to display the spatial arrangement of seeds and vascular tissue of soft fruit.

Near-infrared reflectance

Near-infrared reflectance (NIR), can been used for measuring moisture content using light-emitting diodes. These operate at water-absorbing wavelengths and may find application in rapid on-line determinations of soil moisture, milk and feed composition (Cox 1988). An NIR spectrum analyser has also been used successfully to test the taste of rice and the results correlated well ($r = 0.926$) with those from a taste panel (Hosaka 1987). NIR has been studied in relation to measuring the internal qualities of fruit. Correlations between sugar content of apples, peaches, pineapples and mangoes and NIR measurement have been shown (Kouno *et al.* 1993a). NIR measurement was achieved using a Nireco Model 6500 near-infrared spectrophotometer by placing the fruit so that the light beam on the surface was at right-angles to the fruit surface and covered with a black cloth to avoid the influence of external light. Four places around the equator of the fruit were selected and the NIR beam was irradiated at 2 nm intervals from 400–2500 nm on to the fruit and the average absorbance was measured (Kouno *et al.* 1993b). Fruit firmness, acidity and soluble solids content were measured directly afterwards on the same fruit and adequate correlations were shown between the NIR measurement and soluble solids content for both mangoes (multiple regression coefficient 0.954) and pineapples (multiple regression coefficient 0.825). Results for acidity and firmness were also encouraging for mangoes with multiple regression coefficients of 0.856 for acidity, 0.949 for the firmness of the unpeeled fruit and 0.920 for the peeled fruit. For pineapples the multiple regression coefficients were 0.686 for acidity, 0.460 for firmness of unpeeled fruit and 0.568 for peeled fruit (Kouno *et al.* 1993b). Near-infrared techniques have been used in the measurement of moisture content in crops (Williams and Norris 1987) and have been shown to be able to measure the sugar content of fruit non-destructively (Chen and Sun 1991). NIR spectral data gave good correlations with nitrogen and calcium concentrations of pear fruit peel. It may therefore be possible to obtain information on the actual percentage of fruits with undesirable mineral concentrations, andnon-destructively segregate desirable and undesirable fruits before they are placed into store (Righetti and Curtis 1989). Peirs *et al.* (2000) showed for Jonagold, Golden Delicious, Elstar, Cox's Orange Pippin and Boskoop apples that it is possible to use NIR spectroscopy as a non-destructive technique for measuring internal apple quality.

Radiation

Both X-rays and gamma-rays have been used to assess quality and maturity characteristics of fresh produce. A lettuce harvester was developed which used X-rays to determine which heads were sufficiently mature for harvesting (Lenker and Adrian 1971). Garrett and Talley (1970) used gamma-rays for the same purpose. The basis of these tests depends on the rate of transmission of the rays through the lettuce, since this depends on the density of the head that increases as the lettuce matures. X-rays can also be used to detect internal disorders of crops such as hollow heart in potatoes (Finney and Norris 1973), split pit in peaches (Bowers *et al.* 1988) and granulation in oranges (Johnson 1985). X-rays were used in prototype potato harvesters to differentiate between the tubers and extraneous matter such as clods of earth and stones (Whitney 1993). The equipment was mounted behind the chain lifter, but was found to be impractical because the height of fall of the tubers was too great.

Physiological

For fruit, which pass through a distinct climacteric rise in respiration during ripening, it may be possible to sample the fruit, keep it at a relatively high temperature (for apples in Britain a temperature of about 20°C would be appropriate) and measure its respiration rate. By doing this it may be possible to predict the number of days the fruit would have taken if left on the tree to commence the climacteric rise. Respiration rate is calculated by measuring their uptake of oxygen or the output from the fruit of such gases as carbon dioxide, ethylene or other organic volatile compounds associated with ripening. Recasens *et al.* (1989) measured the concentration of ethylene in the core of apples at different times during maturation and found that for many cultivars the commercial harvest date was 10–15 days after the initiation of the ethylene increase. However, there are problems of applying this method in practice. Testoni and Eccher-Zerbini (1989) found high variability in the ethylene content between fruits and also poor correlation between internal ethylene and other maturity indices such as skin colour, firmness, titratable acidity and soluble solids. In previous studies on the respiration rate of apples it was found that for the cultivars studied, which were Cortland, Delicious, Golden Delicious and McIntosh, there was no definite point on the climacteric curve associated with harvest date (Blanpied 1960).

3
Harvesting and handling methods

Introduction

This chapter deals with the ways in which crops are removed from their parent plant and transported from the field. The method used should take into account:

- the delicacy of the crop
- the importance of speed during and directly after harvesting
- the economy of labour in the operation
- the need for the harvesting method to fulfil the market requirements.

Crop damage

Where harvesting and handling operations are not carried out with sufficient care and attention, damage can be done to the crop that may have repercussions during subsequent marketing and storage operations. These include:

- shortening of the potential maximum crop storage life due to increased respiration or ethylene biosynthesis
- increased levels or microorganism infection through damaged areas
- increases in some physiological disorders.

The types of injury that can be inflicted on the crop include cuts, scuffs and bruises. Compression, impact or vibration can cause bruises.

Ragni (1997) showed that in Italy most defects to apples were caused during harvesting and transport to the packhouse, rather than by maltreatment during grading and packaging. It was found that the grading and weighing/packaging machines did not cause damage to the fruits to a level that would be noticed by the consumer. However, overturning the boxes of fruits could cause bruising of the pulp, noticeable by the consumers. Jarimopas and Therdwongworakul (1994) found that mechanical damage to mangoes due to the washing machine only and the combination of washing and disease control machines were 0.6% and 1.5%, respectively. Only 0.4% of control mangoes were damaged. The washing machine consisted of a rolling conveyor with sprayers and brushes. The disease control machine was a 360 litre steel bin with rotating plates to submerge mangoes in 500 μg litre^{-1} benomyl solution at 55°C for 5 minutes. Peaches suffered the most severe impacts when dropping on to belt conveyors, into the bottom of bins, and being placed manually on rotating surfaces (Ragni *et al.* 2001). Damage caused by three falls of different peach cultivars was compared and Rich May showed 29% damage, Big Top 24%, Ruby Rich 15% and Flaminia only 5%.

Various inexpensive devices could minimize impact damage, such as the use of chutes to avoid fruit falling on to a belt conveyor immediately above a support bar and the use of soft plastic 'cushions' at the bottom of bins. The use of such devices would allow some fruit such as peaches to be picked at a riper stage and so enhance the flavour (Ragni *et al.* 2001).

Cuts

These are self-explanatory, where the crop has come into contact with a sharp object that can penetrate its surface. They can be inflicted where knives are used during harvesting or in opening boxes of packed fruits or vegetables. They have also been observed in badly made nailed wooded field boxes or inappropriately stapled fibreboard cartons. It is also important that harvesters should have short fingernails.

Scuffing

This occurs when fruits or vegetables are caused to move across a hard, usually rough, surface so that the cuticle and layers of cells are scraped away by abrasion. In comparisons with different types of postharvest damage inflicted on citrus fruits, Tariq (1999) found that scuffing caused the greatest reduction in postharvest life.

Compression bruising

Where the downward force on the crop is above a threshold level it can be bruised. This damage may also be a function of time, especially where the pressure is close to the threshold value. It may be as a result of overfilling boxes and then stacking the boxes so that the crop in the lower boxes supports the weight. It is also commonly found in the lower layers of potatoes and onions in bulk stores. In any species of crop there may be differences between cultivars in their susceptibility to compression damage. It may also be related to their moisture content: the higher their moisture content: the more they are susceptible, which may be related to preharvest cultural effects. Compared with bulk storage, box storage reduced the average weight loss of stored potatoes by 16%, the average quality index of subcutaneous tissue discoloration by 30% and the percentage of pressure spots with a depth >1.5 mm by a factor of 15 (Scheer *et al.* 2000). Hironaka *et al.* (2001) applied a static load of 196 newton (20 kg force) to the potato cultivar Norin-ichigo during storage at 7°C for 60 days. They found that static loading increased the glucose, fructose and reducing sugar contents, and also the invertase activity during storage.

To overcome the problem it may be necessary to redesign boxes or ensure that they are not over-filled. In crops that are bulk stored a certain level of compression bruising may be accepted to maximize the capacity of the stores for economic reasons.

Impact bruising

This results either from the crop being dropped or from something hitting it. The damage might be obvious on the surface of the crop or it might be internal. Internal blackening of potatoes is a common form of this latter effect, which is temperature sensitive and occurs more frequently when the tuber temperature is below 8°C or above 20°C (John Love, personal communication). Even when the crop is protected within a fibreboard box, impact damage can occur if the box is dropped or it is over-filled. Work carried out on onion bulbs showed that if they were dropped on to a hard surface they could be damaged even if the fall was as little as 30 cm (Thompson *et al.* 1972). In a study of handling of yams (Thompson 1972a), it was shown that subjecting the tubers to impacts commonly experienced during handling resulted in greatly increased losses during subsequent storage. The contribution to the total amount of subcutaneous tissue discoloration in potatoes was 16% for harvesting through bin-filling, 22% for storage, 27% for shovelling through truck loading and 35% for truck unloading through packaging (Scheer *et al.* 2000).

Mexican exports of Hass avocados to European and Japanese markets are affected by the presence of dark spots on the fruit caused by the fungus *Colletotrichum gloeosporioides* and also as a consequence of bruising during harvest and postharvest (Zamora-Magdaleno *et al.* 2001). The dark spots were associated with oxidation of polyphenols deposited in the cell walls of the dead cells and in intercellular spaces.

Bruised and non-bruised tomato fruit were placed individually inside the electronic nose-sampling vessel and the 12 conducting polymer sensors were lowered into the vessel and exposed to the volatiles given off by the fruit. The electronic nose proved to be a useful tool to identify and classify non-destructively tomato fruit exposed to mechanical injuries (Moretti *et al.* 2000). To avoid impact damage, crops must be handled with great care. This is particularly the case where they are harvested and handled mechanically. Chutes and cushioned pads should be used to break the fall of the crop and distances of fall should be kept to a minimum.

Vibration bruising

This occurs when crops are being transported, especially in lorries. It is common where the crop is packed loosely in the lorry or even in boxes and is largely the result of the crop moving and impacting on each other or the walls of the lorry or box. It can result in an increase in the respiration rate of the crop in addiiton to surface bruising. To minimize the effect, the crop needs to be packed tightly to reduce its movement. Equipment is available in packhouses to 'tight fill' boxes. Internal dividers can be used on accurately size-graded produce to reduce mutual impacts. To reduce vibration injury, specially sprung lorries can be used for transporting soft fruit, or bubble polyethylene film liners at the bottom of crates can be used for grapes (John Love, personal communication). Ragni *et al.* (2001) showed that in trays of peaches vibrations were more intense at the top of a stack than in those at the bottom.

Harvesting

Harvesting operations are frequently carried out in traditional ways by hand, but machines and even chemical sprays are used as aids or methods to speed the harvesting and field handling operations.

Hand harvesting of fruits

For soft fruit such as strawberries and raspberries, which are borne on low-growing plants, harvesting is carried out by breaking the fruit stalk and putting them into a suitable container (see Figure 12.99). This may be the box or punnet into which the fruit is to be taken directly to market or they may be placed in a container which is taken to a packing centre for grading and transfer to a consumer pack. A bag that fits over the hand can help to speed picking and reduce damage to these delicate fruit (Figure 3.1).

Fruits that are borne on trees, such as apples, mangoes, citrus fruits and avocados, are more difficult to harvest. Traditionally the harvester would carry a ladder and use that to reach the fruit (see Figure 12.72). This is very time consuming and various ways have been developed to speed up the operation. A long pole with a bag at the end together with some device for cutting or breaking the fruit stalk is commonly used (Figure 3.2). Picking platforms are available which enable the harvester to be towed around the orchard from tree to tree so the platform can be raised and lowered, enabling the fruit to be picked (Figure 3.3). The way in which the fruit is actually removed from the tree can be important. An example is that in California grapefruit are cut from the tree using clippers, which is assumed to reduce fungal infections, whereas in Florida they are harvested by twisting and pulling the fruit to break the fruit stalk, where it has been shown that there is a higher incidence of rotting when fruit stalks are cut. This effect was shown for papaya, where fruits harvested by cutting the fruit stalk had a lower incidence of rotting during subsequent storage than fruit harvested by twisting and pulling it from the plant (Thompson and Lee 1971).

Figure 3.1 Harvesting aid which fits over the hand and can be used for soft fruit such as raspberries.

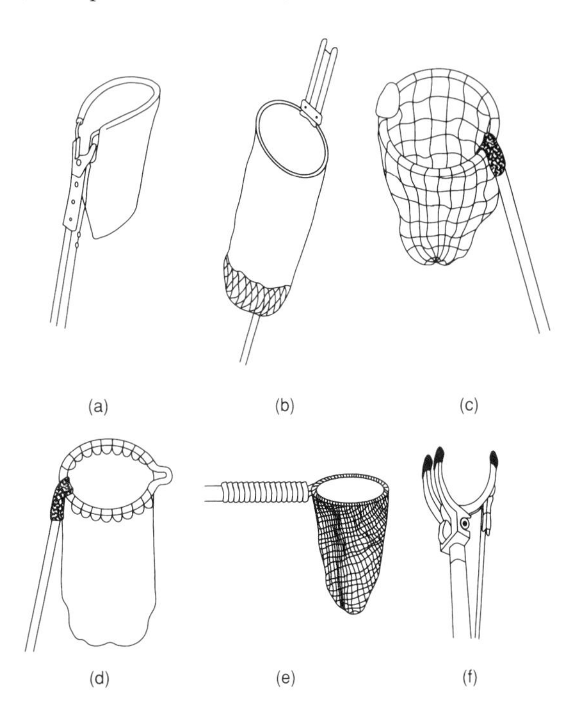

Figure 3.2 Bags and other fitments used on picking poles which are used as a harvesting aid for fruit on tall trees. (a) Pole with clippers attached, from Thailand, (b) For harvesting papaya in Thailand. (c) Hand-woven bag with cutting edge opposite the pole attachment. (d) Canvas bag with cutting notch. (e) Mango harvester from the Philippines. (f) Apple harvester from the UK. Sources: Bautista (1990), Kitmoja and Kader (1993).

Figure 3.3 A platform used to facilitate fruit harvesting from tall trees in the UK.

Hand harvesting of vegetables

Low-growing vegetables are harvested in the same way as described for low-growing fruit. With root crops these must often be dug from the soil, usually by inserting a garden fork, or other similar tool, below the crop and prising it upwards. Other root crops such as beetroot and radish can be pulled from the soil. Great care must be exercised to ensure that the tool is well below the crop so that it is not damaged.

Mechanical harvesting of fruit

Very little fruit destined for the fresh market is harvested by machines because the damage likely to be caused to them could result in rapid deterioration in quality during the marketing chain. Zamora-Magdaleno *et al.* (2001) found that damage to avocados was more likely when crops are harvested mechanically. bit destined for processing may be harvested mechanically, but it is usually important to process it very soon after harvesting otherwise it can deteriorate. Oranges for juice extraction may be removed from the trees by powerful wind machines being dragged through the orchard followed by a device for collecting the fruit from the ground. The grass beneath the trees is usually mown before harvesting and these devices usually have nylon brushes that sweep up the fruit and transfer it to containers. However, they will also collect other things on the surface of the grass, which can be a problem if sheep have been grazed in the orchard. Tree shakers can also be used, which are attached to the tree trunk and violently shake the tree to dislodge the fruit. Canvas sheets mounted on a frame may be placed beneath the tree to break the fall of the fruit and thus reduce damage levels. The canvas slopes away from the tree so that the fruit rolls gently away, preventing bruising caused by other fruit falling from the tree (see Figure 12.84, in the colour plates). These 'shake and catch' methods of harvesting can result in considerable damage to fruits. Miller *et al.* (1973) reported a damage level of 51% for Golden Delicious and Berlage and Langmo (1979) reported a damage level of 64%. These fruit were grown on standard trees and the damage level was assessed as fruits not conforming to the 'Extra Fancy' grade.

In order to reduce the difficulty of dislodging the fruit from the tree, chemical sprays have been recommended. These encourage the formation of the natural abscission layer on the fruit stalk and should be applied a few days prior to harvesting. Ethrel, abscisic acid and cycloheximide have all been shown to be effective but are not permitted for use in all countries. Hitchcock (1973) showed the effectiveness of Ethrel treatment on gooseberries and the importance of concentration (see Figure 12.60). In practical use the time between application of Ethrel and fruit fall was affected by fruit maturity and weather conditions. Ethrel can reduce the shelf life of fruits and is not allowed by some supermarkets, especially on tomatoes (John Love, personal communication 1994). Grapes and soft fruits for processing, such as blackcurrants, may be harvested by tractor-mounted machines which have combing fingers which are run up the stems pulling, off the fruit bunches, and also a high proportion of the leaves (see Figure 12.40). The fruit are removed from the plants by using small-amplitude, high-frequency vibration by the rubber fingers being mounted on a vertical drum that rotates at the same velocity as the harvester moves forward. The fruit may subsequently be separated from the leaves by blowers and the juice extracted. In other cases the fruit is taken straight from the field and passed through a press and filter system. An important factor to be considered with mechanical harvesting is to breed cultivars whose fruit all mature at the same time. Work on the development of mechanical harvesters for strawberries was carried out at Silsoe College (Hitchcock 1973; Williams 1975) where the berries were detached by tensile force with a shaker unit and collected on a sledge. With these mechanical harvesters for strawberries and raspberries (see Figure 12.100) the harvested fruit needs to be graded before it is processed or marketed. Fruit trees or bushes can be specially trained to facilitate mechanical harvesting. Problems with raspberry harvesters is that fruit matures over a period of 20–40 days and requires 5–10 pickings or more if each fruit is to be picked in prime condition (Ramsey 1985). To overcome this problem, a machine was developed in Scotland that shook the canes at such a frequency that only the ripe fruit fell off. This was coupled with an effective catching device. Apple trees grown on a hedgerow system could be harvested with combing fingers giving 85% Grade 1 fruit, although a significant proportion of the fruit remained on the trees (LeFlufy 1983).

In the meadow orchard system of growing apples, peaches, babaco and guavas, the whole of the tree is harvested just above the ground each year or every other year. The fruit are then separated from the stems in machines in the packhouse (Mohammed *et al.* 1984). Using this method, trees are planted very close together with densities in the order of 10 000 trees hectare^{-1} compared with conventional densities of 87–125 trees hectare^{-1} (Alper *et al.* 1980). The possibility of planting babaco at 1 × 1.25 m giving 8000 plants ha^{-1} and a potential yield of 320 tonnes ha^{-1} was discussed by van Oosten (1986).

Other planting and training methods for tree to facilitate mechanical harvesting have been described, including the Lincoln Canopy that was developed in New Zealand. Trees are spaced 3 × 4.5 m and the trees are trained on wires supported on T frames (1.5 m high × 2.4 m wide × 15 m apart). The fruits are detached from these trees by a lightweight impactor located below the canopy that strikes upwards while advancing between each stroke. The detached fruit are caught on a polyethylene-padded conveyor (Diener and Fridley 1983).

A prototype hand-held mango harvester was described by Salih and Ruhni (1993). It consisted of an adaptation of the commonly used pole with a bag attached. The new design had a cutter bar attached at the end of the pole above the bag that was operated from a small electric motor powered by a 12-volt battery.

Mechanical harvesting of vegetables

Mechanical aids to harvesting vegetables such as lettuce, cauliflowers and cabbages involve cutting the vegetables by hand and placing them on a conveyer on a mobile packing station that is slowly conveyed across the field (Figure 3.4). The vegetables are packed on site and transferred to a vehicle for direct transport to a precooler or directly to market (Figure 3.5). Tissue damage during harvesting can stimulate enzymatic action resulting in the production of off-flavours, so they should be processed as quickly as possible after harvest. Mechanical harvesting of green beans may lead to discoloration of the broken ends of the beans,

Figure 3.4 Tractor-mounted conveyer belt to facilitate hand harvesting with a trailer used for field packing.

Figure 3.5 Details of the trailer used for field packing cauliflowers. See also Figure 12.48.

especially if there is a delay in processing them (Sistrunk *et al.* 1989). Brussels sprout harvesters have been developed which perform similar operations to pea viners, harvesting the whole plant and cutting the sprouts from the stems. In both of these cases special processing cultivars have been bred where all the pea pods or brussels sprouts mature at exactly the same time. Brussels sprouts may be stopped to ensure more uniform development (John Love, personal communication 1994). In root crops mechanized harvesters remove the crop from the ground either by undercutting it or, in the case of some potato harvesters, inverting soil in the ridges in which the crop is grown (see Figure 12.81). Moldboard ploughs, subsoilers and other tillage tools used in root crop harvesting may require a lot of power to remove them from the soil. Using vibrating digger blades can reduce the power required, which varies with the amplitude of the vibration (Kang *et al.* 1993). Digger blade performance was better in separating tubers from the soil with high-amplitude vibration and tractor traction was insufficient at low amplitudes.

Field transport

After the crop has been removed from the plant it should be taken to market, a packhouse or a store. Various systems and packing materials have been developed for this purpose. The selection of type and design is related to protection of the produce, convenience of handling and cost effectiveness. Sometimes the crop is harvested into one type of container and is then transferred to another type for transport from the field. An example is the picking boxes used for apples that are worn around the harvester's neck. Fruit may be transferred from these to a pallet box to be taken and stacked in the storage chamber. Types of package and the material from which they are made vary considerably.

Bananas are particularly delicate and easily damaged fruit. Exposure of harvested bananas to carelessness during loading, unloading and overloading, insufficient and poor cushioning and poor truck suspension systems and road conditions seems to be the factors causing physical damage and stress in the in Philippines (Alvindia *et al.* 2000). Transporting bananas to the packhouse has been the subject of considerable research. Overhead cableways, specially padded boxes and padded trailers have all been used. These are reviewed in detail in Chapter 12. Thompson and Lee (1971) showed that packing papaya fruit in specially padded boxes for field transport reduced postharvest losses compared with traditional methods.

4
Precooling

Introduction

To achieve maximum storage life for many crops or to reduce losses during their marketable life, it is essential to keep them at the most appropriate temperature which is usually just above that which will cause chilling or freezing injury. Chilling injury occurs when crops develop temperature-associated physiological disorders or abnormalities when exposed to temperatures above those which would cause them to freeze. They may not produce symptoms until the crop is exposed to higher temperatures. Chilling injury may be apparent as (Thompson 1971):

- failure to ripen in climacteric fruit
- different forms of external abnormalities
- internal discoloration or predisposition to microorganism infection.

Crop susceptibility to chilling injury is influenced by factors such as exposure time, crop cultivar and the conditions in which the crop was grown. Intermittent warming and temperature preconditioning have been shown to reduce the effects of chilling injury in certain crops (Wang 1982). The exact mechanism by which chilling injury affects the crop has still not been fully determined. It has been shown to be concerned with loss of membrane integrity and ion leakage from cells and changes in enzyme activity (Wang 1982), but exactly why some crops are susceptible and some resistant is still a major research topic. Freezing injury, on the other hand, is caused when ice crystals form within the cells of fruits and vegetables. Since these cells contain soluble material, the freezing temperature is below the freezing point of pure water at normal pressure, that is, lower than 0°C. Freezing injury can occur at temperatures above those at which the cells will freeze because, according to Burton (1982), initial ice formation will be in the more dilute liquid on the surface of the cell walls.

To maximize the effect of precooling, the crop should be brought to the optimum temperature as quickly as possible after harvest. Precooling is not always necessary since some crops are not very perishable and rapid cooling would make no difference to their marketable condition or storage life. In many cases, such as curing root crops, drying onions and quailing oranges, there are definite storage advantages to exposing the crop to high temperature after harvest and not cooling them quickly. Also for some crops precooling would be too expensive and therefore not cost effective. In Brazil, Menezes *et al.* (1998) exposed two melon hybrid cultivars, Gold Mine and Agroflora 646, to solar radiation for 0, 1, 2, 3 or 4 hours between 10.00 and 15.00 hours directly after harvest and noted the effect on subsequent storage. They found that leaving harvested fruits in the sun for those periods had no effects on quality or storage life. Precooling always needs to be combined with a cold chain, which means that they are transferred to storage at the optimum temperature directly after precooling.

Lallu and Webb (1997) found that for kiwifruit it was more economical to construct a cool store with cooling capability than to combine a precooler with a holding store. The establishment cost of a cool store with cooling capability was approximately 40% less than that of the combination of a precooler and holding store.

Heat removal

The rate of precooling of crops depends on:

- the difference in temperature between the crop and cooling medium
- accessibility of the cooling medium to the crop
- the nature of the cooling medium
- the velocity of the cooling medium
- the rate of transfer of heat from the crop to the cooling medium.

Heat can be removed from the crop in one of three ways:

- conduction
- convection
- radiation.

In precooling heat removal is usually achieved by conduction, that is, the direct movement of heat from one object or substance to another by direct contact. If two objects or substances are touching and they are at different temperatures, heat will always move from the warmer to the cooler. If a crop has a relative by large mass and a small surface area, the crop will be cooled more slowly than if it had a smaller mass and or a larger surface area, because in the former the heat from the inside of the crop has to move to the surface of the crop before it is transferred.

Exposing crops to the sun will heat them above the air temperature by radiation. This temperature increase can be over 10°C (Rickard *et al.* 1978). If a crop is harvested in the early morning, it will be cooler since the sun has not warmed the air or the crop. The crop will also have a lower level of metabolic heat since, generally, the warmer the crop the higher is its metabolic rate. Another factor that affects cooling is the packaging material used for the crop. If the cooling medium is air, the boxes of produce should have ventilation holes that ensure good circulation through the crop. These holes may be placed to coincide with an air stream as in forced-air cooling. For hydrocooling and top icing, the box used must withstand direct contact with water without disintegrating. Plastic boxes are suitable for this purpose, or fibreboard cartons that have been waxed or treated with some other substance that will render them waterproof. Where heat is removed by evaporative cooling, as it is in vacuum precooling, then the crop must not be sealed in moisture-proof film such as polyethylene bags. If they are packed in bags then the bags must have adequate perforations to allow the water vapour to escape.

Several methods of precooling are used commercially and the selection of the most suitable method depends on the crop, the marketable or storage life required and the economics, that is, the added value to the crop compared with the cost of precooling. A comparison of different methods is given in Table 12.3. Different methods include placing ice over the produce, passing cold air around the produce, immersing them in water or reducing the air pressure around them. The different methods give widely different cooling times.

Precooling methods

Top icing

This has been used for many decades and is also called contact icing. It is commonly applied to boxes of produce by placing a layer of crushed ice directly on top of the crop. The ice melts and the cold water runs down through the crop and cools it. It can also be applied as an ice slurry that is hosed on to the top of the crop from a tank. A typical slurry is made from 60% finely crushed ice, 40% water and usually with 0.1% sodium chloride to lower the melting point of the ice, although the water:ice ratio may vary from 1:1–1:4. Ice slurries give greater contact between product and ice than does top icing, and therefore should result in quicker cooling. The main use of top icing is for road transport and can be applied shortly after harvest to, for example, field-packed lettuce or broccoli, to begin precooling as the crop is harvested.

Room cooling

This method simply involves placing the crop in a cold store. This may be the same cold store where the crop is to be stored for long periods or it may be to hold the crop before it enters the marketing chain to facilitate the accumulation of sufficient produce to send to market. The type of room used for this may vary, but generally it consists of a refrigeration unit across which cold air is passed from a fan. The air circulation may be such that the air is blown across the top of the room and falls through the crop by convection. It may also be blown in at the bottom of the store, in which case some type of plenum chamber is used. The drawback to these methods is the length of time it can take for the temperature of the crop to be reduced. The main advantage is its cost because no special facilities or equipment are required. Simple non-refrigerated rooms have been successfully used for precooling. These have the advantage that they can be erected and used in the field where no electricity is available. Examples are evaporative coolers and tractor-mounted ice bank coolers.

Forced-air cooling

The principle of this type of precooling is to place the crop in the cold room but to arrange the airflow pattern so that it is forced directly through the crop. By doing this the heat given out from the surface of the crop is carried away in the stream of cold air, thus setting up a temperature gradient, hence cooling the crop more quickly.

To achieve this rapid cooling, the cooled air may be forced into a plenum chamber and the boxes of crops placed at exits to the plenum chamber. A common type of precooler that uses this method is the letterbox system. The plenum chamber takes up one wall of the cold store, with the wall of the chamber fitted with slots that are closed with removable covers. These slots coincide with pallet bases so that the cooled air is directed into the base. Horizontal exits from the base are blocked off and the base of the pallet is slatted so that the air is forced up through it, up through the produce and recycled back through the cooling unit (Figure 4.1).

A variation of this method is to have the plenum chamber set into the floor of the cold room so that when the produce is stacked in the cold room air is directed up through it. It is essential to ensure that as much air as possible is blown directly through the produce so all possible escape channels must be blocked off with dunnage or inflatable bags. It is also important to achieve an even distribution of air through the crop to ensure uniform cooling. The speed at which the air is blown through the crop also affects the cooling time. In a comparison between the effects of two airflow rates on the half cooling time of single apple fruits, Hall (1972) found it to be 1.25 hours in an air velocity of 40 m min^{-1} and 0.5 hours at 400 m min^{-1} (see Table 12.3).

An efficient forced-air cooling system can pass air through the crop of a very high velocity, which can lead to desiccation of the crop. To reduce this effect, various methods of humidifying the cooling air have been devised. One of these was the 'ice bank cooler' (Lindsey and Neale 1977; Neale *et al.* 1981). In conventional cold stores, air is blown over cooling coils, which are metal pipes of various designs through which a cooled liquid is passed. Where rapid cooling is necessary the surface area of the pipes must be large and the temperature of the cooling liquid must be considerably lower than the air temperature to achieve good heat exchange. This can result in moisture condensing or freezing on the surface of the coils, which reduces their efficiency, but perhaps even more important, it can increase the speed with which the crop is being desiccated. The ice bank cooler has the cooling coils immersed in water so that the water freezes building up ice around them. Water is then pumped over the ice and sprayed in a fine mist in a counter current of air entering the plenum chamber. The air passes through a filter to take out liquid water particles from it and is then passed through the crop (Figure 4.2). The effect is that the air is both cooled and humidified. Humidity is close to 100% r.h. in a well constructed system and will result in minimum desiccation of the crop during cooling. For precooling in Vietnam, forced-air cooling using an ice bank was recommended (Herregods *et al.* 1995). Elansari *et al.* (2000) described a wet deck precooling system for grapes used prior to refrigeration based on an ice bank.

Portable ice bank coolers have been constructed. These are small (about 1 tonne capacity) insulated trailers that can be towed into the field and driven from the power take off of a tractor. The ice bank is built up overnight using the mains electricity and the water

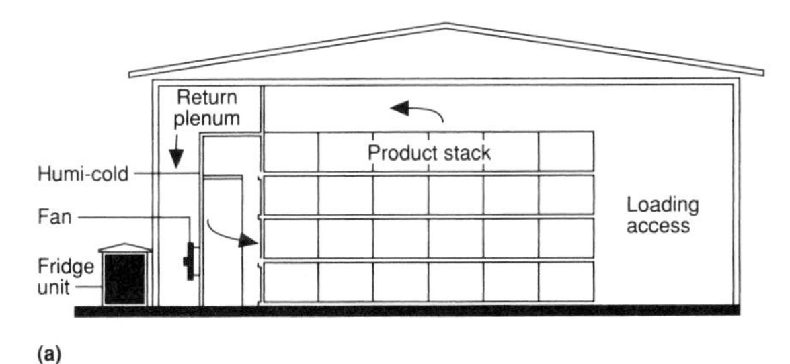

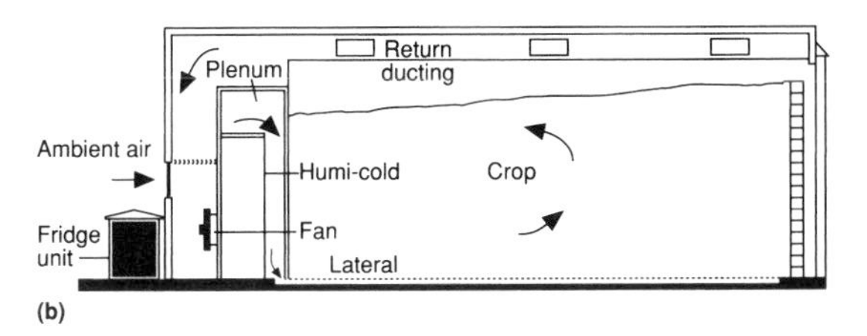

Figure 4.1 A forced-air high-humidity cooler. (a) using the letterbox ventilation system for produce stored in boxes and (b) for bulk stored crops. Source: Positive Ventilation Limited, Yaxley, Peterborough, UK.

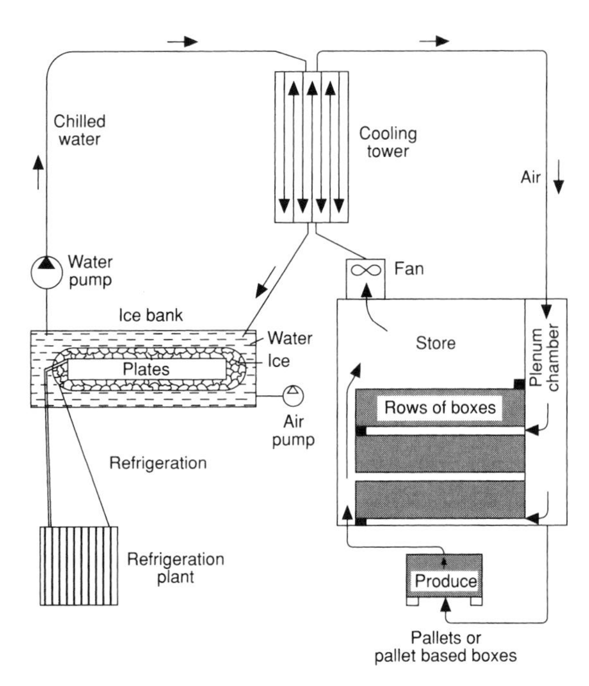

Figure 4.2 Schematic representation of an ice bank cooler. Source: Neale *et al.* (1981)

pump and air circulation fan are driven from the engine of the tractor in the field. It is useful for crops such as strawberries where it may be important to begin cooling directly after harvest.

Other systems have been developed to provide high-velocity, high-humidity air for rapid precooling. These are marketed under trade names such as 'Humifresh,' 'Humi-cold' and 'Airspray.'

Hydrocooling

The transmission of heat from a solid to a liquid is faster than the transmission of heat from a solid to a gas. Therefore, cooling crops by contact with cooled water is much quicker than transfer of heat from the crop to air. Hydrocooling can therefore be very quick and result in no loss of weight of the crop during the precooling operation. If the crop is simply immersed in iced water, the water in direct contact with the crop will heat up and the rate of cooling will be slow. To achieve maximum effect the cooled water must constantly be passed over the crop. Submersing the crop in cold water that is constantly being circulated through a heat exchanger can do this. Where crops are being transported around a packhouse in water, the transport water can also act as a hydrocooler.

A common design for a hydrocooler is to transport the crop on a perforated conveyer belt over a tank of water. Water is pumped from the tank over cooling coils, or blocks of ice, and allowed to fall on the crop and then through to the tank below. This system has the added advantage that the speed of the conveyer can be adjusted to the time required to cool the crop. Hydrocooling has the advantage over other precooling methods that it can help clean the produce. However, the water can become contaminated with microorganisms that can result in increased levels of spoilage during subsequent storage or marketing. Chlorine may be added to the water if this is a problem.

Hydrocooling is unsuitable for many crops. Capsicums that were hydrocooled had a higher incidence of rot during subsequent storage than those that were not, even when chlorine was added to the water (Hughes *et al.* 1981). This was due to water being trapped between the fruit and the calyx. Cooling time depends on the size of the produce. Asparagus spears, being long and narrow, can be hydrocooled in about 2 minutes, but capsicums, which are large and globular, took about 10 minutes to cool (Figure 4.3). Tomatoes, melons and leafy vegetables are among the crops that can be successfully hydrocooled. With Kesar mangoes Kapse and Katrodia (1997) found lower weight loss and firmer fruits during storage after being hydrocooled to 12°C compared with fruits hydrocooled to 16°C or non-hydrocooled fruits. Stem end rot caused by *Botryodiplodia theobromae* and anthracnose caused by *Colletotrichum gloeosporioides* did not occur for up to 13 days of storage when fruits had been precooled. The longest mean shelf-life of 14.3 days was in fruits that had been precooled at 12°C, followed by those precooled to 16°C (12.3 days) and non-precooled fruits (10.2 days). For some crops that are stored for long periods of time hydrocooling may have little or no effect. For example, after 9 months of storage of Gala apples at 0.5°C and 97% r.h. with 1% O_2 and 3% CO_2, hydrocooling at harvest had no effect on firmness, acidity, peel colour, the incidence of internal breakdown, cracked fruit, scald and rot (Brackmann and Balz 2000).

A family of standard size, reusable and recyclable plastic containers was developed by Hui *et al.* (1999) to handle produce better and facilitate precooling. Results indicate that in order to provide an even distribution of water during hydrocooling, the total bottom openings should account for 5% of the total

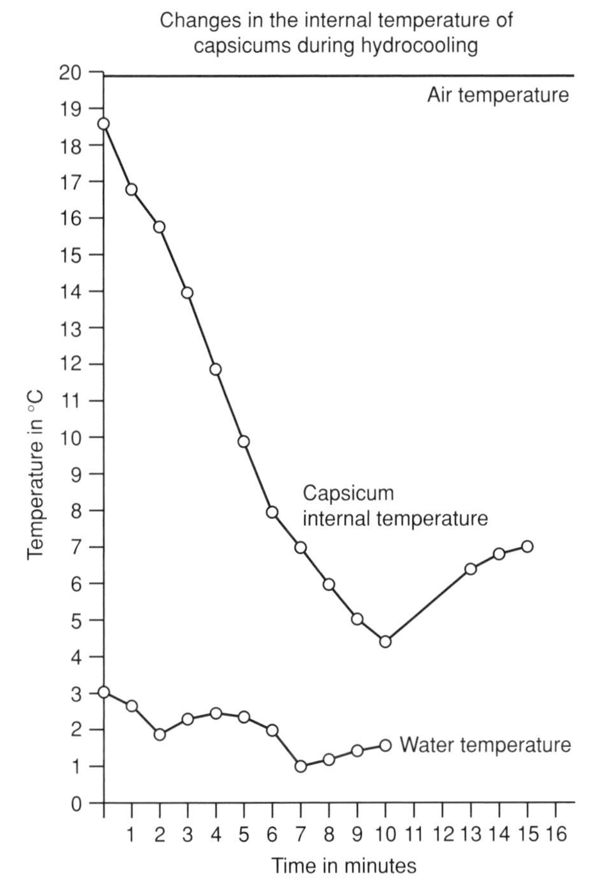

Figure 4.3 Changes in the internal temperature of capsicums during hydrocooling. Source: Seymour *et al.* (1981)

bottom surface for containers having side wall openings located at a height of 10 mm or more, measured from the bottom of the container. Furthermore, this percentage of the total bottom surface could be reduced to 3.5% for containers having side wall openings located over 20 mm from the container bottom.

Vacuum cooling

Cooling is achieved by the latent heat of vaporization rather than conduction. At normal air pressure of 760 mm of mercury water will boil at 100°C. As air pressure is reduced so the boiling point of water is reduced, and at 4.6 mm of mercury water boils at 0°C. For every 5 or 6°C reduction in temperature, under these conditions, Barger (1961) found that the crop would lose about 1% in weight. The speed and effectiveness of cooling are related to the ratio between the mass of the crop and its surface area (Table 4.1), so it is particularly suitable for leaf crops such as lettuce. In forced-air cooling the air passes over the surface of the crop cooling the outside, while the inside is cooled by heat transfer from inside to the outside of the crop. In contrast, with vacuum cooling of leaf vegetables such as lettuce, the reduced pressure is exactly the same around the leaves in the centre as it is around the leaves on the outside. This means that the cooling is very even and quick throughout the crop. Where there is a low ratio between mass and surface area or there is an effective barrier to water loss from the crop surface, vacuum cooling can be slow. Also crops such as tomatoes, which have a relatively thick wax cuticle, are not suitable for vacuum cooling.

Vacuum coolers have to be strongly constructed to withstand the vacuum (Figure 4.4) and therefore tend to be expensive to construct. They are made from heavy duty steel and are usually cylindrical in shape, to withstand the low pressure. The vacuum is usually achieve by a vacuum pump attached to the cylinder and the moisture from the crop is condensed on cooling coils situated on the outlet from the cylinder to the pump. In spite of the high cost, they are probably used for most lettuce marketed in Europe and the USA because the method is so quick.

Loss of water during marketing of crops not only affects their value where they are sold by weight, but

Table 4.1 Product temperature after 25–30 minutes vacuum cooling at 4.0–4.6 mmHg with a condenser temperature of −1.7–0°C. The original product temperature was 20–22.2°C (source: adapted from Barger 1963)

Crop	Temperature after 25–30 minutes (°C)
Courgettes	18
Potatoes	18
Potatoes (peeled)	15.5
Carrot roots	14
Snap beans	12
Cauliflower	10.5
Cabbage	7
Asparagus	7
Celery	7
Artichokes	6
Peas	6
Broccoli	5.5
Brussels sprouts	4.5
Corn	4.5
Carrot tops	2
Lettuce	1
Green onions	1

Figure 4.4 A vacuum cooler in operation at a storage and packaging company in The Netherlands.

Figure 4.5 The inside of a vacuum cooler with water spray facilities used for precooling lettuce in California where it is called 'Hydrovac'.

can also affect their value because of loss of quality. In order to reduce this problem with vacuum cooling, the crop may be sprayed with water before loading it into the vacuum cooler. Special vacuum coolers have been developed called 'Hydrovac' coolers, which have a built-in water spray inside the cooler (Figure 4.5), which is activated towards the end of the cooling operation. Martinez and Artes (1999) found that the levels of both pink rib and heart leaf injury of winter-harvested Coolguard iceberg lettuce was reduced by vacuum cooling after storage for 2 weeks at 2°C and then held for 2.5 days at 12°C for shelf-life compared with non-precooled lettuce.

Results of trials with vacuum cooling of lettuce by Rennie *et al.* (2001) showed that changing the pressure reduction rate had no effect on the mass loss, temperature reduction per percent mass loss or the temperature difference between various locations. The significance of these findings is that vacuum coolers with smaller vacuum pumps (slower pressure reduction rate) could be designed without any changes to the cooling characteristics of lettuce within the pressure reduction rates used in the study.

5 Packaging

Introduction

The function of a package is primarily to contain and protect the produce. With fresh fruits and vegetables there are often two levels of packaging. The first is the pack in which the produce is offered to the consumer. The second is the pack that contains the consumer pack and is used to transport the product to the retail market. The size of the package is therefore important; the consumer pack should be designed in terms of the amount the market or customer requires in a single unit. In some circumstances the overall package size may be dictated by what a person can reasonably lift or carry. At least one British supermarket has a maximum weight for the overall package in order to protect their staff from having to handle heavy weights.

The length of journey, the environmental conditions, the type of handling and any hazard influence packaging in relation to protection of the produce. The cost and cost effectiveness of the package and whether it is easy to assemble, fill and close are primary considerations. The use of standard pallets is increasing and the package may have to be designed to fit these or to fit a standard refrigerated container. Fresh fruit and vegetables being living organisms give out heat and gases that can be detrimental if allowed to accumulate in the package so they may need to be ventilated. Certain types of packaging can be used to extend the storage life of crops, such as using plastic films to modify the atmosphere around the crop, or to protect it from infection or infestation. Root crops can be packed into material such as coir dust, peat (Figure 5.1) or even sawdust to protect them and provide a humid environment to preserve the crop (Thompson *et al.* 1973d). The package may also help in presentation of the crop to enhance its value or help its sale. There are often legal requirements for consumer or wholesale packs where information on the origin, type, and grade of the product need to be displayed on the outside of the pack.

Most fruits and vegetables are harvested and placed in a container for transport to a packhouse or directly into a store. In many cases the crop is packed into a container in the field, and it stays in the same container right through to the wholesale market or even the retail market. The advantages of this are that the crop is handled only once so the potential for causing mechanical injury to it is reduced. Also, handling takes time and labour, both of which can be expensive. The construction, operation and maintenance of a packhouse can likewise be expensive and direct field packing can eliminate the necessity for having a packhouse. In Europe fruit and vegetable packhouses must comply with the provisions of European Council Directive 93/43/EEC.

Figure 5.1 Sweet potatoes packed for export in a wooden box with peat. The peat keeps a moist environment around the crop and delays deterioration. This is also used on early potatoes, yams and cassava.

Types of packaging

No packaging

This is unusual because it is normally time consuming to handle each fruit or vegetable individually and it leaves them unprotected from handling damage. In many developing countries, fruit such as melons, plantains and pumpkins and root crops such as yams and cassava are just stacked into the backs of lorries for transport to the market. A certain level of damage in inflicted on the crop (see Figure 12.108), but this is tolerated because it is perceived that it would be less cost effective to use packaging to reduce the damage.

In many cases, banana bunches are carried from the field balanced on people's heads or shoulders, but bruising and neck damage to the banana fingers are normally a consequence. Wooden trays padded with plastic foam (see Figure 12.26) on which the bunch is placed will reduce this damage (Stamford *et al.* 1971). Mechanical means are also used to transport banana bunches from the field, consisting of overhead cableways running throughout the banana plantation (Figure 5.2). These cableways have miniature trolleys which run on the cables from which the banana bunches are suspended. The trolleys are connected together in groups to form train that is then towed together to a packhouse where the cables converge. A labourer, an animal or a small tractor may pull the trains.

Second-hand containers

Harvesting produce into cartons, wooden boxes or metal cans that have been used for other commodities is common practice for small-scale farmers in many developing countries. The Sudan imports tomato paste in 10 kg cans. After the pulp has been used they are reused by local farmers for harvesting fresh tomatoes into, and became the standard measure on local markets. Second-hand containers must be clean, especially where they are used for products that are to be consumed unwashed or uncooked because of the contamination risk (John Love, personal communication 1994).

Bags and sacks

Many crops are harvested into bags that may be made from a variety of materials such as paper, polyethylene film, sisal, hessian (Figure 5.3) or woven polypropylene. This provides a relatively cheap method but gives little protection to the crop from handling and transport damage. They are commonly used for crops such as potatoes, onions, cassava and pumpkins, where some damage occurs to the fruit, but the extra cost of packing them in such a way that there would be no damage would be uneconomical. For high-risk products woven baskets and sacks are not acceptable in many countries because of the risk of contamination (John Love, personal communication 1994). Crops such as sugar peas, mange tout or fine beans destined for export from East Africa should not be harvested into woven natural fibre bags because of contamination (Chris Anstey, personal communication 1994). The amount packed in each bag is important. If a bag is too heavy the contents are more likely to be damaged. An example is cassava in Colombia that is packed into sacks containing about 75 kg. These are carried on a man's back to a lorry or from a lorry to a wholesale outlet. Because the sacks are so heavy the carrier almost invariably drops the bag, which damages the contents.

Sacks are traditionally made from fibres that are extracted from plants, cleaned, dried, spun and woven into sheets. Commonly used fibres are jute (*Corchorus capsularia* gives the best quality, but *C. olitorius*

Figure 5.2 Overhead cableway for transporting bananas from the field to the packhouse in Machala, Ecuador.

Figure 5.3 Hessian sacks used for transporting pumpkins from Jamaica to Britain.

is also used) and sisal (*Agave sisalana*). The former is a native of India and the fibre is harvested from the inner bark. The sacks are locally called gunny bags. The fibres are fine and shiny but are injured by exposure to water. For non-returnable transport, sacks are made from plain woven jute of 250 g m^{-2} or less. Sisal is a native of Central America and the fibre is extracted from the leaves. It is more water resistant than jute but cannot be spun as finely. Fibres from kennel (*Hibiscus cannabis*), hemp (*Cannabis sativa*) and abaci (*Musa textiles*) have all been used to make sacks for handling of crops. Woven sacks provide good ventilation but give almost no protection against mechanical injury.

Paper bags are made almost exclusively from natural Kraft, which is a strong brown paper, made from wood pulp. It is prepared by a sulphite process and is not bleached. Semi-bleached and fully bleached Kraft, which are off-white and white in colour, respectively, are not commonly used for fresh produce. To make bags more resistant to water two sheets may be stuck together with bitumen, making what is called 'union Kraft.' Coating one side of the Kraft with extruded low-density polyethylene film can make a more effective water-resistant bag. Many materials can be mixed with the wood pulp during the manufacture of the Kraft to give specific properties. These include various woven and extruded plastics and natural fibres. In some cases a transparent panel may be set into the bag to enhance the presentation of the crop. Paper quality is specified on weight per unit areas, which is called grammage in the trade. For fresh produce, Kraft of about 70–100 g m^{-2} is commonly used often as multiwalled bags for such crops as potatoes. Paper bags may be closed with ties by sewing or by gluing. For potatoes the bags usually contain 5, 10 or 25 kg. Paper bags are made by cutting the sheets to the appropriate size, folding them and sewing up two sides, leaving one side open through which to load the crop.

Woven baskets

Baskets are traditional containers into which crops are placed at harvest (Figure 5.4). They have been used in most countries and are of a variety of designs. They are usually made from split bamboo, rattan or palm leaves that are woven to form traditional-shaped containers and are usually conical in shape, or if they are squarish or rectangular they tend to have sloping sides and rounded corners because of the way in which they are made. Construction is normally by hand and often provides an important craft industry at the village level in many less developed countries. Baskets provide very good ventilation for crops, but because the materials from which they are made are rough, they can often damage the crop by abrasion when the baskets of produce are being handled and transported. Cushioning material such as paper or leaves is often used to line baskets to reduce abrasion, but this also reduces ventilation.

Baskets may be made from cheap materials, but these have poor stacking strength and when placed on top of each other the weight of the baskets is taken by the produce and not by the basket, which may lead to damage to the crop due to compression. Baskets are also difficult to stack because of their rounded, often irregular shape. This may also result in poor space utilization compared with the squared corners in boxes. Baskets that are reused for crop transport are difficult to keep clean, and may be a source of insect or micro-organism contamination.

Wooden field boxes

Bruce boxes were made from thin pieces of wood bound together with wire and were of two sizes, the bushel with a volume of 2200 cubic inches and the half-bushel box. They were developed in the USA where fruit were put into them at harvest for transport to wholesale and retail markets. They had the advantage that they could be packed flat when empty and were inexpensive so they could be non-returnable. However, they provided limited protection from mechanical damage during transport. The use of wires,

Figure 5.4 Woven basket used for field packing of a variety of crops. Here they are being used for cauliflowers in the Cameron Highlands in Malaysia.

staples or other obstacles to recycling is discouraged and the Bruce box has almost been phased out.

Rigid wooden boxes of various capacities are commonly used to transport produce from the field to the packhouse. These should be of a size that can be easily lifted, when full, by one person. They are commonly made from timber that has been sawn to the appropriate size and shape, but hardboard and plywood are also used. The boxes are usually assembled by nailing them together, but staples or glue are also used. The amount of ventilation can be arranged easily during the construction of the box. Wooden boxes can be expensive and heavy, which affects their use in crop handling. To make their use economic it is often important to use the same box over many journeys. Being relatively heavy they are not suitable for packing crops that are to be transported by airfreight. Most boxes are rigid and have good stacking properties. However, they are often made from cheap, unplaned wood whose rough surface can cause damage to the crop during transport. Also, the way in which they are constructed, especially when nailed, can damage the crop. Many wooden boxes are made by hand, which can result in slight variations in size and shape that in turn can make them difficult to stack on lorries or in storerooms. Wood can take up moisture and if not properly cured it can lose moisture and become distorted. Non-standard sizes can make them difficult to stack on lorries or in stores, and unplaned inner surfaces or protruding nails can damage produce. Field boxes are made from thin pieces of wood, widely spaced so that they are light in weight and cheap to make. These may be used for crops such as cabbages for transport from the field to the market. Wooden boxes are not acceptable to many UK supermarkets (John Love, personal communication 1994) because they may harbour microorganisms.

Plastic field boxes

These are recyclable and are used many times because they are strong and durable. Many are designed in such a way that they can nest inside each other when empty, to facilitate transport, and stack one on top of another when full (Figure 5.5). Others have metal bars that facilitate the nest stack design (Figure 5.6). Plastic boxes are strong, have good stacking strength and, if correctly made, a smooth surface. This means that they give good protection to the produce during handling and transport. They are completely water resistant and therefore easy to keep clean. They tend to be expensive and therefore each box has to be used many times to make them economic. If a suitable system can be devised for returning the boxes to the field from the market, plastic boxes can provide a very economical system for crop handling.

For supermarkets they must be washed with potable water before they are reused, and lining boxes with old newspaper, grass, straw, etc., is not acceptable for high-risk produce (John Love, personal communication 1994). Some plastic boxes produced in large quantities can be used for single journeys for high-value crops. In Thailand, fresh litchis and longans are packed in such boxes for export to Singapore (Figure 5.7). Expanded polystyrene boxes have been used for many years for non-returnable transport of crops such as watercress (Figure 5.8).

Most plastic boxes used for crops are made by injection moulding. This involves subjecting the

Figure 5.5 Plastic field box of the nest/stack type; the photograph was taken in Sri Lanka but the design is common throughout the world.

Figure 5.6 A moving-bar nest/stack returnable plastic field box packed with mangosteens in Thailand.

molten plastic to very high pressures to force it into the appropriate mould. This means that the equipment for making the boxes is very expensive since it has to be sufficiently robust to withstand conditions of high temperature and pressure. The cost of the plastic from which the boxes are made is relatively cheap, so this type of box can be produced much more cheaply where very large quantities of the same design and dimensions are required. The materials used are high-density polyethylene, polypropylene or poly(vinyl chloride). Polyethylene has higher impact strength, a lower rate of degradation when exposed to ultraviolet radiation and a slightly higher density than polypropylene. Both plastics will deteriorate under field conditions and ultraviolet protection chemicals should be added to the plastic during manufacture to reduce the rate at which they become brittle.

Figure 5.7 Longans packed in non-returnable plastic boxes in the field in Thailand, awaiting export to Hong Kong and Singapore.

Figure 5.8 Cress packed in expanded polystyrene boxes for local marketing in the UK.

Biodegradable plastic material was developed as a response to environmental concerns. One such material is polyhydroxybutyrate, which is a thermoplastic polymer marketed as 'Biopol,' which can easily be broken down by soil bacteria. It is made by fermentation of sugars by a naturally occurring soil bacterium that produces it and stores it as an energy source. It can be used for injection moulding, can withstand temperatures of up to 100°C and most closely resembles polypropylene (R.P. Pearson, unpublished report 1991). Another biodegradable material is marketed as 'CornPak.' It was developed in 1990 at the University of Illinois and is marketed by Arcola Grain Products Inc. It is made from ground puffed corn with soybean lecithin and trace amounts of newspaper that is formed into 19 mm diameter spheres. The marketing company claim that in the USA over 35 000 tonnes of Styrofoam pellets were used each year by the packaging industry, which is equivalent to 20 million bushels of corn if it were replaced by CornPak.

Collapsible plastic boxes are available which have the advantage of being packed flat when empty, thus taking up little space when being returned to the field. They have a shorter life than other types of plastic boxes and it takes a little time to assemble them. A system based on collapsible plastic boxes was developed by IFCO (see Figure 12.35) and had been used in the fruit and vegetable packing industry.

Expanded polystyrene is used for fresh produce (Figure 5.8). They are light and have good stacking strength but are brittle and easily damaged. They have good insulation properties, which means that they can protect cooled crops from rapid rises in temperature when they are removed from a cold room to an environment where the temperature is high. However, they will retain respiratory heat of the crop and fruit must be adequately precooled before the boxes are lidded, and they are also not ideal for the UK because they are not recyclable and tend to blow away when empty (John Love, personal communication 1994).

Pallet boxes

The size of pallet boxes can vary but they are most commonly based on the standard size for a European pallet of 1 × 1.2 m and the are usually about 0.5 m high. These have a capacity of about 500 kg depending on the crop that they contain. They are made from wood or plastic and are used for a whole range of crops. The produce is commonly loaded into them in the field and they are then transported directly to the store or packhouse.

Fibre-board boxes

These are made from either laminated fibreboard or, more commonly, corrugated fibreboard. They may be used for transporting produce from the field to the packhouse, in which case the same carton may be used on several occasions. A study showed that in certain circumstance using corrugated cartons for transporting yams from the field to the packhouse could be economic because of reduced injury to the tubers even when the cartons were used only once (Thompson 1972a). However, this type of box, when used in the field, is normally for produce that will go in the same pack throughout the marketing chain. A study of field packing of papaya fruit showed considerably less damage to the fruit and less postharvest disease than fruits packed in packhouses (Thompson and Lee 1971). Using cartons directly in the field is not acceptable to the supermarket sector because they may get dirty and contaminated in the fields. Boxes are also made of mixtures of fibreboard and plywood or hardboard. The reasons for these mixtures are either to reduce the cost of the box or to make it lighter and more durable.

Fibreboard boxes are made from either wood pulp or recycled paper, although other materials have been used, e.g. popcorn (Daily Telegraph, 31 October 1990), but this increased packaging costs by 40%. The result, however, was an edible box but it had only a short shelf-life. Other non-wood pulps have been made from bagasse, bamboo, straw, etc. The softwood pulp, usually from *Pinus* spp., is formed into sheets that are called 'Kraft' (from the German word for strength). The Kraft can be described as virgin where the fibres are processed straight from the tree, or recycled where they have been obtained from waste paper, fibreboard boxes, etc. The American Paper Institute defines virgin Kraft as that which contain at least 80% new fibre, that is, not more than 20% recycled fibre. The Kraft is then made up into either solid fibreboard or corrugated fibreboard.

Other facing materials include the following:

- Jute, also called test and cylinder, which is usually made from reprocessed used fibreboard boxes. It may be made in layers of different composition or it may be homogenized.
- Schrenz, also called chip or bogus, which is made from recycled fibre of whatever type might be available, even used newsprint. It is naturally greyish in colour and inferior in strength characteristics.
- Laminates, which are two-ply sheets, often one layer of Kraft and one layer of Schrenz.

Solid fibreboard is not commonly used for fresh produce packing but it can provide an attractive container. It is made from layers of Kraft that are glued together to produce the required thickness of board. Specifications of solid fibreboard are based on the total board thickness, which varies in the range from about 0.8–3 mm, or weight of board per unit area, which varies from about 500–2000 g m^{-2} for fresh produce boxes. The Kraft used for the manufacture of solid fibreboard is usually between 60 and 160 g m^{-2} with a white Kraft outer facing which gives an attractive appearance. Water-insoluble adhesives and resin additives are used for solid fibreboard boxes that are to be used for fresh produce to increase their strength when stored in high-humidity conditions. Coating with wax, nitrocellulose or polyethylene increases their resistance to water penetration.

Corrugated fibreboard boxes are more commonly used for fresh produce because they are, weight for weight, stronger than solid fibreboard boxes. The basic corrugated fibreboard boxes are made from three layers of Kraft; two outer layers which are called 'facings' or 'liner boards' and an inner layer which is called the 'fluting structure' or the 'corrugating medium.' This is called single wall board. Double wall board is made from three facings and two flutings. One factor that affects the strength of the board is the fluting contour, which is defined on the height of each flute and the number of flutes (Table 5.1).

The characteristics of board made from the different flute structures can vary considerably. A-flute board is more susceptible to manufacturing and handling damage, but theoretically it should have good column crush strength and therefore good stacking strength. However, with commercially produced board, because of its fragility, its stacking strength is rarely better than that of C-flute. A-flute has the greatest cushioning effect of all the types of corrugated board. B-flute has the highest flat crush and is the most resistant to damage during manufacture and handling. It has the narrowest width and therefore poor compression strength and poor stacking characteristics. C-flute has properties

Table 5.1 Number and height of flutes used in the construction of fibreboard used for fresh fruit and vegetables

Flute contour	Flutes per linear metre	Flute height (mm)
A	108–128	4.76
B	154–174	2.38
C	128–148	3.57

between those of A and B but historically it was developed after them, hence its alphabetical place. It was developed in the hope of combining the top to bottom compression strength of A-flute and the high flat crush and greater control over fabrication of B-flute. All three flutes are still in use, particularly B and C. For double wall board, AB, BA, AC, CA, BC and CB combinations are all used. Combinations of two identical flutes such as AA, BB or CC are uncommon. There is another flute, called either D- or E-flute, which has 305 corrugations per metre and a flute height of about 1.19 mm, but it is not normally used for fresh produce.

The material used for the manufacture of the fluting needs to have good stiffness and formability. The highest quality fluting material is made from 'semi-chemical,' which consists of hardwood fibres made by a neutral sulphite or comparable process that is partly chemical and partly mechanical because of the difficulty in defibring hardwoods. The appearance and other characteristics may vary depending on the species of hardwood used. Other fluting media include bogas, Kraft and straw. Bogus is based on recycled Kraft waste and if properly made can be of good quality. Kraft can be used where high tear and puncture resistance is required, but it does not corrugate well. Straw is little used since it produces an inferior fibre. Unbleached virgin coniferous Kraft is most appropriate for liner materials for boxes to be used for fresh produce. It has a low rate of moisture uptake compared with liners made from recycled material, high tearing resistance and good stiffness. Where recycled materials have to be used, a substance increase of 50% or more may be necessary to achieve a similar strength to virgin Kraft.

The characteristics of the fibreboard boxes are also greatly affected by the weight of the board used. For Kraft liner board these normally vary between 125 and 450 g m^{-2}, for jute between 210 and 90 g m^{-2} and for semi-chemical fluting between 112 and 180 g m^{-2}, 112–127 g m^{-2} for normal use and 150–180 g m^{-2} for heavy-duty applications. Single wall board is usually made from liners of similar weight. Specifications for fibreboard are governed by specific standards, e.g. Rule 41 of the Uniform Freight Classification of the USA. Other countries also have their own specifications, e.g. Fibreboard Case Specifications BS 1133(7) in the UK and AFNOR NF Q 12-008 in France. There are a whole series of standard tests for fibreboard and fibreboard boxes, such as vertical impact test, horizontal impact test, transit vibration test and bursting strength.

Fibreboard boxes easily take up moisture from the atmosphere. This affects the strength of the boxes and therefore their ability to protect and contain the crop. To make the boxes more resistant to water penetration, waxes can be applied to them. These are commonly blends of microcrystalline paraffin wax with additives such as poly(vinyl acetate) and latexes. Paraffins by themselves are unsuitable because their melting points are too low and they fracture easily. Coatings may be applied during fabrication of the fibreboard. Sufficient coating must be applied to ensure the board is sufficiently water resistant, but not too much as this may inhibit the effect of the adhesive used in the assembly of the board. Coatings may also be applied to the board or the box.

The design of boxes made from fibreboard takes into account the need to protect the crop, display it to its best advantage and supply it in units that are required or preferred by the market. These designs have evolved over several decades and can be described by code numbers in accordance with the International Fibreboard Case Code (Export Packaging Note Number 19, International Trade Centre, United Nations Centre for Trade and Development and the General Agreement on Tariffs and Trade) (ITC 1992). The system was devised by Johan Selin in 1958 and allocates a series of letters to a box or case design. The first two letters in the code describe the basic type of box as follows:

- 02 slotted type boxes
- 03 telescopic type boxes
- 04 folder type boxes and trays
- 05 slide type boxes
- 06 rigid type boxes
- 07 ready glued cases
- 09 interior fitments such as liners, pads and partitions.

Package recycling

Packaging is very expensive and its disposal after use can be a major source of pollution. Packaging creates a waste problem. New laws are being enacted in many countries on disposal of packaging. In Germany a company was set up as long ago as 1990, called Duales System Deutschland GmbH, to establish a household-orientated system for collecting used packaging for recycling in a way that is easy for people to use and is environmentally friendly. Over 400 companies have joined the scheme, which permits them to put the Duales System Deutschland GmbH logo on their package. The logo is called *der grüne Punkt* (green dot). Only packaging that can be recycled after collection is allowed to carry the *der grüne Punkt* symbol. Companies are licensed to use the symbol, for which they must pay a fee. Since then many recycling schemes have been developed throughout the world. In Britain most local authorities run a voluntary recycling scheme for household waste.

Modified atmosphere packaging

Modified atmosphere packaging can be defined as an alteration in the composition of gases in and around fresh produce by respiration and transpiration when such commodities are sealed in plastic films (G.E. Hobson, personal communication 1997). For fresh fruits and vegetables this is commonly achieved by packing them in plastic films. Plastic film bags are made from the by-products of the mineral oil refining industry. Different researchers and commercial packers use different units to describe film thickness (see Appendix).

Church and Parsons (1995) reviewed developments in modified atmosphere packaging. These included:

- lowpermeability films
- high-permeability films
- O_2 scavenging technology
- CO_2 scavengers and emitters
- ethylene absorbers
- ethanol vapour generators
- tray-ready and bulk modified atmosphere packaging systems
- easy opening and resealing systems
- leak detection
- time–temperature indicators
- gas indicators
- combination treatments
- predictive/mathematical modelling.

The surface of the bags is slippery, which can make them very difficult to stack, especially large bags. Low-density polyethylene is commonly produced in this manner for the ubiquitous 'polythene' bag so frequently used in fresh produce packaging. The permeability of plastic films to gases and water vapour will vary with the type and thickness of the plastic that is used. Generally, the acceptable time during which fruit can remain in a modified atmosphere varies with cultivars and storage conditions, but it is usually 2–4 weeks, which meets the requirement of export shipment and marketable life.

Packing crops into plastic film bags can lead to a build-up of water vapour and the respiratory gases ethylene and CO_2 and a reduction in O_2. The effects of these gas changes may have beneficial or detrimental effects on the crop. The beneficial effects, which can result in packing crops in plastic films, have led to this method being referred to as modified atmosphere packaging (Kader *et al.* 1989a). Modified atmosphere storage can also refer to sealing fruit and vegetables in a gas-impermeable container where respiratory gases change the atmosphere over time.

If a crop is in equilibrium with its environment, the rate of gas exchange will be the same in both directions. Hence if the concentration of the respiration gases around the outside of the crop is changed by surrounding it with plastic film, this will change the concentrations of the gases inside the crop. The concentration of these gases will then vary with characteristics such as the type and thickness of the film used and innate crop characteristics such as the mass of produce inside the bag, type of crop, temperature, maturity and activity of microorganisms.

Given an appropriate gaseous atmosphere, modified atmosphere packaging can be used to extend the postharvest life of crops in the same way as controlled atmosphere storage for example with mushrooms (Lopez-Briones *et al.* 1993). Another example was given by Pala *et al.* (1994). They studied storage in sealed low-density polyethylene film of various thicknesses at 8°C and 88–92% r.h. on green peppers and found that non-wrapped peppers had a shelf-life of 10 days compared with 29 days for those sealed in 70 nm low-density polyethylene (LDPE) film (Table 5.2). This method of packaging also gave good retention of sensory characteristics.

Batu and Thompson (1994) stored tomatoes at both the mature green and the pink stage of maturity either not wrapped or sealed in polyethylene films of 20, 30 or 50 μm thickness at either 13 or 20°C for 60 days. All pink tomatoes that were not wrapped were overripe and soft after 30 days at 13°C and after 10–13 days at 20°C. Pink tomatoes in polymeric films were still firm after 60 days stored at either 13 or 20°C although those in 20 μm film were slightly softer than those in the other two films. Green tomatoes sealed in 20 and 30 μm film reached their reddest colour after 40 days at 20°C and 30 days at 13°C. Fruit in 50 μm film still had not reached their maximum red colour even after 60 days at both temperatures. All green tomatoes sealed in polymeric film were very firm even after 60 days of storage at 13 and 20°C. Unwrapped tomatoes remained acceptably firm for about 50 days at 13°C and 20 days at 20°C.

The effects of polyethylene film wraps on the postharvest life of fruits may also be related to moisture conservation around the fruit and the change in the CO_2 and O_2 contents. This was shown by Thompson *et al.* (1972a) for plantains, where fruit stored in moist coir or perforated polyethylene film bags had a longer storage life than fruits that had been stored unwrapped. However, there was an added effect when fruits were stored in non-perforated bags, which was presumably due to the effects of the changes in the CO_2 and O_2 levels (Table 5.3). Hence the positive effects of storage of fresh preclimacteric fruits in sealed plastic films may be, in certain cases, a combination of its effects on the CO_2 and O_2 contents within the fruit and the maintenance of a high moisture content. The effect of moisture content is more likely to be a reduction in stress of the fruit that may be caused by a rapid rate of water loss in non-wrapped fruit. This in turn may result in increased ethylene production to internal threshold levels that can initiate ripening.

Table 5.2 Number of days during which green pepper fruit retained either good or acceptable quality during storage at 8°C and 88–92% r.h. (after Pala *et al.* 1994)

	Days in storage	Sensory score[a]				Storage time (days)	
		A[b]	C[c]	T[d]	F[e]	Good quality	Acceptable quality
Before storage	0	9	9	9	9		
Non-wrapped control	15	4.3	4.6	2.7	4.2	7	10
20 μm LDPE	22	5.7	5.5	5.7	5.3	10	22
30 μm LDPE	20	5.8	6.0	6.0	4.5	10	20
50 μm LDPE	20	6.0	5.0	6.0	5.0	15	20
70 μm LDPE	29	7.8	7.8	7.6	7.8	27	29
100 μm LDPE	27	5.5	5.6	5.7	5.5	10	27

[a]1–9 where 1–3.9 = non acceptable, 4–6.9 = acceptable, 7–9 = good.
[b]Appearance.
[c]Colour.
[d]Texture.
[e]Taste and flavour.

Table 5.3 Effects of wrapping and packing material on ripening and weight loss of plantains stored at tropical ambient conditions of 26–34°C and 52–87% r.h. (source: Thompson *et al.* 1972a)

Packing material	Days to ripeness	Weight loss at ripeness (%)
Not wrapped	15.8	17.0
Paper	18.9	17.9
Moist coir fibre	27.2	(3.5)[a]
Perforated polyethylene	26.5	7.2
Polyethylene	36.1	2.6
LSD ($P = 0.05$)	7.28	2.81

[a] The fruit actually gained in weight.

Table 5.4 Effects of packaging material on the quality of yam (*Dioscorea trifida*) after 64 days at 20–29°C and 46–62% r.h. Fungal score was 0 = no surface fungal growth, 5 = tuber surface entirely covered with fungi. Necrotic tissue was estimated by cutting the tuber into two length ways and measuring the area of necrosis and expressing it as a percentage of the total cut surface

Package type	Weight loss (%)	Fungal score	Necrotic tissue (%)
Paper bags	26.3	0.2	5
Sealed 0.03 mm thick polyethylene bags with 0.15% of the area as holes	15.7	0.2	7
Sealed 0.03 mm thickpolyethylene bags	5.4	0.4	4

Hansen (1975) showed that Jonagold apples kept in cold stores at 0.5, 2.5 or 3.5°C retained satisfactory organoleptic quality only until the end of January even at the lowest temperature. However, the apples could be stored at 3.5°C in polyethylene bags (with a CO_2 content of 7%) until April. In controlled atmosphere stores (6% CO_2 with 3% O_2 at 2.5°C) fruit quality was satisfactory to the end of May. Under all conditions apples of this cultivar showed no rotting, CO_2 damage or physiological disorders.

Modified atmosphere packaging does not always have a beneficial effect on the postharvest life of fresh fruits and vegetables. Storing yam in polyethylene bags was shown to reduce weight loss, but to have little effect on surface fungal infections and internal browning of tissue (Table 5.4). Geeson (1989) also showed that film packages with lower water vapour transmission properties could encourage rotting of the tomatoes.

Modified atmosphere packaging is used to slow the deterioration of prepared fruit and vegetables and this technique is often referred to as minimal processing. An example of this technique is described by Lee and Lee (1996), who used low-density polyethylene film of 27 mm thickness for the preservation of a mixture of carrot, cucumber, garlic and green peppers, where the steady-state atmosphere at 10°C inside the bags was 5.5–5.7% CO_2 and 2.0–2.1% O_2.

Film types

The actual concentration of gases in the fruit will also be affected to a limited degree by the amount of space between the fruit and the plastic film, but mainly by the permeability of the film. Several different plastics are used for this purpose (Table 5.5).

A range of high-shrink multilayer high-speed machineable shrink films made from polyolefin is available. Polyethylene is also used for shrink film packaging.

Film permeability

Respiring fresh fruits and vegetables sealed in plastic films will cause the atmosphere to change, in particular O_2 levels to be depleted and CO_2 levels to be increased. The transmission of CO_2 and O_2 though plastic films will vary with film type, but generally films are four to six times more permeable to CO_2 than to O_2. However, Barmore (1987) indicated that the relationship between CO_2 and O_2 permeability and that of water vapour is not so simple. Variation in transmission of water vapour can therefore be achieved to some extent, independently of transmission of CO_2 and O_2 using such techniques as producing multilayer films by coextrusion or applying adhesives between the layers. This permeability is invariably restricting and often

Table 5.5 A selection of plastic films which have been used for fruit and vegetable packaging

Film	Abbreviation
Cellulose acetate	CA
Ethylene–vinyl acetate copolymers	EVAL
Ethylene–vinyl alcohol copolymers	EVOH
High-density polyethylene	HDPE
Ionomer	–
Linear low-density polyethylene	LLDPE
Low-density polyethylene	LDPE
Medium-density polyethylene	MDPE
Oriented polypropylene	OPP
Polybutylene	PB
Poly(ethylene terephthalate)	PET
Polyolefin	–
Polypropylene	PP
Polystyrene	PS
Polyvinylbutyral	PVB
Poly(vinyl chloride)	PVC
Poly(vinylidene chloride)	PVDC

Table 5.6 Effects of number and size of perforation in 3 lb (1.36 kg) 150 gauge polyethylene film bags of Yellow Globe onions on the relative humidity in the bags, rooting of the bulbs and weight loss after 14 days at 24°C (Hardenburg 1955)

Perforations				
Number	Size (mm)	r.h. in bag (%)	Onions rooted (%)	Weight loss (%)
0	–	98	71	0.5
36	1.6	88	59	0.7
40	3.2	84	40	1.4
8	6.4	–	24	1.8
16	6.4	54	17	2.5
32	6.4	51	4	2.5
0[a]	–	54	0	3.4

[a]Kraft paper with film window.

holes are punctured in the bags to improve ventilation (Table 5.6).

The permeability of films to gases (including water vapour) varies with the type of material from which they are made, temperature, in some cases humidity, the accumulation and concentration of the gas and the thickness of the material. Polyethylene film, as indicated above, is available in several forms with different properties. Schlimme and Rooney (1994) showed that there was a range of permeabilities that could be obtained from films with basically the same specifications. A selection of these is summarized as follows:

1. **Low-density polyethylene** is produced by polymerization of ethylene monomer, which produces a branched-chain polymer with a molecular weight of 14 000–1 400 000 and a density ranging from 0.910–0.935 g cm^{-3}. Low-density polyethylene film has the following specifications: 3900–13 000 O_2 cm^3 m^{-2} day^{-1}, 7700–77 000 CO_2 cm^3 m^{-2} day^{-1}, at 1 atmosphere for 0.0254 mm thick at 22–25°C at various or unreported relative humidity and 6–23.2 g water vapour m^{-2} day^{-1} at 37.8°C and 90% r.h. (Schlimme and Rooney 1994).
2. **High-density polyethylene** has 75–90% crystalline structure with an ordered linear arrangement of the molecules with little branching and a molecular weight of 90 000–175 000. It has a typical density range of 0.995 to 0.970 g cm^{-3} and has greater tensile strength, stiffness and hardness than low-density polyethylene. High-density polyethylene film has the following specifications: 520–4 000 O_2 cm^3 m^{-2} day^{-1}, 3900–10 000 CO_2 cm^3 m^{-2} day^{-1}, at 1 atmosphere for 0.0254 mm thick at 22–25°C at various or unreported relative humidity and 4–10 g water vapour m^{-2} day^{-1} at 37.8°C and 90% r.h. (Schlimme and Rooney 1994).
3. **Medium-density polyethylene** has a density in the range 0.926–0.940 g cm^{-3}, 2600–8293 O_2 cm^3 m^{-2} day^{-1}, 7700–38 750 CO_2 cm^3 m^{-2} day^{-1}, at 1 atmosphere for 0.0254 mm thick at 22–25°C at various or unreported relative humidity and 8–15 g water vapour m^{-2} day^{-1} at 37.8°C and 90% r.h. (Schlimme and Rooney 1994).
4. **Linear low-density polyethylene** combines the properties of low-density polyethylene film and high-density polyethylene film, giving a more crystalline structure than low-density polyethylene film but with a controlled number of branches which make it tougher and suitable for heat sealing. It is made from ethylene with butene, hexene or octene, with the last two co-monomers giving enhanced impact resistance and tear strength. Permeability was given as 7000–9300 O_2 cm^3 m^{-2} day^{-1}, at 1 atmosphere for 0.0254 mm thick at 22–25°C at various or unreported relative humidity and 16–31 g water vapour m^{-2} day^{-1} at 37.8°C and 90% r.h. (Schlimme and Rooney 1994).

Film permeability to gases is by active diffusion where the gas molecules dissolve in the film matrix and diffuse through in response to the concentration gradient (Kester and Fennema 1986). An equation to describe film permeability was given by Crank (1975) as follows:

$$P = \frac{Jx}{A(p_1 - p_2)}$$

where

J = volumetric rate of gas flow through the film at steady state
x = thickness of film
A = area of permeable surface
p_1 = gas partial pressure on side 1 of the film
p_2 = gas partial pressure on side 2 of the film ($p_1 > p_2$).

Gas flushing

The levels of CO_2 and O_2 can take some time to change inside the modified atmosphere pack. In order to speed this process, the pack can be flushed with nitrogen to reduce rapidly the O_2 or the atmosphere can be flushed with an appropriate mixture of CO_2, O_2 and nitrogen. In other cases the pack can be connected to a vacuum pump to remove the air so that the respiratory gases can change within the pack more quickly. Gas flushing is more important for non-respiring products such as meat or fish, but it can profitably be used with fresh fruits and vegetables. In work described by Aharoni *et al.* (1973), yellowing and decay of leaves were reduced when the lettuce cultivar Hazera Yellow was pre-packed in closed polyethylene bags in which the O_2 concentration was reduced by flushing with nitrogen. Similar but less effective results were obtained when the lettuce was pre-packed in closed polyethylene bags not flushed with nitrogen, or when open bags were placed in polyethylene-lined cartons. Andre *et al.* (1980a) showed that fungal development during storage could be prevented in asparagus spears by packing them in polyethylene bags with silicon elastomer windows and flushing with 30% CO_2 for 24 hours or by maintaining a CO_2 concentration of 5–10%.

The changes in the atmosphere inside the sealed plastic film bag depend on the characteristics of the material used to make the package, the environment inside and outside the package and the respiration of the produce it contains. This was illustrated in experiments described by Zagory (1990), which showed the effects of flushing chilli peppers stored in plastic film with a mixture containing 10% CO_2 and 1% O_2 compared with no gas flushing (Figure 5.9).

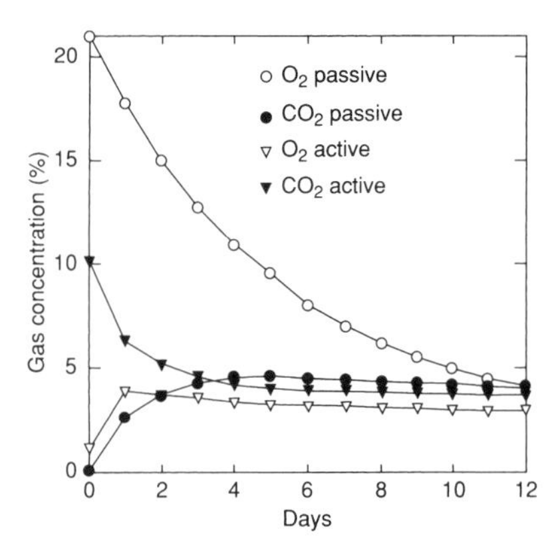

Figure 5.9 Simulation of a passive versus active modification of modified-atmosphere packaging of 'Anaheim' chilli peppers at 10°C in Cryovac SSD-310. Source: Zagory (1990).

Modelling

Ben Yehoshua *et al.* (1995) reviewed modified atmosphere packaging under the following headings: modified humidity packaging with an example of the effects on bell peppers, interactive and microporous films, effects of perforations in the film, and modelling of packaging. Day (1994) also reviewed the concept of mathematical modelling of modified atmosphere packaging for minimally processed fruit and vegetables. Evelo (1995) showed that the package volume did not affect the steady-state gas concentrations inside a modified atmosphere package but did affect the non-steady-state conditions. A modified atmosphere packaging model was developed using a systems-oriented approach that allowed the selection of a respiration model according to the available data. Temperature dependence was explicitly incorporated into the model. The model was suitable for assessing optimal modified atmosphere packaging in realistic distribution chains. O_2 and CO_2 concentrations inside plastic film packages containing mango cultivar Nam Dok Mai fruits were modelled by Boon-Long *et al.* (1994). Parameters for poly(vinyl chloride), polyethylene and polypropylene films were included. A method of determining respiration rate from the time history of the gas concentrations inside the film package instead of from direct measurements was devised. Satisfactory results were obtained in practical experiments carried out to verify the model.

Many other mathematical models have been developed and published (e.g. Cameron *et al.* 1989; Lee *et al.* 1991; Lopez-Briones *et al.* 1993), which can help in predicting the atmosphere around fresh produce sealed in plastic film bags. The last authors suggested the following model:

$$\frac{1}{x} = \frac{\hat{A}m}{x_0 S} \cdot \frac{1}{K} + \frac{1}{x_0}$$

where

x = O_2 concentration in the pouches (%)
x_0 = initial O_2 concentration within the pouches (%)
$\hat{A}$ = proportionality between respiration rate and O_2 concentration including the effect of temperature (ml g^{-1} day^{-1} atm^{-1})
K = O_2 diffusion coefficient of the film (includes effects of temperature) (ml m^{-2} day^{-1} atm^{-1})
S = surface area for gas exchange (m^2)
m = weight of plant tissue (g).

Lee *et al.* (1991) obtained a set of differential equations representing the mathematical model for the modified atmosphere system. In this case, the rate of reaction (r) is given by:

$$r = V_m C_1 / [K_M + C_1(1 + C_2/K_1)]$$

where

C_1 = substrate concentration (which is O_2 concentration in the case of respiration rate)
V_m, K_M = parameters of the classical Michaelis–Menten kinetics, V_m being the maximal rate of enzymatic reaction and K_M the Michaelis constant
C_2 = inhibitor concentration (CO_2 concentration in the case of respiration)
K_1 = constant of equilibrium between the enzyme–substrate–inhibitor complex and free inhibitor.

Combining this equation with Fick's law for O_2 and CO_2 permeation, Lee *et al.*. (1991) estimated the parameters of this model (V_m, K_M, and K_1) from experimental data and then performed numerical calculations of the equations.

Strawberries (cultivars Pajaro and Selva) were used to test a model describing gas transport through micro-perforated polypropylene films and fruit respiration involved in modified atmosphere packaging by Renault *et al.* (1994). Some experiments were conducted with empty packs initially filled with either 100% nitrogen or 100% O_2. The simulations agreed very well with experiments only if areas of approximately half the actual areas replaced the cross-sectional area of the micro-perforations in order to account for the resistance of air around the perforations. It was also possible to fit the model to gas concentration changes in packs filled with strawberries, although deviations were encountered due to contamination of strawberries by fungi. The model was used to quantify the consequences of the variability of pack properties, number of micro-perforations per pack and cross-sectional area of these perforations on equilibrium gas concentrations and to define minimum homogeneity requirements for modified atmosphere packaging.

Quantity of product and the gas content

The quantity of produce inside the sealed plastic film bag has been shown to affect the equilibrium gas content (Table 5.6), but the levels of CO_2 and O_2 do not always follow what would be predicted from permeability data and respiration load of the crop. Zagory (1990) showed that there was a negative linear relationship between the weight of fresh chillies in a sealed Cryovac SSD-310 film package between CO_2 and O_2 levels and CO_2 levels were reduced with increasing

Table 5.6 Effects of the amount of asparagus spears sealed inside different film types on the equilibrium CO_2 and O_2 contents at 20°C (Lill and Corrigan 1996)

Plastic film type	Permeability at 20°C (ml m^{-3} atm^{-1} day)	Product load (g)	CO_2 (%)	O_2 (%)
W.R. Grace RD 106	23 200 CO_2, 10 200 O_2	100	2.5	9.7
W.R. Grace RD 106	23 200 CO_2, 10 200 O_2	150	103.2	6.2
W.R. Grace RD 106	23 200 CO_2, 10 200 O_2	200	3.5	4.1
Van Leer Packaging Ltd	33 200 CO_2, 13 300 O_2	100	3.6	4.5
Van Leer Packaging Ltd	33 200 CO_2, 13 300 O_2	200	4.2	3.6
Van Leer Packaging Ltd	33 200 CO_2, 13 300 O_2	250	4.6	2.0
Chequer Systems Ltd	45 300 CO_2, 16 400 O_2	100	3.4	15.4
Chequer Systems Ltd	45 300 CO_2, 16 400 O_2	250	3.9	16.4
Chequer Systems Ltd	45 300 CO_2, 16 400 O_2	3000	6.1	11.3

product in the bag whereas O_2 levels were proportionately higher.

The number of fruit packed in each plastic bag can alter the effect of modified atmosphere packaging. An example is that plantains packed with six fruits in a bag ripened in 14.6 days compared with 18.5 days when fruits were packed individually when stored at 26–34°C (Thompson *et al.* 1972a).

Zagory (1990) also demonstrated the relationship between the weight of produce in a package (20 cm × 30 cm) and its O_2 and CO_2 contents for one type of plastic film. The relationship between O_2 and CO_2 levels was linear (Figure 5.10), but varied considerably with a four-fold variation in produce weight, which illustrates the importance of varying just one factor.

Perforation

Punching holes in the plastic can maintain a high humidity around the produce, but it may be less effective in delaying fruit ripening (Table 5.2) because it does not have the same effect on the CO_2 and O_2 contents of the atmosphere inside the bag. The holes may be very small and in these cases they are commonly referred to as micro-perforations.

Various studies have been conducted on the effects of packing various produce at different temperatures on the internal gas content. Fruits of the pear cultivars Okusankichi and Imamuraaki decayed most if placed in unsealed 0.05 mm plastic bags and least in sealed bags with five pin holes (bags with 10 holes showed more decay). Weight loss of fruit after 7 months of storage was 8–9% and less than 1% in unsealed and sealed bags, respectively. In sealed bags, the CO_2 concentration reached 1.9% after 2 months of storage (Son *et al.* 1983). Hughes *et al.* (1981) showed that capsicums sealed in various plastic films had a higher percentage of marketable fruit than those stored in air.

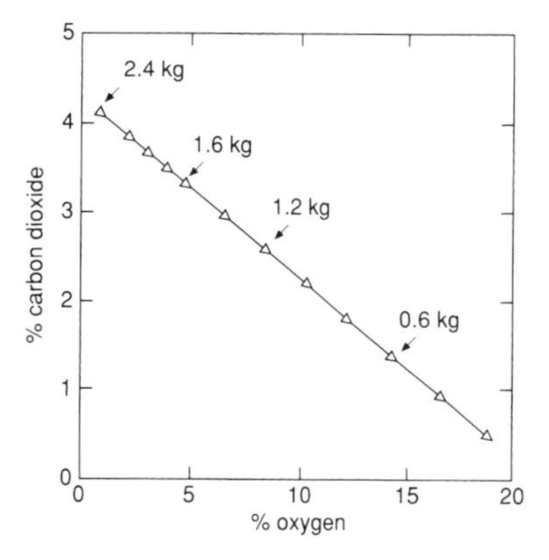

Figure 5.10 Equilibrium gas concentrations in modified atmosphere package as a function of product weight for 'Anaheim' chilli peppers in Cryovac SSD-310. Source: Zagory (1990).

Absorbents

'Active packaging' of fresh produce has been carried out for many years. This system usually involves the inclusion of a desiccant or O_2 absorber within or as part of the packaging material. These are mainly used to control insect damage, mould growth, rancidity and discoloration in a range of perishable food products such as meat, herbs, grains, beans, spices and dairy products. One such product is marketed as 'Ageless ' and uses iron reactions to absorb O_2 from the atmosphere (Abe 1990).

Ascorbic acid-based sachets (which also generate CO_2) and catechol-based sachets are also used as O_2 absorbers. The former is marketed as Ageless G or Toppan C or as Vitalon GMA (when combined with iron), and the latter as Tamotsu. Mineral powders are incorporated into some films for fresh produce, particularly in Japan (Table 5.7).

It was claimed that among other things these films can remove or absorb ethylene, excess CO_2 and water and be used for broccoli, cucumber, lotus root, kiwifruit, tomato, sweetcorn and cut flowers (Industrial Material Research 1989, quoted by Abe 1990). Eun *et al.* (1997) described silver-coated ceramic coatings used on low-density polyethylene film for the storage of enoki mushrooms. Coated films extended the storage life of the mushrooms from 10 days for those stored in non-coated low-density polyethylene film to 14 days for those in silver-coated ceramic films.

Ethylene absorbents can be included inside modified atmosphere packages and the following examples illustrate some of their uses. A major source of deterioration of limes during marketing is their rapid weight loss that can give the skins a hard texture and an unattractive appearance. Packing limes in sealed polyethylene film bags inside cartons resulted in a weight loss of only 1.3% in 5 days, but all the fruits degreened more rapidly than those which were packed just in cartons, where the weight loss was 13.8% (Thompson *et al.* 1974d). However, this degreening effect could be countered by including an ethylene absorbent in the bags (see Figure 12.68). Scott and Wills (1974) described experiments in which compounds that absorb ethylene, CO_2 and water were

Table 5.7 Commercially available films in Japan (source: Abe 1990)

Trade name	Manufacturer	Compound	Application
FH film	Thermo	Ohya stone/PE	Broccoli
BF film	BF Distribution Research	Coral sand/PE	Home use
Shupack V	Asahi Kasei	Synthetic zeolite	–
Uniace	Idemitsu Petro chemical	Silica gel mineral	Sweetcorn
Nack fresh	Nippon Unicar	Cristobalite	broccoli and sweetcorn
Zeomic	Shinanen	Ag zeolite/PE	–

packed with cultivar William's Bon Chrétien (Bartlet) pears held in sealed polyethylene bags. After storage at –1°C for 18 weeks in air or in bags with a CO_2 absorbent (calcium hydroxide) the fruit was in poor condition externally but free from brown heart. All other fruit was stored in the presence of about 5% CO_2 and was externally in excellent condition but was affected by brown heart. The ethylene absorbent potassium permanganate reduced the mean level of ethylene from 395–1.5 μg l^{-1} and reduced brown heart from 68–36%. The presence of calcium chloride tended to increase brown heart. The storage treatments also affected the levels of six other organic volatiles in the fruit.

Proprietary products such as Ethysorb, Purafil and Sorbilen are available which can be placed inside the crop store or even inside the actual package containing the crop. They are made by impregnating an active alumina carrier (Al_2O_3) with a saturated solution of potassium permanganate and then drying it. The carrier is usually formed into small granules; the smaller the granules the larger is the surface area and therefore the quicker their absorbing characteristics. Any molecule of ethylene in the package atmosphere which comes into contact with the granule will be oxidized so a larger surface area is an advantage. The oxidizing reaction is not reversible and the granules change colour from purple to brown, which indicates that they need replacing. Strop (1992) studied the effects of Ethysorb sachets in polyethylene film bags containing broccoli. She found that the ethylene content in the bags after 10 days at 0°C was 0.423 μg $litre^{-1}$ for those without Ethysorb and 0.198 μg $litre^{-1}$ for those with. The rate of ethylene removal from packages using this material is affected by humidity (Lidster *et al.* 1985). At the high humidities found in modified atmosphere packages the rate of ethylene removal of potassium permanganate was shown to be reduced. Investigations were conducted to study the effect of the ethylene absorber Sorbilen on the storage quality of cabbages, carrots, cucumbers, tomatoes, grapes, cut roses and carnations. The inclusion of Sorbilen decreased the storage decay of cabbages, carrots, cucumbers, tomatoes and grapes and the accumulation of the mycotoxic patulin, and contributed to better preservation of ascorbic acid and sugars in all the products (Lavrik *et al.* 2000).

CO_2 absorbents or a permeable window can be used with modified atmosphere packaging to prevent atmospheres developing which could damage the crop. The concentration of the various respiratory gases, O_2, CO_2 and ethylene, within the crop is governed by various factors including the gas exchange equation that is a modification of Fick's law:

$$-ds/dt = (C_{in} - C_{out})DR$$

where

$-ds/dt$ = rate of gas transport out of the crop
C_{in} = concentration of gas within the crop
C_{out} = concentration of gas outside the crop
D = gaseous diffusion coefficient in air
R = a constant specific to that particular crop.

Companies are marketing films for which they claim can greatly increase the storage life of packed fruit and vegetables. These are marketed with trade names such as 'Maxifresh,' 'Gelpack,' and 'Xtend.'

When an ethylene absorbent (potassium permanganate) was included in the tube the increase in storage life was three to four times compared with unwrapped fruit. Fruits sealed in polyethylene tubes with ethylene absorbents could be stored for 6 weeks at 20°C or 28°C and 16 weeks at 13°C (Satyan *et al.* 1992).

The investigations on the effects of modified atmosphere conditions by Ali Azizan (1988) (quoted by Abdullah and Pantastico 1990) showed that the total soluble solids, total titratable acidity and pH in Pisang Mas bananas did not change during storage in modified atmospheres at ambient temperature. However, Wills (1990) mentioned that an unsuitable selection of packaging material can still accelerate the ripening of fruits or enhance CO_2 injury when the ethylene accumulates over a certain period.

Adjustable diffusion leaks

A simple method of controlled atmosphere storage was developed for strawberries (Anon 1920). The gaseous atmospheres containing reduced amounts of O_2 and moderate amounts of CO_2 were obtained by keeping the fruit in a closed vessel fitted with an adjustable diffusion leak.

Marcellin (1973) described the use for storing vegetables of polyethylene bags with silicone rubber panels which allow a certain amount of gas exchange. Good atmosphere control within the bags was obtained for globe artichokes and asparagus at 0°C and green peppers at 12–13°C when the optimum size of the bag and of the silicon gas-exchange panel had been determined. Gas exchange for carrots and turnips at 1–2°C was hampered by condensation on the inner surface of the panel. The storage life of all the vegetables was extended compared with storage in air, but losses due to rotting were high for green peppers that seemed unsuitable for this type of storage. Rukavishnikov *et al.* (1984) described 'controlled atmosphere storage' under a 150–300 μm polyethylene film, with windows of membranes selective for O_2 and CO_2 permeability made of polyvinyltrimethylsilane or silicon–organic polymers. Apples and pears stored at 1–4°C under the covers had 93 and 94% sound fruit, respectively, after 6–7 months, whereas with ordinary storage at similar temperatures over 50% losses occurred after 5 months.

Hong *et al.* (1983) described the storage behaviour of fruits of *Citrus unshiu* at 3–8°C in plastic films with or without a silicone window. After 110–115 days, 80.8–81.9% of fruits stored with the silicone window were healthy with good coloration of the peel and excellent flavour, whereas for those stored without a silicone window only 59.4–76.8% were healthy with poor coloration of the peel and poor flavour. Controls stored for 90 days had 67% healthy fruit with poor quality and shrivelling of the peel, calyx browning and a high rate of moisture loss. The size of the silicone window was 20–25 $cm^2\ kg^{-1}$ fruit giving less than 3% CO_2 and less than 10% O_2.

The suitability of the silicone membrane system for controlled atmosphere storage of Winter Green cabbage was studied by Gariepy *et al.* (1984) using small experimental chambers. Three different controlled atmosphere starting techniques were then evaluated using a cabbage cultivar imported from Florida. In each case, there was a control with the product stored under a regular atmosphere at the same temperature. The silicone membrane system maintained controlled atmosphere storage conditions of 3.5–5% CO_2 and 1.5–3% O_2, where as 5% and 3%, respectively, were expected. After 198 days of storage, the total mass loss was 14% under controlled atmosphere storage compared with 40% under a regular atmosphere. The three methods used to achieve the controlled atmosphere storage conditions did not have any significant effect on the storability of cabbage, but all maintained 3–4% CO_2 and 2–3% O_2, whereas 5% and 3%, respectively, were expected. In both experiments, cabbage stored under controlled atmosphere storage showed better retention of colour, fresher appearance, firmer texture and lower total mass losses than those stored under a regular atmosphere.

Raghavan *et al.* (1982) used a silicone membrane system for storing carrots, celery, rutabagas and cabbage. Carrots were stored for 52 weeks and celery, rutabagas and cabbage for 16 weeks. The required higher CO_2 levels were achieved within the membrane system compared with normal air composition in a standard cold room. Design calculations for selection of the membrane area are also presented in the paper.

Conclusions

Modified atmosphere packaging of fruit and vegetables has been used commercially for decades and has been the subject of enormous quantities of research. However, it has never quite achieved what appeared to be its full potential. In theory it should be possible to replace the expensive controlled atmosphere storage techniques used increasingly by the industry with modified atmosphere packaging. The reason why it has not been done is related to the control that is necessary to achieve optimum results. With modified atmosphere packaging the levels or gases cannot be controlled precisely and also where there are temperature fluctuations this can have major effects on the internal gas content. It can therefore give unpredictable results that are not acceptable in modern marketing and storage situations.

6
Postharvest treatments

Introduction

The storage or marketable life of crops can be extended by various treatments applied to them after harvesting. The most important of these is temperature management, including the cold chain where the temperature of the crop is reduced rapidly directly after harvesting to stabilize the crop, and then it is maintained under these conditions until it reaches the consumer. This is dealt with in Chapter 4. Exposing crops to brief periods of high or low temperature after harvest can also have beneficial effects, particularly in pest and disease control. A variety of chemicals are applied to crops after harvest to control diseases, delay or prevent sprouting or affect the crop's metabolism. Strict legislation controls their use to protect the consumer. Waxes and other coating materials are applied to crops, but again their use is controlled by law. Crops may be exposed to certain gases such as ethylene or in other circumstances precautions have to be taken to protect the crops from undesirable gases.

Minerals

During growth, the crop absorbs mineral nutrients from the soil. The amount taken up depends on various factors, including the amount available in the soil. If a certain nutrient is available in low quantities it can lead to deficiencies in the crop, which in turn can result in a loss of storage life or quality. Nutrient imbalance can be affected by other nutrients. If there is an excess amount of one nutrient in the soil it may be taken up in preference to another, causing imbalance within the crop. It is obviously better to grow the crop on soils that are not deficient in that particular nutrient, but this is not always possible. The usual solution to this problem is to apply the nutrient to the growing crop either as a soil fertilizer or a foliar spray, but postharvest application can also be beneficial. The postharvest application of calcium can be used for apples to reduce the development of physiological disorders occurring during storage, particularly bitter pit (Sharples *et al.* 1979). Postharvest application of calcium has also been shown to inhibit ripening of tomatoes (Wills and Tirmazi 1979), avocados (Wills and Tirmazi 1982) and mangoes (Wills and Tirmazi 1981). In these three cases the calcium was applied by vacuum infiltration (250 mmHg) with calcium chloride at concentrations within the range 1–4%; higher concentrations could result in skin damage. These treatments were found to delay ripening significantly without affecting fruit quality. Vacuum (32 kPa) and pressure infiltration (115 kPa) of Kensington Pride mangoes with 2–8% calcium chloride solution resulted in delayed softening of fruit during storage at 20°C of 8–12 days compared with untreated fruit (Yuen *et al.* 1993). However, peel colour remained partially green when fruits were ripe and some peel injury occurred in treated fruit. Cling film (19 μm thick), shrink film (17 μm thick) or polyethylene bags (50 μm thick) appeared to be as effective as calcium infiltration in delaying ripening without peel injury and undesirable retention of excessive green colour in the ripe fruit (Yuen *et al.* 1993). Drake and Spayd (1983) pressure infiltrated

Golden Delicious apples with calcium chloride (3% at 3 pounds per square inch for 8 minutes) before storage for 5 months at 1°C. These fruit were firmer and more acid than untreated fruit stored for the same period. Calcium infiltration was shown reduce chilling injury and increase disease resistance in stored fruit (Yuen 1993). The addition of lecithin (phosphatidylcholine) to the postharvest application of calcium can enhance its effect in controlling bitter pit in apples (Sharples *et al.* 1979). At 18°C apple fruits treated with calcium (4% calcium chloride) had a slightly lower respiration rate, but this effect was greatly enhance when lecithin (1%) was added. At 3°C lecithin-treated apples had reduced ethylene production but there was no effect on carbon dioxide production (Watkins *et al.* 1982). Duque and Arrabaca (1999) found evidence that suggested that the role of postharvest calcium in fruit ripening is not related to respiration.

Astringency removal

Astringency occurs in many fruit owing to the presence of tannins. The latter can impart an unpleasant flavour to the fruit and are often associated with immature fruit. In bananas tannins polymerize as the fruit ripens and they lose their astringency (Von Loesecke 1949). Treatment of persimmon with high levels of carbon dioxide can be used to reduce astringency. Storage of persimmons in 4% CO_2 at –1°C for about 2 weeks before removal from storage followed by 6–18 hours in 90% carbon dioxide at 17°C removed astringency from fruits (Hardenburg *et al.* 1990). Placing fruit in a PVC film tent containing 80–90% CO_2 for 24 hours reduced astringency (Kitagawa and Glucina 1984). Constant-temperature, short-duration treatment (20–25°C and 90–95% carbon dioxide) of the cultivar Hiratanenashi showed that astringency disappeared 3–4 days after removal (Matsuo *et al.* 1976). Treating persimmon fruits with alcohol has been known for over 100 years in Japan to reduce their astringency . Spraying fruit with 35–40% ethanol at the rate of 150–200 ml per 15 kg of fruit was shown to have removed astringency 10 days later in ambient storage. This treatment took longer to remove the astringency than the carbon dioxide treatment, but the fruit were considered of better quality (Kitagawa and Glucina 1984).

Antioxidants

Storage disorders such as scald in apples can be controlled by a pre-storage treatment with an antioxidant. Ethoxyquin (1,2-dihydro-2,2,4-trimethylquinolin-6-yl ether) marketed as 'Stop-Scald,' or DPA (diphenylamine), marketed as 'No Scald' or 'Coraza,' should be applied directly to the fruit within a week of harvesting. Sive and Resnizky (1989) described the successful application of DPA and ethoxyquin to Granny Smith, Delicious and Golden Delicious apples in cold stores, by thermal fogging and thermonebulization. Different cultivars of apples respond differently to these treatments; for example, only DPA is effective in controlling scald on the Delicious cultivars. In the USA the Government approved the postharvest application of these two chemicals to apples with maximum residues of 3 μg $litre^{-1}$ for ethoxyquin and 10 μg $litre^{-1}$ for DPA (Hardenburg and Anderson 1962). Residue levels of DPA in apples were found to vary depending on the application method and the position of the fruit in the pallet box (Table 6.1). There are restrictions on the sale of treated fruits in several countries and local legislation should be consulted before they are applied.

Sprout suppressants

Many plants produce natural vegetative storage organs. These may be the edible parts of the crop that are harvested and stored or marketed. They are often natural organs of perennation that have a dormant period and then regrow. Botanically they may be root tubers, stem tubers, corms, bulbs, swollen roots or a combination of more than one structure. As natural storage organs, many of them can be stored for protracted periods of time. The limiting factor in storage is usually when the dormancy period ends and the

Table 6.1 Residues of DPA in Granny Smith apples 1–2 months after treatment (source: Chapon, *et al.* 1987, quoted by Papadopoulou-Mourkidou 1991)

Method of application	Position of apples in the pallet box	(ppm)
Drenching	Top	1.25
	Middle	1.30
	Bottom	2.15
Thermofogging	Top	4.25
	Middle	4.65
	Bottom	9.60

organ regrows. Even after the dormancy period has ended the crop might not sprout if the conditions, especially temperature, are not conducive.

Sprout suppression in onion bulbs is achieved by preharvest application with maleic hydrazide. This is applied to the leaves of the crop about 2 weeks before harvesting so that the chemical can be translocated into the middle of the bulb in the meristematic tissue where sprouting is initiated.

Sprouting of potatoes is suppressed at 5°C and below, and in yams no sprouting was observed during 5 months of storage at 13°C, but tubers sprouted during that period at temperatures of 15°C and above. In potatoes the natural dormancy period varies between cultivars. With some dormant storage organs sprouting can be suppressed to extend the storage life by applying growth-regulating chemicals to the crop. In bulbs, such as onions, this is not possible because the meristematic region where sprouting occurs is deep inside the bulb and difficult to treat with chemicals postharvest. In potatoes a range of chemicals have been shown to suppress sprouting (Burton 1989). The most commonly used is CIPC, also called chloropropham, which is vaporized and applied as a thermal fog to the stored crop. A granular formulation is also available for application to tubers as they are loaded into store. This can impede the curing process and result in increased levels of deep skin spot infections. IPC (propham) is also used either by itself or in a formulation with CIPC.

CIPC residue concentrations (MRL) from the Environmental Protection Agency in the USA were revised in 1996 to 30 mg kg^{-1} (Kleinkopf *et al.* 1997). Lentza-Rizos and Balokas (2001) found that the concentrations of CIPC in individual potato tubers were in the following ranges:

- 1.8–7.6 mg kg^{-1} 10 days post-application (mean 3.8 mg kg^{-1})
- 0.7–4.0 mg kg^{-1} 28 days post-application (mean 2.9 mg kg^{-1})
- 0.8–3.8 mg kg^{-1} 65 days post-application (mean 2.2 mg kg^{-1})

and in composite samples:

- 4.3–6.1 mg kg^{-1} 10 days post-application (mean 4.9 mg kg^{-1})
- 3.1–4.2 mg kg^{-1} 28 days post-application (mean 3.8 mg kg^{-1})
- 2.6–3.2 mg kg^{-1} 65 days post-application (mean 2.9 mg kg^{-1}).

Peeling removed 91–98% of the total residue; washing reduced residues by 33–47%. Detectable residues were found in boiled potatoes and the boiling water, and in French fries and the frying oil (Lentza-Rizos and Balokas 2001).

Sprout suppression can also be achieved by irradiating crops such as onions, potatoes and yam tubers. Although the law may permit this, there is consumer resistance and economic factors against its use and it is rarely applied commercially.

Fruit coating

Fruit are dipped or sprayed with a range of materials postharvest to improve their appearance or delay deterioration. A USA patent was taken out on a product (Ukai *et al.* 1976) that could be used to coat fruit, which upon drying left a membrane on the surface that would controllably suppress their respiration rate. The end effect was to extent the storage life of the fruit. The composition of the coating material was a water-soluble high polymer such as a polysaccharide and a hydrophobic substance selected from a group consisting of hydrophobic solids and hydrophobic and non-volatile liquids such as natural waxes. After application to fruit, the dried coating consisted of fine continuous micovoids (Ukai *et al.* 1976).

Tal Prolong

Banks (1984) described a coating called Tal Prolong, which consisted of sucrose esters of fatty acids and carboxymethylcellulose. It was claimed that it could delay the ripening of bananas. He postulated that the effect was due to the restriction of gas exchange between the fruit and its surrounding atmosphere. This led to a build-up of CO_2 and a depletion of O_2, thus causing an effect achieved in controlled atmosphere storage. Bancroft (1989) showed some reduction in levels of disease on stored apples coated with Tal Prolong.

Semperfresh

Suhaila *et al.* (1992) found that guavas coated with a material similar to Tal Prolong, called Semperfresh, and stored for 2 months at 10°C were in better condition than untreated fruit, but coating fruit with palm oil was more effective, and cheaper, than Semperfresh. Banana crowns coated with Semperfresh also showed delayed development of crown rot caused by infection with *Colletotrichium musae*. However, when fruits were inoculated with *C. musae* after harvest the results were inconsistent (Thompson *et al.* 1992). The control of crown rot could be enhanced by the inclusion of organic acids in the coating material (Al-Zaemey *et al.* 1993). Kerbel *et al.* (1989) showed that for Granny Smith apples coated with Semperfresh (0.75% and 1.5%) and stored at 0°C for 4 or 5 months ripening was delayed 'somewhat' compared with fruit with no coating. The treatments did not reduce levels of bitter pit or superficial scald (Kerbel *et al.* 1989). The company who manufactured and marketed Semperfresh was Surface Systems International. In 1992 and 1993 they brought out a range of different formulations for different purposes. Ban-seel was a liquid formulation to protect and extend the storage life of bananas and plantains. Nu-coat Flo was for citrus fruits. Brilloshine C was to protect, shine and extend the shelf-life of citrus fruit. Brilloshine L was to protect, shine and extend the shelf-life of apples, avocados, melons and other non-citrus fruit. Surface Systems International has also developed a formulation that could be applied to potatoes to control sprouting during storage.

Chitosan

N,O-Carboxymethylchitosan can be used to produce films for coating fruit that are selectively permeable to gases such as CO_2, O_2 and ethylene (Hayes 1986; Davies *et al.* 1988). Golden Delicious apples coated with this chemical (marketed as Nutri-Save) were kept in storage at 0°C for 6 months and were of superior quality to those stored without coating (Davies *et al.* 1988). Papaya fruits coated in 1.5% Nutri-Save showed no significant difference in weight loss, abnormal browning and limited colour development compared with untreated fruits during storage at 16°C (Maharaj and Sankat 1990). All the fruits were of the cultivar Tainung Number 1 and were harvested at the colour break stage and placed in hot water at 48°C for 20 minutes and benomyl (1.5 g $litre^{-1}$) at 52°C for 2 minutes.

Chitosan has been successfully used as a postharvest coating of peppers and cucumbers to reduce water loss and maintain their quality (El-Ghaouth *et al.* 1991). Chitosan at either 5 or 10 mg ml^{-1} reduced the incidence of brown rot (caused by *Monilinia fructicola*) on peaches. Chitosan-treated peaches were also firmer and had higher titratable acidity and ascorbic acid content than control peaches (Li and Yu 2001). Jiang and Li (2001) suggested that the application of a chitosan coating to longans could extend their postharvest life, maintaining quality and, to some extent, controlling decay

In Granny Smith apples which had been coated with either 1.5% Semperfresh or 1.5% Nutri-Save, Kader (1988) found that the internal carbon dioxide level was lower and the internal oxygen level was higher in the Semperfresh-coated fruit than those which had been coated with Nutri-Save. He also showed that coating with Nutri-Save reduced ethylene production and fruit respiration rate to a larger degree than Semperfresh coating. From these results he concluded that Nutri-Save at 1.5% forms a surface barrier on the fruit that was less permeable to these gases than 1.5% Semperfresh.

Vapor Gard

Another fruit coating, marketed as 'Vapor Gard', is an anti-transpirant. Its effects on the postharvest life of Harumanis mangoes were studied by Lazan *et al.* (1990). The effects of coating the fruit in a 1.3% solution was to reduce water loss, retard firmness decrease, reduce the decrease in ascorbic acid content, inhibit malic enzyme activity and increase polygalacturonase activity compared with untreated fruit.

Biofresh

Biofresh is also a sucrose fatty acid ester. Elstar and Shampion apples were dipped in 1% Biofresh for 20 seconds and kept in cold storage for 3 months by Xuan and Streif (2000). The dipped fruits were firmer, had reduced weight loss and ethylene production, retained titratable acidity and had a lower respiration rate than the untreated fruits.

Trehalose

A sugar which has been tested as a coating material for food preservation is trehalose (Roser and Colaco 1993; A. H. Abboudi, unpublished results 1993). Trehalose is a disaccharide made up of two glucose molecules that are linked by their reducing sugar carbon. This makes trehalose very stable reducing sugar that is chemically almost inert and biologically non-toxic. These properties make it a useful preservative for biological molecules and food. The use of trehalose in preservation of dehydrated material is covered by European Patent 0 297 887 B1 (22 April 1992).

Antiethylene

Ethylene is synthesized in plant material. In climacteric fruit copious amounts may be produced. The amount of ethylene synthesized can also depend on harvest maturity, handling method, storage temperature and gaseous content around the fruit or vegetables (see also Chapter 7). Exposure of harvested plant material to 1-MCP (1-methylcyclopropene) has been shown to reduce ethylene production of fruits, vegetables and flowers. De Wild (2001) even suggested that short-term controlled atmosphere storage might be replaced by 1-MCP treatment.

1-MCP is available as a commercial preparation called Ethylbloc powder. Ethylene production, respiration and colour changes in Fuji apples were all inhibited following 1-MCP treatment at 0.45 mmol m^{-3} (Fan and Mattheis 1999). Watkins *et al.* (2000) showed it reduced superficial scald incidence, and accumulations of α-farnesene and conjugated trienols during air storage, but that its efficacy on stored apples was affected by cultivar and storage conditions. Rupasinghe *et al.* (2000) also showed that the contents of α-farnesene and its putative superficial scald-causing catabolite, conjugated triene alcohol, in the skin were reduced 60–98% by 1-MCP and the incidence of superficial scald suppressed by 30% in McIntosh and 90% in Delicious. The threshold concentration of 1-MCP inhibiting *de novo* ethylene production and action was 1 ml $litre^{-1}$. 1-MCP-treated McIntosh and Delicious apples were significantly firmer than untreated apples following cold storage and post-storage exposure to 20°C for 7–14 days (Rupasinghe *et al.* 2000). De Wild (2001) made a single application of 1-MCP directly after harvesting to Jonagold and Golden Delicious apples, followed by 6 months of storage at 1°C under controlled atmosphere conditions of 1.2% O_2 and 4% CO_2 or cold storage at 1°C followed by a shelf-life of 7 days at 18°C. No loss of firmness was found after the use of 1-MCP, and yellow discoloration of Jonagold decreased. Cold-stored apples showed greater firmness loss than controlled atmosphere stored apples, unless treated with 1-MCP. Treated apples did not show firmness losses during shelf-life at 18°C.

With Cavendish bananas harvested at the mature-green stage, treatment with 1-MCP delayed peel colour change and fruit softening, extended shelf-life and reduced respiration rate and ethylene production (Jiang *et al.* 1999). Ripening was delayed when fruits were exposed to 0.01–1.0 μl l^{-1} 1-MCP for 24 hours and increasing concentrations of 1-MCP were generally more effective for longer periods of time. Similar results were obtained with fruits sealed in 0.03 mm thick polyethylene bags containing 1-MCP at either 0.5 or 1.0 ml l^{-1}, but delays in ripening were longer at about 58 days.

Ku and Wills (1999) showed that 1-MCP markedly extended the storage life of Green Belt broccoli through a delay in the onset of yellowing during storage at both 20 and 5°C and in development of rotting at 5°C. The beneficial effects at both temperatures were dependent on the concentration of 1-MCP and treatment time.

Salicylic acid

Salicylic acid 2-hydroxybenzoic acid, [$C_6H_4(OH)COOH$] has been used for medical purposes for over a century. It is extracted from the willow (*Salix* spp.) and also occurs in the fruit of *Ribes*, *Fragaria* and *Rubus* (Hulme 1970). It has been investigated for its effects on plant material. For example, immersing peach fruit in 0.10 g l^{-1} salicylic acid and storage at room temperature resulted in a delay in the peak of ethylene production, a reduction in electrolyte leakage and an initial reduction in polyphenol oxidase activity compared with untreated fruits (Han *et al.* 2000). Despite its importance in medicine it is not clear whether it would ever be approved for postharvest application to fruit and vegetables.

Curing

Many root crops have a cork layer over the surface that is called a periderm. This serves as a protection against microorganism infections and excessive water loss. This layer can be broken or damaged during harvesting and handling operations so curing is essentially a wound-healing operation to replace the damaged periderm and is achieved by exposure to an appropriate temperature at high humidity for a period of time. Curing is applied to fruits, especially citrus fruits. The mechanism is different to that root for crops but it effectively heals wounds and reduces disease levels. Specific conditions are given under each crop in Chapter 12. Drying is also carried out to aid preservation of bulb crops such as bulb onions and garlic. This does not involve drying the crop to an even low moisture content as in dehydration, but only drying the outer layers.

Lignification of injuries is an important component of defence against postharvest diseases. The effect of elicitor treatment on accumulation of lignin at 30°C was evaluated using storage tissues of four plants. Root tissues of daikon (radishes), turnip and sweet potato all exhibited increased lignification in response to elicitation with pectinase, chitosan or a yeast extract. Acorn squash (*Cucurbita pepo*) tissue responded only to pectinase. Squash tissue elicited and then held at 28°C for 18 hours developed considerable resistance to infection by *Penicillium italicum*. The lignification process was complete within 24 hours on all four plant tissues (Stange and McDonald 1999).

Hot water treatment

Crops may be immersed in hot water or brushed with hot water before storage or marketing to control diseases. The principal of this treatment is that the disease fungi are actively growing on fruit. After harvest the fruit are not growing. Cells that are actively dividing are more susceptible to damage and therefore it is possible to find a time/temperature regime that will kill the fungi and thus control the disease without damaging the fruit. It may also be necessary to include a fungicide in the water to achieve complete control. Additionally it has been used on fruit to control insect infestation. Hot water treatment has been applied to various crops, including mangoes, papaya, citrus fruits, blueberries and sweet potatoes. Details are included under individual crops in Chapter 12.

Vapour heat treatment

This treatment was developed to control infections of fruit flies in fruit. It consists of stacking the boxes of fruit in a room that is heated and humidified by the injection of steam. The exposure time is judged by placing a temperature probe in the centre of the fruit or alongside the seed. The temperature and exposure time are adjusted to kill all stages of the insects (eggs, larvae, pupa and adult) but not damage the fruit. The most difficult stage to control by both vapour heat treatment and hot water treatment is the larval stage, because it tends to be further from the fruit surface and thus exposed to high temperatures for shorter periods (Jacobi *et al.* 1993). Hydrothermic quarantine treatment at 46.5°C for 90 minutes for Tommy Atkins and Kent mangoes was described by Lizana and Ochagavia (1997) for fruit exported from Chile to the USA. Vapour heat treatment can cause injury to some fruits. McGuire (1991), Esquerra and Lizada (1991) and Coates *et al.* (1993) reported that vapour heat treatment can also be used to control fungal diseases.

Degreening

The colour of many fruits is governed by the presence in the skin of carotenoids and xanthophylls, which give the fruit orange or yellow colours, and chlorophylls, which give them a green colour. The change from green to yellow or orange, associated with maturation and ripening of many fruit, may involve pigment synthesis but in many cases it is simply the breakdown of chlorophyll that has been masking the orange or yellow pigments (Seymour 1986). Oranges are often degreened after harvest by exposing them to ethylene gas under controlled temperature and humidity. Application of ethylene may be by an initial injection of the correct level or by continuously trickling it into the room. For citrus fruits there are various recommendations, e.g. those by Winston (1955), Eaks (1977) and McCornack and Wardowski (1977), which are given in detail in Chapter 12.

7
Storage

Store management and organization

Organization of storage varies with the scale of the operation. On-farm crop storage is frequently carried out. With small farmers the quantities stored are often very small and therefore sophisticated facilities may not be appropriate. In such cases, if crops need to be stored it may be more economic for groups of farmers to form a storage cooperative where the group combine to purchase and run central storage facilities. This is particularly important where complex systems are used such as controlled atmosphere stores. The expensive electronic equipment that is used to sample and regulate the store atmosphere is a relatively fixed cost for small or large stores. The facilities and organization of such an enterprise can allow for purchasing and marketing at more favourable rates. This centralized type of organization would require separate and specialized management.

Storage of fruits and vegetables is practised for various reasons. It is part of orderly marketing, where the storage period is usually short, to allow for accumulation of sufficient produce by a grower or group of growers to send to market. It may be stored in wholesale markets during the period when it is being sold. Also it may be stored when the price at a particular time is low to await an increase in price. Certain crops are stored for long periods of time to extend the duration of their availability. In this latter case the crop is grown purposely for long-term storage, and even specific cultivars are grown. However, long-term storage can be expensive and require a high level of technical knowledge of the crop. The factors which need to be taken into account before embarking on crop storage are:

- knowledge of the appropriate storage conditions
- cultivar or variety of crop suitable for storage
- appropriate storage facilities available
- suitable management available.

The cost of storage can be very high and the following costs need to be taken into account:

- erection of storage structure
- maintenance of structure
- cost of loading and unloading crop
- possible cost of electricity
- possible cost of storage containers
- depreciation on capital.

Fruits and vegetables are living organisms. Their condition and marketable life are affected by such things as temperature, humidity, the composition of the atmosphere which surrounds them, the level of damage that has been inflicted on them and the type and degree of infection with microorganisms. They will deteriorate during storage through:

- loss of moisture
- loss of stored energy, e.g. carbohydrates
- loss of other foods, e.g. vitamins
- physical losses through pest and disease attack
- loss in quality from physiological disorders.

All these factors should be taken into account before a crop storage enterprise is undertaken. The alternatives to storage should also be examined. These alternatives may include:

- importing the crop from another region where it can be grown off-season
- growing a different cultivar which may produce a crop out of season
- changing the method of cultivation to extend the production season or convincing the consumer to eat a crop that is in season.

If finally it is decided to go ahead with storage, the crop must increase in price during the storage period to offset the costs (Figure 7.1). In many cases the retail price of a crop is relatively stable throughout the year. In a study in Zambia (Thompson 1971b), it was shown that the price of potatoes fluctuated by less than 3% in the Copperbelt and by about 7% in Lusaka, and onions by about 7% in both places. In a study of onions, the break-even price by month of controlled atmosphere storage decreased as the capacity level of the storage facility increased and as the rate of interest decreased (Hancock and Epperson 1990). If consumers acculated a large inventory of onions or the demand was truly seasonal, the latent demand for onions over the technically feasible storage period may be insufficient to cover storage costs.

The situation may be complicated in many situations in less dveloped countries by the need to preserve foreign exchange and perhaps provide work for labour in the consuming country. Such factors may override purely economic considerations and crops may be stored which could be imported more economically. This principle must hold true in purely commercial situations whatever type of storage is contemplated. In many market situations crops will be stored when their supply is in excess of demand. If such an enterprise is undertaken there must be a reasonable assumption that the market price will rise to pay for storage costs, but with many crops, especially in less developed countries, the stored crop may be competing with freshly harvested crop which may enjoy a premium market price. There are cases where a certain amount of a crop will be stored to restrict the market supply, and subsequently released over a period of time. These subsequent prices may not be higher but the result may have overall economic advantages. This presupposes that the organization carrying out storage is on a sufficient scale to influence prices, as in the case of a national marketing board.

In some cases fruit and vegetables destined to be processed are first stored to even out the supply to the factory. Often the objective is to provide a constant supply of raw material for the factory for as long a time as possible because labour force and machinery are fixed costs. Often this type of storage is different from storage of crops destined for the fresh market. This can be true both technically and from the point of view of the economic factors which are used to justify storage. Technically the crop may be stored at temperatures which would cause chilling injury when removed to higher temperatures for the fresh market. This is because crops for processing are usually processed directly from the store. If a crop is to be processed by dehydration it may not be a disadvantage for it to lose water during storage. If this is the case, the crop for processing may be stored at low relative humidity. From the economic aspect, the processed crop may have a market all the year round but the harvest season may be only a few weeks. The crop that is processed directly after harvest may be very profitable whereas that which is processed after long periods of storage may make a loss. However, on balance the enterprise is profitable and the processor can supply the market and therefore the processor retains market share.

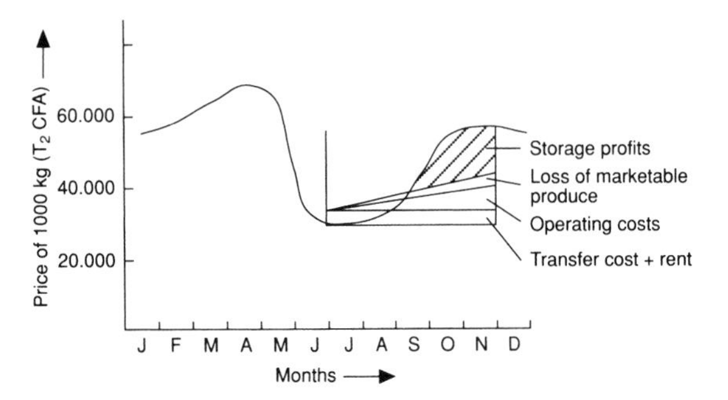

Figure 7.1 Prices of Irish potatoes at Kumbo market in the Cameroon and storage costs and profits of one storage season per year. Source: Toet (1982).

Store design and method

Human beings have stored crops for centuries and techniques have been developed in different societies involving varying methods and structures, some of which survive today. Stores can be grouped into those that do not require refrigeration, which are dealt with in this section, and those that do, which are dealt with later in the chapter and under separate crop headings in Chapter 12. Types of simple stores are given in the following.

In situ

In situ storage effectively means delaying the harvest of the crop until it is required. This can be done in certain cases with root crops, but it does mean that the land where the crop was grown remains occupied and a new crop cannot be planted there. The crop may also be exposed to pest and disease attack. In cassava it was shown that delayed harvest can result in reduced acceptability and starch content and the risk of pre-harvest losses (Rickard and Coursey 1981). In colder climates the crop may be exposed to the detrimental effects of freezing or chilling injury.

Burying

Root crops can be packed into material such as coir dust, peat (Figure 5.1) or even sawdust to protect them and provide a humid environment to preserve the crop (Thompson, *et al.* 1973). CIP (1992) compared the storage of root crops for 2 months in tropical ambient conditions buried in various materials. It was found that brick kiln soil gave lower storage losses (Table 7.1). In India, potatoes are traditionally stored buried in sand. Maintaining high humidity around fruit such as plantains can help to keep fruit in the preclimacteric stage so that where fruit were stored in moist coir dust they could remain green and preclimacteric for over 20 days in Jamaican ambient conditions (Thompson *et al.* 1972a).

Pits

Placing the crop in a pit or trench is a traditional way of storage used in many countries. The pit or trench is dug at the edges of the field where the crop has been grown. If the field slopes, it is important that the pits or trenches are placed at a high point in the field, especially in regions of high rainfall. The pit or trench may be lined with straw or other organic material, filled with the crop to be stored and covered by a layer of organic material and then a layer of soil. Sometimes wooden boards are placed on the soil surface before the soil is put on. The lack of ventilation may cause problems with rotting and ventilation trenches may be dug down to the base of the store. Ventilation holes may also be left at the top of the store covered with straw in such a way as to allow air to pass out but no rain to penetrate. The thickness of the soil covering will depend on the climate. In very cold conditions soil up to 25 cm may be required to protect the crop from frost. In very hot conditions a thick soil or organic material covering may also be necessary to prevent large temperature rises during the day. In European countries many root crops have been traditionally stored in this way, and even apples. In the Sudan potatoes are stored in pits with a heavy straw covering for insulation. Labour requirements for construction, loading and unloading are very high and storage periods are normally short, otherwise there are high crop losses. Storage of taro corms in pits lined with leaves or plastic film extended their storage life by about 4 weeks compared with the traditional method of storing them in heaps in the shade in the South Pacific islands (Gollifer and Booth 1973).

Table 7.1 The effects of packaging material on storage losses of potatoes after 2 months in ambient conditions. (source: CIP 1992)

Medium	Weight loss (%)[a]	Rotting loss (%)[a]
Sand	15.4b	20.0a
Brick-kiln soil	8.7c	8.6b
Dried soil	12.7b	9.8b
Wood ash	19.8a	6.2b
Rice husks	20.7a	18.0a

[a]Figures followed by the same letter were not significantly different ($p = 0.05$).

Clamps

These have been used as a traditional method of storing potatoes and as recently as 1963 some 48% of potatoes stored in Britain were stored in clamps (Figure 7.2). Since then the proportion has decreased

Figure 7.2 Traditional potato clamp. Source: photo library FAO Rome.

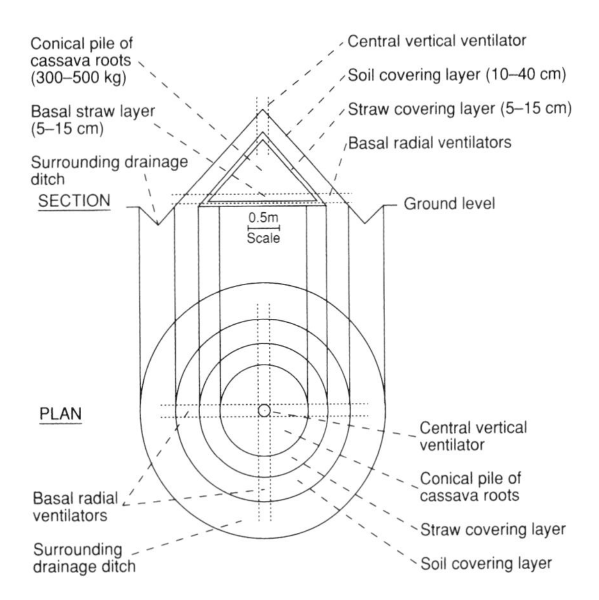

Figure 7.3 Design of cassava storage clamp. Source: Booth (1975).

rapidly and the method is uncommon today. Clamps were also developed for cassava storage in Colombia (Figure 7.3), where successful storage for over 8 weeks was reported (Booth 1977).

The design of potato clamps in Britain varied between regions and a common design was as follows. An area of land at the side of the field was selected that was not subject to water logging. The width of the clamp usually ranges from 1–2.5 m, with any suitable length. The dimensions are marked out and the potatoes piled on the ground in an elongated conical heap. Sometimes straw is laid on the soil before the potatoes. The central height of the heap of potatoes depends on their angle of repose, and is about one third its width. Straw is then laid on the heap of potatoes in overlapping layers starting at the bottom and working upwards. At the top straw is bent over the ridge so that rain will tend to run off the structure. The straw thickness should be 15–25 cm when compressed. After about 2 weeks the clamp is covered with soil to a depth of 15–20 cm, but this may vary depending on the climate. Potatoes stored in clamps in hot climates should be of 1.5 m maximum width with a ventilating duct at ground level down the centre (Figure 7.4). Extra straw covering may be needed in hot climates (CIP 1981). Cassava stored in clamps for 1 month have 75% of the original weight in marketable condition, with additional losses after up to 3 months of storage usually being small (Booth 1977).

Windbreaks

This is a traditional way of storing onions in Britain and farmers claim they can be stored in this way for up to 6 months. Windbreaks are constructed by driving wooden stakes into the ground in two parallel rows about 1 m apart. The stakes should be slightly sloping outwards and should be about 2 m high. A wooden platform is built between the stakes about 30 cm from the ground, often made from upturned wooden boxes. Chicken wire is fixed between the stakes and across the two ends of the windbreak. The onion bulbs are then loaded into the cage and covered with 15–20 cm of straw. On top of the straw a polyethylene film sheet (about 500 gauge) is placed and weighted down with stones to protect it from the rain (Figure 7.5). The windbreak should be sited with its longer axis at right-angles to the prevailing wind.

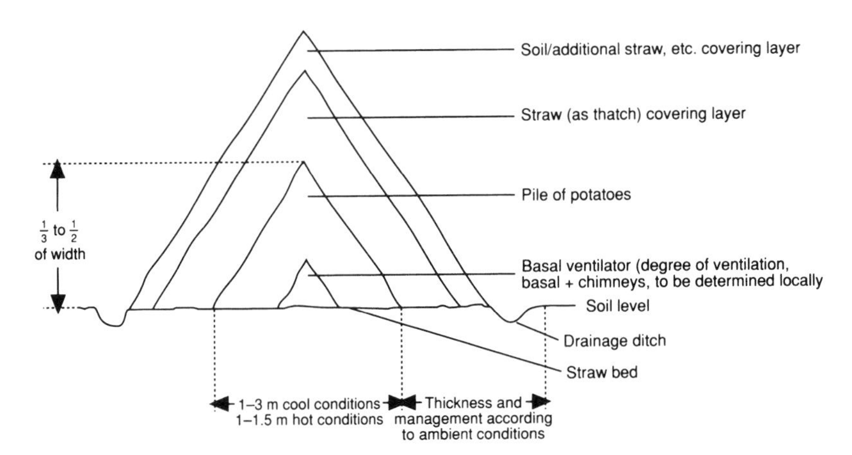

Figure 7.4 Details of the construction of a potato clamp. Source: CIP (1981).

Figure 7.5 Simple windbreak field onion store in the UK.

Figure 7.6 Yam barn in Nigeria showing the construction using live branches that produce leaves to provide a shaded humid environment.

Cellars

These are underground or partly underground rooms, often below a house. This location provides good insulation, which means they are cool in warm ambient conditions and protected from excessively low temperatures in cold conditions. In Britain they were used for storing crops such as apples, cabbages, onions and potatoes during the winter. The crops were usually spread out thinly on shelves to ensure good air circulation. Onions and garlic would be plaited into bunches and suspended from hooks, a method that is commonly used throughout the world. In the mountainous regions of Nepal cellars are dug into the sides of slopes for storing apples. In that area night temperatures are low and the hillside provides insulation to reduce the temperature rise during the day.

Barns

Farmers build various simple structures for crop storage. In Nigeria yams are stored in specially constructed barns (Figures 7.6 and 7.7). The traditional barn consists of a framework of raffia palm poles supported by erect stems of *Newbouldia laevis*, with the live plants giving a canopy to provide shade and shelter. The design is described in detail in Chapter 12 under white yam, *Dioscorea alata*. Yams are stored in slatted wooden trays in the coastal region of Cameroon (Lyonga 1985). Ezeh (1992) described a modified design for a yam barn. It was a rectangular building with walls made of wire mesh and supported by timber pillars $10.2 \times 10.2 \times 3.0$ m with a cement floor and the roof made of corrugated aluminium sheets; the ceiling was constructed with bamboo sticks and raffia mats. Results showed that investing in this improved storage barn in Nigeria was economically worthwhile for both long and short term. In the Sudan onions are stored in barns called 'rakubas' (see Table 12.41) that are made from either mud walls with a thatched roof (Figure 7.8) or palm leaves (Figure 7.9).

Evaporative coolers

When water evaporates from the liquid to the vapour phase, it requires energy. This principle can be used to cool stores by passing the air that is introduced into the store first through a pad of water. The degree of cooling depends on the original humidity of the air and the efficiency of the evaporating surface. If the

Figure 7.7 Detail of how yam tubers are tied to the upright poles in a Nigerian yam barn.

Figure 7.8 Traditional mud-walled and thatched onion store in the Sudan.

Figure 7.9 A traditional palm leaf store, called a rakuba, being used for onions in the Sudan. Source: Professor Sulafa Khalid Musa.

ambient air has low humidity and this is humidified to close to 100% r.h. then a large reduction in temperature can be achieved. This can provide cool, moist conditions in the store. Both active and passive evaporative cooling systems are used. In passive systems the cooling pads are placed over the entrances to the store and kept moist. These entrances should be in a position so that the prevailing wind can blow through them. In active stores air is drawn into the store by a fan through a moist pad. The pad (Figure 7.10) can be kept moist by constantly pumping water over it. This latter type is much more efficient in cooling but does require an electricity supply. A comprehensive review of different designs of evaporative coolers was given by Bautista (1990), including their use during road transport and on small boats.

In Nigeria, the shelf-life of carrots stored in a brick wall evaporative cooler was extended by 4–20 days relative to ambient storage (Babarinsa *et al.* 1997). In India, Fuglie *et al.* (2000) found that cold stores were the only viable alternative for long-term (5–8 months) storage of table and seed potatoes. However, evaporative cool stores could significantly reduce losses in farm stores for shorter-term storage, but their higher capital costs, compared with farmers' traditional methods, limit their adoption prospects among Indian potato farmers. The two evaporative cooler structures tested by Acedo (1997), one with jute sack walling and the other with a rice hull wall insert, maintained temperatures 1–6°C lower and relative humidities 10–20% higher than ambient.

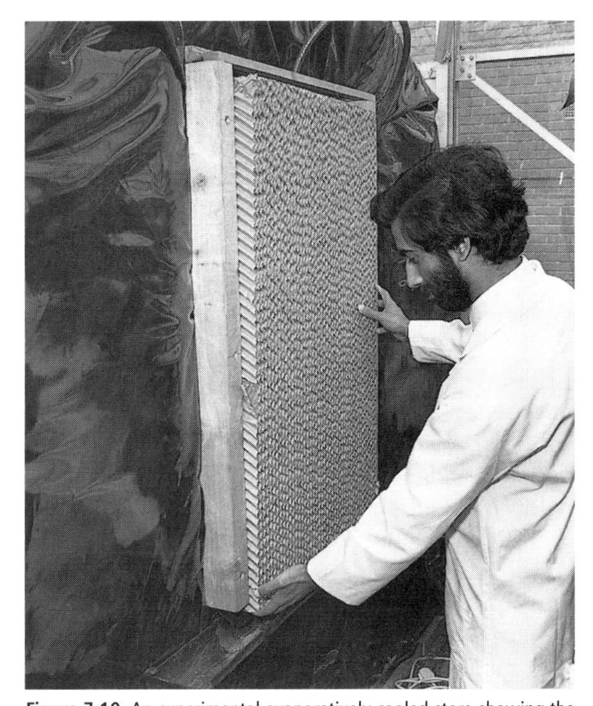

Figure 7.10 An experimental evaporatively cooled store showing the constantly moistened Celdec pad through which the air is passed to cool the insulated store. Source: Mr A.J. Hilton.

Night ventilation

In hot climates, the variation between day and night temperatures can be used to keep stores cool. The store should be well insulated and the crop is placed inside. A fan is built into the store that is switched on when the outside temperature is lower than the temperature within the store. This is at night and when the temperatures have equalized the fan is switched off. A differential thermostat that constantly compares the air temperature outside the store with the internal store temperature can control the fan. In a study on such a night-ventilated store, Skultab and Thompson (1992) showed that onions could be maintained at an even temperature very close to the minimum night temperature. The design that was used is shown in Figure 7.11. A time switch can also be used which would be set, for example, to switch the fan on for 2 hours each day between 5.0 and 7.0 a.m. An onion store was built and tested in the Sudan which was similar to that described by Skultab and Thompson (1992).

Passive night-ventilated stores have been designed which rely on opening the store doors at nights to admit the cold air and closing them during the day. Convective night-ventilated stores have been developed which use heated rocks to draw air currents through the store at night (Bishop 1992). These two methods require constant manual attention.

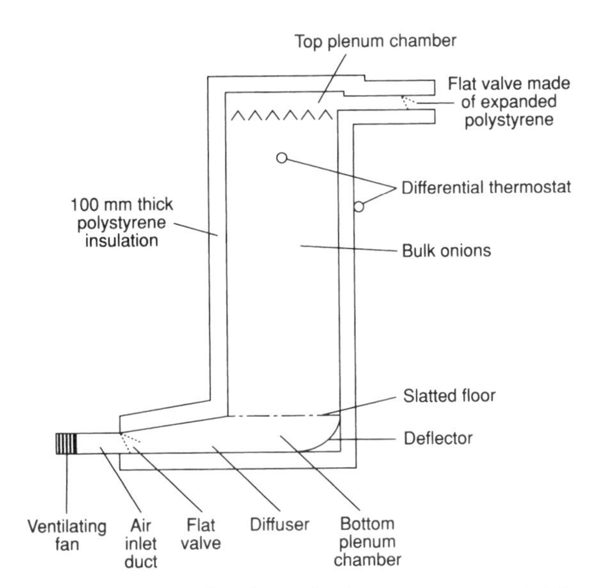

Figure 7.11 Diagram of a night ventilated onion store. Source: Skultab and Thompson (1992).

Refrigerated storage

Generally, the lower the storage temperature for fruit and vegetables down to their freezing point, the longer is the storage life. The increase in the rate of deterioration is related to the metabolic processes of the crop. Within the physiological temperature range of the crop the respiration rate increases exponentially with temperature, so that for every 10°C rise in temperature the increase in metabolism is of the order of two- to three-fold. This is called the Q_{10} value and is predicted by the van't Hoff rule. Many crops suffer from chilling injury at temperatures well above their freezing point, therefore it is essential to define clearly the temperature range over which the prediction applies. This varies not only between different crop species, but also with different varieties of the same species. In some crops, storage at or just above their freezing point can greatly extend their marketable life compared with storage at 0°C. This has been shown for crops such as pears and broccoli. In Japan, this low-temperature technology was developed in fruit and vegetable marketing to maintain a cold chain for produce of between 0 and –3°C right through the distribution network, even to the consumer's refrigerator (Anon 1988).

There is a relationship between temperature and respiration rates, but for some crops the relationship is not linear. Burton (1962) showed that for potatoes the respiration actually increased at lower temperatures (Figure 7.12). With onions, sprouting and therefore respiration of bulbs decreased with increasing temperature over the range 15–25°C (Karmarkar and Joshi 1941). Many crops suffer from chilling injury at temperatures well above that at which they will freeze.

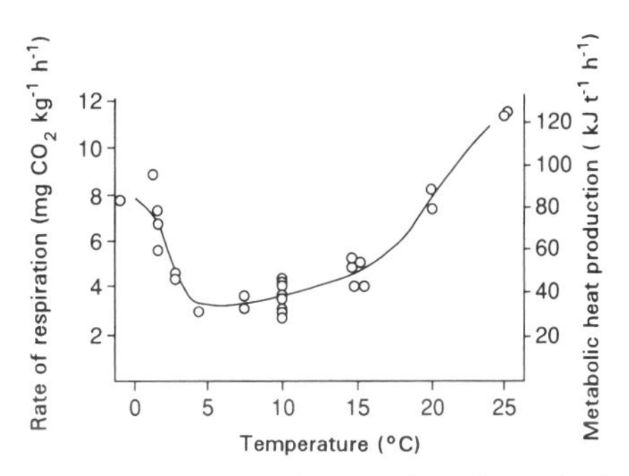

Figure 7.12 Respiration rates of potato tubers of several cultivars at different temperatures. Source: Burton (1989).

Refrigeration equipment

The most effective way of achieving accurate and constant temperature control is through refrigeration of an insulated structure. Refrigeration equipment works by having pipes containing a liquid or gas (called the refrigerant) inside the store. These pipes pass out of the store; the liquid is cooled and passed into the store to reduce the air temperature of the store as it contacts the cooled pipes. A simple refrigeration unit consists of an evaporator, a compressor, a condenser and an expansion valve. The evaporator is the pipe that contains the refrigerant mostly as a liquid at low temperature and low pressure, and is the part of the system that is inside the store. The evaporator causing the refrigerant to boil absorbs heat. The vapour is drawn along the pipe through the compressor, which is a pump that compresses the gas into a hot, high-pressure vapour. This is pumped to the condenser, where the gas is cooled by passing it through a radiator. The radiator is usually a network of pipes open to the atmosphere. The high-pressure liquid is passed through a series of small-bore pipes that slows the flow of liquid so that a high pressure builds up. The liquid then passes through an expansion valve that controls the flow of refrigerant and reduces its pressure. This reduction in pressure results in a reduction in temperature, causing some of the refrigerant to vaporize. This cooled mixture of vapour and liquid refrigerant passes into the evaporator, so completing the refrigeration cycle. In most stores a fan passes the store air over the coiled pipes containing the refrigerant, which helps to cool the air quickly and distribute it evenly throughout the store. The fan gives out heat, which is undesirable, and studies of methods of eliminating the fan in cold stores have been undertaken (Robinson 1990). The basis of these studies was to use convective currents caused in the store by heat given out by the crop and the cooled air from the coils of refrigerant pipes to maintain an even temperature throughout the store. The amount of heat absorbed by the cooling coils is related principally to the temperature of the refrigerant that they contain and the surface area of the coils.

Humidity control

In order to maintain the quality of stored fruit and vegetables, they are normally kept in humid conditions. For most crops, the higher the humidity the better. However, if the humidity is too high in a refrigerated store water may condense on the crop surface and increase rotting. In some cases low humidity can increase rotting (Thompson 1972). Also in studies where kiwifruit were stored at 0°C with 40–59%, 65–80% or 92–97% r.h. by Bautista Banos *et al.* (2000), the overall general pattern showed that with an increase in humidity infection levels decreased. The temperature differential between the evaporator refrigerant and the store air determines the moisture content of the store air (Table 7.2).

If the refrigerant temperature is low compared with the store air temperature, then water will condense on the evaporator. If the refrigerant is below 0°C ice will be formed over the cooling coils, making them less efficient in cooling the store air. This removal of moisture from the store air reduces its relative humidity and increases its vapour pressure deficit. This means that the stored crop will loose water more quickly, which can adversely affect its marketability. In order to reduce crop desiccation, the refrigerant temperature should be kept close to the store air temperature. However, with respiratory heat from the crop, temperature leakage through the insulation and doors and heat generated by fans ,it may not be possible to maintain the store temperature. There remains the possibility of increasing the area of the cooling surface and adding devices such as fins to the cooling pipes or coiling them into spirals. In a study on the humidity of the air in an apple cold store design in New Zealand, Amos *et al.* (1993) found that the evaporated surface area and the occurrence of precooling within the store were the design and operational factors having the greatest effect on the relative humidity of the air. Door protection and management and floor insulation were the next most significant effects on relative humidity, but close stacking of storage bins could have the reverse effect.

Table 7.2 Approximate effects on store humidity of the temperature differential between the evaporator refrigerant and the store air

Temperature differential (°C)		
Natural convection	Air circulation by a fan	Relative humidity (%)
7–8	4–6	91–95
8–9	6–7	86–90
9–10	7–8	81–85
10–11	8–9	76–80
11–12	9–10	70–75

A study on evaporator coil refrigerant temperature, room humidity, cool-down and mass loss rates was carried out by Hellickson *et al.* (1995) in commercial 1200 bin apple storage rooms. Rooms in which the evaporator coil ammonia temperatures were dictated by cooling demand required a significantly longer time to achieve desired humidity levels than rooms in which evaporator coil temperatures were controlled by a computer. The overall mass loss rate of apples stored in a room in which the cooling load was dictated by refrigerant temperature was higher than in a room in which refrigerant temperature was maintained at approximately 1°C during the fruit cool-down period.

Another device is to have secondary cooling so that the cooling coils do not come into direct contact with the store air. One such system is the 'jacketed store.' These stores have a metal inner wall inside the store's insulation with the cooling pipes cooling the air in that space. This means that a low temperature can be maintained in the cooling pipes without causing crop desiccation and the whole wall of the store becomes the cooling surface. Ice bank cooling is also a method of secondary cooling, where the refrigerant pipes are immersed in a tank of water so that the water is frozen. The ice is then used to cool water and the water is converted to a fine mist that is used to cool and humidify the store air (Lindsey and Neale 1977; Neale *et al.* 1981). See the section on forced-air precooling in Chapter 4.

A whole range of humidifying devices can also be used to replace the moisture in the air which has been condensed out on the cooling coils of refrigeration units. These include spinning disc humidifiers, where water is forced at high velocity on to a rapidly spinning disc. The water is broken down into tiny droplets that are fed into the air circulation system of the store. Sonic humidifiers utilize energy to detach tiny water droplets from a water surface that are fed into the stores' air circulation system.

Solar-driven refrigerated stores

Since part of the refrigeration cycle requires the input of energy, there has been interest in tropical countries in using the sun to provide this energy. One such project was developed in the Sudan by technical cooperation between the Sudanese and Dutch Governments (Anon 1990). The store had a single stage ammonia/water absorption refrigerator with 13 kW peak cooling power, which was designed to keep 10 tonnes of agricultural products at a minimum temperature of 5°C. The 50 m^3 store was constructed and tested on bananas, grapefruit, onions and potatoes. The project in the Sudan proved not to be cost effective when compared with a conventional cold store operated from mains electricity. The price of electricity would have to increase by a factor of 10 before the solar-driven cold store could compete with grid-driven compressor systems.

Solar voltaic cells have been available for decades, but again suffer from the same problem as described above, that is, there are cheaper sources of energy. They have had some limited practical use in developing countries where they have been used to power small, portable refrigerators for keeping medicines in remote areas where there is no mains electricity. However, Umar (1998) claimed that solar cooling systems can be used to preserve food in rural locations in tropical countries either by the production and distribution of ice or by the provision of large cold stores for community use.

Controlled atmosphere stores

Introduction

Controlled atmosphere is still mainly applied to stored apples, but studies of other fruit and vegetables have shown that it has wide applicability. Commercial practice can be basically the same as that described in the early part of the 20th century. Factors to be taken into account with controlled atmosphere storage include:

1. It is expensive, therefore only the best fruit should be stored.
2. If there is a choice, small fruit should be stored, as it has been found with apples that they store better than large fruit.
3. Fruit should be placed in storage as soon as possible after harvest and in any case within a day.
4. The store should be closed and cooled each evening during loading.
5. Loading should be as quick as possible and certainly within 1 week.
6. For apples, at whatever subsequent temperature they are to be stored, the fruit should be initially cooled to 0°C.
7. Only one type of crop should be stored in the same room and preferably only one cultivar.

History

The effects of gases on harvested crops have probably been known for centuries. In Eastern countries, fruits were taken to temples, where incense was burned, to improve ripening. In Britain, crops were stored in pits which would have restricted ventilation and thus affected the O_2 and CO_2 levels in the pit. The earliest documented scientific study was by J.E. Bernard in 1819 in France, who showed that harvested fruit absorbed O_2 and gave out CO_2. He also showed that fruit stored in atmospheres containing no O_2 did not ripen, but if they were held for only a short period in zero O_2 and then placed in air they continued to ripen. These experiments showed that storage in zero O_2 gave a life of 28 days for peaches, prunes and apricots and up to 3 months for apples and pears. In 1856, B. Nice built a commercial cold store using ice to keep it below 1°C. A decade later he experimented with modifying the CO_2 and O_2 in the store by making it air tight. This was achieved by lining the store with metal sheets, thickly painting the edges of the metal and having tightly fitted doors. It was claimed that 4000 bushels of apples were kept in good condition in the store for 11 months, and that the O_2 level in the store was so low that a flame would not burn. J. Fulton in 1907 observed that fruit could be damaged where large amounts of CO_2 were present in the store. R.W. Thatcher in 1915 experimented with apples sealed in boxes containing different levels of gases and concluded that CO_2 greatly inhibited ripening. West (1916) described the effects of different concentrations of CO_2 on the respiration of seeds. Kidd and West (1917) carried out comprehensive studies on CO_2 and O_2 on fruit and vegetables and called the technique gas storage (Kidd and West 1927). The modern application of the manipulation of gases in crop stores on a commercial basis can be traced to their work. In the early 1940s, the term gas storage was replaced by controlled atmosphere storage and it has been the subject of an enormous number of studies, in spite of which it is still not known precisely why it works. A current hypothesis is that the success of controlled atmosphere storage is the result of a stress response, but the mechanism is still unknown (J.R. Stow, personal communication 2002). The effect that O_2 and CO_2 have on crops varies with factors such as the species and cultivar of crop, the concentration of the gases, the crop temperature and its stage of maturity or ripeness. There are also interactive effects of the two gases, so that the effect of the CO_2 and O_2 in extending the storage life of a crop may be increased when they are combined.

Oxygen reduction

The crop is loaded into an insulated storeroom whose walls have been made gas tight. The time taken for the levels of these two gases to reach the optimum (especially for the O_2 to fall from the 21% in normal air) can reduce the maximum storage life of the crop. It is therefore common to fill the store with the crop, seal the store and inject nitrogen gas until the O_2 has reached the required level, and then maintain it in the way described below. The nitrogen may be obtained from large liquid nitrogen cylinders. Malcolm and Gerdts (1995) in a review described three main methods used to produce nitrogen gas used for controlled atmosphere applications. These were cryogenic distillation, a pressure swing adsorption system and membrane systems. Hollow-fibre technology is also used to generate nitrogen to inject into the store (Figure 7.13). The walls of these fibres are differentially permeable to O_2 and nitrogen. Compressed air is introduced into these fibres and by varying the pressure it is possible to regulate the purity of the nitrogen coming out of the equipment and produce an output of almost pure nitrogen. Another type of nitrogen generator is called 'pressure swing adsorption.' This has an air compressor which passes the compressed air through a molecular sieve that traps the O_2 in the air and allows the nitrogen to pass through. It is a dual-circuit system so that when one circuit is providing nitrogen for the store, the other

Figure 7.13 Nitrogen generators in use at a commercial controlled atmosphere store in the UK.

circuit is being renewed. The O_2 content at the output of the equipment will vary with the throughput. For a small machine with a throughput of 5 $nm^3 h^{-1}$ the O_2 content will be 0.1%, at 10 $nm^3 h^{-1}$ O_2 it will be 1% and at 13 $nm^3 h^{-1}$ it will be 2% (Thompson 1996). In China a carbon molecular sieve nitrogen generator has been developed (Shan-Tao and Liang 1989). This was made from fine coal powder which was refined and formed to provide apertures similar to the size of a molecule (3.17 Å) but smaller than an O_2 molecule (3.7 Å). When air was passed through the carbon molecular sieve under high pressure, the O_2 molecules were absorbed on the carbon molecular sieve and the air passing through had an enriched nitrogen level. A pressurized water CO_2 scrubber was described by Vigneault and Raghavan (1991) which could reduce the time required to reduce the O_2 content in a controlled atmosphere store from 417–140 hours.

Carbon dioxide enrichment

For crops which have only a very short marketable life, such as strawberries, the CO_2 level may also be increased to the required level by direct injection of CO_2 from a pressurized gas cylinder. This is particularly important where controlled atmospheres are used in transport.

Controlled atmosphere controls

The temperature in the store room is controlled by mechanical refrigeration and the composition of the atmosphere is constantly analysed for CO_2 and O_2 levels. When the O_2 has reached the level required for the particular crop being stored it is maintained at that level by frequently introducing fresh air from outside the store. Usually tolerance limits are set at, say ±0.1%, so that when the O_2 level is 0.9% air is vented until it reaches 1.1%. The same applies to CO_2. When a predetermined level is reached the atmosphere is passed through a chemical that removes CO_2 and then back into the store. This is called 'active scrubbing.' Alternatively, the CO_2 absorbing chemical may be placed inside the store where it can keep the level low (usually about 1%); this is called 'passive scrubbing.' This method of controlled atmosphere store is referred to as 'product generated,' since the gas levels are produced by the crop's respiration. The time taken for the levels of these two gases to reach the optimum (especially for the O_2 to fall from the 21% in normal air) can reduce the maximum storage life of the crop. It is therefore common to fill the store with the crop, seal the store and inject nitrogen gas until the O_2 has reached the required level and then maintain it in the way described above.

Controlled atmosphere storage on aroma and flavour

Controlled atmosphere storage increases the postharvest life of crops and there has been concern that extended storage may adversely affect the eating qualities (Blythman 1996). Many experiments have been carried out to test the effects of controlled atmosphere storage on fruit and vegetable quality and the results have been variable. Aroma volatiles in Golden Delicious apples were suppressed during storage at 1°C for up to 10 months in atmospheres containing high levels of CO_2 (3%) or low levels of O_2 (1 or 3%) compared with storage in air (Brackmann 1989). The measurement of the organic volatile production of the apples was over a 10 day period at 20°C after removal from the cold store. In McIntosh and Cortland apples most organic volatile compounds were produced at lower rates during ripening after controlled atmosphere storage than those produced from fruits ripened immediately after harvest (Yahia 1989). Storage of pears in controlled atmosphere conditions did not have a deleterious effect on fruit flavour compared with those stored in air and were generally considered better (Meheriuk 1989a). However, controlled atmosphere storage generally resulted in significantly higher acid levels in fruit (Table 7.3).

In studies of litchis, fruit were packed in punnets overwrapped with plastic film with 10% CO_2 or vacuum packed and stored for 28 days at 1°C followed by 3 days shelf-life at 20°C. It was found that their taste and flavour was unacceptable, whereas they remained acceptable for non-wrapped fruits (Ahrens and Milne 1993).

Oxygen effects

Many chemical reactions in crops are catalysed by enzymes and the reactions require molecular O_2. Therefore, as O_2 levels in plant cells fall, the rates of chemical reactions decrease and metabolism is reduced. This effect tends to be accelerated at very low levels of O_2. If the O_2 level in the cells is too low there may be undesirable changes in the chemicals that contribute to the flavour and aroma of the crop. At very

Table 7.3 Effect of controlled atmosphere storage on the acidity and firmness of Spartlett pears stored at 0°C and their taste after 7 days of subsequent ripening at 20°C (source: Meheriuk 1989a)

Storage atmosphere		Storage time (days)	Acidity (malic acid, mg permg 100 ml)[a]	Taste (score 1= like, 9 = dislike)	Firmness (Newton)[a]
CO_2 (%)	O_2 (%)				
0	21	60	450b	6.1	62a
5	3	60	513a	7.6	61ab
2	2	60	478ab	7.6	60b
0	21	120	382b	4.3	55b
5	3	120	488a	6.7	58a
2	2	120	489a	6.9	58a
0	21	150	342b	2.9	51b
5	3	150	451a	7.7	58a
2	2	150	455a	7.1	58a
0	21	180	318b	–	40b
5	3	180	475a	–	62a
2	2	180	468a	–	61a

[a]Figures followed by the same letter were not significantly different ($p = 0.05$) by Duncan's multiple range test.

low levels of O_2 the tricarboxylic acid cycle is inhibited but the glycolytic pathway may continue. This results in less efficient energy production during respiration since there is insufficient O_2 to metabolize stored carbohydrates to water and CO_2. Instead, the blocking of the glycolytic pathway results in a build-up of acetaldehyde and ethanol, which are toxic to the cells if allowed to accumulate. The amount of O_2 in the cells is proportional to:

- the O_2 level surrounding the crop
- the permeability of the crop surface
- the diffusion of O_2 through the crop
- temperature of the crop
- innate metabolic rate of the crop.

Robinson *et al.* (1975) studied the respiration rate of a variety of crops at different temperatures stored in air or in 3% O_2. Respiration was generally lower when crops were stored in 3% O_2 than when they were stored in air, but there was some interaction between O_2 and temperature in that the effect of the reduced O_2 level was more marked at higher storage temperatures.

The effect that low O_2 has on the crop also depends on the duration of exposure. Ke and Kader (1989) showed that fruits such as pears, stone fruit, blueberries and strawberries stored at between 0 and 5°C for up to 10 days would tolerate O_2 levels of between 0.25 and 1%, whereas apples could tolerate these levels for longer periods. Detrimental effects of low O_2 on the crop include the accumulation of ethanol and acetaldehyde to concentrations that can result in off-flavours (Ke and Kader 1989). The level of O_2 at which anaerobic respiration may occur in plant cells may be as low as 0.2%, but the gradient of O_2 concentration between the store atmosphere and the cells of the crop requires the maintenance of a higher level around the crop (Kader 1986). The gradient is affected by the ease of gas penetration through the cells and cuticle or periderm of the crop and the utilization of O_2 by the crop. In modern controlled atmosphere apple stores where they are stored at 1% O_2, an alcohol detector is fitted which sounds an alarm if ethanol fumes are detected in the store. This enables the store operator to increase the O_2 level and no damage should have been done to the fruit. The detector technology is based on that used by the police to detect alcohol fumes on the breath of motorists. It has been shown that low O_2 levels in stores can affect the crop in several ways (Richardson and Meherink 1982). These are summarized in Table 7.4.

Table 7.4 The effects of oxygen on postharvest responses of fruit, vegetables and flowers (various sources)

Reduced respiration rate
Reduced substrate oxidation
Delayed ripening of climacteric fruit
Prolonged storage life
Delayed breakdown of chlorophyll
Reduced rate of production of ethylene
Changed fatty acid synthesis
Reduced degradation rate of soluble pectins
Formation of undesirable flavours and odours
Altered texture
Development of physiological disorders

Carbon dioxide effects

The effect of CO_2 in extending the storage life of crops appears to be on reduction in respiration of the crop. Knee (1973) showed that CO_2 could inhibit an enzyme (succinate dehydrogenase) in the tricarboxylic acid which is part of the crop's respiratory pathway. The atmosphere inside the tissue of the crop will equalize with the atmosphere in the store if it is at the same temperature and pressure. Where the level of CO_2 in the store is increased this will therefore increase its levels within the crop tissue (Table 7.5).

It has been suggested that high levels of CO_2 in stores can compete with ethylene for binding sites in fruits (Burg and Burg 1967). They showed that in the presence of 10% CO_2 the biological activity of 1% ethylene was abolished. Yang (1985) showed that CO_2 accumulation in the intercellular spaces of fruit acts as an ethylene antagonist.

Physiological disorders in fruit associated with excess CO_2 levels may be associated with this disruption of the respiratory pathway leading to an accumulation in the crop cells of alcohol and acetaldehyde. The sensitivity of crops to CO_2 vary widely. This may be related to the permeability of the crop tissue to gas exchange (intercellular spaces, cuticle thickness and composition, presence of stomata, lenticels and hydrothodes and the nature of the cell tissue), which could affect the level of CO_2 in the crop cells.

Intermittent exposure of crops to high levels of CO_2 may have beneficial effects. Marcellin and Chaves (1983) showed that exposing avocado fruit to 20% CO_2 for 2 days each week during storage at 12°C delayed their senescence and reduced decay compared with fruit stored for 7 days a week in air. They also showed that the same CO_2 treatment of fruit stored at 4°C reduced the symptoms of chilling injury.

High levels of CO_2 (10–15%) can control grey mould in grapes for limited periods. Prolonged exposure to these levels can damage the fruit, but up to 2 weeks of exposure did not have a detrimental effect (Kader 1989). Strawberries stored in high concentrations of CO_2 had reduced levels of decay (Table 7.6) without damaging the fruit.

Exposing grapefruit to 10–20% CO_2 can reduce chilling injury during subsequent storage at 7–10°C (Hatton and Cubbage 1977). Exposure to 60–90% CO_2 for 24 hours at 17–20°C removed astringency in persimmon (Ben Arie and Guelfat-Reich 1976).

Physiological disorders associated with high CO_2 include those listed in Table 7.7.

Table 7.5 The effects of CO_2 on postharvest responses of fruit, vegetables and flowers (various sources)

Decreased synthetic reactions in climacteric fruit
Delaying the initiation of ripening
Inhibition of some enzymic reactions
Decreased production of some organic volatiles
Modified metabolism of some organic acids
Reducing the rate of breakdown of pectic substances
Inhibition of chlorophyll breakdown
Production of 'off-flavours'
Induction of physiological disorders
Retarded fungal growth on the crop
Inhibition of the effect of ethylene
Changes in sugar content (potatoes)
Effects on sprouting (potatoes)
Inhibition of postharvest development
Retention of tenderness
Decreased discoloration levels

Table 7.6 The effects of carbon dioxide concentration during storage for 3 days at 5°C on the levels of decay in strawberries and the development of decay when removed to ambient conditions (source: Harris and Harvey 1973)

	CO_2 level (%)			
Storage	0	10	20	30
3 days of storage at 5°C	11	5	2	1
+ 1 day at 15°C in air	35	9	5	4
+ 2 days at 15°C in air	64	26	11	8

Table 7.7 Physiological disorders associated with high CO_2

Crop	Disorder	Reference
Apples	Brown heart	Kidd and West (1923)
Apples	Core flush	Lau (1983)
Apples	Low temperature breakdown	Fidler (1968)
Broad bean	Pitting	Tomkins (1965)
Broccoli	Accelerated softening	Lipton and Harris (1974)
Broccoli	Off-flavours	Lipton and Harris (1974)
Cabbage	Internal browning	Isenberg and Sayles (1969)
Capsicums	Internal browning	Morris and Kader (1977)
Kiwifruit	Off-flavours	Harman and McDonald (1983)
Lettuce	Brown stain	Stewart and Uota (1971)
Mushrooms	Cap discoloration	Smith (1965)
Potato	Curing inhibition	Butchbaker *et al.* (1967)
Spinach	Off-flavours	McGill *et al.* (1966)
Strawberry	Off-flavours	Couey and Wells (1970)
Tomato	Uneven ripening	Morris and Kadar (1977)
Tomato	Surface blemishes	Tomkins (1965)

Nitrogen effects

Placing fruit and vegetables in total nitrogen results in anaerobic respiration, but exposure for short periods can have beneficial effects. Treatment of tomato fruits prior to storage in atmospheres of total nitrogen were shown to retard their ripening (Kelly and Saltveit 1988), but they could be ripened normally when removed to a normal atmosphere. Similar effects were shown with avocados by Pesis *et al.* (1993).

Carbon monoxide

Carbon monoxide is a colourless, odourless gas that is flammable and explosive in air at concentrations between 12.5% and 74.2%. It is extremely toxic to animals with haemoglobin and exposure by human beings to 0.1% for 1 hour can cause unconsciousness and for 4 hours death. However, it can have beneficial effects in crop storage. If added to controlled atmosphere stores, with levels of O_2 between 2% and 5%, carbon monoxide can inhibit discoloration of lettuce on the cut butts or from mechanical damage on the leaves. Concentrations in the range 1–5% carbon monoxide were effective, but the effect was lost when the lettuce was removed to normal air at 10°C. The respiration rate of lettuce was reduced during a 10 day storage period at 2.5°C when carbon monoxide was added to the store. A combination of 4% O_2, 2% CO_2 and 5% CO was shown to be optimum in delaying ripening and maintaining good quality of mature green tomatoes stored at 12.8°C after being subsequently ripened at 20°C (Morris *et al.* 1981). In peppers and tomatoes the level of chilling injury symptoms could be reduced, but not eliminated, when carbon monoxide was added to the store.

The addition of carbon monoxide in a store can partially reduce the detrimental effects of ethylene in causing russet spotting on the leaves (Kader 1987). The reduction in respiration rate of vegetative tissue in the presence of 1–10% carbon monoxide was reversed in climacteric fruit. In tomatoes carbon monoxide was found to stimulate respiration rate, but this effect could be minimized by reducing the level of O_2 in the store to below 5% (Kader 1987).

Carbon monoxide has fungistatic properties, especially when combined with low O_2 (Kadar 1983). *Botrytis cinerea* on strawberries was reduced where the carbon monoxide level was maintained at 5% or higher in the presence of 5% O_2 or lower. Decay in mature green tomatoes stored at 12.8°C was reduced when 5% carbon monoxide was included in the storage atmosphere (Morris *et al.* 1981). Carbon monoxide is an analogue of ethylene and it was shown to initiate ripening of bananas when the fruit were exposed to concentrations of 0.1% for 24 hours at 20°C.

Ozone

Ozone (O_3) is a powerful oxidizing agent which has a penetrating odour and is unstable at room temperature. It is a highly toxic gas to human beings and exposure to concentrations in excess of 0.1 μl litre^{-1} should be avoided. Information on its use on fruits and vegetables is often contradictory. It has been shown that fungal growth and rotting can be controlled in some vegetables and fruit exposed to 1–2 μl litre^{-1}. Its effectiveness is increased with high humidity in the store. However, there is evidence that it is phytotoxic and peaches were damaged by exposure to 1 μl litre^{-1} O_3 and lettuce and strawberries to 0.5 μl litre^{-1} O_3. The rate of decomposition to O_2 is rapid.

Valencia oranges were exposed continuously to 0.3 ± 0.05 μl litre^{-1} (volume/volume) O_3 at 5°C for 4 weeks. Eureka lemons were exposed to an intermittent day–night ozone cycle with 0.3 ± 0.01 μl litre^{-1} O_3 only at night, in a commercial cold storage room at 4.5°C for 9 weeks. Also both oranges and lemons were continuously exposed to 1.0 ± 0.05 μl litre^{-1} O_3 at 10°C in an export container for 2 weeks. Exposure to O_3 did not reduce the final incidence of green or blue mould (*Penicillium digitatum* and *P. italicum*) on inoculated fruit, although the incidence of both diseases was delayed by about 1 week and infections developed more slowly under O_3. Sporulation was prevented or reduced by O_3 without noticeable phytotoxicity to the fruit (Palou *et al.* 2001). O_3 treatment of mushrooms at 100 mg h^{-1} for up to 25 minutes prior to packaging PVC plastic film caused an increase in external browning rate and a reduction of the internal browning rate during 7 days of subsequent storage at 5, 15 or 25°C (Escriche *et al.* 2001). The O_3 treatment had no significant differences in terms of texture, maturity index and weight loss of mushrooms.

Ethylene (ethene)

Ethylene (C_2H_4), which was discovered in 1796, was originally called olefiant gas. It should more correctly be called ethene but modern commercial use favours

Table 7.8 Sensitivity (concentration of ethylene causing half maximum inhibition of growth of etiolated pea seedlings at 20°C) of pea seedlings to ethylene in different levels of oxygen, carbon dioxide and nitrogen (source: Burg and Burg 1967)

Nitrogen (%)	Oxygen (%)	Carbon dioxide (%)	Ethylene sensitivity (μl C_2H_4 $litre^{-1}$)
99.3	0.7	0	0.6
97.8	2.2	0	0.3
82.0	18.0	0	0.14
80.2	18.0	1.8	0.3
74.9	18.0	7.1	0.6

the term ethylene. It can be produced by heating a mixture of one part of ethanol with two parts of sulphuric acid. It is a colourless gas with a sweetish odour and taste, has asphyxiant and anaesthetic properties and is flammable. Physical characteristics include:

- molecular weight 28.05
- vapour pressure (9.9°C) 5.1 bar(g)
- specific volume (20°C, 1 atm) 861.5 ml g^{-1}
- boiling point (1 atm) −103.7°C
- freezing point (1 atm) −169.5°C
- density of gas (at 0°C, 1 atm) 1.26 g ml^{-1}
- flammable limits in air 3.1–32% v/v
- autoignition temperature 543°C
- specific gravity (air = 1) 0.974.

Ethylene was known to have physiological effects on crops, but Gane (1934) first identified it as a volatile chemical produced by ripening apples. It was assumed that ethylene was produced only in climacteric fruits during the ripening phase, but with the development of chromatographic analytical techniques it is clear that all crops are able to produce ethylene under certain conditions (Hulme 1971; Roberts and Tucker 1985). Microorganisms that attack stored crops can also produce ethylene.

The effects of introducing ethylene to the store on climacteric fruits is to initiate them to ripen if the:

- concentration of the gas is above the required threshold
- temperature of the fruit is within a certain range
- exposure time is sufficient
- level of other gases in the store is conducive.

Ethylene may also hasten the ripening processes of climacteric fruit after they have been initiated to ripen. It can also affect non-climacteric fruit, flowers and all types of vegetables. It is synthesized in plant tissue and its biosynthesis in fruit is related to O_2 levels. The rate of ethylene production by apples was shown to be about half in an atmosphere of 2.5% O_2 compared with fruit stored in air (Burg 1962). CO_2 levels in store affected ethylene biosynthesis. It was shown that increased levels of CO_2 in controlled atmosphere stores containing apples reduced ethylene levels (Tomkins and Meigh 1968). The sensitivity of plant tissue to ethylene has been shown to be affected by O_2 and CO_2 levels in the store (Table 7.8). The effects of exogenous ethylene on the tissue of plants are variable, but it has been shown to affect the rate of respiration of crops (Table 7.9).

Ethylene removal from stores

The control of internal concentrations of ethylene in crops may ultimately be limited by the resistance of the crop to diffusion rather than its removal from the atmosphere surrounding the crop (Dover 1989). However, Stow (1989) showed that the ethylene concentration in the core of apples was reduced by lowering the ethylene concentration in the containers in which the fruit were stored. There are various ways in which ethylene can be removed from stores (Knee

Table 7.9 Effects of ethylene on respiration rate of selected crops (source: Solomos and Biale 1975)

	Respiration rate (μl O_2 g^{-1} h^{-1})	
Crop	Control	Ethylene
Apple	6	16
Avocado	35	150
Beet	11	22
Carrot	12	20
Cherimoya	35	160
Grapefruit	11	30
Lemon	7	16
Potato	3	14
Rutabaga	9	18
Sweet potato	18	22

et al. 1985; Gorini *et al.* 1989). These involve either a chemical reaction or absorption.

Absorption

Molecular sieves and activated carbon can hold organic molecules such as ethylene. When fresh air is passed through these substances the molecules are released. This means that they can be used in a two-stage system where the store air is being passed through the substance to absorb the ethylene while the other stage is being cleared by the passage of fresh air. After an appropriate period, the two stages are reversed. Hydrated aluminium silicate can also be used to absorb ethylene. This has a complex lattice structure forming a honeycomb network which can be lined with anions which will loosely hold any cations with which they come into contact. A natural zeolite called clinoptilolite is used. This has the following structure:

$$\underset{\text{Exchangeable cations}}{(Na_2K_2Ca)_3} \quad \underset{\text{Structural cations}}{(Ai_9Si_{30}O_{72}).\ 24H_2O}$$

Activated charcoal, which is also called activated carbon, is used for absorbing carbon dioxide from stores, but it will also absorb ethylene. Baumann (1989) described a simple scrubber system which could be used in stores to remove both gases using activated charcoal. A chart was also presented which showed the amount of activated charcoal required in relation to the carbon dioxide levels required and ethylene output of the fruit in the store.

Reaction

Ethylene can be oxidized at room temperatures when it comes into contact with potassium permanganate. Proprietary products are available which can be placed inside the crop store or even inside the actual package containing the crop. The oxidizing reaction is not reversible and the granules change colour from purple to brown to indicate that they need replacing. In trials with the proprietary ethylene absorbent Purafil in apple stores, Blanpied (1989) found variable efficiency in ethylene removal (Table 7.10). The rate of removal of ethylene with another proprietary ethylene absorbent, Ethysorb, was shown to be affected by the absorption system used and the concentration of the ethylene within the store (Table 7.11).

Stow (1989a) showed an interaction between ethylene levels in apple stores and the concentration of O_2. He also showed that an ethylene absorbent (1 kg of vermiculite soaked with 2 litres of a saturated solution of potassium permanganate) reduced the ethylene concentration in the store at all oxygen levels studied (Figure 7.14). See also Chapter 5 where the use of ethylene absorbants in modified atmosphere packaging is discussed more fully.

Table 7.10 Ethylene removal efficiency by chemisorption with Purafil ES in a controlled atmosphere store containing 200 tons of apples (source: Blanpied 1989)

Days in storage	Ethylene removal efficiency[a]	Ethylene concentration in store (ppm)
1	0.73	2.6
2	0.46	1.1
3	0.42	1.2
7	0.38	1.7
11	0.21	3.3
16	0.07	4.4

[a] 475 m^3 $hour^{-1}$ through 85 kg of Purafil beads.

Table 7.11 The rate of removal of ethylene in ppm with Ethysorb (source: Dover and Bubb undated)

	Rate of ethylene removal g Ethysorb $hour^{-1}$	
Concentration of ethylene (ppm)	Closed static system	Closed circulating system
0.01	0.023	0.028
0.10	0.230	0.280
1.00	2.30	2.800

Catalytic converters

These also remove ethylene by chemical reaction. Air from the store is passed through a device where it is heated to over 200°C in the presence of an appropriate catalyst, usually platinum (Wojciechowski 1989). Under these conditions the ethylene in the air is oxidized to carbon dioxide and water. It requires an energy input of 30–80 W per cubic metre of purified air, so it is a high energy-consuming method. However, with suitable heat exchangers it is possible to make the method more energy efficient. One such device, called the 'Swingtherm', reduces energy consumption to 7–14 W m^{-3}. Another ethylene-converting device, the 'Swingcat', was marketed by Tubamet AG of Vaduz in Liechtenstein in 1993.

Ozone scrubbers

Ozone is a powerful oxidizing agent and reacts with ethylene to produce carbon dioxide and water. It can be easily generated from oxygen by ultraviolet radia-

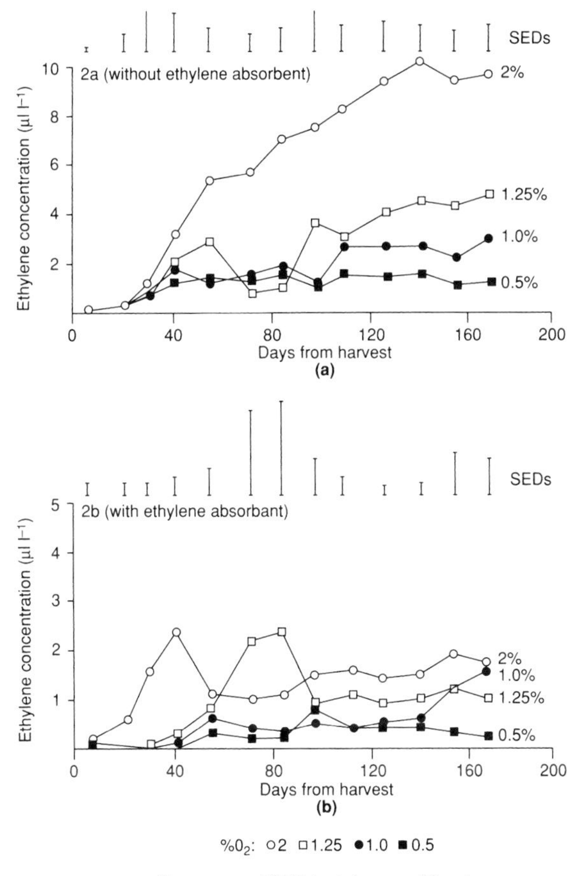

Figure 7.14 Effects of ethylene absorbents on the ethylene levels in controlled atmosphere stores with different oxygen levels. Source: Stow (1989a).

tion or electrical discharge. Scott *et al.* (1971) and Scott and Wills (1973) described the development of a reaction chamber which could be used in stores to remove ethylene. It consists of an ultraviolet lamp giving out radiation at 184 and 254 nm which produce both ozone and atomic oxygen. The store atmosphere is passed through the reaction chamber with a fan and any ethylene that it contains is rapidly oxidized. The outlet of the reaction chamber contains rusty steel wool which will react with any excess O_3 and prevent it from entering the store because it could be toxic to crops or to workers in the store.

Hypobaric storage

Store designs which enable them to be kept under a vacuum have been used for controlled atmosphere storage. The reduction in pressure reduces the partial pressure of oxygen and thus its availability to the crop in the store. The reduction in the partial pressure of the oxygen is proportional to the reduction in pressure. However, in crop stores the humidity must be kept high and this water vapour in the store atmosphere has to be taken into account when calculating the partial pressure of oxygen in the store. To do this, the relative humidity must be measured and, from a psychometric chart, the vapour pressure deficit can be calculated. This is then included in the following equation:

$$\frac{P_1 - VPD \times 21}{P_0} = \text{partial pressure of } O_2 \text{ in the store}$$

where

P_0 = outside pressure at normal temperature (760 mmHg)

P_1 = pressure inside the store

VPD = vapour pressure deficit inside the store.

Since the crop in the hypobaric store is constantly respiring, it is essential that the store atmosphere is constantly being changed. This is achieved by a vacuum pump evacuating the air from the store. The store atmosphere is constantly being replenished from the outside. The air inlet and the air evacuation from the store are balanced in such a way as to achieve the required low pressure within the store.

There are two important considerations in developing and applying this technology to crop storage. The first is that the store needs to be designed to withstand low pressures without imploding. The second is that the reduced pressure inside the store can result in rapid water loss from the crop. To overcome the first, stores have to be strongly constructed of thick steel plate with a curved interior. For the second the air being introduced into the store must be saturated (100% r.h.); if it is less than this, serious dehydration of the crop can occur. The control of the oxygen level in the store can be very accurately and easily achieved and simply measured by measuring the pressure inside the store with a vacuum gauge. This method also has the advantage of constantly removing ethylene gas from the store, which prevents it building up to levels which could be detrimental to the crop. The effects of hypobaric storage on fruits and vegetables has been reviewed by Salunkhe and Wu (1975) and by Burg (1975), showing considerable extension in the storage life of a wide range of crops when it is combined with refrigeration compared with refrigeration alone. In other work, these extensions to storage life under hypobaric conditions have not been confirmed (Hughes *et al.* 1981). The hypobaric-stored capsicums also had a significantly higher weight loss during storage than the air-stored product. Hypobaric storage of bananas was shown to increase their storage life. When bananas were stored at 14°C their storage life was 30 days at 760 mmHg, but when the pressure was reduced to 80 or 150 mmHg the fruit remained unripe for 120 days (Apelbaum *et al.* 1977). When these fruits were subsequently ripened they were said to be of very good texture, aroma and taste.

A more recent system of hypobaric storage controls water loss from produce without humidifying the inlet air with heated water (Burg 1993). This is achieved by slowing the evacuation rate of air from the storage chamber to a level where water evaporated from the produce by respiratory heat exceeds the amount of water required to saturate the incoming air. Using this technique with roses stored at 2°C and 3.33 $\times 10^3$ N m^{-2}, Burg (1993) found that flowers stored for 21 days with or without humidification at a flow rate of 80–160 cm^3 min^{-1} lost no significant vase life compared with fresh flowers.

The effectiveness of short hypobaric treatments against postharvest rots was investigated by Romanazzi *et al.* (2001), who found that it reduced fungal infections in sweet cherries, strawberries and table grapes.

8
Disease control

Introduction

Postharvest diseases of fruit and vegetables are caused by infections with fungi and bacteria. Virus infections can cause diseases, which manifest themselves postharvest, but these are rare. Fungi and bacteria can exist as either parasites or saprophytes. The former feed from living matter and the latter from dead organic matter. Parasites can be obligate or facultative. Obligate parasites can only survive on living hosts, generally only attack a limited range of hosts and are very difficult or impossible to grow on artificial culture media. Facultative parasites can adapt to live on dead organic matter, often have a wide host range and can easily be cultured on artificial media. Most fungi require an acidic (pH 2.5–6) environment in whereas to grow and develop whereas bacteria thrive best in neutral conditions and only a few species can grow at levels below pH 4.5. Bacteria therefore do not usually infect fruit, which are acidic, normally only vegetables and flowers. Fungi may be unicellular or multicellular in the case of yeasts, but most fungi are made up of hyphae that are multicellular filamentous thread-like structures, which can form a dense, fluffy mass, called mycelium. Some information is given on diseases in the sections on individual crops in Chapter 12, but more comprehensive references are Snowdon (1990, 1991) and Waller (2001).

Legislation

There is a strong aversion in many consumers to the treatment of fruits and vegetables with chemical pesticides. This has resulted in increasing legislation on their use, especially after harvest. Chemical pesticide residues in fruits and vegetables are restricted and controlled by legislation. There is a United Nations body set up to provide a scientific basis for legislation called the World Health Organization (WHO) Expert Group on Pesticide Residues with the Food and Agriculture Organization (FAO) Panel of Experts on Pesticide Residues (JMPR). In Europe the DG VI (Agriculture) of the European Commission has the responsibility of harmonizing chemical pesticide residues in food. With the introduction of the Single European Market on 1 January 1993 a first list of 55 chemicals was produced that could be applied to foods. The list is constantly being updated and it contains the maximum residue levels (MRLs) for different chemicals on different crops. It provides Directives for MRLs and these are communicated in Directives issued in the Official Journal of the European Communities. Where no MRL exists for a certain crop and chemical it is subject to a default level (LOD). European Union legislation is effectively removing most postharvest fungicides and insecticides from use on fruits and vegetables. This means that for a wide range of tropical crops there are only a few permitted compounds and for some crops none. For producers in developing countries it is important to try every possible means of non-chemical protection to avoid problems on entry to the export markets.

COLEACP in Paris has been given the task of developing an action programme to the European Commission. This has three main purposes:

1. To set up a communication programme to raise awareness of the European Union harmonization programme for pesticide regulations and to provide accurate, regularly updated and easily accessible information in a user-friendly format on MRLs for tropical and minor crops produced in Asian, Caribbean and Pacific (ACP) countries and a database.
2. Where MRLs have been set by default at the LOD, to search for available data on essential crop/chemical combinations in order to submit applications for the setting of practical MRL for GAP use in these crops. Where data are missing or inadequate, trials will be set up to establish the missing data.
3. To set up mechanisms (and work with other organizations where possible) to build the capability of ACP exporters and exporter associations to develop sustainable systems of crop production, minimizing pesticide use through integrated crop management and integrated pest management (ICM/IPM) to comply with European Union (and US) regulations, and also to protect the health of farm workers and preserve the environment.

Microorganism control

A large number of chemicals are applied to crops to control fungi that cause diseases. Laws to protect the consumer strictly govern their use, and these laws may vary between countries, although most countries conform to the Food and Agriculture Organization of the United Nations and the World Health Organization regulations.

Some chemicals are permitted to be applied to crops only before harvest and the interval of time that must take place between application and harvesting is usually specified. Others are also allowed to be applied postharvest, usually with strict regulations on the maximum permitted residue, the MRL that can remain in the crop. Residue levels of chemical fungicides in a crop are related to the concentration of the chemical used, but also to the time of the crop in storage and the formulation of the fungicide (Table 8.1).

Fungi are constantly changing, and where a particular chemical fungicide is constantly used strains of the fungus may develop which are tolerant to that chemical. This commonly occurs with the benzimidazole group of fungicides. Benomyl, which is a member of the benzimidazole group, was used to control postharvest diseases of yams (Thompson *et al.* 1977). However, during commercial application, a rot caused by infection with *Penicillium sclerotigenum* was frequently observed on the tubers. In *in vitro* tests this organism was found to be tolerant to benomyl (Figure 8.1). This tolerance was confirmed in in vitro tests, but the organism was highly susceptible to the fungicide imazalil, which gave good control of the disease (Table 8.2).

The application of some chemicals can actually lead to an increase in the level of disease. An example is the postharvest application of dichloran to control postharvest diseases of yams (Thompson *et al.* 1977). The increase (Figure 8.2) was probably due to synergy

Table 8.1 Residue levels in μg litre^{-1} after storage at 21°C on the skin and in the pulp of citrus fruits of benomyl applied at 0.5 g litre^{-1} in water or in oil emulsion (Papadopoulou-Mourkidou 1991)

Storage period (days)	Water	Oil
Residue in the skin		
1	0.17	0.83
8	0.14	0.52
21	0.01	0.13
Residue in the pulp		
1	Not detectable	0.01
8	0.01	0.03
21	0.05	0.14

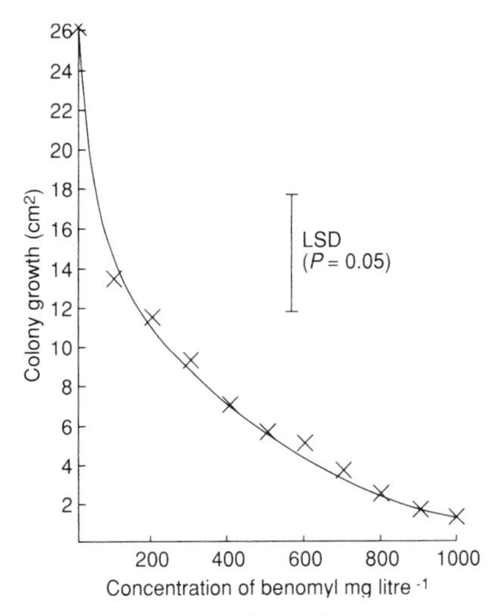

Figure 8.1 Colony size of *Penicillium sclerotigenum* Yamamoto on potato dextrose agar containing different concentrations of benomyl after 14 days' growth at 25°C. Source: Plumbley *et al.* (1984).

Table 8.2 Effects of benomyl and imazalil on the growth (in cm^2) of a benomyl tolerant strain of *Penicillium sclerotigenum* on yam tubers (*Dioscorea cayenensis*) during storage at 20°C (source: Plumbley *et al.* 1984)

Treatment	Storage time in (days) 7	14	21	28
Control	1.10	2.20	3.19	3.66
Benomyl (500 ppm)	0.83	1.87	2.05	3.49
Benomyl (1000 ppm)	0.60	1.34	2.35	2.99
Imazalil (500 ppm)	0	0	0	0
LSD (p = 0.05)	0.56			

between the various organisms that were infecting the tuber.

Control of bacteria by chemicals is not normally necessary and where it is practised it is usually achieved by dipping or washing the crop in a solution containing active chlorine at 75–125 μg $litre^{-1}$. Household bleach is often used as the source of chlorine. Abu Baker and Abdul-Karim (1994) showed that dipping sapota fruit in the fungicides benomyl, captafol or captan reduced not only the fungal population on the fruit but also the bacterial population. Most of these chemicals are toxic to human beings and therefore great care must be taken during handling the chemicals, especially before they are diluted, and appropriate protective clothing must be worn.

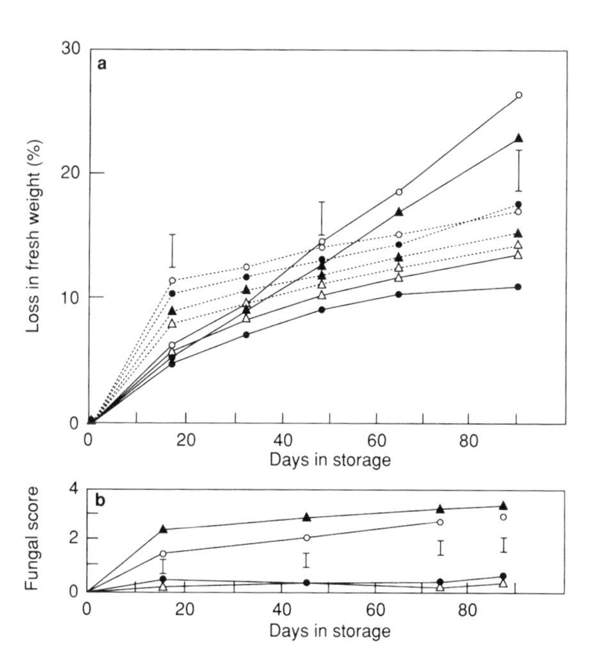

Figure 8.2 Effects of fungicidal treatment on (a) weight loss and (b) the level of surface fungi on store yams (*Dioscorea rotundata*). Source: Thompson *et al.* (1977).

Chemical application methods

The methods of application include:

- Dipping or spraying the crop in a solution or suspension of the chemical; this may be in hot water to enhance control. The crop may be passed below a shower of the diluted chemical. This is called 'cascade' application.
- Applying the chemical as a spray but producing very fine particles and then charging them to give a fine, even coat of chemical over the crop.
- Applying the chemical as a dust, usually the active chemical diluted with an inert powder such as talc.
- Volatile chemicals can be applied as a vapour that is circulated in a confined space containing the crop. Some chemicals need to be heated to aid circulation, and this process is called 'thermal fogging.' In other cases chemicals which have a high thermal stability can be burnt and the smoke introduced into the store.
- The chemical may be absorbed on a pad made of suitable material such as absorbent paper. These pads are placed over cut surfaces such as the cut crown of a hand of bananas. In this case the pad absorbs latex from the cut surface, which also helps to keep the pad in position.

Dipping

The crop or part of the crop is immersed in water containing an appropriate concentration of a chemical that is toxic to the fungi which are known to cause disease on that crop. However, the chemical must either be not toxic to the crop or have acceptable levels of toxicity. The residue of the chemical remaining in the crop after treatment must be at a level that will not endanger public health. In order to improve the effectiveness of these dips, additives may be included in the formulation. These include wetting agents, which reduce the surface tension and allow a better coating of the chemical on the crop, and acids, such as citric acid, which lower the pH of the fungicide and can make it more effective. In some cases the area to be protected can be targeted. In pineapples infection of the fruit is commonly through the cut fruit stalk. Dipping this area directly after it has been cut is normally sufficient to control postharvest disease. This allows for reduced levels of fungicide application, which may thus leave a lower residue and costs less in chemical used. The effectiveness of the fungicidal treatment may be enhanced by heating the water into which the fruit are dipped. In the control of mango anthracnose, postharvest dips in fungicides in cold water rarely reduced levels of infection, but when fungicide was applied in hot water complete control could be achieved. The optimum recommended conditions varied between different research worker, but 500 μg litre^{-1} of benomyl in water at 55°C for 5 minutes appeared generally effective without damaging the fruit.

Sprays

Spraying fruit and vegetables with a diluted fungicidal spray postharvest may be more effective than dipping. This is because where the crop is dipped the fungicide concentration may be reduced if the crop has been washed and is still wet. Also, fungicide may be preferentially removed from the mixture when the crop is removed. Additionally, many fungicidal chemicals are formulated in such a way that they are not in solution, but in a fine suspension. This can result in a concentration gradient from the top the bottom of the tank tunless the suspension is frequently agitated. This is less likely to occur with sprays, but it is important with sprays to ensure that the crop is evenly coated with the chemical.

Knapsack sprays (see Figure 12.31) are sometimes used postharvest, but more frequently the spray is built into a grading line in a packhouse. In this case the crop is passed over rollers which constantly revolve the fruit to ensure that all sides are evenly exposed to the spray. The cascade applicator is a modification of this method where the fungicide is applied as a curtain of liquid under which the fruit pass. This method is good for bananas where the cut crowns are placed upwards, because this is the area where the fungi are likely to attack the fruit (Thompson and Burden 1995). It is less useful in fruits such as citrus where it is important to have an even coating of chemical all over the fruit surface.

Electrostatic sprays

Breaking up the pesticide solution into fine droplets and then giving them an electric charge was first developed for field spays. The main advantages of this system are that they give increased uniformity of application, enabling the application rates to be reduced without loss of biological activity (Coffee 1980; Law 1982). The principle on which they work is that the particles all have the same electrical charge and thus repel each other. They are attracted to earth, which in this case is the crop, and can form a thin, even coat. This method was demonstrated as having a postharvest application by Cayley *et al.* (1987), who developed it for postharvest applications of various fungicides on potatoes. The tubers were passed below an electrostatic sprayer on a roller table. Subsequent work by Anthony (1991) demonstrated its effectiveness in controlling crown rot on bananas with low levels of fungicide. Application methods for bananas were similar to those described by Cayley *et al.* (1987).

Dusting

Various dusts can be applied to crops postharvest to control postharvest diseases. Wood ash is commonly used or sometimes limes (calcium hydroxide) for disease control in stored yams (Coursey 1967; Ogali *et al.* 1991; Thompson 1992). Fungicidal chemicals diluted in an inert carrier, such as talc, are available for postharvest use anddust formulations have been used on potatoes as they are being loaded into storeThe advantage of dusts over some sprays is that the crop is still dry after application. Some vegetables may be affected by bacteria when wet, so the result of the application of water diluted fungicides could be an increase in bacterial diseases. The disadvantage of dusts is that it is difficult to achieve complete coverage over the surface of the crop.

Fumigation

Various fumigants have been applied to crops, postharvest, for various purposes. Their use is strictly controlled by legislation that is constantly changing. Sulphur dioxide is used for controlling postharvest diseases of grapes (Pentzer and Asbury 1934; Ryall and Harvey 1959; Couey and Uota 1961), litchis (Jurd 1964; Sauco and Menini 1989; Underhill *et al.* 1992; La-Ongsri *et al.* 1993; Milne 1993; Tongdee 1993) and snap beans (Henderson and Buescher 1977). Acetaldehyde fumigation can be also be used on grapes (Avissar *et al.* 1989).

Paper pads or paper wraps, impregnated with diphenyl fungicide, were commonly applied to citrus fruit. The chemical vaporized slowly, protecting the fruit from fungal infection.

Sprout suppressants can be applied to root crops in store by fumigation. Tecnazene was reported to be applied as a fumigant in bulk stores, but because of the technical difficulties of application, contractors normally apply it (Burden and Wills 1989). 2-Aminobutane (2-AB) could be used as a fumigant in stored potatoes (Graham and Hamilton 1970). Fumigating oranges with 2-AB was also recommended for postharvest disease control (Eckert 1969).

Chemical pads

Paper pads impregnated with a fungicidal chemical were developed for the banana industry in the Windward Islands. These are called crown pads and they are used to prevent fungal infections on the cut crowns of fruit. Normally postharvest disease control is achieved by dipping or spraying the fruit with fungicide when it is taken to the packhouse. However, when bananas are dehanded in the field and packed directly into export cartons, it poses problems of how to protect the cut surface of the crown of the hand from infection. Applying these pads to the cut surface solved this. The pads are made from several layers of soft paper previously soaked in a fungicide (often thiabendazole) and then dried. The pad also absorbs latex given out from the cut surface and prevents it staining the banana fingers. Potassium aluminium sulphate may be added to the pads, which helps to coagulate the latex. It was claimed that this system contributed to an increase of 35% in exportable quantities (Hope-Mason 1984). The crown pads proved very effective in controlling crown rot (Genay 1990) and the fungicide residue levels in the banana are likely to be small (Johanson *et al.* 1989). However, the pads were difficult to apply using the technique described by the manufacturer and in Genay's experiments they were secured by rubber bands to ensure complete and even contact between the cut crowns and the pads. In observations by the author of commercial packs, this even contact between crowns and pads is not always achieved so these excellent results would not always be achieved in commercial practice (Table 8.3). This may be reflected in the fact that crown pads are not now used as a method of crown rot control in the major banana-producing countries. A British company, Harcloth of Bury, Lancashire, have patented the crown pad system on bananas.

Non-fungicidal methods of disease control

Considerable interest exists in controlling postharvest diseases and pests of crops without using chemical fungicides. There are several alternative strategies some of these are dealt with in the following.

Yeasts

It has been known for many years that some microorganisms found on the surfaces of growing plants can attack potential fungal pathogens. Some yeasts have been shown to be effective. Aspire is a commercial formulation of the yeast *Candida oleophila* registered for postharvest application to citrus for the control of green mould (*Penicillium digitatum*). Its mode of action appears to be that it competes with the pathogen for nutrients released by injuries. The yeast *C. guilliermondii* was shown to be antagonistic towards *Penicillium* spp. (McGuire 1993) and could be successfully incorporated into the commercial citrus waxes FMC 705, FMC 214 and Nature Seal where it

Table 8.3 Effects of chemical application method on the level of crown rot disease on bananas, where 0 = no infection and 7 is crown completely covered (source: Genay 1990)[a]

	Level of infection		Hands with pedicel rots (%)	Commercially marketable hands (%)	
Days in store	5	10	10	5	10
Microstat	2.4b	4.9a	66b	88bc	29a
Cascade	2.8b	4.8a	57ab	71ab	17a
Crown pads	0.8a	3.9a	36a	100c	63b
Untreated	3.06	5.3a	70b	67a	20a

[a]Figures followed by the same letter were not significantly different ($p = 0.05$).

was shown to survive for 2 months at 12°C. In trials by Brown *et al.* (2000) with Hamlin and Valencia oranges, colonization by *C. oleophila* of puncture-related injuries that either encompassed oil glands or individually ruptured glands was achieved within 1–2 days at 21°C. Ruptured oil glands were colonized more effectively if treated 7 hours after injury rather than immediately, since peel oil was toxic to *C. oleophila* but not to *P. digitatum* spores. Colonization of puncture injuries by the yeast was comparable after 2 days at 21 and 30°C, but no colonization occurred at 13°C. Growth rates of the yeast were similar in waxed and non-waxed fruits. *C. oleophila* colonized punctures more uniformly than individually damaged oil glands, and provided more effective control of *P. digitatum* originating at punctures than at oil gland injuries. Incubating treated fruits at 30°C for 2 days before storage at 21°C enhanced the control of *P. digitatum.*

Arras and Arru (1999) also evaluated the fungitoxic activity of antagonist yeasts against postharvest citrus fruit diseases. Trials were first performed in the laboratory using the yeasts *Pichia guilliermondii* (strain 5A), *Candida oleophila* (strain 13L), *Rhodotorula glutinis* (strain 21A) and Aspire. *P. guilliermondii* 5A strongly inhibited *P. italicum* on artificially wounded Oroblanco (a hybrid between pummelo and grapefruit) fruits with Aspire coming second. More than 200 yeasts were selectively isolated from microbial populations on the surface of different fruits by Lima *et al.* (1998) and tested against various postharvest fruit pathogenic fungi. At 20°C, the antagonists significantly reduced rot incidence and showed a wide range of activity on different host–pathogen combinations. Two of the yeasts grew in culture at temperatures ranging from 0–35°C. In assays performed *in vitro* at 24°C, the antagonists showed low sensitivity towards several fungicides common on fruits and vegetables. Antagonistic epiphytic yeasts were isolated from the surface of mandarin orange in Turkey and grapefruit in Israel by Kinay *et al.* (1998). Among the mandarin orange isolates, three were effective in inhibiting green mould. Among the grapefruit isolates, several yeasts showed high efficacy against green and blue mould and sour rot caused by *Geotrichum candidum.* At the wound site, the population of yeasts increased gradually and maintained a high concentration during one week of storage at 20°C. Teixido *et al.* (1999) found that preharvest application of the antagonistic yeast *Candida sake* on Golden Delicious apples was less effective against Penicillium rot *Penicillium expansum* than postharvest treatment. No advantages in biocontrol were observed when cold-stored apples were treated with the yeast antagonist both pre- and postharvest.

The inclusion of a *Bacillus* species or a rose-coloured yeast isolated from lemon leaves in either carnauba and polyethylene wax on citrus fruits was not effective in postharvest disease control (Visintin *et al.* 2000).

Other fungi

The yeast-like fungus *Aureobasidium pullulans* (isolate LS-30) displayed antagonistic activity against *Botrytis cinerea, Penicillium expansum, Rhizopus stolonifer* and *Aspergillus niger* on Italia grapes, and *B. cinerea* and *P. expansum* on Royal Gala apples (Castoria *et al.* 2001). Yam tubers that were sprayed with a conidiospore suspension of T*richoderma viride* in potato dextrose broth showed a large reduction in the range and number of microflora, including pathogens, on the tuber surface during 5 months of storage in a traditional yam barn (Okigbo and Ikediugwu 2001).

Bacteria

Several types of bacteria that attack fungi have been isolated from plants. *Bacillus* spp. on their own or combined with a fungicide could be used to control postharvest diseases of mangoes, avocados and litchis (Korsten *et al.* 1993). *B. subtilis* and *B. licheniformis* isolated from mango leaves and applied to fruit as a warm water dip controlled anthracnose levels. The effects were enhanced by the inclusion of either benomyl or prochloraz in the dip. *Bacillus* spp. isolated from leaves and fruit of avocados were more effective in controlling anthracnose and stem end rot of avocados when applied as a postharvest dip than prochloraz applied in the same way. The combination of *B. subtilis* and prochloraz was more effective than when they were applied separately. *B. stearotermophilus, B. megaterium* and *B. licheniform*is isolated from the phylloplane of litchis were more effective in reducing fruit browning and postharvest decay of litchis than benomyl when the treatments were applied as warm water dips (Korsten *et al.* 1993). Good results have been found in the control of anthracnose in mangoes and avocados with *Bacillus subtilis.* While not giving complete control it may have an application in reducing the levels of chemical fungicide application (Korsten *et al.*

1993). *In vitro* studies of *Bacillus stearotermophilus*, *B. magaterium* and *B. licheniformis* isolated from litchi trees showed that they effectively inhibited growth in 11 postharvest pathogens of litchi fruit (Korsten *et al.* 1993a). Postharvest dips with these antagonists were shown to be more effective in controlling decay, postharvest infection and fruit browning than a warm water plus benomyl treatment (Korsten *et al.* 1993a).

Pseudomonas syringae strains ESC-10 and ESC-11 produce syringomycin and control green and blue moulds of citrus caused by *Penicillium digitatum* and *P. italicum*, respectively (Bull *et al.* 1998). Isolates of *Pseudomonas syringae* NSA-6 and MA-4 reduced brown rot *Monilinia fructicola* to 28% and 73%, respectively, from 98% in the inoculated control after 5 days of incubation at 22°C. Both isolates reduced rhizopus rot *Rhizopus stolonifer* to 5% and 8% from 53% in the inoculated control after 5 days of incubation. Isolates MA-4 and NSA-6 suppressed brown rot from 63–30% and from 95–71–81%, respectively, after 3 and 4 days of incubation at 22°C. The use of 0.5% calcium chloride in the soak suspension significantly improved the activity of *P. syringae*, but the use of 'peach wax' (Decco 282) increased brown rot incidence and negated the beneficial effect of calcium chloride (Zhou *et al.* 1999).

Pseudomonas sp. isolates, associated with the skin of banana inhibited germination and appressoria formation *in vitro* of conidia of *Colletotrichum musae* (de Costa and Subasinghe 1998). It also inhibited the growth of *Fusarium* spp., *Botryodiplodia* spp. and *Ceratocystis paradoxa* by 30–42%. Crown rot was significantly reduced using the antagonistic bacteria as a postharvest dip.

Bacillus thuringiensis has been used for controlling insects for many decades. Its postharvest use in controlling tuber moth (*Phthorimaea operculella*) in stored potatoes was decribed by Das *et al.* (1998).

Ultraviolet light

When air is passed through UV light some of the oxygen molecules are broken down for an instant to atomic oxygen and ozone. This can have the effect of destroying pathogens when the fruit or vegetable is passed through the light. The overall results on strawberries from investigations by Nigro *et al.* (2000) indicated that treatment with low UV-C doses produced a reduction in postharvest decay related to induced resistance mechanisms. Gonzalez-Aguilar *et al.* (2001) also found that UV-C irradiation for 10 minutes of ripe Tommy Atkins mangoes was effective in suppressing decay and maintaining firmness during storage. However, it appears to be only partially effective in practice and it was concluded by Stevens *et al.* (1997) that UV-C and biological control agents could be used together as an alternative to chemical control of some storage diseases. Also relatively low levels of ozone are toxic to fruit and vegetables, although no UV damage was observed on treated fruits after storage by Gonzalez-Aguilar *et al.* (2001). See also Chapter 7 in the section on ozone.

Organic volatiles

Some organic volatile compounds produced by fruit are thought to have an effect on fungi. The inhibitory effect of (*E*)-2-hexenal, a volatile component of many fruits, especially after wounding, was studied for its potential for controlling mould development on fruits during storage by Archbold *et al.* (2000). A 100 ml volume of the compound was placed in 1 litre low-density film-wrapped clamshell containers with 150 g of either the blackberry cultivar Chester Thornless or the grape cultivar Flame Seedless and stored at 2°C for 7 days. Following removal of the over-wrapped film and chemical from the containers and transfer to 20°C, mould development was reduced at 20°C following 14 days of exposure to (*E*)-2-hexenal in 2°C storage.

Essential oils

Essential oils may provide alternatives and supplements to conventional antimicrobial additives in foods. Elgayyar *et al.* (2001) evaluated essential oils extracted from anise, angelica, basil, carrot, celery, cardamom, coriander, dill weed, fennel, oregano, parsley and rosemary against *Listeria monocytogenes*, *Staphylococcus aureus*, *Escherichia coli* O157:H7, *Yersinia enterocolitica*, *Pseudomonas aeruginosa*, *Lactobacillus plantarum*, *Aspergillus niger*, *Geotrichum* and *Rhodotorula*. Oregano essential oil showed the greatest inhibition. Coriander and basil were also highly inhibitory to *E. coli* O157:H7 and to the other bacteria and fungi tested. Anise oil had little inhibitory effect on bacteria but it was highly inhibitory to the fungi. There was no inhibition of either fungi or bacteria with carrot oil.

Arras and Usai (2001) in *in vitro* studies found that *Thymus capitatus* essential oil at 250 μl litre^{-1} had fungitoxic activity against *Penicillium digitatum, P. italicum, Botrytis cinerea* and *Alternaria citri*. The fungitoxic activity of *T. capitatus* essential oils at 75, 150 and 250 ppm on healthy orange fruits, inoculated with *P. digitatum* by spraying and placed in 10 litre desiccators, was weak at atmospheric pressure (3–10% inhibition at all three concentrations). In vacuum conditions (0.5 bar), conidial mortality on the exocarp was high at 90–97% at all three concentrations. These data proved not to be statistically different from treatments with thiabendazole at 2000 ppm.

Carbon dioxide

In a review of the fungistatic effects of high CO_2, Kader (1997) indicated that levels of 15–20% could retard decay incidence on cherry, blackberry, blueberry, raspberry, strawberry, fig and grape. In celery stored at 8°C disease suppression was greatest in atmospheres of 7.5–30% CO_2 with 1.5% O_2, but there was only a slight reduction in 4–16% CO_2 with 1.5% O_2 or in 1.5–6% O_2 alone (Reyes 1988). Pretreatment of Italia grapes with 20% CO_2 for 48 hours before storage reduced decay during subsequent storage at 0°C and 90–95% r.h. for 30 days in both types of grape. It had a marked effect during the first 10 days of cold storage on organic grapes (Massignan *et al.* 1999). Summer Red nectarines tolerated an air enriched with 15% CO_2 atmosphere for 16 days at 5°C. Development of brown rot decay in fruits inoculated with *Monilinia fructicola* 24 hours before storage was arrested. After 3 days of ripening in air at 20°C, the progression of brown rot disease was rapid in all inoculated nectarines, demonstrating the fungistatic effect of CO_2 (Ahmadi *et al.* 1999). Eris and Akbudak (2001) found that peaches stored at 0°C and 90% r.h. had lower levels of decay in 5% CO_2 and 2% O_2 than in air. Control of infections by the anthracnose and crown rot pathogen of bananas by exposure to 15% CO_2 or 1% O_2 resulted in a reduction in growth of the pathogen but not complete inhibition (Al-Zaemey *et al.* 1994). Also that level of CO_2 is toxic to the fruits, so this does not appear to be an alternative to fungicides. See also Chapter 7 in the section on controlled atmosphere storage.

Oxygen

Reports on the effects of reduced O_2 levels in storage on the development of diseases of fruit and vegetables have shown mixed results, but some positive effects have been reported. Parsons *et al.* (1974) showed considerable reduction in disease levels on tomatoes in controlled atmosphere storage, with most of the effect coming from the low O_2 levels and little additional effect from increased CO_2 levels (Table 8.4).

Organic Ziv bananas pretreated with a 2% O_2 atmosphere at 20°C for either 48 or 72 hours immediately after harvest was effective in reducing crown rot decay after simulated transport, ethylene ripening and shelf-life, but less so than 0.2% thiabendazole treatment (Pesis *et al.* 2001). See also Chapter 7 in the section on controlled atmosphere storage.

Nitrous oxide

Qadir and Hashinaga (2001) showed that fumigation and a mixture of 80% N_2O and 20% O_2 delayed the appearance of disease and reduced the lesion growth rate on fruit. This response to N_2O was dose and time dependent. This suppression of decay by N_2O treatment was thought to be a direct inhibitory effect on fungal growth rate and/or increased resistance of host tissue. The fruits and diseases tested included Fuji apples, inoculated with *Alternaria alternata* and *Penicillium expansum*, Toyonoka strawberries with *Botrytis cinerea, Fusarium oxysporum* f. sp. *fragariae* and *Rhizopus stolonifer*, Satsuma mandarin with *Geotrichum candidum*, Momotaro tomato with *F. oxysporum* f. sp. *lycopersici*, Fuyu persimmon with *Colletotrichum acutatum* and seedling guava with *R. stolonifer*.

Table 8.4 Effects of controlled atmosphere storage conditions on the decay levels (%) of tomatoes harvested at the green mature stage (Parsons *et al.* 1974)

Storage atmosphere	After removal from 6 weeks at 13°C	Plus 1 week at 15–21°C	Plus 2 weeks at 15–21°C
Air (control)	65.6	93.3	98.6
0% CO_2 with 3% O_2	2.2	4.4	16.7
3% CO_2 with 3% O_2	3.3	5.6	12.2
5% CO_2 with 3% O_2	5.0	9.4	13.9

Curing

Many root crops have a cork layer over the surface that is called a periderm. This serves as a protection against microorganism infections and excessive water loss. This layer can be broken or damaged during harvesting and handling operations so curing is essentially a wound-healing operation to replace the damaged periderm. Sealing citrus fruits in plastic film bags and then exposing them to 34–36°C for three days resulted in the inhibition of *Penicillium digitatum* infection and reduced decay and blemishes through lignification and an increase in antifungal chemicals in the peel of the fruit without deleterious effects on the fruit (Ben Yehoshua *et al.* 1989a, b). See also individual crops in Chapter 12.

Sodium carbonate

Immersion of lemons in a 3% solution of sodium carbonate plus hot water treatment at 52°C was as effective as imazalil treatment against *Penicillium digitatum* (Lanza *et al.* 1998).

Jasmonates

Methyl jasmonate is a plant growth regulator. There is evidence that jasmonates can also affect microorganisms. Droby *et al.* (1999) showed that postharvest application of jasmonic acid and methyl jasmonate reduced decay caused by the green mould *Penicillium digitatum* after either natural or artificial inoculation of Marsh Seedless grapefruit. The most effective concentration of jasmonates for reducing decay in cold-stored fruits or after artificial inoculation of wounded fruit at 24°C was 10.5 mol $litre^{-1}$. They suggested that the effect was indirectly by enhancing the natural resistance of the fruits to *P. digitatum* at high and low temperatures. Exposure of cultures of *Aspergillus flavus* to methyl jasmonate vapour by Goodrich-Tanrikulu *et al.* (1995) inhibited aflatoxin production. The amount of aflatoxin produced depended on the timing of the exposure.

There has been considerable research on the physiological properties of jasmonates, for example in mangoes where Gonzalez-Aguilar *et al.* (2000) showed that treatment with methyl jasmonate may prevent chilling injury symptoms without altering the ripening process. Droby *et al.* (1999) found that jasmonic acid and methyl jasmonate effectively reduced chilling injury in Marsh Seedless grapefruit after cold storage. Meir *et al.* (1996) suggested that methyl jasmonate might mediate the plant's natural response to chilling stress, and by its application might provide a simple means to reduce chilling injury in chilling-susceptible commodities such as avocado, grapefruit and capsicums.

Acibenzolar

Acibenzolar (*S*-methyl benzo[1,2,3]thiadiazole-7-carbothioate) is a chemical activator of induced systemic acquired resistance. When it was applied at 0.25–2.0 mg active ingredient ml^{-1} to harvested strawberry fruit by Terry and Joyce (2000), it delayed the development of grey mould disease caused by *Botrytis cinerea* during storage at 5°C by about 2 days.

Acidification

Thompson *et al.* (1992) found that in *in vitro* studies both the germination and growth of *Colletotrichium musae* were greatly reduced at pH 3 and 15°C. Al-Zaemey *et al.* (1993) studied the effects of fruit coating and organic acids on *in vitro* control of *C. musae*, but although reductions in the growth of the fungus were observed there was no complete control.

Coatings

Banks (1984, 1989) described a fruit coating called Tal Prolong, which consisted of sucrose esters of fatty acids and carboxymethylcellulose, and showed some reduction in levels of disease on apples coated with Tal Prolong compared with untreated fruit. Banana crowns coated with Semperfresh, which is similar to Tal Prolong and is also made up of sucrose esters of fatty acids and carboxymethylcellulose, showed delayed development of crown rot caused by infection with *Colletotrichium musae*. This effect could be enhanced by the inclusion of organic acids in the coating material (Al-Zaemey *et al.* 1993). See also the section on fruit coatings in Chapter 6.

9
Safety

Introduction

Fungi and bacteria will infect fresh fruit and vegetables especially during prolonged storage. In some cases the infection may be entirely superficial but would still make the crop either entirely unmarketable or at least reduce its commercial value. In other cases they can cause it to rot completely. Some of these organisms can produce toxic metabolic products. Those produced by fungi are called mycotoxins; those produced by bacteria are called by the name of the organism that produces them.

Church and Parsons (1995) have dealt with legislation related to safety of fruit and vegetables in a detailed review. There is considerable legislation related to food sold for human consumption. In the UK, for example, the Food Safety Act 1990 states that it is an offence to sell or supply food for human consumption if it does not meet food safety requirements. In the late 1990s the British Government set up a Food Standards Agency, one of whose concerns is food safety.

The use of modified atmosphere packaging has health and safety implications. One factor that should be taken into account is that the gases in the atmosphere could possibly have a stimulating effect on microorganisms. Farber (1991) stated that while modified atmosphere packaged foods have become increasingly more common in North America, research on the microbiological safety of these foods was still lacking. The growth of aerobic microorganisms is generally optimum at about 21% O_2 and falls off sharply with reduced O_2 levels whereas generally for anaerobic microorganisms their optimum growth is at 0% O_2 and falls as the O_2 level increases (Day 1996). With many modified atmospheres containing increased levels of CO_2, the aerobic spoilage organisms, which usually warn consumers of spoilage, are inhibited, whereas the growth of pathogens may be allowed or even stimulated, which raises safety issues (Farber 1991). Hotchkiss and Banco (1992) stated that extending the shelf-life of refrigerated foods might increase microbial risks in modified atmosphere packaged produce in at least three ways:

- increasing the time in which food remains edible increases the time in which even slow-growing pathogens can develop or produce toxin
- retarding the development of competing spoilage organisms
- packaging of respiring produce could alter the atmosphere so that pathogen growth is stimulated.

In the USA, the Food and Drug Administration's Bacteriological Analytical Manual sets out culture-based methods for detecting the presence of dangerous bacteria in food. Shearer *et al.* (2001) tested a polymerase chain reaction-based detection system for its sensitivity in detecting *Salmonella enteritidis*, *Escherichia coli* O157:H7, *Listeria monocytogenes* and other *Listeria* species. The technique was tested on fresh alfalfa sprouts, green peppers, parsley, white cabbage, radishes, onions, carrots, mushrooms, leaf lettuce, tomatoes, strawberries, cantaloupe, mango, apples and oranges. Generally the method was more sensitive than culture-based methods, but neither method was successful with low levels on any of the crops. However, this method allowed the detection of *S. enteritidis*,

E. coli O157:H7 and *L. monocytogenes* at least 2 days earlier than the conventional culture methods.

Micotoxins

Many microorganisms produce secondary metabolic products that are toxic. Over 150 species of fungi have been shown to be capable of producing mycotoxins and new ones are constantly being found.

Aflatoxin

The most common of these mycotoxins is aflatoxin, which is produced by *Aspergillus flavus* and *A. parasitic* and has been shown to infect dozens of food products, including groundnut kernels, coconut, wheat, rice, flour, dry beans and some leafy foods. Aflotoxins are highly toxic and carcinogenic. The presence of the fatty acid hydroperoxides, which can form in plant material either preharvest under stress or postharvest under improper storage conditions, correlates with high levels of aflatoxin production (Goodrich-Tanrikulu *et al.* 1995). Mycotoxins can also be found in fresh fruit although it is much less common than in dried products. Singh *et al.* (2000) found *A. flavus* infection in Indian jujube fruits (*Ziziphus mauritiana*) associated with decay and that approximately 86% of the *A. flavus* isolates were toxic. Exposure of cultures of *A. flavus* to methyl jasmonate vapour was shown to inhibit toxic production. The amount of aflatoxin produced depended on the timing of the exposure (Goodrich-Tanrikulu *et al.* 1995).

Patulin

The Guardian newspaper of 11 February 1993 reported that there was considerable public concern in the United Kingdom when apple juice in a supermarket was found to be contaminated with levels of the mycotoxin patulin [4-hydroxy-4*H*-furo[3,2-*c*]pyran-2(6*H*)-one]. This toxin has carcinogenic properties and has a maximum permitted level of 50 μg litre^{-1} in fruit juices. Its presence results from apples or pears that have been contaminated with *Penicillim patulum*, *P. expansum*, *P. urticae*, *Aspergillus clavatus* or *Byssochlamys nivea* and stored for too long before being processed. The toxin was found to occur in apple juice at levels 500–2500 μg litre^{-1} if it was made from rotting apples (Steiner *et al.* 1999).

In a market survey in South Africa by Leggott and Shephard (2001), eight of 31 fruit juice samples had patulin with concentrations ranging between 5 and 45 μg litre^{-1} with a mean of 10 μg litre^{-1}. Of six whole fruit products, two samples were contaminated with 10 μg litre^{-1} of patulin and of 10 infant fruit juices, six samples had patulin concentrations ranging between 5 and 20 μg litre^{-1}, but infant fruit purées showed no detectable patulin contamination. In the Cote d'Ivoire, eight out of 44 samples of fruit juice tested by Ake *et al.* (2001) were found to contain patulin, but a level of over 50 μg litre^{-1} was found in only one traditionally manufactured sample. In 11 Australian apple and pear juices tested by Steiner *et al.* (1999), the patulin content was below the limit of 50 μg litre^{-1}, in the apple juices the patulin content was more than 20 μg litre^{-1} while in the pear juices patulin could not be detected.

Lavrik *et al.* (2000) showed that Sorbilen, which is an ethylene absorber, decreased storage decay and accumulation of the patulin.

Edible mushrooms

In addition to the fungi that infect fruit and vegetables, it is well known that a number of other fungi produce various types of poisons. The toxic mushrooms, or toadstools, may be as large as some of the edible mushrooms, and some of the most poisonous species closely resemble edible species. Toxicity obviously depends on species, but can also be affected by the environment and growing conditions. Poisonous mushrooms are mostly members of the class Basidiomycetes, although some are Ascomycetes, such as the poisonous false morel (*Gyromitra esculenta*). The jack-o-lantern (*Clitocybe illudens*) is an orange–yellow fungus of woods and stumps and glows in the dark, but superficially resembles the delicious edible chanterelle (*Cantharellus cibarius*). *Lactarius* has milky or bluish juice and this genus contains the edible *L. deliciosus* and also several poisonous species. Most deaths attributed to mushroom poisoning result from eating members of the genus *Amanita*, especially the destroying angel, *A. virosa*, or the death cap, *A. phalloides*. Sterry (1995) gives more details of toxic species, including many photographs and drawings.

Other species that are poisonous, but may look like edible mushrooms, include:

- *Inocybe patouillardii* — *Cortinarius orellanus*
- *Inocybe fastigiata* — *Entoloma sinuatum*
- *Clitocybe dealbata* — *Hypholoma fasciculare.*

Bocchi *et al.* (1995) reported that in the 5 years 1989, 1992, 1993, 1994 and 1995 there were 20–49 human cases of fungal poisoning seen annually at a hospital in Parma in Italy. The species responsible were:

- *Agaricus hortensis* (2 cases)
- *Agaricus romagnesii* (2 cases)
- *Agaricus xanthoderma* (2 cases)
- *Agrocybe aegerita* (1 case)
- *Amanita phalloides* (17 cases)
- *Armillaria mellea* (19 cases)
- *Boletus edulis* (9 cases)
- *Boletus satanas* (6 cases)
- *Calocybe gambosa* (1 case)
- *Clitocybe candicantes* (4 cases)
- *Clitocybe nebularis* (9 cases)
- *Entoloma lividum* (33 cases)
- *Hygrophorus penarius* (1 case)
- *Hypholoma sublateritium* (2 cases)
- *Leccinum scabra* (4 cases)
- *Lepiota brunneoincarnata* (2 cases)
- *Lepiota lilacea* (1 case)
- *Leucoagaricus leucothites* (2 cases)
- *Macrolepiota procera* (1 case)
- *Omphalotus olearius* (17 cases)
- *Pleurotus ostreatus* (1 case)
- *Ramaria formosa* (4 cases)
- *Russula olivacea* (5 cases)
- *Xerocomus chrysenteron* (1 case).

The reasons why the toxicity varied could have been that with some of these fungi it they were eaten in poor condition or without adequate cooking (Bocchi *et al.* 1995).

Bacterial toxins

Escherichia coli

The coliform bacterium *E. coli* O157:H7 produces toxins and has the ability to grow on salad vegetables. This was demonstrated in a study by Abdul Raouf *et al.* (1993) where raw salad vegetables were subjected to minimal processing and storage conditions simulating those routinely used in commercial practice. Behrsing *et al.* (2000) found that packaging vegetables in an atmosphere containing 3% O_2 and 97% nitrogen had no apparent effect on populations of *E. coli* O157:H7, psychrotrophs or mesophiles. Populations of viable *E. coli* O157:H7 declined on vegetables stored at 5°C and increased on those stored for 14 days at both 12 and 21°C. The most rapid increases in populations of *E. coli* O157:H7 occurred on lettuce stored at 21°C. They found that an unknown factor or factors associated with carrots may inhibit the growth of *E. coli* O157:H7. The reduction in pH of vegetables was correlated with initial increases in populations of *E. coli* O157:H7 and other naturally occurring microflora. Eventual decreases in *E. coli* O157:H7 in samples stored at 21°C were attributed to the toxic effect of accumulated acids. Changes in the visual appearance of vegetables were not influenced substantially by growth of *E. coli* O157:H7, which is a potential danger to consumers.

E. coli O157:H7 is also a potential problem on fresh fruit. Dingman (2000) found that four of five apple cultivars tested (Golden Delicious, Red Delicious, McIntosh, Macoun, and Melrose) inoculated with *E. coli* O157:H7 promoted the growth of the bacterium in bruised tissue independent of the degree of ripeness or whether they were harvested from the tree or dropped fruit. When fruit was stored for 1 month at 4°C prior to inoculation with *E. coli* O157:H7, all five cultivars supported growth of the bacterium. Yu *et al.* (2001) found that two strains of *E. coli* O157:H7 survived externally and internally on stored strawberries at 23°C for 24 hours and at 10, 5, and –20°C for 3 days. The bacteria inside the fruit either survived as well as or better than bacteria on the surface.

In order to control infections, hygiene and carefully handling must be paramount. Other supplementary methods have been studied. A coating on Ruby Red grapefruit and Valencia oranges prepared from a shellac formulation with morpholine to achieve pH 9 was toxic to *E. aerogenes* and *E. coli* on storage at 13°C. The addition of the preservative paraben to the basic shellac had further inhibitory effects (McGuire and Hagenmaier 2001). Dipping inoculated fresh lettuce leaves and broccoli florets in hypochlorite solutions of 50 or 100 mg $litre^{-1}$ for up to 2 minutes was not effective in eliminating *E. coli* populations, although they significantly reduced the *E. coli* counts compared with those inoculated and not dipped (Behrsing *et al.* 2000). Yu *et al.* (2001) found that 3% hydrogen peroxide was the most effective chemical treatment in reducing the bacterial population, although other work has shown that this treatment can be phytotoxic. In other work, innoculated apples, oranges and tomatoes that had been submerged in sterile deionized water containing 1.5% lactic acid plus 1.5% hydrogen peroxide for 15 minutes at 40°C had reduced bacterial pathogens compared with those in deionized water only. Furthermore, substantial populations of the pathogens survived in the control wash water, whereas no *E. coli* O157:H7, *S. enteritidis* or *L. monocytogenes* cells were detected in the chemical treatment solution. The sensory and qualitative characteristics of apples treated with the chemical wash solution were not adversely affected by the treatment (Venkitanarayanan *et al.* 2000).

Salmonella

The bacterium *Salmonella typhimurium* produces toxins. Samples were taken from healthy and decayed portions of 341 fruits and vegetables collected in local supermarkets in the USA. Suspected *Salmonella* occurred in 20.2% of healthy and in 26.4% of decayed portions. The fungi *Alternaria* spp. and *Botrytis* spp caused two-thirds of the fungal infections causing the decay. In a similar analysis of 121 samples with mechanical injuries, in which about two-thirds were gouges, cuts and bruises, there were no significant differences in *Salmonella* incidence between injured and non-injured portions. Of 332 suspected *Salmonella* randomly isolated from healthy and decayed or injured portions, 5.1% were confirmed as *Salmonella* by physiological and serological testing (Wells and Butterfield 1999).

Botulism

In the USA, Draughon and Mundt (1988) investigated 34 cases of acid food botulism, of which 17 involved home-canned tomatoes. They concluded that obviously mouldy tomatoes should not be used in preparing juices and home-canned tomatoes, since mouldy tissue had a high pH and may harbour *Clostridium botulinum* and botulinal toxin. Church and Parsons (1995) mentioned that there was a theoretical potential fatal toxigenesis through infections by *C. botulinum* in the depleted O_2 atmospheres in modified atmosphere packed fresh vegetables. It was claimed that this toxigenesis has not been demonstrated in vegetable products without some sensory indication, that is, they have an 'off' taste (Zagory and Kader 1988, quoted by Church and Parsons 1995). However, Roy *et al.* (1995) showed that the optimum in-package O_2 concentration for suppressing cap opening of fresh mushrooms was 6% and that lower O_2 concentrations in storage are not recommended because they could promote growth and toxin production by *C. botulinum*. Sugiyama and Yang (1975) showed that with mushrooms prepacked in plastic film, *C. botulinum* not only grew on the mushrooms but also produced toxins.

Betts (1996), in a review of hazards related to modified atmosphere packaging of food, indicated that vacuum packing of shredded lettuce had been implicated in a botulinum poisoning outbreak. Toxin was not formed when the inoculated packages were kept at temperatures as low as 6°C. The results indicate that fresh vacuum-packaged enoki mushrooms do not present a botulism hazard when cultured aseptically and stored refrigerated. In the USA, botulinal toxin was not detected in 148 packages from 14 independent lots by Malizio and Johnson (1991) when spores were added to the packages. Spoilage was evident prior to toxin formation. Hauschild (1989) found that some strains of *C. botulinum* could grow at temperatures as low as 3.5°C. CO_2 levels of over 20% can retard bacterial growth such as *Erwinia carotovora* on potatoes. The degree of retardation increases with increasing concentrations of the gas, but at these high levels *C. botulinum* may survive (Daniels *et al.* 1985).

Measures to reduce the microbiological contamination of ready-to-use fruits and vegetables were recommended by Hguyen-The (1991) and included good manufacturing practices, disinfections of the product, standards and specifications to control the finished product, i.e. a use-by date that limits the shelf-life to 1 week and a storage temperature of 4°C. These will reduce the multiplication of microorganisms during transport and retail.

Listerosis

The food-borne human pathogen *Listeria monocytogenes*, which is a Gram-positive bacterium, can cause the disease listerosis, which normally has only mild influenza-like effects on healthy people, but can be fatal to pregnant women, the elderly and those suffering from chronic diseases. The bacterium can grow well in the low acid conditions found in vegetables. Problems of listerosis are not commonly associated with stored fruit and vegetables. However, in a survey by Heisick *et al.* (1989) for the US Food and Drugs Administration, *L. monocytogenes* was found in

21% of the samples of potatoes, 14% of radishes and 2% or less of cucumbers and cabbages. Berrang *et al.* (1989) found that it grew well on asparagus, broccoli and cauliflower in storage at 15°C, and it even grew on asparagus in storage at 5°C. They also found that under controlled atmosphere conditions *L. monocytogenes* grew just as well as in air, but since vegetables can be stored for longer periods under these conditions this can result in a longer time for the bacteria to grow. Conway *et al.* (2000) found that *L. monocytogenes* survived and its population increased on Delicious apple slices at 10 or 20°C in air or controlled atmospheres of 0.5% O_2 and 15% CO_2, but did not grow at 5°C.

Shigellosis

This is a disease caused by infection with the bacteria *Shigella* spp. It is also called bacterial dysentery and the symptoms are diarrhoea, fever and abdominal pains. Contamination has been shown to be from many sources, including shredded lettuce (Davis *et al.* 1988).

Aeromonas hydrophila

A. hydrophila is a Gram-negative bacterium that is widely spread in nature. Callister and Agger (1987) found it on virtually every type of vegetable they analysed from a grocery store. They also showed that it could grow on vegetables at temperatures of 5°C or even lower. The bacteria can cause diarrhoea, which is usually mild, but more severe cases have been reported. In experiments with 'ready-to-use' lettuce and mixed salads, Guerzoni *et al.* (1996) found that the inclusion of red chicory was an inhibiting ingredient to *A. hydrophila*. They also found an antagonistic effect of *Lactobacillus plantarum* against *A. hydrophila*, but the presence of *L. plantarum* appeared to negate the inhibiting effect of red chicory.

Safety in controlled atmosphere stores

Oxygen and carbon dioxide

Low O_2 and high CO_2 can have a direct lethal effect on human beings working in those atmospheres. In Britain the Health and Safety Executive (HSE 1991) showed that work in confined spaces, such as controlled atmosphere stores, could be potentially dangerous and entry must be strictly controlled, preferably through some permit system. Also, it is recommended that anyone entering such an area should have emergency breathing apparatus and also proper training and instruction in the precautions. Stringent procedures need to be in place, including a person on watch outside and the formulation of a rescue plan. It was also indicated (Anon 1974) that when a store is sealed anyone entering it must wear breathing apparatus. Warning notices should be placed at all entrances to controlled atmosphere stores and an alarm switch located near the door inside the chamber in the event that someone may be shut in. There should be a release mechanism so that the door can be opened from the inside. When the produce is to be unloaded from a store, the main doors should be opened and the circulating fan run at full speed for at least 1 hour before unloading is commenced. In the UK the areas around the store chamber must be kept free of impedimenta in compliance with the appropriate Agricultural Safety Regulations.

O_2 levels in the atmosphere of 12–16% can affect human muscular coordination, increase respiration rate and affect their ability to think clearly. At lower levels vomiting and further impediment to coordination and thinking can occur. At levels below 6% human beings rapidly lose consciousness, and breathing and heartbeats stop (Bishop 1996). The limits for CO_2 levels in rooms for human occupation was quoted by Bishop (1996) from the Health and Safety Executive publication EH40-95, *Occupational Exposure Limits*, as 0.5% CO_2 for continuous exposure and 1.5% CO_2 for 10 minutes exposure.

Great care needs to be taken when using modified atmosphere packaging containing 70 or 80% O_2, or even higher levels, because of potential explosions.

Ethylene

Ethylene is used in fruit ripening rooms. It is a colourless gas with a sweetish odour and taste, has asphyxiate and anaesthetic properties and is flammable (see also the section on ethylene in Chapter 7). Its flammable limits in air are 3.1–32% volume for volume and its autoignition temperature is 543°C. Care must be taken when the gas is used for fruit ripening to ensure those levels in the atmosphere do not reach 3.1%. As added precautions all electrical fitting must be of a 'spark-free' type and warning notices relating to smoking and fire hazards must be displayed around the rooms.

Toxicity of packaging material

Another safety issue is the possibility of the plastic films being used in modified atmosphere packaging being toxic. Schlimme and Rooney (1994) reviewed the possibility of constituents of the polymeric film used in modified atmosphere packaging migrating to the food that they contain. They showed that it is unlikely that the polymerized constituents would be transferred to the food because of their high molecular weight and insolubility in water. All films used can contain some non-polymerized constituents (monomers) that could be transferred to the food and the Food and Drugs Administration in the USA and also the European Community have regulations related to these 'indirect additives.' The film manufacturer must therefore establish the toxicity and extraction behaviour of the constituents with specified food simulants.

Packhouse safety

Clarke (1996) reviewed safety and hygiene of the workforce and suggested the following. In some countries there are strict laws to control the health and safety of all workers and these must, of course, be adhered to. In the absence of such laws one could use the principles laid down here as a guide.

> 'Clothing should be provided that is suitable for each task. Most workers would be engaged on the inspection or packing which would require simply an overall, a cap or hat and probably rubber gloves if preferred. The overall would often be of the coat type with a zip or button front. Trousers or skirts would often be open to personal choice although some employers may well provide trousers for work in what is often a cool, draughty and elevated position. Shoes would also commonly be provided with a firm, non-slip sole and sturdy foot support. Boots would not normally be necessary but the lighter, ladies fashion shoes and high heels, etc., must clearly be avoided.
>
> The start-up of all machines is a procedure which can lead to danger if not done to a set order. First a warning hooter should sound, giving everybody ample chance to stand clear of the machines, and then after a few seconds of the hooter the whole line should start up. This could, however, lead to unnecessary overloads on the electricity supply system as all motors would reach peak demand at once. It is just as accepable to start all motors in reverse order with the the last machines in the production line starting up first and the others in progression going back down the line to the first machines. The first motors are likely to be loaders and feeding belts so it is important that all the line is fully operational before any crop arrives.
>
> In case of danger on any part of the process line there should again be close adherence to the regulations that govern all electrical machines for health and safety at work. For example, "Every electrical motor shall be controlled by an efficient switch for starting and stopping so placed to be easily worked by the person in charge of the motor. In every place in which machines are being driven by any electric motor there shall be a means at hand for either switching off or stopping the motor/machine if necessary to prevent danger." This would normally be achieved by the provision of a large red button placed at convenient positions all along the process line. All other regulations such as insulation should be taken care of by the machine manufacturer.
>
> Walkways and forklift truck ways should be clearly marked on the floor in bright yellow paint so that both workforce and visitors are not likely to stray into the danger zone. Forklift trucks should be equipped with a hooter to warn anyone in danger and should be powered by gas to avoid fumes.'

10
Fruit ripening conditions

Introduction

Fruits can be classified into two groups, climacteric and non-climacteric. The former can be defined as fruit that can be ripened after harvest and the latter as fruit that do not ripen after harvest. The term climacteric was first applied to fruit ripening in the mid-1920s by Kidd and West (1927). They observed an increase in the respiration rate as measured by the production of CO_2 by Bramley's Seedling apples around the time of the normal commercial harvest. Biale and Barcus (1970) published measurements of the respiration rate of some selected fruit and classified them into climacteric, non-climacteric and indeterminate on the basis of their respiration rate. They classified banana, biriba, breadfruit, mango, papaya and soursop as climacteric, cacao, cashew and guava as non-climacteric and jackfruit, jambo, passionfruit and sapote as indeterminate.

Other fruit which have been shown to display typical climacteric respirational patterns include annonas, apricot, avocado, blueberry, cantaloupe melon, durian, feijoa, kiwifruit peach, persimmon, plum, pome fruits, and watermelon. Other fruit that can be classified as non-climacteric include aubergine, blackberry, capsicum, cherry, citrus fruits, cucumber, grape, litchi, olive, pineapple, pomegranate, raspberry, strawberry and tamarillo. In some cases there is conflict of opinion on classification. For example, fig and melon are classified as climacteric by Wills *et al.* (1989), but Biale (1960) classified them both as non-climacteric. Pratt (1971) showed that both Honeydew muskmelon and PMR-45 cantaloupe were climacteric. Biale and Barcus (1970) classified passionfruit as indeterminate whereas Akamine *et al.* (1957) classified it as climacteric. Grierson (1993), commenting on the classification of fruits into climacteric and non-climacteric, said that parts of citrus fruits are climacteric and other parts of the same fruit are non-climacteric.

The respiration rate of non-climacteric fruit and vegetables tends to decrease progressively during development. Wills (1998) found in experiments with strawberries, oranges, lettuces, beans, Chinese cabbage, bak choi (*Brassica chinensis*), choi sum (*B. parachinensis*) and gai lan (*B. alboglabra*) exposed to air containing 10–0.005 μl $litre^{-1}$ of ethylene at either 0–2.5 or 20°C that the storage life of all produce was linearly extended with a logarithmic reduction in ethylene concentration across the whole concentration range. The ethylene level found to accumulate around produce in markets was in the range 0.06–1.45 μl $litre^{-1}$, suggesting that premature ageing occurred in all non-climacteric crops during normal marketing. There appears to be no benign level of ethylene so that any reduction in ethylene concentration should help in maintaining the quality of non-climacteric fruits and vegetables. In non-climacteric fruit such as strawberry there was little change in ethylene production during ripening (Manning 1993). However, he showed that auxin produced by the achenes could affect the colour (anthocyanin production) of the ripening fruit. Removal of the achenes from developing strawberry fruit resulted in quicker colouring of the area where they were removed.

Climacteric fruits are often harvested in an immature state and ripened postharvest. Most climacteric fruit species will ripen normally on the tree after they have reached full maturity. However, in the case of avocado, which is a climacteric fruit, they do not normally ripen until they are harvested and can remain in a mature but unripe condition on the tree until picked (Seymour and Tucker 1993).

Ripening involves a number of physical and chemical changes that occur in fruit. These normally occur when the fruit has developed to what is referred to as full maturity. However, immature fruit may be harvested and exposed to certain postharvest conditions (temperature, gas content in the atmosphere, humidity) that are conducive to ripening. The changes that can occur during ripening may be independent of each other. However, under the correct circumstances these processes are initiated together and proceed together, producing an acceptably ripe fruit. In other circumstances changes such as reduction in acidity or change in colour may occur at a different rate to other processes producing a fruit which is fully ripe in all other aspects except that one.

Initiation of ripening occurs when a threshold level of ethylene is reached in the cells of the fruit. This occurs naturally within the fruit at a point during maturation. It can occur earlier if the fruit undergoes stress, e.g. if a disease-causing microorganism attacks the fruit or the tree on which it is growing. Ripening can also be initiated by exposing the fruit to exogenous ethylene so that the threshold level infiltrates into the cells of the fruit. As fruits approach maturity they become more sensitive to ethylene-initiated ripening (Knee *et al.* 1987). Duque and Arrabaca (1999) found that the respiratory climacteric and the induction of alternative oxidase were tightly and temporally linked.

Changes during fruit ripening

Colour

The most obvious change in many fruits during ripening is their external colour. In tomatoes chlorophyll levels are progressively broken down into phytol. It was observed that in cherry tomatoes the total chlorophyll level was reduced from 5490 μg per 100 g fresh weight in green fruit to 119 μg per 100 g fresh weight in dark-red fruit (Laval-Martin *et al.* 1975). Concurrent with this degradation process, lycopene, carotenes and xanthophylls are synthesized, giving the fruit its characteristic colour, usually red (Grierson and Kader 1986). Total caroteniods in cherry tomatoes increased from 3297 μg per 100 g fresh weight in green fruit to 11 694 μg 100 g fresh weight in dark-red fruit (Laval-Martin *et al.* 1975). This colour development occurs in both the pulp and the flesh of the tomato. The optimum temperature for colour development is 24°C; at 30°C and above lycopene is not formed.

The pigments in the peel of bananas and plantains are chlorophylls, carotenoids and xanthophylls. The change in colour of ripening fruits is associated with the breakdown of chlorophylls with carotenoid levels remaining relatively constant (Seymour 1985; Montenegro 1988) (Figure 10.1). Cavendish banana cultivars can fail to degreen completely when they are ripened at 25°C and above (Seymour *et al.* 1987; Semple and Thompson 1988). This can result in bananas, which are ripe in every other respect, remaining green. The higher the temperature, the more obvious is the effect. It is one cause of the physiological disorder of Cavendish bananas called 'pulpa crema' or 'yellow pulp.' This occurs when banana fruit are initiated to ripen on the plant, but because the ambient temperature is above 25°C the pulp ripens but the chlorophyll in the skin is not fully broken down. With plantains it was shown that complete chlorophyll destruction can occur even at 35°C (Seymour 1985; Seymour *et al.* 1987). Studies on why this effect of temperature on degreening of Cavendish bananas occurs failed to reach a definitive conclusion (Black-

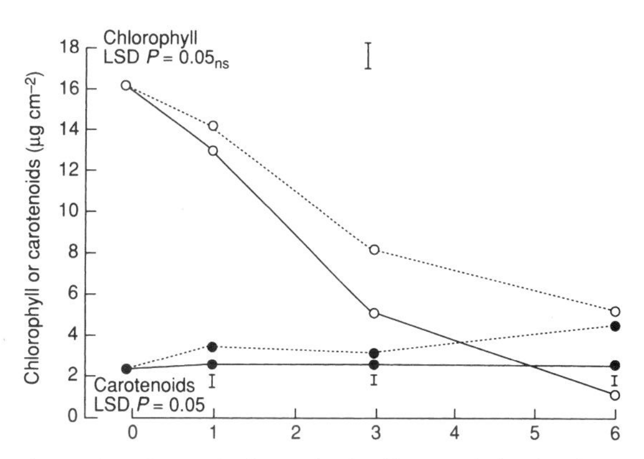

Figure 10.1 Changes in pigment levels of bananas during ripening at either 20°C or 35.2°C. Fruit were exposed to 1000 μl l^{-1} of ethylene before being ripened in air. Source: Seymour (1985).

bourn *et al.* 1990), although there was some indication that it was related to the thylakoid ultrastructure of the chloroplasts (Seymour 1985; Blackbourn *et al.* 1990).

Colour changes during ripening of many fruit, including bananas, have been used as a rough guide to the stage of ripeness in many banana cultivars. It is commonly used commercially in the form of colour-matching charts based on the degree of peel yellowing where colour index 1 is allocated to fruit that are dark green and colour index 6 is for fruit which are fully yellow (Von Loesecke 1949; Lizada *et al.* 1990). Examples of these colour charts are supplied by commercial banana companies and can also be found in Stover and Simmonds (1987), Abdullah and Pantastico (1990) and Thompson (1996). Standard colour charts exists for a variety of fruits for example tomatoes (see Figure 12.107, in the colour plates).

Advanced stages of banana ripeness are characterized by the appearance of brown flecks or spots on the skin. Conflicting reports, however, dispute the colour index at which these superficial spots appear. For triploid bananas of the same cultivar, Miem (1980) reported colour index 5.5, whereas Ng and Mendoza (1980) found that they occur at colour index 3.5–4.

Texture

Fruits normally soften progressively during ripening (Figure 10.2). Instrumental measurement of fruit texture needs careful application and interpretation to ensure they are relevant to organoleptic perception of texture (Batu and Thompson 1993). Although exact biochemical mechanisms have not yet been fully established, it is believed that softening is largely due to the breakdown of starch and other non-pectic polysaccharides in the pulp, thereby reducing cellular rigidity (Lizada *et al.* 1990). Softening of bananas during ripening appears to be associated with two or three processes (Smith 1989). The first of these is the breakdown of starch to form sugars since starch granules could have a structural function in the cells. The second is the breakdown of the cell walls due to the solublization of pectic substances and even the breakdown of cellulose. Smith (1989) found evidence of increased activity of cellulase during banana ripening. The possible third is the movement of water from the peel of the banana to its pulp during ripening. This process could affect the turgidity of the skin, which would be enhanced by transpirational losses. This change in the moisture status of the fruit also contributes to the ease with which the peel can be detached from the pulp. In studies of Japanese pears, Sirisomboon *et al.* (2000) found that the skin contributed 70–80% of the firmness of the fruit.

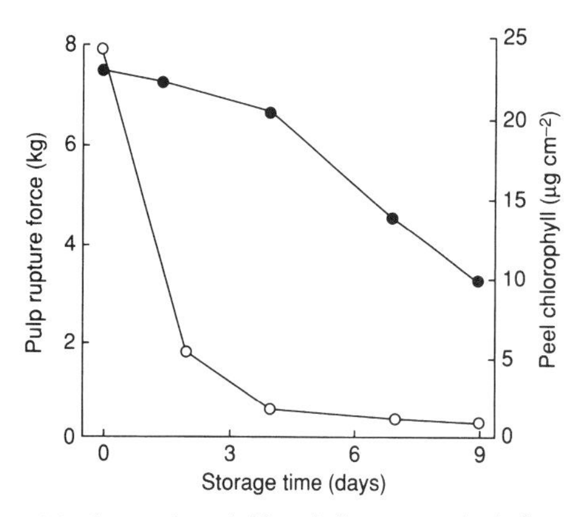

Figure 10.2 Changes in peel chlorophyll content and pulp firmness of Keitt mangoes during storage at 22°C. Source: Medlicott and Thompson (1985).

During ripening there are major changes in the pectic polymers in the cell walls (Tucker and Grierson 1987). Neutral sugars, mainly galactose in most fruit but also some loss of arabinose, which are major components of the cell wall neutral pectin, are lost. There are also losses of acidic pectin. The solubility of these polyuronides increases and in several cases they have been shown to be progressively depolymerized (Tucker and Grierson 1987). Sirisomboon *et al.* (2000) also found that the solubilization of non-soluble pectin to water-soluble pectin appeared to influence the texture of Japanese pears.

Stow and Genge (1993) measured the cell wall strength of apples using osmotic techniques. They found that cell walls do not weaken during fruit softening and suggested that softening results from loss of cell-to-cell cohesion. The soluble pectin content of apples did not correlate with fruit firmness, which casts doubt on the hypothesis that softening results from loss of pectin from the middle lamella of cells (Stow and Genge 1993). The major effect of ethylene removal from stored apples is a delay in the onset of softening (Dover and Stow 1993). They also suggested that removal of ethylene from the store could slow softening once it has started. Cellulase is involved in fruit softening during ripening of avocado fruits. At least 11 multiple forms were identified (Kanellis and Kalaitzis 1992).

Genetic engineering has produced fruit that do not soften normally, but due to restrictions on products of biotechnology in the European Union and other countries, the market was restricted.

As bananas and plantains ripen, the ratio of the mass of the pulp to the mass of the peel increases and the peel becomes progressively easier to detach from the pulp. This pulp to peel ratio can be a measure of the fruit's ripeness.

Carbohydrates

During the developmental stage of climacteric fruit there is a general increase in starch content. The most striking chemical change during ripening is the hydrolysis of starch to simple sugars. Starch occurs in the plastids of unripe avocados at levels of about 12 mg g^{-1} on a dry weight basis, but this reduces to undetectable levels in ripe fruit (Pesis *et al.* 1976). In kiwifruit starch was shown to be hydrolysed to glucose and fructose and to a lesser degree to sucrose during ripening (Okuse and Ryugo 1981). Medlicott and Thompson (1985) and Medlicott *et al.* (1986) showed that during ripening of mangoes the starch content was completely hydrolysed to sugar during ripening. The major sugars were identified as glucose, fructose and sucrose, with the last being in the largest proportion (Figure 10.3). The reduction in the level of sucrose towards the end of the storage period was probably due to it being used by the fruit for metabolic activity after all the starch had been hydrolysed. In dessert bananas, ripening involves a reduction in starch content from around 15–25% to less than 5% in the ripe pulp, coupled with a rise of similar magnitude in total sugars (Barnell 1943; Desai and Deshpande 1975; Lizada *et al.* 1990). During the early part of ripening sucrose is the predominant sugar, but in the later stage glucose and fructose predominate (Barnell 1943).

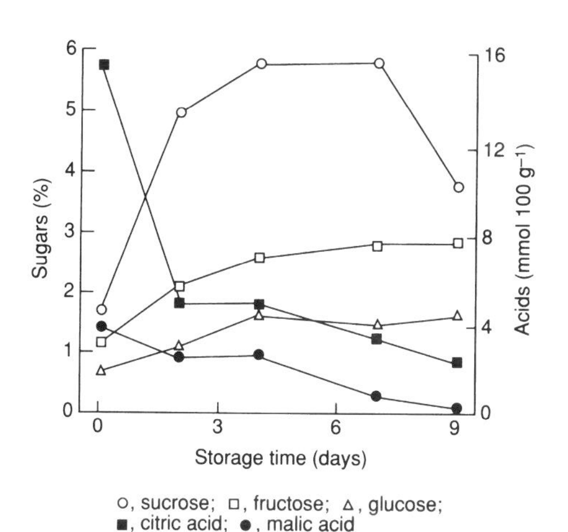

Figure 10.3 Changes in the sugar and acid contents of Keitt mangoes during storage at 22°C. Source: Medlicott and Thompson (1985).

The proportion of the different sugars is related to the stage in the respiratory climacteric of the fruit (Hubbard *et al.* 1990). It was also shown that starch is broken down to sucrose by the action of sucrose phosphate synthetase and non-reducing sugars from sucrose by acid hydrolysis. The onset of the starch–sugar conversion has been shown to be influenced by harvest maturity, with more mature fruits responding earlier. These changes have been demonstrated in both triploid (*Musa* AAA) (Madamba *et al.* 1977) and diploid (*Musa* AA) (Montenegro 1988) bananas. In bananas the breakdown of starch is usually completed during ripening, but in plantains this breakdown is not complete even when the fruit is fully yellow and soft (George 1981). In Apple bananas (*Musa* AA) the fruit still has residual starch when it is fully yellow and needed to be ripened further before being eaten (Wei and Thompson 1993).

Acids

Although the development of sweetness is important, organic acids also influence the overall fruit flavour. Acids help form the desirable sugar–acid balance necessary for a pleasant taste. The acidity of fruits generally decreases during ripening. Organic acids present in fruit vary with different fruit, with malate and citrate being the most common (Ulrich 1970; Ali Azizan 1988).

Bananas and plantains, like most other fruits, are acidic in with a pulp pH below 4.5 (Von Loesecke 1949). Palmer (1971) showed that the main acids in bananas were citric, malic and oxalic acids and the levels of these acids normally increase during ripening. Malic acid is the major organic acid in apples and pears but some apples contain quantities of citric acid and some pears quantities of quinic acid (Ulrich 1970). Titratable acidity in bananas increased during ripening at 20°C and then decreased (Desai and Deshpande 1975); for Dwarf Cavendish it was 1.95 meq. per 100 g fresh weight at harvest and reached a peak of 5.15. Medlicott and Thompson (1985) showed that in Keitt mangoes the principal acids were citric and malic acids and that there was a large decrease in citric acid dur-

Table 10.1 The change in ascorbic acid content (mg per 100 g fresh weight of the pulp of three varieties of bananas in storage at 20°C (source: Desai and Deshpande 1975)

Variety	At harvest	After storage for 3 weeks
Dwarf Cavendish	0.25	3.11
Rasabale	5.10	12.45

ing ripening, but only a small decrease in malic acid (Figure 10.3). The Ascobic acid content of three banana varieties (Dwarf Cavendish, Rasabale, Rajabale) increased during ripening at 20°C for 21 days (Table 10.1) and then decreased slightly to the end of the experiment (35 days).

Phenolic compounds

Bananas and plantain fruits can contain high levels of phenolic compounds, especially in the peel (Von Loesecke 1949). Phenolics such as tannins are polymerized to insoluble compounds, resulting in a reduction in astringency in the ripe banana fruit (Lizada *et al.* 1990). Tannins are perhaps the most important phenolic from the point of view of fruit utilization because they can give fruit an astringent taste. As fruit ripens their astringency becomes lower, which seems to be associated with a change in structure of the tannins, rather than a reduction in their levels, in that they form polymers (Von Loesecke 1949; Palmer 1971). Phenolics are common in many fruits and are responsible for the oxidative browning reaction when the pulp of fruit (especially immature fruit) is cut. The enzyme polyphenol oxidase is responsible for this reaction (Palmer 1971). In Carabao mango fruit there was a progressive decrease in total phenolic content during ripening, which was associated with loss in astringency (Tirtosoekotjo 1984). Levels of polyphenols, particularly the high molecular weight polyphenols, in guava were shown to decrease as the fruit matures (Itoo *et al.* 1987). Young guava fruit contained 620 mg per 100 g fresh weight, of which 65% were condensed tannins (Itoo *et al.* 1987). In persimmon the soluble tannin content in fruit varies between cultivars and in astringent cultivars it is particularly high in immature fruit (Ito 1971).

The latex which spurts from the broken pedicel of immature mango fruits is structurally similar to phenolic allergens found in other genera of Anacardiaceae. The main component of mango latex was identified as 5-[2-(*Z*)-heptadecenyl]resorcinol (Bandyopadhyay *et al.* 1985).

Flavour and aroma

Flavour is a subtle and complex perception combining taste, smell and texture or mouth feel. Ripening usually brings about an increase in simple sugars to give sweetness, a decrease in organic acids and phenolics to minimize astringency and an increase in volatiles to produce the characteristic flavour (Pantastico 1975). The characteristic aroma of ripe fruit is due to the production of a complex mixture of individual volatile components.

In apples and pears, butyl ethanoate, 2-methylbutyl ethanoate and hexyl ethanoate are typical flavour and aroma compounds that are synthesized during ripening (Dimick and Hoskins 1983). Terpenoid compounds such as linalool and its epoxide and farnesene have been shown to develop in some apple cultivars (Dimick and Hoskins 1983). Controlled atmosphere storage of apples in either 2% O_2 and 98% nitrogen or 2% O_2, 5% CO_2 and 93% nitrogen showed that few organic volatile compounds were produced during the storage period (Hatfield and Patterson 1974). Even when the fruit were removed from storage they did not synthesize normal amounts of esters during ripening.

At least 350 flavour and aroma compounds have been shown to occur in ripe bananas and their synthesis generally occurs relatively late in the climacteric, with acetate and butyrate esters accounting for about 70% of the total volatile emanations (Tressl and Drawert 1973). McCarthy *et al.* (1963) claimed that amyl and butyl esters gave bananas their distinctive flavour and aroma and a fruity flavour and aroma, respectively. The exact relationship between the chemistry and the flavour and aroma of bananas has not yet been fully determined. However, in addition to the volatile compounds above, other esters, and also aldehydes, alcohols and ketones, have been associated with fruit flavour and their production rates can increase during banana ripening (Tressl and Jennings 1972).

Over 80 volatile aroma compounds have been identified in guava fruit (Askar *et al.* 1986). The production of volatile chemicals has been shown to change during maturation of guava. As the fruit matures the production of isobutanol, butanol and sesquiterpenes decreases and in mature and ripe fruit ethyl acetate, ethyl caproate, ethyl caprylate and *cis*-hexenyl acetate levels are high (Askar *et al.* 1986).

More than 400 substances have been shown to contribute to the odour of tomatoes, but no single compound or simple combination of these compounds has the typical smell of ripe fruit (Hobson and Grierson 1993).

Toxicity

Toxins may exist in unripe fruit that reduce as the fruit ripens. Tomatoes at the green stage of maturity contain a toxic alkaloid called solanine that decreases during ripening (Laval-Martin *et al.* 1975). Ackee fruit contain the toxin hypoglycin in the arils, which gradually reduces as the fruit matures (Larson *et al.* 1994).

Controlled atmosphere storage on ripening

The level of CO_2 and O_2 in the surrounding environment of climacteric fruit can affect their ripening rate. Controlled atmosphere storage has been shown to suppress the production of ethylene by fruit. The biosynthesis of ethylene in ripening fruit was shown in early work by Gane (1934) to cease in the absence of O_2. Ethylene concentrations in the core of apples were generally progressively lower with reduced O_2 concentrations in the store over the range of 0.5–2% O_2 (Stow 1989a). Wang (1990) reviewed the literature on the effects of CO_2 and O_2 on the activity of enzymes associated with fruit ripening, and cited many examples of the activity of these enzymes being reduced in controlled atmosphere storage. This is presumably due, at least partly, to many of these enzymes requiring O_2 for their activity.

Quazi and Freebairn (1970) showed that high CO_2 and low O_2 delayed the high production of ethylene associated with the initiation of ripening in bananas, but the application of exogenous ethylene was shown to reverse this effect. Wade (1974) showed that bananas could be ripened in atmospheres of reduced O_2, even as low as 1%, but the peel failed to degreen, which resulted in ripe fruit which were still green. Similar effects were shown at O_2 levels as high as 15%. Since the degreening process in Cavendish bananas is entirely due to chlorophyll degradation (Seymour *et al.* 1987), the controlled atmosphere storage treatment was presumably due to suppression of this process. Hesselman and Freebairn (1969) showed that ripening of bananas, which had already been initiated to ripen by ethylene, was slowed in low O_2 atmospheres.

Goodenough and Thomas (1980, 1981) also showed suppression of degreening of fruits during ripening, in this case with tomatoes ripened in 5% CO_2 combined with 5% O_2. Their work, however, showed that this was due to a combination of suppression of chlorophyll degradation and the suppression of the synthesis of carotenoids, lycopene and xanthophyll. Jeffery *et al.* (1984) also showed that lycopene synthesis was suppressed in tomatoes stored in 6% CO_2 combined with 6% O_2.

Ethylene biosynthesis was studied in Jonathan apple fruits stored at 0°C under controlled atmosphere storage conditions with raised CO_2 concentrations (0–20%) and low (3%) and high (15%) O_2 concentrations. Fruits were removed from storage after 3, 5 and 7 months. Internal ethylene concentration, 1-aminocyclopropane 1-carboxylic acid (ACC) levels and ethylene-forming enzyme activity were determined in fruits immediately after removal from storage and after holding at 20°C for 1 week. Increasing CO_2 concentration from 0–20% at both low and high O_2 concentrations inhibited ethylene production by fruits. ACC levels were similarly reduced by increasing CO_2 concentrations even at low O_2; low O_2 enhanced ACC accumulation but only in the absence of CO_2. Ethylene-forming enzyme activity was stimulated by CO_2 up to 10% but was inhibited by 20% CO_2 at both O_2 concentrations. The inhibition of ethylene production by CO_2 may, therefore, be attributed to its inhibitory effect on ACC synthase activity (Levin *et al.* 1992). In other work, storing preclimacteric fruits of apple cultivars Barnack Beauty and Wagner at 20°C for 5 days in 0% O_2 with 1% CO_2 (99% nitrogen) or in air containing 15% CO_2 inhibited ethylene production and reduced ACC concentration and ACC synthase activity compared with storage in normal atmospheres (Lange *et al.* 1993).

Low O_2 and increased CO_2 concentrations were shown to have an effect on the decrease in ethylene sensitivity of Elstar apple fruits and Scania carnation flowers during controlled atmosphere storage (Woltering *et al.* 1994). Storage of kiwifruit in 2–5% O_2 with 0–4% CO_2 reduced ethylene production and ACC oxidase activity (Wang *et al.* 1994). Ethylene production was lower in Mission figs stored at 15–20% CO_2 concentrations compared with those kept in air (Colelli *et al.* 1991).

Design of ripening rooms

The primary requirements for ripening rooms are that they should:

- have a good temperature control system
- have good and effective air circulation
- be gas tight
- have a good system for introducing fresh air.

Air circulation around the fruit is important to prevent local accumulations of the CO_2 given out by the fruit and to ensure good contact between the fruit and the ethylene gas applied to initiate ripening. In the case of bananas and some other types of fruit the boxes are lined with polyethylene film and usually transported to ripening rooms stacked on pallets. Several systems are used. One is to remove each box from the pallet, pull back the plastic film and re-stack them on pallets so that there is a space between boxes. In other cases the pallets are stacked so that one hand hole in each box is facing outwards. As the fruit are loaded into the rooms the plastic just inside each hand hold is torn to facilitate air exchange. This is especially important where fruit have been vacuum packed.

Air circulation systems are usually largely convectional. Air is blown through the cooler and then across the top of the store just below the ceiling. The cooled air falls by convection through the boxes of fruit and is taken up at floor level for recirculation. Many modern ripening rooms have air channels in the floor through which air is circulated at high pressure. This forces it up through the pallets of boxes and should give better air circulation. Special devices such as inflatable air bags placed between pallets are now used to ensure better air circulation and, therefore, more even fruit ripening.

Good ventilation to enable fresh air to be introduced is very important for successful fruit ripening. During the period of initiation of ripening, which is usually 24 hours, no fresh air is introduced to the rooms. This is the period when ethylene is introduced. Directly after this period the rooms must be thoroughly ventilated. In a study of bananas in Ecuador (Thompson and Seymour 1984), it was found that the CO_2 level in ripening rooms had increased to 7% during this 24 hour initiation period. Even with a good fan extraction system it took 40 minutes of ventilation to bring the CO_2 levels to below 1%. This ventilation with fresh air must be repeated every 24 hours during subsequent ripening to prevent levels of CO_2 becoming too high. If rooms are not frequently ventilated ripening can be delayed, or abnormal ripening can occur.

The rooms need to be gas tight in order to ensure that threshold levels of ethylene are maintained around the fruit during the initiation period. The most common place where leakage occurs is around the doors. It is crucial, therefore, that special gas-tight doors are fitted and that they have suitable rubber gaskets. These must be inspected regularly to ensure that they have not been damaged. Gas can also be lost through the walls of ripening rooms. Commonly these are metal-lined on the inside with mastic between joints to ensure no gas can pass through them. Gas-tight paint can be used on walls. All holes through the walls for plumbing and electrical fittings must be blocked with mastic.

During the 1990s, there was an increasing demand for all the fruit being offered for sale in a supermarket to be of exactly the same stage of ripeness so that it has an acceptable and predictable shelf-life. This led to the development of a system called 'pressure ripening,' which is used mainly for bananas but is applicable to any climacteric fruit (see Chapter 12 under Bananas for more details). The system is the same as that described above, but it also involves direction of the circulating air so that it is channelled through the boxes of fruit. This results in the ethylene being in contact equally with all the fruit in the room. At the same time the CO_2 is not allowed to concentrate around the fruit (see Figure 12.39).

It is advisable to have high humidity of 90–95% r.h. in ripening rooms. To this end, many rooms are fitted with some humidification device such as a spinning disc humidifier. However, if the rooms are full of fruit and the refrigerant in the cooling coils used to maintain the room temperature is regulated to within a few degrees of the required room temperature, then this should be sufficient to keep the humidity high. More information on humidification is given in Chapter 7, and safety aspects of ripening rooms are deal with in Chapter 9.

Ethylene on ripening

The change in physiology of climacteric fruit from maturation to ripening is initiated when cellular quantities of ethylene reach a threshold level (Hoffman and Yang 1980). The metabolic pathway within the cells of the plant that result in ethylene production is complex. This involves a series of steps culminating in the synthesis of SAM (*S*-adenosylmethionine), which is converted to ACC (1-aminocyclopropane-1-carboxylic acid) by the action of ACC synthase. ACC is the immediate precursor of ethylene biosynthesis in plants. ACC oxidase is required to convert ACC to ethylene (Hamilton *et al.* 1990). It is a labile enzyme and is sensitive to oxygen and can be inhibited at temperatures above 35°C and under anaerobic conditions. It was suggested that high levels of CO_2 in stores could compete with ethylene for binding sites in fruits (Burg and Burg 1967). They showed that in the presence of 10% CO_2 the biological activity of 1% ethylene was abolished. Yang (1985) also showed that CO_2 accumulation in the intercellular spaces of fruit acts as an ethylene antagonist.

Little ethylene, or ethylene precursors, has been detected in maturing fruits and it is not clear exactly what causes its rapid production and, thus, the beginning of the climacteric rise. However, the application of ethylene to fruit that are not yet fully mature can initiate the climacteric and the endogenous production of ethylene for example in bananas (Biale 1950). In bananas, the principal source of ethylene is its biosynthesis in the fruit pulp (Ke and Tsai 1988). This increase in ethylene synthesis is followed by changes in the fruits' physiology, texture and composition associated with ripening as described above including the typical climacteric rise in respiration.

Threshold levels of ethylene will be reached naturally at fruit maturity, but can also arise by the fruit being put under stress during production by water shortage or infection by disease-causing organisms or by mechanical damage to the fruit during harvesting and handling operations. Exposing fruit postharvest to low humidity may also result in sufficient stress to the fruit to initiate ripening. This was shown by Thompson *et al.* (1974a) for plantains. The threshold level can also be achieved by exposing the fruit, for a sufficient period, to a threshold concentration of ethylene in a gas-tight room. This is the principal used in most commercial fruit ripening rooms.

Sources of ethylene

Ethylene application methods

Detailed information on ethylene is covered in Chapter 7 and ripening requirements are dealt with for many crops in Chapter 12. With bananas, very low concentrations of ethylene are sufficient for mature fruit at 14–19°C (the temperatures commonly used commercially). These are in the range of 1–10 parts per million ($\mu l\ litre^{-1}$) for 24 hours. However, in commercial practice 1000 $\mu l\ litre^{-1}$ is commonly used to ensure ripening. This is partly because many ripening rooms are not fully gas tight and the concentration may be rapidly reduced through leakage. Giant Cavendish bananas from various commercial sources in the Caribbean and Latin America were all successfully initiated to ripen by exposure to 10 $\mu l\ litre^{-1}$ of ethylene for 24 hours at 19°C (Thompson and Seymour 1982). The same paper recorded that 1000 $\mu l\ litre^{-1}$ of acetylene for 24 hours at 19°C was required to achieve the same ripening initiation. This requirement of higher concentrations of acetylene than ethylene to achieve a similar biological effect had previously been described by Burg and Burg (1967).

There are several sources of ethylene that can be used in fruit ripening and degreening. The source and the method selected for applying ethylene to fruit depend on cost, convenience and safety factors.

Liquid

Compounds that decompose in or on the crop to release ethylene can have the advantage of easy application. Etacelasil [2-chloroethyltris(ethoxymethoxy) silane] or ACC (1-aminocyclopropane-1-carboxylic acid), which is the immediate precursor of ethylene biosynthesis in plants, have not been used practically for the application of ethylene to crops. However, 2-chloroethylphosphonic acid, which is commonly called 'Ethrel' or 'Ethephon,' has been used as a source of ethylene for decades and a voluminous literature exists on its application and effects. These include initiation of flowering in pineapples, anti-lodging agent in cereals, stimulation of latex flow in rubber, enhanced degreening of citrus fruits and initiation of ripening in climacteric fruit. Ethrel is hydrolysed in plant tissue to produce ethylene, phosphate and chloride:

$$Cl{-}CH_2{-}CH_2{-}PHO_3^- \xrightarrow{OH^-} CH_2 = CH_2 + PH_2O_4^- + Cl^-$$

Ethylene can also be released from Ethrel by mixing it with a base such as sodium hydroxide. Ethrel 'C' will release 93 g from 1 litre or 74.4 litres of ethylene gas per litre of Ethrel. It has been used in this way to initiate fruit to ripen by placing containers of Ethrel in a gas-tight room containing the fruit and then adding the base to the containers. It is a simple, effective method of initiating fruit ripening, but tends to be expensive compared with other methods. Degreening of citrus fruits can be achieved by dipping fruit in 1000 µl litre^{-1} Ethrel, which had the same effect as exposing them to 50 µl litre^{-1} ethylene gas for 24 hours (Amchem Information Sheet 46). Direct dipping of crops which are to be eaten, in Ethrel) may be governed by food legislation in some countries.

In commercial ripening rooms in the Yemen, buckets of sodium hydroxide were placed throughout the ripening rooms. When all these were in place, measured amounts of Ethrel were added. This gave an instant release of ethylene gas into the room. This is a simple, effective method, but tends to be very expensive. A comparison was made between using this method and initiating ripening with calcium carbide. The Ethrel method was found to be about 50 times more expensive (Thompson 1985).

Large gas cylinders

Ethylene is available in large steel cylinders where it is stored under pressure (Figure 10.4). Typical cylinders are number 1, which is 1520 mm high and 230 mm in diameter, and number 3, which is 940 mm high and 140 mm in diameter. The former contains 15 kg (12.9 m^3) of ethylene and the latter 3 kg (2.6 m^3). Because ethylene is highly flammable, the use of large cylinders of the pure gas is discouraged. In order to allow some margin for error it is usually used diluted with nitrogen. Typical mixtures are 95% nitrogen and 5% ethylene or 95.5% nitrogen and 4.5% ethylene. The method of application is to meter the gas into the ripening room containing the fruit through a pipe. The volume of the room should have been previously calculated and the volume of ethylene introduced calculated with a flow meter and a stop-watch.

Figure 10.4 Small local banana ripening room in Brazil showing large ethylene cylinders connected to an injection system.

Small gas cylinders

These are steel cylinders, also called lecture tubes, and commonly contain 35 litres of ethylene (Figure 10.5). Two types are available. One type has a cover which, when it is punctured, releases all the gas inside. The second type can be fitted with a metering device to allow for slow and controlled release of the gas. The former is the type commonly used for initiating fruit to ripen commercially. The way it is applied is to calculate the volume of the ripening room and the release the gas from the correct number of cylinders to achieve the correct concentration of ethylene required for ripening or degreening.

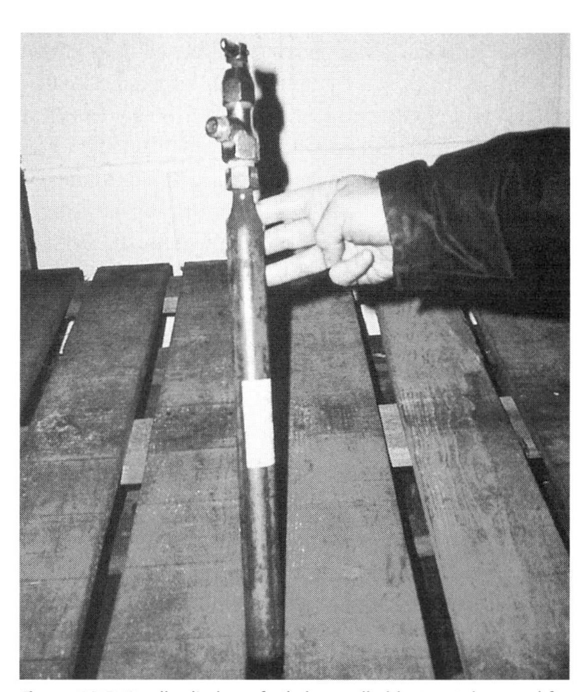

Figure 10.5 Small cylinders of ethylene called lecture tubes used for ripening of fruit.

Ethylene generators

These are devices that are placed in ripening rooms. A liquid is poured into them and they are connected to an electrical power source, and they produce ethylene over a protracted period (Figure 10.6). The manufacturers of these generators do not provide information on exactly what the liquid is which they supply for use in the generators or the process by which the ethylene is generated. A possible way of generating ethylene would be to heat ethanol in a controlled way in the presence of a copper catalyst. Care should be exercised in doing this because of the flammability of the alcohol.

The way in which generators are used is to calculate the volume of the store and place the correct number of generators in the store to provide the required ethylene concentration. This method has the advantage of supplying ethylene to the store over 16 hours rather than applying it in one dose from cylinders. This means that there is a better chance of achieving the desired ripening or degreening effect where there are problems with the room being perfectly gas tight. Gull (1981) used this method for ripening tomatoes. Blankinsop and Sisler (1991) reported that effluent produced by the catalytic generators had a distinctly different odour than that produced by ethylene from cylinders. When they compared the two methods for ripening tomatoes they found inconclusive results in that taste panellists could detect a difference between the fruit ripened by the two methods, but they did not express a preference for either treatment. There tended to be less variation in colour between fruits ripened with ethylene from cylinders compared to those ripened from ethylene generators.

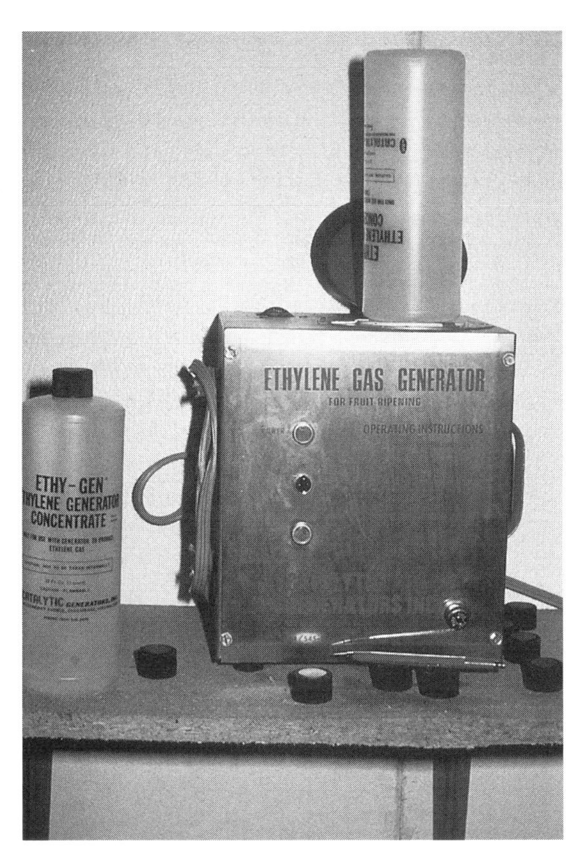

Figure 10.6 Commercially available ethylene generator of the type commonly used to initiate fruit ripening showing the bottles of liquid which are poured into the generator.

Alternative gases to ethylene

Other gases have been shown to initiate ripening of fruit. All are considerably less effective than ethylene (Burg and Burg 1967) and a prerequisite for effectiveness appears to be that their chemical structure must contain an unsaturated bond, that is, they must have a double or triple bond between carbon atoms.

Acetylene

The most commonly used alternative is acetylene. It is produced in less developed countries throughout the world using calcium carbide because it is cheaper than ethylene sources and easier to apply in simple ripening rooms. Calcium carbide is a by-product of the iron and steel industry and the material available contains impurities. Technical grade calcium carbide of regular chip size 4–7 mm conforming to British Standard BS 642 (1965) is the type most commonly used. If it were pure calcium carbide then 1 kg would produce 300 litres of acetylene gas. The gas is released when the calcium carbide is exposed to moisture. The reaction can be violent, so the way it is commonly applied is to wrap small amounts (just a few grams) in twists of newspaper and put these among the bananas to be ripened. The high humidity reacts with the calcium carbide, giving a slow release of acetylene. Where large quantities of acetylene are required quickly, the small amounts of calcium carbide can be dropped carefully into large buckets of water.

Great care must be exercised and the operator must wear protective clothing, including a protective face

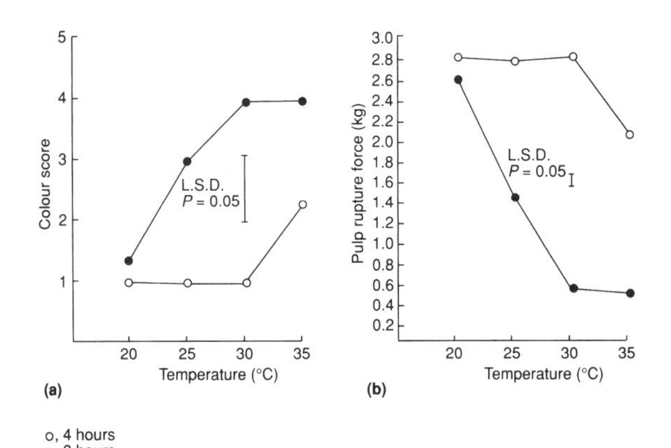

Figure 10.7 Effects of exposure for either 4 or 8 hours to 1 ml l^{-1} acetylene at various temperatures on the ripeness of bananas after 6 days of storage at 20°C. (a) Peel colour score; (b) pulp rupture force. Source: Smith *et al.* (1986).

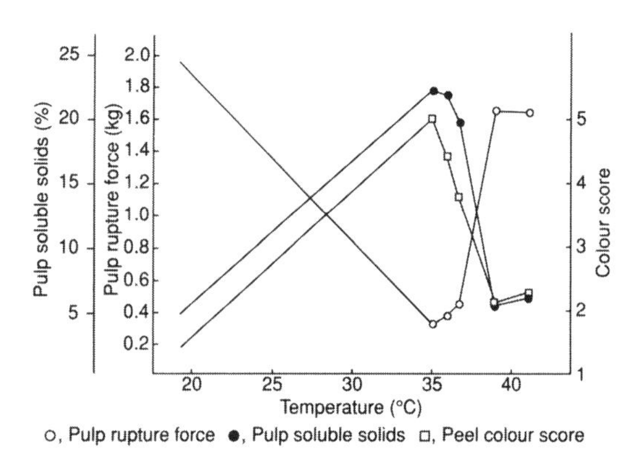

Figure 10.8 Effects of exposure to acetylene at 1 ml l^{-1} for different temperatures on the ripening of bananas at 20°C for 6 days. Source: Smith *et al.* (1986).

mask, and leave the area immediately. Hazard warnings should be put up and flames, cigarettes or electrical fittings that could cause a spark must be eliminated from the area.

Acetylene has been shown to be effective in initiating ripening of bananas. Thompson and Seymour (1982) showed that bananas were initiated to ripen when exposed to 0.1 or 1 ml litre^{-1} at 18°C for 24 hours. Smith *et al.* (1986) showed that the effects of acetylene on bananas was proportional to temperature and exposure time (Figures 10.7 and 10.8). Exposure of fruit to 1 ml litre^{-1} of acetylene for 4 hours did not initiate ripening at between 20 and 30°C but it did initiate ripening at 35°C. With 8 hours of exposure fruit were not initiated at 20°C, were partially initiated at 25°C and were fully initiated at 30°C. Acedo and Bautista (1993) showed that Latundan (*Musa* ABB) and Saba (*Musa* BBB) bananas showed that exposure to 5 g of calcium carbide for 1 day effectively enhanced ripening. Latundan harvested at full maturity required a lower level of calcium carbide then Saba harvested at the full three-quarter stage. In mangoes it was found that at 25°C and a 24-hour exposure time then at least 1 ml litre^{-1} of acetylene or 0.01 ml litre^{-1} of ethylene was required to initiate ripening (Medlicott *et al.* 1987b). A 1-day treatment of bananas with *Gliricidia* sp. or rain tree leaves at 5% of fruit weight enhanced ripening of Saba but was less effective than the calcium carbide treatment (Acedo and Bautista 1993).

Simple methods

Smoke

Another simple method of initiating ripening is to light a smoky fire in the ripening room. This can produce various gases, including acetylene, ethylene and carbon monoxide, which will initiate ripening. It is used in many developing countries and a study in the Sudan showed that it appeared to be effective, but there was considerable variation in the speed of ripening throughout the room.

Damage

Wounding the banana bunch stalks or even the fruit may produce ethylene in response to the wound. Bautista (1990) described a simple technique used in Southeast Asia where a stick is inserted into the stalk of Jackfruit, which not only makes a convenient handle for carrying the fruit but also initiates it to ripen. She also mentioned other methods, including cutting, scraping or 'pinching' papaya, chico or avocado, which can hasten ripening.

Fruit generation

Fruit that are ripening and thus giving out ethylene can be placed in an air-tight room with green fruit. A continuous system could be worked out for commercial application of this method. However, the room would need to be frequently ventilated to ensure there was no build-up of CO_2, which is known to inhibit the effect of ethylene.

11
Marketing and transport

Marketing

Dixie (1989) gave the following two definitions of marketing: (1) 'The series of services involved in moving a product (or commodity) from the point of production to the point of consumption' and (2) 'Marketing involves finding out what your customers want and supplying it to them at a profit.' Marketing of perishable crops becomes increasingly important as the standard of living of consumers increases, because they require higher, more consistent quality and also fruit and vegetables when they want them, not necessarily when they are in season.

In many countries, the production of fresh fruit and vegetables is currently lower than the market requirement. These normally result in a seller's market where the farmer can sell all the crops that are grown, and there may be little incentive to supply high-quality crops on the market. This happens in less developed countries. In other societies crop production, or potential crop production, is greater than the demand. The effect of this, in European markets for example, has not necessarily reduced the price of the crops as might have been expected, but increased the quality of the crops being offered to the consumer. It is not unknown for fruit and vegetables to be destroyed in order to maintain market prices. This can even happen after the crop has been harvested and packed for the market. In industrialized countries the farmer is increasingly studying the market to determine its requirements. An example is organic fruit and vegetables that usually command a premium market price. However, their higher cost of production has resulted in production not always being economic and there are groups lobbying for government subsidies.

Where markets are not well regulated, this has led to fluctuating supplies of crops such as vegetables. This is because the farmer may see that a particular crop is commanding a high price on the market because of under-supply. While the farmer changes his production to this crop, other farmers see the same opportunity and also change, resulting in an over-supply. This can lead to low prices and farmers stopping growing that crop, leading to a shortage, and the cycle repeats itself. This has led to the establishment of commodity marketing boards to regulate supply to the perceived demand. There are several cases where these marketing boards have had major effects on produce marketing, e.g. the New Zealand Apple and Pear Board and Agrexco in Israel, but many others have had limited success in the perishable crop sector.

The flow of fruit and vegetables from the producer to the consumer is not simply governed by the market forces of supply and demand. These forces are tempered, modified and controlled by import duties, quotas and other trade barriers. The General Agreement on Tariffs and Trade (GATT) was first implemented in 1948 as a mechanism to promote free and fair trade between its member states and several rounds of negotiations were carried out in the following years and rules that governed international trade were formulated. The Uruguay round of GATT began in 1986 and was signed in April 1994 in Marrakech. This opened up trade throughout the world, including that in fruit and vegetables. The Uruguay round of GATT led to the creation of the World Trade Organization (WTO) to reform and police international trade agreements. This put an onus on the more than 100 states that signed the agreement to reform both their internal and external trade policies to open up international trade. However, the rules are complicated and allow for states and trading blocks to subsidise production and impose tariffs on imports to protect their own farmers.

There are many cases where trade is not a 'level playing field' and it is often to the detriment to farmers in developing countries and the benefit of farmers in the more wealthy countries. Such liberalization of trade can also lead to specific difficulties for some producer countries. There were riots in India against the WTO rules. The banana industry in the French West Indies, the Canary Islands and the African Caribbean and Pacific (ACP) countries have had preferential access to European Union countries. Many of these ACP countries have vulnerable economies, which are dependent on banana exports, and because of factors such as comparatively high labour costs, or lower capital investment they are unable to compete with other countries on price. The WTO ruling against these trade preferences between ACP countries and the European Union has been a potential source of problems in these fragile economies. In the Windward Islands the industry is being rationalized by measures which include reducing the numbers of loading ports for the fruit, in an effort to make them more competitive before 2005 when the current marketing arrangements with the EU are planned to be terminated.

Marketing systems can be considered on the basis of their structure, conduct and performance. Structure relates to their relative size, number of institutions and the various functions of the participants. Conduct is described by the level of control by government and the level of competitiveness of market participants regarding the product, price and promotional strategies. The performance can be measured in several ways, but the most common one is by its efficiency. Profits, turnover, innovation and research and development may measure efficiency. These three factors are interdependent in that each will affect the way the market is structured and the way business is conducted, and both in turn will affect its performance.

Many perishable crops are seasonal and their value, or selling price, varies throughout the year. Orderly marketing can be achieved in certain cases by storing the crop from the harvest season to the season of scarcity. Many different marketing systems have been used in many developing and industrial countries. However, the problem remains of getting fresh fruit and vegetables from the producer to the consumer with the minimum loss of quality or quantity, at the time the consumer desires them and at minimum cost. All marketing systems have some flaw in them, and many have been started and have had to close after periods of operation. All these systems are constantly changing and evolving to suit changing market and political situations.

Marketing systems

Direct marketing

The simplest system is where the farmer sells directly to the consumer. For this the consumer may call directly at the farm. In many countries this has led to the development of farm shops where the farmer harvests the crop and offers it for sale. This is often supplemented by purchasing crops from wholesale markets to increase the range of crops offered. 'Pick-your-own' is where the farmer grows the crop and customers harvest it themselves and pay the farmer for the quantity they have harvested.

Middlemen

Middlemen may visit the farm to purchase from the farmer. They will then resell the produce to a wholesaler or retailer. In many situations the middlemen have their own labour force, which relieves the farmer from having to harvest the crop: he simply sells the crop by area or weight harvested. An example of this is the Kissan Company in Bangalore in India. They are major processors of mangoes, and they have a team of workers whose function is to go from plantation to plantation harvesting fruit that is then sent directly to the factory. In many less developed countries the farmer harvests the crop, transports it to market and then sells it. In many cases these sales are at the retail level and it can take the farmer a considerable amount of time to sell only a small quantity.

Direct to retailer

In industrial countries, many farmers, particularly those on a large scale, sell direct to supermarkets. This is commonly on a contract basis and can be a very convenient way of marketing for the producer or importer. The supermarkets, however, have stringent quality requirements and times of delivery. Only the best organized and efficient farmers who are forward thinking are consistently successful in these markets. In the UK, with over-production of many crops, some supermarket suppliers fail.

Wholesale markets

Specialist wholesale markets have been established in towns and cities throughout the world to assist marketing of fresh fruits and vegetables. Many of these have been in existence for centuries. These consist of groups of middlemen who purchase produce, often from the producer, the producer's cooperative or importer (sometime through brokers) and sell the produce to a retailer. The situation is usually more complicated with wholesalers selling to secondary wholesalers or to exporters (particularly in some developing countries) or directly to the consumer. Fresh produce is mainly highly perishable and has a fragmented production base, which makes it extremely difficult to standardize quality and production specifications (Henderson 1993). Wholesaling of fresh produce should minimize the number of transactions and reduce the stocks required and be involved with price discovery, minimizing wastage and assembly, stockholding, financing, standardization and breaking bulk (Sturgess 1987). The demand for fresh produce is relatively inelastic in that large price fluctuations create only small fluctuations in demand. This can result in prices being unstable and the need for buyers and sellers to have frequent and close contact (Henderson 1993). In the United Kingdom at the Covent Garden market, the average markup for wholesaling services was about 18% of the value of primary production at the farm gate or the dockside (Sturgess 1987).

Retail markets

Retail outlets vary from carts on the streets, through established open markets (Figure 11.1, in the colour plates) and small greengrocer's shops to multiple retailers called supermarkets (Figure 11.2, in the colour plates). Throughout the world this diversity of retail outlets has been retained, but the multiple retail sector has constantly increased its market share. This is due to various factors, including the consistent quality of produce and convenience of purchasing the weekly shop in one place. Retail stores were traditionally supplied by the retailer going to the wholesale market early in the morning and purchasing the items required. With the multiple sector, various systems have been put in place to obtain supplies. These usually involve the produce being delivered to a distribution centre where crops are held and sometimes graded and sent to individual retail outlets. In Britain there has been considerable debate about supermarkets obtaining locally produced fruit and vegetables. Some progress has been achieved but they often have difficulties in maintaining consistent supplies of sufficient quality.

State marketing

In the past in some communist countries, the state was responsible for the marketing of fruits and vegetables. This system prevailed in the People's Democratic Republic of Yemen before unification with the Yemen Arab Republic. The result was that harvesting and handling systems were very inefficient and delivered poor-quality produce to the consumer. In China the marketing of fresh produce was entirely in the hands of the state until the mid-1980s. After that time wholesale and retail markets were allowed to evolve. It is interesting to observe that this system involved middlemen with lorries going to the countryside and bringing the produce to places in the towns and cities where the produce was offered for sale. In 1986, the Dazhongsi Farm and Sideline Products Wholesale Market was established in Beijing. This provided a site in which wholesale trade could take place and by 1993 it had 10 000 customers and total annual sales of 58 billion kg. It is the largest of the four wholesale markets in Beijing and is operated by the local government and charges 2% on sales to meet its operating costs. It also pays an additional 3% in tax to the Government.

Dutch auctions

The Centraal Bureau van de Tuinbouwveilingen in the Netherlands operates a special type of wholesaling. These wholesale auctions are owned by growers who produce, harvest, grade, prepare, pack and sent the produce to the auction (Hill and Selassie 1993). Auction staff are responsible for quality and assembling the produce into suitable quantities for sale. The lots of produce are offered for sale by the movement of the arm of a clock that moves from a high price to progressively lower prices. Prospective purchasers can stop the clock at any point and then become the owner of that produce at the price on the clock at the time it was stopped. Computerization has been developed in controlling the auction clock and linking the clock with the auction accounts department and distribution. Clocks from several auctions can be displayed together with information on supply and demand (Crawford and Selassie 1993). The marketing system in the Netherlands often leads to high wastage levels of good-quality produce. This is done to keep prices high so that when some lots do not achieve a predetermined minimum price they are taken off the market and destroyed. A mean price is taken for all the produce of that type which has been sold at that time and the owners of the produce which has been destroyed receive the same proportional share as those whose produce was sold.

Panels

Many organizations market their produce through panels. These are common in the UK and are fruit- and vegetable-importing companies who are responsible for marketing and distribution of a product or group of products from one source. For a company to be a panel member it must reach the standards required by the supplier. An example of this would be the citrus panel which is supplied by Outspan. Outspan would invite a company to join their panel which they feel had sufficiently high standards, distribution network and contacts within the industry. The panel member then has the responsibility to supply retailers. In most cases the supermarket sector of the fruit and vegetable retailing industry deals through panellists rather than directly with the marketing board. The supermarket may prefer to work in this way because it provides flexibility so that if the produce does not achieve the required standard it can easily be returned to the panel member to be retailed elsewhere or otherwise disposed of. The panellists receive a commission for the services that they supply. This system does extend the marketing chain and may change or be modified to meet competition in a changing market situation.

Cooperative marketing associations

Cooperative marketing is where a group of farmers associate to market their produce together. There are many examples of successful marketing cooperatives. East Kent Packers in England was formed by a group of local apple and pear producers and the company is still owned by the grower members. They provide very sophisticated controlled atmosphere storage facilities and grading and packing equipment. Members grow and harvest their crops and send them to East Kent Packers, where they are either graded, packed and sent to market or placed in storage for later marketing. The stores and packhouse are used for other crops on a contract basis when they are not required for members' fruit, which provides additional income for members. Growers' cooperatives have the advantages for small- to medium-sized farmers of increasing the quantity of produce to be marketed, thus improving continuity of supply, bargaining position and the employment of special marketing techniques and staff. In some cases marketing and related activities are supported by statutory bodies such as marketing boards or export boards.

Fair trade

In spite of the creation of the World Trade Organization, there is much international trade that exits to the benefit of developed countries at the expense of less developed countries; 20% of the world's population had 82 times the income of the bottom 20%. There are many organizations that have been set up to address this imbalance of trade and income. Fair Trade is a non-government organization (NGO) that is over 40 years old. It has a charity background but commercial companies are also involved. It publishes a set of values in which trade takes place. Other objectives are to give value for money and good quality, so there is something in it for consumers. Its objective is sustainability 'in it for the long term' so that it provides a stable situation for the grower.

The main product with which it is involved is coffee marketed as Café Direct. Fair Trade is a movement not a brand, but becoming a brand is their aim in UK, especially with Café Direct. Café Direct has been in existence for over 10 years and is the sixth largest supplier of coffee in Britain. It has suppliers of 1.5 million growers and their families. It comes from a higher cost base and therefore it charges a premium for its brand. Much of this is passed to the grower, who either invests the premium into his or her own co-operative or they can use it to improve their standard of living. In the UK and Japan (where the People Tree brand is trendy and cool and good with young people) many people are conscious of it but in Spain and the USA they are not.

Fair Trade has become increasingly concerned with the marketing of bananas and supports a cooperative in Ghana. A fruit sticker is attached to each cluster of fruit and a premium price is charged to the consumer. Most of this is passed on to the producer so those small farmers receive a better income than they would if they marketed their fruit in other ways.

Market analysis

There are two basic systems that can be applied to the marketing of fresh fruit and vegetables. The first is simply to market what farmers can produce. This system is very common in developing countries and usually is well founded and has evolved over generations. Supply and demand for the crops are not always in balance and there can be large seasonal fluctuations in both quantities available on the market and the price received for the crop. In extreme situations it may not be worth harvesting the crop. Many ways of overcoming this situation have been developed, including storage to link periods of excess supply and market promotion to increase the demand at that time. This marketing system can be a major problem where new markets are being developed. This is particularly important where export development or a processing industry is planned. The alternative is market analysis. This means looking at the proposed market and producing for that market. This can involve producing specific varieties of specific crops to a defined quality and predetermined schedule. It usually also involves management and postharvest practices which are required by the market. One analytical procedure, which can be used to assess the market potential, is SWOT analysis (strengths, weaknesses, opportunities and threats). The stages are to look first at the planned business to determine what it can do or be developed to do best. When this has been decided, the difficulties that are likely to be encountered in the proposed business must be estimated and ways in which these weaknesses can be overcome must be formulated. The market must be studied in detail to determine current supply and demand factors by studying previous market data and picking up possible trends of what the market might be in the future. This will also include determining any legislation governing standards and which varieties of crop receive the best prices and at what time of the year.

Market situations, especially on export markets, are constantly changing and it is important to know what competitors, or potential competitors, are doing or planning to do. This should be taken into account when planning the business. Crop production is subject to environmental conditions and sometimes pest and disease infestation and in developing countries the supply of materials that are necessary for production and postharvest operations.

Retaining the crop in optimum condition throughout the marketing chain also becomes increasingly important in ensuring that marketing is profitable. A major contributor to this is the means of transporting the crop from the producer to the consumer.

Branding

At one time branding seemed just to be related to tough-looking men in leather using a red hot iron on cattle, but the concept has now become a major occupation of clever people putting a logo or name into the mind of people. The subject of branding of fresh fruits was reviewed by Best (1993). In its simplest form a brand can be defined as 'a name, term, symbol, design or all of these intended to identify and differentiate goods and services one from another' (Kotler 1991). However, according to Davies (1992), brands are more than just a 'label to differentiate' and are in fact 'a complex system that represents a variety of ideas and attributes separate and distinct from the product, each capable of undergoing change independently of each other.' Murphy (1992) defined a brand as 'a trademark which, through careful management, skilful promotion and wide use, comes in the mind of the consumers, to embrace a particular and appealing set of values and attributes, both tangible and intangible. It is therefore much more than the product itself; it is much more than merely a label. To the consumer it represents a whole host of attributes and a credible guarantee of quality and origin. To the brand owner it is in effect, an annuity, a guarantee of future cash flows.' He identifies the brand as the output of a management process which invests resources to develop assets which are increasingly being recognized on company balance sheets and being assessed for their profit-earning capability.

Increasingly the brand is seen not as a thing which people involved in marketing do to their product, but as a relationship that they develop with their customers. de Chernatony and McDonald (1992) suggested that successful brands are 'an identifiable product, service, person or place, augmented in such a way that the buyer or user perceives relevant unique added values which match their needs most closely. Furthermore, its success results from being able to sustain these values in the face of competition.' The success of brands stems from the management of the entire marketing mix to create an augmented commodity form of a product which has meaning to the consumer beyond the functional product. Although brands are most commonly associated with products, and in particular fast-moving consumer goods, they can be developed for a wide range of market offers, including services, countries and people.

These definitions trace the development of the understanding of the brand over time, and reflect the changing emphasis of branding from physical products to services such as financial services. Their emphasis on the non-functional aspects of branding appears to be particularly relevant to the fresh fruit industry.

Consumers are aware of brands in the fresh produce industry. In a survey, reported by Best (1993), prompt awareness by people in the UK for brands in the citrus industry were:

- Jaffa 95%
- Outspan 93%
- Sunkist 54%
- Morocco 30%
- Florida 24%.

An understanding of the parts which comprise a brand is paramount to understanding why some 'products get commodity prices, giving them low margins, and other brands get the commodity price plus what ever the brand is worth beyond the product' (King 1984). Levitt (1986) described the brand as having several components which might be applied to a fresh fruit brand, as follows:

1. Core attributes that meet the basic needs of the consumer. For example, an apple.
2. Generic attributes that exist in the commodity form of the product, delivering the functional attributes meet the basic need of consumers. For example, the Cox's Orange Pippin variety, ripe, blemish free, graded, labelled, available in stores.
3. Expected attributes where the generic form is tailored to needs of specific customers. For example, Cox's Orange Pippin grown in England, tasting like English Cox's Orange Pippin, available in stores servicing the targeted consumer segment, labelled English Cox Apples.
4. Augmented attributes meeting consumers' non-functional needs. For example, assurances of taste and quality of traditional Cox's Orange Pippin, reassurances that English Cox's Orange Pippin can only be grown in England, and that English Cox's Orange Pippin represent an English way of life.
5. Potential attributes that are the limitless enhancements of the brand, often intangible and emotional, which underlie its extended life. For example, upgrading of the cultivar to include new features, extension of the Cox's Orange Pippin image to other products such as jams, and juices, keeping adjustments in promotions to keep in touch with perceived English self-image of the main consumer segment.

National transport

This is the transfer of produce from the farm to the place where it is to be offered for sale. The system varies in both less developed countries and the industrialized countries from simply packing the produce in some kind of vehicle for transport, to having comprehensive environmental control around the crop. The reasons for selecting a particular type of transport may be related to:

- special characteristics of the crop
- its value
- what is available
- what is most appropriate.

Where crops are moved short distances, quickly from the field to the market or directly to the consumer, it is usually unnecessary for most crops to need any temperature management or other controlled conditions. This simple kind of transport can appear cheap but it may result in high crop losses due to physical damage. In a study of banana transport in the Sudan it was shown that bunches packed in lorries padded with banana leaves and transported from the field to the ripening rooms had extremely high levels of damage. Packing the bananas in plastic boxes for transport greatly reduced these losses (Silvis *et al.* 1976).

In a study of sweet potatoes by Tomlins *et al.* (2000) in Tanzania, the handling and transport system resulted in up to 20% and 86% of roots with severe breaks and skinning injury, respectively and a reductions in market value of up to 13%. The most severe impacts occurred during unloading and loading from road vehicles and ships. However, skinning injury and broken roots were correlated with a large number of minor impacts.

International trade

From the Middle Ages, small quantities of dried spices, fruits and beverage crops were traded internationally, mainly from the East to Europe, but these were relatively small quantities and were expensive. International trade in fresh fruit and vegetables was scarce. In the late 18th century, citrus fruits were traded throughout Europe, but were mainly from orchards in Spain and Portugal. At that time the human populations in Europe were comparatively small and mainly involved in some aspect of agriculture or the service sector, and there was little incentive for international trade. In the 19th century, the industrial revolution stimulated manufacture and the trade in raw materials to service the industry. This was particularly so in Britain, where the population rose from a fairly stable level of some 9 million in 1801 to over 36 million by 1911. This stimulated the demand for food and attracted imports. Initially these were mainly from other European countries, but with the development of steam ships they increasingly came from further afield. Dietary habits also changed and the market for fruit and vegetables increased. This demand was supplied by local producers, and led to the development of such technologies as producing crops in heated glasshouses, but also to an increasing extent by imports. This was especially stimulated when the British Government abolished the import duty on fruit in 1860. Citrus fruits were the main import, but apples, cherries, grapes, melons, pears and plums were also traded. Pineapples were exported from the West Indies to Britain from 1842 in schooners. In order to supply out-of-season fruit, shipments of apples, grapes and other fruits were being made from South Africa to Britain from 1888, with mixed results. The first export shipment of bananas was from Jamaica and was reported to be 500 stems to Boston in 1866, which took 14 days, and the fruit were said to have been sold at a profit (Sealy *et al.* 1984). There was no indication of temperature control during this or subsequent shipments in the latter part of the 19th century. In 1879, both steam ships and schooners were being used to transport fruit from the Caribbean Islands and Central America to the USA.

International sea transport of perishable foodstuffs using refrigeration began in the latter part of the 19th century. The use of cooling for transport began in 1873 or 1874 using ice through which air was circulated by fans on shipments of meat from the USA to Britain, and this developed into a regular trade in 1877 when mechanical refrigeration was used on the ships. The first records of refrigerated shipment of fresh fruits and vegetables were in 1880 from Italy to Britain and included pears and peas. In 1901 the vessel Port Morant carried 23 000 stems of bananas from Jamaica to Britain using mechanical refrigerated holds for the company Elder and Fyffes (Sinclair 1988).

Cold chain

Cold chain transport, where the crop is precooled directly after harvest and kept at a constant temperature throughout the marketing chain, is being increasingly practised in both industrial and non-industrial countries. The system involves a high capital expenditure but ensures a high level of quality maintenance and a reduction in physical losses of crops. In Britain the major retailers (supermarkets) have distribution centres where growers deliver their crops. The supermarket will have agreed price, quantity and quality with the producer. They will also specify the precise time at which it must be delivered and the temperature of the crop at the time of delivery. A well run cold chain is therefore essential in order to achieve these requirements.

Transport by sea

For export, fresh fruit and vegetables are largely transported in temperature-controlled cargo space in either 'break bulk' or in containers such as 'reefers,' porthole or ventilated types. In trial shipments of

Table 11.1 Costs (other than labour) and returns in Trinidad and Tobago dollars of a shipment of mangoes of the cultivar Julie shipped from Trinidad to Britain at 13°C taking 10 days (source: Thompson 1971a)

Expenditure		
Cost of fruit per carton	1.60	
Cost of carton	0.71	
Refrigerated sea transport	1.34	
Total		3.65
Income		
Grade 1	53%	12.24
Grade 1	19%	3.96
Grade 2	19%	3.60
Wastage	9%	0.00
Total		19.80

mangoes from the West Indies to Britain it was shown that the fruit was in good condition on arrival and was profitably sold (Table 11.1).

Refrigerated containers were first introduced in the 1930s but it was only in the 1950s that large numbers were transported on ships. Their use was from the West Coast of the USA to Hawaii. In the late 1960s the porthole container was introduced. Different operators developed different specifications and sizes of the various types of container. This was particularly evident in their length, which varied from 17, 20, 24, 35–40 feet long. This variation prevented interchange of containers between different operating lines. Currently only a small part of the world supply of containers deviate from the standard 20 and 40 foot lengths recommended by the International Standards Organization in 1967. The first purpose-built container ship was completed in 1969 that had a capacity of 1300 20-foot equivalent units and during succeeding years they have increased in capacity. Numbers have also increased; for example, over the period 1970–1988 the world container fleet has increased from 195 000–2 869 000 20-foot equivalent units.

Break bulk

Break bulk refers to a system of transport where individual boxes or pallets of produce are stacked directly in the hold of the ship. Bananas account for about 35% of the fresh produce transported in this way. Most produce is palletized before shipping and maintained in this condition throughout the marketing chain; in some cases this is right up to the retail outlet. Palletization reduces handling and therefore labour costs and damage to produce. The standard international pallet is 1000 × 1200 mm on which produce is commonly stacked to about 2100 mm high. The holds of ships are commonly 2200 mm high and reefers the same or higher, so that pallet loads can fit in with an adequate clearance to allow for air circulation. Boxes should be secured on the pallet with edging strips so that they do not move during transport. During winter when the seas are rough there is a considerable increase in the level of damage to boxes and to the fruit they contain. Dunnage or inflatable bags may be placed between pallets to improve air circulation and temperature control.

Reefer containers

A large and increasing amount of fresh fruit and vegetables is transported by sea freight refrigerated (reefer) containers (Thompson and Stenning 1994). Reefers are insulated containers which have their own individual refrigeration units They are built to International Standards Organization specification, which specifies that they must be sufficiently strong so that they are capable of holding 30 tonnes and withstand a vertical load of up to nine containers stacked above it. These containers have about 70 mm of insulation in the walls, generally polyurethane foam. A new insulated container will have a heat leakage of about 22 W K^{-1}, but some containers designed solely for the carriage of fresh fruit will have thinner insulation with a heat leakage of around 35 W K^{-1} (Heap 1989). The foam almost always contains low-conductivity halocarbons to improve performance. Insulation efficiency reduces with time, due to loss of halocarbons and moisture ingress, by about 3–5% per year (Heap 1989). Sizes vary but commonly used international sizes are given in Table 11.2.

Operating costs of containers are usually much higher than those of break bulk, but apparently they can also be lower since they depend on the journey (Table 11.3). Bananas are the most important fruit that is transported by sea freight and almost all international trade in bananas is still carried out using break bulk. This suggests that the figures in Table 11.3 might have changed. The trade from Martinique and Guadeloupe in the Caribbean is heavily subsidized through the EU Common Agricultural Policy and there they use containers for bananas. The relatively new industry from the port of Tema in Ghana uses con-

Table 11.2 Sizes and capacities of reefer containers (source: Seaco Reefers)

Type	External dimensions	Internal dimensions
20 foot (RM2)		
Length	6096 mm	5501 mm
Width	2438 mm	2264 mm
Height	2591 mm	2253 mm
Capacity	28.06 m^2	
Tare	3068 kg	
Maximum payload	21 932 kg	
ISO payload	17 252 kg	
40 foot (RM4)		
Length	12 192 mm	11 638 mm
Width	2438 mm	2264 mm
Height	2591 mm	2253 mm
Capacity	59.81 m^2	
Tare	4510 kg	
Maximum payload	27 990 kg	
ISO payload	25 970 kg	
40 foot (RM5)		
Length	12 192 mm	11 638 mm
Width	2438 mm	2264 mm
Height	2896 mm	2557 mm
Capacity	68.03 m^2	
Tare	4620 kg	
Maximum payload	27 880 kg	
ISO payload	25 860 kg	

tainers, but this reflects the scale of operation and the difficulty in securing break bulk transport. For the growers in Ghana it means that the only way the industry can be viable is by obtaining subsidized prices mainly through the Fair Trade Organization.

Porthole containers

Porthole containers, which are also called Conair containers, are insulated containers that do not have their own refrigeration unit. Instead they have two 10 cm diameter ports which can be connected to an external refrigeration system via a flexible hose. They are used in ships where the containers are plugged into the ship's refrigeration system. They may also be plugged into a unit at the port to cool the produce while awaiting transport. The advantages of this type of system are that the rental costs on the containers are lower and the units can be cooled more quickly because of the larger refrigeration capacity from the ship's system. Disadvantages are that special facilities are required both at the port and on the ship. Refrigeration can only be started when the container has reached the port and not at the packhouse where it may be loaded. With reefer containers a diesel generator may be fitted or attached to the container so that its contents may be cooled directly after it has been loaded.

Table 11.3 A comparison between the c.i.f. costs of international transport of bananas communicated to UNCTAD in 1981 in US $ per carton (source: Sinclair 1988)

Journey	Break bulk	Containerized
Central America to North America	5.64	5.95
Central America to Europe	7.63	7.41
French West Indies to France	8.42	8.23

Ventilated containers

Ventilated containers are standard metal containers that have no insulation or refrigeration unit but have some kind of ventilation system. This may simply be square apertures along each of the long sides with downward pointing fixed louvres. A row is situated near the top and another row near the bottom down each of the long sides of the container (Figure 11.3). Ventilation is passive using natural convection currents. A more elaborate system has extractor fans installed high up on the doors of the container. These usually have the same ventilation of the passive system but with just one row

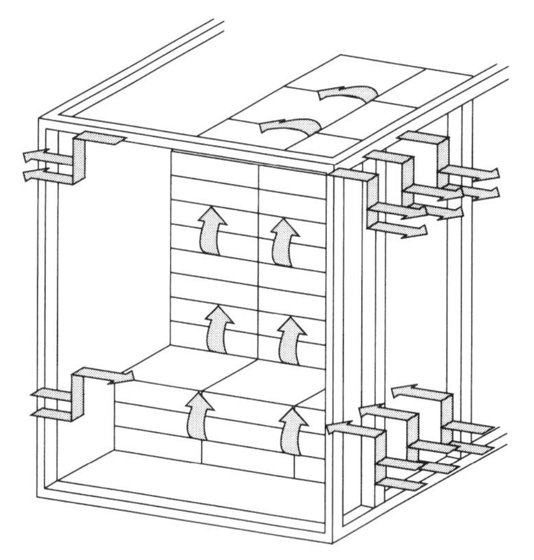

Figure 11.3 Ventilated container showing the airflow movement through the cargo. The ventilator covers have been removed from the end side panels.

of apertures towards the bottom of the container on each of the long sides. The attraction of these types of container is their low rental cost compared with reefers. SeaVent used a passively ventilated container filled with 17.5t of green coffee in extremes of temperature and humidity in a simulated trial of the conditions that would be encountered during transport from the tropics to Europe. It was found that the coffee was in perfect condition, and this method is now being used on a large commercial scale throughout the world. It is also used for transport of crops such as potatoes and onions even for long distances such as from Australia to the UK.

Modified atmosphere containers

Modified atmosphere containers have been used for several decades with varying degrees of success. Transfresh of Salinas in California build containers which are designed to be gas tight and which have an aluminium track fitted around the door to which is fitted a plastic curtain to eliminate leakage through the door. After loading, the container is sealed and the required gaseous atmosphere is injected through purging ports.

Controlled atmosphere containers

Controlling the levels of some of the gases in reefer containers has been used for many years to increase the marketable life of fruits and vegetables during transport (Figure 11.4). Many commercial systems have been manufactured. Champion (1986) reviewed the state of the art of controlled atmosphere transport as it

Figure 11.4 Packed controlled atmosphere reefer container plugged in to the mains electricity awaiting loading on the dockside in Castries, St Lucia, in 1998.

existed at that time. He listed the companies who cooperated in the production of the paper, which gives some indication of the large number of companies who were involved in the commercial application of this technology, as follows:

- AgriTech Corporation, USA
- CA (Container) Systems, UK
- Finsam International, Norway
- Franz Welz, Austria
- Fresh Box Container, West Germany
- Industrial Research, the Netherlands
- Synergen, UK
- TEM, USA
- Transfresh Corporation, USA
- Transfresh Pacific, New Zealand.

Many of those companies have since gone out of business and others have been started. Champion (1986) also defined the difference between controlled atmosphere containers and modified atmosphere containers. The latter has the appropriate mixture injected into the sealed container at the beginning of the journey with no subsequent control, which means that in containers being used to transport fresh fruit and vegetables the atmosphere will constantly change during transport. Controlled atmosphere containers have some mechanism for measuring the changes in gases and adjusting them to a pre-set level. Dohring (1997) stated that the world fleet has increased four-fold since 1993 and in 1997 consisted of 38 000 reefer containers with only about 1000 providing control of humidity CO_2 and O_2, that is, controlled atmosphere containers. A stylized diagram of a typical controlled atmosphere container is given in Figure 11.5.

Dohring (1997) claimed that 'avocados, stone fruit, pears, mangoes, asparagus and tangerine made up over 70% of container volumes in recent years. Lower value commodities, e.g. lettuce, broccoli, bananas and apples, make up a greater percentage of the overall global produce trade volumes but cannot absorb the added CA costs in most markets.' Previously Harman (1988) had suggested the use of controlled atmosphere containers for transport and storage for the New Zealand fruit industry, but there has been only limited uptake. Lizana *et al.* (1997) found that long-distance sea transport for 35 days of Hass avocado fruits from Chile to the USA was feasible in refrigerated controlled atmosphere containers. They found that the best controlled atmosphere storage combination for Hass

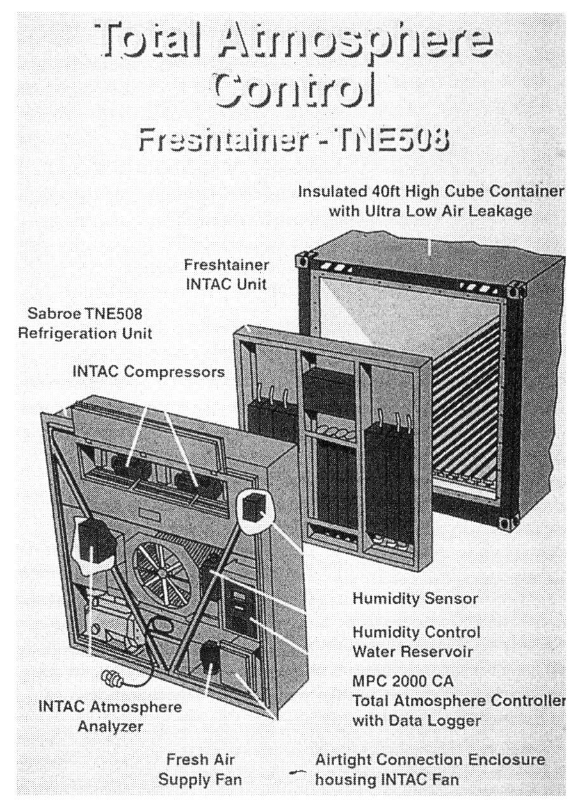

Figure 11.5 Diagram of a controlled atmosphere container (Freshtainer).

avocado fruits was at 6°C and 90% r.h. with 5% CO_2 and 2% O_2.

Gas sealing

The degree of control over the gases in a container is affected by how gas tight the container is; some early systems had a leakage rate of 5 $m^3 h^{-1}$ or more, but with current systems it can be below 1 $m^3 h^{-1}$ (Garrett 1992). Much of the air leakage is through the door and fitting plastic curtains inside the door could reduce the leakage, but they were difficult to fit and maintain in practice. A system introduced in 1993 has a single door instead of the double doors of reefers containers, which are easier to make gas tight. Other controlled atmosphere containers are fitted with a rail from which a plastic curtain is fitted to make the container more gas tight.

Gas generation

Controlled atmosphere containers are gas tight, so the respiration rate of the fruit will eventually produce lower O_2 and higher CO_2 levels. Ilinski *et al.* (2000) found for the apple cultivars Antonovka, Martovskoe, Severni Sinap and Renet Chernenko that the CO_2 concentration in controlled atmosphere containers reached 15–19% and O_2 was reduced to 0.4 to 0.8% within 3–3.5 days. This may be too slow and therefore the systems used to generate the atmosphere in the containers fall into three categories (Garrett 1992):

1. The gases that are required to control the atmosphere are carried with the container in either a liquid or solid form.
2. Membrane technology is used to generate the gases by separation.
3. The gases are generated in the container and recycled with pressure absorption technology and swing absorption technology.

The first method involves injecting nitrogen into the container to reduce the level of O_2 with often some enhancement of CO_2 (Anon 1987). It was claimed that such a system could carry certain cooled produce for 21 days compared with an earlier model, using nitrogen injection only, which could be used only on journeys not exceeding 1 week. The gases were carried in the compressed liquid form in steel cylinders at the front of the container, with access from the outside. O_2 levels were maintained by injection of nitrogen if the leakage into the container was greater than the utilization of O_2 through respiration by the stored crop. If the respiration of the crop was high the O_2 could be replenished by ventilation.

In containers which use membrane technology, the CO_2 is generated by the respiration of the crop and nitrogen is injected to reduce the O_2 level. The nitrogen is produced by passing the air through fine porous tubes, made from polysulphones or polyamides, at a pressure of about 5–6 bar (see Figure 7.13). These will divert most of the oxygen through the tube walls leaving mainly nitrogen, which is injected into the store (Sharples 1989).

A controlled atmosphere reefer container, which has controls that can give more precise control over the gaseous atmosphere, was introduced in 1993. The containers use ventilation to control O_2 levels and a patented molecular sieve to control CO_2. The molecular sieve will also absorb ethylene and has two distinct circuits which are switched at predetermined intervals so that while one circuit is absorbing, the other is being regenerated. The regeneration of the molecular sieve beds can be achieved when they are warmed to 100°C to drive off the CO_2 and ethylene. This system of regeneration is referred to as temperature swing where the gases are absorbed at low temperature and released at high temperature. Regeneration can also be achieved by reducing the pressure around the molecular sieve, which is called pressure swing. During the regeneration cycle the trapped gases are usually ventilated to the outside, but they can be directed back into the container if this is required. The levels of gas, temperature and humidity within the container are all controlled by a computer which is an integral part of the container. It monitors the levels of oxygen from a paramagnetic analyser and the CO_2 from an infrared gas analyser and adjusts the levels to those which have been pre-set in the computer (Figure 11.6).

Lallu *et al.* (1997) described experiments on the transport of kiwifruit in containers where the atmosphere was controlled by either nitrogen flushing and 'Purafil' to absorb ethylene or lime and 'Purafil'. The former maintained CO_2 levels of approximately 1% and O_2 at 2–2.5% whereas in the latter the CO_2 levels were 3–4% and O_2 levels increased steadily to 10%. A control container was included in the shipments and on arrival the ethylene levels in the three containers were less than 0.02 μl $litre^{-1}$.

Champion (1986) mentions the 'Tectrol' gas sealing specifications. Simulated commercial export of mangoes using the 'Transfresh' system of controlled atmosphere container technology was described by Ferrar (1988). Avocados and bananas were stored in two 'Freshtainer' controlled atmosphere containers, 40 feet long and controlled by microprocessors. The set conditions for avocados were 7.4 ± 0.5°C for 8 days followed by 5.5 ± 0.5°C for 7 days, with 2 ± 0.5% O_2 and a maximum of 10 ± 0.5% CO_2. Those for bananas were 12.7 ± 0.5°C for 8 days followed by 13.5 ± 0.5°C for 11 days, with 2.0 ± 0.5% O_2 and a maximum of 7 ± 0.5% CO_2. Temperatures and container atmospheres were continually monitored and fruit quality was assessed after the predetermined storage period. The results confirmed that the containers were capable of very accurate control within the specified conditions. Also, the fruit quality was better than that for controls, which were avocados held in a cold store for 2 weeks at 5.5°C in a normal atmosphere and bananas held at 13.5 ± 0.5°C in an insulated container (Eksteen and Truter 1989).

Figure 11.6 Controlled atmosphere reefer container controls. Source: Tim Bach of Cronos.

Peacock (1988) described a controlled atmosphere transport experiment. Mango fruits were harvested from three commercial sites in Queensland when the content of total soluble solids was judged to be 13–15%. Fruit were de-stalked in two of the three sources and washed, dipped for 5 minutes in 500 μg $litre^{-1}$ benomyl at 52°C, cooled and sprayed with prochloraz (200 μg $litre^{-1}$), dried and sorted, and finally size graded and packed in waxed fibreboard cartons. Some cartons contained polypropylene inserts, which cupped the fruits, other fruits were packed on an absorbent pad, but the majority were packed on expanded polystyrene netting. After packing, fruits were transported by road to a precooling room overnight (11°C). Pulp temperatures were 18–19°C the following morning; after 36 hours of transport to Brisbane and overnight holding in a conventional cool room, the fruits were placed in a controlled atmosphere shipping container. At loading, the pulp temperature was 12°C. Fruits were stored in the container at 13°C with 5% O_2 and 1% CO_2 for 18 days.

A UK company, Cronos Containers, supply inserts, called the Cronos Controlled Atmosphere System (manufactured under licence from BOC), which can convert a standard reefer container to a controlled atmosphere container. The installation may be permanent or temporary, and is self-contained, taking about 3 hours the first time. If the equipment has already been installed in a container and is subsequently dismantled, then reinstallation can be achieved in about 1 hour. The complete units measure $2 \times 2 \times 0.2$ m, which means that 50 of them can be fitted into a 40 foot dry container for transport. This facilitates management of the system. It also means that they take up little of the cargo space when fitted into the container, only 0.8 m^3. The unit operates alongside the container's refrigeration system and is capable of controlling, maintaining and recording the levels of O_2, CO_2 and humidity to the levels and tolerances pre-set into a programmable controller. Ethylene can also be removed from the container by scrubbing. This level of control is greater than for any comparable controlled atmosphere storage system, increasing shipping range and enhancing the quality of fresh fruit, vegetables, flowers, fish, meat, poultry and similar products. The system is easily attached to the container floor and bulkhead, and takes power from the existing reefer equipment with minimal alteration to the reefer container. The design and manufacture are robust to allow operation in the marine environments that will be encountered in typical use. The system fits most modern reefers and is easy to install, set up, use and maintain. A menu-driven programmable controller provides the interface to the operator, who simply has to pre-set the required percentages of each gas to levels appropriate for the product in transit. The controls are located on the front external wall of the container, and once set up a display will indicate the measured levels of O_2, CO_2 and relative humidity. The system consists of a rectangular aluminium mainframe on to which the various sub-components are mounted (Figure 11.6). The compressor is located at the top left of the mainframe and is driven by an integral electric motor supplied from the control box. Air is extracted from the container and pressurized (up to 4 bar) before passing through the remainder of the system. A pressure relief valve is incorporated in the compressor along with inlet filters. Note also that a bleed supply of external air is ducted to the compressor and is taken via the manifold with a filter mounted external to the container. The compressed air is cooled prior to passing through the remainder of the system. A series of coils wound around the outside of the air cooler radiated heat back into the container. The compressed air then passes into this component, which removes the pressure pulses produced by the compressor and provides a stable air supply. The water trap passing into the controlled atmosphere storage system then removes moisture. Water from this component is ducted into the water reservoir and used to increase the humidity when required. Two activated alumina drier beds are used in this equipment, each located beneath one of the nitrogen and CO_2 beds. Control valves are used to route the air through parts of the system as required by the conditioning process. Mesh filters are fitted to the outlet, which vents nitrogen and carbon dioxide back into the container, and to the outlet which vents oxygen to the exterior of the container. Nitrogen and CO_2 beds are located above the drier sieve beds and contain zeolite for the absorption of nitrogen and CO_2.

Another type of reefer container that has controls that can give a more precise control over the gaseous atmosphere was introduced in 1993. The specifications are given in Table 11.4. The containers use ventilation to control oxygen levels and a patented molecular sieve to control CO_2. The molecular sieve will also absorb ethylene and has two distinct circuits which are

Table 11.4 Specifications for a refrigerated controlled atmosphere container (Freshtainer INTAC 401)

	External dimensions	Internal dimensions	Door
Length	12 192 mm	11 400 mm	
Width	2438 mm	2280 mm	2262 mm
Height	2895 mm	2562 mm	2519 mm
Height	2895 mm	2562 mm	2519 mm
Capacity	66.6 m		
Tare	5446 kg		
Maximum payload	24 554 kg		

- Temperature range at 38°C ambient is –25 to +29°C (±0.25°C)
- Oxygen down to 1% (+1% or –0.5%) up to 20 l h^{-1} removal
- Carbon dioxide 0–80% (+0.5% or –1%) up to 180 l h^{-1} removal
- Humidity 60–98% (±5%)
- Ethylene removal rate 120 l h^{-1} (11.25 mg h^{-1})
- Water recycled to maintain high humidity

switched at predetermined intervals so that while one circuit is absorbing, the other is being regenerated. The regeneration of the molecular sieve beds can be achieved when they are warmed to 100°C to drive off the carbon dioxide and ethylene. This system of regeneration is referred to as temperature swing where the gases are absorbed at low temperature and released at high temperature. Regeneration can also be achieved by reducing the pressure around the molecular sieve, which is called pressure swing. During the regeneration cycle the trapped gases are usually ventilated to the outside, but they can be directed back into the container if this is required. The levels of gas, temperature and humidity within the container are all controlled by a computer that is an integral part of the container. It monitors the levels of oxygen from a paramagnetic analyser and the carbon dioxide from an infrared gas analyser and adjusts the levels to those which have been preset in the computer.

Figure 11.7 Typical airfreight container.

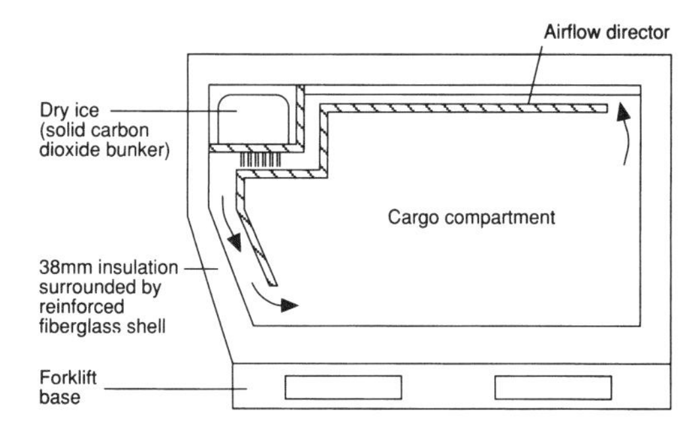

Figure 11.8 Diagram of a vertical section through a controlled atmosphere airfreight container.

International transport by airfreight

International transport of fresh fruit and vegetables is constantly being questioned in the news media, especially in relation to airfreight transport. This is mainly in terms of the high environmental costs of using fossil fuels for aeroplanes to provide consumers with crops all the year round, which in the past were seasonal. Most of this export trade is from less developed countries to industrialized countries and can be a major source of income for producing countries. This is a very involved subject and not one for a technical work such as this.

For export, fruit and vegetables are usually packed into cartons which in turn are packed into either closed containers, igloos or net-covered pallets which are specially designed to fit into a specific type of aircraft. The pallets or containers are loaded with the produce and then taken to the aircraft for loading to speed the process. Containers come in various shapes and sizes (Figures 11.7 and 11.8) which are designed to maximize the use of the interior of the aircraft.

Akinaga and Kohda (1993) measured the conditions to which airfreight cargo was exposed to during transport. They showed that most aircraft have pressurized cargo compartments which are maintained at low humidity (20–40% r.h.) and the same temperature as the passenger compartment. Okras loaded into a freight container without precooling showed a progressive decrease in temperature and increase in humidity during the flight. The increase in humidity inside the container was considered to be due to the poor ventilation characteristics of the container (Akinaga and Kohda 1993). In another shipment observed by Akinaga and Kohda (1993), the carbon dioxide and ethylene concentrations inside a unit load device lined with PVC film for waterproofing containing chrysanthemum flowers were shown to increase (Figure 11.9). However, the maximum level of CO_2 was

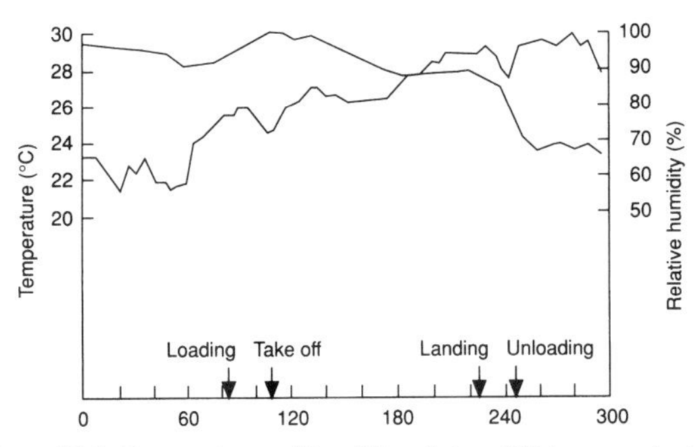

Figure 11.9 Temperature and humidity of okras (*Hibiscus esculentus*) during airfreight transport. Source: Akinga and Kohda (1993).

Table 11.5 Controlled atmosphere containers used for airfreight (source: Envirotainer Worldwide)

Type	Internal capacity (m^2)	Tare weight (kg)	Maximum gross weight (kg)	Dry ice capacity (kg)
LD3	3.5	190	1591	56
LD7/9	8.6	499	6033	56
LD5/11	6.0	408	3175	90

only about 0.5% and ethylene at 0.23 μl $litre^{-1}$ and was unlikely to have any effect on the flowers.

Containers that are used on aircraft can have facilities that provide some control over temperature and gases. Reid (2001) recommended a cold chain including refrigerated air containers to reduce losses that occur during shipping and handling of cut flowers. One company which produces and has patented such equipment is Envirotainer Worldwide in the USA (Sinco 1985). It produces three sizes of insulated containers which are cooled by a fan circulating air across dry-ice (solid CO_2) and then through the product in the container. The container has up to 57 kg of dry-ice in a compartment that is separated from the cargo compartment. There are no moving parts in the container and air circulation is by convection. Temperature in the cargo compartment is regulated by airflow controls. The sizes are given in Table 11.5.

Lufthansa have the LD3 airfreight container with a payload of 1350 kg and LD7 with a payload of 4755 kg (Sinclair 1988). They are insulated with Alu-foam panels affixed to a pallet and air is passed over the dry-ice by means of a battery-operated fan which can maintain the temperature within the container at the selected range for the entire transport time.

Air cargo is charged by weight. However, if the volume is greater than 366 in^3 kg^{-1} (6 litres kg^{-1}), a formula may be applied by the airline to increase the basic charge. The transport cost is a major item in marketing of fresh produce. Examples of costs for beans and cut flowers exported from Kenya to the UK and the Netherlands, respectively, were given by Jaffee (1993) (Table 11.6) and for limes from Sudan to the UK by Thompson *et al.* (1974d) (Table 11.7).

The limes used by Thompson *et al.* (1974d) were the local cultivar, Baladi. It had a good strong flavour but was generally smaller than the market required, which was greater than 42 mm diameter. In studies it was found that the distribution of fruit sizes in a commercial harvest was:

- >42 mm diameter: 7.5% with a mean weight of 43.8 g
- 38–42 mm diameter: 25.2% with a mean weight of 33.7 g
- <38 mm diameter: 67.3% with a mean weight of 22.6 g.

Airfreight charges are normally considerably higher than sea freight (Table 11.8).

Temperature monitoring

Monitoring the storage conditions of the crop during transport is very important. Reefer containers are fitted with chart recorders that monitor the temperature of the air before and after it passes over the cooling coils. In modern containers an integral computer often carries out this function. In addition, portable

Table 11.6 Distribution of sales revenue, expressed as a percentage of the total, between Kenya and the UK for French beans (April 1985) and Kenya and the Netherlands for cut flowers (source: Jaffee 1993)

	Fresh french beans (%)	Cut beans (%)
F.O.B. Kenya	25.6	19.2
Freight and handling	21.3	12.8
Importer's costs and profits	12.7	6.4
Wholesaler's costs and profits	10.4	10.6
Retailer's costs and profits	30.0	50.0

Table 11.7 Costs and returns (in £ sterling) for an airfreight shipment of limes based on one carton containing 153 fruits weighing 4.13 kg (source: Thompson *et al.* 1974d)

	Expense	Total
Purchase of fruit	0.38	
Carton	0.12	
Airfreight from Khartoum to London	0.84	
Customs and agents charges	0.45	
Total expenditure (excluding labour)		£1.79
Receipts		£2.69

Table 11.8 Comparison of the variable costs of transporting avocados or mangoes from the Kenyan highlands to Germany either by airfreight or by 20 foot reefer container by sea freight. All figures have been converted to US $ and the study was made in June 1986

	Airfreight 2 tonnes	Sea freight 9 tonnes
Transport: field to packhouse	9	41
Transport: packhouse to airport	7	
Transport: packhouse to seaport		260
Stowing in container		33
Export clearance at airport	500	
Export clearance at seaport		33
Handling at seaport		24
Export fees	13	59
Air freight	1374	
Sea freight		3000
Total	1903	3417
Costs c.i.f. per kilogram	0.95	0.38

recorders are often placed in the boxes of produce in order to monitor the actual conditions close to the produce. These are often contained in sealed units, which are sometimes used as evidence where there is a dispute as to the temperature to which the crop has been exposed. Several companies make these recorders, perhaps the best known being Ryan Instruments Inc. The recorders have a temperature sensor based on a bimetallic strip attached to a pressure-sensitive roll of paper which passes below a pointer attached to the bimetallic strip and records a trace of the temperature over time. A British company, Green PC, developed a smaller device (about 55 mm long × 30 mm diameter) called 'Tiny Talk' (Figure 11.10). The temperature

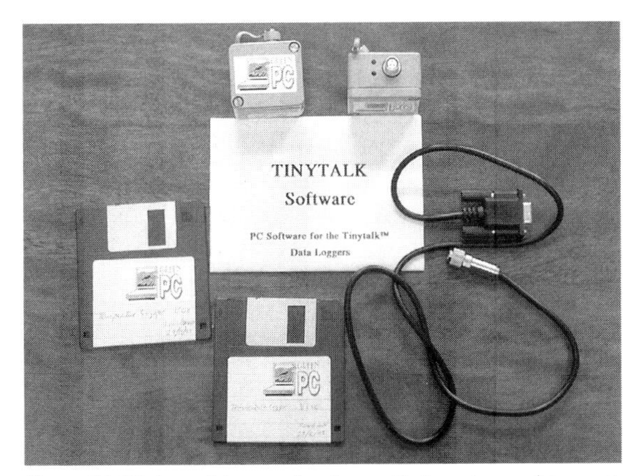

Figure 11.10 A small temperature logger that can be used to monitor the condition inside boxes of fruit and vegetables during transport. Source: Dr J.R. Robinson.

is monitored and recorded on a chip and can be downloaded to a computer to give a graph of the temperature over time. Tiny Talk has the following characteristics (Robinson 1993):

- accurate time-stamped temperature measurement to ±0.3°C
- recording temperature range of –40 to +75°C
- data points with user configuration from 15 minutes to 1 year duration
- water-resistant diecast enclosure
- simple and fast recovery of data
- automatic graphic results or printout on any personal computer.

This design of recorder has now completely replaced Ryan recorders. They have become reasonably cheap and flexible with the provision for downloading the data to a computer.

International quarantine

In order to prevent the transmission of crop pests and diseases between countries, safeguards and restrictions are placed on the international movement of fresh produce. National laws may prevent the import of fresh produce from designated areas or require that they are treated in such a way as to eliminate certain pests or diseases.

Fruit flies are a major concern, particularly in fruit exported to countries such as Japan, the USA and Australia. Important fruit fly genera are (Vijaysegaran 1993):

- *Anastrepha* – South America, Central America and the West Indies
- *Ceratitis* – Africa, but has spread throughout the world except Asia
- *Rhagoletis* – South America, Central America, North America and Europe
- *Dacus* – mostly Africa
- *Bactocera* – Asia, Australia and South Pacific.

Fruits which are susceptible to insect attack are allowed into Japan from certain specified countries provided a specified disinfestation treatment is applied (Table 11.9).

Since the establishment of the single market in the European Union in January 1993, health requirements for plant material entering the Union have been

Table 11.9 Fruits given special entry permits by the Japanese Government (source: Kitagawa 1993)

Country or area	Fruit	Cultivar	Designated pest	Disinfestation method
Australia	Orange		Mediterranean fruit fly, Queensland fruit fly	Cold treatment (1°C, 16 days)
Australia	Lemon		Mediterranean fruit fly, Queensland fruit fly	Cold treatment (1°C, 14 days)
Canada	Cherry		Codling moth	Methyl bromide fumigation
Chile	Grape		Mediterranean fruit fly	Cold treatment (0°C, 12 days)
China	Melon		Melon fly	(Pest-controlled areas)
Israel	Orange		Mediterranean fruit fly	Cold treatment (0.5°C, 14 days)
Israel	Grapefruit		Mediterranean fruit fly	Cold treatment (0.5°C, 13 days)
Israel	Sweety		Mediterranean fruit fly	Cold treatment (1.5°C, 16 days)
New Zealand	Cherry		Codling moth	Methyl bromide fumigation
New Zealand	Nectarine		Codling moth	Methyl bromide fumigation
New Zealand	Apple		Codling moth	Methyl bromide fumigation + cold treatment (0.5°C, 25 days)
Philippines	Mango	Carabao	Oriental fruit fly, melon fly	Vapour heat treatment (46°C, 10 minutes)
Spain	Lemon		Mediterranean fruit fly	Cold treatment (2°C, 16 days)
South Africa	Orange		Mediterranean fruit fly	Cold treatment (–0.6°C, 12 days)
South Africa	Lemon		Mediterranean fruit fly	Cold treatment (–0.6°C, 12 days)
South Africa	Grapefruit		Mediterranean fruit fly	Cold treatment (–0.6°C, 12 days)
Swaziland	Orange		Mediterranean fruit fly	Cold treatment (–0.6°C, 12 days)
Swaziland	Lemon		Mediterranean fruit fly	Cold treatment (–0.6°C, 12 days)
Swaziland	Grapefruit		Mediterranean fruit fly	Cold treatment (–0.6°C, 12 days)
Taiwan	Orange		Oriental fruit fly	Cold treatment (1°C, 14 days)
Taiwan	Mango	Irwin, Haden	Oriental fruit fly, melon fly	Vapour heat treatment (46.5°C, 30 minutes)
Taiwan	Litchi		Oriental fruit fly	Vapour heat treatment (46.2°C, 20 minutes) + cold treatment (2°C, 42 hours)
Thailand	Mango	Nang Klarngwun	Oriental fruit fly, melon fly	Vapour heat treatment (46.5°C, 10 minutes)
Thailand	Mango	Nam Dorkmai, Pimsen dang, Rad	Oriental fruit fly, melon fly	Vapour heat treatment (47°C, 20 minutes)
USA	Nectarine		Codling moth	Methyl bromide fumigation
USA	Walnut		Codling moth	Methyl bromide fumigation
Hawaii	Papaya	Solo	Mediterranean fruit fly or melon fly	Vapour heat treatment (47.2°C)
Washington	Cherry		Codling moth	Methyl bromide fumigation
Oregon	Cherry		Codling moth	Methyl bromide fumigation
California	Cherry		Codling moth	Methyl bromide fumigation

Table 11.10 Fruit that required a phytosanitary certificate to enter the European Union in 1993

Scientific name	Common name	Origin
Anona L.	Custard apple	Third countries[a]
Citrus L. and *hybrids*	Orange, lemon, lime, etc.	Third countries
Cydonia Mill.	Quince	Non-European countries
Diospyros L.	Persimmon, date plum	Non-European countries
Fortunella and hybrids	Kumquat	Third countries
Malus	Apple	Non-European countries
Mangifera	Mango	Non-European countries
Passiflora	Passionfruit	Non-European countries
Poncirus and hybrids	Ornamental citrus	Third countries
Psidium	Guava	Non-European countries
Pyrus	Pear	Non-European countries
Ribes	Gooseberry, redcurrant, blackcurrant	Non-European countries
Rubus	Blackberry, raspberry, dewberry, loganberry	Third countries
Syzygium	Clove, jambolan, rose Apple	Non-European countries
Vaccinium	Cranberry, blueberry	Non-European countries
Zea mays	Maize (seeds)	Third countries

[a] Third countries are countries other than those in the European Union.

harmonized. The general rule is that all plant material intended for planting from third countries (countries outside the EC), except seeds and aquarium plants, need a phytosanitary certificate. In addition, a certificate is required for certain fruit entering the EU (Table 11.10), but these are constantly being revised and the local EU office should be consulted before shipments are made.

Chemical pesticide residues in fruits and vegetables are restricted and controlled by legislation. See also Chapter 8 in the section on legislation. The work of DG VI (Agriculture) of the EC has the responsibility of harmonizing chemical pesticide residues in food. With the introduction of the single market on 1 January 1993, a first list of 55 chemicals was produced which contained their maximum residue levels (MRLs). The European Union provides Directives for MRLs and these are communicated in Directives issued in the Official Journal of the European Communities, e.g. Council Directive 93/58/EEC of 29 June 1993.

12
Postharvest technology of fruits and vegetables

Introduction

Recommended storage conditions can vary considerably even for the same species. Therefore, to use this information commercially it may be necessary to experiment. Some of the information is conflicting and in some cases the sources quoted may have used the same original data as other sources. Storage recommendations for individual crops in this chapter are mainly given in chronological order. In some cases original recommendations have been superseded and it may be that the latest reference may be the most reliable where there is a conflict of recommendations. Also, some of the sources use only common names for the crops and since the same common name may be used for very different crops, this may also be a source of confusion and error.

It is impossible to be absolutely definite about storage conditions because even though there is a recommendation that a crop should be stored at a certain temperature, humidity and even gaseous environment and under those conditions they will have a certain storage life, this can be affected by many other factors. These are especially related to factors such as the condition in which the crop was grown, the harvest maturity, the crop cultivar and the conditions in which the crop has been kept between harvest and being placed in store. Precooling, other crops in the store and contamination by microorganisms also affect the storage life of the crop. Most of the recommendations refer to the actual maximum storage life under the conditions specified so that the crop is still in a condition to allow it to pass through normal marketing channels. In some cases shelf-life is given at marketing temperatures. Some data given on shelf-life at room temperature or under ambient conditions may also be contentious, especially when from practical experience the reader knows that they are kept in a fruit bowl at home regularly for longer periods and are still fit for consumption.

Where detailed studies of storage conditions have been made over long periods of time, it is known that these may have to be varied from year to year, often based on the experience of the store manager. An example is with apples, where the condition of the fruit may be monitored during storage and the storage conditions modified as a result of the monitoring.

Some general recommendations can be made where specific data are not available by extrapolating from a similar crop. Generally leaf vegetables and herbs can be stored at 0°C since chilling injury does not occur on most species. There is an increasing market for wild and unusual types of mushrooms and they are becoming more common in greengroceries and supermarkets. From the literature, chilling injury does not appear to be a problem and so the fruiting bodies can be stored at 0°C unless specific experience indicates that it results in injury.

Acerolas

Malpighia glabra, *M. emarginata*, Malpighiaceae

Botany

Acerola is a fleshy drupe, which has a bright red skin when ripe and has three lobes each containing a triangular seed. They can be up to 3 cm in diameter and are extremely rich in ascorbic acid with up to 4 g per 100 g of the pulp (Purseglove 1968). There is some confusion in the literature between Acerolas and Barbados or West Indian cherry (*M. punicifolia*, *M. glabra*). Probably they all require similar postharvest conditions.

Harvesting

They are normally harvested when their skins are pinkish orange to light red. Harvesting is by hand although tree shakers (see Figure 12.84) have been tried (Pantastico 1975). Fruits harvested green had the highest ascorbic acid levels, but their eating quality was not as desirable as those of ripe fruits (Santos *et al.* 1999).

Storage

Cold storage at 4–7°C with 60% r.h. maintained the fruit quality but they still became soft. Freezing was more efficient than cold storage in reducing ascorbic acid losses and changes in the level of acidity (Santos *et al.* 1999). Other refrigerated storage recommendations are as follows:

- 0°C and 85–90% r.h. for 8 weeks (Pantastico 1975)
- 0°C and 85–90% r.h. for 50–58 days (SeaLand 1991).

Ackee, akee

Bligha sapida, Sapindaceae.

Botany

The fruit is a capsule and is bright red. It is ovoid in shape, about 6 cm long and 3 cm in diameter. It is a native of West Africa, but is part of the national dish of Jamaica together with salt fish. The edible portion is the aril, which can be eaten raw but is more commonly cooked and is yellow and resembles scrambled egg both in appearance and taste. The genus is named after Captain Bligh of the British ship HMS *Bounty*.

Harvesting

It should be harvested when fully mature, otherwise it may be toxic. The fruit is green as it develops on the tree and then turns yellow, then red as it matures. When the fruit is fully mature on the tree it splits open along three sutures revealing its three black seeds on yellow arils (Figure 12.1, in the colour plates). The arils are the edible portion of the fruit; care must be taken to remove the pink or purplish membrane near to the seed as this can be toxic. Hypoglycin [L-2-amino-3-(methylenecyclopropyl)propionic acid] is the toxic ingredient of the unripe fruit. A case study involving a 27-year-old Jamaican man was reported by Larson *et al.* (1994). The man developed jaundice, pruritus, intermittent diarrhoea and abdominal pain following chronic ingestion of canned ackee fruits. Liver function tests returned to normal and his symptoms were completely resolved following cessation of fruit intake.

Storage

No information on its postharvest life could be found in the literature.

Ripening

As indicated above, they should be harvested when the fruit is fully red and it splits along the three sutures. However, companies who process the fruit in Jamaica have been seen to lay the fully red fruit on wire trays in ambient temperatures of about 28–30°C to allow for good air circulation until they split, and then can them. This practice is potentially dangerous.

African breadfruit

Treculia africana, Moraceae.

Botany

It is a tree that produces fruit some 45 cm in diameter, which contain edible seeds that are boiled, roasted or made into flour (Purseglove 1968). It should not be confused with the breadfruit *Artocarpus altilis*, which belongs to the same family.

Storage

Nwaiwu *et al.* (1995) found that the microorganisms associated with biodeterioration of pods were 72% fungi (*Aspergillus niger*, *Rhizopus stolonifer* and *Botryodiplodia theobromae*) and 28% bacteria (*Erwinia* spp.). No bacteria were isolated from seeds only from the pods. Storage at 5°C reduced spoilage from 100–0% and prestorage treatment with 15% sodium chloride solution reduced spoilage during subsequent storage from 100–5%.

Amelanchier

Amelanchier alnifolia, Rosaceae

Botany

The fruits are black when ripe and are borne on a medium-sized shrub. It is a native of Western North America.

Controlled atmosphere storage

The cultivars Pembina, Smoky, Northline and Thiessen were stored at 0.5°C in 2, 10 and 21% O_2 factorially combined with either 0.035 or 5% CO_2 for 56 days by Rogiers and Knowles (2000). Storage of all four cultivars was improved by controlled atmospheres compared with those stored in air. The 5% CO_2 atmosphere combined with 21 or 10% O_2 was most effective at minimizing losses in fruit soluble solids, anthocyanin, firmness and fresh weight. Fungal infection by *Botrytis cinerea* occurred on fruits stored in air after 8 weeks but was eliminated on those stored with 5% CO_2 at all O_2 concentrations. Ethanol did not exceed 0.03% in fruits stored in any of the atmospheres.

Amaranth

Amaranthus hybridus, Amaranthaceae.

Botany

This is grown mainly as a leafy vegetable. Some other species of *Amaranthus* are grown for their grains, e.g. *A. caudatus*, *A. cruentus*, *A. hypochondriacus* and *A. leucocarpus*. The last species was important to the Aztecs in Mexico. The Emperor Montezuma was said to receive an annual tribute of 200 000 bushels of amaranth seed (Purseglove 1968).

Harvesting

This is done when individual leaves are still young and tender.

Storage

The only specific storage recommendation was 0°C and 95–100% r.h. for 10–14 days (SeaLand 1991). Various other methods of preservation have been described. Sun drying of the amaranth leaves resulted in reductions in the levels of ascorbic acid and β-carotene of 87.4 and 16.3%, respectively. After storage of the dried leaves for 3 months, the retention of ascorbic acid was 42 mg per 100 g at 22°C, 40 mg per 100 g at 28°C and 36 mg per 100 g at room temperature (30–32°C) and for β-carotene 32.4 mg per 100 g at 22°C, 29.5 mg per 100 g at 28°C and 27.4 mg per 100 g at room temperature on a dry matter basis (Mziray *et al.* 2000).

American grapes

Vitus labrusca, *V. raparia*, *V. rupestris* and hybrids, Vitaceae.

Botany

Most commercial cultivars of grapes are from the species *V. vinifera*, even those grown in the USA, and most harvesting and handling practices are the same for all species and are dealt with under the heading of *V. vinifera*. The fruit is also a berry borne in racemes or bunches and they are not climacteric.

Disease control

American species are very sensitive to sulphur dioxide, a treatment commonly used on *V. vinifera* to control mould, and it is not recommended for American species because they will suffer damage (Anon 1968; Lutz and Hardenburg 1968).

Storage

Specific refrigerated storage recommendations for American Grapes are as follows:

- –0.6 to 0°C and 85–90% r.h. for 4 weeks (Phillips and Armstrong 1967)
- –1 to 0°C and 85–90% r.h. for 3–4 weeks (Concord type) (Anon 1967)
- –0.6 to 0°C and 85% r.h. for 4–8 weeks (Lutz and Hardenburg 1968)
- –0.6 to 0°C and 85% r.h. for 3–5 weeks for Warden, 3–6 weeks for Niagra and Moore, 4–7 weeks for Concord and Delaware, 5–8 weeks for Catawba (Anon 1968).

Anise, anis

Pimpinella anisum, Umbelliferae.

Botany

The edible portion is the leaf petiole, which can be used in salads. The seeds are used for flavouring liquors such as aguardiente in Colombia. The essential oil, aniasol, may be extracted and used as a perfume or medicine.

Pest control

The use of 60% CO_2 to disinfest anise in a 5-day exposure in a 4 m^3 stainless-steel fumatorium and 30 m^3 fumigation chamber at temperatures ranging between 28 and 35°C resulted in complete control of pupal and adult stages of *Stegobium paniceum* and *Lasioderma serricorne* (Hashem 2000).

Storage

The only reference to storage is 0°C and 90–95% r.h. for 14 to 21 days (SeaLand 1991).

Añus, cubios

Tropaeolum tuberosum, Tropaeolaceae.

Botany

It produces conical or ellipsoidal tubers that are up to 15 cm long and 6 cm in diameter (Kay 1987). They are usually greyish-white with the furrows tinged purple, but can be various colours.

Harvesting

This is usually done some 7 months after planting and it is carried out by digging them by hand.

Storage

Tubers have a storage life at ambient temperatures for up to 6 months (Kay 1987).

Apples

Malus domestica, Rosaceae.

Botany

The fruit is a pome and is classified as climacteric. Their freezing point is about –2.8 to 2.5°C (Wright 1942).

Texture

Landfald (1966) showed that fruit softened during storage even at 0°C (Table 12.1). Stow and Genge (1993) suggested that softening results from loss of cell-to-cell cohesion. Dover and Stow (1993) found that the major effect of ethylene removal from stored apples is to delay in the onset of softening and that removal of ethylene from the store could slow the rate of softening once it has started.

Acidity

Malic acid is the major organic acid in apples and pears but some apples contain quantities of citric acid and some pears quantities of quinic acid (Ulrich 1970).

Flavour and Aroma

These are related to organic volatiles and are dealt with in more detail in Chapter 10. An experiment measured organic volatile production of Golden Delicious over a 10 day period at 20°C after removal from the cold store. It was found that aroma volatiles were suppressed during storage at 1°C for up to 10 months in atmospheres containing 3% CO_2 or 1–3% O_2 compared with storage in air (Brackmann 1989). Yahia (1989) found that in McIntosh and Cortland most organic volatile compounds were produced at lower rates during ripening after controlled atmosphere storage than those produced from fruits ripened immediately after harvest. In experiments on the storage of Filippa, Ingrid Marie and Lobo, Landfald (1966) found that temperatures in the range 4–12°C did not affect their flavour score, but those which had been stored at 0°C had an inferior flavour.

Table 12.1 Effects of temperature and duration of storage on the firmness of apples in kg cm^{-2} (initial value 9.2). Results are the means of three cultivars (source: Landfald 1966)

	Temperature (°C)			
Days in storage	0	4	8	12
30	8.9	7.6	6.6	6.3
60	7.7	6.7	5.9	5.7
90	7.2	6.3	5.7	5.5
120	6.7	5.9	5.5	5.3

Table 12.2 Effects of temperature and duration of storage on the ground colour score (0–10) of apples. Results are the means of three cultivars (source: Landfald 1966)

	Temperature (°C)			
Days in storage	0	4	8	12
30	2.3	3.1	4.5	6.5
60	2.8	4.3	6.3	8.4
90	3.7	5.1	7.8	9.5
120	4.0	5.3	8.3	9.7

Colour

The ground colour of fruits can change during storage (Landfald 1966), which is mainly due to the breakdown of chlorophyll (Table 12.2).

Preharvest factors

Link (1980) showed that high rates of application of nitrogen fertilizer to apple trees could adversely affect the flavour of the fruit. For good storage quality Cox's Orange Pippin it was found that they required the following composition, on a dry matter basis for storage until December at 3.5°C:

- N 50–70%
- P 11% minimum
- K 130–160%
- Mg 5%
- Ca 5%.

For storage until March they required 4°C with 2% O_2 and less than 1% CO_2 and 4.5% (Sharples 1980).

Daminozide

Daminozide applied to trees can improve storage (Sharples 1967), but it has been withdrawn from the market in several countries (John Love, personal communication, quoted by Thompson 1996). More information is included in Chapter 1.

Harvest maturity

Selection of the correct maturity for harvesting is often a matter of experience on the part of the grower. They should be harvested just as they begin to ripen for good quality and a good storage and marketable life. There are a variety of techniques that are used or could be used to increase the precision. In order to determine the maturity of fruit various techniques have been developed and these are discussed in the following. All the tests, except the 'colour' test and the 'time' test, involve harvesting representative samples. These are reviewed in detail Chapter 2.

Colour

Skin colour can change during maturation and charts are used for some cultivars of apple, but it is not a reliable method since the changes tend to be subtle.

Time

The time between flowering and fruit being ready for harvesting may be fairly constant, which gives an approximate guide to when fruit should be harvested.

Starch

Starch is converted to sugar as harvest time is approached and the assessment of starch content with the starch/iodine test (Figure 12.2) (Cockburn and Sharples 1979). Studies using this technique on apples gave inconsistent results in England, but in some other countries, for example in Turkey, it works well on apples and excellent charts have been produced by Dr Umit Ertan, Dr Sozar Ozelkok and Dr Kenan Kaynas at the Ataturk Horticultural Research Institute in Yalova for individual cultivars.

Firmness

Pressure testers were first developed for apples (Magness and Taylor 1925) and are currently available in various forms (see Figure 12.85).

Vibration and acoustic tests

Equipment that puts vibration energy into a fruit, and measures its response to this input, has been tested but not used commercially.

Figure 12.2 The starch/iodine test used to determine the maturity of pears. Photograph: Dr R.O. Sharples.

Near-infrared reflectance

Correlations between sugar content and NIR measurements have been shown (Kouno *et al.* 1993a) but NMR is not used commercially.

Nuclear magnetic resonance

NMR date have also been shown to correlate well with the sugar content of apples (Cho and Krutz 1989), but NIR is currently not used commercially.

Harvest methods

Apples were traditionally grafted on rootstocks that resulted in tall trees, which made them difficult to harvest. Traditionally, the harvester would carry a ladder and use that to reach the fruit (Figure 12.3). A picking box could be used and worn around the harvester's neck. Fruit may be transferred from these to a pallet box to be taken and stacked in the storage chamber or taken to the packhouse. Types of package and the material from which they are made vary considerably. Hand harvesting is very time consuming and various ways have been developed to speed up the operation. A long pole with a bag at the end together with some device for cutting or breaking the fruit stalk has been used. Picking platforms are available that enable the harvester to be towed around the orchard from tree to tree so the platform can be raised and lowered, enabling the fruit to be picked (Figure 12.4).

Apple trees grown on a hedgerow system could be harvested with combing fingers giving 85% Grade 1 fruit, although a significant proportion of the fruit remained on the trees (LeFlufy 1983). In the meadow orchard system the whole of the tree is harvested just above the ground every other year.

Figure 12.3 Ladder used for apple harvesting.

Prestorage treatments

A variety of chemicals have been applied to apples before storage, including calcium or lecithin to prevent bitterpit, *N*, *N*-dimethylaminosuccinamic

Figure 12.4 A platform used to facilitate fruit harvesting from tall trees in the UK.

acid, diphenylamine, Nutri-Save, Semperfresh 1-MCP and various chemical fungicides. The postharvest application of calcium can also be used for apples to reduce the development of physiological disorders occurring during storage. Drake and Spayd (1983) pressure infiltrated Golden Delicious apples with calcium chloride at 3% at 3 lb in^{-2} for 8 minutes, before storage for 5 months at 1°C. These fruit were firmer and more acid than untreated fruit stored for the same period of storage.

Packhouses

Finney (1970) discussed sorting apples on the basis of firmness and much work has taken place since then, but it is still not used in commercial packhouses. Clarke (1996) described a size-grading machine where there are two cables along which the crop runs, made from high-tensile, polyethylene-coated steel. When the crop reaches the prescribed point, ejector bars controlled by solenoid valves and operated by compressed air cylinders push it into the chute. The machine is very gentle to the crop and high drops are unnecessary, although padded chutes can be provided.

Chen *et al.* (1992) discussed vibrational techniques for grading fruit and identified many of the factors that influence the acoustic responses of apples. However, it is difficult to test the fruit without any physical contact. The main approach was to impart a band of frequencies into the fruit by a contacting emitter and then to pick up the transmitted signal frequencies using a detector that rests briefly on the surface of the fruit. This would most commonly be a piezoelectric surface or an accelerometer. Rapid analysis of the acoustic resonance signal by fast Fourier transformation means that the whole sorting process can be carried out fairly rapidly but not yet at commercially attractive rates for most fruits. Experimental machines do exist, however, which often take up to 1 second per fruit, which is too long (Kouno *et al.* 1993a–c).

Physiological disorders

A more comprehensive discussion of the subject can be found in Fidler (1973) and Snowdon (1990). The principal disorders are as follows:

Superficial scald

Scald is a physiological disorder that can develop on apples and pears during storage and has been associated with ethylene levels in the store, which results in the skin turning brown. Scald susceptibility in Delicious apples was found to be strain dependent. Although storage in 0.7% O_2 effectively reduced scald in Starking and Harrold Red fruits picked over a wide range of maturity stages, it did not adequately reduce scald in Starkrimson fruits after 8 months of storage (Lau and Yastremski 1993). To control scald, Van der Merwe (1996) suggested storage at –0.5°C and 0–1% CO_2 with 1.5% O_2 for 7 months. Granny Smith apples were stored at –0.5°C for 6 months in normal atmosphere, for 9 months in 1.5% O_2 and 0% CO_2 and after storage they were ripened at 20°C for 7 days before evaluation. In a normal atmosphere, all control fruits developed superficial scald, but only a few in controlled atmosphere storage developed scald (Van Eeden *et al.* 1992). Scald can also be controlled by a prestorage treatment with an antioxidant. Ethoxyquin (1,2-dihydro-2,2,4-trimethylquinoline-6-yl ether), marketed as 'Stop-Scald,' or DPA (diphenylamine), marketed as 'No Scald' or 'Coraza,' should be applied directly to the fruit within 1 week of harvesting. In the USA, the Government approved the postharvest application of these two chemicals to apples with maximum residues of 3 µg $litre^{-1}$ for ethoxyquin and 10 µg $litre^{-1}$ for DPA (Hardenburg and Anderson 1962). Residue levels of DPA in apples were found to vary depending on the application method and the position of the fruit in the pallet box. There are restrictions on the sale of treated fruits in several countries and local legislation should be consulted before they are applied. Different cultivars of apples respond differently to these treatments; for example, only DPA is effective in controlling scald on the Delicious cultivars. Successful application of DPA, ethoxyquin and captan to Granny Smith, Delicious and Golden Delicious apples in cold stores, by thermal fogging and thermonebulization, was described by Sive and Resnizky (1989).

Ilinski *et al.* (2000) found that exposure of apples to 19–21% CO_2 for 2–4 days followed by controlled atmosphere storage resulted in good scald control whereas a short exposure of 1–5 hours had no such effect. After controlled atmosphere storage for 250 days, fruit firmness, acidity and overall sensory acceptability of fruits were similar in treatments with initial O_2 reduction by respiration compared with traditional atmosphere establishment. Fidler *et al.* (1973) also showed that scald development in storage could be related to the CO_2 and O_2 levels but that the relationship varied between cultivars (Table 12.3).

Table 12.3 Effects of controlled atmosphere storage conditions during storage at 3.5°C for 5–7 months on the development of superficial scald in three apple cultivars (Fidler *et al.* 1973)

Storage conditions		Cultivar		
CO_2 (%)	O_2 (%)	Wagener	Bramley's Seedling	Edward VII
0	21	100	–	–
0	6	–	89	75
0	5	100	–	–
0	4	–	85	62
0	3	9	30	43
0	2.5	–	17	43
0	2	0	–	–
8	13		24	3

Core flush

This disorder is also called brown core and has been described on several cultivars, where it develops during storage as a brown or pink discoloration of the core while the flesh remains firm. It has been associated with CO_2 injury but may also be related to chilling injury and senescent breakdown. The effects of O_2 shock treatment on physiological disorders have also been described. Johnson and Ertan (1983) found that at 4°C in 1% O_2 the fruits were free of core flush and other physiological disorders and their quality was markedly better than those stored in 2% O_2. Keeping apples in 0% O_2 for the first 10 days of storage was shown to prevent core flush in early-picked Jonathan from highly affected orchards. Wang (1990) reviewed the effects of CO_2 and concluded that core flush was due to exposure to high levels of CO_2 at low storage temperatures. No damage due to anaerobic respiration was observed in any of the treatments (Resnizky and Sive 1991).

Bitterpit

Bitterpit (Figure 12.5, in the colour plates) is principally associated with calcium deficiency during the period of fruit growth and may by detectable at harvest or sometimes only after protracted periods of storage. The addition of lecithin (phosphatidylcholine) to the postharvest application of calcium can enhance its effect in controlling bitter pit (Sharples *et al.* 1979). At 18°C apples treated with 4% calcium chloride had a slightly lower respiration rate, but this effect was greatly enhanced when lecithin (1%) was added. At 3°C lecithin-treated apples had reduced ethylene production but there was no effect on CO_2 production (Watkins *et al.* 1982).

Disease control

It has been common practice for many years to drench apples with a fungicide just before they are loaded into store. However, legislation on the postharvest use of fungicides is being increasingly restricted. Spraying the crop during growth can reduce postharvest infections. In Britain, single sprays with 0.025% benomyl in either June, July or August to apple trees controlled rots, caused by infection with *Gloeosporium* spp., which developed in subsequent storage from September onwards at 3.3°C in fruit from unsprayed trees. The observations were made over two seasons and the fungicide was applied at 1120 litres hectare^{-1} (Edney *et al.* 1977). With the pressure against the use of postharvest fungicides, various biological control methods have proved to be successful (Teixido *et al.* 1999). See also Chapter 8 for more details.

Fruit coatings

Coating fruit with Tal Prolong (Bancroft 1989), Semperfresh (Kerbel *et al.* 1989) or chitosan (Davies *et al.* 1988) has been reported. See the section on fruit coatings in Chapter 6.

Precooling

Precooling is not a common commercial practice with apples but there is some information in the literature, which gives different recommendations. MAFF (undated) found that placing the fruit in a conventional cold storage was the most acceptable, but forced air cooling with high humidity was suitable, hydrocooling less suitable and vacuum cooling unsuitable. However, hydrocooling was reported to be the most effective (Table 12.4) since it rapidly cooled the fruit and had no detrimental effects on their quality.

Table 12.4 Effects of precooling method on the half-cooling time of apples packed loosely in 18 kg boxes (source: Hall 1972)

Cooling method	Half cooling time
Conventional cool room	12 h
Tunnel with air velocity of 200–400 m min^{-1}	4 h
Jet cooling with air velocity of 740 m min^{-1}	45 min
Hydrocooling (loose fruit)	20 min

Chilling injury

Some cultivars suffer from chilling injury at temperatures above 0°C, and the environment in which they are grown can also affect susceptibility. Symptoms include a brown discoloration of the cortex with streaks of darker brown in the vascular region and the tissue remaining moist (Wilkinson 1970). In Ellison's Orange and some other cultivars, symptoms can take the form of ribbon-shaped brown patches on the skin and extending 2–3 mm into the cortex, often called 'ribbon scald' (Wilkinson 1970).

Simple stores

In Britain and many other European countries before refrigeration storage was developed, cellars were traditionally used for storage during the winter. The apples were usually spread out thinly on shelves to ensure good air circulation. In the mountainous regions of Nepal caves are dug into the sides of slopes for storing apples and fitted with a wooden door. The doors may be left open at night to ensure that the temperature in the store remains low. In such a climate the humidity is often low, which can lead to desiccation of the fruit. To relieve this, the floor of the store can be kept wet (Thompson 1978).

Low oxygen storage

Ke and Kader (1991) showed that apples at temperatures between 0 and 5°C would tolerate oxygen levels of between 0.25 and 1% for over 10 days. In modern controlled atmosphere apple stores, where they are stored at levels as low as 1% oxygen for long periods, an alcohol detector is fitted which sounds an alarm if ethanol fumes are detected in the store. This enables the store operator to increase the oxygen level and no damage should have been done to the fruit.

Storage and controlled atmosphere storage conditions

A large commercial store in the UK uses 3.5°C with 1% CO_2 and 1.2% O_2 for Cox's Orange Pippin. To achieve this, the store is sealed directly after loading is completed and the temperature is reduced as quickly as possible. When it is below 5°C nitrogen is injected to reduce the oxygen level to about 3%. The O_2 level is allowed to decline to about 2% for 7 days through fruit respiration and then progressively to 1.2% over the next 7 days. Bramley Seedling apples were treated in a similar way with the final conditions being 3.5–4°C with 5% CO_2 and 1% O_2 (East Kent Packers Limited, personal communication 1994). General recommended storage conditions that were not specific to a particular cultivar were given as 0–1°C with 1–2% CO_2 and 2–3% O_2 (Kader 1985). Kader (1992) revised these recommendations to 0–1°C with 1–5% CO_2 with 1–3% O_2. Another general recommendation was given by SeaLand (1991) as –1.1°C and 90–95% r.h. for 90–240 days for non-chilling sensitive fruit and 4.4°C and 90–95% r.h. for 40–45 days for chilling sensitive fruit. Storage temperature is very important as it affects the maximum time the apples can be stored. This includes chilling injury and senescent breakdown. Fruit colour and firmness were also affected by storage temperature (Tables 12.1 and 12.2).

Specific storage recommendations for major cultivars are described in the following.

Allington Pippin

Fidler (1970) recommended 3–3.5°C with 3% CO_2 and 3% O_2.

Aroma

Storage trials of Aroma apples at 2°C and 88% r.h. in air, or 2% O_2 with 2% CO_2 or 3% O_2 with 3% CO_2 for 115, 142 and 169 days were described by Haffner (1993). One set of samples analysed immediately and a second set of samples analysed after 1 week in a cold store and a third set after 1 week in a cold store followed by 1 week in ambient conditions of 18–20°C and 25–30% r.h. to simulate shelf-life. Weight loss was 0.6% per month for controlled atmosphere storage compared with 1.5% in cold storage. Controlled atmosphere storage controlled *Gloeosporium album* [*Pezicula alba*] and *G. perennans* [*P. malicorticis*]. Fruits undergoing simulated shelf-life experiments showed a severe loss in quality after storage with damage from *Gloeosporium* rot, wilting and over-ripeness. Controlled atmosphere storage delayed the maturity process, controlled weight loss and resulted in better eating quality during shelf-life.

Boskoop

Meheriuk (1993) recommended 3–5°C with 0.5–2% CO_2 and 1.2–2.3% O_2 for 5–7 months. Koelet (1992) advocated 3–3.5°C with 0.5–1.5% CO_2 and 2–2.2% O_2. Hansen (1977) suggested a temperature of not lower than 4°C with 3% CO_2 and 3% O_2. Schaik (1994) described experiments where fruits of the cultivar Schone van Boskoop (also called Belle de Boskoop) from several orchards were held in controlled atmosphere storage with a combined scrubber/separator or with a traditional lime scrubber, in 0.7% CO_2 and 1.2% O_2, at 4.5°C, followed by holding at 15°C for 1 week or at 20°C for 2 weeks. Ethylene concentration during storage with a scrubber/separator rose to only 20 μl $litre^{-1}$, compared with up to 1000 μl $litre^{-1}$ in a cell without a scrubber. The average weight loss, rots (mostly *Gloeosporium* spp.), flesh browning and core flush were also slightly lower in the scrubber/separator cell, but the incidence of bitterpit was slightly greater. Using a scrubber/separator did not affect fruit firmness and colour. Herregods (1993) and Meheriuk (1993) summarized general recommendations from a variety of countries as follows:

	°C	CO_2%	O_2%
Belgium	3–3.5	<1	2–2.2
Denmark	3.5	1.5–2	3–3.5
France	3–5	0.5–0.8	1.5
Germany (Saxony)	3	1.1–1.3	1.3–1.5
Germany (Westphalia)	4–5	2	2
Netherlands	4–5	$\ll 1$	1.2
Switzerland	4	2–3	2–3

Bowden's seedling

Fidler (1970) recommended 3.5°C with 5% CO_2 and 3% O_2.

Braeburn

A temperature of 0°C and 96% r.h. with 1% O_2 and 3% CO_2 for 8 months plus 7 days at 31°C shelf-life with no physiological disorders was recommended by Brackmann and Waclawovsky (2000). However, >1.5% CO_2 with <1.5% O_2 could result in a higher incidence of flesh and core browning (Rabus and Streif 2000). In Canada, storage at 0°C in 1.2 or 1.5% O_2 with 1.0 or 1.2% CO_2 retained firmness and acidity better than storage in air, but fruit had more core browning and superficial scald (Lau *et al.* 1993; Elgar *et al.* 1998). Growing conditions were shown to influence these disorders, so following a cool growing season it was recommended that Braeburn be harvested at starch index values between 2.5 and 3.0 (0–9 scale) and stored in air storage at 0°C to avoid the risks of scald, browning disorder and internal cavities. The fruits could be stored in <1% CO_2 (preferably close to 0.1%) and >1.5% O_2 after warm seasons (Lau *et al.* 1993). Clark and Burmeister (1999) found that during storage at 0.5°C, with 2% O_2 and 7% CO_2 there was 1.5% brown tissue after 2 weeks, increasing to 15.9% and 21.3% after 3 and 4 weeks, respectively. It was suggested that they should be stored for 2 weeks in air followed by 1% CO_2 and 3% O_2, all at 0°C (Elgar *et al.* 1998). This is because they found that browning disorder appeared to develop during the first 2 weeks of storage, and storage in air at 0°C prior to controlled atmosphere storage decreased the incidence and severity of the disorder. Development of browning induced in Braeburn fruits by a damaging CO_2 concentration could be detected with magnetic resonance imaging (Clark and Burmeister 1999).

Bramley's seedling

Fidler and Mann (1972) recommended 3–4°C with 8–10% CO_2 and about 11–13% O_2. Wilkinson and Sharples (1973) recommended 3.3–4.4°C with 8–10% CO_2 and no scrubber for 8 months. Sharples and Stow (1986) recommended 4–4.5°C with 8–10% CO_2 where no scrubber is used and 4–4.5°C with 6% CO_2 and 2% O_2 where a scrubber is used. The following conditions were said to be used in their commercial controlled atmosphere stores by East Kent Packers Limited: 3.5–4°C with 5% CO_2 and 1% O_2 (East Kent Packers Limited, personal communication 1994). Johnson (1994, personal communication 1997) recommended the following:

Temperature (°C)	CO_2 (%)	O_2 (%)	Storage time (weeks)
4–4.5	8–10	11–13	39
4–4.5	6	2	39
4–4.5	5	1	44

Cacanska Pozna

Jankovic and Drobnjak (1994) stored Cacanska Pozna at 1°C and 85–90% r.h., either in a controlled atmosphere (7% CO_2 with 7% O_2) or under normal atmospheric conditions. Weight losses and decay were negligible in controlled atmospheres (0–0.78%). Fruits stored in controlled atmosphere stores exhibited no physiological disorders.

Charles Ross

Fidler (1970) recommended 3.5°C with 5% CO_2 and 3% O_2. Sharples and Stow (1986) recommended 3.5–4°C and 5% CO_2 for fruit stored without a scrubber and 5% CO_2 with 3% O_2 for those stored with a scrubber.

Cortland

There was a recommendation of 3.3°C with 2% then 5% CO_2 and 3% O_2 for 5–6 months (Anon 1968). Meheriuk (1993) recommended 0–3°C with 5% CO_2 and 2–3.5% O_2 for 4–6 months. Herregods (1993) and Meheriuk (1993) summarized general recommendations from a variety of countries as follows:

	Temperature (°C)	CO_2 (%)	O_2 (%)
Canada (Nova Scotia)	3	4.5	2.5
Canada (Nova Scotia)	3	1.5	1.5
Canada (Quebec)	3	5	2.5–3
USA (Massachusetts)	0	5	3

Cox's orange pippin

Fidler and Mann (1972) and Wilkinson and Sharples (1973) recommended 3.3–3.9°C with 5% CO_2 and 3% O_2 for 5 months, but where core flush is a problem then 0–1% CO_2 with 1.8–2.5% O_2 for 6 months was recommended. Stoll (1972) suggested 4°C with 1–2% CO_2 and 3% O_2. Sharples and Stow (1986) advocated 3.5–4°C with 5% CO_2 where no scrubber is used and 3.5–4°C with 5% CO_2 and 3% O_2 (for storage to mid February), <1% CO_2 with 2% O_2 (for storage to late March), <1% CO_2 with 1.25% O_2 (for storage to late April) and <1% CO_2 with 1% O_2 (for storage to early May) where a scrubber is used. East Kent Packers Limited used 3.5°C with 1% CO_2 and 1.2% O_2 for Cox's Orange Pippin in their commercial stores (East Kent Packers Limited, personal communication 1994). Koelet (1992) recommended 3–3.5°C with 0.5–1.5% CO_2 and 2–2.2% O_2. Meheriuk (1993) recommended 3–4.5°C with 1–2% CO_2 (sic) and 1–3.5% O_2 for 5–6 months. Johnson (1994, personal communication 1997) recommended the following:

Temperature (°C)	CO_2 (%)	O_2 (%)	Storage time (weeks)
3.5–4	5	3	20
3.5–4	<1	2	23
3.5–4	<1	1.2	28
3.5–4	<1	1	30

Herregods (1993) and Meheriuk (1993) summarized general recommendations from a variety of countries as follows:

	Temperature (°C)	CO_2 (%)	O_2 (%)
Belgium	3–3.5	<1	2–2.2
Denmark	3.5	1.5–2	2.5–3.5
UK	4–4.5	<1	1.25
France	3–4	<1	1.2–1.5
Germany (Saxony)	3	1.7–1.9	1.3–1.5
Germany (Westphalia)	4	2	1–2
Netherlands	4	⩽1	1.2
New Zealand	3	2	2
Switzerland	4	2–3	2–3

Crispin

Sharples and Stow (1986) recommended 3.5–4°C with 8% CO_2 where no scrubber is used, but stated that there was a risk of the physiological disorder of core flush at 8% CO_2.

Delicious

See also the sections on Red Delicious, Starking and Golden Delicious. Drake (1993) described the effects of a 10 day delay in harvesting date and/or a 5-, 10- or 15-day delay in the start of controlled atmosphere storage (1% O_2 and 1% CO_2) on Delicious fruits (Bisbee, Red Chief and Oregon Spur strains). Delayed harvest increased red colour of the skin at harvest in Bisbee and Red Chief, but not Oregon Spur; soluble solids content and size also increased, but up to 12% of firmness was lost, depending on strain. Immediate establishment of controlled atmosphere conditions after harvest resulted in good quality fruits after 9 months of storage, but reduced quality was evident when controlled atmosphere establishment was delayed by 5 days (the interim period being spent in refrigerated storage at 1°C). Longer delays did not result in greater quality loss. Oregon Spur had a redder colour at harvest and after storage than the other two strains. Sensory panel profiles were unable to distinguish between strains, harvest dates or delays in the time of controlled atmosphere establishment.

Mattheis *et al.* (1991) stored Delicious fruit in 0.5% O_2 plus 0.2% CO_2 at 1°C for 30 days and found that they developed high concentrations of ethanol and acetaldehyde. Scald susceptibility in Delicious apples was found to be strain dependent. Whereas storage in 0.7% O_2 effectively reduced scald in Starking and Harrold Red fruits picked over a wide range of maturity stages, it did not adequately reduce scald in Starkrimson fruits after 8 months of storage (Lau and Yastremski 1993).

Discovery

Discovery has been reported to have poor storage qualities and Sharples and Stow (1986) recommended 3–3.5°C with <1% CO_2 and 2% O_2 for 7 weeks.

Edward VII

Fidler (1970) recommended 3.5–4°C with 8–10% CO_2 with information supplied only for stores with no scrubber and Wilkinson and Sharples (1973) recommended 3.3–4.4°C with 8–10% CO_2 and no scrubber for 8 months. Sharples and Stow (1986) also recommended 3.5–4°C and 8% CO_2 for fruit stored without a scrubber.

Egremont Russeta

Wilkinson and Sharples (1973) recommended 3.3°C with 5% CO_2 and 3% O_2 for 5 months. Fidler (1970) recommended 3.3–4.4°C with 7–8% CO_2 with information supplied only for stores with no scrubber.

Ellison's Orange

Fidler (1970) and Sharples and Stow (1986) recommended 4–4.5°C and 5% CO_2 with 3% O_2 for fruit stored with a scrubber or less than 1% CO_2 with 2% O_2.

Elstar

Sharples and Stow (1986) recommended 1–1.5°C <1% CO_2 with 2% O_2 for those stored with a scrubber. Meheriuk (1993) recommended 1–3°C with 1–3% CO_2 and 1.2–3.5% O_2 for 5–6 months. In a trial with Elstar apples stored in 1 or 3% O_2, the ethylene concentration in the lower O_2 concentration remained less than 1 μl $litre^{-1}$ after 8 months. After holding at 20°C for 1 week, fruit firmness declined only slightly and taste was considered good. In the higher O_2 concentration, however, the ethylene concentration rose rapidly to almost 100 μl $litre^{-1}$ and fruit hardness declined markedly after holding at 20°C (Schaik 1994). Elstar fruits were stored in controlled atmospheres of 1 or 5% O_2 with 0.5% CO_2 or 3% O_2 with 0.5 or 3.0% CO_2 at 1.5°C. An ethylene scrubber was used continuously in the low-ethylene cabinets and at intervals in other cabinets to remove excess ethylene produced by the fruits. Fruit firmness was measured 1 week after removal from storage and holding at 20°C. Low O_2 concentrations in addition to low temperature further decreased ethylene sensitivity in apples. Increased CO_2 concentrations did not affect fruit firmness when fruits were stored under low ethylene but were clearly beneficial when ethylene was present in the storage atmosphere (Woltering *et al.* 1994). Schouten (1997) compared storage at 1–2°C in either 2.5% CO_2 with 1.2% O_2 or <0.5% CO_2 with 0.3–0.7% O_2. The fruit stored in 0.5% CO_2 with 0.3–0.7% O_2 were shown to be of better quality both directly after storage and after a 10 day shelf-life period in air at 18°C. Herregods (1993) and Meheriuk (1993) summarized general recommendations from a variety of countries as follows:

	Temperature (°C)	CO_2 (%)	O_2 (%)
Belgium	1	2	2
Canada (Nova Scotia)	0	4.5	2.5
Canada (Nova Scotia)	0	1.5	1.5
Denmark	2–3	2.5–3	3–3.5
UK	1–1.5	<1	2
France	1	1–2	1.5
Germany (Saxony)	2	1.7–1.9	1.3–1.5
Germany (Westphalia)	2–3	3	2
Netherlands	1–2	2.5	1.2
Slovenia	1	3	1.5

Empire

Meheriuk (1993) recommended 0–2°C with 1–2.5% CO_2 and 1–2.5% O_2 for 5–7 months. Herregods (1993) and Meheriuk (1993) summarized general recommendations from a variety of countries as follows:

	Temperature (°C)	CO_2 (%)	O_2 (%)
Canada (British Columbia)	0	1.5	1.5
Canada (Ontario)	0	2.5	2.5
Canada (Ontario)	0	1	1
USA (Michigan)	0	3	1.5
USA (New York)	0	2–2.5	1.8–2
USA (Pennsylvania)	–0.5–0.5	0–2.5	1.3–1.5

Fiesta

Sharples and Stow (1986) recommended 1–1.5°C and <1% CO_2 with 2% O_2 for those stored with a scrubber.

Fortune

Fidler (1970) recommended 3.5–4°C with 7–8% CO_2 for stores with no scrubber or either 5% CO_2 and 3% O_2 or <1% CO_2 and 2% O_2 for stores fitted with a scrubber.

Fuji

Meheriuk (1993) recommended 0–2°C with 0.7–2% CO_2 and 1–2.5% O_2 for 7–8 months. Fuji were stored at 0°C in a controlled atmosphere (1.5% O_2 with than <0.5% CO_2). Fruits were removed from cold storage after 3, 5, 7 and 9 months and assessed for storage life and sensory acceptability on return to air at 20°C after 0, 4, 11, 18 and 25 days. Harvest maturity was not critical for Fuji apples, which maintained a good level of consumer acceptability for up to 9 months in storage and 25 days at 20°C in air (Jobling *et al.* 1993). Work in Australia showed that fruits store well in 2% O_2 and 1% CO_2 at 0°C (Tugwell and Chvyl 1995). Herregods (1993) and Meheriuk (1993) summarized general recommendations from a variety of countries as follows:

	Temperature (°C)	CO_2 (%)	O_2 (%)
Australia (South)	0	1	2
Australia (Victoria)	0	2	2–2.5
Brazil	1.5–2	0.7–1.2	1.5–2
France	0–1	1–2	2–2.5
Japan	0	1	2
USA (Washington)	0	1–2	1–2

Gala

Gala were stored at 0°C in 1.5% O_2 with <0.5% CO_2 for up to 9 months (Jobling *et al.* 1993). Johnson (1994) recommended 3.5–4°C in <1% CO_2 with 2% O_2 for 23 weeks. However, more recent work has shown that Gala grown in the UK can be stored at 1.5°C in O_2 concentrations as low as 1% with CO_2 concentrations of 2.5–5% (Stow 1996). Meheriuk (1993) recommended –0.5 to 3°C with 1–5% CO_2 and 1–2.5% O_2 for 5–6 months. Herregods (1993) and Meheriuk (1993) summarized general recommendations from a variety of countries as follows:

	Temperature (°C)	CO_2 (%)	O_2 (%)
Australia (South)	0	1	2
Australia (Victoria)	0	1	1.5–2
Brazil	1–2	2.5–3	1.6–2
Canada (British Columbia)	0	1.5	1.2
Canada (Nova Scotia)	0	4.5	2.5
Canada (Nova Scotia)	0	1.5	1.5
Canada (Ontario)	0	2.5	2.5
France	0–2	1.5–2	1–2
Germany (Westphalia)	1–2	3–5	1–2
Netherlands	1	1	1.2
Israel	0	2	0.8–1
New Zealand	0.5	2	2
South Africa	–0.5	2	2
Switzerland	0	2	2
USA (Washington)	0	1–2	1–2

Gloster

Sharples and Stow (1986) recommended 1.5–2°C with <1% CO_2 and 2% O_2 for Gloster 69. Hansen (1977) advocated 0–1°C with 3% CO_2 and 2.5% O_2. Meheriuk (1993) suggested 0.5–2°C with 1–3% CO_2 and 1–3% O_2 for 6–8 months. Herregods (1993) and Meheriuk (1993) summarized general recommendations from a variety of countries as follows:

	Temperature (°C)	CO_2 (%)	O_2 (%)
Belgium	0.8	2	2
Canada (Nova Scotia)	0	4.5	2.5
Canada (Nova Scotia)	0	1.5	1.5
Denmark	0	2.5–3	3–3.5
France	1–2	1–1.5	1.5
Germany (Saxony)	2	1.7–1.9	1.3–1.5
Germany (Westphalia)	1–2	2	2
Netherlands	1	3	1.2
Slovenia	1	3	1
Slovenia	1	3	3
Spain	0.5	2	3
Switzerland	2–4	3–4	2–3

Golden Delicious

Stoll (1972) recommended 2.5°C with 5% CO_2 and 3% O_2. Koelet (1992) recommended 0.5–1°C with 1–3% CO_2 and 2–2.2% O_2. Meheriuk (1993) recommended –0.5 to 2°C with 1.5–4% CO_2 and 1–2.5% O_2 for 8–10 months. Golden Delicious were stored at 3% CO_2 with 21% O_2, 3% CO_2 with 3% O_2, 3% CO_2 with 1% O_2, 1% CO_2 with 3% O_2, 1% CO_2 with 1% O_2 and 1% CO_2 with 21% O_2. All controlled atmosphere storage combinations suppressed volatile production except for fruits stored in 1% CO_2 with 3% O_2 in which volatile production was high (Harb *et al.* 1994). The effect of ultra-low O_2 (3% CO_2 with 1% O_2) and low ethylene (1 μl $litre^{-1}$) storage on the production of ethylene, CO_2, volatile aroma compounds and fatty acids of Golden Delicious apples harvested at preclimacteric and climacteric stages were evaluated during 8 months of storage and during the following 10 days at 20°C in air. 1-Aminocyclopropane-1-carboxylic acid oxidase activity, increase in ethylene and CO_2 and production of aroma volatiles and fatty acids were lower during the 10 days after storage following ultra-low O_2 storage than following non-controlled atmosphere cold storage. The reduction of all these parameters was more pronounced in fruits harvested at the preclimacteric stage than in those harvested when climacteric. Increasing CO_2 concentrations from 1–3% under ultra-low O_2 conditions (1% O_2) intensified the inhibition of respiration and the production of ethylene, aroma and fatty acids. On the other hand, removal of ethylene from the store atmosphere only slightly affected CO_2 and aroma production by apples (Brackmann *et al.* 1995). Van der Merwe (1996) recommended –0.5°C and 1.5% CO_2 with 1.5% O_2 for 9 months. Sharples and Stow (1986) recommended 1.5–3.5°C with 5% CO_2 and 3% O_2.

In experiments conducted over 4 years at 1% O_2 with 1% CO_2 or 1% O_2 with 6% CO_2, fruit rot, scald and flesh browning were prevented to a large extent (Strempfl *et al.* 1991). Starking Delicious and Golden Delicious can be stored for up to 9 months without appreciable loss of quality at 0°C in 3% CO_2 with 3% O_2 (Ertan *et al.* 1992).

Golden Delicious in storage at 0°C had a threshold for peel injury at 15% CO_2 with 3% O_2, with little difference between 3% and 5% O_2 levels at 20% CO_2. The development of superficial scald (which did not occur at 3% O_2) diminished at 15% O_2 with increasing CO_2 and was completely inhibited by 20% CO_2. Fruit softening was inhibited with increasing CO_2 levels in a similar manner at both 3% and 5% O_2 levels, even though the higher O_2 level enhanced softening. No softening occurred at 20% CO_2 regardless of O_2 level both during storage and subsequent ripening. This could be related to the total inhibition of ethylene evolution, which occurred at 20% CO_2 at both O_2 levels. Reducing the O_2 level predominantly inhibited chlorophyll degradation. Titratable acidity was highest when CO_2 was increased to 10% at 15% O_2 and to 5% at 3% O_2 (Ben Arie *et al.* 1993).

Controlled atmosphere storage treatments suppressed aroma production compared with storage at 1°C. The greatest reduction was found under ultra-low O_2 (1%) and high CO_2 (3%) conditions. A partial recovery of aroma production was observed when controlled atmosphere stored fruits were subsequently stored for 14 days under cold storage conditions (Brackmann *et al.* 1993). There was a recommendation of –1.1 to 0°C with 1–2% CO_2 and 2–3% O_2 (Anon 1968). De Wild (2001) recommended 1°C with 1.2% O_2 and 4% CO_2. Herregods (1993) and Meheriuk (1993) summarized general recommendations from a variety of countries as follows:

	Temperature (°C)	CO_2 (%)	O_2 (%)
Australia (South)	0	2	2
Australia (Victoria)	0	1	1.5
Australia (Victoria)	0	5	2
Belgium	0.5	2	2
Brazil	1–1.5	3–4.5	1.5–2.5
Canada (British Columbia)	0	1.5	1–1.2
Canada (Ontario)	0	2.5	2.5
China	5	4–8	2–4
France	0–2	2–3	1–1.5
Germany (Saxony)	2	1.7–1.9	1.3–1.5
Germany (Westphalia)	1–2	3–5	1–2
Netherlands	1	4	1.2
Israel	–0.5	2	1–1.5
Italy	0.5	2	1.5
Slovenia	1	3	1
Slovenia	0	3	1
South Africa	–0.5	1.5	1.5
Spain	0.5	2–4	3
Switzerland	2	5	2–3
Switzerland	2	4	2–3
USA (New York)	0	2–3	1.8–2
USA (New York)	0	2–3	1.5
USA (Pennsylvania)	–0.5–0.5	0–0.3	1.3–2.3
USA (Washington)	1	<3	1–1.5

Storing Golden Delicious infested with eggs of *Rhagoletis pomonella* in a 100% nitrogen atmosphere at 20 ± 1°C resulted in 100% mortality after 8 days but apples developed detectable off-flavour and it was suggested that the treatment might be useful for cultivars less susceptible to anoxia (Ali Niazee *et al.* 1989).

Granny Smith

Meheriuk (1993) recommended –0.5 to 2°C with 0.8–5% CO_2 and 0.8–2.5% O_2 for 8–10 months. Van der Merwe (1996) suggested –0.5°C and 0–1% CO_2

with 1.5% O_2 for 7 months. Granny Smith apples were stored at –0.5°C for 6 months in a normal atmosphere and for 9 months in 1.5% O_2 with 0% CO_2. After storage the fruits were ripened at 20°C for 7 days before evaluation for superficial scald. In a normal atmosphere, all control fruits developed scald. In controlled atmosphere storage only a few apples developed scald (Van Eeden *et al.* 1992).

In South Tyrol (Italy), controlled atmosphere storage using ultra-low O_2 regimes (0.9–1.1% O_2) compared with conventional levels of 2.5–3.0% O_2 controlled scald on Granny Smith fruits for about 150 days. No accumulated alcoholic flavour or any external or internal visual symptoms of low O_2 injury, either during or after storage, were observed (Nardin and Sass 1994).

Granny Smith apples that were kept at 46°C for 12 hours, 42° for 24 hours or 38°C for 72 or 96 hours before storage at 0°C for 8 months in 2–3% O_2 with 5% CO_2 were firmer at the end of storage and had a higher °Brix:acid ratio and a lower incidence of superficial scald than fruits not heat treated. Prestorage regimes of 46°C for 24 hours or 42°C for 48 hours resulted in fruit damage after storage (Klein and Lurie 1992).

After 6.5 months of storage at –0.5°C, the average percentage of Beurre d'Anjou pears and Granny Smith apples with symptoms of scald ranged from zero in fruits stored in 1.0% O_2 with 0% CO_2 or 1.5% O_2 with 0.5% CO_2 to 2% in fruits stored in 2.5% O_2 with 0.8% CO_2 and 100% in control fruits stored in air. Herregods (1993) and Meheriuk (1993) summarized general recommendations from a variety of countries as follows:

	Temperature (°C)	CO_2 (%)	O_2 (%)
Australia (South)	0	1	2
France	0–2	0.8–1	0.8–1.2
Germany (Westphalia)	1–2	3	1–2
Israel	–0.5	5	1–1.5
Italy	0–2	1–3	2–3
New Zealand	0.5	2	2
Slovenia	1	3	1
South Africa	0–0.5	0–1	1.5
Spain	1	4–5	2.5
USA (Pennysylvania)	–0.5–0.5	0–4	1.3–2
USA (Washington)	1	<1	1

Howgate Wonder

Sharples and Stow (1986) recommended 3–4°C with 8–10% CO_2 with data only available for controlled atmosphere storage with no scrubbing.

Idared

Idared was stored at 1°C and 85–90% r.h., either in a controlled atmosphere (7% CO_2, 7% O_2) or under normal atmospheric conditions. Weight loss and decay were negligible in controlled atmospheres (0.0–0.78%). Fruits in controlled atmosphere storage exhibited no physiological disorder (Jankovic and Drobnjak 1994). Meheriuk (1993) recommended 0–4°C with <1–3% CO_2 and 1–2.5% O_2 for 5–7 months. Hansen (1977) recommended a temperature of not lower than 3°C with 3% CO_2 and 3% O_2. Sharples and Stow (1986) recommended 3.5–4.5°C with 5% CO_2 and 3% O_2 or <1% CO_2 and 1.25% O_2 for 37 weeks. In order to maintain fruit at a satisfactory level of firmness, Stow (1996a) suggested storage in 1% O_2 with <1% CO_2 for no more than 13 weeks. Herregods (1993) and Meheriuk (1993) summarized general recommendations from a variety of countries as follows:

	Temperature (°C)	CO_2 (%)	O_2 (%)
Belgium	1	2	2
Canada (Ontario)	0	2.5	2.5
UK	3.5–4	8	–
UK	3.5–4	<1	1.25
France	2–4	1.8–2.2	1.4–1.6
Germany (Westphalia)	3–4	3	1–2
Slovenia	1	3	3
Spain	2	3	3
Switzerland	4	4	2–3
USA (Michigan)	0	3	1.5
USA (New York)	0	2–3	1.8–2
USA (New York)	0	2–3	1.5
USA (Pennsylvania)	–0.5–0.5	0–4	2–3

Ingrid Marie

Fidler (1970) recommended 1.5–3.5°C with 6–8% CO_2 for stores with no scrubber or <1% CO_2 and 3% O_2 for stores fitted with a scrubber.

James Grieve

Fidler (1970) and Sharples and Stow (1986) recommended 3.5–4°C with 6% CO_2 and 5% O_2 or <1% CO_2 and 3% O_2.

Jonagold

Koelet (1992) recommended, for what he refers to as 'Jonah Gold,' 0.5–1°C with 1–3% CO_2 and 2–2.2% v. Meheriuk (1993) recommended 0–2°C with 1–3% CO_2 and 1–3% O_2 for 5–7 months. In experiments with Jonagold stored at 5°C and 95% r.h. for 9 months it was shown that the lower the O_2 concentration (over the range 17.0% O_2 with 4.0% CO_2, 1.0% O_2 with 1.0% CO_2 or 0.7% O_2 with 0.7% CO_2), the slower were the decreases in firmness and titratable acidity and the slower was the change of skin colour. No alcohol flavour or physiological disorders developed in fruits stored at 0.7% O_2 and this concentration was recommended by Goffings *et al.* (1994). Hansen (1977) recommended 0–1°C with 6% CO_2 and 2.5% O_2 Awad *et al.* (1993) recommended 1°C for 6 months. Sharples and Stow (1986) recommended 1.5–2°C with <1% CO_2 and 2% O_2 or <1% CO_2 and 1.25% O_2. Johnson (1994) showed that a slight difference in temperature at the same controlled atmosphere conditions could have an affect on maximum storage life as follows:

Temperature (°C)	CO_2 (%)	O_2 (%)	Storage time (weeks)
1.5–2	<1	1.5–2	32
1.25	<1	1.5–2	36

Jonagold were stored at 1°C and 85–90% r.h., either in a controlled atmosphere (7% CO_2, 7% O_2) or under normal atmospheric conditions. Weight losses and decay were negligible in controlled atmospheres (0–0.78%). Fruits stored in controlled atmosphere storage exhibited no physiological disorder (Jankovic and Drobnjak 1994). After storage at 3% O_2 with 1% CO_2 followed by 3 weeks of shelf-life at 20°C, scald was 23%. At 1.5% O_2 with 1% CO_2 and 3% O_2 with 5% CO_2, scald was 2% and 6%, respectively, while at 1.5% O_2 with 5% CO_2 no scald was observed. This shows that in Jonagold both O_2 and CO_2 affect the occurrence of scald. At high O_2 and low CO_2 the level of scald (39%) was significantly higher in early harvested apples compared with apples from a normal harvest date (13% scald) (Awad *et al.* 1993). Jonagold apples were stored at 0°C in air or in a controlled atmosphere. Controlled atmosphere storage at 0°C 1.5% O_2 with 1.5% CO_2 for 6 months significantly reduced the loss of acidity and firmness, decreased production of volatile compounds by half, but did not influence total soluble solids content (Girard and Lau 1995). De Wild (2001) recommended 1°C with 1.2% O_2 and 4% CO_2. Herregods (1993) and Meheriuk (1993) summarized general recommendations from a variety of countries as follows:

	Temperature (°C)	CO_2 (%)	O_2 (%)
Belgium	0.8	1	2
Canada (British Columbia)	0	1.5	1.2
Canada (Nova Scotia)	0	4.5	2.5
Canada (Nova Scotia)	0	1.5	1.5
Denmark	1.5–2	<1	2
UK	1.5–2	8	–
UK	1.5–2	<1	2
France	0–1	2.5–3	1.5
Germany (Saxony)	2	1.7–1.9	1.3–1.5
Germany (Westphalia)	1–2	3–6	1–2
Netherlands	1	4	1–2
Slovenia	1	3	3
Slovenia	1	6	3
Slovenia	1	3	1.2
Spain	2	3–4	3
Switzerland	2	4	2
USA (New York)	0	2–3	1.8–2
USA (New York)	0	2–3	1.5

Jonathan

There was a recommendation of 0°C with 3–5% CO_2 and 3% O_2 for 5–6 months (Anon 1968). Stoll (1972) advocated 4°C with 3–4% CO_2 and 3–4% O_2. Meheriuk (1993) suggested –0.5 to 4°C with 1–3% CO_2 and 1–3% O_2 for 5–7 months. Fidler (1970), Fidler and Mann (1972), and Sharples and Stow (1986) recommended 4–4.5°C with 6% CO_2 and 3% O_2. With Jonathan harvested at the optimum time, the loss of hardness after storage until March or June was less in 0.9% O_2 and 4.5% CO_2 than 1.2% O_2 and 4.5% CO_2 directly after storage or after holding for 1 week at 20°C but not after 2 weeks. With later harvested fruits, O_2 concentration did not affect fruit hardness. However, flesh browning was greater in fruits stored in 0.9% O_2 than in 1.2% O_2 (Schaik *et al.* 1994). Jonathan was shown to be sensitive to CO_2 injury, which appeared both externally and internally. The threshold for peel injury was 10% CO_2 and it occurred more severely and earlier at 15% O_2 than at 3% O_2 with little difference between the two O_2 levels at 20% CO_2. The development of superficial scald (which did not occur at 3% O_2) diminished at 15% O_2 with increasing CO_2 and was completely inhibited by 20% CO_2. Fruit softening was inhibited with increasing CO_2 levels in a similar manner at both O_2 levels, even though the higher O_2 level enhanced softening. No softening occurred at 20% CO_2 regardless of O_2 level during both storage and subsequent ripening. This could be related to the total inhibition of ethylene evolution which occurred at 20% CO_2 at both O_2 levels. Chlorophyll degradation was predominantly inhibited by increasing the level of CO_2 (Ben Arie *et al.* 1993). Herregods (1993) and Meheriuk (1993) summarized general recommendations from a variety of countries as follows:

	Temperature (°C)	CO_2 (%)	O_2 (%)
Australia (South)	0	3	3
Australia (South)	0	1	3
Australia (Victoria)	2	1	1.5
China	2	2–3	7–10
Germany (Westphalia)	3–4	3	1–2
Israel	–0.5	5	1–1.5
Slovenia	3	3	3
Slovenia	3	3	1.5
Switzerland	3	3–4	2–3
USA (Michigan)	0	3	1.5

Jupiter

Sharples and Stow (1986) recommended 3–3.5°C with <1% CO_2 and 2% O_2.

Karmijn

Hansen (1977) recommended a temperature of not lower than 4°C with 3% CO_2 and 3% O_2.

Katy (Katja)

Sharples and Stow (1986) recommended 3–3.5°C with <1% CO_2 and 2% O_2.

Kent

Sharples and Stow (1986) recommended 3.5–4°C with 8–10% CO_2 and about 11–13% O_2 or <1% CO_2 and 2% O_2.

King Edward VII

Fidler and Mann (1972) recommended 3–4°C with 8–10% CO_2 and about 11–13% O_2.

Lady Williams

Lady Williams were stored at 0°C in a controlled atmosphere of 1.5% O_2 with <0.5% CO_2. Fruits were removed from storage after 3, 5, 7 and 9 months and assessed for storage life and sensory acceptability on return to air at 20°C after 0, 4, 11, 18 and 25 days. Lady Williams is a late-maturing cultivar and the level of acidity (which decreased with time) is an important quality parameter for fruits of this cultivar (Jobling *et al.* 1993). Work in Australia showed that fruits can be stored for 6–8 months in 2% O_2 and 1% CO_2 at 0°C (Tugwell and Chvyl 1995).

Laxton's Fortune

Sharples and Stow (1986) recommended 3–3.5°C with 5% CO_2 and 3% O_2 or <1% CO_2 and 2% O_2 using a scrubber and 7–8% CO_2 for stores without a scrubber.

Laxton's Superb

Fidler (1970), Wilkinson and Sharples (1973) and Sharples and Stow (1986) recommended 3.5–4.5°C with 7–8% CO_2 and 3% O_2 for 6 months.

Lord Derby

Sharples and Stow (1986) recommended 3.5–4°C with 8–10% CO_2 with data only available for stores without a scrubber.

Lord Lambourne

Sharples and Stow (1986) recommended 3.5–4°C with 8% CO_2 with data only available for stores without a scrubber.

McIntosh

There was a recommendation of 3.3°C with 2–5% CO_2 with 3% O_2 for 6–8 months (Anon 1968). Meheriuk (1993) recommended 2–4°C with 1–5% CO_2 and 1.5–3% O_2 for 5–7 months. Sharples and Stow (1986) recommended 3.5–4°C with 8–10% CO_2, with data only available for stores without a scrubber. Sharples and Stow (1986) recommended 3.5–4°C with less than 1% CO_2 and 2% O_2. Hansen (1977) recommended for McIntosh-Rogers a temperature not lower than 3°C with 6% CO_2 and 2.5% O_2. Herregods (1993) and Meheriuk (1993) summarized general recommendations from a variety of countries as follows:

	Temperature (°C)	CO_2 (%)	O_2 (%)
Canada (British Columbia)	3	4.5	2.5
Canada (British Columbia)	3	5	2.5
Canada (British Columbia)	1.7	5	2.5
Canada (Nova Scotia)	3	4.5	2.5
Canada (Nova Scotia)	3	1.5	1.5
Canada (Ontario)	3	5	2.5–3
Canada (Ontario)	3	1	1.5
Canada (Quebec)	3	5	2.5
Germany (Westphalia)	3–4	3–5	2
USA (Massachusetts)	1–2	5	3
USA (Michigan)	3	3	1.5
USA (New York)	2	3–5	2–2.5
USA (New York)	2	3–5	4
USA (Pennsylvania)	1.1–1.7	0–4	3–4

Melrose

Melrose were stored at 1°C and 85–90% r.h., either in a controlled atmosphere (7% CO_2 with 7% O_2) or under normal atmospheric conditions. Weight losses and decay were negligible in controlled atmospheres (0–0.78%). Fruits stored in controlled atmospheres exhibited no physiological disorders whereas bitterpit was observed under normal atmospheric conditions (Jankovic and Drobnjak 1994). Hansen (1977) recommended 0–1°C with 6% CO_2 and 2.5% O_2. Meheriuk (1993) recommended 0–3°C with 2–5% CO_2 and 1.2–3% O_2 for 5–7 months. Herregods (1993) and Meheriuk (1993) summarized general recommendations from a variety of countries as follows:

	Temperature (°C)	CO_2 (%)	O_2 (%)
Belgium	2	2	2–2.2
France	0–3	3–5	2–3
Germany (Westphalia)	2–3	3	1–2
Slovenia	1	3	3
Slovenia	1	3	1.2

Merton Worcester

Fidler (1970) recommended 3–3.5°C with 7–8% CO_2 for stores with no scrubber or 5% CO_2 and 3% O_2 for stores with a scrubber.

Michaelmas Red

Fidler (1970) recommended 1–3.5°C with 6–8% CO_2 for stores with no scrubber or 5% CO_2 and 3% O_2 for stores with a scrubber.

Monarch

Fidler (1970) recommended 0.5–1°C with 7–8% CO_2 for stores with no scrubber or 5% CO_2 and 3% O_2 for stores with a scrubber.

Morgenduft

In Italy, controlled atmosphere storage using ultra-low O_2 regimes (0.9–1.1% O_2 compared with conventional levels of 2.5–3.0% O_2) controlled scald on Morgenduft fruits for 210 days. Fruits did not develop any alcoholic flavour or any external or internal visual symptoms of low O_2 injury, either during or after storage (Nardin and Sass 1994).

Mutsu

Fidler (1970) recommended 1°C with 8% CO_2 with data supplied only for stores not fitted with a scrubber. Hansen (1977) recommended 0–1°C with 6% CO_2 and 2.5% O_2. Meheriuk (1993) recommended 0–2°C with 1–3% CO_2 and 1–3% O_2 for 6–8 months. Herregods (1993) and Meheriuk (1993) summarized general recommendations from a variety of countries as follows:

	Temperature (°C)	CO_2 (%)	O_2 (%)
Australia (Victoria)	1	1	1.5
Australia (Victoria)	0	3	3
Denmark	0–2	3–5	3
Germany (Westphalia)	1–2	3–5	1–2
Japan	0	1	2
Slovenia	1	3	1
USA (Michigan)	0	3	1.5
USA (Pennsylvania)	–0.5–0.5	0–2.5	1.3–1.5

Newton Wonder

Controlled atmosphere storage was not recommended, only air storage at 1.1°C for 6 months (Fidler 1970; Wilkinson and Sharples 1973).

Norfolk Royal

Fidler (1970) recommended 3.5°C with 5% CO_2 and 3% O_2.

Northern Spy

There was a recommendation of 0°C with 2–3% CO_2 and 3% O_2 (Anon 1968).

Pacific Rose

Johnston *et al.* (2001) recommended 0°C.

Pink Lady

Work in Australia showed that fruits can be stored for up to 9 months in 2% O_2 and 1% CO_2 at 0°C without developing scald (Tugwell and Chvyl 1995).

Red delicious

In South Tyrol, Italy, controlled atmosphere storage, using ultra-low O_2 regimes (0.9–1.1% O_2 compared with conventional levels of 2.5–3.0% O_2), controlled scald on Red Delicious fruits for about 150 days without the development of alcoholic flavour or any external or internal visual symptoms of low O_2 injury, either during or after storage (Nardin and Sass 1994). Red Delicious fruits stored for up to 7 months in 1–5% O_2 and 2–6% CO_2 had superficial scald levels of <3% (Xue *et al.* 1991). Van der Merwe (1996) recommended –0.5°C and 1.5% CO_2 with 1.5% O_2 for 9 months. Fidler (1970) and Sharples and Stow (1986) recommended 0–1°C with either 5% CO_2 and 3% O_2 or <1% CO_2 and 3% O_2. Meheriuk (1993) recommended –0.5 to 1°C with 1–3% CO_2 and 1–2.5% O_2 for 8–10 months. Herregods (1993) and Meheriuk (1993) summarized general recommendations from a variety of countries as follows:

	Temperature (°C)	CO_2 (%)	O_2 (%)
Australia (South)	0	1	2
Australia (Victoria)	0	1.5	1.5–2
Canada (British Columbia)	0–1	1	1.2–1.5
Canada (British Columbia)	0–1	1	0.7
Canada (Ontario)	0	2.5	2.5
Canada (Ontario)	0	1	1
France	0–1	1.8–2.2	1.5
Germany (Westphalia)	1–2	3	1–2
Israel	–0.5	2	1–1.5
Italy	0.5	1.5	1.5
New Zealand	0.5	2	2
South Africa	–0.5	1.5	1.5
Spain	0	2–4	3
USA (Massachusetts)	0	5	3
USA (New York)	0	2–3	1.8–2
USA (Pennsylvania)	–0.5–0.5	0–0.3	1.3–2.3
USA (Washington)	0	<2	1–1.5

Ribstone Pippin

Fidler (1970) recommended 4–4.5°C with 5% CO_2 and 3% O_2.

Rome Beauty

There was a recommendation of –1.1 to 0°C with 2–3% CO_2 and 3% O_2 (Anon 1968).

Royal Gala

Work in Australia showed that Fruits of Royal Gala can be stored for up to 5 months in 2% and 1% CO_2 at 0°C (Tugwell and Chvyl 1995), but Stow (1996) in the UK found that there was a large loss in flavour after 3 months storage. Johnston *et al.* (2001) recommended 0°C.

Spartan

Fidler (1970) recommended –1 to 0°C with 5–8% CO_2 for stores with no scrubber or <1% CO_2 and 2% O_2 for stores with a scrubber. Sharples and Stow (1986) recommended 1.5–2°C with 6% CO_2 and 2% O_2 for 42 weeks. Meheriuk (1993) recommended 0–2°C with 1–3% CO_2 and 1.5–3% O_2 for 6–8 months. Herregods (1993) and Meheriuk (1993) summarized general recommendations from a variety of countries as follows:

	Temperature (°C)	CO_2 (%)	O_2 (%)
Canada (British Columbia)	0	1.5	1.2–1.5
Canada (Ontario)	0	2.5	2.5
Canada (Quebec)	0	5	2.5–3
Denmark	0–2	2–3	2–3
UK	1.5–2	<1	2
UK	1.5–2	6	2
Switzerland	2	3	2–3
USA (New York)	0	2–3	1.8–2
USA (Pennsylvania)	–0.5–0.5	0–1	2–3

Starking

Starking were stored at –0.5°C for 4 months in a normal atmosphere, followed by 7 months in 1.5% O_2 and 2% CO_2. After storage the fruits were ripened at 20°C for 7 days before evaluation for superficial scald. Starking fruits stored in controlled atmospheres were similar during 1988 and 1989 (Van Eeden *et al.* 1992). Starking Delicious could be stored for up to 9 months without appreciable loss of quality at 0°C in 3% CO_2 with 3% O_2 (Ertan *et al.* 1992). Sfakiotakis *et al.* (1993) stored Starking Delicious fruit under various conditions including 2.5% O_2 with 2.5% CO_2 and ultra-low O_2 (with 1% O_2 and 1% CO_2) storage all at 0.5°C and 90% r.h. They showed that these conditions generally retarded fruit softening and markedly reduced scald development; 1% O_2 with 1% CO_2 was the most promising treatment for reducing fruit softening. Low O_2 without CO_2 was equally effective in extending fruit storage life, while 2.5% O_2 with 2.5% CO_2 storage gave moderate results. Sensory evaluation of fruits stored for 7–8 months in 2.5% O_2 with 2.5% CO_2 and 1% O_2 with 1% CO_2 gave high scores without any low O_2 injury or alcoholic taste. During storage, the ethylene scrubber used (Ethysorb) was effective in reducing the ethylene concentration in the gas phase below 0.3 μl $litre^{-1}$ in the 2.5% O_2 with 2.5% CO_2 and ultra-low O_2 treatments and below 1.6 μl $litre^{-1}$ in air storage. However, internal fruit ethylene concentrations were found to be above the physiological levels capable of inducing ripening, even in the ultra-low O_2 treatments. Calcium treatment delayed fruit flesh softening in Starking Delicious apples (Guan *et al.* 1998).

Stayman

There was a recommendation of –1.1 to 0°C with 2–3% CO_2 and 3% O_2 (Anon 1968).

Sundowner

Work in Australia showed that fruits can be stored in 2% O_2 and 1% CO_2 at 0°C with no adverse effects developing when fruit were stored below 1°C (Tugwell and Chvyl 1995).

Sunset

Sharples and Stow (1986) recommended 3–3.5°C with 8% CO_2, with data only presented for stores without a scrubber.

Suntan

Sharples and Stow (1986) recommended 3–3.5°C with either 5% CO_2 and 3% O_2 or less than 1% CO_2 and 2% O_2.

Tydeman's Late Orange

Sharples and Stow (1986) recommended 3.5–4°C with either 5% CO_2 and 3% O_2 or less than 1% CO_2 and 2% O_2.

Undine

Hansen (1977) recommended a temperature of not under 2°C with 3% CO_2 and 2.5% O_2.

Virginia Gold

Kamath *et al.* (1992) recommended 2.2°C with 2.5% O_2 and 2% CO_2 for up to 8 months. Under these condition fruits were firmer than those held in normal cold storage, they did not shrivel even when stored without polyethylene box liners and soft scald was eliminated.

Winston

Sharples and Stow (1986) recommended 1.5–3.5°C with 6–8% CO_2, with data only presented for stores without a scrubber.

Worcester Pearmain

Fidler (1970) recommended 0.5–1°C with 7–8% CO_2 for stores with no scrubber or 5% CO_2 and 3% O_2 for stores with a scrubber. Wilkinson and Sharples (1973) suggested 0.6–1.1°C with 5% CO_2 and 3% O_2 for 6 months. Sharples and Stow (1986) advocated 0.5–1°C and 7–8% CO_2 for fruit stored without a scrubber and 5% CO_2 with 3% O_2 for those stored with a scrubber.

Ethylene

The rate of ethylene production by apples was shown to be about half in an atmosphere of 2.5% O_2 compared with fruit stored in air (Burg 1962). Ethylene biosynthesis was shown to be affected by CO_2 levels in store. Increased levels of CO_2 reduced ethylene levels in controlled atmosphere stores containing apples (Tomkins and Meigh 1968). In Cox's Orange Pippin apples stored at 3.3°C, softening was hastened with ethylene levels of 1 μl $litre^{-1}$ in the store (Knee 1976). However, Stow (1989) showed that lowering the ethylene concentration in the containers in which the fruit were stored reduced the ethylene concentration in the core of apples. Ethylene concentrations in the core of apples were generally progressively lower with reduced O_2 concentrations in the store over the range of 0.5–2% oxygen (Stow 1989a).

Apricot

Prunus armeniaca, Rosaceae.

Botany

The fruit is a drupe and is classified as climacteric. The skin varies in colour with different cultivars from pale yellow to deep orange with red freckles.

Harvesting

If harvested immature they lack the full flavour of tree-ripened fruit (Fidler 1963; Lutz and Hardenburg 1968). Skin colour is commonly used to judge harvest maturity. Fan *et al.* (2000) defined maturity stage 1 as light green, partially turning to a straw colour, and maturity stage 2 as straw colour on most of the fruit surface. In South Africa the Deciduous Fruit Board produced a colour chart, AP.1. Jooste and Taylor (1999) recommended that the cultivar Bebeco be harvested between values of 4 and 6 on that chart as the primary indicator of harvest maturity. They also found that flesh firmness as measured by a penetrometer of between 8.0–11.5 kg could be used as a very reliable secondary indicator. Soluble solids content of the juice is also used (Botondi *et al.* 2000). Softening was used to determine fruit harvest maturity by Claypool (1959) and Rood (1957). As the fruit matures and ripens the chlorophyll content reduces so that light transmission properties of fruit can be used to measure objectively chlorophyll content and therefore their ripeness. The wavelength used for apricots was 380–730 nm and the correlation coefficient with organoleptic measurements was in the range 0.84–0.93 (Yeatman *et al.* 1965; Czbaffey 1984, 1985).

Prestorage treatments

Perfection apricot fruits were treated with 1 $\mu l\ l^{-1}$ 1-MCP for 4 hours at 20°C, then stored at 0 or 20°C. The onset of ethylene production was delayed, respiration rate was reduced and firmness and titratable acidity better retained following the 1-MCP treatment and storage at both temperatures. It also delayed the production of volatile alcohols and esters during ripening at 20°C and they had less colour change than controls (Fan *et al.* 2000).

Storage

Their shelf-life in a simulated room temperature of 20°C and 60% r.h. was shown to be only 1–2 days (Mercantilia 1989). Refrigerated storage recommendations are as follows:

- –0.5 to 0°C and 85–95% r.h. for 1–2 weeks (Phillips and Armstrong 1967 and Hardenburg *et al.* 1990)
- 0°C and 90% r.h. for 1–2 weeks (Mercantilia 1989)
- –1 to 0°C and 90–95% r.h. for 1–4 weeks (Snowdon 1990)
- –0.5°C and 90–95% r.h. for 7–14 days (SeaLand 1991)
- 0°C (Fan *et al.* 2000).

Controlled atmosphere storage

J.E. Bernard in 1819 in France (quoted by Thompson 1998) showed that apricots could be successfully stored in atmospheres containing no O_2 for up to 28 days, then removed to air for ripening. Controlled atmosphere storage can lead to an increase in internal browning in some cultivars (Hardenburg *et al.* 1990). A 1% O_2 atmosphere maintained acceptable firmness and colour during storage at 15°C for fruit harvested with 13–14 °Brix. Early harvested fruits, with 9–10 °Brix, did not benefit from low O_2 (1% or 2%) storage because they had not reached the optimal soluble solids content whereas for late harvest fruit (13–14 °Brix) the use of 1% O_2 at a higher temperature (15°C) than that used commercially can be an alternative to low temperature (5°C) as a shipping treatment or short-term storage (Botondi *et al.* 2000). Other recomendations include:

- 2.5–3% CO_2 with 2–3% O_2 at 0°C for Blenheim (Royal) retained their flavour better than air stored fruit when they were subsequently canned (Hardenburg *et al.* 1990)
- 2–3% CO_2 with 2–3% O_2 (SeaLand 1991)
- 0–5°C with 2–3% CO_2 and 2–3% O_2 but not to be used commercially (Kader 1985, 1989).

Hypobaric storage

Haard and Salunkhe (1975) found that storage life could be extended from 53 days in cold storage to 90 days in cold storage combined with reduced pressure of 102 mmHg. They found that hypobaric storage delayed carotinoid production, but after storage carotinoid, sugar and acid levels were the same as those in cold storage alone.

Ripening

Optimum ripening can be achieved at 18–24°C (Hardenburg *et al.* 1990). Fruits that were harvested while they were still green and kept for 3 days at 19°C with 1000 $\mu l\ litre^{-1}$ ethylene for the first 24–48 hours lacked the aroma and flavour of fruits which were left on the trees for a further 6 or 7 days to ripen naturally (Fideghelli *et al.* 1967).

Arracacha, Peruvian parsnip

Arracacia xanthorrhiza, Umbelliferae.

Botany

The plant produces spindle-shaped, smooth-skinned, fleshy tuberous roots, several of which are borne as lateral projections from the central rootstock. They are light brown in colour with white flesh and are usually small, but can weigh up to 3 kg (Kay 1987). Popenoe 1992) stated that 'it produces a carrot-like root with a delicate flavour of cabbage, celery and roasted chestnuts and can be boiled, fried or added to stews.' They are a staple root crop in parts of South and Central America, particularly Colombia, Ecuador and Peru.

Disease control

Mainly fungal and bacterial rots and desiccation, leading to high market wastage, caused rapid postharvest deterioration. Washing roots increased rotting and the addition of chlorine to the washing water was ineffective. Dipping roots after harvest in benomyl, with or without maneb, reduced but did not eradicate rotting. Individual wrapping of roots in plastic cling or shrink films reduced rotting and desiccation (Thompson 1981).

Storage

At room temperature in Bogatá (about 25°C and 40% r.h.) they turned black in 3 days (Amezquita Garcia 1973). Refrigerated storage recommendations are as follows:

- 3°C for 1 month (Kay 1987)
- 10°C and 90% r.h. for 1 month with a small weight loss and slight fungal infection (Amezquita Garcia 1973)
- 10°C and 90% r.h. (Lutz and Hardenburg 1968)
- 25°C and 40% r.h. for <3–4 days (Thompson 1981).

Modified atmosphere packaging

17–20°C and 68–70% r.h. for 7 days if wrapped in plastic cling film or plastic shrink film (Thompson 1981).

Transport

Burton (1970) studied shipments from Puerto Rico to Chicago at 13.5°C taking 9 days. On arrival most were in poor condition, but the sound tuberous roots were stored for a further 30 days at 15°C and 65% r.h. After this time their weight loss over the 30 days was 15% and 12% had decayed. Treatment before storage with 2500 μl litre^{-1} chlorine plus wax reduced these losses to 1% and 6%, respectively.

Arrowroot

Maranta arundinacea, Marantaceae.

Botany

The plant produces large, fleshy, cylindrical, subterranean rhizomes up to 2.5 cm in diameter and up to 25 cm long, which are covered in overlapping scale leaves. They are largely used for starch extraction. The starch they produce is easily digested and therefore in demand for infants and invalids. The rhizomes can also be eaten after boiling or roasting (Kay 1987).

Harvesting

Harvest maturity is some 10–12 months after planting when the leaves begin to wilt and die down and the stems lodge (Purseglove 1975). In St Vincent, harvesting is usually by hand, but modified potato harvesters have been used in both St Vincent and Australia.

Storage

Storage is usually achieved by delaying the harvest, but the rhizomes may become fibrous and some of the starch is eventually converted to sugar when they sprout (Purseglove 1975). There is little information on storage life except that they deteriorate rapidly after harvest; the cultivar Banana must be processed within 2 days and the cultivar Creole within 7 days of harvesting (Kay 1987).

Asian pears, Japanese pears, nashi

Pyrus pyrifolia, P. serotina, P. ussuriensis variety sinensis, Rosaceae.

Botany

This is closely related to the European pear *P. communis* and probably has similar postharvest characteristics. The fruit is a pome and tends to be spherical rather than pear-shaped. They have a golden russet brown skin and crisp white juicy and somewhat granular flesh. Sirisomboon *et al.* (2000) indicated that the fruit were firm, sweet and juicy, bruise more easily the European pears and ripen on the tree rather than during storage.

Harvesting

In Japan commercial harvesting starts around late August to early September some 70 days after pollination.

Curing

Fruits of Niitaka were harvested at commercial harvest and cured at ambient temperature of about 18°C for 5, 10 or 15 days followed by stepwise cooling until equilibrium at 0.5°C. Skin blackening, which is a physiological problem with storage of this cultivar, was observed only in fruits stored without curing. Curing treatments did not significantly alter other fruit quality characteristics during storage but affected the physical properties of the fruit's epidermis (Park *et al.* 1999).

Storage

Refrigerated storage recommendations are as follows:

- 2°C for 6 months for Early Gold (Zagory *et al.* 1989)
- 1.1°C and 90–95% r.h. for 150–180 days (SeaLand 1991)
- 0.5°C (Park *et al.* 1999).

Controlled atmosphere storage

Low-O_2 atmospheres reduced yellow colour development, compared with air. The cultivar 20th Century was shown to be sensitive to CO_2 concentrations above 1% when exposed for longer than 4 months or to 5% CO_2 when exposed for more than 1 month (Kader quoted by Zagory *et al.* 1989). The cultivar Shinko became prone to internal browning, perhaps due to CO_2 injury, after the third month in storage (Zagory *et al.* 1989). An experiment at 2°C in 1, 2 or 3% O_2 of Early Gold and Shinko showed no clear benefit compared to storage in air (Zagory *et al.* 1989). Kader (1989) recommended 0–5°C with 0–1% CO_2 and 2–4% O_2, which, however, had limited commercial use. Meheriuk (1993) indicated that controlled atmosphere storage of Nashi was still in an experimental stage but gave provisional recommendations of 2°C with 1–5% CO_2 and 3% O_2 for about 3–5 months and that longer storage periods may result in internal browning. Meheriuk also stated that some cultivars are subject to CO_2 injury but indicated that this is only when they are stored in concentrations of over 4%.

Ethylene

Their shelf-life was found to be determined by their level of ethylene production. Relatively high levels of ethylene reduce storage potential and fruit quality (Itai *et al.* 1999). However, Inaba *et al.* (1989) found that there was no change in respiration rate from exposure to ethylene at 100 μl $litre^{-1}$ for 24 hours at any temperature between 5 and 35°C.

Asian spinach

Brassica spp., *Chrysanthemum* spp., *Ipomea* spp.

Botany

There are various species and varieties, especially of *Brassica*, that are grown for their leaves and have similar postharvest requirements. These include:

- Chinese Cabbage, *B. pekinensis*
- Chinese Kale, Jie Lan, Gai Lan, *B. alboglabra* (Figure 12.7)
- Chinese Mustard, Jie Cai, *B. juncea* (Figure 12.8)
- Choi Sum, Cai Xin, *B. parachinensis*, *B. rapa* var. *parachinensis* (Figure 12.9)
- Mibuna, Mizuna, *B. rapa* var. *nipposinica*
- Pak Choi, Bai Cai, *B. chinensis*, *B. chinensis* var. *pekinensis*, *B. rapa* var. *chinensis* (Figure 12.10)
- Tatsoi, *B. rapa* var. *rosularis*
- Tong Hao, *Chrysanthemum spatiosum* (Figure 12.11).
- Water Spinach, Tong Cai, *Ipomea reptans*, *I. aquatica*.

Storage and controlled atmosphere storage

Generally, storage is optimum for Asian brassicas at 0°C and humidity as close to 100% r.h. as possible. O'Hare *et al.* (2000) found that at retail temperatures

Figure 12.7 Chinese kale, *Brassica alboglabra*. Source: Dr Wei Yuqing.

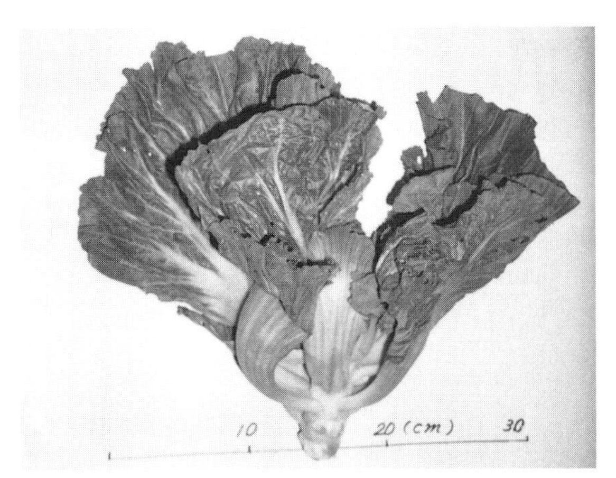

Figure 12.8 Chinese mustard, *Brassica juncea*. Source: Dr Wei Yuqing.

Figure 12.9 Choi sum, *Brassica parachinensis*, *B. rapa* var. *parachinensis*. Source: Dr Wei Yuqing.

the shelf-life of these vegetables can be severely curtailed because of leaf yellowing. They used controlled atmosphere storage to reduce losses as follows:

- Chinese Mustard, 10°C with 2% O_2 and 5% CO_2 increased shelf-life by 153% compared with cold storage alone
- Choy Sum (Choi Sum), 10°C with 2% O_2 and 5% CO_2 increased shelf-life by 105% compared with cold storage alone
- Mibuna, 10°C with 2% O_2 and 5% CO_2 increased shelf-life by 86% compared with cold storage alone. Mibuna, which was the least tolerant of CO_2, achieved a 70% increase in shelf-life 2% O_2 alone
- Mizuna, 10°C with 2% O_2 and 5% CO_2 increased shelf-life by 40% compared with cold storage alone
- Tatsoi, 10°C with 2% O_2 and 5% CO_2 increased shelf-life by 153% compared with cold storage alone.

Figure 12.10 Pak choi, *Brassica chinensis*, *B. chinensis* var. *pekinensis*, *B. rapa* var. *chinensis*. Source: Dr Wei Yuqing.

Figure 12.11 Tong hao, *Chrysanthemum spatiosum*. Source: Dr Wei Yuqing.

Asparagus

Asparagus officinalis, Liliaceae.

Botany

The part that is eaten is the young shoot, which is called the spear (Figure 12.12) and which arises from the rootstock as it begins to grow in spring after over-winter dormancy. It will freeze at about –1.4 to 1.1°C (Wright 1942) or –0.89 to 0.61°C (Weichmann 1987).

Harvesting

If not harvested the spears continue to elongate into long, spindly stems and produce fibres for support and eventually they produce leaves. Spears should be harvested before fibres have been produced so that they are crisp and tender. Villanueva Suarez *et al.* (1999) found that the most important postharvest changes in the spears corresponded to increases in xylose and glucose from insoluble dietary fibre (from 2.62–5.14 and from 7.50–10.14 g per 100 g dry matter after 21 days of storage, respectively) and decreases in galactose (from 1.43–0.93 g per 100 g) from soluble dietary fibre. In practice they are normally harvested when the shoots are 10–15 cm above ground by cutting them off at ground level. During the harvest season this is usually every 1–3 days. They are often harvested with a section of tough white butt to protect them from mechanical injury during handling. For certain markets they are harvested with a special knife when the tips are just beginning to emerge from the soil. The spears are cut off about 20 cm below ground. In the latter case the asparagus is white.

Figure 12.12 Evaluating the quality of freshly harvested asparagus.

Handling

In order to preserve their freshness directly after harvesting, the spears can be placed upright in field boxes, which may include a layer of damp moss at the bottom. Spears must be handled gently. In experiments Lallu *et al.* (2000) found that impact on apical tissues after drops from 0, 50, 100 or 150 mm resulted in 0, 34, 36 and 64% tip rot, respectively, after 5 days at 20°C and 93–95% r.h.

Hot water treatment

Immersing spears in water at 47.5°C for 2–5 minutes and cooling as soon as possible after the heat treatment prevented curvature during 7 of days storage at 10°C. Treatment at a higher temperature and/or for longer periods resulted in an unacceptable decline in overall appearance. Treatment temperature and duration needed to be adjusted for spear diameter; small diameter spears required a shorter exposure time or lower temperature (Paull and Chen 1999).

Chilling injury

Lutz and Hardenburg (1968) reported chilling injury after storage at 0°C for more than 10 days.

Precooling

Villanueva-Suarez *et al.* (1999) showed that the speed at which fibre formation processes occur was proportional to temperature, so rapid cooling was recommended. In a study where spears were held at 18°C and 50% r.h. for different periods until they lost 0, 2, 4, 6 or 8% in weight, then stored at 0.5°C it was found that spears which had lost 8% in weight were unsaleable after 14 days of storage, whereas those that had lost 0% in weight were still saleable after 28 days (Karup 1990). In a comparison of different precooling methods, overall quality was marginally higher and weight loss significantly less in hydrocooled spears than in forced-air or passively cooled spears. It was therefore recommended that spears are hydrocooled or forced-air cooled within 4–12 hours of harvest (Lallu *et al.* 2000). A.K. Thompson (unpublished data), working in a processing factory in Korea, found that they could be successfully hydrocooled within 2 minutes. Vacuum cooling was less successful. After 25–30 minutes of vacuum cooling at 4.0–4.6 mmHg with a condenser temperature of –1.7 to 0°C, the spears' temperature fell from 20–22°C to 7°C (Barger 1963).

Storage

For the cultivar Limbras 10, spears were harvested at intervals during a 24-hour cycle. They exhibited a clear diurnal pattern in postharvest shelf-life, with spears harvested at 0200 hours lasting 1.1 days longer at 20°C than spears harvested at 1400 hours (Lill and Borst 2001). Spears may be stored with their butts in water. This technique has been used in processing plants for overnight storage. Their shelf-life at 20°C was said to be only 1 day (Mercantilia 1989). During storage where the temperature fluctuated from the average 6°C by ±5, 3 or ±1°C every 12 or 6 hours, Ito and Nakamura (1985) found that the spears deteriorated increasingly rapidly with increasing amplitude of fluctuation. Refrigerated storage recommendations are as follows:

- 0.6°C and 90% r.h. (Wardlaw 1937)
- 0–2°C and 80% r.h. for 21–50 days (Wardlaw 1937)
- 0°C and 95–98% r.h. for a maximum of 1 week (Wardlaw 1937)
- 0–0.5°C and 85–95% r.h. for 2–4 weeks (Anon 1967)
- 0–1.7°C and 95% r.h. for 3 weeks (Phillips and Armstrong 1967; Anon 1968)
- 0°C and 95% r.h. for up to 10 days (Shipway 1968)
- 0°C and 95% r.h. for 3–4 weeks (Pantastico 1975)
- 4–5°C and high humidity for up to 21 days (Tindall 1983)
- 0°C and 90–95% r.h. for 14 days (Mercantilia 1989)
- 4°C for 7 days (Mercantilia 1989)
- 8°C for 4 days (Mercantilia 1989)
- 2.2°C and 90–95% r.h. for 14–21 days (SeaLand 1991)
- 0°C and 95–100% r.h. for 14–21 days (white) (SeaLand 1991)
- 0–2°C and 95–100% r.h. for 2–4 weeks (Snowdon 1991).

Controlled atmosphere storage

Respiration rate was strongly affected by temperature, but also by the O_2 level in the storage atmosphere (Table 12.5). Also, the effect of reduced O_2 was greater the higher was the storage temperature.

Table 12.5 Effects of temperature and reduced O_2 level on the respiration rate in CO_2 production in mg kg^{-1} h^{-1} of asparagus spears (Robinson *et al.* 1975)

	Temperature (°C)				
	0	5	10	15	20
In air	28	44	63	105	127
In 3% O_2	25		45		75

Storage recommendations were as follows:

- 4°C with 10–14% CO_2 and high humidity. At higher temperatures these levels of CO_2 could be injurious, but levels of 5% CO_2 had no effects, nor did reduced O_2 (J.R. Stow, personal communication 1977)
- 0–5°C with 5–10% CO_2 and 21% O_2, which had a good effect but limited commercial use (Kader 1985 and 1992)
- 1–5°C with 10–14% CO_2 and 21% O_2, which had only a slight effect (Saltveit 1989)
- 0°C with 12% CO_2 or at 5°C or just above at 7% CO_2 retarded decay and toughening and brief exposure to 20% CO_2 can reduce soft rot at the butt end (Hardenburg *et al.* 1990)
- 0°C with 5–10% CO_2 and 20% O_2 (SeaLand 1991).

Modified atmosphere packaging

Villanueva-Suarez *et al.* (1999) found that the changes observed in the fibre of the non-packed asparagus stored in air at 20°C were more rapid and pronounced than those in polyethylene bags in either air or a modified atmosphere of 15% O_2, 10% CO_2 and 75% nitrogen. Losses of glucose, uronic acids, arabinose and galactose were greater in packed spears stored in a modified atmosphere than in packed spears stored in air.

Atemoyas

Annona squamosa × *A. cherimolia*, Annonaceae.

Botany

It is a climacteric fruit, which is a cross between the cherimoya and the sugar apple. The fruit is a fleshy syncarp formed by the fusion of the pistils and receptacle with custard-like sweet pulp containing many hard seeds. In the cultivar African Pride stored at 20°C from pre- to post-climacteric stages, the major changes during ripening were a continuous decrease in starch, a continuous increase in fructose and glucose, an increase in sucrose to a maximum at the climacteric, an increase in malic acid early in the climacteric rise and a decrease in ascorbic acid after the climacteric. Eating quality was optimal 2 days after the climacteric peak (Wills *et al.* 1984).

Splitting

Paull (1996) found that fruits tended to split during maturation. He found that in the cultivars African Pride and Gefner splitting coincided with a 3-fold increase in soluble solids, which resulted in osmotic and turgor changes during ripening leading to the movement of water from the skin, and possibly the receptacle, to the flesh. The increase in receptacle diameter increased the stress on the flesh and skin, leading to fruit splitting.

Chilling injury

Storage at 0, 5 or 10°C all resulted in chilling injury (Wills *et al.* 1984). Batten (1990) found that fruits stored at 4°C and high humidity for more than 2 days developed symptoms of chilling injury, which were blackening of skin and browning of flesh, although even after 5 days at 4°C fruits that were then ripened at 20°C had a very good flavour.

Storage

Storage recommendations are:

- 25°C for 4 days (Wills *et al.* 1984)
- 20°C for 5 days (Wills *et al.* 1984)
- 15°C for 8 days (Wills *et al.* 1984)
- 8°C for 1 week for the cultivars Africa Pride, Pink's Mammoth, Neilsen and QAS (Brown and Scott 1985)
- 12°C or 15–16°C for 2 weeks for the cultivars Africa Pride, Pink's Mammoth, Neilsen and QAS (Brown and Scott 1985)
- 12°C for 6 days followed by ripening for 3 days at 20°C (Batten 1990)
- 12.8°C and 85–90% r.h. for 28–42 days (SeaLand 1991).

Ripening

At 20°C the ripening time was about 5 days for freshly harvested fruit (Batten 1990). Wills *et al.* (1984) also recommended 20°C for 5 days for optimum eating quality. Fruits ripened more quickly and with good flavour at 28°C, whereas ripening was slower and the quality impaired at 32°C (Batten 1990).

Aubergines, egg plants

Solanum melongena, Solanaceae.

Botany

The fruit is a pendant berry, which can be ovoid to oblong in shape. Most cultivars are dark purple, although white, yellow and variegated varieties are grown. Aubergine is classified as a non-climacteric fruit and will freeze at about –1.0 to 0.7°C (Wright 1942) or –0.94 to 0.78°C (Weichmann 1987). Inaba *et al.* (1989) found that there was increased respiration in response to ethylene at 100 μl $litre^{-1}$ for 24 hours at 20–35°C, but the effect was much reduced at lower temperatures.

Quality standards

The Organization for Economic Cooperation and Development (OECD) published a brochure of standards for aubergines in 1987.

Harvesting

The fruit are harvested when they reach a size that the market requires, but if this is delayed too long they can become seedy, dull and fibrous (Pantastico 1975).

Chilling injury

Chilling injury was reported after 32 days storage at 7.2°C in the form of internal browning (Wardlaw 1937). Chilling injury can result in surface scald, browning, pitting and excessive decay that may not be apparent until the fruit are removed from storage (Wardlaw 1937; Lutz and Hardenburg 1968). The symptoms can be confused with CO_2 injury.

Storage

Their shelf-life at a simulated room temperature of 20°C and 60% r.h. was shown to be only 3–4 days (Mercantilia 1989). Refrigerated storage recommendations are as follows:

- 0–4.4°C for 2–4 weeks (Platenius *et al.* 1933)
- 7.2–10°C and 80–85% r.h. (Wardlaw 1937)
- 10–15.6°C and 80–85% r.h. for 10 days (Wardlaw 1937)

- 7.2–10°C and 90% r.h. for 1 weeks (Lutz and Hardenburg 1968)
- 10°C and 75% r.h. for 5–8 days with 3.1–3.6% loss (Heydendorff and Dobreanu 1972)
- 10–12.8°C and 92% r.h. for 2–3 weeks with 9.6% loss (Pantastico 1975)
- 8.3–10°C and 85–90% r.h. for 4 weeks with 17.7% loss (Pantastico 1975)
- 10–13°C and 90–95% r.h. for 10–14 days with a weight loss of about 10% (Tindall 1983)
- 8–10°C and 90–95% r.h. for 10–14 days (Mercantilia 1989)
- 8–12°C and 90–95% r.h. for 7–10 days (Hardenburg *et al.* 1990)
- 10°C and 90–95% r.h. for 10–14 days (SeaLand 1991)
- 8–12°C and 90–95% r.h. for 1–2 weeks (Snowdon 1991).

Controlled atmosphere storage

Sayed *et al.* (2000) measured the respiration of aubergines in CO_2 and O_2 at different concentrations. They found that in 0% CO_2 the respiration rate was 60 mg CO_2 kg^{-1} h^{-1} whereas in 6.7% CO_2 it was 30 mg CO_2 kg^{-1} h^{-1}. Also in 20.6% O_2 it was 60 mg CO_2 kg^{-1} h^{-1}, whereas in 1.7% O_2 it was only 28 mg CO_2 kg^{-1} h^{-1}. This indicates the controlled atmosphere storage could be beneficial to aubergines. High CO_2 in the storage atmosphere can cause surface scald browning, pitting and excessive decay; these symptoms are similar to those caused by chilling injury (Wardlaw 1937). Storage in 0% CO_2 with 3–5% O_2 was recommended by SeaLand (1991).

Ethylene

Levels of decay and fruit stalk abscission were increased in the presence of ethylene (Schouten and Stork 1977).

Avocados, alligator pear, midshipmen's butter

Persea americana, Lauraceae

Botany

Fruit is produced on a small tree that is native to the West Indies and Central America. There are three races, all belonging to *P. americana*. High oil content of the fruit characterizes the Mexican and Guatemalan races and their ability to withstand low storage temperatures (Condit 1919), whereas the West Indian race has low oil content is intolerant to low storage temperatures and is often considered to have a better flavour. Thompson *et al.* (1971) and Purseglove (1968) indicated the following:

Race	Oil content (%)	Refrigeration tolerance
Mexican	Up to 30	Most cold tolerant
Guatemalan	8–15	Less cold tolerant
West Indian	3–10	Highly susceptible to cold

It is classified as a climacteric fruit. Biale and Young (1962) showed that at both 5 and 30°C no climacteric rise in respiration occurred in a Mexican and Guatemalan hybrid, but a climacteric occurred at all the intervening temperatures. It will freeze at about –2.8 to 2.6°C (Wright 1942).

Preharvest factors

In some work in South Africa (Kohne *et al.* 1992), it was shown that the cultivar Fuerte grown on different seedling rootstocks showed a large variation in both yield and quality of fruit (Table 12.6).

Harvesting

The oil content of the fruit may be used to determine the harvest maturity of avocados. The Agricultural Code of California specifies '... avocados, at the time of picking and at all times thereafter, shall contain not less than 8% of oil by weight of the avocado excluding the skin and seed.' Eaks (1990) in California found that during development on the tree the oil content in Fuerte increased to just below 20% then levelled off, and in Hass there was a progressive increase throughout development to just over 15% before levelling off. During subsequent storage of the fruit at temperatures

Table 12.6 Effect of clonal rootstock on the yield and quality of Hass avocados (source: Kohne *et al.* 1992)

Rootstock	Yield (kg per tree)	Fruit internally clean (%)
Thomas	92.7	96
Duke 7	62.1	100
G 755	12.1	100
D 9	7.4	64
Barr Duke	3.1	70

in the range 0–20°C the oil content remained constant. Oil content in avocados is related to moisture content and in South Africa fruit for export should have the following minimum moisture content:

Cultivar	Moisture content (%)
Fuerte	80
Pinkerton	80
Hass	77
Ryan	77
Edranol	74

Blumenfeld (1993) also showed that there was a good correlation between taste and oil content and dry matter and oil content for avocados grown in Israel and also recommended that rate of dry matter accumulation could be used to predict optimum harvest time. However, these methods of stipulating harvest maturity have little relevance to avocados grown in the tropics, for two reasons. First, it is based on a sampling technique where it is assumed that the sample of fruit on which the oil or dry matter analysis has been taken is representative of the whole field. In the subtropics, for example in California, South Africa and Israel, there are distinct seasons and flowering of avocados occurs after a cold season and the trees tend to flower and thus set fruit over a short period of time. Fruit of the same variety in one orchard will have fruit, which therefore mature at about the same time, and so a representative sample can be taken. In the tropics, the flowering period, even on the same tree, is a much more protracted and so there is a wide range of fruit maturities. It is rarely possible, therefore, to obtain a representative sample. There may even be fully mature fruit and flowers on the same tree at the same time. Second, those grown in the subtropics tend to be of the Mexican or Guatemalan races or a cross between the two, whereas in the tropics the West Indian race of avocados is often more common and this race has fruit which have a much lower oil content even when fully mature.

In South Africa, the early-maturing fruit tended to have better quality and postharvest characteristics than those that matured later. Milne (1993) showed, in studies over two seasons, where early-picked fruit had an average of over 80% 'clean fruit' whereas late-picked fruit had an average of just over 50%. This quality effect may have an effect on storage to extend the season of availability because the fruit that may be required for storage are usually those of the poorest quality.

Nuclear magnetic resonance has also been shown to correlate well with oil content in avocados (McCarthy *et al.* 1989), but the technology is expensive and no record could be found of its commercial use. Self *et al.* (1993) have shown that there is a relationship between ultrasonic velocity and the ripeness of avocados. Acoustic and vibration test equipment puts vibration energy into a fruit, and measures its response to this input (Figure 12.13). Good correlations have been found using the second resonant frequency to determine avocado maturity (Darko 1984; Punchoo 1988). A sensor system for automatic non-destructive sorting of firm (tree-ripe) avocado fruits used vibrational excitement of one side of the fruit, while measuring the transmitted vibration energy on the other. Special signal processing hardware and software were developed for computing several alternative firmness indexes,

Figure 12.13 A device that was developed by a company in Israel to be used by customers at retail outlets to test the ripeness of avocados. It is based on surface, which puts energy into the fruit and measures the fruit's response to that energy.

which were highly correlated with values obtained by the standard destructive piercing force test method. Optimal classification algorithms were developed whereby fruit could be classified into two or three firmness grades (Peleg *et al.* 1990).

Disease control

Colletotrichum gloeosporiodes, which causes anthracnose, and a lenticular rot caused by *Dothiorella gregaria* can infect the fruit in the field and develop postharvest (Pantastico 1975). However, fungal infections were found to be associated only with area of primary damage, except where fruits were over-ripe where the most common organism was *Botryodiplodia theobromae* (Thompson *et al.* 1971). Infections can also occur postharvest through the cut stalk (Pantastico 1975). *Bacillus* spp. on their own or combined with a fungicide could be used to control postharvest diseases of avocados (Korsten *et al.* 1993). *Bacillus* spp. isolated from leaves and fruit of avocados were more effective in controlling anthracnose and stem end rot of avocados when applied as a postharvest dip than prochloraz applied in the same way. The combination of *B. subtilis* and prochloraz was more effective than when they were applied separately (Korsten *et al.* 1993).

Prestorage treatments

Postharvest application of calcium has also been shown to inhibit ripening of avocados (Wills and Tirmazi 1982). The calcium was applied by vacuum infiltration (250 mmHg) with calcium chloride at concentrations within the range 1–4%; higher concentrations could result in skin damage. These treatments were found to delay ripening significantly without affecting fruit quality. Treatment of fruits prior to storage in atmospheres of total nitrogen was also shown to retard the ripening of avocados (Pesis *et al.* 1993).

Chilling injury

Certain cultivars were reported to be susceptible to chilling injury when stored below 13°C (Salunkhe and Desai 1984). Symptoms of chilling injury include pitting, browning of pulp near the seed or in the tissue midway between the seed and the skin, failure to soften when transferred to a higher temperature, off-flavour, vascular strands and development of a brownish appearance. In other work chilling injury was shown to occur at 10–11.1°C for cultivars of the West Indian race and 4.4–6.1°C for the Mexican and Guatemalan races (Pantastico 1975). In a study of avocados from trees propagated from seed in Grenada, Thompson *et al.* (1971) showed that there was considerable variation in storage life and chilling injury symptoms between fruit from different trees (Table 12.7). It meant that even at 7°C some 77% of the fruit ripened without showing symptoms of chilling injury and at 13°C some 27% actually suffered from chilling injury. It is therefore difficult to generalize, as did Pantastico. However, it does seem clear that the Mexican and Guatemalan races are less susceptible to chilling injury than the West Indian race.

Controlled atmosphere storage can affect susceptibility to chilling injury. There were less chilling injury symptoms in Booth 8, Lula and Taylor after refrigerated storage in controlled atmospheres than in refrigerated storage in air (Haard and Salunkhe 1975; Pantastico 1975). Corrales-Garcia (1997) found that Hass stored at 2 or 5°C for 30 days in air, 5% CO_2 with 5% O_2 or 15% CO_2 with 2% O_2 had higher chilling injury of fruits stored in air that the fruits in controlled atmosphere storage, especially those stored in 15% CO_2 with 2% O_2. Spalding and Reeder (1975) showed that storage of fruits at either 0% CO_2 with 2% O_2 or 10% CO_2 with 21% O_2 fruit had less chilling injury and less anthracnose during storage at 7°C than fruits stored in air. Intermittent exposure of the cultivar Hass to 20% CO_2 increased their storage life at 12°C and reduced chilling injury during storage at 4°C compared with those stored in air at the same temperatures (Marcellin and Chevez 1983).

Avocados of the Fuerte cultivar will ripen normally at temperatures between 9 and 24°C, but in the presence of 100 μl $litre^{-1}$ ethylene chilling injury occurred

Table 12.7 Effects of storage temperature on time from harvest to softening and chilling injury symptoms of West Indian seedling avocados (source: Thompson *et al.* 1971)

Storage temperature (°C)	Mean days to softening	% Affected by chilling injury after storage
7	16	23
10	13	16
13	16	27
18	9	0
27	8	0

at 12°C (Lee and Young 1984). Chaplin *et al.* (1983) showed increased susceptibility to chilling injury in Haas in the presence of ethylene. Application of calcium to avocados can reduce their susceptibility to chilling injury during subsequent storage (Hofman and Smith 1993).

Fruits stored for 4–10 weeks at 2°C had reduced severity of chilling injury symptoms and percentage of injured fruits after they had been dipped for 30 seconds in methyl jasmonate at 2.5 μM for Fuerte and Hass and 10 μM for Etinger (Meir *et al.* 1996).

Storage and controlled atmosphere storage

General recommendations for storage are:

- 3% CO_2 with 10% O_2 (Stahl and Cain (1940)
- 5–10°C and 90% r.h. for 2–4 weeks (Anon 1967)
- 8–12°C and 85–90% r.h. for 2–4 weeks (unripe) (Kader 1985)
- 5–8°C and 85–90% r.h. for 1–2 weeks (ripe) (Kader 1985)
- 5% CO_2 and a minimum of 3% O_2 (Fellows 1988)
- 20°C for 11 days (Saucedo Veloz *et al.* 1991)
- 5–13°C with 3–10% CO_2 and 2–5% O_2 (Kader 1992)
- 5°C and 85% r.h. with 9% CO_2 and 2% O_2 (Lawton 1996)
- 10–13°C with 3–10% CO_2 and 2–5%O_2 (Bishop 1996)
- 5.5°C with 10% CO_2 with 2% O_2 for up to 4 weeks (Van der Merwe 1996)
- 7–13°C and 90–95% r.h. with 2–5% O_2 and 3–10% CO_2 for 2–4 weeks (Burden 1997).

High CO_2 treatment (20%) applied for 7 days continuously at the beginning of storage or for one day a week throughout the storage period at either 2 or 5°C maintained fruit texture and delayed ripening compared with fruit stored throughout in air (Saucedo Veloz *et al.* 1991). Storage in total nitrogen or total CO_2, however, caused irreversible injury to the fruit (Stahl and Cain 1940). Intermittent exposure of crops to high levels of CO_2 may have beneficial effects. Marcellin and Chaves (1983) showed that exposing avocado fruit to 20% CO_2 for 2 days each week in storage at 12°C delayed their senescence and reduced decay compared with fruit stored for 7 days a week in air.

Recommendations for specific cultivars, varieties or production region are given below.

Anaheim

- 10% CO_2 with 6% O_2 at 7°C for 38 days compared with only 12 days in air at the same temperature (Bleinroth *et al.* 1977).

Booth 1

- 4.5°C and 85–90% r.h. for 2–4 weeks for unripe fruits (Snowdon 1990).

Booth 8

- 4.4°C and 85–90% r.h. for 1 month or longer (Lutz and Hardenburg 1968)
- 4.5°C and 85–90% r.h. for 2–4 weeks for unripe fruits (Snowdon 1990).

California grown fruit

- 3.3°C and 85–90% r.h. for 14–28 days (SeaLand 1991)
- 3–10% CO_2 with 2–5% O_2 (SeaLand 1991).

Ettinger

- 5.5°C and 85–90% r.h. for 3–4 weeks for unripe fruits (Snowdon 1990).

Fuchs

- 12.8°C and 85–90% r.h. for 2 weeks (Lutz and Hardenburg 1968)
- 12°C and 2% O_2 with 10% CO_2 for 3–4 weeks (Spalding and Reeder 1975)
- 10–13°C and 85–90% r.h. for 2 weeks for unripe fruits (Snowdon 1990).

Fuerte

- 7.2°C for 2 months in 3 or 4–5% CO_2 with 3 or 4–5% O_2 which was more than a month longer than they could be stored at the same temperature in air (Overholster 1928)
- 7.2°C for 2 months in 4–5% CO_2 with 4–5% O_2 (Wardlaw 1937)
- 5–6.7°C had double or triple the storage life in 3–5% CO_2 with 3–5% O_2 than in air (Pantastico 1975)
- 10% CO_2 with 6% O_2 at 7°C for 38 days compared with only 12 days in air at the same temperature (Bleinroth *et al.* 1977)

- 5.5–8°C and 85–90% r.h. for 3–4 weeks for unripe fruits (Snowdon 1990)
- 2–5°C and 85–90% r.h. for 1–2 weeks for ripe fruit (Snowdon 1990).

Guatemalan varieties

- 5.6–7.2°C and 85–90% r.h. for 4 weeks with 10% loss (Pantastico 1975)
- 4°C (Salunkhe and Desai 1984).

Gwen

Lizana *et al.* (1993) described storage at 6°C and 90% r.h. in controlled atmosphere conditions of either 5% CO_2 with 2% O_2, 5% CO_2 with 5% O_2, 10% CO_2 with 2% O_2, 5% CO_2 with 2% O_2, 10% CO_2 with 5% O_2 or 0.03% CO_2 with 21% O_2 (control) for 35 days followed by 5 days at 6°C in normal air and 5 days at 18°C to simulate shelf-life. All fruit in controlled atmosphere storage retained their firmness and were of excellent quality regardless of treatment whereas the control fruit were very soft and unmarketable after 35 days of storage at 6°C.

Hass

- 5.5–8°C and 85–90% r.h. for 3–4 weeks for unripe fruits (Snowdon 1990)
- 2–5°C and 85–90% r.h. for 1–2 weeks for ripe fruit (Snowdon 1990)
- 5°C for 6 weeks followed by 4 days at 20°C resulted in excessive softening, browning and storage rots whereas 2°C for 6 weeks followed by 4 days at 20°C resulted in fruit of better quality (Saucedo Veloz *et al.* 1991).

Lizana and Figuero (1997) compared several controlled atmosphere storage conditions and found that 6°C with 5% CO_2 and 2% O_2 was optimum. Under these conditions they could be stored for at least 35 days and then kept at the same temperature in air for a further 15 days and still be in good condition.

Corrales-Garcia (1997) found that fruit stored at 2 or 5°C for 30 days in air, 5% CO_2 with 5% O_2 or 15% CO_2 with 2% O_2 had higher chilling injury for fruits stored in air that the fruits in controlled atmosphere storage, especially those stored in 15% CO_2 with 2% O_2. In experiments described by Meir *et al.* (1993) at 5°C, increasing CO_2 levels over the range of 0.5, 1, 3 or 8% and O_2 levels of 3 or 21% inhibited ripening (on the basis of fruit firmness and peel colour change). The most effective CO_2 concentration was 8%, with either 3 or 21% O_2; with a combination of 3% O_2 and 8% CO_2 fruits could be stored for 9 weeks. After controlled atmosphere storage, fruits ripened normally and underwent typical peel colour changes. Some injury to the fruit peel was observed, probably attributable to low O_2 concentrations. Peel damage was seen in 10% of fruits held in 3% O_2 with either 3 or 8% CO_2 but it was too slight to affect their marketability.

Lula

- 4–7°C and 98–100% r.h. with 10% CO_2 and 2% O_2 for 8 weeks (Spalding and Reeder 1972)
- 10°C with 9% CO_2 and 1% O_2 for 60 days (Pantastico 1975)
- 7.2°C with 10% CO_2 and 1% O_2 for 40 days (Pantastico 1975)
- 4.4°C and 85–90% r.h. for 1 month or longer (Lutz and Hardenburg 1968)
- 4.5°C and 85–90% r.h. for 4–5 weeks (Snowdon 1990).

Controlled atmosphere storage at 3–5% CO_2 and 3–5% O_2 delayed fruit softening and 9% CO_2 with 1% O_2 at 10°C kept them in an acceptable eating quality and appearance for 60 days (Hardenburg *et al.* 1990).

Mexican varieties

- 8°C (Salunkhe and Desai 1984).

Pollock

- 12.8°C and 85–90% r.h. for 2 weeks (Lutz and Hardenburg 1968).
- 10–13°C and 85–90% r.h. for 2 weeks for unripe fruits (Snowdon 1990).

Subtropical grown fruit

- 7°C and 90% r.h. for 2–4 weeks (Mercantilia 1989).

Taylor

- 4.5°C and 85–90% r.h. for 2–4 weeks for unripe fruits (Snowdon 1990).

Tropical grown fruit

- 13°C (Mercantilia 1989)
- 20°C and 60% r.h. for 2–7 days (Mercantilia 1989)
- 10°C and 85–90% r.h. for 14–28 days (SeaLand 1991)
- 3–10% CO_2 with 2–5% O_2 (SeaLand 1991).

Waldrin

- 13°C (Mercantilia 1989)
- 20°C and 60% r.h. for 2–7 days (Mercantilia 1989)
- 10°C and 85–90% r.h. for 14–28 days (SeaLand 1991)
- 10°C and 3–10% CO_2 with 2–5% O_2 (SeaLand 1991).

West Indian fruit

- 12.8°C and 85–90% r.h. for 2 weeks (Lutz and Hardenburg 1968)
- 15°C for 2 weeks for msot West Indian cultivars (Thomson *et al.* 1971)
- 12.8°C and 85–90% r.h. for 2 weeks with 6.3% loss (Pantastico 1975).

Modified atmosphere packaging

Storage life was extended by 3–8 days at various temperatures by sealing individual fruit in polyethylene film bags (Haard and Salunkhe 1975). Fuerte fruit sealed individually in 0.025 mm thick polyethylene film bags for 23 days at 14–17°C ripened normally on subsequent removal to higher temperatures (Aharoni *et al.* 1968). Levels of gases inside the bags after 23 days storage were 8% CO_2 and 5% O_2 (Aharoni *et al.* 1968). Thompson *et al.* (1971) showed that sealing various seedling varieties of West Indian avocados in polyethylene film bags greatly reduced fruit softening during storage at various temperatures (Table 12.8).

Other recommendations include 5°C in 30 µm thick polyethylene film bags for Hass (Meir *et al.* 1997) or 4–7.5°C in 0.05 mm polyethylene bags for Fuerte (Scott and Chaplin 1978).

Ripening

Seymour and Tucker (1993) reported that avocado fruit do not normally ripen until they are harvested and can remain in a mature but unripe condition on the tree until picked. Changes during ripening are softening, starch breakdown and flavour development.

Table 12.8 Effects of storage temperature and wrapping treatment (0.031 mm thick polyethylene film bags compared with unwrapped fruit) on the number of days to softening of avocados (Thompson *et al.* 1971)

	7°C	13°C	27°C
Not wrapped	32	19	8
Sealed film bags	38	27	11

Cellulase was involved in fruit softening during ripening of Hass (Kanellis and Kalaitzis 1992). Starch occurs in the plastids of unripe avocados at levels of about 12 mg g^{-1} on a dry weight basis, but this reduced to undetectable levels in ripe fruit (Pesis *et al.* 1976). During the ripening of Fuerte the oil content remained constant (Eaks 1990). Recommended ripening conditions were 18–21°C with exposure to 10 µl $litre^{-1}$ of ethylene for 24–72 hours (Wills *et al.* 1989). Hatton *et al.* (1965a) recommended 15.5°C as the optimum temperature for ripening Florida avocados.

Babacos

Carica × *heilbornii* nm *pentagona*, Caricaceae.

Botany

It is a hybrid between the mountain papaya (*C. pubescens*) and *C. stipulata*. The fruit is a fleshy berry, oblong in shape with grooves down the sides, blunt at the stem end and pointed at the distal end. It is characteristically green and waxy as it develops and ripens to a yellow–orange coloured skin. The flesh is pale orange, succulent and juicy with a faint strawberry aroma and a low soluble solids content of about 6% when ripe (Moreton and Maclead 1990). The flavour is of a rather bland papaya (*C. papaya*). It is closely related to the papaya and the fruit has a similar appearance so it is likely to be climacteric. The milky sap from the fruit can cause skin irritation in some people. It is a native of Ecuador, but is grown commercially in New Zealand.

Harvesting

The fruit turns from green to yellow during ripening (Figure 12.14, in the colour plates). If the fruit is harvested when it is still green it may be possible to develop the fruit colour after harvest but not all the full flavour characteristic.

Storage

Refrigerated storage recommendations are as follows:

- 6°C for 5 weeks with a subsequent shelf-life of 1 week at 20° C (Harman 1983)
- 10°C and 90% r.h. for 5–6 weeks (green) (Snowdon 1990)

- 7°C and 90% r.h. for 4 weeks (turning) (Snowdon 1990)
- 4°C for 2 weeks with 1 week subsequently at 20°C, but blotchy ripening can appear (Mencarelli *et al.* 1990)
- 8°C for up to 3 weeks with 1 week at 20°C, but fruits were very soft (Mencarelli *et al.* 1990)
- 7.2°C and 85–90% r.h. for 7–21 days (SeaLand 1991).

Bamboo shoots

Bambusa oldhami, B. vulgaris, Phyllostachys pubescens, Graminineae.

Botany

The edible parts are the thick shoots that have just emerged from the soil in bamboo clumps and which, if left would produce a new cane (Figure 12.15). They are soft and are boiled before eating. At 1–2°C, the respiration rate was about 0.7 μl CO_2 kg^{-1} h^{-1} (Kleinhenz *et al.* 2000).

Figure 12.15 Harvesting bamboo shoots in China. Source: Dr Wei Yuqing.

Storage

Microbial decay appeared to be the primary of cause of postharvest deterioration since bacterial, fungal and coliform microorganisms were present on external surfaces and internal tissues of shoots (Kleinhenz *et al.* 2000). Shoots stored at 5°C and about 80% r.h. had a lower fibre content, less discoloration of the cut ends and were of better quality than those stored at 30°C and about 80% r.h. (Chen *et al.* 1989). At 8°C discoloration and fungal growth restricted the shelf-life of bamboo shoots to not more than 10 days (Kleinhenz *et al.* 2000).

Modified atmosphere packaging

Under traditional storage at ambient temperature of 20–25°C without packaging, high transpirational weight loss limited shelf-life to 1 day, but with low-density polyethylene film, shelf-life was extended to about 6 days (Kleinhenz *et al.* 2000). The shelf-life of shoots in macro-perforated low-density polyethylene film bags (8.9% area perforated) was 17 days. In contrast, accumulation of condensate in low-permeable materials, low-density polyethylene film bags and particularly heat-sealed PVC film reduced shelf-life to 21 and 14 days, respectively, due to loss of visual packaging quality. The most suitable packaging materials were thin, micro-perforated (45 μm, 0.01% perforation) low-density polyethylene film bags and non-perforated low-density polyethylene film (10.5 μm thick). This reduced the total weight loss of bamboo shoots to about 5% after 28 days, minimized condensation within packages and extended the shelf-life to, and probably beyond, 28 days (Kleinhenz *et al.* 2000).

Banana passionfruit

Passiflora mollissima, P. antioquiensis, P. van-volxemii, Passifloraceae.

Botany

The name banana passionfruit is used for two species, both of which produce yellow-skinned fruit about 7 cm long. *P. mollissima* grows wild in the high Andes and has large pink flowers (Figure 12.16, in the colour plates). It has been introduced to New Zealand and Hawaii. *P. antioquiensis* is synonymous with *P. van-volxemii*, has bright red flowers and is a native of the higher altitudes of Colombia. Tellez *et al.* (1999) found that fruits of two new cultivars, Ruizquin 1 and Ruizquin

2, of *P. mollissima* showed climacteric behaviour. The edible portion is the aromatic pulp around the seeds.

Harvesting

The fruit changes from green to yellow as it matures and this change can be used to determine when to harvest. Harvesting is by hand.

Disease control

Damage due to *Colletotrichum* sp. was minimal, whereas *Botrytis* sp. caused slight damage. Growing plants were sprayed with the fungicide thiabendazole (0.5 ml litre^{-1}) and grapefruit seed extract (3 ml litre^{-1}) during 5 months with spraying intervals of 7, 14 and 21 days from the appearance of floral buds until the occurrence of green mature fruits. These treatments significantly reduced damage by *Colletotrichum* sp., while *Botrytis* sp. was not detected (Caceres *et al.* 1998).

Storage

Storage at 10–12°C and 87–89% r.h. increased the storage life by 50% compared with ambient storage at 24–25°C and 75–77% r.h. Losses of fruit in cold storage were 2.8% during 30 days (Collazos *et al.* 1984). Caceres *et al.* (1998) stored green fruit, without disinfecting, in a cold room with free ventilation at 13°C and 90% r.h. for 20 days. Storage at 8°C resulted in a slower loss of weight and firmness, a lower respiration rate, a higher content of total soluble solids, a lower pH value of the juice and a higher percentage of acidity compared with storage at room temperature of about 20°C (Tellez *et al.* 1999).

Modified atmosphere packaging

Fruit storage in 25 × 75 cm polyethylene bags 0.025, 0.0375 or 0.05 mm thick and with 0, 6 or 12 perforations of 0.5 cm in diameter per bag was investigated. Bags of 0.05 mm thick with 6 perforations were the most suitable (Collazos *et al.* 1984).

Bananas

Musa, Musaceae.

Botany

Banana is a climacteric parthenocarpic berry. It does not conform fully to the Linnean system of plant classification and they are therefore usually referred to by their ploidy and how much *M. balbisiana* and *M. acuminata* have contributed (Stover and Simmonds 1987). For example, all the Cavendish types and Gros Michel would be referred to as *Musa* AAA since they are triploids of *M. acuminata* and Apple banana is *Musa* AAB since two thirds of its chromosomes come from *M. acuminata* and one third from *M. balbisiana*. The most common types grown for international trade are the various mutations of Cavendish. With the Cavendish types grown at tropical temperatures, ripening only occurs in the pulp and at the usual temperatures in the lowland tropics the skins would remain green while the pulp ripens, becomes soft and changes colour to the typical cream colour of ripe bananas. The yellow colour of the skin of ripe Cavendish bananas is due to the presence of carotenoids, which remain at constant levels throughout the ripening process. The change in colour from green to yellow is entirely due to the breakdown of the green pigment chlorophyll. The presence of chlorophyll in the skin masks the yellow colour of the carotenoids (Blackbourn *et al.* 1990). Seymour *et al.* (1987, 1987a) and Semple and Thompson (1988) showed that as the ripening temperature increases over 20°C progressively less chlorophyll breaks down during the ripening process. At 25°C the amount of chlorophyll present in the peel of fully ripe bananas means that they look unripe from the outside, but the pulp is a cream colour, and is soft and sweet just like ripe fruit. This effect is often referred to as a physiological disorder called '*pulpa crema*' in Latin America and 'yellow pulp' in the Caribbean. Wainwright and Hughes (1989) in their studies of *pulpa crema* indicated that the disorder was associated the plants being under stress while being grown in adverse conditions. The mature pulp will freeze at about –3.6 to 3.0°C (Wright 1942).

Preharvest factors

Black Sigatoka is a leaf spot or black leaf streak disease caused by infection with the fungus *Mycosphaerella fijiensis*, which in endemic to most banana-exporting countries. It does not infect the fruit. Its effect is to damage or kill leaves. This reduces the photosynthetic area of the plant, which can lead to a reduction in yield. This leaf damage in turn causes stress to the plants and a reduction in the green life of the harvested fruit. It has been found that fruit of infected plants behave as though they were physiologically 1–2 weeks older than those of uninfected plants of the same age. In trials reported by Turner (1997), reducing leaf number from 12–7 during fruit growth did not affect bunch weight, but reduced green life by 6 days. With fewer leaves there was no further reduction in green life, but bunch weight was reduced by 8%. Ramsey *et al.* (1990) found that plants with less than five viable leaves (that is, five leaves not greatly affected by Black Sigatoka) at harvest produced lighter bunches, due to smaller fingers. All bunches from plants with fewer than four leaves were field-ripe, that is, suffering from *pulpa crema.*

Figure 12.17 Placing a protective plastic sleeve on the bunch shortly after emergence.

Harvest maturity

In order to maximize yield and have sufficient shelf-life for marketing it is necessary to be precise in determining harvest maturity. The actual age of the bunch harvested varies between countries, whether the fruit is to be marketed locally or exported, if exported how long the journey will be and if exported whether Banavac or controlled atmosphere storage will be used. For international trade it is important to be very precise in order to reduce wastage from fruit becoming over-ripe and to maintain quality. There are several ways that are used to determine optimum harvest time.

Bunch age

To measure bunch age, every week a plastic bag is placed over each banana bunch that is newly emerged with at least three to four hands exposed (Figure 12.17) and tied with a ribbon, the colour of which is changed every week so it is easy to tell the age of the bunch (Figure 12.18). The coloured ribbons used to identify the age of the bunch are according to a set calendar so that all growers in one area or even one country use the same colour for the same week. The age is usually between 9 and 13 weeks old. The primary reason for the plastic bag is to protect the bunch from insects, scuffing, sand blasting and bruises, and there is some evidence that it also accelerates its growth.

In India, the cultivar Giant Governor was harvested at 10-day intervals from 90–150 days after opening of the third hand in March to August and September to February. It was found that dry matter and starch contents were highest between 90 and 100 days and thereafter declined (Dhua *et al.* 1992).

Figure 12.18 The protective plastic sleeve is tied in a knot at the bottom but a gap is left to allow water to escape and a coloured ribbon is tied on to identify the age of the bunch.

Calliper grade

Calliper grade, which measures the width on an individual finger on a bunch, is also used. As bunches get older, the fingers become thicker and therefore measuring their width is a measure of their age. Simple callipers are used, which show the maximum and minimum acceptable width for fruit. Several types of tool are used; one is a hand-held plastic type (Figure 12.19) and another consists of pieces of strong steel wire held on a stick (Figure 12.20).

There is variation in calliper grade across the bunch so measurement is always on the middle finger of the second hand from the top of the bunch (Figure 12.21).

The calliper grade used will vary between countries and whether the fruit is to be marketed locally or exported, but it is usually within the range 37–45 thirty-seconds of an inch. Calliper grade is related to the class which is commonly used in the banana industry (Table 12.9).

The way in which the system works in practice is that fruit with a calliper grade of over 46 should not be packed for export. Bunches that are 10 or 12 weeks old and which have reached a calliper grade of 43–45 must be harvested. For export normally three ages of banana bunches are harvested in a given week, according to the instructions issued by the exporting company. Bunches of two of the youngest ages (say 10- and 11-week-old fruit) can be harvested if they have reached the minimum calliper grade (finger diameter). The oldest bunches (say 12 weeks old) are harvested regardless of calliper grade. If these bunches have less than the minimum calliper grade (often 39), they should be rejected for export and marketed locally.

Figure 12.19 Plastic banana callipers with the maximum acceptable finger diameter at one end and the minimum at the other.

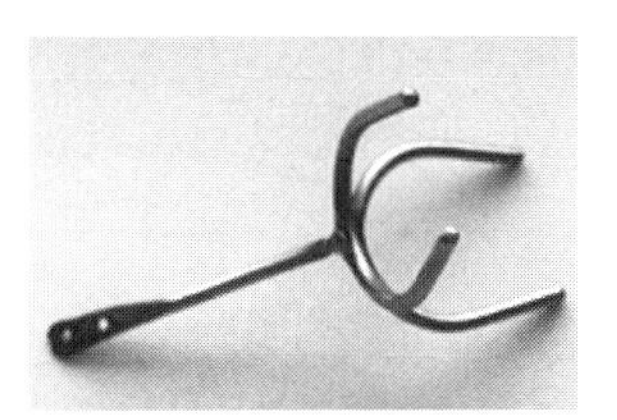

Figure 12.20 Metal banana callipers with the maximum acceptable finger diameter at one end and the minimum at the other. These are fixed to a pole so that the diameter can be measured from the ground.

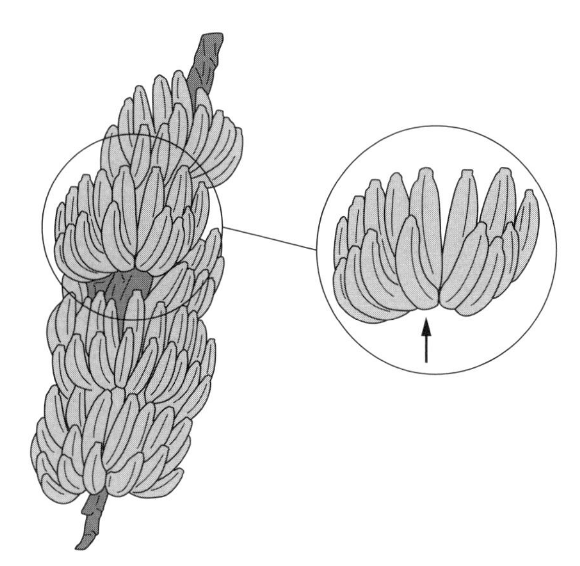

Figure 12.21 The finger that is tested with callipers is the middle finger on the second hand from the top. Source: WIBDECO, Windward Islands.

If bananas are allowed to mature fully before harvest and harvesting is shortly after rainfall or irrigation, the fruit can easily split during handling operations, allowing microorganism infection and postharvest rotting. The tetraploid cultivar Goldfinger harvested at the light ¾ full stage (36–39 mm grades) had an adequate green life (Seberry *et al.* 1998).

Variation in harvest maturity

There is always a variation in harvest maturity between individual clusters of bananas in any commercial

Table 12.9 The relationship between the different methods of expressing commercial calliper grade

Class	Calliper grade		
	Grade[a]	Grade[b]	Metric[c]
Light 3/4	37	5	29
	38	6	30
	39	7	31
	40	8	32
3/4	41	9	33
	42	10	33
	43	11	34
	44	12	35
Full	45	13	36
	46	14	37

[a] Thirty-seconds of an inch.
[b] Thirty-seconds of an inch above 1 inch.
[c] Millimetres.

shipment. This is because estimation of harvest maturity is recording how old each bunch is and combining this with a measurement of the diameter of predetermined fingers on the bunch (calliper grade). In some cases the experienced harvester also takes the shape of the fruit (called angularity) into account. There is an art to this and it does depend on the skill of the harvester. Also within a bunch there is always some gradation in age of fruit from the top to the bottom (Ahmed *et al.* 2001), simply because of the way they develop.

Harvest maturity on green life

Bunch harvest maturity affects their green life, that is, the time before they will initiate to ripen. Montoya *et al.* (1984) found that the older the bunch, the shorter is its green life. They found that this relationship was stronger than the relationship between bunch calliper grade and green life. They reported a green life of over 50 days for 13-week-old fruit at 13°C. Peacock (1975) found that for each week that bananas were harvested earlier than normal harvest age, the green life increased by 3–5 days, while bunch weight decreased by about 10%. Marriott *et al.* (1979) estimated a 0.4-day reduction in green life of bananas per day increase in bunch age. Srikul and Turner (1995) concluded that the green life of bananas declined exponentially as the fruit grows. Individual bananas weighing 90 g had a green life of 90 days, bananas weighing 140 g had 33 days and bananas weighing 190 g had 15 days.

Flavour

Harvest maturity appeared to have little effect on organoleptic properties, although Wei and Thompson (1993) reported that fruit harvested more immature tended to have a slightly firmer texture when fully ripe (Table 12.10). However, Ahmed *et al.* (2002) found very strong evidence that for Robusta, which is also a Cavendish type, the fruit had much better organoleptic properties the more mature they were when harvested (Table 12.11). The cultivar was also shown to affect organoleptic properties since Yellow Banana, which is a *Musa* AA type, had higher organoleptic characteristics than the standard Cavendish type (Table 12.10).

Table 12 10 Effects of harvest maturity of the cultivars Yellow banana (*Musa* AA) and Giant Cavendish (*Musa* AAA) on the organoleptic properties of the ripe fruit (source: Wei and Thompson 1993)

	Harvest maturity		Cultivar		
	Immature	Mature[a]	Yellow banana	Cavendish	CV (%)
Sweetness	3.19	3.39ns	3.72	2.86	8.3
Flavour	3.32	3.58ns	4.04	2.86	10.0
Texture	4.02	3.56	4.17	3.41	7.2

[a] ns = not significantly different ($p = 0.05$), otherwise = significant, $p = 0.01$ and $p = 0.001$ respectively.

Table 12.11 Effect of harvest maturity (time after anthesis) and hand position on the taste panel scores of Robusta fruit after exposure to exogenous ethylene and ripening at 13°C (source: Ahmed *et al.* 2002)

Time after anthesis (weeks)	Position of hand on bunch														
	Top	Mid.	Low	Top	Mid.	Low	Top	Mid.	Low	Top	Mid.	Low	Top	Mid.	Low
	Flavour			Sweetness			Astringency			Off-odours			Acceptance		
14	3.0	3.0	2.3	3.4	3.4	2.6	2.0	2.2	2.3	1.0	1.1	1.2	2.6	2.1	1.5
12	2.5	2.1	1.5	3.1	3.0	2.4	2.3	2.4	2.4	1.2	1.1	2.1	2.5	2.4	1.2
10	1.9	1.6	1.3	3.0	2.3	1.9	2.5	2.6	3.0	1.7	2.0	2.0	2.0	1.9	1.1
LSD ($P = 0.05$)	0.37			0.29			0.35			0.40			2.4		
CV (%)	17.3			19.1			14.3			26.1			12.1		

Harvesting

Harvesting should always start at one end of the field moving to the other, and checking for harvestable bunches.

Traditional harvesting methods

The usual method of harvesting bananas, still commonly used in some producing countries, is to cut the bunch from the plant and carry it on the head or shoulder to the edge of the field to await transport to the packing station. The bunches are stacked, either horizontally or vertically, by the roadside, sometimes left exposed to the sun, and then loaded on to vehicles and carried to the packing station.

With tall types, the method of harvesting the bunch is partly to cut through the pseudostem of the plant about 2 m above ground; this allows the plant to bend over under the weight of the bunch, which is then supported by the carrier while the harvester cuts through the bunch stalk about 30 cm above the top hand. With short types such as Dwarf Cavendish bananas, the bunch stalk is cut directly through and the bunch carried away.

This method of harvesting requires multiple handling of the bunches of fruit before they are dehanded at the packing station, and can result in both bananas and plantains suffering considerable damage. A number of methods are used to transport bunches from the field to packing stations, including wrapping individual bunches in foam mats. Of the methods tried, carrying the bunches from the field on foam-padded trays and transporting them stacked vertically on the distal end in padded vehicles, have given some reduction in damage. No matter what the methods of harvesting and transport, bunches which are repeatedly handled during stacking, loading and unloading still suffer considerable damage.

Field dehanding

One method that is employed commercially is as follows: when a bunch that is ready for harvest, two clean intact banana leaves are cut and placed flat on the ground with their midribs facing upwards. A slight cut in the upper part of the banana pseudostem is made and the bunch gently pulled down to a suitable height (Figure 12.22). The individual hands are cut off, starting from the bottom of the bunch and working upwards with a sharp knife, retaining a portion of stem with each hand, holding three fingers of the cluster

Figure 12.22 Harvesting the bunches by partially cutting the pseudo stem, then pulling down the bunch. Source: WIBDECO, Windward Islands.

firmly but gently as the hand is cut off. A knife is placed so that it rests on the crown and then pulled downwards through the stem to give a clean cut. The cut should be so that the movement of the knife is away from the harvester to prevent injury. The crown is trimmed evenly and the two 'wing' fingers are cut off, together with any damaged fingers. The hand should be held downward by its crown to prevent latex dripping on to the fruit. The cut crowns should be placed on the leaf midrib so the latex drains away. The hands are cut into clusters, usually of 4–8 fingers, and the crowns trimmed neatly. Only fruit that are of the quality required should be selected. Contact between the soil and the cluster, especially the crown, should be avoided as this can result in infection. The clusters should be left for about 10 minutes to drain, avoiding longer delays between clustering and transport to the packing shed, because this again can facilitate infection.

Plastic packing trays are used in which is placed at least a double layer of polyethylene film sleeving at the bottom (Figure 12.24), ensuring that the film covers the edges of the box. The hands are placed, crown downwards, on the film (Figure 12.25) in rows (normally five rows), inter-laying each row with at least a

Figure 12.24 In field dehanding, the clusters are placed crown downwards on padded trays.

Figure 12.25 The trays are filled tightly with fruit with layers of plastic between each layer.

double layer of polyethylene film between each row of clusters. It should be ensured that the film is clean and not contaminated with latex and the clusters are not placed on top of each other in the tray. The trays are carried to the packing shed (Figure 12.26), always leaving them in the shade. When the clusters are taken from the trays, toppling should be avoided as they may be bruised.

In building and siting the packing shed, thought must be given to the movement of the sun during the day and the direction of the wind and rain. Leaky roofs should be repaired immediately. Shed workers, the fruit or the empty or packed boxes should not be exposed to the sun or rain.

Overhead cableways

The most successful method of minimizing damage during the movement of bananas or plantains from the point of harvesting to the packing station has been where the bunches of fruit have remained suspended vertically by the proximal end of the main bunch stalk throughout. This principle has been applied for several decades in areas of large-scale banana production in Latin America and elsewhere by the use of overhead cableways (see Figure 5.2). The cableway involves high capital investment to install and maintain, and its viability depends on the use of a high-yield production system, often including irrigation. The

Figure 12.26 Carrying trays of clusters from the field to the packhouse in the Windward Islands.

ideal area for the installation of a cableway system is said to be 250 hectares, but some large-scale producers operate on the basis of 400-hectare units.

The overhead 'cable' is, in fact, a single-strand high-tensile steel wire suspended about 2.25 m above ground from arches fabricated from galvanized pipe. The cableway serves as a runway for roller conveyor hooks, from which the bunches of fruit are suspended (Subra 1971). The rolling hooks carrying the bunches from the field to packing station are made up into trains totalling up to 30; they are joined together in groups by 1.2 m long spacing rods, which reduce damage to the fruit by preventing the swinging bunches from striking each other. The trains of bunches can be pulled manually or by a small tractor. Some plantations have used a suspended motor hanging on the cableway to haul the bunches, but this method is not universally used, possibly because there is heavy wear on the driving pulleys and the cable. The layout of the cableway is based on a regular-shaped planting of bananas or plantains with a centrally located packing station. It involves a central primary cableway running the length of the planting with secondary lateral branches serving the whole plot. These secondary branches are ideally spaced 100 m apart so that harvesters need carry the bunches no more than 50 m to hang them on the rolling hooks.

For successful operation, the cableway must be included in the original planning of the plantation, so that its layout can be coordinated with other features such as drainage ditches, irrigation, roads or railways and packing station location. The topography of the land is of primary importance; a slope of less than 2 in 1000 (0.2%) is indicated, especially for the lateral branches, where it will be difficult to restrain the train of rolling hooks for the bunches to be attached.

The cableway remains in place for the life of the planting, but with such large monoculture areas there are usually accumulated pest and disease problems, which require an integrated control programme.

A version of this type of cableway, including modifications to recover bananas from hillsides, was designed for use in the Windward Islands (Kemp and Matthews, 1977).

Hillside wireways

A system designed specifically for recovering bananas from hillside plantings was set up in Jamaica before the Second World War, and has also been used successfully in Australia (Barnes 1943). It is an endless wire system, with the banana bunches suspended from hooks attached to the wire rope. The wire runs around two pulleys located on platforms at the top and bottom of the hillside respectively, with a simple braking mechanism to control the rate of descent.

The wireway is operated by gravity; the weight of the bunch moving down returns the empty hook on the other side to the top of the slope. Although this system is effective in moving bananas down to the roadside, the bunches may still have to be carried some way along the hillside and the problem of multiple handling of the bunches from there to the packing house still exists. Possibly it would be more effective if the bananas were dehanded and packed, using crown pads, at the lower end of the wire way.

Padded trailers

In some countries, including Martinique and Guadeloupe, bunches are placed on padded shelves in trailers which are placed alongside the plantation. The padding used is polyurethane foam and bunches are carefully harvested and foam pads placed between each hand to protect it from rubbing. The bunches are then carried to the trailer and secured on to a shelf. When the trailer is full it is hooked on to a tractor and driven slowly to the nearby packhouse.

In a study in Jamaica, in which bunches of bananas were stacked horizontally in lorries padded with banana leaves, it was found that almost all damage caused to the fruit occurred during the loading and unloading operations. Even when the vehicles were driven over unmade or pot-holed roads, the damage was only slightly increased (Jamaica Banana Board, unpublished report 1971).

Packhouse operations

Dehanding fruit in the field and packing them directly into fibreboard boxes for export eliminates the need for a packing station, and has the advantage of saving the capital and operating costs of such a facility. The system was used in the Windward Islands until the 1990s; however, it was discontinued and all the systems currently used involve the use of a packing station or packhouse. This may vary in design and construction from a simple pole and thatch structure to a purpose-designed permanent building. The operations required for preparing and packing bananas for export will be the same irrespective of the type of structure. They include dehanding, washing, selection and grading, fungicide application and packing.

Dehanding

In many packing systems, fruit arrive at the packing station as bunches. Where a cableway is installed the bunches are brought right to the dehanding site still suspended from the rolling hook on which they were placed when harvested. If they have been brought by other means they should not be stacked on the ground, but immediately hung from an overhead structure at the correct height. In all cases the dehanding site should be close to the end of the wash tank. Usually the bunch is hung from the proximal end, with the largest hands uppermost.

When dehanding, the crown should be cut close to the main stem of the bunch (Figure 12.27). It should be evenly cut, leaving as much as possible of the crown attached to the hand, otherwise its outer fingers may be detached during subsequent handling. The design of the knife used for dehanding varies considerably in different countries. A curved knife, with the inside curve sharpened, is usual but in some countries a curved chisel type of knife is preferred (Figure 12.28). In all cases the knife must be very sharp to give a clean, smooth cut in a single movement. As the bunches of bananas or plantains are dehanded the hands must be placed immediately in the wash tank.

Figure 12.27 Fruit being dehanded and placed in a delatexing tank in a packhouse in Ecuador.

Figure 12.28 A chisel type of dehanding tool being used in a packhouse in Ecuador.

Washing and grading

As soon as the hands of fruit have been removed from the bunch, they are placed in the wash tank to remove dirt and latex, which exudes from the cut surface of the crown. There should be a flow of water through the tank to avoid the accumulation of dirt and fungal spores, which may infect the crown (Figure 12.29). In some sites this flow is achieved by diverting water from a stream through the tank; alternatively, pumps can be used to keep up the flow of fresh water. If there is no continuous fresh water supply, chlorine (at 100 mg litre^{-1} active chlorine) and/or alum (10 g litre^{-1}) are normally added to recirculated water to remove latex and destroy microorganisms (Figure 12.30). Normally, the flow of water is from the dehanding end of the tank, so that the hands of bananas or plantains move along to the far end, where workers select and grade them prior to fungicide application. Separate washing and de-latexing tanks are sometimes used, in which case the dehanded fruit is washed for 4 minutes and

Figure 12.29 Delatexing tank in a packhouse in Ecuador.

Figure 12.30 Dipping clusters in a tank of fungicide before packing in St Lucia.

de-latexed for 10 minutes. More commonly a single tank is used, in which the hands are left for 15–20 minutes to remove latex. Latex has no physiologically detrimental properties, but if left on the surface of the fruit causes unsightly staining, which can affect the fruits' market value.

Quality inspection

At the packing plant, an inspector checks for minimum finger length and calliper grade, as well as different types of damages, and marks the hands or whole bunches that have to be rejected. The selection procedure continues at the water tanks of the packing plants.

Diseases

Since the introduction of the shipment of bananas as hands packed in fibreboard boxes, the principal postharvest disease problem has been crown rot. This develops from the infection of the cut surface of the crown by pathogenic fungi. If not treated with fungicide these infections develop during ripening in the importing country and cause decay of the crown tissue, which may spread into the fruit stalk or even the fruit itself during marketing.

Several different fungi have been found to be associated with crown rot (Lucezic *et al.*, 1967; Griffee and Burden, 1976), the most common being the banana anthracnose fungus *Colletotrichum musae*, which often occurs in mixed infections with *Fusarium semitectum* (Knight *et al.*, 1977). *C. musae* also infects other injuries on bananas, particularly those on the ridges of angular fruit, causing decay which spreads into the pulp during ripening. This fungus gets its name from the ability to cause latent infections of banana fruit. These infections occur at any time during the development of the fruit in the field, and usually only develop as the fruit ripens, causing round lesions known as anthracnose. This was a problem when bananas were shipped as bunches with prolonged shipping times, or when ripened at temperatures above 18°C. It is rarely seen in hands packed in boxes.

Several other fungi cause postharvest diseases of banana fruit. They include:

- *Botryodiplodia theobromae* — Stalk and fruit rot
- *Ceratocystis paradoxa* — Stem end rot
- *Pyricularia grisea* — Pitting disease
- *Verticillium dahliae* — Cigar end rot.

Most of these diseases are only occasionally serious where infection levels are high and favourable conditions occur; none are as widespread as crown rot. Pitting disease, *Magnapothe grisea*, however, is a serious field problem in some production areas. It can also be serious on fruit after harvest owing to the development of latent infections that are not controlled by postharvest treatments (Meredith 1963; Stover 1972).

All postharvest disease organisms are widespread in the field, growing and sporulating on decaying banana flowers, bracts and leaves. The spores are blown by wind or splashed by rain on to the fruit, and are also carried on the bunches to contaminate the packing station environment, including the washing water.

Disease control

Since most postharvest diseases of bananas originate in the field from infected plant trash, plantation hygiene and a regular disease control programme will play a major part in reducing the inoculum's load. In addition, it is important to avoid carrying banana trash to the packing house. Dead flower remains are a prolific source of fungal pathogen spores; on many commercial plantings, the flower remains are removed when the plastic covers are placed on the developing bunches in the field. In other systems they are left until harvest, but in that case they should be removed before they are taken into the packhouse.

In addition to these precautions, it is now customary to use a fungicide treatment on the dehanded bananas before they are packed for export (Burden 1969). Bananas taken to packing houses on the bunch are liable to suffer more handling damage and are subject to infection of injuries on the body of the fruit as well as of the cut surface of the crown. This makes it necessary to treat the whole hand with fungicide.

The most effective postharvest control of crown rot is provided is by treatment with the benzimidazole group of fungicides, those most commonly used being thiabendazole (2-thiazol-4-ylbenzimidazole) and benomyl [methyl–1-(butylcarbamoyl)benzimidazol–1-yl carbamate] (Griffee and Pinegar, 1974). Thiabendazole is available in the form of wettable powders and emulsifiable concentrates and benomyl as a wettable powder. Concentrations used should be in accordance with local official recommendations; importing countries have statutory regulations concerning the residue levels permitted on fresh produce. *In vitro* resistance of certain crown rot fungi to benzimidazole fungicides has occurred. In the event of field resistance development, other fungicides can be used, e.g. imazalil [1-(allyloxy-2,4-dichlorophenethyl)imidazole]. In the Philippines, thiophanate methyl [1,2-di(3-methoxycarbonyl-2-thioureido) benzene] is used to control crown rot on export bananas. New legislation exists in the EU, USA and other countries which seeks to restrict all postharvest application of chemicals to fresh fruit and vegetables; however, in practice, most bananas for international trade are treated with either thiabendazole or imazalil.

Fungicide application is made after the hands of fruit have been washed, but they must first be drained of excess water, especially when dipping in small amounts of fungicide suspension, otherwise the water adhering to the fruit may dilute the fungicide below its effective concentration. Since good coverage of the crown is necessary, it is usual to stack the washed hands in perforated plastic trays to drain, followed by fungicide application. The fungicide suspension can be applied by different methods according to the scale of operations and resources available. No matter what application method is used, it is essential to keep the fungicide agitated to prevent the active material from settling out. Also, if large amounts of fruits are to be treated, it will be necessary to replenish the fungicide from time to time to replace that lost by 'dragout'.

Figure 12.31 Spraying the crowns of bananas with a fungicide just prior to packing in St Lucia.

For small-scale operations, the fungicide can be applied by dipping each hand in a bowl of fungicide or spraying the upturned hands in a tray using a hand-operated knapsack sprayer (Figure 12.31). Rubber gloves must always be worn to protect the skin from the fungicide.

In large-scale packing stations, fungicide is applied mechanically to run-off, by passing the hands in trays under either a spray or cascade of fungicide (Figure 12.32). A problem associated with high-pressure spraying of wettable powder suspensions is the tendency for the spray nozzles to block, or for their orifices to wear owing to abrasion by the wettable powder. These problems are avoided by the use of a simple cascade applicator (Burden and Griffee 1974; Kemp and Matthews 1974). This uses a low-pressure, high-volume pump to deliver a cascade of fungicide (about 50 litres per minute) over the hands of bananas as they are carried down a gravity roller conveyor (Figure 12.33). The excess fungicide is returned to the reservoir and recirculated. Here again the volume and concentration of fungicide must be restored from time to time when large quantities of fruit are treated.

Recent studies have been carried out on the possibility of applying fine charged particles of the diluted fungicide, to improve its distribution and effectiveness and reduce potential residue levels. The fungicide is applied over a roller conveyor as in the cascade applicator, but using a fine spray of particles with uniform electrical charges, which repel each other. The

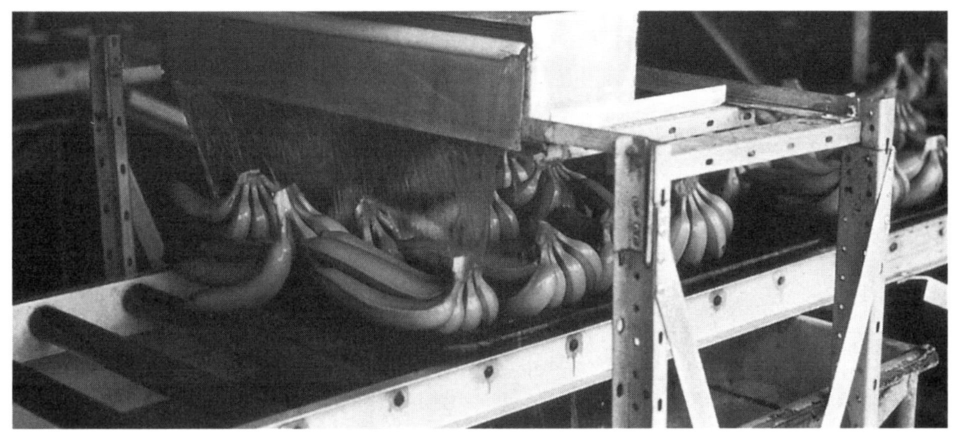

Figure 12.32 A cascade fungicide applicator in a in a packhouse in Ecuador.

bananas are earthed so that the particles are attracted to them. The hands are stacked in trays so that the crowns, which are uppermost, receive the greatest amount of spray (Anthony 1991).

Alternatives to fungicides

Treating the freshly cut crowns with citric acid is used on many organic bananas to control crown rot (Al-Zaemey *et al.* 1993, 1994). Benomyl application could be reduced by half when applied concurrently with pressure infiltration (1.03×10^5 Pa for 2 minutes) of both citric acid and acetic acid to *Musa* AAB 'Embul' (Perera and Karunaratne 2001).

Packaging

The final handling of the bananas in the packing station involves packing the hands in the boxes in which they are to be marketed. Since the 1960s, virtually all bananas exported outside their production region have been packed in fibreboard boxes holding an average weight of about 14 kg or, more commonly, 18 kg each (28 or 40 lb) depending on market preference (Figure 12.34). The boxes are usually of the fully telescopic slotted type, often with a divider to improve the stacking strength. They are assembled by gluing or stapling, which is normally done by hand, but machines are available. The box dimensions used vary,

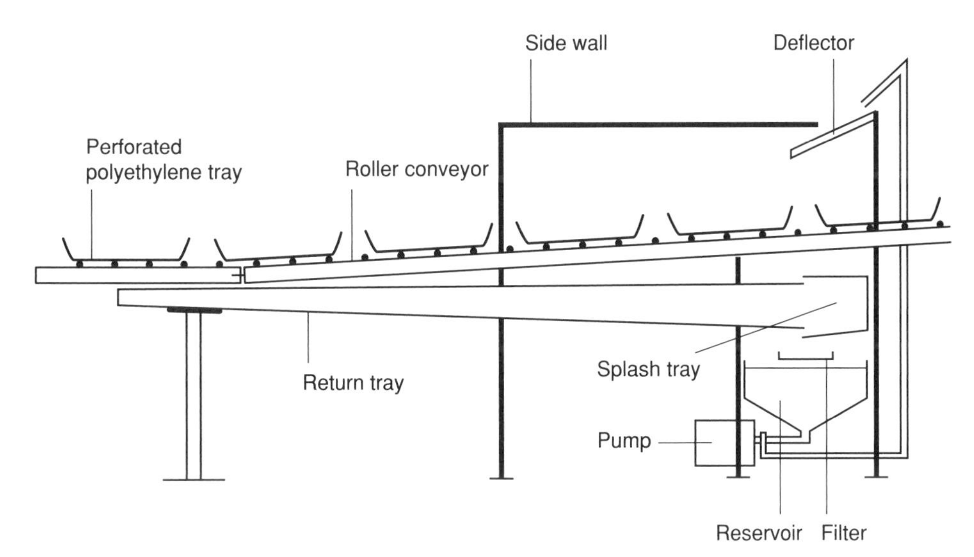

Figure 12.33 Diagram of a cascade fungicide applicator. Source: National Institute of Agricultural Engineering, Silsoe, UK.

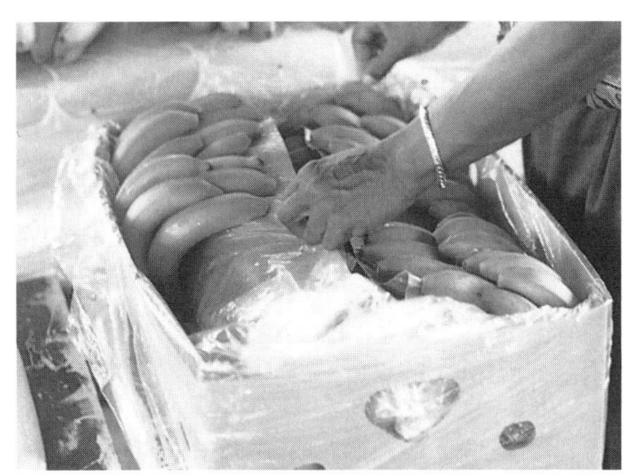

Figure 12.34 Fruit being packed in a 40 lb box inside a plastic sheet for export from the Windward Islands.

Figure 12.35 Reusable collapsible plastic box from the IFCO design used in banana transport trials from the Windward Islands to Britain.

but now tend to conform to international standard modular dimensions for palletized handling. Boxes must be very strong to withstand stacking pressures, which can be as much as 250 kg in some break-bulk stacks.

It is essential that the box design should allow for good ventilation of the contents. Ventilation holes should be as large as permissible and located to give free air flow through the contents of close stacked boxes. This will maintain an even temperature throughout the fruit during refrigerated shipment. However, ventilation holes can affect the stacking strength of fibreboard boxes, so a compromise should be reached to maximise ventilation and stacking strength.

Most exporters use polyethylene film liners to help reduce moisture loss from the bananas and provide some protection from damage during handling and transport. This procedure varies with different producers and the distance the bananas must be transported. Some fold sheets of polyethylene around the fruit, others pack all the box contents in a sealed bag under partial vacuum and a few use individual small bags, each containing a single hand of fruit. Vacuum packing is commonly used where the transport time is long because it can modify the carbon dioxide and oxygen levels around the fruit, which is said to extend their green life. Banavac is a patented method of vacuum packing.

Polyethylene film packing has also been used to delay ripening of boxed bananas during transport, as an alternative to refrigeration. The hands of green fruit, sealed in polythene bags with an ethylene absorbent (potassium permanganate), remained in the green, hard condition for up to 18 days at ambient temperatures, during transport from North Queensland to Auckland, New Zealand (Scott *et al.* 1971).

The bananas should be packed in a regular pattern in the box in such a way that the hands of fruit do not move and damage each other when the box is handled. The pack should be full and tight enough to prevent the contents moving, but not so full that the bananas are damaged by pressure when the boxes are stacked.

Trials have been carried out with returnable plastic boxes. The system used is called IFCO (Figure 12.35). The boxes will collapse and pack flat for return to the producer country after they have been used. A reliable washing procedure is also necessary to ensure that the fruit are not contaminated. One of the incentives to use such a box is the European Union's decision to make an increasing amount of packaging material reusable. In unpublished trials it was found that there was an increase in chilling injury (underpeel discoloration) in fruit packed in IFCO boxes.

Transport

The packed boxes are usually loaded on to a lorry or pickup (Figure 12.36) and taken to a ship where individual boxes are placed directly into refrigerated holds. This method is called 'break bulk.' In some cases the boxes are loaded on to pallets at the packhouses and the pallets transported and loaded on to ships (Figure 12.37). Palletization means that fewer boxes can be loaded into ships' holds. Refrigerated containers (reefers) are used, the banana boxes being loaded

Figure 12.36 Packed boxes, outside a field packhouse, loaded ready for transport to the ship in the Windward Islands.

Figure 12.38 Refrigerated vehicle for transporting fruit in Britain.

directly into 20 or 40 foot insulated containers at the packhouse and then transported to the ship by road. The containers transported by sea to the importing country and then by road directly to the ripening rooms. In some cases the containers are loaded (stuffed) at the docks of the exporting country. Reefers are used in some countries for export bananas, but break bulk shipment (Figure 12.38) is cheaper and is usually preferred where reefer ships are available.

Figure 12.37 Fruit palletized at the packhouse and loaded into a lorry for transport to the ship in the Windward Islands.

Green bananas (that is, preclimacteric) are shipped under refrigeration to prevent the initiation of ripening before they arrive at their destination. Stacking boxes of fruit in ships' holds or containers must be done with great care to ensure adequate ventilation to all boxes. This is particularly important where palletization is used because with a solid stack of boxes on a pallet the air will tend to go around it, leaving some of the boxes inadequately cooled. Techniques to ensure that air is directed through the boxes include stapling strips of fibreboard between pallets, or even having inflatable bags jammed around the pallets.

Postharvest treatments

In experiments with Cavendish bananas harvested at the mature-green stage, treatment with the anti-ethylene compound 1-methylcyclopropene (1-MCP) at 0.01–1.0 μl $litre^{-1}$ for 24 hours delayed peel colour change and fruit softening, and extended shelf-life and suppressed respiration and ethylene evolution. Similar results were obtained with fruits sealed in polyethylene bags (0.03 mm thick) containing 1-MCP at various concentrations, but longer delays in ripening, about 58 days, were achieved (Jiang *et al.* 1999).

Chilling injury

Bananas, in common with other tropical fruits, are subject to chilling damage at temperatures well above freezing. At temperatures at, or below, about 12°C the green fruit develop a dull, often grey, skin colour, starch is no longer converted to sugar and they subsequently fail to ripen properly (Wardlaw, *et al.* 1939). They eventually become black and decay. To avoid this problem, bananas are shipped under a controlled temperature of 13–14°C, and are provided with regular changes of fresh air throughout the voyage to avoid damage due to the accumulation of carbon dioxide.

Precooling

The ship's hold is commonly precooled prior to loading the boxes into the ship in most exporting countries. This is done both to test that the refrigeration equipment is working adequately and to cool the bananas more quickly. The practice has been used in the Caribbean, but it was found that the temperature in the hold rose quickly to ambient during loading. Also, workers were reluctant to move in and out of the cold rooms, believing that it adversely affects their health. The ship's refrigeration has capacity for rapidly cooling the fruit, which do not appear to be adversely affected.

Precooling the bananas before loading has been discussed by importers and producers. There is a case for it, but it is usually considered to be too expensive and therefore not justified on the basis of cost-benefit analysis.

It is desirable, however, that when bananas are shipped in reefer containers with integral or 'clip-on' refrigeration units, both the boxes of fruit and the containers should be pre-cooled and loaded very quickly, because many of the containers' refrigeration units are designed only to maintain cool temperatures, not for the rapid removal of field heat.

Storage

General refrigerated storage and transport recommendations are 13.3–14.4°C and 90–95% r.h. for all cultivars except Gros Michel, which is 11.7°C (Hardenburg *et al.* 1990), or 14.4°C and 85–95% r.h. for 7–28 days (SeaLand 1991). For specific types or stages of ripeness, the following were recommended.

Cavendish

- 12.8–14.4 with 85–90% r.h. (Pantastico 1975)
- 13°C and 85–90% r.h. for 10–20 days (green) (Snowdon 1990).

Dwarf Cavendish

- 11.5–11.7°C (Wardlaw 1937; Anon 1968)
- 12.2°C from Trinidad to Britain (Simmonds 1966)
- 11.4°C from West Africa to Europe (Simmonds 1966).

Green bananas

- 12–14°C and 90–95% r.h. for 2–3 weeks (Mercantilia 1989).

Gros Michel

- 11.5–11.7°C (Wardlaw 1937; Anon 1968)
- 11.7°C from tropical America and West Africa to Europe and USA (Simmonds 1966)
- 12.2°C from Trinidad to Britain (Simmonds 1966)
- 13°C and 85–90% r.h. for 10–20 days (green) (Mercantilia 1989).

Lacatan

- 14–14.4°C (Wardlaw 1937; Anon 1968)
- 13.1°C from Jamaica to Britain (Simmonds 1966)
- 12.2°C from Trinidad to Britain (Simmonds 1966)
- 13°C (Purseglove 1975)
- 12.8–15.6°C for 4 weeks with 85–90% r.h. (Pantastico 1975)
- 13–15°C and 85–90% r.h. for 1 month (green) (Snowdon 1990).

Latundan

- 14.4–15.6°C for 3–4 weeks with 85–90% r.h. (Pantastico 1975).

Robusta, Poyo

- 11.1–11.7°C from Samoa to New Zealand (Simmonds 1966)
- 13°C and 85–90% r.h. for 10–20 days (green) (Snowdon 1990).

Valery

- 13°C (Purseglove 1975)
- 13°C and 85–90% r.h. for 10–20 days (green) (Snowdon 1990).

Yellow bananas

- 13°C for 3–6 days (Mercantilia 1989)
- 20°C for 1–2 days (Mercantilia 1989)
- 13–15°C and 85–90% r.h. for 5–10 days (coloured) (Snowdon 1990).

Controlled atmosphere storage

The benefits of reduced O_2 and increased CO_2 were very good in that they both delayed ripening and high CO_2 additionally has been claimed to reduce chilling injury symptoms. CO_2, above certain concentrations, can be toxic to bananas. Symptoms of CO_2 toxicity were softening of green fruit, internal browning and an undesirable texture and flavour. Sarananda and Wilson Wijeratnam (1997) found that fruit stored at 14°C in 1–5% O_2 had lower levels of crown rot than those stored in air. This fungistatic effect continued even during ripening in air at 25°C. They also found that CO_2 levels of 5 and 10% actually increased rotting levels during storage at 14°C.

Many of the reports in the literature do not include which type or variety of banana is being referred to. Typical storage conditions for unspecified bananas were given as:

- 12–16°C with 2–5% CO_2 and 2–5% O_2 by Bishop (1996)
- Smock *et al.* (1967) showed that at 15°C with O_2 levels of 2% and CO_2 of 8% bananas, can be stored for 3 weeks compared to the fruits which had green life for only 2 days at room temperature
- 5% or lower levels of CO_2 in combination with 2% O_2 (Woodruff 1969)
- 5% CO_2 with 4% O_2 which can extend the shelf-life by 2–3 times (Hardenburg *et al.* (1990)
- Fellows (1988) recommended a maximum of 5% CO_2
- it was reported that in experiments with 25 controlled atmosphere storage combinations 0.5, 1.5, 4.5, 13.5 and 21% O_2 combined with 0, 1, 2, 4 and 8% CO_2 all at 20°C the optimum controlled atmosphere storage conditions appeared to be 1.5–2.5% O_2 combined with possibly 7–10% CO_2 (Anon 1978); under these conditions there was an extension of the green life of the fruit of about six times
- 12–15°C with 2–5% CO_2 and 2–5% O_2 (Kader 1985, 1992)
- 12–16°C with 2–5% CO_2 and 2–5% O_2 (Kader 1989)
- 2–5% CO_2 with 2–5% O_2 (SeaLand 1991)
- 14°C with a range of 12–16°C with 2–5% CO_2 and 2–5% O_2 (Kader 1993)
- 11.5°C with 7% CO_2 with 2% O_2 for 6–8 weeks for South Africa (Van der Merwe 1996)
- 14–16°C, 95% r.h., 2–5% CO_2 and 2–5% O_2 (Lawton 1996)
- 13–14°C and 90–95% r.h. with 2–5% O_2 and 2–5% CO_2 for 1–4 weeks (Burden 1997).

In a study in Southeast Asia it was found that banana cultivars differed in their response to controlled atmosphere storage conditions. Generally, the respiration rate of fruit was slowed down with increased CO_2 and the decreased O_2 in the surrounding atmosphere. As a result of reduced respiratory activity ripening were slowed (Abdullah and Pantastico 1990). Abdullah and Pantastico (1990) also observed that Lakatan (Musa AA) could tolerate O_2 levels as low as 2% at 15°C if the CO_2 level was around 8%. A similar response to controlled atmosphere storage was shown on Bungulan (Musa AAA), which were stored at 5–8% CO_2 and 3% O_2 (Calara 1969; Pantastico 1976). Castillo *et al.* (1967) recommended that Latundan (Musa AAB) should be stored at 10–13.5% O_2. The general recommendation for storage of Dwarf Cavendish (Musa AAA) was 2% O_2 and 6–8% CO_2 at 15°C (Pantastico and Akamine 1975). Abdul Rahman *et al.* (1997) indicated that anaerobic respiration occurred in Berangan at 12°C in 0.5% O_2 but not in 2% O_2. They also showed that after storage for 6 weeks in both 2% O_2 and 5% O_2 fruit had a lower respiration rate at 25°C than those which had been stored for the same period in air.

In Australia, McGlasson and Wills (1972) stored bananas in 3% O_2 and 5% CO_2 for 182 days at 20°C and the fruit still ripened normally when removed to air. Some preliminary work in Australia on Williams, a Cavendish type (Musa AAA), suggested that concentrations around 1.5–2.5% O_2 and 7–10% CO_2 at 20°C were the optimum controlled atmosphere storage conditions. Under these conditions, an extension of preclimacteric period of about six times was achieved compared with that in normal air. Dillon *et al.* (1989) found that fruits stored in 5% O_2 had lower activities of some enzymes during storage and ripened more slowly than fruits stored in air.

Wills *et al.* (1990) described another type of controlled atmosphere storage treatment. Cavendish bananas were stored in a total nitrogen atmosphere at 20°C for 3 days soon after harvest. These fruit took about 27 days to ripen in subsequent air storage, compared with untreated fruit, which ripened after about 19 days. Parsons *et al.* (1964) found that bananas could be stored satisfactorily for several days at 15.6°C in an atmosphere of 99–100% nitrogen. However, this treatment should only be applied to high-quality fruit without serious skin damage and with short periods of treatment, otherwise fruits failed to develop a full yellow colour when subsequently ripened even though the flesh softened normally. Parsons *et al.* (1964) observed that Cavendish fruit stored in $<1\%$ O_2 failed to ripen normally, developed off-flavour, a dull yellow to brown skin with a 'flaky' grey pulp. Wilson (1976) also mentioned that fermentation takes place when bananas are ripened at 15.5°C in an atmosphere containing 1% O_2.

Controlled atmosphere storage of ripening fruits

Controlled atmosphere stored ripening fruits in 3% O_2 for 15 days at 14°C and 95% r.h. had higher firmness and lower °Brix values for the first 3 days after ethylene treatment than fruits stored in air at the same temperature. These differences disappeared as fruits reached commercial retail-store ripeness. No significant differences were found in any of the colour parameters (Madrid and Lopez-Lee 1998). Storage of ripening fruits in 2% O_2 also suppressed respiration rate, maintained the green peel colour and increased ethanol production compared with storage in air. However, the ethanol content of fruit flesh was lower than in fruits stored in 0 and 1% O_2 atmospheres, and remained constant during storage (Imahori 1998).

Modified atmosphere packaging

For long-distance transport of bananas, a system was developed and patented by the United Fruit Company called 'Banavac' (Badran 1969). The system uses polyethylene film bags, 0.04 mm thick, in which the fruit are packed to a nominal 18.14 kg net and a vacuum is applied and the bags are sealed. Typical gas contents, that were claimed to develop in the bags through fruit respiration during transport at 13–14°C, were about 5% CO_2 and 2% O_2. Polyethylene film is about six times more permeable to O_2 than to CO_2 (Fuchs and Temkin-Gorodeiski 1971), so it is difficult to see how the method could be so precise. In later work, Satyan *et al.* (1992) stored preclimacteric bananas in 0.1 mm thick polyethylene tubes at 13, 20 and 28°C and found that their storage life was increased 2–3 times compared with unwrapped fruit at the same temperature. Scott *et al.* (1971) found that storage life in polyethylene bags was extended for 6 days longer than non-wrapped fruit at 20°C. Tongdee (1988) found that the green life of Kluai Khai bananas could be maintained for more than 45 days in polyethylene bags at 13°C. According to Shorter *et al.* (1987), the storage life of banana can be increased five times when they are stored in plastic film (where the gas content stabilized at about 2% O_2 and 5% CO_2) with an ethylene absorber compared with fruit stored without wraps. In later work, Turner (1997) found that bananas had an extended green life of 22–55 days stored at 13°C and 18–40 days at 17°C. However, Wills (1990) mentioned an unsuitable selection in packaging materials can enhance CO_2 injury or actually accelerate the ripening of fruits when ethylene is accumulated over a certain period. If the film was insufficiently permeable, packaging Apple banana in polyethylene film and storage at 13–14°C accumulated levels of CO_2, which could prove toxic to the fruit (Wei and Thompson 1993). In this latter study, the symptoms of CO_2 injury were observed as pulp browning when CO_2 levels were between about 5% and 14%.

Potassium permanganate

At 18°C, Fuchs and Temkin-Gorodeiski 1971 found that the oxidizing agent, potassium permanganate, that breaks down ethylene, placed inside 0.03 mm thick sealed polyethylene film bags each containing 5–6 kg of Dwarf Cavendish delayed ripening. Non-wrapped fruit became ripe after 7 days whereas fruit in sealed polyethylene bags with or without potassium permanganate were still green after 14 days. Initiating the fruit to ripen before sealing it in the bags resulted in CO_2 levels of up to 30% and O_2 levels down to 0.5%. In the latter cases the fruits failed to degreen.

Scott *et al.* (1970) stored Williams in 0.038 mm thick polyethylene film, some containing potassium permanganate and some with calcium hydroxide to absorb CO_2, and some with neither and some with both. In some cases CO_2 reached high levels, but this might have been due to the fruits having been initiated to ripen before being placed in the bags. Again, this could have accounted for the high levels of ethylene in some of the bags. The inclusion of potassium permanganate invariably kept ethylene levels low. There was high variability between fruit from the same treatment but generally the fruits sealed in polyethylene bags remained unripe for longer (Table 12.12). Fruits

Table 12.12 Effects of absorbents (potassium permanganate and calcium hydroxide) on the level of gases in 0.038 mm thick polyethylene film bags contained bananas stored at 20°C (source: Scott *et al.* 1970)

Absorbent	Ethylene (μl $litre^{-1}$)	CO_2 (%)	O_2 (%)
None	0.06–8.5	2.8–20.7	1.6–10.0
$Ca(OH)_2$	0.06–8.9	1.3–9.5	1.2–16.3
$KMnO_4$	0.004–0.93	3.2–10.0	1.2–8.5
$Ca(OH)_2$ + $KMnO_4$	0.005–0.078	1.3–5.5	1.9–11.2

that were firm to touch when the bags were opened ripened with normal flavour and texture. Fruit that were obviously soft also had collapsed pulp.

Hypobaric storage

Storage at sub-atmospheric pressure slowed the rate of the ripening processes in bananas, thus prolonging their storage life. This effect was inversely related to the pressure; fruit stored at atmospheric pressure ripened after 30 days, whereas at 150 or 80 mmHg they remained unripe after 120 days. No injuries that could be attributed to the sub-atmospheric pressure were observed and the fruit that were ripened after hypobaric storage had very good texture, aroma and taste (Burg 1975). A combination of 150 mmHg and one air exchange every 2 hours was found to create beneficial conditions for storage of high-quality banana fruit for up to 120 days (Apelbaum *et al.* 1977).

Ripening

Von Loesecke (1947) reviewed the ripening practices that were used at that time, normally on whole bunches. In the UK, 15.6–18.3°C was used in summer and in winter 23.9°C for West Indian fruit (Gros Michel) and 21.1°C for those from the Canary Islands (Dwarf Cavendish). These temperatures were maintained for 12 or 24 hours then reduced to 15.6–16.7°C according to the condition of the fruit. If the fruit did not 'break colour' within 3 days, the hands were cut off and ripened in boxes at 18.3–21.1°C. In the USA ripening was generally at 16.7–20°C. However, this depended on the maturity and condition of the fruit. For 'hard green fruit,' 22.8–25.6°C was used with high humidity for 12 hours. High humidity, of 90–95% r.h., was used during the early stages of ripening, dropping to 75–85% r.h. as the fruit ripened. The use of ethylene in ripening bananas was first described in the 1920s using 1 part in 1000 parts of air. It is normally used commercially, because ethylene produces uniformity in ripening and accelerates slow ripening fruit.

Currently in Britain bananas are commonly ripened at 16°C, whereas in some other European countries 18°C is preferred. The amount of gas required to initiate bananas to ripen depends on their stage of maturity at harvest, the pulp temperature of the fruit and the time of exposure of the fruit to the gas. Generally, very low concentrations of ethylene are sufficient for mature fruit at 14–19°C. These are in the range of 1–10 ppm (μl $litre^{-1}$) for 24 hours. However, in commercial practice 1000 μl $litre^{-1}$ is commonly used to ensure ripening. Giant Cavendish bananas from various commercial sources in the Caribbean and Latin America were all successfully initiated to ripen by exposure to 10 μl $litre^{-1}$ ethylene for 24 hours at 19°C (Thompson and Seymour, 1982). From the 1990s there has been an increasing demand for all the fruit being offered for sale in a supermarket to be of exactly the same stage of ripeness so that it has an acceptable and predictable shelf-life. This has led to the development of a system called 'pressure ripening' (Figure 12.39). The system involves the circulating air in the ripening room being channelled through boxes of fruit so that ethylene gas, which initiates ripening, is in contact equally with all the fruit in the room. At the same time, the CO_2, which can impede ripening initiation, is not allowed to concentrate around the fruit. A temperature

Figure 12.39 Pressure ripening room for bananas in Britain.

of 18–21°C using ethylene at 10 μl $litre^{-1}$ for 24 hours in 85–90% r.h. was recommended by Wills *et al.* (1989).

Rahman *et al.* (1995a) found that the yellowness of the fruits was not affected by temperatures over the range 14–25°C, but fruits ripened at 14–18°C retained more peel green colour when fully ripe than those at the higher temperatures. They also found that fruits ripened at lower temperatures were firmer in texture, and 20°C was found to be the best ripening temperature with regard to flavour. Ripening of hard green preclimacteric fruit can take anything from 4–10 days, depending on temperature (Table 12.13).

Ripening Goldfinger (*Musa* AAAA) at 16°C with more than 96% r.h., applying 1 μl $litre^{-1}$ ethylene gave fruits of the best appearance and colour. There was a lower incidence of ripe-fruit rots in Goldfinger than Williams (Cavendish, *Musa* AAA) and Lady Finger (*Musa* AAB), but Goldfinger fruits softened more rapidly than those of the other two cultivars (Seberry 1998).

At 20°C, Valery ripened to an edible condition at 95–100, 70–75 or 30–35% r.h. following a normal climacteric pattern of respiration; the high relative humidity accelerated the climacteric by several days (Lizana and Lizana 1975).

Ripening-initiated Valery held in gas mixtures containing 10 μl $litre^{-1}$ ethylene, 2–10% O_2 and 0–10% CO_2, ripened at a slightly lower rate than in air and had comparable quality when ripe. Fruit ripened very slowly and had low quality, however, when the O_2 concentration was reduced to 1% (Liu 1978). Ethylene at 10–100 μl $litre^{-1}$ in air inhibited the development of superficial senescent spots on ripe bananas, cultivar Valery. The ethylene did not affect the respiration rate but accelerated softening of the partially ripe or ripe bananas at 21°C (Liu 1976). Ripening mature green fruits for 24 hours at 20°C with 200 μl $litre^{-1}$ ethylene accelerated ethanol formation and increased the acetaldehyde and ethanol contents of flesh tissue. When the fruits were subsequently stored in 0 or 1% O_2 atmospheres, the respiration rate was suppressed and the skin colour remained green, but ethanol production increased compared with those stored in air. Ethanol accumulation tended to cause an off-flavour, especially at 0% O_2.

Barbados cherries, West Indian cherries

Malpighia punicifolia, *M. glabra*, Malpighiaceae.

Botany

The fruit is a slightly triangular shaped drupe, and contains three triangular ridged seeds. See also Acerola (*M. glabra*, *M. emarginata*).

Storage

The only specific information on postharvest life is 0°C and 85–90% r.h. for 49–56 days (SeaLand 1991).

Beefsteak fungus

Fistulina hepatica.

Botany

It is a bracket fungus up to about 30 cm across and 7 cm thick and feels soft, moist and sticky. When cultivated in China the mature basidiocarps weighed 60–80 g and the pilei were 10–16 cm in diameter (Huang *et al.* 1996). It is tongue or kidney shaped and blood red in colour, darkening as it ages with red flesh that oozes a red fluid when cut. It contains tannic acid and therefore should be soaked before cooking. It has whitish pores turning dark red with reddish brown

Table 12.13 Temperature effects on speed of ripening of preclimacteric fruit with 90–95% r.h. Ethylene injected at 1000 μl $litre^{-1}$ for the first 24 hours (Source: Geest, undated)

	Ripening temperature (°C)									
Days to ripen	Day 1	Day 2	Day 3	Day 4	Day 5	Day 6	Day 7	Day 8	Day 9	Day 10
4	17.8	16.7	15.6	13.3						
5	17.8	15.6	15.6	14.4	13.9					
6	17.8	15.6	14.4	14.4	13.3	13.3				
7	16.7	15.6	14.4	14.4	13.3	13.3	13.9			
10	15.6	14.4	14.4	14.4	13.9	13.9	13.9	13.9	13.9	13.9

spores. It grows parasitically on deciduous trees, particularly oak, usually near the base. Infection causes brown heart rot, resulting in the wood being stained. Studies by Krempl (1989) of infected *Quercus robur* and *Q. petraea* concluded that the discoloured oak is not reduced in strength, and that the effect of the infection is just the stain, which is merely an aesthetic defect. However, infection can be lethal to the tree.

Storage

No information could be found on its postharvest life, but from the literature it would appear that all types of edible mushrooms have similar storage requirements and, in the absence of other information, the recommendations for the cultivated mushroom (*Agaricus bispsorus*) could be followed.

Beetroots, red beet

Beta vulgaris variety *conditiva*, Chenopodiaceae.

Botany

The edible portion is the swollen root, which is dark red in colour. It will freeze at about –2.9 to 2.7°C (Wright 1942) or –1.67 to 1.06°C for beetroots and –0.94 to 0.39°C for beet tops (Weichmann 1987).

Harvesting

They should be harvested when they are a suitable size for the market, but while they are still young, tender and free from fibres. They are sometimes harvested and marketed while still immature with their tops still attached.

Precooling

Forced air cooling with high humidity was the most suitable, hydrocooling was suitable, air cooling in a conventional cold store was less suitable and vacuum cooling was generally unsuitable (MAFF undated).

Chilling injury

Symptoms are a darker coloured flesh that oozes red droplets when cut.

Storage

Fidler (1963) pointed out that in 1963 clamp or pit storage was often used over winter in temperate countries, but if the ambient temperature rose too high the beets might sprout. In the UK less rot was found on beets in storage at 4.4 or 2.2°C than at 0°C (Fidler 1963). Storing in bulk bins lined with 0.125 mm polyethylene film was recommended by Shipway (1978). In other work, using plastic liners in bins was actually better than refrigerated storage (Table 12.14). At room temperature of 20°C and 60% r.h. it was reported in Mercantilia (1989) that they could be kept for about 1 week. Recommendations for refrigerated storage are as follows:

- 0–1.7°C and 79% r.h. (Wardlaw 1937)
- 0°C and 90–95% r.h. for 1–3 months (Phillips and Armstrong 1967)
- 0–3.3°C and 90–95% r.h. for 6–8 months (Shipway 1968)
- 0°C and 95% r.h. for 3–5 months (Anon 1968; Hardenburg *et al.* 1990)
- 0–1.7°C and 90–95% r.h. for 8–14 weeks with 33% loss (Pantastico 1975)
- 0°C and 90–95% r.h. for 55–90 days for bunched beets (Tindall 1983)
- 4°C and 95–98% r.h. for 6 months (Mercantilia 1989)
- 8°C for 4 months (Mercantilia 1989)
- 12°C for 2 months (Mercantilia 1989)
- 0°C and 95–100% r.h. for 90–150 days (beetroot) (SeaLand 1991)
- 2.8°C and 95–98% r.h. for 120–180 days (red beet) (SeaLand 1991)
- 0–1°C and 95–100% r.h. for 3–8 months (Snowdon 1991)

Table 12.14 Storage of red beet in different systems over a 6 month period (Weichmann 1987)

Cooling system	Storage conditions	Plastic liners	Wastage (%)	Water loss (%)
Ambient air	2–12°C, 90% r.h.	None	2.3	12.9
	2–12°C, 90% r.h.	Yes	0.7	2.7
Refrigeration	0–1°C, 95% r.h.	None	2.3	6.3
	0–1°C, 95% r.h.	Yes	3.5	0.9

Table 12.15 Effects of temperature and reduced O_2 level on the respiration rate (CO_2 production in mg kg^{-1} h^{-1}) of beetroot (Robinson *et al.* 1975)

	Temperature (°C)				
	0	5	10	15	20
Mature beetroot without leaves					
In air	4	7	11	17	19
In 3% O_2	6		7		10
Bunching with leaves					
In air	11	14	22	25	40
In 3% O_2	7		14		32

- 0°C and 90–95% r.h. for 10–14 days for bunched beets (SeaLand 1991)
- 0–1°C and 95–100% r.h. for 2–3 weeks for bunched beets (Snowdon 1991).

Controlled atmosphere storage

Controlled atmosphere storage was said to have a slight to no effect (SeaLand 1991) and atmospheres containing over 5% CO_2 can damage the beet (Shipway 1978). This is partly confirmed by the findings of Robinson *et al.* (1975), who showed that at 0°C the respiration rate was higher in low O_2 than in air (Table 12.15). However, for beetroot that were stored with their leaves still attached, storage in low O_2 reduced the respiration rate.

Belle apples, Jamaican honeysuckle, water lemons

Passiflora laurifolia, Passifloraceae

Botany

The vine, which has quadrangular stems like *P. quadrangularis*, bears oval to ellipsoidal fruit, which are capped with three large green bracts, up to 7 cm long with orange to yellow soft skin and a central cavity filled with seeds and juicy pulp. It is a native of Central America, the north-eastern part of South America and the West Indies, and has been widely distributed throughout the tropics.

Storage

No information on its postharvest life could be found.

Biriba, wild soursop

Rollinia spp., Annonaceae

Botany

There are some 50 species of *Rollinia*. Biale and Barcus (1970), in experiments with Amazonian fruits, showed that *R. orthopetala* was a climacteric fruit. In other work, biriba is identified as *R. mucosa* and *R. deliciosa*. *Rollinia* spp. are small- to medium-sized trees. The fruits are heart shaped and can be up to 15 cm in diameter. The flesh, in which is embedded black seed about 1 cm long, is soft, sweet and viscous when ripe. Forty-two volatile components were isolated from the pulp of *R. mucosa*, grown in Cuba, of which the major volatiles were α-pinene, β-pinene and β-caryophyllene (Pino 2000).

Extracts

R. mucosa, *R. papilionella*, *R. emarginata*, *R. emarginata*, *R. ulei* and *R. deliciosa* are acetogenin-producing plants. The Annonaceous acetogenins are a series of apparently polyketide-derived fatty acid derivatives that possess tetrahydrofuran rings and a methylated γ-lactone (sometimes rearranged to a methyl keto-lactone) with various hydroxy, acetoxy and/or ketoxy groups along the hydrocarbon chain. They exhibit a broad range of potent biological activities, including cytotoxic, antitumour, antimalarial, antimicrobial, immunosuppressant, antifeedant and pesticidal (Rupprecht *et al.* 1990).

Storage

No information on its postharvest life could be found.

Bitter gourd, pepino, kerela, bitter cucumber, balsam pear

Momordica charantia, Cucurbitaceae.

Botany

The fruit is a pepo up to 25 cm long, which is ribbed with numerous tubercles containing bitter flesh with many brown seeds up to 15 mm long with scarlet arils. The carotenoid content in the aril increased considerably during fruit ripening, lycopene being the main component, reaching 64.75 mg per 100 g fresh weight when fully ripe. Both chlorophyll a and b contents decreased significantly in the fruits stored at higher temperatures compared with those at lower temperatures (Tan *et al.* 1999). The tender shoots and leaves are used as spinach and the seeds as a condiment.

Harvesting

This is usually based on size and is about 15–20 days after flowering.

Storage

Their shelf-life in simulated room temperature of 20°C and 60% r.h. was shown to be about 3–5 days (Mercantilia 1989). Refrigerated storage recommendations are as follows:

- 0.6–1.7°C and 85–90% r.h. for 4 weeks (Pantastico 1975)
- 1–2°C and 85–90% r.h. for 20–30 days (Tindall 1983)
- 10°C and 90% r.h. for 2–3 weeks (Mercantilia 1989)
- 11.7°C and 85–90% r.h. for 14–21 days (SeaLand 1991).

Modified atmosphere packaging

In experiments, the fruit harvested at horticultural maturity were packaged in polyethylene film bags and stored at 1, 10, 20 or 30°C. The lutein and cryptoxanthin contents in the pulp increased during storage at 20 and 30°C, but increased only slightly at 1°C and tended to decrease during storage and 10°C. The pattern of changes of carotenoids in the aril was similar to that in the pulp (Tan *et al.* 1999).

Bitter yam, cluster yam

Dioscorea dumetorum, Dioscoreaceae.

Botany

The tubers are small and borne in clusters that may be fused together. The flesh may be from white to yellow. The yellow ones are likely to contain the highly toxic alkaloid dehydrodioscorine, but it can also occur in white-fleshed tubers; however, cultivars have been selected for low toxicity (Purseglove 1975).

Harvesting

This is usually 8–10 months after planting and they are dug by hand.

Detoxification

The tubers are sliced and soaked in frequent changes of water, usually with the addition of common salt.

Disease control

Aspergillus niger and *Penicillium oxalicum* were the most common fungi on stored tubers with the optimum temperature for growth of 26–30°C (Ogundana *et al.* 1970), but no growth was observed on undamaged tubers.

Storage

Kay (1987) mentioned that the tubers do not store well and become hard and inedible about 4 weeks after harvesting.

Blackberries, brambles

Rubus ulmifolius, *R. fructicosus* and other species, Rosaceae.

Botany

The fruit is a collection of drupes, which are classified as non-climacteric. Hulme (1971) observed that the fruit continued to change in colour during the first 24 hours after harvesting. Its freezing point was given as –2.2 to 1.4°C by Wright (1942).

Harvesting

They are normally harvested by hand with the receptacle in place when the drupes are fat, juicy and shiny. When fully mature they have a soluble solids content of about 9%, an acid content of 0.68–1.84% w/w and a sugar:acid ratio of 2.8 (Hulme 1971).

Precooling

Forced air cooling with high humidity was the most acceptable method, air cooling in a conventional cold store was suitable and hydrocooling and vacuum cooling less suitable (MAFF undated).

Table 12.16 Effects of temperature and reduced O_2 level on the respiration rate (CO_2 production in mg kg^{-1} h^{-1}) of Bedford Giant blackberries (Robinson *et al.* 1975)

	Temperature (°C)				
	0	5	10	15	20
In air	22	33	62	75	155
In 3% O_2	15		50		125

Storage

Refrigerated storage recommendations are as follows:

- –0.6 to 0°C and 85–90% r.h. for 5–7 days (Phillips and Armstrong 1967)
- –1 to 0°C and 90% r.h. for 5–7 days (Anon 1967)
- –0.6 to 0°C and 90–95% r.h. for 2–3 days (Lutz and Hardenburg 1968)
- –0°C and 90% r.h. for 10 days for Bedford Giant (Robinson *et al.* 1975)
- –0.5 and 90–95% r.h. for 2–3 days (SeaLand 1991).

Controlled atmosphere storage

At atmosphere containing 20–40% CO_2 can be used to maintain the quality of machine harvested blackberries for processing during short-term storage at 20°C (Hardenburg *et al.* 1990). They retain their flavour well if they are cooled rapidly and stored in an atmosphere containing up to 40% CO_2 for 2 days (Hulme 1971). Kader (1989) recommended 0–5°C with 10–15% CO_2 and 5–10% O_2. The respiration rate was strongly affected by temperature, but also by the O_2 level in the storage atmosphere (Table 12.16); however, the effect of reduced O_2 was similar at all the storage temperatures.

Blackcurrants

Ribes nigrum, Grossulariaceae.

Botany

The fruits are small and contain many small seeds, and when ripe they turn shiny and black. When mature they have a total sugar content of about 7.9% and a total solids content of 19.7%, made up of 13.8% soluble solids and 5.9% insoluble solids, and a sugar:acid ratio of 2.1 (Hulme 1971). They are borne in clusters called 'strigs' and the fruit matures in order along the strigs with the first one nearest the pedicel and the last being the terminal fruit. They are natives of Europe. They have become popular and a major crop in Britain since the blackcurrant juice drink Ribena was developed in the 1960s.

Harvesting

It is normal to harvest the whole cluster of fruit together, even though they will contain under-mature fruit (Hulme 1971). They can be picked by hand, but with the great demand for blackcurrant juice they are mainly machine harvested. The machines are usually tractor-mounted systems that have combing fingers that are run up the stems, pulling off the fruit bunches, and also a high proportion of the leaves (Figure 12.40). This is achieved by small-amplitude, high-frequency vibration by the rubber fingers that are mounted on a vertical drum that rotates at the same velocity as the

Figure 12.40 Mechanical harvester used for blackcurrants for processing into juice in the UK. Source: Professor H.D. Tindall.

harvester moves forward. The fruit may subsequently be separated from the leaves by blowers and the juice extracted. In other cases the fruit is taken straight from the field and passed through a press and filter system to remove leaves and other material. An important factor to be considered with mechanical harvesting is to breed cultivars whose fruit all mature at the same time (Hitchcock 1973; Williams 1975).

Precooling

Forced-air cooling with high humidity was the most acceptable, air cooling in a conventional cold store was suitable and hydrocooling and vacuum cooling less suitable (MAFF undated).

Storage

Hulme (1971) found that fruit held for 24 hours at ambient temperatures continued to darken in colour and lost about 3% in weight and 7% in ascorbic acid content on a fresh weight basis. Refrigerated storage recommendations are as follows:

- –1 to 0°C and 90% r.h. for 1–2 weeks (Anon 1967)
- –0.6 to 0°C and 90–95% r.h. for 1–2 weeks (Lutz and Hardenburg 1968)
- –0.5 to 0°C and 90–95% r.h. for up to 4 weeks (Hardenburg *et al.* 1990)
- –0°C and 90–95% r.h. for 1–2 weeks (Snowdon 1990)
- –0.5°C and 90–95% r.h. for 7–14 days (SeaLand 1991).

Controlled atmosphere storage

Respiration rate was reduced at all temperatures tested in 3% O_2 in the storage atmosphere, but particularly at the highest temperature (Table 12.17).

Smith (1957), quoted by Hulme (1971), recommended 2°C with 50% CO_2 for 1 week followed by 25% CO_2 for the remainder of the storage period. Stoll (1972) recommended 2–4°C with 40–50% CO_2 and 5–6% O_2. After storage at 18.3°C with CO_2 levels of 40% for 5 days the fruit had a subsequent shelf-life of 2 days, with 2–3% of the fruit being unmarketable owing to rotting (Wilkinson 1972). Skrzynski (1990) described experiments where fruits of the cultivar Roodknop were held at 6–8°C for 24 hours and then transferred to 2°C for storage for 4 weeks in one of the following:

- 20% CO_2 with 3% O_2
- 20% CO_2 with 3% O_2 for 14 days then 5% CO_2 with 3% O_2
- 10% CO_2 with 3% O_2
- 5% CO_2 with 3% O_2
- ambient air.

The best retention of total ascorbic acid content was obtained in both treatments with 20% CO_2. In years with favourable weather preceding the harvest, storage in 20% CO_2 completely controlled the occurrence of moulds caused mainly by *Botrytis*, *Mucor* and *Rhizopus* species. For juice manufacture, storage of fruit at 2°C with 50% CO_2 for 7 days followed by a further 3 weeks with 25% CO_2 was recommended by Smith (1957). There was an accumulation of alcohol and acetaldehyde but the juice quality was not affected. Fruit stored in CO_2 levels of 40% at 18.3°C for 5 days had a subsequent shelf-life of 2 days, with 2–3% of the fruit being unmarketable owing to rotting (Wilkinson 1972). Fruits of the cultivar Rosenthal were stored at 1°C in either a controlled atmosphere of 10, 20 or 30% CO_2 all with 2% O_2 or a high CO_2 environment of 10, 20 or 30% CO_2 all with >15% O_2. The optimum CO_2 concentration was found to be 20%. Ethanol accumulation was higher under controlled atmosphere storage than higher CO_2 environment conditions. Fruits could be stored for 3–4 weeks under controlled atmosphere storage or high CO_2 environment conditions, compared with 1 week for fruits stored at 1°C in a normal atmosphere (Agar *et al.* 1991).

Table 12.17 Effects of temperature and reduced O_2 level on the respiration rate (CO_2 production in mg kg^{-1} h^{-1}) of blackcurrants (Robinson *et al.* 1975)

	Temperature (°C)				
	0	5	10	15	20
In air	16	27	39	90	130
In 3% O_2	12		30		74

Black radish

Raphanus sativus variety niger, Cruciferae.

Botany

The edible part is the tuberous root, which normally has white flesh.

Storage

The limiting factor to storage is usually when they sprout. Their shelf-life at a simulated room temperature of 20°C and 60% r.h. was shown to be about 1 week (Mercantilia 1989). Refrigerated storage recommendations are as follows:

- 0°C and 90–95% r.h. for 2–4 months (Kay 1987)
- 0°C and 90–95% r.h. for 4 months (Mercantilia 1989)
- 4°C for 2 months (Mercantilia 1989)
- 8°C for 1 month (Mercantilia 1989)
- 0°C and 90–95% r.h. for 60–120 days (SeaLand 1991).

Black sapotes

Diospyros ebenaster, Ebenaceae.

Botany

The fruits are olive green when mature, with a soft sweet pulp that is dark brown in colour. It is a native of Mexico.

Storage

The factor limiting storage was softening (Nerd and Mizrahi 1993). The shelf-life at 20°C and 85% r.h. was 13 days with 11–16% water loss (Nerd and Mizrahi 1993). Refrigerated storage recommendations are as follows:

- 12.8°C and 85–90% r.h. for 14–21 days (SeaLand 1991)
- 10°C and 85% r.h. for 29 days with 11–16% water loss (Nerd and Mizrahi 1993).

Modified atmosphere packaging

Fruits wrapped in plastic film had 40–50% longer storage life at 20°C than fruits not wrapped (Nerd and Mizrahi 1993).

Blueberries, bilberries, whortleberries

Vaccinium corymbosum, Vacciniaceae.

Botany

The fruit is a berry and is classified as climacteric. They are usually deep purple or red to almost black.

Harvesting

They should be harvested when they are fully coloured and handling should be minimized as they are very delicate and damage easily and the bloom on the fruit surface can be destroyed. The berries are very small and have to be picked by hand, which is very laborious. Mixtures of soft ripe and hard ripe fruits should be avoided as the over-ripe ones can speed deterioration (Lutz and Hardenburg 1968).

Over the row, self-propelled vibrating machines have been developed for harvesting blueberries (Nelson 1966). These machines are fitted with collecting pans and conveyor belts and are capable of harvesting 1.5 acres of fruit per hour. Machine-picked blueberries had a harvesting cost of only 76% of that of hand-harvested fruit, but 50% of the crop was lost during mechanical harvesting (Liner 1971). In subsequent work, the cost of harvesting and sorting machine-harvested blueberries for processing was 44% of the cost of hand harvesting and packaging fruit for the fresh market (Safley 1985). Blueberry bushes, which are regularly mechanically harvested, yielded 6–42% less fruit than bushes that had been hand picked (Mainland 1971). Machine harvesting blueberries caused 50 times more damage to canes than hand harvesting and machine-harvested fruit were 10–30% softer and developed 11–41% more decay than hand-harvested fruit (Mainland *et al.* 1975). Mechanically harvested blueberries could be maintained in good condition at 20°C for up to 2 days in high CO_2 atmospheres (Morris *et al.* 1981).

Hot water treatment

A 3 minute immersion of blueberries in water at 46–55°C reduced subsequent decay by as much as 90%; 40°C was ineffective and 55°C injured the fruit (Burton *et al.* 1974).

Storage

Berries stored at 4.4°C or above gradually developed a tough texture (Lutz and Hardenburg 1968). The cultivar Bluecrop stored at 12°C had high scores for taste, aroma, bitter flavour and flavour intensity and had a high percentage titratable acidity. Fruits stored at 4°C scored high for sweet taste, colour (more blue), acidic taste, blueberry flavour, firmness, crispness and juiciness (Rosenfeld *et al.* 1999). Their shelf-life at a simulated room temperature of 20°C and 60% r.h. was shown to be only 1–2 days (Mercantilia 1989). Refrigerated storage recommendations are as follows:

- –0.6 to 0°C and 90–95% r.h. for 2 weeks or with some loss of quality for 4–6 weeks (Anon 1968; Lutz and Hardenburg 1968)
- –0.5 to 0°C and 90–95% r.h. for about 2 weeks (Hardenburg *et al.* 1990)
- 0°C and 90% r.h. for 10–14 days (Mercantilia 1989)
- 0°C and 90–95% r.h. for 2–4 weeks (Snowdon 1990)
- 0°C and 90–95% r.h. for 21–28 days (Red Whortleberry) (SeaLand 1991)
- –0.5 and 90–95% r.h. for 10–18 days (Blueberry) (SeaLand 1991).

Controlled atmosphere storage

The storage of fresh blueberries for 7–14 days at 2°C in an atmosphere of 15% CO_2 delayed their decay by 3 days after they had been returned to ambient temperature compared with storage in air at the same temperature (Ceponis and Cappellini 1983). They also showed that storage in 2% O_2 had no added effect over the CO_2 treatment. Kader (1989) recommended 0–5°C with 15–20% CO_2 and 5–10% O_2 for optimum storage. Ellis (1995) recommended 0.5°C and 90–95% r.h. with 10% O_2 and 10% CO_2 for 'medium-term' storage.

Modified atmosphere packaging

Sealing fruit in polyethylene film box liners resulted in maintaining their turgidity and reducing weight loss during storage at 0°C, but off-flavours developed when they were stored for about 6 weeks (Lutz and Hardenburg 1968). This seems a very long time for storage in view on the information on controlled atmosphere storage.

Blewit, field blewit

Lepista saeva.

Botany

It is a gill fungus with a cap diameter of up to 10 cm and a height of up to 8 cm, and has a mealy smell and a mild flavour. The cap is domed as it emerges and gradually flattens and may become depressed in the centre with a wavy margin and has a dry smooth surface. It is pale brown, becoming darker towards the centre. The gills are buff with pink spores. The stalk has a very bulbous base and is white, sometimes streaked and lined with an intense lilac colour. It is commonly found in grassy areas and deciduous woods and may be found growing in large circles in autumn. It is said to cause stomach upsets in some people, but otherwise it is delicious.

Storage

No information could be found on its postharvest life, but from the literature it would appear that all types of edible mushrooms have similar storage requirements and the recommendations for the cultivated mushroom (*Agaricus bispsorus*) could be followed.

Bottle gourds, white flowered gourds

Lagenaria siceraria, Cucurbitaceae

Botany

The fruit is a pepo which can grow up to 1 m long, with a hard skin with edible or bitter flesh containing many white seeds, which can be up to 2 cm long. The chlorophyll content decreased during fruit development, but the carotenoid content increased (Bhatnagar and Sharma 1994).

Harvesting

Fruit length and diameter increased continuously, but especially up to 28 and 20 days after anthesis, respectively, in the cultivars Pusa Summer and Prolific Long, which were edible, tender and green from 12–24 days after anthesis (Bhatnagar and Sharma 1994).

Storage

They had a shelf-life of about 10 days at a room temperature of 26–30°C and 54–68% r.h. (Waskar *et al.* 1999). A refrigerated storage recommendation was 7.2°C and 85–90% r.h. for 4–6 weeks with 3.2% loss (Pantastico 1975).

Modified atmosphere packaging

Non-packed fruits of the cultivar Samrat could be stored for up to 18 days when packed in 100 gauge polyethylene bags with 2% vents. Fruits packed in polyethylene bags and stored in a cool chamber at 20.1–21.2°C and 90–95% r.h. had a storage life of 28 days (Waskar *et al.* 1999).

Boysenberries

Rubus sp., Rosaceae.

Botany

It is a tri-specific hybrid between blackberry, loganberry and raspberry, but the receptacle is small and there are few seeds. The fruit is a collection of drupes and is classified as non-climacteric.

Harvesting

They are normally harvested by hand with the receptacle in place. When fully mature they have a total solids content of about 13 or 14%, an acid content of about 1.5% and a sugar:acid ratio of 3.5 (Hulme 1971).

Storage

The only storage recommendation was –0.6 to 0°C and 85–90% r.h. for 2–4 days (Phillips and Armstrong 1967).

Breadfruits

Artocarpus altilis, Moraceae.

Botany

It is a climacteric fruit reaching a peak in respiration rate of 3 ml CO_2 kg^{-1} h^{-1} at 20°C (Biale and Barcus 1970). They are globose to oblong in shape and up to 30 cm in diameter (Figure 12.41). The fruits are syncarps that are seedless and are produced on large trees.

Figure 12.41 Breadfruit growing in Jamaica with the typical surface latex that indicates it is ready for harvesting.

Figure 12.42 Breadnut growing in Brazil.

It is a basic starch staple in the Pacific islands and transporting plants to the West Indies was the reason for the voyage of HMS Bounty that ended in the famous mutiny. The same species also produces a seeded form, which is called breadnut or chataigne, and the seeds have a flavour similar to chestnuts (Figure 12.42). The African breadfruit (*Treculia africana*) is grown in south-eastern Nigeria for its edible seeds (Nwaiwu *et al.* 1995).

Harvesting

In Jamaica they recognize two harvest maturities, 'young' and 'fit.' Young is characterized by its light green skin, almost complete absence of external latex and the fruit segments closely packed, indicating that they are not fully grown. Fit is characterized by a darker green skin, with some browning and external dried latex and the segments not closely packed. Fit is preferred in Jamaica and is cooked by roasting or baking. Young is cooked by boiling. In a comparison of harvesting methods, Thompson *et al.* (1974c) found that the traditional method of detaching the fruit from the tree with a forked stick and allowing them to fall to the ground, then transporting them to the market in hessian sacks, was satisfactory. Some of the fruit were damaged, but those which did not split had a similar quality to those which were caught and not allowed to hit the ground.

Chilling injury

Storage at either 2.5 or 7°C resulted in the skin of the fruit turning from green to dull brown in 2–3 days. When these fruits were removed to a higher temperature they lost weight rapidly and softening occurred in small circular areas over the surface, increasing in number and size with time (Thompson *et al.* 1974).

Storage

It is used as a starch staple vegetable, so that as soon as it begins to ripen it becomes soft and starch is converted to sugar and it is unacceptable. Bernardin *et al.* (1994) claimed that browning of the peel was the limiting factor in storage. In tropical ambient conditions they had a very short storage life. At 26 or 28°C its storage life was only 2 or 3 days (Thompson *et al.* 1974c; Maharaj and Sankat 1990a). Bernardin *et al.* (1994) found that their shelf-life at 16°C could be extended from 4–9 days by immersing them in 2% calcium chloride solution for at least 3 hours directly after harvesting. In certain rural areas of Jamaica it is the practice to store them underwater at ambient temperature. It is believed to extend their storage life, which was confirmed by Thompson *et al.* (1974a), who found that they did not begin to soften until after 14 days at 28°C, at which time they split, having absorbed too much water. Refrigerated storage recommendations are as follows:

- 12.5°C and 92–98% r.h. for about 5 days (Thompson *et al.* 1974c)
- 13.3°C and 85–90% r.h. for 14–40 days (SeaLand 1991)
- 13°C and 95% r.h. for 1–3 weeks (Snowdon 1990).

Modified atmosphere packaging

A temperature of 12.5°C in polyethylene bags for up to 13 days was recommended by Thompson *et al.* (1974c). Wrapping combined with refrigeration at 16°C markedly reduced external skin browning and there were high chlorophyll levels after 10 days of storage. This treatment also delayed fruit ripening as evidenced by changes in texture, total soluble solids and starch content (Samsoondar *et al.* 2000). Passam *et al.* (1981) recommended 14°C for up to 10 days for the cultivar Whiteheart when enclosed in polyethylene film. The inclusion of an ethylene-oxidizing agent (potassium permanganate) within the polyethylene bag did not significantly improve storage (Passam *et al.* 1981).

Transport

A shipment from Java to the Netherlands at 3°C was referred to in Wardlaw (1937), but there was no information on their condition on arrival. In the 1970s, commercial sea shipments were made in refrigerated containers from Jamaica to Britain at 13°C taking some 10 days. The fruits were precooled directly after harvest and packed individually in polyethylene bags and arrived in Britain in good condition. However, the importer eventually stopped the shipments because the exporter found it difficult to ensure precooling and fruits were often left unprotected during the day awaiting collection. This meant many of the fruit were soft and ripe on arrival in Britain and therefore unmarketable.

Broad beans, horse beans, Windsor beans

Vicia faba, Leguminosae.

Botany

The edible portion is the flat, pale green or white seeds that are borne in pods or legumes. They have been cultivated from ancient times; they originated in the Near East and are now cultivated throughout both tropical and temperate countries.

Harvesting

Harvest maturity is between 90 and 220 days after sowing, depending on the cultivar and growing conditions. They are commonly marketed still in their pods. Maturity is when the pods are well filled, but while they are still dark green in colour and 'snap.' They are harvested by hand and are usually picked two or three times from the same plant. They may also be allowed to mature fully to produce dry beans, in which case they are machine harvested.

Storage

Refrigerated storage recommendations are as follows:

- 0–1°C and 85–95% r.h. for 2–3 weeks (Anon 1967)
- 2–4°C and 85–90% r.h. for 10–14 days (Lutz and Hardenburg 1968)
- 0–1°C and 95–100% r.h. for 2–3 weeks (Snowdon 1991).

Table 12.18 Effects of temperature and reduced O_2 level on the respiration rate (CO_2 production in mg kg^{-1} h^{-1}) of broad beans (Robinson *et al.* 1975)

	Temperature (°C)				
	0	5	10	15	20
In air	35	52	87	120	145
In 3% O_2	40		55		80

Controlled atmosphere storage

Although no specific information could be found on controlled atmosphere storage, Tomkins (1965) found that physiological disorders associated with high CO_2 included pitting. Also, it was found that low levels of O_2 resulted in a reduced respiration rate at all the temperatures tested (Table 12.18) so controlled atmosphere storage may be useful.

Broccoli, calabrese

Brassica oleracea variety *italica*, Cruciferae.

Botany

The edible portion is the terminal and axillary green heads and buds. They will freeze at about –1.8 to 1.4°C (Wright 1942) or –1.17 to 0.39°C (Weichmann 1987). Inaba *et al.* (1989) found that there was increased respiration in response to ethylene at 100 μl $litre^{-1}$ for 24 hours at 20–35°C, but the effect was much reduced at lower temperatures.

Harvesting

Harvesting is just before the flower buds open, which are cut when they are 20–25 cm long with the axillary shoots cut at the same time (Pantastico 1975). If harvesting is delayed, yellow flowers appear and they have a tough texture. Also, they can have an unpleasant flavour if they are over-developed. However, this is not likely to occur until heads are overmature and unmarketable (John Love, personal communication).

Precooling

Forced-air cooling with high humidity was the most suitable method, vacuum cooling was also suitable and air cooling in a conventional cold store and hydrocooling were less suitable (MAFF undated). After 25–30 minutes of vacuum cooling at 4.0–4.6 mmHg with a condenser temperature of –1.7 to 0°C the broccoli temperature fell from 20–22.2°C to 5.5°C (Barger 1963). Top icing was also commonly carried out (Phillips and Armstrong 1967), either in the field directly after harvest or after packing in the packhouse.

Carbon dioxide damage

Physiological disorders associated with high CO_2 include accelerated softening and off-flavours (Lipton and Harris 1974).

Storage

Their shelf-life at a simulated room temperature of 20°C and 60% r.h. was shown to be only 1–2 days (Mercantilia 1989). After protracted storage the leaves discolour, the buds drop off and the tissue becomes soft; however, refrigerated storage recommendations are as follows:

- 0°C and high humidity of about 90% (Wardlaw 1937)
- 0°C and 95% r.h. for 7–10 days, with their shelf-life reduced by half for every 5.5°C rise in temperature (Tindall 1983)
- 0°C and 90–95% r.h. for 7 days (Phillips and Armstrong 1967)
- 0°C and 90% r.h. for 1–2 weeks (Mercantilia 1989)
- 0°C and 95–100% r.h. for 10–14 days (Hardenburg *et al.* 1990)
- 0°C and 90–95% r.h. for 10–14 days (SeaLand 1991)
- 0–1°C and 95–100% r.h. for 1–2 weeks (Snowdon 1991).

Controlled atmosphere storage

Respiration rate was strongly affected by temperature, but also by the O_2 level in the storage atmosphere (Table 12.19). Also, the effect of reduced O_2 was greater the higher the storage temperature.

Table 12.19 Effects of temperature and reduced O_2 level on the respiration rate (CO_2 production in mg kg^{-1} h^{-1}) of sprouting broccoli (Robinson *et al.* 1975)

	Temperature (°C)				
	0	5	10	15	20
In air	77	120	170	275	425
In 3% O_2	65		115		215

Shelf-life extension at 10°C could be achieved with storage in 2–3% O_2 with 4–6% CO_2 (Ballantyne 1987). Other storage recommendations were as follows:

- 0°C and up to 10% CO_2 and 1% O_2 (Lipton and Harris 1974; McDonald 1985; Deschened *et al.* 1990)
- 0–5°C with 5–10% CO_2 and 1–2% O_2, which had a good effect but was of limited commercial use (Kader 1985, 1992)
- 0–5°C with 5–10% CO_2 and 1–2% O_2, which had a high level of effect (Saltveit 1989)
- 0°C with 5–10% CO_2 with 1–2% O_2 (SeaLand 1991).

Modified atmosphere packaging

Packaging in low-density polyethylene film or polybutadiene film at 0°C retained their freshness, but packaging with thicker film and subsequent storage at higher temperature induced the production of acetaldehyde, ethanol and acetic acid, more so from stem tissues than from florets (Chachin *et al.* 1999). Mineral powders incorporated into films can be used for broccoli (Industrial Material Research 1989, quoted by Abe 1990).

Brussels sprouts

Brassica oleracea variety gemmifera, Cruciferae.

Botany

The edible portion is the vegetative buds that are borne in the leaf axils of the rigid stem. Their freezing point range was given as –1.28 to 0.83°C (Weichmann 1987).

Harvesting

Harvesting is carried out when the buds are mature and tight. They can be harvested by pulling and twisting by hand or being cut with a sharp knife. Brussels sprout harvesters have been developed which perform similar operations to pea viners; harvesting the whole plant and cutting the sprouts from the stems. Special processing cultivars have been bred where all the Brussels sprouts mature at exactly the same time, so facilitating mechanical harvesting. Brussels sprouts may be stopped, by removing the terminal growing point, to ensure more uniform development (John Love, personal communication 1994, quoted by Thompson 1996).

Curing

Curing appeared not to have any affect. Heat treatment of the sprouts with moist air at 40, 45, 50 or 55°C for 0, 30, 60 or 90 minutes had no significant effects on the rate of senescence or storage quality (Wang 2000).

Precooling

Forced-air cooling with high humidity was the most acceptable method, vacuum cooling was suitable, air cooling in a conventional cold store and hydrocooling were less suitable (MAFF undated).

Storage

Prolonged storage resulted in the loss of the bright green colour, yellowing of the leaves, wilting, discoloration of the cut stems and decay (Lutz and Hardenburg 1968; Shipway 1968). Their shelf-life at a simulated room temperature of 20°C and 60% r.h. was shown to be only 1 day (Mercantilia 1989). Refrigerated storage recommendations are as follows:

- 0–1.1°C (Wardlaw 1937)
- 0°C and 90–95% r.h. for 3–4 weeks (Phillips and Armstrong 1967)
- 0–1°C and 95% r.h. for 7 days (Shipway 1968)
- 0°C and 90–95% r.h. for 3–5 weeks (Lutz and Hardenburg 1968)
- 0–1.7°C and 90–95% r.h. for 4–6 weeks (Pantastico 1975)
- –1°C and 90–95% r.h. for 20 days (Mercantilia 1989)
- 0°C for 14 days (Mercantilia 1989)
- 2°C for 8 days (Mercantilia 1989)
- 4°C for 5 days (Mercantilia 1989)
- 8°C for 3 days (Mercantilia 1989)
- 0°C and 95–100% r.h. for 3–5 weeks (Hardenburg *et al.* 1990)
- 0–1°C and 95–100% r.h. for 3–5 weeks (Snowdon 1991)
- 0°C and 90–95% r.h. for 21–35 days (SeaLand 1991).

Controlled atmosphere storage

Storage recommendations were as follows:

- 2.5, 5 or 7% CO_2 were all found to be suitable by Pantastico (1975)
- Kader (1985) and Saltveit (1989) recommended 0–5°C with 5–7% CO_2 and 1–2% O_2 which had a good effect on storage but was of no commercial use

Table 12.20 Effects of temperature and reduced O_2 level on the respiration rate (CO_2 production in mg kg^{-1} h^{-1}) of Brussels sprouts (Robinson *et al.* 1975)

	Temperature (°C)				
	0	5	10	15	20
In air	17	30	50	75	90
In 3% O_2	14		35		70

- Kader (1989) recommended 0–5°C with 5–10% CO_2 and 1–2% O_2 which had an excellent potential benefit, but again limited commercial use
- 5–7.5% CO_2 and 2.5–5% O_2 helped to maintain quality at 5 or 10°C, but at 0°C with O_2 levels below 1% internal discoloration can occur (Hardenburg *et al.* 1990)
- 0°C with 5–7% CO_2 and 1–2% O_2 (SeaLand 1991).

The respiration rate was strongly affected by temperature, but also by the O_2 level in the storage atmosphere. The effect of reduced O_2 was similar over the storage temperature range studied (Table 12.20).

Modified atmosphere packaging

Lutz and Hardenburg 1968 recommended packaging in plastic film to reduce moisture loss.

Ethylene

Exposure to ethylene can result in the rapid breakdown of chlorophyll (Richardson and Meheriuk 1982). It can also cause elongation of the internodes. This growth can open up a tight Brussels sprout giving an effect that the trade calls 'blown,' and which results in them having little commercial value.

Cabbages

Brassica oleracea, variety capitata, Cruciferae.

Botany

The edible portion is the leaves and staminal bud, which is commonly called the heart. Some cultivars do not produce a heart, e.g. spring cabbage grown in Britain. Cabbages will freeze at about –0.5 to 0.4°C (Wright 1942) or –1.22 to 0.17°C (Weichmann 1987). Inaba *et al.* (1989) found that there was an increased respiration rate in response to ethylene at 100 μl $litre^{-1}$ for 24 hours at 20–35°C, but the effect was much reduced at lower temperatures.

Harvesting

Harvest maturity is when the cabbage is fully developed so that it feels firm and solid to the touch. They are harvested by cutting the stalk just below the bottom leaves with a sharp knife. The outer leaves are trimmed, usually removing 3–6, together with any diseased, damaged or necrotic leaves. Any heads that are soft, immature, diseased or damaged by caterpillars should be discarded at harvest. Mechanical aids to harvesting involve cutting it by hand and placing it on a conveyor on a mobile packing station, in the same way as cauliflowers (see Figures 3.5, 3.6 and 12.48), which is slowly conveyed across the field (Holt and Sharp 1989).

Physiological disorders

These were reviewed by Shipway (1978):

- Black speck is where there are minute dark spots throughout the head, often associated with the stomata.
- Internal tip burn is where the margins of inside leaves turn brown, but the outer leaves look normal.
- Necrotic spot is where there are oval sunken spots a few mm across often grouped around the midribs.
- Pepper spot is where tiny black spots occur on the areas between the veins, which can increase during storage.

Physiological disorders associated with high CO_2 include internal browning (Isenberg and Sayles 1969).

Precooling

It is not normally precooled, but in an experiment with vacuum cooling after 25–30 minutes at 4.0–4.6 mmHg with a condenser temperature of –1.7 to 0°C, the cabbage temperature fell from 20–22°C to 7°C (Barger 1963). Forced-air cooling with high humidity was the most acceptable method for head cabbage, air cooling in a conventional cold store was suitable and hydrocooling and vacuum cooling less suitable. For leaf cabbage, forced-air cooling with high humidity and vacuum cooling were the most acceptable (MAFF undated).

Storage

They can be bulk stored for up to 3.5–4 m deep (Shipway 1978). Their shelf-life at a simulated room temperature of 20°C and 60% r.h. was shown to be about 20 days for Savoy (Mercantilia 1989). Refrigerated storage recommendations are as follows:

- 0°C for 3 months with 10% weight loss (Platenius *et al.* 1933)
- 0°C and 90–95% r.h. for 3–6 weeks for early cultivars or –1 to 2.2°C and 90% r.h. for others (Wardlaw 1937)
- 0°C for 2–4 months (Fidler 1963)
- 0°C and 85–90% r.h. for 2–6 months (Anon 1967)
- 0°C and 95% r.h. for up to 8 months with 1–1.5% weight loss per month and a total wastage including trimming of 20–25% (Shipway 1968)
- 1–2°C and 90–95% r.h. for up to 7 months with 4.5% weight loss per month (Sohonen 1969)
- 0–1.7°C and 92–95% r.h. for 4–6 weeks (wet season) or 12 weeks (dry season) with losses of 10% for the latter (Pantastico 1975)
- 0°C and 90–95% r.h. for 200 days (white) (Mercantilia 1989)
- 2°C for 100 days (Mercantilia 1989)
- 4°C for 75 days (Mercantilia 1989)
- 8°C for 50 days (Mercantilia 1989)
- –2°C and 90–95% r.h. for 120 days for Savoy (Mercantilia 1989)
- 0°C for 40 days for Savoy (Mercantilia 1989)
- 3°C for 20 days for Savoy (Mercantilia 1989)
- 20°C for about 5 days for Savoy (Mercantilia 1989)
- 0°C and 95–100% r.h. for 90–180 days (SeaLand 1991)
- 0–1°C and 95–100% r.h. for 3 months (green) (Snowdon 1991)
- 0–1°C and 95–100% r.h. for 6–7 months (white) (Snowdon 1991).

Controlled atmosphere storage

Respiration rate was strongly affected by temperature and to a lesser extent by the O_2 level in the storage atmosphere. Also, the effect of reduced O_2 was similar throughout the storage temperature range studied (Table 12.21). In other work the respiration rate was given as 2.7 mg CO_2 kg^{-1} h^{-1} for freshly harvested head cabbage (Weichmann 1987).

Table 12.21 Effects of temperature and reduced O_2 level on the respiration rate (CO_2 production in mg kg^{-1} h^{-1}) of Primo cabbage (Robinson *et al.* 1975)

	Temperature (°C)				
	0	5	10	15	20
In air	11	26	30	37	40
In 3% O_2	8		15		30

Storage recommendations were as follows:

- 0°C with 2.5–5% CO_2 and 5% O_2 for 5 months for Danish cultivars (Isenburg and Sayles 1969)
- 0–5°C with 3–6% CO_2 and 2–3% O_2 which had a high level of effect (Saltveit 1989).
- 2.5–5% CO_2 with 2.5–5% O_2 at 0°C (Hardenburg *et al.* 1990)
- Kader (1985) recommended 0–5°C with 5–7% CO_2 and 3–5% O_2, 0–5°C with 3–6% CO_2 and 2–3% O_2 (Kader 1992), which had a good effect and was of some commercial use for long-term storage of certain cultivars
- 5–7% CO_2 with 3–5% O_2 for green, red or Savoy (SeaLand 1991).

Caimetos, star apples

Chrysophyllum cainito, Sapotaceae.

Botany

The fruit are globose and up to 10 cm in diameter with a green or purplish tough skin containing an unpleasant-tasting latex. The pulp is white and sweet in which are embedded the small, hard, brown, glossy seeds. When the fruit is cut in cross-section the centre has a star-like appearance. It is a native of Central America. Adesina and Aina (1990) referred to the African star apple as *C. albidum*.

Harvesting

Harvest maturity is based on the colour change of the skin. Harvesting of half-mature fruit was recommended by Wardlaw (1937), then curing in a well ventilated room for 2 days before storage for up to 3 weeks at 0°C. This treatment apparently resulted in chilling injury to the fruits but it did not noticeably affect the flavour.

Storage

Adesina and Aina (1990) found that storage of the African star apple (*C. albidum*) at 5°C resulted in chilling injury and decay after 3 weeks, However, refrigerated storage recommendations for *C. cainito* are as follows:

- 2.8–5.6°C and 90% r.h. for 3 weeks (Pantastico 1975)
- 3.3°C and 90% r.h. for 21 days (SeaLand 1991).

Calamondnis, Philippine limes

Citrus mitis, Rutaceae.

Botany

The fruit is a berry or hesperidium, and being a citrus is non-climacteric. It has a bright orange skin when mature and is 2–3 cm in diameter. The peel is thin and easily removed and the segments are juicy with a unique flavour.

Storage

Refrigerated storage recommendations are as follows:

- 8.9–10°C and 90% r.h. for 2 weeks with 6.5% weight loss (Pantastico 1975)
- 8.9°C and 90% r.h. for 14 days (SeaLand 1991).

Canistel, egg fruit

Pouteria campechiana, *P. lucuma*, *Lucuma nervosa*, Sapotaceae.

Botany

The tree is evergreen, attractive and fruitful on a range of soils from clay to limestone and tolerant of drought. The fruit are orange–yellow when mature and pointed, globose or ovoid in shape and 5–12 cm long. The sweet pulp is also orange and mealy in texture and has up to three dark brown shiny seeds. It is usually eaten raw. It is a native of northern South America and it grows wild and under cultivation in southern Mexico, Central America, the West Indies, the Bahamas and, to a limited extent, in southern Florida (Morton 1983).

Pest control

At 1°C, 14 days would be needed to achieve quarantine security to control Caribbean fruit fly (*Anastrepha suspensa*), and at 3°C a minimum of 15 days would be required. Hot water treatment at 46°C for 90 minutes or at 48°C for 65 minutes resulted in the development of dark blotches on the peel and a 2–3 mm thick layer under the peel that did not soften (Hallman 1995).

Storage

The fruits were said to keep well (Morton 1983) and storage at 1 or 3°C for 17 days did not cause appreciable loss of fruit quality compared with fruit stored at 10°C (Hallman 1995). In contrast, SeaLand (1991) recommended 12.8°C and 85–90% r.h. for 21 days.

Cape gooseberries, physalis, Peruvian cherry

Physalis peruviana, Solanaceae.

Botany

The fruit is a berry that is yellow when mature and is globular in shape and up to 2 cm in diameter. The fruit is enclosed in a brown bladder-like calyx (Purseglove 1968). It is a native of South America but has become an important fruit in South Africa, hence one of its common names.

Harvesting

Optimum storage was when fruits were harvested at the maturity stage with green pedicels (Lizana and Espina 1991). Fruits were harvested by Sarkar *et al.* (1993) 45 days (mature green), 50 days (ripening) or 55 days (fully ripe) post-anthesis. They found that the mature green fruits retained much more acidity during storage, whereas ripening fruits retained appreciable amounts of ascorbic acid during storage. For long-term storage they found that fruits should be harvested at the ripening stage.

Diseases

Fungal infection by *Alternaria*, *Botrytis*, *Cladosporium* and *Penicillium* spp. occurred in storage, especially at 7°C (Lizana and Espina 1991).

Storage

Their shelf-life at a simulated room temperature of 20°C and 60% r.h. was shown to be 1–2 weeks

(Mercantilia 1989). Refrigerated storage recommendations are as follows:

- 14°C and 80% r.h. for 1–2 months (Mercantilia 1989)
- 14°C and 80% r.h. (Snowdon 1990)
- 4°C and 90–95% r.h. for 72 days for fruits with the calyx attached (Fischer *et al.* 1990)
- 12.2°C and 80% r.h. (SeaLand 1991)
- 0°C for 33 days before keeping them at 7 days at 18°C (Lizana and Espina 1991).

Capsicums, sweet peppers, bell peppers

Capsicum annum variety *grossum*, Solanaceae.

Botany

The fruit is an indehiscent many-seeded berry and is conical or globose in shape, inflated with a basal depression. The fruit is classified as non-climacteric. They have a mild flavour and are green as they develop and change colour to various shades or red or yellow as they mature. Green peppers will freeze at about –1.2 to 0.9°C (Wright 1942) or –0.94 to 0.61°C (Weichmann 1987). Inaba *et al.* (1989) found that there was increased respiration in response to ethylene at 100 μl $litre^{-1}$ for 24 hours at 20–35°C, but the effect was much reduced at lower temperatures.

Harvesting

The increase in size as they are growing is frequently used to determine when they should be harvested, which may simply be related to the market requirement. Most fruit are still green at harvest and therefore not physiologically mature. However, fully mature fruit, those which are red or yellow, are also harvested and it was found that fruit of this maturity lost less weight during storage than immature fruit, that is, those which were still green (Lutz and Hardenburg 1968). Allowing fruits to mature fully before harvesting, that is, leaving them longer on the plants, results in lower yields per plant.

Coatings

Chitosan has been successfully used as a postharvest coating of peppers to reduce water loss and maintain their quality (El-Ghaouth *et al.* 1991).

Diseases

In a survey of nine capsicum crops over four seasons the major cause of wastage was soft rots caused by *Botrytis cinerea* and *Rhizoctonia carotae* (Geeson *et al.* 1983, 1988). Infection levels of *B. cinerea* were related to weight loss, that is, the fruit with a high weight loss were more susceptible (Tronsmo 1989). Storage above 10°C encourages bacterial soft rots (Lutz and Hardenburg 1968).

Physiological disorders

Physiological disorders associated with high CO_2 include internal browning (Morris and Kadar 1977). Sunscald resulted in light-coloured soft areas on the skin. The disorder may be latent and develop during storage or marketing as water-soaked areas eventually becoming dry and brown.

Precooling

Capsicums that were hydrocooled, which took about 10 minutes (see Figure 4.3), had a higher incidence of rots during subsequent storage than those that were not hydrocooled, even when chlorine was added to the water (Hughes *et al.* 1981). This was due to water being trapped between the fruit and the calyx. Forced-air cooling with high humidity was the most acceptable method, air cooling in a conventional cold store was suitable and hydrocooling and vacuum cooling less suitable (MAFF undated).

Chilling injury

Chilling injury occurred in 5 days at 0°C as surface pitting followed by decay, which only became manifest when fruits were transferred to room temperatures (Phillips and Armstrong 1967). Chilling injury has also been reported to occur within a few hours after removal from at storage temperatures below 7.2°C as pitting and surface discoloration near the calyx (Anon 1968).

Red bell pepper fruits of the cultivar Maor that had been dipped in methyl jasmonate at 10 μM for 30 seconds then stored for 4–10 weeks at 2°C had reduced severity of chilling injury symptoms and a lower percentage of injured fruits compared to those not treated (Meir *et al.* 1996).

Storage

A major factor that influences storage life is weight loss, which results in shrivelling and loss of their crisp texture. Their shelf-life at a simulated room temperature of 20°C and 60% r.h. was shown to be only 2 days

(Mercantilia 1989). In spite of the clear evidence of chilling injury reported above, several researches make storage recommendations at very low temperatures. It is difficult to justify these and it may be that the research did not include observations of the fruit after removal from storage. However, refrigerated storage recommendations are as follows:

- 10°C and 16 days or 4.4°C for 28 days or 0°C for 40 days (Platenius *et al.* 1933)
- 0°C and 90% r.h. for 32 days (Wardlaw 1937)
- 0°C and 90–95% r.h. for 1–3 weeks (Wardlaw 1937)
- 7–10°C and 85–90% r.h. for 8–10 days (1, Phillips and Armstrong 1967)
- 7.2–10°C and 90–95% r.h. for 2–3 weeks (Lutz and Hardenburg 1968)
- 10°C and 75% r.h. for 5–8 days with 1.9–2.2% loss (Heydendorff and Dobreanu 1972)
- 21°C and 65% r.h. for 2 days with 1.4–2.2% loss (Heydendorff and Dobreanu 1972)
- 7.2°C and 85–90% r.h. for 3–5 weeks with 7.1% loss (green) (Pantastico 1975)
- 5.6–7.2°C and 90–95% r.h. for 2 weeks (ripe) (Pantastico 1975)
- 5–7°C and 95% r.h. for up to 14 days (Tindall 1983)
- 0.5°C and 97–98% r.h. for 40 weeks in an ice bank store (Geeson *et al.* 1988)
- 2–2.5°C and 90–95% r.h. in a conventional refrigerated store for a few months before becoming flaccid (Geeson *et al.* 1983, 1988)
- 15°C (Umiecka 1989)
- 8°C and 90–95% r.h. for 14 days (Mercantilia 1989)
- 5 or 10°C for 10 days (Mercantilia 1989)
- 4 or 14°C for 6 days (Mercantilia 1989)
- 1°C for 2 days (Mercantilia 1989)
- 7–13°C and 90–95% r.h. for 2–3 weeks (Hardenburg *et al.* 1990)
- 7–10°C and 95–100% r.h. for 1–2 weeks (Snowdon 1991)
- 10°C and 90–95% r.h. for 12–18 days (SeaLand 1991).

Controlled atmosphere storage

Storage in low O_2 had more effect on reducing respiration rate at temperatures higher than at 0°C (Table 12.22).

High levels of CO_2 could result in internal browning (Morris and Kadar 1977). Saltveit (1989) recommended 8–12°C with 0% CO_2 and 2–5% O_2, which had only a slight effect. Otma (1989) showed that at 8°C and >97% r.h. the optimum conditions for storage were 2% CO_2 with 4% O_2. Storage with O_2 levels of 3–5% retarded the respiration rate but high CO_2 reduced the loss of green colour but also resulted in calyx discoloration (Hardenburg *et al.* 1990). SeaLand (1991) recommended 0–3% CO_2 with 3–5% O_2. Kader (1985 and 1992) recommended 8–12°C with 0% CO_2 and 3–5% O_2, which had a fair effect but was of limited commercial use. The maximum storage time of California Wonder at 8.9°C in air was 22 days, but in 10% CO_2 with 5% O_2 it was 38 days (Pantastico 1975).

Table 12.22 Effects of temperature and reduced O_2 level on the respiration rate (CO_2 production in mg kg^{-1} h^{-1}) of green peppers (Robinson *et al.* 1975)

	Temperature (°C)				
	0	5	10	15	20
In air	8	11	20	22	35
In 3% O_2	9		14		17

Modified atmosphere packaging

Lutz and Hardenburg (1968) found that fruits in perforated polyethylene film bags in cold storage had double the storage life of those stored not in bags. Hughes *et al.* (1981) showed that those stored sealed in various plastic films had a higher percentage of marketable fruit than those stored in air (Table 12.23).

Table 12.23 Effects of controlled atmosphere storage, plastic film wraps and hypobaric storage on the mean percentage of sound fruit after 20 days at 8.8°C followed by 7 days at 20°C (source: Hughes *et al.* 1981)

Storage	Sound fruit (%)		
Air control	43		
Controlled atmosphere storage			
	CO_2 (%)	O_2 (%)	
	0	2	42
	3	2	39
	6	2	21
	9	2	29
Plastic wrap			
Sarenwrap	67		
Vitafilm PWSS	57		
Polyethylene Pe301	69		
VF 71	63		
Hypobaric Storage (mmHg)			
152	46		
76	36		
38	42		

Lee and Lee (1996) used low-density polyethylene film of 27 μm thickness for the preservation of a mixture of carrot, cucumber, garlic and green peppers. The steady-state atmosphere at 10°C inside the bags was 5.5–5.7% CO_2 and 2.0–2.1% O_2.

Hypobaric storage

Capsicums stored at 8.8°C at either 150, 76 or 38 mmHg did not have an increased subsequent storage life compared with those stored in air under the same conditions (Hughes *et al.* 1981) (Table 12.23). The hypobaric-stored capsicums also had a significantly higher weight loss during storage than those stored at normal pressure.

Transport

Transport life could be extended in controlled atmospheres of 8.9°C with 2–8% CO_2 and 4–8% O_2 (Pantastico 1975).

Carambola, star fruit

Averrhoa carambola, Oxalidaceae.

Botany

They are oval or elliptical fleshy berries with distinct ridges giving them a star shape in cross-section and are up to about 15 cm long (Figure 12.43, in the colour plates). They vary in weight from <80 g to >200 g (Campbell 1989). The skin is shiny green as it develops and a bright yellow when fully mature. The flesh is crisp, juicy and aromatic and contains many small seeds. From experiments, O'Hare (1993) showed that they appeared to be non-climacteric fruits.

Harvesting

Commercial harvesting is at colour break (Figure 12.44) to reduce susceptibility to mechanical injury during handling (O'Hare 1993). Fruits harvested 10 or 11 weeks after fruit set could be stored for about 2 weeks at 5 or 10°C before disease began to develop, but conversely fruits harvested 12 or 13 weeks after fruit set could be stored successfully for 4 weeks at 5 or 10°C (Osman and Mustaffa 1996). Fruit development in Florida takes 60–70 days to reach harvest maturity. Fruits harvested too early tend to brown severely during storage (Campbell 1989). Harvesting is carried out entirely by hand to reduce physical damage to the fruits (Osman and Mustaffa 1996).

Figure 12.44 Carambola fruit from two harvests.

Pest control

They require quarantine treatment for control of the Caribbean fruit fly (*Anastrepha suspensa*) prior to shipment to the USA and some other countries. Low-dose irradiation (1.0 kGy) was shown to be effective for protecting against fruit flies, but carambolas are susceptible to irradiation-induced peel injury. In experiments, low-dose gamma irradiation treatment generally reduced fruit quality, but packaging fruit in clamshell polystyrene containers rather than conventional fibreboard boxes, prior to treatment, mitigated the effects on quality. Use of clamshell containers reduced peel pitting, stem-end breakdown, shrivelling and weight loss after storage at 5 or 7°C for 14 days. Fruits held in clamshell containers were also firmer, with slightly less green peel and had lower total soluble solids contents than fruits stored in fibreboard boxes, but their flavour was not as good (Miller and McDonald 1998). Miller and McDonald (2000) investigated hot water and vapour heat treatments for fruit fly control.

Precooling

Miller and McDonald (2000) found that cooling the fruits with iced water greatly reduced fruit quality compared with ambient air, ambient water and refrigerated air treatments. Refrigerated air treatment at the targeted storage temperature of 10°C resulted in the least fruit damage. Refrigerated air, ambient air and ambient water treatments resulted in similar degradation, including peel pitting, bronzing, decay and weight loss. However, ambient air fruits were the least firm and ambient water fruits had the lowest flavour quality. Carambolas cooled with refrigerated air had the least injury compared with fruits cooled with the other methods.

Chilling injury

Chilling injury symptoms appeared after 5 weeks of storage at 5°C on unripe fruit but not in ripe fruits (Wan and Lam 1984).

Storage

Sugar levels remained constant during storage, although fruits will continue to lose chlorophyll and synthesize carotenoids after harvest. Acidity can decline during storage, and this is often undesirable as it can be associated with a bland flavour. No storage decay was detected in ripe and unripe fruits held at 5°C for 12 weeks (Wan and Lam 1984). Their shelf-life at a simulated room temperature of 20°C and 60% r.h. was shown to be 4–5 days (Mercantilia 1989). Refrigerated storage recommendations are as follows:

- 5°C for 3 weeks with browing, including stem-end browning, being the main limiting factor in storage (Campbell 1989)
- 6°C and 90% r.h. for 3 weeks (Mercantilia 1989)
- 10°C for the fresh market fruits and 5°C for processing (Sankat and Balkissoon 1990).
- 5–10°C and 90% r.h. for 3–4 weeks (Snowdon 1990)
- 8.9°C and 85–90% r.h. for 21–28 days (SeaLand 1991)
- 5°C for 6 weeks (O'Hare 1993)
- 10°C and 85–88% r.h. (Osman and Mustaffa 1996).

Controlled atmosphere storage

Storage at 7°C and 85–95% r.h. with either 2.2% O_2 and 8.2% CO_2 or 4.2% O_2 and 8% CO_2 resulted in low losses of about 1.2% in 1 month and fruit retained a bright yellow colour. Controlled atmosphere stored fruit also had good retention of firmness, °Brix and acidity compared with fruits stored in air (Renel and Thompson 1994).

Modified atmosphere packaging

Unripe fruits in sealed 0.04 mm thick polyethylene bags remained green during storage at 5°C but turned yellow after the bags were opened and the fruit exposed to 20°C for 9 days (Wan and Lam 1984).

Carrots

Daucus carota subspecies *sativus*, Umbelliferae.

Botany

The edible portion is the modified swollen taproot. It will freeze at about –1.4 to 1.3°C (Wright 1942) or –1.83 to 1.39°C (Weichmann 1987). Inaba *et al.* (1989) found that there was increased respiration in response to ethylene at 100 μl litre^{-1} for 24 hours at 20–35°C, but the effect was much reduced at lower temperatures. The definitions for carrots and sweet potatoes were changed in the European Union Jam and Similar Products Regulations. The regulation stated that from January 1991, carrots and sweet potatoes are 'sometimes fruit and not vegetables.' The reason given for this was that in Portugal they may be used in the manufacture of jam and European Union regulations specify that jams can only be made from real fruit. Carrots and sweet potatoes therefore had to be classified as fruits when used for jam and as vegetables when used for other purposes.

Preharvest factors

Splitting during growth can affect postharvest losses. The incidence of damage in carrots was shown to be affected by the total amount of irrigation and the time when it was applied. Heavy irrigation during the first 90 days after drilling resulted in up to 20% growth splitting, while minimal irrigation for the first 120 days followed by heavy irrigation resulted in virtually split-free carrots with a better skin colour and finish and only a small reduction in yield (McGarry 1993).

Harvesting

They are usually harvested when they reach a size that is acceptable to the market. In Britain they may be stored in the ground and harvesting them as required (Fidler 1963).

Precooling

Hydrocooling was the most suitable method, forced-air cooling with high humidity was also suitable, air cooling in a conventional cold store was less suitable and vacuum cooling was generally unsuitable (MAFF undated). However, with carrots marketed with their tops attached vacuum cooling was also suitable. After 25–30 minutes of vacuum cooling at 4.0–4.6 mmHg with a condenser temperature of –1.7 to 0°C the temperature of the carrot tops fell to 2°C from 20–22.2°C (Barger 1963).

Storage

Hydrolysis of sucrose, which affects flavour and disease, was lowest at 0°C compared with higher temperatures (Hasselbrink 1927). They used to be stored in field clamps in a similar way to potatoes (Fidler 1963, Lutz and Hardenburg 1968). Storage in bulk containers of 0.5–1 tonne capacity was recommended, but storage in bags can restrict the airflow passed the carrots (Shipway 1978). Their shelf-life at a simulated room temperature of 20°C and 60% r.h. was shown to be only 7 days for topped mature carrots (Mercantilia 1989). Refrigerated storage recommendations are as follows:

Bunched immature with tops attached

- 0°C and 90–95% r.h. for 4 weeks with 25% loss (Pantastico 1975)
- 0°C and 95% r.h. for 10–14 days (Anon 1968)
- 0°C and 95–100% r.h. for 21 days with about 15% weight loss (Tindall 1983)
- 0°C and 95–100% r.h. for 2 weeks (Hardenburg *et al.* 1990)
- 0–1°C and 95–100% r.h. for 2–4 weeks (Snowdon 1991).

Topped immature

- 0–1°C and 95–100% r.h. for 4–6 weeks (Snowdon 1991).

Topped mature

- 0–1.7°C for 22 weeks with 7% weight loss (Hasselbrink 1927)
- 0°C and over 90% r.h. for 5 months with 10% weight loss (Wardlaw 1937)
- 0–1.1°C and 79% r.h. (Wardlaw 1937)
- 0°C and 95% r.h. for 20–24 weeks with 20–35% loss (Pantastico 1975)
- –1 to 1°C and 90–95% r.h. for 6–7 months (Shipway 1968)
- 0°C and 90–95% r.h. for 4–5 months (Lutz and Hardenburg 1968)
- 0°C and 95% r.h. for 4–5 months (Anon 1968)
- 0°C and 95–100% r.h. for 100–150 days or longer with 20–35% weight loss (Tindall 1983)
- 0°C and 90–95% r.h. for 180 days (Mercantilia 1989)
- 2°C for 100 days (Mercantilia 1989)
- 8°C for 50 days (Mercantilia 1989)
- 0–1°C and 98–100% r.h. for 7–9 months (Hardenburg *et al.* 1990)
- 0°C and 95–100% r.h. for 28–180 days (SeaLand 1991)
- 0–1°C and 95–100% r.h. for 4–8 months (Snowdon 1991).

Controlled atmosphere storage

Respiration rate was strongly affected by temperature, but also by the O_2 level in the storage atmosphere (Table 12.24). The effect of reduced O_2 was about the same at all the storage temperatures.

Storage at 2°C in 1–2% O_2 for 6 months was reported to have been successful (Platenius 1934). However, increased decay was reported at 0°C and 95% r.h. in atmospheres of 6% CO_2 with 3% O_2 compared with storage in air (Pantastico 1975). Also, controlled atmosphere storage was not recommended by Hardenburg *et al.* (1990) since carrots stored in 0–1°C and 98–100% r.h. with 5–10% CO_2 and 2.5–6% O_2 had more mould growth and rotting than those stored in air. Fidler (1963) found that high levels of CO_2 in the storage atmosphere could give the roots a bitter flavour.

Modified atmosphere packaging

Lee and Lee (1996) used low-density polyethylene film of 27 μm thickness for the preservation of a mixture of carrot, cucumber, garlic and green peppers. The

Table 12.24 Effects of temperature and reduced O_2 level on the respiration rate (CO_2 production in mg kg^{-1} h^{-1}) of carrots (Robinson *et al.* 1975)

	Temperature (°C)				
	0	5	10	15	20
In air	13	17	19	24	33
In 3% O_2	7		11		25

steady-state atmosphere at 10°C inside the bags was 5.5–5.7% CO_2 and 2.0–2.1% O_2.

Ethylene

Ethylene in the storage atmosphere was shown to increase their respiration rate (Sarkar and Phan 1979). Carrots exposed to ethylene levels as low as 1% can produce isocoumarin, which gives them a bitter flavour (Fidler 1963; Chalutz *et al.* 1969). Isocoumarin content was shown to increase with increasing concentration of ethylene over the range 0.5–50 μl $litre^{-1}$ (Lafuente *et al.* 1989). Exposure to 0.1 μl $litre^{-1}$ ethylene for 30 days at 5°C resulted in little isocoumarin formation whereas exposure to 0.5 μl $litre^{-1}$ for 14 days resulted in over 20 mg per 100 g in the peel (Lafuente *et al.* 1989).

Cashew apples

Anacardium occidentale, Anacardiaceae.

Botany

The main product of the tree is the nut that produces the edible kernel, the cashew nut. Cashew nut shell oil is also extracted from the shell of the nut. The portion of the fruit which is called the apple is actually the pedicel or swollen fruit stalk. The nut on the end of the pedicel is actually the true fruit. The apple was classified as a non-climacteric fruit by Biale and Barcus (1970). Pratt and Mendoza (1980) also showed that the pattern of ethylene production confirmed that the cashew apple was non-climacteric. The cashew apple grows very slowly until the nut matures (Figure 12.45, in the colour plates). It then grows very rapidly and ripens (Thompson 1968). Removal of the nut from the receptacle initiated rapid growth of the cashew apple and earlier maturation.

Prestorage treatments

The lowest incidence of microbial decay and weight loss occurred in cashew apples treated with 1% mustard oil, in which the storage limit was 6 days (i.e. some fruits still not rotten), compared with only 2 days for controls that were dipped in distilled water. In all treatments, the fruit soluble solids content peaked after about 2 days of storage, and then declined, and the fruit ascorbic acid content decreased throughout storage (Narayan *et al.* 1993).

Storage

Refrigerated storage recommendations are as follows:

- 0–1.7°C and 85–90% r.h. for 5 weeks with 22% loss (Singh and Mathur 1963; Pantastico 1975)
- 0°C and 85–90% r.h. for up to 35 days (SeaLand 1991).

Modified atmosphere packaging

Narayan *et al.* (1993) suceaafully stored the cultivar VTH-30 in perforated white polyethylene bags at room temperature of about 28°C and 66–75% r.h.

Cassava, monioc, tapioca, yuca

Manihot esculenta, Euphorbiaceae.

Botany

The edible portion is the swollen root or root tuber as it is sometimes called. Its origin is in Latin America where it has been cultivated for thousands of years. It now forms a starch staple throughout the tropics. They develop by secondary thickening of adventitious roots close to the stem. There are normally up to 10 tubers per plant, which are cylindrical or tapering, up to 15 cm in diameter and up to 1 m long (Ingram and Humphries 1972; Kay 1987). They have a brown outer skin above a white or pink inner skin, with white flesh. Hydrocyanic acid can be present in the roots, but this varies with cultivar. For consumption as a vegetable, after boiling or roasting, 'sweet' clones are used, e.g. the cultivar Chirosa Armenia in Colombia.

Vascular streaking

Roots can deteriorate within a day or so of harvesting owing to a physiological disorder called vascular streaking (Thompson and Arango 1977). The vascular bundles turn a bluish black colour and can be seen as points across the root when it is cut at right-angles (Figure 12.46). A low susceptibility to vascular streaking was correlated with a low dry matter content and a high sugar content (van Oirschot *et al.* 2000). Methods of control were reviewed by Beeching (1998) and included curing and cutting off the stem 20–30 cm above the ground 2–3 weeks before harvesting. In experiments, six cassava cultivars were harvested 0–39 days after pruning by van Oirschot *et al.* (2000). After pruning, the susceptibility for all cultivars tested was drastically

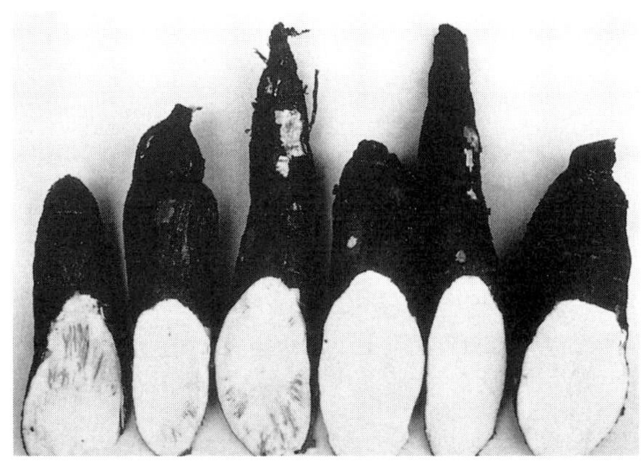

Figure 12.46 Cassava roots cut to reveal vascular streaking on the three left-hand side ones. Source: Dr R.H. Booth.

reduced, reaching a minimum of around 25% of the original value for a pruning–harvest interval of up to 25 days. With a longer interval the plants slowly developed a new leaf canopy, normal assimilation recommenced and starch content increased. Storage for 9 days at 5°C inhibited vascular streaking but its incidence increased with increasing temperature over the range of 10–25°C (Akhimienho 1999). The sugar content, i.e. the sugar:starch ratio, of cassava roots is positively correlated with their resistance to postharvest physiological deterioration (van Oirschot *et al.* 2000).

Harvesting

In Trinidad, Wickham and Wilson (1988) harvested roots the cultivars White Stick, Black Stick and M Col.22 from 11-month-old plants. However, harvesting can be delayed for several months. Roots left unharvested in the field maintained acceptable cooking quality for 20 weeks after the normal harvest time (Wickham and Wilson 1988). Because of its shape, cassava is difficult to harvest without damaging the swollen roots. The roots are commonly just pulled from the ground by the stem but this can result in damage, especially when the soil is dry and hard, resulting in rapid postharvest deterioration. In some cases the soil is carefully dug away from each swollen root individually to avoid damaging them.

Cassava harvesters for processing are common but the roots must be processed quickly after harvest to avoid deterioration. In the 1970s, the Centro Instituto de Agricultura Tropical in Colombia developed a cassava harvester for the fresh market. It consisted of equipment towed behind a tractor that gripped the stem and at the same time undercut the roots so that they were pulled from the soil. In tests it worked well on flat land when the soil was moist, but it was ineffective on sloping, dry soils. Odigboh (1991) described a cassava harvester that could remove the crop from the ground with minimum damage to the roots when they were grown in ridges. It operated at a theoretical rate of 0.25–0.4 ha h^{-1} depending on the soil type and field conditions. It had two gangs of reciprocating power take-off drivers that dug two opposite sides of the ridge from the furrow bottom to remove the cassava roots.

Handling

In Colombia, cassava is packed into sacks containing 75 kg. Because these sacks are so heavy to carry they are borne on workers' backs. When loading lorries they are dropped with some considerable force from the workers' backs to the bed of the lorry. A simple solution would be to reduce the pack size to say 25 kg. This, however, proved unacceptable to the trade and other methods of reducing wastage had to be investigated (Thompson 1978). Twenty-five years later the same method is being used with similar levels of damage.

Curing

Some work in Colombia (Booth 1975) indicated that cassava could be successfully cured by exposure to

Table 12.25 Effects of time and temperature at 95% r.h. on the number of layers of cells that developed during curing of 2 mm deep cuts at 25 or 35°C in cassava roots (source: Akhimienho 1999)

	Days after damage									
	1		2		4		8		10	
	25°C	35°C	25°C	35°C	25°C	35°C	25°C	35°C	25°C	35°C
Lignified	0	0	1.5	1	2.5	2.5	4.5	4	4.5	4
Suberized	0	0	0	0	0.5	0.5	1	1	3	3

25–40°C and 80–85% r.h. Suberization was said to occur in 1–4 days with periderm 3–5 days later. In a subsequent study by Akhimienho (1999), it was shown that curing of tissue damaged by cutting was related to the depth of cut. Shallow wounds 2 mm deep (Table 12.25) could be cured, but wounds 8–10 mm deep showed no curing over the 10 day period.

Detoxification

The roots can contain the glucoside linamarin, which is converted to hydrogen cyanide by the enzyme linamarase. The levels of cyanide in 28 cassava cultivars were determined in Fiji and ranged from 1.4–12.1 mg per 100 g of fresh peeled tuber (Aalbersberg and Limalevu 1991), which seems high. Removal of the cyanogenic glucosides that occur especially in the clones grouped as 'bitter cassava' is important since they are highly toxic. Grating followed by fermentation and sometimes heating on a flat plate normally remove them. Gari, a staple fermented meal in West Africa, is prepared in this way. Washing in constant running water or repeated boiling in changes of water is also used.

Waxing

Coating roots with paraffin wax (Figure 12.47) kept fresh cassava in good condition and prevented the development of vascular streaking for 1–2 months at room temperature in Bogotá in Colombia (Young *et al.* 1971). This coating treatment is currently applied to roots exported from Costa Rica where the following procedure was successfully developed:

1. Harvest the roots with care to avoid damage.
2. Wash thoroughly directly after harvest to remove all the surface soil.
3. Dry in an oven at 50°C for 2 hours.
4. Immerse the roots, while still hot, in paraffin wax at 62°C (at 64°C and over fumes are produced by the wax that can detrimentally affect the workers).
5. Allow to cool.
6. Transport in boxes at 12°C.

If the roots are immersed in paraffin wax without the 2 hour drying, the wax does not bind closely with the surface of the root and it may crack and the wax will be dislodged during subsequent handling.

Figure 12.47 Waxed cassava in UK imported from Costa Rica.

Storage structures

In situ storage effectively means delaying the harvest of the crop until it is required, but it does mean that the land where the crop was grown remains occupied and a new crop cannot be planted there. The crop may also be exposed to pest and disease attack. It was shown that delayed harvest could result in reduced acceptability and starch content and the risk of preharvest losses (Rickard and Coursey 1981). In Uganda, fresh cassava was stored mainly by reburying or placing it in water for a few days (Ameny 1990).

Clamps were also developed for cassava storage in Colombia (see Figure 7.3), where successful storage for over 8 weeks was reported (Booth 1977). Extra straw covering may be needed in hot climates (CIP 1981). Cassava stored in clamps for 1 month had 75% of the original weight in marketable condition with additional losses on storage for up to 3 months being usually only small (Booth 1977).

Wickham and Wilson (1988) dipped roots in benomyl fungicicde and stored them in moist storage in polyethylne-lined boxes filled with sand and/or sawdust. Vascular streaking occurred if roots were removed from moist storage and placed under ambient conditions, unless moist storage was continued for 11–14 weeks depending on cultivar. Callus, adventitious roots and secondary adventitious roots grew at the distal end of all stored cultivars. Some roots developed a spongy appearance and large cavities under moist storage and especially on removal to ambient conditions. Cooking quality was maintained in all stored roots for the first 6 weeks but a considerable decline in quality was evident after 8–11 weeks depending on cultivar.

Storage conditions

Reilly *et al.* (2001) showed that the rapid postharvest deterioration was related to oxidative processes. In a comparison between highly susceptible and less susceptible cultivars, they suggested that high levels of catalase activity may play a role in delaying the deterioration response. Their shelf-life at a simulated room temperature of 20°C and 60% r.h. was shown to be 2–4 weeks (Mercantilia 1989). Refrigerated storage recommendations are as follows:

- 1.7–3.3°C and 80–90% r.h. for 2 weeks with 11% loss, but they also had internal browning and mould growth (Singh and Mathur 1953a)
- 13–14°C for 9 days (Puerto Rico) (Burton 1970)
- 3°C gave losses of 6–7% per week (Ingram and Humphries 1972)
- 0°C and 85% r.h. for 23 weeks (Pantastico 1975)
- 0–2°C and 85–90% r.h. for several months (Tindall 1983)
- 1°C for 90% r.h. for 5–6 months (Mercantilia 1989)
- 0–2°C and 85–90% r.h. for 5–6 months (Snowdon 1991)
- 10°C and 90–95 for 10–14 days (SeaLand 1991).

Controlled atmosphere storage

Akhimienho (1999) found that storage in 0% O_2 with 40–65% CO_2 and the balance nitrogen at all temperatures within the range 5–35°C inhibited both vascular streaking and mould growth over the 10-day experimental period. She also found that 21% O_2 with 10–20% CO_2 also inhibited vascular streaking and mould growth, but low O_2 storage (1 or 5%) had no apparent effect.

Modified atmosphere packaging

Averre (1971) found that wrapping the roots in moist paper and then sealing them in polyethylene film bags prevented vascular streaking during an 8-day storage period at either 10 or 40°C with only slight development at 25°C. Dipping the roots in a fungicide (benomyl or thiabendazole) and packing them in plastic film directly after harvest (Thompson and Arango 1977) reduced vascular streaking during storage for 8 days at 22–24°C compared with roots stored unwrapped. However, the use of the fungicides benomyl and thiabendazole is now restricted in many countries.

Cauliflower

Brassica oleracea variety *botrytis*, Cruciferae.

Botany

The edible portion is the curd, which is made up of the abortive flowers on hypertrophied branches. They will freeze at about −1.2 to 1.1°C (Wright 1942) or −1.17 to 0.94°C (Weichmann 1987). Inaba *et al.* (1989) found that there was increased respiration in response to ethylene at 100 μl $litre^{-1}$ for 24 hours at 20–35°C, but the effect was much reduced at lower temperatures.

Harvesting

Harvesting should take place when the curds have reached an acceptable size, but before the flower stalks elongate and become discoloured, ricy or leafy. If harvesting is delayed they can have a tough texture and an unpleasant flavour when they are over developed. However, this is not likely to occur until heads are considerably overmature and unmarketable (John Love, personal communication).

The plants are cut just below the lowest leaf and they should be transported to the store or packhouse without pruning the outer leaves as they protect the curd from damage. In practice, the outer leaves are cut to just above the curd and they are loaded into crates for transport to the packhouse or packed directly into boxes for marketing. Mechanical aids to harvesting involve cutting it by hand and placing it on a conveyor on a mobile packing station, which is slowly conveyed across the field (Figure 12.48) (Holt and Sharp 1989).

Figure 12.48 Cauliflower harvester in operation in the UK. See also Figures 3.4 and 3.5.

Precooling

Forced-air cooling with high humidity was the most suitable method, vacuum cooling, air cooling in a conventional cold store and hydrocooling were all less suitable (MAFF undated).

Storage

Storage is commonly in 0.5 tonne slatted boxes, which are covered with polyethylene film after cooling to maintain humidity. If they lose excess moisture (more than about 5%), they can have a wilted appearance and the curds may be rubbery. Their shelf-life at a simulated room temperature of 20°C and 60% r.h. was shown to be only 2 days (Mercantilia 1989). Refrigerated storage recommendations are as follows:

- −1°C and 90% r.h. for 3 months (Wardlaw 1937)
- 0–1°C and 85–90% r.h. for 3–6 weeks (Anon 1967)
- 0°C and 90–95% r.h. for 2–4 weeks (Phillips and Armstrong 1967; Anon 1968; Hardenburg *et al.* 1990)
- 1°C and 90–95% r.h. for up to 3 weeks (Shipway 1968)
- 0–1.7°C and 85–90% r.h. for 7 weeks with 30.4% loss for Snowball (Pantastico 1975)
- 0–1°C and 95% r.h. for up to 28 days with about 30% weight loss (Tindall 1983)
- 4°C for approximately 7 days (Tindall 1983)
- 0°C and 90–95% r.h. for 42 days (Mercantilia 1989)
- 2°C for 32 days (Mercantilia 1989)
- 4°C for 18 days (Mercantilia 1989)
- 8°C for 8 days (Mercantilia 1989)
- 0–1°C and 85–90% r.h. for 2–3 weeks (Hardenburg *et al.* 1990)
- 0°C and 90–95% r.h. for 20–30 days (SeaLand 1991)
- 0–1°C and 95–100% r.h. for 2–4 weeks (Snowdon 1991).

Controlled atmosphere storage

Respiration rate was strongly affected by temperature, but also by the O_2 level in the storage atmosphere (Table 12.26), but the effect of reduced O_2 was greater at 20°C.

Table 12.26 Effects of temperature and reduced O_2 level on the respiration rate (CO_2 production in mg kg^{-1} h^{-1}) of April Glory cauliflowers (Robinson *et al.* 1975)

	Temperature (°C)				
	0	5	10	15	20
In air	20	34	45	67	126
In 3% O_2	14		45		60

Considerable work has been done in the Netherlands (originally in the Sprenger Institute) since the 1950s. This work has shown very little benefit of controlled atmosphere storage, but storage at 0–1°C and >95% r.h. with 5% CO_2 and 3% O_2 gave a better external appearance than cold sorage alone but had no effect on curd quality (Mertens and Tranggono 1989). The respiration rate was decreased in 3% O_2 compared with storage in air (Romo-Parada *et al.* 1989). Other recommendations are:

- 0°C with 10% CO_2 and 10% O_2 (Wardlaw 1937)
- 10% CO_2 and 11% O_2 for 5 weeks (Smith 1952)
- 4.4 or 10°C with 5% CO_2, but some injury was evident after the curds were cooked (Ryall and Lipton 1972)
- 0°C and 100% r.h. 3% O_2 and either 2.5 or 5% CO_2 gave products still acceptable and marketable after 7 weeks of storage (Romo-Parada *et al.* 1989)
- 0°C and 100% r.h. 3% O_2 and 10% CO_2 resulted in yellowing and leakage of the curd after 7 weeks of storage (Romo-Parada *et al.* 1989)
- 1°C with 2.5% CO_2 and 1% O_2 for 71–75 days or in the same atmospheres but at 5°C for 45 days (Adamicki 1989).
- 5% or more CO_2 and 2% or less O_2 injured the curds and did not extend their storage life (Hardenburg *et al.* 1990)
- 0°C and 2–5% CO_2 with 2–5% O_2 (SeaLand 1991)
- 0–5°C with 2–5% CO_2 and 2–5% O_2 had a fair effect but was of no commercial use (Kader 1985, Kader 1992).

Modified atmosphere packaging

Plastic films or cellophane wraps can be used to protect the curds during storage and transport. The cultivar Siria was grown in Spain in winter with three harvesting periods of January, March and April by Artes and Martinez (1999). They were subjected to simulated marketing trials of 1 week at 1.5°C followed by 2.5 days at 20°C packed in different films [14.5 μm poly(vinyl chloride), 11, 15 or 20.5 μm low-density polyethylene or 11.5 μm microwavable low-density polyethylene]. After the shelf-life simulation, the best results were obtained in the 11.5 μm low-density polyethylene film. The gas composition was about 16% O_2 and 2% CO_2 during cold storage, and about 11% O_2

and 3.5% CO_2 during retail sale simulation. The overall quality, yellowing and browning of the head and *Alternaria* spp. development were at similar levels among the films studied. Weight loss was considerably lower for low-density polyethylene films than for the poly(vinyl chloride) film.

Cauliflower fungus

Sparassis crispa, Basidiomycetes

Botany

S. crispa is an edible fungus that is cultivated in Japan. It is comprised of densely packed wrinkled lobes that arise from the central rooting base and, as the common name suggests, resembles a compact cauliflower or even a bath sponge. They can be up to 40–50 cm in diameter and are greyish to tan in colour, which darkens with age, and the texture ranges from crisp to tough. They have an aromatic smell. It is a bracket fungus, which infects trees, particularly larch and Douglas fir, and can cause them to rot. In Europe it can be found growing from conifer stumps in autumn.

Harvesting

They should be harvested when they are still young, at which stage they are good to eat.

Storage

No information could be found on its postharvest life, but from the literature it would appear that all types of edible mushrooms have similar storage requirements and the recommendations for the cultivated mushroom (*Agaricus bispsorus*) could be followed.

Celeriac, turnip rooted celery

Apium graveolens variety rapaceum, Umbelliferae.

Botany

The edible portion is the swollen base of the stem, which superficially resembles a turnip but is not actually part of the root as the turnip is (Figure 12.49). It is about 10 cm in diameter, irregular in shape and subglobose, with a brown skin and white flesh tasting of celery.

Figure 12.49 Celeriac at a wholesale market in Britain.

Harvesting

Harvesting is carried out when it has reached an acceptable size. In Northern Europe this is in the autumn. The tops are usually twisted off or cut off at harvest.

Storage

It can be stored for short periods of time in slatted crates in cool sheds (Lutz and Hardenburg 1968). Their shelf-life at a simulated room temperature of 20°C and 60% r.h. was shown to be 7 days (Mercantilia 1989). Refrigerated storage recommendations are as follows:

- 0–1°C and 85–90% r.h. for 3–5 months (Anon 1967)
- 0°C and 90–95% r.h. for 3–4 months (Lutz and Hardenburg 1968)
- 0°C and 90–95% r.h. for 160 days (Mercantilia 1989)
- 2°C for 120 days (Mercantilia 1989)
- 4°C for 90 days (Mercantilia 1989)
- 8°C for 50 days (Mercantilia 1989)
- 0°C and 97–99% r.h. for 6–8 months (Hardenburg *et al.* 1990)
- 0°C and 95–100% r.h. for 180–240 days (SeaLand 1991)
- 0–1°C and 95–100% r.h. for 3–5 months (Snowdon 1991).

Controlled atmosphere storage

Storage recommendations are as follows:

- 0–5°C with 2–3% CO_2 and 2–4% O_2 had only a slight effect (Saltveit 1989)
- 5–7% CO_2 and low O_2 increased decay during 5 months of storage and therefore controlled atmosphere storage was not recommended (Hardenburg *et al.* 1990).

SeaLand (1991) showed that controlled atmosphere storage had a slight to no effect.

Celery

Apium graveolens variety *dulce*, Umbelliferae.

Botany

The edible portion is the leaf petioles. They are often ridged up with soil as they grow to keep the bottoms of the petioles white, a process that is called 'blanching.' They will freeze at about –1.3 to 1.1°C (Wright 1942) or –0.67 to 0.22°C (Weichmann 1987).

Harvesting

Harvesting is carried out when they reach a size acceptable to the market, but if this is delayed too long they become tough and fibrous. This is because the parenchyma cells break down, leaving open spaces and producing a pithy texture, and collenchyma fibres are produced, giving a tough stringy texture (Wardlaw 1937). Also, excessive water loss during storage can cause them to become tough (Lutz and Hardenburg 1968).

Precooling

Vacuum cooling was the most suitable method, forced-air cooling with high humidity and hydrocooling were also suitable and air cooling in a conventional cold store was less suitable (MAFF undated). After 25–30 minutes of vacuum cooling at 4.0–4.6 mmHg with a condenser temperature of –1.7 to 0°C the petiole temperature fell from 20–22°C to 7°C (Barger 1963). Wardlaw (1937), Phillips and Armstrong (1967) and Lutz and Hardenburg (1968) all recommended hydrocooling as soon as possible after harvest.

Chilling injury

Chilling injury was reported to occur during storage at –0.3°C as a loosened epidermis that can be detected by twisting the stalk (Wardlaw 1937).

Storage

Shelf-life at 20°C was given as only 1 day by Mercantilia (1989). Refrigerated storage recommendations are as follows:

- –0.3 to 0°C and 95–98% r.h. (Wardlaw 1937)
- 0–1.1°C, 2–4°C, 0.6–1.7°C, 0°C all at 90% r.h. (Wardlaw 1937)
- 0°C and 95% r.h. for up to 3 months if cooled rapidly after harvest (Phillips and Armstrong 1967)
- 0°C and 95% r.h. for 8–10 weeks (Anon 1967)
- 0°C and 90–95% r.h. for 2–3 months (Lutz and Hardenburg 1968)
- –0.5 to 1.5°C and 90–95% r.h. for 12–14 weeks (Shipway 1968)
- 0°C and 90–95% r.h. for 1–2 months (Amezquita 1973)
- –0.6 to 1°C and 92–95% r.h. for 8 weeks with 15.2% loss (Pantastico 1975)
- 0°C and 90–95% r.h. for up to 21 days (Tindall 1983)
- 4°C for a maximum of 14 days with 15% weight loss (Tindall 1983)
- 0°C and 90–95% r.h. for 28 days (Mercantilia 1989)
- 2°C for 12 days (Mercantilia 1989)
- 8°C for 4 days (Mercantilia 1989)
- 0°C and 98–100% r.h. for 2–3 months (Hardenburg *et al.* 1990)
- 0°C and 90–95% r.h. for 14–28 days (SeaLand 1991)
- 0–1°C and 95–100% r.h. for 1–3 months (Snowdon 1991).

Controlled atmosphere storage

Respiration rate was strongly affected by temperature, but also by the O_2 level in the storage atmosphere, fairly consistently at all the temperatures tested (Table 12.27). Wardlaw (1937) found that 0, 4 or 10°C for 7 days with 25% CO_2 resulted in browning at the base of the petioles, reduced flavour and a tendency for petioles to break away more easily.

Storage recommendations were as follows:

- 1–4% O_2 helped to preserve the petioles and 2.5% CO_2 'may be injurious,' but 9% CO_2 during one months storage 'seemed harmless' (Pantastico 1975)
- Kader (1985) recommended 0–5°C with 0% CO_2 and 2–4% O_2 or 0–5°C with 0–5% CO_2 and 1–4% O_2 (Kader 1992), which had a fair effect but was of limited commercial use

Table 12.27 Effects of temperature and reduced O_2 level on the respiration rate (CO_2 production in mg kg^{-1} h^{-1}) of white celery (Robinson *et al.* 1975)

	Temperature (°C)				
	0	5	10	15	20
In air	7	9	12	23	33
In 3% O_2	5		9		22

- Saltveit (1989) recommended 0–5°C with 3–5% CO_2 and 1–4% O_2, which had only a slight effect
- 5% CO_2 and 3% O_2 at 0°C reduced decay and loss of green colour (Hardenburg *et al.* 1990)
- Reyes (1989) reviewed work on controlled atmosphere storage and concluded that at 0–3°C with 1–4% CO_2 and 1–17.7% O_2 storage could be prolonged for 7 weeks and he specifically referred to his recent work which showed that at 0–1°C and 2.5–7.5% CO_2 with 1.5% O_2, market quality could be maintained for 11 weeks.

SeaLand (1991) recommended 0°C with 2–5% CO_2 and 2–4% O_2.

Modified atmosphere packaging

Perforated polyethylene film liners in crates or cartons can be used to minimize water loss (Lutz and Hardenburg 1968).

Ethylene

Exposing them to ethylene can result in the rapid breakdown of chlorophyll in the tops and an increase in decay (Mack 1927).

Cep, penny bun boletus

Boletus edulis.

Botany

The cap of this edible fungus is up to 20 cm in diameter and 20 cm high and is hemispherical as it emerges, becoming more rounded and flattened as it develops. It tends to have a slippery surface when wet and its colour is a rich chestnut-brown, becoming darker towards the centre, and can have narrow white margins with white flesh and a nutty aroma. It has a thick stalk that is white with a brown-netted pattern tapering from the base. It is common in August and September in deciduous woodland, especially beech woods, but is also found in coniferous woods. At one time it was used commercially for some canned mushroom soup in Britain instead of the cultivated mushroom because of its superior flavour and texture when cooked. However, it ceased to be used because application of the Trade Description Act insisted that it should be called 'Boletus soup,' a name that apparently did not appeal to marketing staff. Several other species, which have Boletus as their common name, are edible, including Bay Boletus (*Boletus badius*), Orange Birch Boletus (*Leccinum versipelle*), Brown Birch Boletus (*Leccinum scabrum*), Larch Boletus (*Suillus grevillei*), and also other species of *Boletus* that have no common name, *B. appendiculatus*, *B. erythropus* and *B. luridus*.

Storage

No information could be found on its postharvest life, but from the literature it would appear that all types of edible mushrooms have similar storage requirements and the recommendations for the cultivated mushroom (*Agaricus bispsorus*) could be followed.

Chanterelles

Cantharellus cibarius, Basidiomycetes.

Botany

It is a bracket fungus. On emergence the cap is a flattened dome with a wavy margin. As it develops it expands and becomes funnel or trumpet shaped, still with a wavy margin. The cap is deep yellow–orange and has a smooth surface. On the underside of the cap it is wrinkled and yellow. The stem tapers towards the base and is also yellow–orange. It is a delicious fungus, which is said to have a distinct smell of apricots when fresh.

Harvesting

They are picked by hand as soon as they are big enough.

Storage

Their shelf-life at a simulated room temperature of 20°C and 60% r.h. was shown to be only 2–3 days (Mercantilia 1989). Refrigerated storage recommendations are as follows:

- 0°C and 90% r.h. for 2 weeks (Mercantilia 1989)
- 0°C and 90–95% r.h. for 14 days (SeaLand 1991).

Chard, spinach beet

Beta vulgaris variety *cicla*, Chenopodiaceae.

Botany

There are cultivars that have been specially selected to produce large quantities of leaf, which is the edible portion, and little swollen root. The flavonoid content of fresh leaves of the cultivar Green was in the range 2.4–3.0 mg g^{-1} fresh weight. The cultivar Yellow contained only flavone C-glycosides (2.1–2.3 mg g^{-1} fresh weight), while the flavonols were not detected. Their ascorbic acid content was between 0.4 and 0.5 mg g^{-1} fresh weight and after domestic storage an 80% loss was observed (Gil *et al.* 1998).

Harvesting

They are harvested by hand by cutting the petioles just above the root. Harvest time is when the leaves are young and tender and before flowering.

Storage

Leaf quality was unacceptable after 3 days storage at 18°C, regardless of humidity level, and at 4°C and 43% r.h. they were unacceptable after 4 days of storage owing to dehydration (Roura *et al.* 2000). Other storage recommendations are as follows:

- 0°C and 85–90% r.h. for 1–2 weeks (Lutz and Hardenburg 1968)
- 0°C and 95–100% r.h. for 10–14 days (SeaLand 1991)
- 8°C and 90–92% r.h. Waskar *et al.* (1998)
- 4°C and 86 or 98% r.h., remained acceptable for 9 days (Roura *et al.* 2000).

Modified atmosphere packaging

Modified atmosphere packaging, with about 7% O_2 and 10% CO_2, had no effect on total flavonoid content after 8 days of storage. The ascorbic acid content decreased, especially in Swiss chard, to reach levels below 50% of the initial content after 8 days of cold storage (Gil *et al.* 1998). Storage life was increased from 1 day at room temperature of 17–30°C with 24–73% r.h. to 6 days in polyethylene bags of 200 gauge with 2% ventilation (Waskar *et al.* 1998).

Chayotes, christophines, chocho

Sechium edule, Cucurbitaceae.

Botany

The edible portion is the berry-like fruit that is called a pepo, of which the outer wall is the receptacle. It is usually pear shaped, up to 20 cm long, with a pale green skin and crisp white flesh containing one central flat seed that is up to 5 cm long (Figure 12.50).

Harvesting

This is carried out when the fruit reach an acceptable size, but before the seed begins to emerge from the apex of the fruit and germinates, because when this occurs the fruit is unpalatable and the flesh can be tough (Figure 12.51). If there is any indication of the fruit splitting then they are too mature to harvest and should be removed from the vine and discarded.

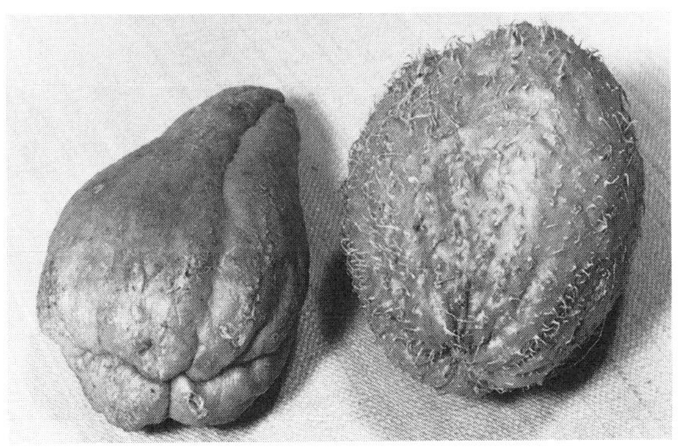

Figure 12.50 Two varieties of chocho; the smooth one on the left is usually the most popular.

Figure 12.51 Chochos imported to the UK from Jamaica showing seed germination, which makes them unmarketable.

Storage

Refrigerated storage recommendations are as follows:

- 7.2°C for 3 weeks (Wardlaw 1937)
- 7.2°C and 85–90% r.h. for 4–6 weeks with 4.9% loss (Pantastico 1975)
- 6°C and 85–90% r.h. for 30 days (Tindall 1983)
- 7.2°C and 85–90% r.h. for 8–10 days (SeaLand 1991)
- 9–11°C and 85–90% r.h. for 4–6 weeks (Snowdon 1991)
- 10°C and 85% r.h. for 10 days followed by 4 days at ambient conditions (Marin-Thiele *et al.* 2000).

Transport

After 9 days of transport at 13–14°C from Puerto Rico to the USA, fruits were generally in good condition with some mould growth, which was identified as *Mycospharella citrullina*. Rotting developed during subsequent marketing to levels which made many of the fruits unmarketable (Burton 1970). In a study of fruit exported to the UK from Jamaica shipped at 13°C for 10 days, 78% were sound, 16% had protruding seeds and 6% had fungal rot (Thompson *et al.* 1979).

Cherimoyas

Annona cherimolia, Annonaceae.

Botany

It is classified as climacteric. The fruit consists of berries fused to a fleshy receptacle, which are heart shaped or conical with a light green skin that may be covered with shallow depressions marking the outlines of the berries, or each berry may end in an abrupt point. The fruits are large and can be up to 15 cm in diameter and weigh up to 2 kg. The flesh is creamy white with a texture of custard when fully mature and contains many dark brown seeds. They are borne on trees that can be up to 6 m high and originated in Peru.

Harvesting

This is done by hand when the fruit begin to soften and give out a characteristic odour. There is a change in the greenness of the skin as they mature, but this is subtle and often difficult to assess.

Storage

Their shelf-life in simulated room temperature of 20°C and 60% r.h. was shown to be 3–4 days (Mercantilia 1989). Refrigerated storage recommendations are as follows:

- 12°C and 90% r.h. for 2–3 weeks (Mercantilia 1989)
- 8–9°C and 90% r.h. for 1–2 weeks (Snowdon 1990)
- 12.8°C and 90–95% r.h. for 14–28 days (SeaLand 1991).

Cherries, sweet cherries

Prunus avium, Rosaceae.

Botany

The fruit is a drupe and is classified as non-climacteric and will freeze at about –4.3 to 3.8°C (Wright 1942). The modern cultivars were selected from the mazzard or gean, which is a tall tree that probably originated in Asia Minor. The mazzard has small dark red to black fruits that have a rich flavour. Modern cultivars are the same colour but can be yellow with a red flush.

Harvesting

Harvest maturity is when the skin turns a bright red colour; delaying harvest until the skin turns dark red results in them having a better flavour, but a shorter marketable life. Also, the red colour makes them attractive to birds and the longer they are left before harvesting the greater will be the loss. Czbaffey (1984, 1985) measured diffuse light transmission in the range 380–730 nm through cherries and found that it was shown to be affected by the level of chlorophyll in the fruit. Chlorophyll decreases during maturation so this could be used as a maturity measurement.

They are harvested by hand. The trees can be tall and therefore pickers often use ladders or mechanical platforms. These latter gave rise to the name 'cherry picker' for the platforms used to service street lighting and pruning trees (see Figure 3.3).

Disease control

A range of fungi have been reported to infect cherry fruits including *Alternaria alternata*, *Botryotinia fuckeliana*, *Botrytis cinerea*, *Cladosporium herbarum*, *Colletotrichum gloeosporiodes*, *Monilina* spp., *Penicillium expansum*, *Rhizopus* spp. and *Stigmina carpophila*. In an experiment, fruit of the cultivar Hedelfingen were inoculated with spores of *Botrytis cinerea*. They were then fumigated with 30 mg litre^{-1} of thymol or acetic acid for 25 min before sealing in modified atmosphere packages and storage at 0°C. After 10 weeks of storage, thymol or acetic acid reduced grey mould rot of *B. cinerea*-inoculated cherries from 36–0.5% or 6%, respectively. However, fruit fumigated with thymol had lower total soluble solids, higher titratable acidity and greater stem browning than acetic acid or non-treated cherries (Chu *et al.* 1999).

Physiological disorders

Scald is a localized translocation of the red pigments and is related to bruising (Hulme 1971). Cracking can occur when fruits are held at high temperature after harvest. Rapid cooling after harvest can be used as a control cracking (Hulme 1971).

Precooling

Air cooling in a conventional cold store was the most suitable method, forced-air cooling with high humidity was also suitable, hydrocooling was less suitable and vacuum cooling was generally considered unsuitable (MAFF undated).

Storage

Refrigerated storage recommendations are as follows:

- –1 to 0°C and 85–90% r.h. for 1–4 weeks (Anon 1967)
- 0°C and 90–95% r.h. for 14 days (Mercantilia 1989)
- 4°C for 9 days (Mercantilia 1989)
- 10°C for 5 days (Mercantilia 1989)
- 20°C for 2 days (Mercantilia 1989)
- –1 to 0.6°C and 90–95% r.h. (Hardenburg *et al.* 1990)
- –1 to 0°C and 90–95% r.h. for 2–3 weeks (Snowdon 1990)
- –1.1°C and 90–95% r.h. for 14–21 days (SeaLand 1991).

Controlled atmosphere storage

Storage recommendations were as follows:

- –1 –0.6°C with 20–25% CO_2 and 0.5–2% O_2 helped to retain fruit firmness, green pedicels and bright fruit colour (Hardenburg *et al.* 1990)
- –1.1 with 20–25% CO_2 with 10–20% O_2 (SeaLand (1991)
- 20–30% CO_2 reduced decay (Haard and Salunkhe 1975)
- 0–5°C with 10–12% CO_2 and 3–10% O_2 (Kader 1985, Kader 1989).

Modified atmosphere packaging

Bing cherries were successfully stored for 2 months at 0°C in Xtend, a plastic film which results in a high internal CO_2 atmosphere. The gas concentration stabilized within 4 days and remained at 7.5% CO_2 and 16% O_2 for 2 months. The main benefits of the Xtend film on cherry quality were reduction of weight loss and maintenance of green pedicel on the fruit. Chinook cherries were also stored successfully for a shorter period of time. Comparison of polyethylene film and Xtend for Chinook cherries showed that green pedicels were maintained better in the Xtend film than in polyethylene bags (Lurie *et al.* 1997).

Hypobaric storage

Hypobaric treatments prolonged the storage life of cherries by 16–33 days. They delayed chlorophyll and starch breakdown, carotenoid formation and the decrease in sugars and titratable acidity (Salunkhe and Wu 1973). The effectiveness of short hypobaric treatments against postharvest rots was investigated by Romanazzi *et al.* (2001). Ferrovia sweet cherries exposed to 0.50 atm for 4 hours had the lowest incidence *Botrytis cinerea*, *Monilinia laxa* and total rots.

Chervil

Anthriscus cerefolium, Umbelliferae.

Botany

It is an annual herb that is grown for its leaves, the flavour of which resembles parsley and fennel, but is more aromatic. They are used in salads and for flavouring soup and are a native of southern Europe and the Levant.

Precooling

Vacuum precooling was recommended by Aharoni *et al.* (1989).

Transport

Aharoni *et al.* (1989) studied freshly harvested chervil under simulated conditions of air transport from Israel to Europe, and also with an actual shipment, during which temperatures fluctuated between 4 and 15°C. Packaging in sealed, polyethylene-lined cartons resulted in a marked retardation of both yellowing and decay. However, sealed film packaging was applicable only if the temperature during transit (probably because of condensation) and storage are well controlled, otherwise perforated polyethylene was better. In other experiments, chlorophyll degradation was delayed effectively when 5% CO_2 in air was applied in a flow-through system.

Chicory, whitloof, radicchio

Cichorium intybus, Compositae.

Botany

The edible portion of the salad chicory cultivars is the leaves, which are grown in the dark from the swollen taproot so that they are blanched (Figure 12.52). The blanching makes them less bitter. In Europe the plants are grown from seed and lifted from the ground in the autumn. The leaves are cut off 2–3 cm above the root and the crowns are covered with soil or sand and placed in heated sheds or greenhouses. The new young growth within the covering is shielded from the light and is therefore blanched. The blanched hearts are used as a cooked vegetable. Curled chicory will freeze at about –0.8 to 0.6°C (Wright 1942). Some cultivars are grown for their large roots, which, when dried, roasted and ground can be blended with coffee or form an ersatz coffee.

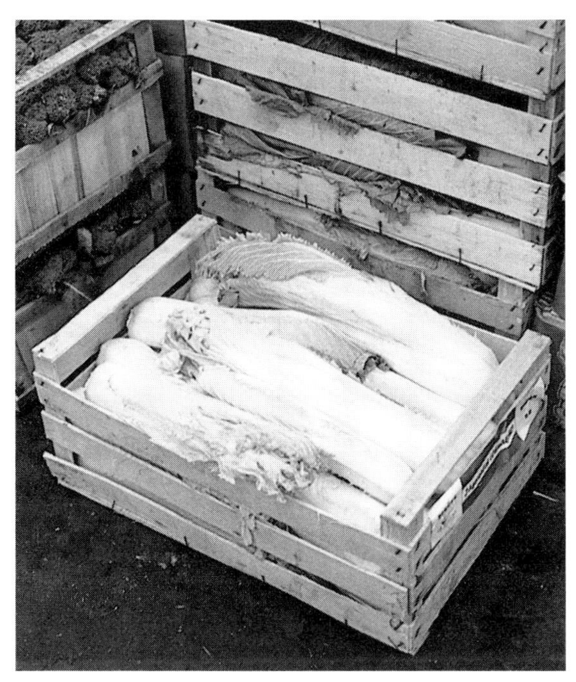

Figure 12.52 Chicory packed in wooden boxes in a wholesale market in Britain.

Physiological disorders

The margins of the leaves may turn yellow and become dry. The disorder is called marginal browning and is thought to be premature senescence caused by poor growing conditions or inappropriate transport or storage (Pantastico 1975).

Precooling

This is recommended to preserve their freshness and is achieved by top icing as soon as possible after harvesting (Wardlaw 1937).

Storage

Their shelf-life at a simulated room temperature of 20°C and 60% r.h. was shown to be only 2 days (Mercantilia 1989). Refrigerated storage recommendations are as follows:

- 0–1°C and 85–95% r.h. for 2–3 weeks (Anon 1967)
- 0°C for 20 days in plastic wraps or bags with the tops left open (Tindall 1983)
- 0°C and 90–95% r.h. for 24 days (Mercantilia 1989)
- 4°C for 12 days (Mercantilia 1989)
- 10°C for 5 days (Mercantilia 1989)
- 0–1°C and 95–100% r.h. for 2–4 weeks (Snowdon 1991)
- 0°C and 95–100% r.h. for 14–28 days (SeaLand 1991)
- 0°C and 95–100% r.h. for 14–21 days for Raddichio (SeaLand 1991).

Controlled atmosphere storage

Atmospheres containing 25% CO_2 caused the central leaves to turn brown (Wardlaw 1937). Storage of Witloof chicory in 4–5% CO_2 with 3–4% O_2 at 0°C delayed greening of the tips in light and delayed opening of the heads (Hardenburg *et al.* 1990). Saltveit (1989) recommended 0–5°C with 4–5% CO_2 and 3–4% O_2 for Witloof chicory, but it had only a slight effect.

Chilles, hot peppers, peppers, cherry peppers, bird chillies

Capsicum annum, *C. frutescens*, Solanaceae.

Botany

The fruit is a berry of extremely variable size, shape and colour. They are very pungent, containing capsaicin ($C_{18}H_{27}NO_3$) in their placenta (Purseglove 1968). It is a bushy plant that is native of tropical America and the Caribbean. The fruits are used fresh to flavour cooking and they are also dried to extend their storage life.

Harvesting

This is based on colour and size and depends on variety and market requirements; some are harvested green and immature whereas others are harvested when they change from green to red, yellow or orange. Capsaicin content was highest in fruit harvested at the initial colouring stage, followed by those in fruits of reddening and greening stages (Park *et al.* 2001). The sugar content was higher in fruits of initial colouring and reddening stages than those of the greening stage. The ascorbic acid content was highest in the reddening fruits followed by the initial colouring fruits and green fruits, while storage losses of ascorbic acid were higher in fruits of the greening stage than those of the colouring and reddening stages (Park *et al.* 2001). During storage, the highest rates of ethylene production and respiration rate were in fruits harvested at the initial colouring stage followed by those fruits at the greening and reddening stages (Park *et al.* 2001).

Disease control

Prasad *et al.* (2000) collected fruit from different parts of India that exhibited varying degrees of decay, discoloration and breaking of the pericarp, detachment of the pedicel and spore dust formation within the fruit. Altogether 67 types of fungi were isolated from the stored chilli fruits, three species of *Phycomycotina*, 11 species of *Ascomycotina* and 53 species of *Deuteromycotina*. *Aspergillus flavus*, *A. terreus*, *A. candidus*, *A. niger*, *A. sclerotiorum*, *Fusarium moniliforme* [*Gibberella fujikuroi*], *F. sporotrichioides*, *Syncephalastrum racemosum*, *Paecilomyces varioti* and *Penicillium corylophilum* were commonly associated with decaying fruits stored in humid regions. Prasad *et al.* (2000) also found that capsaicin content was decreased owing to fungal infection.

Storage

Their shelf-life at a simulated room temperature of 20°C and 60% r.h. was shown to be only 2–3 days (Mercantilia 1989). Refrigerated storage recommendations are as follows:

- 0–4.4°C for fresh fruit (Lutz and Hardenburg 1968)
- 0–10°C with 60–70% r.h. for 6–9 months (Anon 1968; Lutz and Hardenburg 1968)
- 10°C and 90% r.h. for 2–3 weeks (Mercantilia 1989)
- 7–10°C and 90–95% r.h. for 1–3 weeks (Snowdon 1991)
- 10°C and 90–95% r.h. for 14–21 days (SeaLand 1991)
- 5 or 10°C for up to 60 days (Park *et al.* 2001).

Controlled atmosphere storage

Storage recommendations were as follows:

- Kader (1985, 1992) recommended 8–12°C with 0% CO_2 and 3–5% O_2, which had a fair effect but was of no commercial use, but 10–15% CO_2 was beneficial at 5–8°C
- Saltveit (1989) recommended 12°C with 0–5% CO_2 and 3–5% O_2 for the fresh market, but controlled atmospheres had only a slight effect
- For processing Saltveit (1989) recommended 5–10°C with 10–20% CO_2 and 3–5% O_2, which had a moderate effect.
- 10°C and 90–95% r.h. with 0% CO_2 with 3–5% O_2 (SeaLand 1991).

Chinese artichokes

Stachys affinis, Labiatae.

Botany

It is an herbaceous plant that develops many underground stolons, the ends of which swell to form brownish tubers with ridges running horizontally around them to give them a screw-like appearance. It originated in East Asia.

Storage

Their shelf-life at a simulated room temperature of 20°C and 60% r.h. was shown to be only 1–2 days (Mercantilia 1989). Storage recommendations are as follows:

- 0°C and 95% r.h. for 1–2 weeks (Mercantilia 1989)
- 0°C and 90–95% r.h. for 7–14 days (SeaLand 1991).

Chinese bayberries

Myrica rubra, Myricaceae.

Botany

Myrica are aromatic shrubs that bear unisexual flowers and small, greyish fruit. *M. rubra* has bright red fruit and is related to the American Bayberry (*M. pensylvanica*), the Wax Myrtle (*M. cerifera*) and the Sweet Gale or Bog Myrtle (*M. gale*). At 20–22°C, the fruits showed some physiological characters of a respiratory climacteric, such as high rates of respiration and ethylene production. The peroxidase and polygalacturonase activities were closely related to the senescence (Hu *et al.* 2001). In China and Thailand, where they are called 'jiabao,' the berries are dried and treated with salt, sugar and citric acid to make a snack.

Storage

Hu *et al.* (2001) studied storage at 21 ± 1, 11 ± 1 and 1 ± 1°C and did not report any chilling injury, so it seems possible that it can be stored at low temperatures.

Chinese cabbage

Brassica pekinensis, Cruciferae.

Botany

The edible portion is the broad basal shiny leaves that are up to 50 cm long with white midribs. They are formed in a rosette that grows into a compact head. There are many different cultivars, varying from those with long slender heads with broad leaf stalks with the heads open, to those with round heads with crimpled light green leaves that have narrow ribs and closed heads.

Harvesting

Harvesting is carried out when the head is fully formed, but before the flower stem begins to emerge, which is called bolting. They are cut directly below the lowest leaves. Damaged and infected leaves should be removed at harvest and the Chinese cabbage should not be exposed to direct sunlight.

Precooling

Forced-air cooling with high humidity was the most acceptable method, air cooling in a conventional cold store was suitable and hydrocooling and vacuum cooling less suitable (MAFF undated).

Storage

Postharvest losses over a storage period were compared between the traditional method of storing Chinese cabbage in clamps with an improved store where ventilation and air circulation were improved. The results clearly indicate that there are distinct advantages in improved ventilation (Table 12.28).

Their shelf-life at a simulated room temperature of 20°C and 60% r.h. was shown to be only 2 days (Mercantilia 1989). Refrigerated storage recommendations are as follows:

- 0°C for 30–40 days (Tindall 1983)
- 0°C and 90–95% r.h. for 1–2 months (Lutz and Hardenburg 1968)
- 0°C and 90–95% r.h. for 24 days (Mercantilia 1989)
- 4°C for 15 days (Mercantilia 1989)
- 8°C for 10 days (Mercantilia 1989)
- 0°C and 95–100% r.h. for 2–3 months (Hardenburg *et al.* 1990)
- 0°C and 95–100% r.h. for 30–60 days (SeaLand 1991)
- 0–1°C and 95–100% r.h. for 1–2 months (Snowdon 1991).

Table 12.28 Effect of storage method on the postharvest losses (%) of cabbages over 120 days in China (source: Li and Hueng 1992)

Type of loss	Traditional method	Improved storage
Trimming	10–15	10–15
Weight loss	7–12	3–7
Rejects	15–20	5–8
Total	32–47	18–30

Controlled atmosphere storage

Storage recommendations were as follows:

- 1% O_2 was shown to extend the storage life and reduce the decline in ascorbic acid, chlorophyll and sugar content in the outer leaf lamina (Chien and Kramer 1989)
- 0–5°C with 0–5% CO_2 and 1–2% O_2 had only a slight effect (Saltveit 1989)
- 0°C with1% O_2 (Hardenburg *et al.* 1990).

Ethylene

Storage at 10°C for 2 weeks in 100 μl $litre^{-1}$ ethylene compared with ethylene-free air resulted in a higher level of leaf abscission (Wang 1985).

Chinese chives

Allium tuberosum, Alliaceae

Botany

The edible portions are the young, thin, flattish leaves and inflorescences. They have a strong flavour that resembles garlic and are used for seasoning. They probably originated in China, but have spread throughout Asia.

Precooling

Snowdon (1991) mentioned that they have a high respiration rate and therefore it is desirable to precool them possibly with top icing.

Storage

The optimum temperature was approximately 3°C and at that temperature they could be stored for 26 days (Wu and Wu 1998).

Modified atmosphere packaging

Ishii and Okubo (1984) recommended that bundles of about 110 g should be sealed in polyethylene bags with 10 bundles per bag and a bag size of about 30 × 45 cm and about 0.025 mm in thickness. Four bags should be packed vertically in a corrugated box and cooled to 5°C within 1 day. The cooled produce may be transported in insulated containers. Film wrapping was said to be effective in maintaining their freshness and at 3°C losses through decay and flower opening were reduced to <1% after 26 days when polypropylene film wrapping was used (Wu and Wu 1998).

Chinese kale, kale

Brassica alboglabra, Cruciferae.

Botany

The edible part is the leaves, that are large and borne on an erect stem (see Figure 12.7). Most types have smooth leaves, but curly cultivars are cultivated where the leaves are curled and crimped. The kale grown in Europe is *B. oleracea* variety *acephala* and is used mainly as fodder, but is sometimes used as a substitute for cabbage. Respiration rate and ethylene production for Chinese kale reached a peak after 6 days when stored at 20°C (Wu *et al.* 1995).

Curing

The optimal heat treatment with moist air for kale was 45°C for 30 minutes and resulted in higher postharvest quality, delayed yellowing and a lower decline of sugars and organic acids with longer storage at 15°C (Wang 2000).

Storage

In Thailand, they could be stored at room temperature of about 31°C for only two days (Siriphan 1982). During storage at 0°C reducing sugars and soluble protein contents declined in both stems and leaves but increased in the flower buds. Peroxidase activity increased in all parts during storage, and the crude fibre content of the leaves also increased (Wu *et al.* 1995). It is not always clear exactly which kale is being referred to in the literature, but the following are refrigerated storage recommendations:

- 0–1°C and 95% r.h. for 2–4 weeks (Snowdon 1991)
- 0°C and 95–100% r.h. for 10–14 days (SeaLand 1991).

Modified atmosphere packaging

Wrapping in them in wet newspaper and then placing them in plastic bags and storage at 0°C gave the best qualities and longest storage life, which averaged about 16 days (Siriphan 1982).

Chinese pears

Pyrus bertschneideri, Rosaceae.

Botany

The fruit is a pome (Figure 12.53). There is also the Chinese Sand pear, *P. sinensis*, which produces a rather insipid fruit.

Fruit coatings

The cultivars Laiyang Chili and Ya Li were treated with 3, 6 or 9% emulsions of commercial or refined (reduced α-tocopherol levels) plant (soya bean, maize, groundnut, linseed and cottonseed) oils at harvest and stored at 0°C for 6 months (Ju *et al.* 2000). All the oils had similar effects. Ethylene production and respiration rate in fruits treated with 9% oils were lower in early storage and higher in late storage than that in the controls. Oil at 6% reduced internal browning, at 9% it inhibited internal browning completely and at 3% was not effective after 6 months at 0°C and 7 days at 20°C. Plant oil treatment maintained fruit colour, firmness, soluble solids content and titratable acids in a concentration-dependent manner during storage. Internal ethanol was not affected by oil treatment compared with controls, either during storage or 7 days at 20°C. No off-flavour was detected by sensory evaluation in either oil-treated or control fruits.

Figure 12.53 Chinese pears packed for the market in China. Source: Dr Wei Yuqing.

Storage

The only recommendation was 0°C for 6 months (Ju *et al.* 2000).

Chinese radishes, Japanese radishes

Raphanus sativus, variety *Longipinnatus*, Cruciferae.

Botany

The edible portion is the cylindrical white or brown roots, which are long and slender, up to 40 cm long and usually weigh up to 2 kg, but can be as much as 20 kg in some Japanese cultivars. They can be eaten raw, but are most often cooked.

Harvesting

The timing of the harvest is not so critical as salad radishes and they are harvested by digging with a spade or fork.

Storage

Kay (1987) recommended 0°C and 90–95% r.h. for 2–4 months. At higher temperatures storage may be terminated by sprouting, although the naturally occurring campothecin was said to inhibit sprouting, allowing successful storage at 10–20°C (Kay 1987).

Chinese water chestnut, biqi

Eleocharis tuberosa, *E. dolcis* variety *tuberosa*, Cyperaceae.

Botany

It is a root crop and produces two types of rhizome. The first type is produced 6–8 weeks after planting, grows horizontally just below the soil surface, then turns upwards to form daughter plants. The second type occurs later and bends down and produces a single corm at its tip. These corms are flattened and are white when first produced, then become scaly and brown as they mature and up to 4 cm across (Kay 1987). Chen *et al.* (1999) described their culture in China. The principal sugars found in the corms at harvest were sucrose, which was over 90% of the total, glucose and fructose (Kays and Sanchez 1985).

Harvesting

In China they are usually ready for harvest 7–8 months after planting when the green culms (the shoots) have turned brown and have been killed by frost. They are usually dug out by hand, although Kay (1987) indicates that in the USA the soil is carefully lifted onto a mesh screen and worked over with paddles or rubber pads to separate the corms from the soil. They may be harvested semi-mechanically by turning the soil over with a plough and collecting the corms from the surface by hand.

Storage

After harvest, corms rapidly decline in quality owing to desiccation, discoloration and pathogen invasion (Kays and Sanchez 1985). Storage at 4.4°C and 85–90% r.h. for 100–128 days was recommended by SeaLand (1991). Corms for propagation could be dried to 90% of their fresh weight and stored for 6 months in glass bottles at 0–15°C; in this condition they were said to retain 90% viability (Cao *et al.* 1999).

Holding the corms at 1.5°C in aerated aqueous solutions containing 10% sodium chloride at 6 months maintained visual quality, soluble solids and texture, although the concentration of sucrose was reduced while fructose and glucose increased. In aqueous solutions alone storage losses due to desiccation were eliminated, but it resulted in permeation of sound corms with odoriferous metabolites from rotting corms within the container (Kays and Sanchez 1985). The sodium concentration within the tissues increased markedly but did not result in cellular death, and could not be sufficiently removed from unpeeled corms prior to marketing by passive diffusion.

Chinese yams

Dioscorea opposita, *D. batatas*, Dioscoreaceae.

Botany

It is a native of China and is also widely grown in Japan where three forms of tubers are recognized:

- Naga-imo (log and cylindrical)
- Icho-imo (palmer)
- Tsukune-imo (globular).

It is produced only in colder climates in the Far East. Tubers turn brown when cut, which seems to be associated with phenolic amines (Kay 1987).

Harvesting

Harvest maturity is about 6 months after planting. They are dug from the soil, but Naga-imo is difficult to harvest since tubers may reach a depth of 1 m.

Storage

Tubers are consumed mainly directly after harvest. Fully mature tubers can be stored at 5°C for long periods of time and do not turn brown, even when subsequently stored at higher temperatures (Imakawa 1967). They are stored over winter in clamps or cold stores in Japan (Kay 1987). Ventilated storage at about 25°C for up to 60 days can be also be used (Tindall 1983).

Controlled atmosphere storage

Storage in high O_2 tensions or at 5°C reduced browning of tubers (Imakawa 1967).

Chives

Allium schoenoprasum, Alliaceae.

Botany

The edible part is the leaves that have a mild onion flavour and are used to garnish and flavour foods. The plants will withstand frosts and drought and are widely distributed throughout the world.

Harvesting

They tiller and form dense clumps. The leaves are cut close to the ground and brown leaves and flower stalks are discarded before bunching.

Precooling

Vacuum precooling was recommended by Aharoni *et al.* (1989).

Storage

The only information on its storage that could be found was 0–1°C and 95–100% r.h. for 1–2 weeks (Snowdon 1991).

Controlled atmosphere storage

No direct information is available, but chlorophyll degradation was delayed effectively when stored in 5% CO_2 in air applied in a flow-through system (Aharoni *et al.* 1989).

Modified atmosphere packaging

In an experiment, green tops were bunched, 25–30 g per bunch, packed in perforated or non-perforated polythene bags (20 × 25 cm) and stored at 2, 5, 10, 15 or 20°C by Umiecka (1973). The control was kept unpacked. The tops stored better in non-perforated than in perforated bags and the longest satisfactory storage of 14–21 days was in non-perforated bags at 2°C. The non-packed control kept well for 1–2 days at 2°C but deteriorated rapidly at the higher temperatures.

Transport

Aharoni *et al.* (1989) studied freshly harvested chives under simulated conditions of air transport from Israel to Europe, and also with an actual shipment, during which temperatures fluctuated between 4 and 15°C. Packaging in sealed polyethylene-lined cartons resulted in a marked retardation of both yellowing and decay. However, sealed film packaging was applicable only if the temperature during transit and storage was well controlled, otherwise perforated polyethylene was better.

Citron

Citrus medica, Rutaceae.

Botany

The fruit is a berry or hesperidium and is non-climacteric. It is oblong in shape and up to 20 cm long with a bumpy peel that is usually thick. The pulp area is small, green and sour (Purseglove 1968). Its main use is for candied peel. It originated in the area from southern China to India.

Storage

No information on its postharvest life could be found.

Clementines

Citrus reticulata, *C. unshiu*, Rutaceae.

Botany

The fruit is a berry or hesperidium, which is non-climacteric. It is thought to have originated in China and to be similar to the Canton Mandarin. It was selected by Father Clement Rodier in Morocco in the 1890s. According to Davies and Albrigo (1994), the fruit are seedless, with the exception of the variety Monreal. Many different varieties are grown in the Mediterranean region as well as other subtropical and tropical climates.

Harvesting

The juice content increases as they mature on the tree. By taking representative samples of the fruit, extracting the juice in a standard and specified way and then relating the juice volume to the original mass of the fruit, it is possible to specify its maturity. In the USA the minimum juice content at harvest should be 40%.

Degreening

The actual choice of conditions will vary between different situations, particularly whether they have been grown in the tropics or subtropics. The cultivar and the stage of maturity at harvest should also be taken into account. Generally the optimum conditions for degreening are a temperature within the range 25–29°C, humidity as high as possible, generally between 90 and 95% r.h., and ethylene at between 1 and 10 μl $litre^{-1}$ for 24–72 hours.

Storage

Refrigerated storage recommendations are as follows:

- 0–3°C and 85–90% r.h. for 1–2 weeks (Mercantilia 1989)
- 4–5°C and 90% r.h. for 4–6 weeks (Snowdon 1990)
- 4.4°C and 90–95% r.h. for 2–4 weeks (SeaLand 1991).

Cloudberries, baked-appled berries

Rubus chamaemorus, Rosaceae.

Botany

The fruit is a collection of drupes. It produces small, golden-coloured fruit, similar to blackberries (*R. ulmifolius*), and grows wild in boggy areas in cold northern climates, including Arctic Russia and within the Arctic Circle. Because they grow in cold climates the berries are said to mature slowly and develop an intense sweet flavour.

Storage

There are no records of it being grown commercially and no information on its postharvest life.

Coconuts, waternuts, jelly coconuts

Cocos nucifera, Palmae.

Botany

The main edible portion is the nut, which is actually a fibrous drupe containing a thick, brown, dry outside layer consisting of coir, then a hard shell inside which is the white flesh or copra containing the milk. Their main use is the oil extracted from the copra (Purseglove 1975).

Harvesting

After flowering the nut takes a long time to develop. For marketing as fresh nuts they are usually about 11 months old. Nuts are also harvested when they are immature, about 7 months old, when the copra is still a jelly and are called waternuts (Figure 12.54). At this stage the milk has the highest sugar content and the nuts are used for drinking by cutting off the end and drinking the milk directly from the nut. In Thailand, the outsides are trimmed into an attractive pattern and they are sold on the street for passers-by to quench their thirst. In the West Indies and many African countries the waternuts are sold from pickup trucks at the roadside, where the vendor will slice the top off so the milk can be drunk. They will then slice the nut into two and cut a 'spoon' from the husk to that the jelly inside can be scooped up and eaten.

Figure 12.54 Jelly coconut ready for harvesting in Jamaica.

Storage

The white copra is commonly preserved by extraction, shredding and drying. If it is not dried sufficiently, fungi, which can produce toxins, can infect it. The most common of these is aflatoxin, which is mainly produced by *Aspergillus flavus* and *A. parasiticus.* These fungi infect dozens of food products apart from coconuts. Refrigerated storage recommendations for the intact nut are as follows:

- 0–1.7°C and 85–90% r.h. for 1–2 months (Wardlaw 1937)
- 0–1.5°C and 80–85% r.h. for 1–2 months (Hardenburg *et al.* 1990)
- 0°C and 85–90% r.h. for 30–60 days (SeaLand 1991).

Modified atmosphere packaging

Waternuts may be stored in polyethylene film bags or coated with wax to reduce desiccation (Lutz and Hardenburg 1968).

Collards, kale

Brassica oleracea variety *acephala* f. *viridis*, Cruciferae.

Botany

They are erect branched herbs and form a rosette of smooth leaves at the apex of the stem.

Diseases

Operations that involved extended holding times and mechanical handling tended to increase total plate counts of Enterobacteriaceae, yeasts and moulds. Enterobacteriaceae isolated from the frozen product were those normally associated with soil contamination and not normally considered pathogenic to humans (Senter *et al.* 1987).

Curing

Collards cured in moist air at 40°C and high humidity for 60 minutes maintained their leaf structure and had delayed onset of yellowing during subsequent storage, compared with untreated and other heat treatments. When temperature and duration exceeded tolerance levels, heat injury was observed. In some cases of heat injury, tissues maintained their green colour but developed fungal infection (Wang 2000).

Storage

The only recommendation was 0°C and 90–95% r.h. for 10–14 days (SeaLand 1991).

Coriander

Coriandum sativum, Umbelliferae.

Botany

The edible part is the leaves, but they are commonly marketed with their tap root still attached. They are cooked as a vegetable and the young leaves are used for pickles and sauces. The dried fruits are used as a spice, especially to flavour curry and gin.

Precooling

Vacuum precooling was recommended by Aharoni *et al.* (1989).

Storage

Storage life was found to be 1 day at room temperature of 17–30°C with 24–73% r.h. (Waskar *et al.* 1998). Refrigerated storage recommendations are as follows:

- 0°C and 90–95% r.h. for 1 week (Amezquita 1973)
- 0–1.7°C and 90% r.h. for 5 weeks (Pantastico 1975)
- 6–7 days at 8°C and 90–92% r.h. (Waskar *et al.* 1998)
- 4°C for 1 week, although weight loss was approximately 50% (Gomez *et al.* 1999).

Modified atmosphere packaging

Storage in 200 gauge polyethylene bags with 2% ventilation was recommended by Waskar *et al.* (1998); 0°C in modified atmosphere packaging with thicker film gave a longer storage life without detrimental effects (Lee *et al.* 2000). Aharoni *et al.* (1989) recommended packaging in perforated polyethylene-lined cartons.

Ethylene

Yellowing was accelerated during storage in 0.4 μl litre^{-1} ethylene compared with storage in air (Lee *et al.* 2000).

Transport

Aharoni *et al.* (1989) studied freshly harvested coriander under simulated conditions of air transport from Israel to Europe, and also with an actual shipment, during which temperatures fluctuated between 4 and 15°C. Packaging in perforated polyethylene-lined cartons resulted in a marked retardation of both yellowing and decay. In other experiments, chlorophyll degradation was delayed effectively when 5% CO_2 in air was applied in a flow-through system.

Courgettes, summer squash, zucchini, baby marrow

Cucurbita pepo variety *melopepo*, Cucurbitaceae

Botany

The fruit is a pepo and at the normal stage of harvest it contains numerous soft, undeveloped seeds that are white. The cultivars with dark green skin are usually preferred, although there are exceptions, e.g. in the Sudan the local market is mainly for the pale green to yellowish skinned types.

Harvesting

Harvesting is carried out when they reach a size that is required by the market. This is generally about 15 cm long, but some markets prefer them up to 30 cm long and there is a market for 'baby courgettes,' which are 7 or 8 cm long. They are very easily bruised and great care must be taken during harvesting and handling.

Precooling

Forced-air cooling with high humidity was the most suitable, air cooling in a conventional cold store was also suitable and vacuum cooling and hydrocooling were generally unsuitable (MAFF undated).

Storage

In ambient conditions in the Sudan about 50% were considered inedible 2 days after harvest. At room temperature of 20°C and 60% r.h. it was reported in Mercantilia (1989) that they could be kept for 3–5 days. Refrigerated storage recommendations are as follows:

- 2.2–10°C (Wardlaw 1937)
- 5.6°C for several weeks (Wardlaw 1937)
- 0°C and 85–90% r.h. for less than 2 weeks (Phillips and Armstrong 1967)

- 0–4.4°C and 90% r.h. for 4–5 days (Anon 1968, Lutz and Hardenburg 1968)
- 7.2–10°C and 90% r.h. for less than 2 weeks (Lutz and Hardenburg 1968)
- 10°C and 95% r.h. for 7 days (Tindall 1983)
- 10°C and 90% r.h. for 2–3 weeks (Mercantilia 1989)
- 10°C and 90–95% r.h. for 7–14 days for soft-skinned summer types (SeaLand 1991)
- 7.2°C and 90–95% r.h. for 14–21 days (SeaLand 1991)
- 8–10°C and 90–95% r.h. for 1–2 weeks (Snowdon 1991).

Chilling injury

This results in the fruits developing surface pits and turning yellow when removed to higher temperatures. Chilling injury occurred after 10–14 days at 0°C (Lutz and Hardenburg 1968). The use of 1% O_2 reduced chilling injury symptoms at 2.5°C compared with storage in air (Chien and Kramer 1989). The symptoms occurred after 9 days of storage compared with 3 days in air.

Controlled atmosphere storage

Storage in low O_2 was reported to be of little or no value at 5°C (Hardenburg *et al.* 1990), while SeaLand (1991) recommended 5–10% CO_2 with 3–5% O_2.

Cranberries

Vaccinium spp., Vacciniaceae.

Botany

There are two species, the American cranberry, *V. macrocarpon*, which is the most vigorous, and the European cranberry, *V. oxycoccus* which bears smaller fruit. The fruit is borne on small, prostrate, evergreen shrubs and is a bright red berry that will freeze at about –2.9 to 2.4.°C (Wright 1942). Cranberry juice was used as a traditional medicine by North American Indians and subsequently by American sailors to ward off scurvy during long voyages. It is a rich source of ascorbic acid and antioxidants and may have benefits in combating urinary-tract infections and could also have benefits to the cardiovascular system. It was served at the first thanksgiving feast at Cape Cod in 1621 in America.

Preharvest factors

In the USA they are grown in swampy areas and the fields are flooded in winter to protect them from frost. This can be used to advantage in harvesting mechanically.

Harvesting

Harvest maturity is judged on berry size and colour, where almost all the fruit should have turned from green to red before harvesting. In North America, barries are harvested in autumn, but can be left on the plant throughout winter when protected from snow (Hulme 1971). A jump test machine has been developed for measuring the maturity of the fruits postharvest. The berries are bounced and those that are considered just ripe will rebound some 10 cm. Over- and under-ripe fruits will rebound a shorter distance.

In the USA, scoops are used to aid manual harvesting. They have wooden tines spaced about 1 cm apart. These are combed through the bushes, dislodging the berries and depositing them at the back of the scoop. Mechanical harvesters are used mainly for fruits that are to be processed. Many other types have been developed, some of which are used on flooded fields and others on dry fields; the latter are better if the fruit are to be marketed fresh, but they are usually less efficient.

Disease control

An experiment was conducted by dipping fruits in BioSave (a commercial preparation containing *Pseudomonas syringae*) and coating with carnauba wax. The cranberries were then stored at 13°C. Coating with carnauba wax consistently resulted in lower disease incidence than non-coated fruits as follows (the smaller the number the less the disease):

	Weeks in storage		
	4	8	16
Carnauba wax	17	26	42
Non-coated	21	31	56

Two formulations of BioSave gave variable results but the best treatment was BioSave 110 with carnauba wax, which reduced the total decay by 60, 42 and 33%, after 4, 8 and 16 weeks, respectively, compared with the control (Chen *et al.* 2000).

Figure 11.1 A local market in Bedford, UK, where West Indian fruits and vegetables are offered for sale.

Figure 11.2 Supermarket in Beijing, China, showing mainly prepacked produce being offered for sale.

Figure 12.1 Ackee fruit ready for harvesting where the fruit has split and the seeds are exposed.

Figure 12.5 Bitterpit on a Red Delicious apple in Turkey.

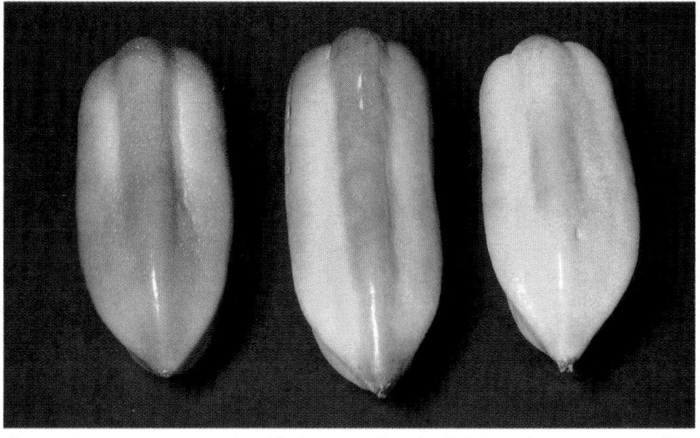

Figure 12.14 Harvest maturity of babaco is when the fruit have begun to turn yellow.

Figure 12.16 Banana passionfruit growing close to Bogota, Colombia.

Figure 12.43 Carambola growing in Malaysia.

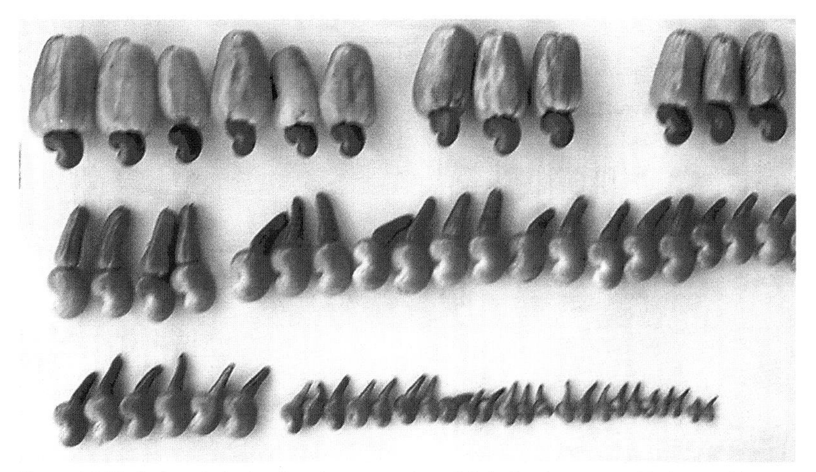

Figure 12.45 Cashew fruit showing the progression of their development, with the nut (the true fruit) developing earlier than the apple (the fruit stalk).

Figure 12.55 Shrink-wrapped cucumber marketed in Britain.

Figure 12.62 Guavas grown in Thailand and packed ready for the market. There are some large cultivars in Thailand that are crisp and sweet when still unripe and can be eaten like dessert apples.

Figure 12.67 Tahiti limes grown in Brazil and packed for export.

Figure 12.76 Turban squash.

Figure 12.84 Oyster mushrooms (*Pleurotus ostreatus*) in a retail pack for a British supermarket.

Figure 12.86 Prickly pear fruit imported into Britain being offered for sale in a wholesale market.

Figure 12.92 Whitecurrants growing in Britain.

Figure 12.93 Salak fruit in a market in Thailand.

Figure 12.97 Fresh sorrel (*Hibiscus sabdariffa*) from the West Indies on the UK market.

Figure 12.104 The best quality of tree tomatoes are those that are harvested when fully red.

Figure 12.106 Tayberries growing in the UK.

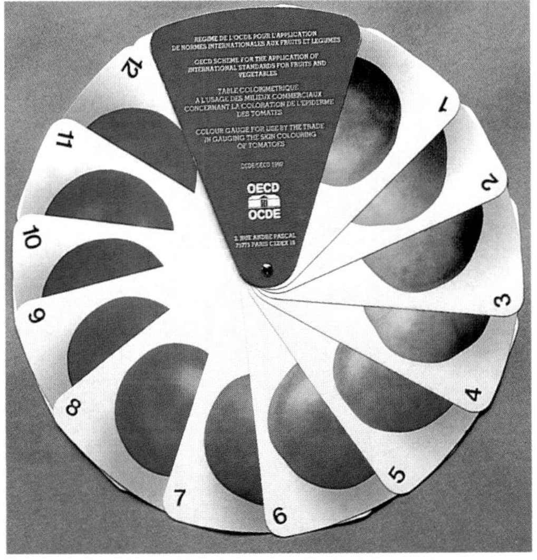

Figure 12.107 Colour chart for use by the trade in gauging the skin colour of tomatoes. Courtesy of OECD.

Storage

Much of the crop was traditionally held in common storage in the USA (Anon 1968; Lutz and Hardenburg 1968) and Canada (Phillips and Armstrong 1967). At room temperature of 20°C and 60% r.h. they could be kept for 10–14 days (Mercantilia 1989). Refrigerated storage recommendations are as follows:

- 2.2–3.3°C for 3–4 months depending on variety (Anon 1968)
- 4.4°C for 6 weeks (Lutz and Hardenburg 1968)
- 2.2–4.4°C and 90–95% r.h. for 3–4 months (Lutz and Hardenburg 1968)
- 4°C and 90% r.h. for 2–4 months (Mercantilia 1989)
- 2–4°C and 90–95% r.h. for 3–4 months (Hardenburg *et al.* 1990)
- 2–4°C and 90–95% r.h. for 2–4 months (Snowdon 1990
- 2.2°C and 90–95% r.h. for 60–120 days (SeaLand 1991).

Chilling injury

Chilling injury occurred in storage below 2.5°C as a change in flesh texture from crisp to rubbery and a loss of natural lustre (Fidler 1963). Lutz and Hardenburg (1968) reported that storage for longer than 4–8 weeks at 0°C may lead to chilling injury.

Controlled atmosphere storage

Kader (1989) recommended 2–5°C with 0–5% CO_2 and 1–2% O_2. However, Hardenburg *et al.* (1990) reported that controlled atmospheres were not successful in increasing their storage life.

Ripening

Poorly coloured fruit can be improved by holding them at 7.2–10°C for a few weeks after harvest (Phillips and Armstong 1967; Lutz and Hardenburg 1968).

Cress, watercress

Nasturtium spp., Cruciferae.

Botany

The edible portion is the leaves and stems that are used in salads. Watercress has been gathered from the wild for centuries in Britain, where it grows along the sides of streams, and it is mostly cultivated in clear running water. There are two main species; Green watercress is *N. officinale*, which is frost susceptible but remains green until the first frosts, and Winter or Brown watercress, which is the hybrid *N. microphyllum* × *officinale*, which turns purple or brown in autumn and is less susceptible to frost damage.

Harvesting

It is cut while it is still tender and young. Yellow and necrotic leaves are discarded and it is commonly tied in bunches and placed in split wooden baskets or plastic trays. Expanded polystyrene boxes have been used for many years for non-returnable transport of trays of watercress (see Figure 5.8).

Precooling

Forced-air cooling with high humidity and vacuum cooling were the most acceptable, air cooling in a conventional cold store was suitable and hydrocooling was said to be not suitable (MAFF undated). This last statement is surprising and also contradicts Fidler (1963), who stated that it is usually hydrocooled. Top icing is also used (Lutz and Hardenburg 1968). Vacuum precooling was recommended by Aharoni *et al.* (1989).

Storage

The factors which limit storage are wilting, yellowing and bacterial rots. Refrigerated storage recommendations are as follows:

- 0.6–1.7°C and 80–82% r.h. (Wardlaw 1937)
- 0–1.7°C and 90–95% r.h. for up to 3 or 4 days (Lutz and Hardenburg 1968)
- 0–2°C for up to 7 days (Tindall 1983)
- 0°C and 95–100% r.h. for 4–7 days (SeaLand 1991)
- 0–1°C and 95–100% r.h. for 3–4 days for watercress (Snowdon 1991)
- 0–1°C and 95–100% r.h. for 1–2 weeks for cress (Snowdon 1991).

Controlled atmosphere storage

No direct information could be found on controlled atmosphere storage, but low O_2 in the store atmosphere was shown to reduce their respiration rate at higher temperatures (Table 12.29). The fact that O_2 had no effect at the optimum storage temperature of 0°C could indicate that it is not an appropriate technique

Table 12.29 Effects of temperature and reduced O_2 level on the respiration rate (CO_2 production in mg kg^{-1} h^{-1}) of watercress (Robinson *et al.* 1975)

	Temperature (°C)				
	0	5	10	15	20
In air	18	36	80	136	207
In 3% O_2	19		72		168

for watercress. In experiments by Aharoni *et al.* (1989), chlorophyll degradation was delayed effectively when 5% CO_2 in air was applied in a flow-through system.

Transport

Aharoni *et al.* (1989) studied freshly harvested watercress under simulated conditions of air transport from Israel to Europe, and also with an actual shipment, during which temperatures fluctuated between 4 and 15°C. Packaging in sealed polyethylene-lined cartons resulted in a marked retardation of both yellowing and decay. However, sealed film packaging was applicable only if the temperature during transit and storage was well controlled, otherwise perforated polyethylene was better.

Cucumber

Cucumis sativus, Cucurbitaceae.

Botany

Cucumber is classified as non-climacteric and is an indehiscent berry called a pepo. It varies in shape from long and cylindrical to almost globular and usually has a dark green skin, but there are yellow cultivars, with whitish green flesh that contains many flat white seeds in the central area. Cucumbers may also be seedless. The cucumber will freeze at about –0.9 to 0.8°C (Wright 1942) or –0.83 to 0.72°C (Weichmann 1987). Inaba *et al.* (1989) found that there was increased respiration in response to ethylene at 100 μl $litre^{-1}$ for 24 hours at 20–35°C, but the effect was much reduced at lower temperatures.

Harvesting

This is carried out when they reach a marketable size, but before they begin to turn yellow and before the seeds develop, as this is associated with a bitter flavour.

Coatings

Chitosan has been successfully used as a postharvest coating of cucumbers to reduce water loss and maintain their quality (El-Ghaouth *et al.* 1991). Wax was sometimes applied to enhance their appearance (Lutz and Hardenburg 1968).

Precooling

Forced-air cooling with high humidity was the most acceptable method, air cooling in a conventional cold store was suitable and hydrocooling was less suitable and vacuum cooling generally unsuitable (MAFF undated).

Storage

Ethanol accumulation was significantly increased by storage for 42 hours at 18°C compared to those stored at 8°C, but storage at 28°C did not cause any further increase than storage at 18°C (Yanez Lopez *et al.* 1998). Refrigerated storage recommendations are as follows:

- 0°C or 4.4°C (Platenius *et al.* 1933)
- 10–15.6°C (Wardlaw 1937)
- 0°C and 85% r.h. (Wardlaw 1937)
- 1.7°C and 85% r.h. for 3–4 weeks (Wardlaw 1937)
- 7.2°C and high humidity (Wardlaw 1937)
- 10°C for 20–25 days (Wardlaw 1937)
- 7–10°C and 90–95% r.h. for up to 2 weeks (Anon 1967)
- 7.2 and 95% r.h. for 10–14 days (Phillips and Armstrong 1967)
- 7.2–10°C and 90–95% r.h. for 10–14 days (Anon 1968; Lutz and Hardenburg 1968)
- 10–11.7°C and 92% r.h. for 2 weeks with 7.2% weight loss (Pantastico 1975)
- 13°C and 95% r.h. for approximately 14 days with a weight loss of 7–10% (Tindall 1983)
- 13°C and 90–95% r.h. for 10 days (Mercantilia 1989)
- 10°C and 90–95% r.h. for 10–14 days (SeaLand 1991)
- 8–11°C and 90–95% r.h. for 1–2 weeks (Snowdon 1991).

Chilling injury

This causes dark-coloured watery blemishes (Wardlaw 1937), pitting, tissue collapse (Lutz and Hardenburg 1968), fungal infection and decay, especially when they have been moved to higher temperatures (Anon 1968; Lutz and Hardenburg 1968; Pantastico 1975). Chilling injury can occur at temperatures below 10°C (Wardlaw 1937; Tindall 1983), 4.4–6.1°C (Pantastico 1975) or 7.2°C for 2 days, which had a slight effect but became severe after 6 days (Lutz and Hardenburg 1968).

Controlled atmosphere storage

Low-temperature storage was shown to result in a low respiration rate with some slight additional effects of reduced levels of O_2 (Table 12.30); 5% O_2 in storage was shown to retard yellowing (Lutz and Hardenburg 1968). Eaks (1956) showed that high CO_2 in the storage atmosphere could have detrimental effects on cucumbers, in that it appeared to increase their susceptibility to chilling injury when they were stored at low temperatures. Short-term exposure to 10 or 20% CO_2 reduced the respiration rate of pickling cucumbers (Pal and Buescher 1993). Other specific storage recommendations were as follows:

- 10°C and 90–95% r.h with 5–7% CO_2 and 3–5% O_2 (SeaLand 1991)
- 8.3°C with 3–5% O_2 gave a slight extension of storage life (Pantastico 1975)
- 5% CO_2 with 5% O_2 extended their storage life (Ryall and Lipton 1972; Pantastico 1975)
- 8–12°C with 0% CO_2 and 3–5% O_2 had a fair effect but was of no commercial use (Kader 1985, 1992)
- 12°C with 0% CO_2 and 1–4% O_2 for the fresh market, and 4°C with 3–5% CO_2 and 3–5% for pickling, both of which had only a slight effect (Saltveit 1989).

Modified atmosphere packaging

In Europe, fruit are commonly marketed in a shrink-wrapped plastic sleeve, which provides a modified atmosphere, to help retain their texture and colour (Figure 12.55, in the colour plates). Lee and Lee (1996) used low-density polyethylene film of 27 μm thickness for the preservation of a mixture of carrot, cucumber, garlic and green peppers. The steady-state atmosphere at 10°C inside the bags was 5.5–5.7% CO_2 and 2.0–2.1% O_2. Mineral powders are incorporated into some films, particularly in Japan, where it was claimed that among other thing these films can remove or absorb ethylene, excess CO_2 and water and can be used for cucumber (Industrial Material Research 1989, quoted by Abe 1990).

Table 12.30 Effects of temperature and reduced O_2 level on the respiration rate (CO_2 production in mg kg^{-1} h^{-1}) of cucumbers (Robinson *et al.* 1975)

	Temperature (°C)				
	0	5	10	15	20
In air	6	8	13	14	15
In 3% O_2	5		8		10

Hypobaric storage

Hypobaric storage at 0.1 atm (76 mmHg) extended the storage life to 7 weeks, compared with 3–4 weeks in cold storage (Bangerth 1974).

Ethylene

Exposing harvested cucumbers to ethylene can result in the rapid breakdown of chlorophyll (Poenicke 1977); 30% CO_2 in the store accelerated ethylene evolution, possibly owing to an early injury response (Pal and Buescher 1993).

Custard apples, bullock's heart

Annona reticulata, Annonaceae.

Botany

It is classified as climacteric with globose to ovoid fruits up to 12 cm in diameter with a smooth skin, which have faint outlines of the fused berries. It is brownish when ripe, often with traces of red or yellow, with a sweet, creamy, white pulp containing many small seeds. It is a native of South America and the Caribbean region. In India and China, *A. squamosa* is sometimes referred to as custard apple (Chen and Chen 1998; Bhadra and Sen 1999; Prasanna *et al.* 2000).

Harvesting

It changes colour from green to brownish as it matures and this and the aroma are used to determine when to harvest. If harvested when it is still green it can have a turpentine flavour, which is said to disappear when it is ripen by exposure to ethylene (Wardlaw 1937).

Storage

Refrigerated storage recommendations are as follows:

- 7–10°C and 85–90% r.h. for 1 week (Lutz and Hardenburg 1968)
- 5°C and 85–90% r.h. for 6 weeks (Pantastico 1975)
- 5–7°C and 85–90% r.h. for 4–6 weeks for Philippino varieties (Snowdon 1990)

- 12 and 85–90% r.h. for West Indian varieties (Snowdon 1990)
- 5°C and 85–90% r.h. for 28–42 days (SeaLand 1991).

Ripening

Wardlaw (1937) recommended exposure to ethylene for improved ripening. At either 0 or 14°C, Wills *et al.* (2001) found that the time to ripen custard apple of the cultivar Pink's Mammoth increased linearly with logarithmic decrease in ethylene concentration over the range <0.005, 0.01, 0.1 1.0 and 10 μl $litre^{-1}$, with the optimum concentration being 10 μl $litre^{-1}$.

Damsons

Prunus damascena, P. insititia, Rosaceae.

Botany

The fruit is a drupe and is probably climacteric. It is very acid but has a strong distinctive flavour and is used mainly for making jam. The tree is commonly used as a rootstock for plums and can be used for peaches and apricots.

Storage

No information on its postharvest life could be found, but it is probably similar to that of the plum, *P. domestica.*

Dates

Phoenix dactylifera, Palmae.

Botany

The edible portion is the fruit, which is a drupe containing a single cylindrical seed surrounded by fleshy pericarp (Figure 12.56). It is usually oblong or ellipsoidal in shape, but sometimes cylindrical or spherical. When fully mature it can be 20–110 mm long and 8–30 mm in diameter (Al Bakr 1972). The date is unusual, in that the food material for the growing embryo is stored as cellulose rather than starch (Dowson 1962). The fruit grow in a bunch, which consists of a main axis with branched stalks up to 1 m long. The skin of the fruit has been described as thin or thick and soft or hard. It is sometime attached to the fruit flesh and in sometimes it is separated from

Figure 12.56 A heavy crop of Khadrawi dates at Kismunt. Source: photo library of FAO, Rome.

the flesh as a little bubble. Shrinkage occurs in the skin when the fruit starts to lose moisture in the late Rutab and Tamr maturity stages (Al Bakr 1972). One characteristic of a good commercial variety is that the skin remains attached to the flesh (Al Bakr 1972). The freezing point of most fruits is usually just below the freezing point of water, but because of their very high sugar content the freezing point of dates is about –16°C.

The date palm can live for up to 100 years and reach up to 24 m to the growing point. The useful height is between 15 and 20 m and the age between 44 and 55 years (Barreveld 1993). Above these two limits the date palm will be dangerous to climb and the yield will also decline. It has been reported to survive temperatures as low as –7°C in Baghdad, –12°C in California and –15.5°C in Texas, and temperatures as high as 50°C. The minimum temperature required for growth is 9°C (Al Bakr 1972). The date palm will flower only when the shade temperature is above 18°C and it will not produce fruit if the temperature is less than 25°C (FAO 1992). It is often mentioned that the date palm likes its 'feet in Heaven and its head in Hell,' as it requires abundant water supply and high temperature for it to grow successfully (Barreveld

1993). They are grown mostly in Islamic countries, which advise the believer to take good care of the date palm, to the point that no other tree has received the same type of attention (Popenoe 1973). Popenoe (1973) stated that it has been in cultivation for more than 7000 years and there are some 1500 varieties of dates in the world. They can be divided into three types:

1. Soft cultivars, which contain mainly invert sugar (glucose and fructose), e.g. Khadrawy, Helawy, Sayer, Saidi, Siwi, Maktoon and Hayany, and are grown in the Middle East and North Africa.
2. Semi-dry cultivars, which contain sucrose e.g. Amri, and Zaholi and Deglet Nour, which is the most widely grown date in California.
3. Dry cultivars, e.g. Sakkoti and Thoory.

There is an increase in respiration in the Rutab stage that Latif (1988) refers to as the climacteric peak, followed by a decline at the end of this stage (Table 12.31), but other than that it is not clear whether the fruit is climacteric or not.

Harvesting

The soluble tannin content of the date decreases with maturity while insoluble ones increase (Al Ogaidi 1986). The speed of this process differs with variety; some loose the astringency at the Khalaal stage and others at the Rutab stage. The stages of maturity at harvest (Table 12.31) are referred to as follows.

Kimiri

The fruit are green. Nearly all cultivars are astringent at this stage. Astringency is caused by a soluble tannin layer below the skin (Al Bakr 1972).

Khalaal

The fruits are fully grown and beginning to change colour to red or yellow. This maturity is rarely used outside the district where they grown, since the fruit are fibrous and usually still astringent. Gupta (1983) showed that a 5–10 minute boiling, depending on variety, could be used to preserved the fruit at this maturity. Excessive boiling caused splitting of fruit and loss of sugar and moisture content.

Rutab

The colour changes to brown or black, the skin becomes loose and the flesh develops a squashy texture. The dates at this stage reach their most acceptable eating quality and are considered to be a fully ripe fruit and are consumed as fresh fruit, but can be stored at 0°C (Barreveld 1993). The length of the Rutab stage on the tree is about 2–4 weeks.

Tamar

They are brown and sufficiently dry to require no further processing. They can be stored at 0–4°C with 70–75% r.h. for about 6 months. Benjamin (1976) suggested that 5°C and 70% r.h. is adequate for the storage for up to 9 months, while –3°C with 70% r.h. was recommended by the FAO (1982) for up to a year.

The harvest time selected depends on climatic factors and consumer preference (Al Bakr 1972). In India, harvesting is commonly carried out at the Khalal stage, after which they are processed into dry dates, which can save them from the spoilage caused by the monsoon rain if they are left to mature further (Gupta 1983). The fruit requires intense heat on the tree to soften, but if it is combined with high humidity the fruit can fall from the bunch at the Rutab stage before it reaches the Tamr stage (FAO 1992).

The traditionally method to harvest fruit is by climbing the tree, cutting the bunch and lowering it to the ground tied to a rope, usually through a pulley system. Picking platforms are available which enable the harvester to be towed around the orchard from tree

Table 12.31 Quality characteristics of some Middle Eastern varieties of dates; respiration rates, at 20°C, are the means for the varieties Zahdi, Diri and Sultani (Latif 1988)

Stage of maturity	Moisture (%)	Reducing sugars	Sucrose (%)	Respiration rate (mg CO_2 kg^{-1} h^{-1})
Kimri	80.2–83.0	33.2–34.0	3.0–4.8	215
Khalaal	56.8–68.0	40.0–63.2	16.0–16.7	16
Rutab	30.8–35.4	58.6–60.0	10.0–24.2	54
Tamr	19.5–26.3	45.3–76.4	7.1–28.4	2

to tree so the platform can be raised and lowered, enabling the fruit to be picked. Cranfield University developed such a machine and tested it in Saudi Arabia in the 1990s. They are expensive and can only be used successfully on flat, firm land where the trees are evenly spaced.

Disease control

Dates with 23% or less moisture content are comparatively safe from microbiological spoilage during storage. In low-temperature storage, the dates with high moisture content retained the soft texture that many consumers prefer. On the other hand, high humidity increased perishability, by increasing the activity of yeasts (Hardenburg 1990). The yeasts *Zygosaccharomyces* spp. are more tolerant of higher sugar concentrations than others found in the dates. Dates that have been attacked by yeasts can be detected by an alcoholic odour (Rygg 1975). Moulds are not as economically important as yeasts. They may cause large losses before and just after the fruit is harvested if rain or high humidity occurs. Species of *Aspergillus*, *Alternaria* and *Penicillium* are the most common moulds found in dates. They can be detected by colour, odour and flavour of the fruit (Rygg 1975). Fennema (1973) recommended a storage temperature of –7°C to prevent microbiological growth.

Physiological disorders

Sugar spotting and darkening are the most important physiological disorder of dates. Sugar spotting is restricted entirely to the invert sugar types of date. Storage below room temperature (±20°C) will reduce the rate of its formation. Soft varieties of dates will have spotting when the moisture percentage is in the high twenties. Higher than 33% or lower than 22% moisture content will prevent spotting (Rygg 1975). Darkening is another type of deterioration that produces some slight off-flavour. In frozen dates the darkening occurs only slightly during storage but rapidly during the thawing period (Mutlak 1984). Mutlak (1984) specified three types of reactions, the enzymatic oxidative browning of simple phenolic compounds, the non-enzymatic oxidative browning of tannin material and the non-enzymatic, non-oxidative browning of reducing sugars. Microwave heating was tried successfully to overcome the darkening effect of the enzymes. Polyphenolase was not active after 1 minute of microwave heating, whereas peroxidase was completely inactive in <1 minute of microwave heating. For fresh dates Rygg (1975) stated that lowering the storage temperature and reducing their moisture content could delay general darkening.

Storage

At room temperature of about 26.7°C and 75% r.h. dates can be stored for 1 month for Deglet Noor (Lutz and Hardenburg 1968). Recommendations for refrigerated storage include:

- 15.6°C for 3 months, 4.4°C for 8 months for Deglet Noor (Lutz and Hardenburg 1968)
- 0°C for 1 year for Deglet Noor (Lutz and Hardenburg 1968)
- –17.8°C for more than 1 year for Deglet Noor (Lutz and Hardenburg 1968)
- 6.7°C and 85–90% r.h. for 2 weeks (Pantastico 1975)
- –3°C with 75–85% r.h. for Angasiah, Sultani, Zahdi, Barban and Um Al Balaliz (Benjamin 1976)
- 0°C and 75% r.h. for Sayer (Benjamin 1976)
- 5°C and 75% r.h. Hilwat Al Jabal for 5 months (Benjamin *et al.* 1985)
- 0°C and 90% r.h. for 1–2 months (Mercantilia 1989)
- 0°C and 85–90% r.h. for 1–2 months (fresh) (Snowdon 1990)
- 0°C or lower and 70–75% r.h. for 12 months (dried) (Snowdon 1990)
- 0°C and 70–75% r.h. for 165–365 days (SeaLand 1991).

Modified atmosphere packaging

Four different packaging materials were used by Butt (1993) for the cultivars Khadrawi and Zaidi at the Khalal stage at 28 ± 2°C for 12 days as follows:

- cardboard boxes without lining
- cardboard boxes with newsprint as lining
- perforated polyethylene bags
- non-perforated polyethylene bags
- the control placed in open trays.

The polyethylene bags whether perforated or not gave significantly better results compared with the other packages and control, with the non-perforated polyethylene bags giving the best quality.

Ripening

Abdul (1988) and Latif (1988) showed that postharvest application of ethylene hastened ripening and its effect increased with increased maturation of the fruit.

Dewberries

Rubus caesius, *R. villosus*, Rosaceae.

Botany

The berries are classified as non-climacteric fruit. The fruit is a collection of drupes, which are smaller and less coherent than blackberries and are black when mature and have a dull surface with a white bloom.

Harvesting

They are normally harvested by hand with the receptacle in place when they have changed colour from red to black and the acid content is about 1% w/w.

Storage

Refrigerated storage recommendations are:

- –0.6–0°C and 85–90% r.h. for 5–7 days (Lutz and Hardenburg 1968)
- 0.5°C and 90–95% r.h. for 2–3 days (SeaLand 1991).

Dill

Anethum graveolens, Umbelliferae.

Botany

It is an annual herb that produces a small, brown fruit called a double achene, which is elliptical and prominently ribbed and about 3 mm long. It is a native of the Mediterranean and western Asia regions and the stems, leaves and particularly the seeds are used as food flavouring. It also has medicinal properties.

Precooling

Vacuum precooling was recommended by Aharoni *et al.* (1989).

Storage

The leaves kept well for 1–2 days at 2°C (Umiecka 1973). Quantitative and qualitative changes were investigated in dill harvested at three stages of maturity and stored for 1–15 days. The essential oil content rose during the first 48 hours of storage and then gradually declined to 21.2% of that in the fresh material on the 15th day. Phellandrene increased and carvone decreased during storage (Zlatev *et al.* 1975).

Controlled atmosphere storage

No direct information could be found, but Aharoni *et al.* (1989) found that chlorophyll degradation was delayed effectively in 5% CO_2 in air.

Modified atmosphere packaging

Green tops were bunched (25–30 g per bunch) packed in perforated or non-perforated polythene bags (20 × 25 cm) and stored at 2, 5, 10, 15 or 20°C. The control was kept unpacked. The tops stored better in non-perforated than in perforated bags and the longest satisfactory storage (9–12 days) was obtained in non-perforated bags at 2°C. The non-packed control kept well for 1–2 days at 2°C but deteriorated rapidly with rising temperature (Umiecka 1973, and 1989).

Transport

Aharoni *et al.* (1989) studied freshly harvested dill under simulated conditions of air transport from Israel to Europe, and also with an actual shipment, during which temperatures fluctuated between 4 and 15°C. Packaging in perforated polyethylene-lined cartons resulted in a marked retardation of both yellowing and decay.

Durians

Durio zibethinus, Bombaceae.

Botany

The fruit is ovoid, foetid and spiny with the walls split into five valves and greenish yellow in colour. They have a creamy-coloured custard-like pulp. They are up to 25 cm in diameter and produce a strong, unpleasant smell which has been described as a combination of over-ripe cheese, rotten onions, turpentine and bad drains, and will taint any other crop with which it is stored (Purseglove 1968). *D. zibethinus*, the most commonly grown durian species, originated in Southeast Asia.

The ripening fruits exhibited a very marked climacteric rise in respiration at 22°C and a parallel increase in ethylene evolution, immediately upon harvest (Tongdee *et al.* 1990). The respiration rate and ethylene production at harvest and the peak climacteric respiration rate were higher in fruits harvested at a more advanced stage.

Harvesting

Purseglove (1968) claimed that it is not considered mature until it has fallen from the tree. However, in practice they are harvested for commercial purposes from the tree before they are fully mature. Vilcosqui and Dury (1997) recommended that fruits are harvested 95–130 days after flowering. The criteria taken into account in judging maturity are the changes in skin colour, which turns from dark to light green as it approaches maturity, then to yellow and sometimes with brown tips. The spines become less stiff as the fruit matures and can be bent more easily. An abscission layer is produced at the base of the fruit stalk, which means that as it matures it is more easily detached from the tree. As the fruit matures it has a more hollow sound when it is tapped. The smell emitted by the fruit indicates that it has begun to ripen. The decision is usually dependent on the skill and experience of the harvester. Images obtained by X-ray, CT and NMR showed that both methods were suitable for detecting differences between immature, mature and over-ripe fruit and internal disorders and rotten pulp in the cultivar Murray, and NMR imaging could be used to identify the physiological disorder called water core (Yantarasri *et al.* 1998).

Coatings

Various coating materials tested by Sriyook el al (1994) were shown to delay dehiscence and ripening. A sucrose fatty acid ester at a concentration of 1% gave the best result in the cultivar Chanee. All coating materials reduced weight loss to 7–14% below that of control fruits. Fruits coated with the sucrose fatty acid ester or 100% apple wax had higher internal CO_2 concentrations than fruits coated with any other coating.

Storage

Purseglove (1968) suggested that they should be eaten immediately after they have fallen from the tree, otherwise they will turn rancid. However, ripe fruits collected after they have fallen to the ground undergo dehiscence within 3 days (Suhaila 1990). Fruits harvested when ripe can be stored in ambient temperature for 4 days (Vilcosqui and Dury 1997). Refrigerated storage conditions have been given as:

- 3.9–5.6°C and 85–90% r.h. for 6–8 weeks (Pantastico 1975)
- 4–5°C and 85–95% r.h. for 30–35 days (Salunkhe and Desai 1984)
- 4–6°C and 85–90% r.h. for 1–2 months (Snowdon 1990)
- 3.9°C and 85–90% r.h. for 42–56 days (SeaLand 1991)
- 15°C for 3 weeks for ripe fruits (Vilcosqui and Dury 1997).

In contrast to some of the above work, Romphophak *et al.* (1997) in Thailand showed that fruits harvested 118 days after anthesis developed chilling injury symptoms at 13.5°C, but could be stored for 15 days at 16.5°C. Fruits harvested 125 days after anthesis could be stored for 15 days at 13.5°C, but for only 10 days at 16.5°C.

Humidity

During storage at 27°C and either 65 or 95% r.h. it was found that there was increased fruit dehiscence at the lower humidity (Sriyook *et al.* 1994).

Chilling injury

Chilling injury symptoms were described as black discoloration on the fruit surface, especially the grooves between the thorns, and occurred after storage at 5°C for 2 weeks (Romphophak and Palakul 1990). Kosiyachinda (1968) described chilling injury symptoms as the rind turning dark brown, and these symptoms occurred in 1.5 days at 1°C, 4 days at 4°C and 14 days at 11°C.

Controlled atmosphere storage

Ripening was inhibited in the cultivars Chanee, Kan Yao and Mon Tong stored in 5–7.5% O_2 at 22°C compared with air, but ripened normally when fresh air was subsequently introduced. O_2 at this level did not affect ripening in fruits harvested at an advanced stage of maturity. Fruits stored in 2% O_2 failed to resume ripening when removed to air. Storage in 10 or 20% CO_2 did not influence the onset of ripening or other quality attributes of ripe fruits and it resulted in little or no reduction in ethylene production (Tongdee *et al.* 1990).

Modified atmosphere packaging

Shrink wrapping of ripe durians, collected after they have fallen to the ground, in a double layer of low-density polyethylene film gave the best results during storage at ambient temperature or at 8°C. At ambient temperature, shrink wrapping prolonged the shelf-life of fruits up to 15 days and percentage dehiscence was low. At 8°C, shrink wrapping prolonged the shelf-life to 8 weeks; at this stage 20% of fruits had dehisced. Shrink wrapping also reduced moisture loss during storage and reduced the release of the characteristic strong durian odour (Suhaila 1990).

Ripening

The ambient conditions where it is grown in South East Asia appear to be satisfactory for ripening. Ketsa and Pangkool (1995) showed that 24–30°C gave satisfactory ripening of the cultivar Chanee. Ketsa and Pangkool (1996) experimented with Chanee fruits harvested 100, 105 or 110 days after full bloom. They were treated with 0, 200 or 400 μl litre^{-1} ethylene at 30–33°C and 60–66% r.h. for 24 hours. Ethylene treatment significantly affected firmness, β-carotene, soluble solids, total sugars and starch content of the pulp 2 days after treatment. Fruit harvested 110 days after full bloom responded to ethylene treatment better than those harvested 100 or 105 days after full bloom.

Ethylene

Ethylene at 100 μl litre^{-1} increased fruit dehiscence during storage at 27°C (Sriyook *et al.* 1994).

Easy peeling citrus fruits

Citrus reticulata, *C. unshiu*, Rutaceae

Storage

Some reviewers give storage for unspecified types of *C. reticulata* and *C. unshiu*, so the following is information on general storage, but see also under Tangerine, Clementine, Manderin and Satsuma. Recommendations include:

- 4.7–7°C and 85–90% r.h. for 3–6 weeks (Anon 1967)
- 0°C and 85–90% r.h. for up to 4 weeks (Lutz and Hardenburg 1968)
- 4°C and 90–95% r.h. for 2–4 weeks (Hardenburg *et al.* 1990)
- 4–5°C and 90% r.h. for 6–8 weeks (Snowdon 1990)
- 4.5–7°C and 90% r.h. for 3–5 weeks for California and Arizona fruit (Snowdon 1990)
- 0–4.5°C and 90% r.h. for 3–5 weeks for Florida and Texas fruit (Snowdon 1990).

Elderberries

Sambucus nigra, Caprifoliaceae.

Botany

There are two edible portions from this shrub: the flowers and the fruit. Both are used for making soft drinks, syrup and wine. The flowers are white and the berries are green, turning black when ripe and borne in clusters. They are natives of Europe, Asia and North America. The name of the genus is derived from a musical instrument called a *sambuca* and in Italy a flute called a *sampogna* is made from the young shoots with the soft pith removed.

Harvesting

For the fruit harvesting is carried out when they have turned black and shiny. Harvesting is by hand, removing the whole cluster by clipping it from the tree.

Storage

It is not produced in commercial orchards, and is mainly collected from the wild, so the only information on postharvest aspects is for the berries and storage was recommended as –0.5°C and 90–95% r.h. for 5–14 days (SeaLand 1991).

Elephant foot yam, elephant yam, suran

Amorphophallus campanulatus, *A. konjac*, *A. paeoniifolius*, *A. rivieri* variety *konjac*, Araceae.

Botany

It is not a true yam, but is more closely related to taro (*Colocasia esculenta*, *Alocasia macrorrhiza*, *Cyrtosperma chamissonis*). It is herbaceous plant that produces a large globose corm that is usually brownish yellow to dull yellow, surrounded by up to 10 cormels. The genus is native to tropical Asia and Africa (Kay 1987). Corms can weigh up to 9 kg.

Harvesting

Kay (1987) found that the growth cycle of the corms is 8–12 months, but it takes three to four seasons for an economically viable crop to be produced. At the end of each growth cycles the leaves wither and die and this is indicative of the time to harvest. They are dug by hand.

Storage

They are carefully cleaned after harvest and stored in heaps, preferably in a well ventilated shed. *A. paeoniifolius* can be stored for several months at 10°C (Kay 1987). Sprouting of *A. paeoniifolius* tubers was induced by storage in diffused light and by presoaking in aqueous solutions of gibberellic acid, Ethrel or thiourea. Thiourea was most effective in breaking dormancy with 92% sprouting in 75 days compared with 18% sprouting in the control. Darkness had an negative effect on sprouting (Kumar *et al.* 1998).

Emblic, Indian gooseberries

Phyllanthus emblica, Euphorbiaceae

Botany

The fruit is globular and 2 or 3 cm in diameter with a thin, pale yellow or greenish skin when mature. The flesh is semi-transparent and acidic and contains a three-celled stone inside which are six seeds. The fruit are produced on medium-sized trees that originated in tropical Asia.

Storage

The only information on postharvest aspects is recommendations for storage at 0–1.7°C and 85–90% r.h. for 8 weeks (Anon 1967; Pantastico 1975).

Endives, escaroles, frisee

Cichorium endiva, Compositae.

Botany

The edible part is the leaves, which are grown like spinach, and the plants are sown and harvested in the open in the same year. In some of the older cultivars the leaves tended to be bitter and were often covered with soil to blanch them to a pale yellow colour before harvesting. The process took 5–20 days depending on the weather conditions. The Curled Endive will freeze at about –0.8 to 0.4°C (Wright 1942) or –0.78 to 0.17°C for Endive and –0.89 to 0.06°C for Escarole (Weichmann 1987).

Storage

At room temperature of 20°C and 60% r.h. they could be kept for about 1.5 days (Mercantilia 1989). Refrigerated storage recommendations are as follows:

- 0°C and 90–95% r.h. for 2–3 weeks (Wardlaw 1937; Phillips and Armstrong 1967; Hardenburg *et al.* 1990)
- 0–1°C and 90–95% r.h. for 2–3 weeks (Endive) (Anon 1967)
- 0°C and 95–100% r.h. for 14–21 days (Endive, Escarole) (SeaLand 1991)
- 2.2°C and 95–98% r.h. for 14–28 days (Belgium Endive) (SeaLand 1991)
- 0°C and 90–95% r.h. for 14 days (Mercantilia 1989)
- 4°C for 8 days (Mercantilia 1989)
- 8°C for 4 days (Mercantilia 1989)
- 0–1°C and 95–100% r.h. for 2–3 weeks (Snowdon 1991).

Controlled atmosphere storage

Controlled atmosphere storage in 25% CO_2 was not recommended as it could cause the central leaves to turn brown (Wardlaw 1937).

Enoki-take mushrooms, winter mushroom, velvet shank

Flammulina velutipes, Hymenomycetes.

Botany

It is an edible gill fungus. As it emerges from the dead wood on which it grows, the cap is conical, but it expands and flattens as it develops. The cap is bright yellow to orange, often darker in the centre when it is wet, and it develops an undulating sticky surface. The gills are yellow and the spores are white. The flesh is yellowish. The stem is tough and orange–brown in colour, darkening, curving and becoming velvety towards the base. They are common in the wild on dead tree stumps throughout Europe and are grown commercially in Japan, the USA and Korea.

Storage

Shelf-life was shown by Minamide *et al.* (1980) to be approximately 14–20 days at 1°C, 10 days at 6°C and 2–3 days at 20°C.

Browning of the pilei, gills and stipes and polyphenol oxidase activity were markedly increased during storage at 20°C (Minamide *et al.* 1980). In storage where the temperature fluctuated from the average 6°C by ±5, ±3 or ±1°C every 12 or 6 hours, Ito and Nakamura. (1985) found that fluctuating temperature had a particularly marked detrimental effect on the quality.

Modified atmosphere packaging

In Japan and the USA they are vacuum-packed in polyethylene film for retail sale (Malizio and Johnson 1991). The effect of poly(vinylidene chloride) coated, oriented nylon, antifogging, wrap or vacuum packing film at 25 or 2°C was investigated. The best treatment for maintaining quality was packing in antifogging film, where they could be stored for 28 days at 2°C (Chi *et al.* 1998). At 0°C in sealed polyethylene bags they were successfully stored for up to 15 days (Chi *et al.* 1996). Low-density polyethylene film 30 μm thick and embedded with silver-coated ceramic extended the storage life from 10 days in low-density polyethylene film without ceramic to 14 days at either 5 or 20°C. After 14 days of storage at 5°C, O_2 and CO_2 concentrations were 7.3% and 8.6%, respectively, in the low-density polyethylene film package without ceramic, and 2.9% and 11.6%, respectively, in the silver-coated ceramic low-density polyethylene film (Eun *et al.* 1997).

Vacuum packaging did not improve their storage life over packing in polyethylene bags (Kang *et al.* 2001). Packaging in polyolefin film bags containing 100 g of mushrooms established an equilibrated atmosphere of 1.7–2.4% O_2 and 4.1–5.6% CO_2 inside the package at 10°C and improved their storage life. The gas permeability of the film was 166 and 731 ml m^{-2} h^{-1} atm^{-1} for O_2 and CO_2, respectively. Temperature fluctuations between 5 and 15°C did not induce a harmful atmosphere inside the polyolefin package, although high temperatures accelerated the quality loss (Kang *et al.* 2001).

Ethylene

Inaba *et al.* (1989) found that there was no response in respiration rate to ethylene at 100 μl $litre^{-1}$ for 24 hours at any temperature between 5 and 35°C.

Fairy ring toadstool

Marasmius oreades

Botany

It is a basidiomycete gill fungus with a cap diameter of up to 6 cm and up to 10 cm high. As the sporophore forms it has a pronounced bulge that flattens and expands with age. The cap is tan coloured and often grooved at the margins, it has white flesh and the gills are brownish and the spores white. It often grows wild in large rings, hence its common name, and is commonly found in grassy places. The stem is tough and pale brown and woolly towards the base. In Japan, Tanesaka (2000) found that the vegetative mycelia, which seemed to be differentiated from the fruit bodies of *Marasmius* spp., formed small colonies on leaf litter. Vetter (1993) found that *M. oreades* was an important protein source with a mean value of 52.8% of its the dry weight.

Storage

No information could be found on its postharvest life, but from the literature it would appear that all types of edible mushrooms have similar storage requirements and the recommendations for the cultivated mushroom (*Agaricus bispsorus*) could be followed.

Feijoas, pineapples guava

Acca sellowiana, Myrtaceae.

Botany

The fruit is classified as climacteric and is oblong in shape and up to 5 cm in diameter. The skin is reddish brown and contains a whitish pulp. When the fruit is cut across there is a four-pointed star shape in the middle that contains small seeds. The flavour of the fruit resembles pineapple, hence one of its common names, combined with strawberry. It originated in Brazil, where it forms a large bush.

Harvesting

As it matures the fruit turns from green to reddish brown and this colour change is used to determine when to harvest. They can be picked when fully grown but still green, and ripened after harvest (Klein and Thorpe 1987).

Prestorage treatments

Pesis and Sass (1994) showed that exposure of fruits of the cultivar Slor to total nitrogen or total CO_2 for 24 hours prior to storage induced the production of aroma volatiles including ethyl acetate and ethyl butyrate, and also acetaldehyde and ethanol.

Storage

At room temperature of 20°C and 60% r.h. the fruit could be kept for 7–10 days (Mercantilia 1989). Refrigerated storage recommendations are as follows:

- 4°C and 90% r.h. for 4 weeks followed by a shelf-life of 5 days at 20°C for the cultivars Apollo, Gemini, Marion and Triumph (Thorp and Klein 1987)
- 10°C and 90% r.h. for 3 weeks (Mercantilia 1989)
- 5°C and 95–100% r.h. for 14–21 days (SeaLand 1991).

Fennel

Foenicullum vulgare, Umbelliferae.

Botany

The young leaves are used as flavouring and the seeds, which contain the essential oils anethole and fenchone, are used medicinally. The taproot is also edible. The swollen leaf bases are used as a vegetable, eaten raw in salads or boiled and have a distinctive anis flavour. It is a native of the Mediterranean region and the Caucasus.

Harvesting

The swollen leaf bases are harvested when they have reached an acceptable size, but are still tender. If they are left too long they become tough and stringy.

Storage

At room temperature of 20°C and 60% r.h. the swollen leaf bases could be kept for 2–3 days (Mercantilia 1989). Refrigerated storage recommendations for the swollen leaf bases are as follows:

- 0°C and 90–95% r.h. for 14–28 days (Mercantilia 1989; SeaLand 1991)
- 0–1°C and 95% r.h. for 1–2 weeks (Snowdon 1991).

Field mushroom

Agaricus campestries

Botany

It is closely related to the cultivated mushroom *A. bisporus* and can be distinguished by the thin ring on the stalk when the cap has opened.

Storage

No information could be found on its postharvest life, but from the literature it would appear that all types of edible mushrooms have similar storage requirements and the recommendations for the cultivated mushroom (*Agaricus bispsorus*) could be followed.

Fig leaf gourds, malibar gourds

Cucurbita ficifolia, Cucurbitaceae

Botany

The fruit is an edible berry, which is oval in shape and up to 20 cm long with a green skin with a white striped or blotched design. The flesh contains numerous large, black seeds that may be used in cookery, as flavouring (Smith 1936; Purseglove 1968), or it can be canned or fermented to produce a beverage. It has been cultivated in the highlands of Central and South America since ancient times.

Storage

The only reference to its postharvest life is that the fruits can be stored for long periods (Smith 1936).

Figs

Ficus carica, Moraceae.

Figure 12.57 Figs packed for local marketing in Mexico.

Botany

The fruit is a synconium that is formed by the swelling of the entire female inflorescence (Figure 12.57). Wills *et al.* (1989) classified it as a climacteric fruit but Biale (1960) classified it as non-climacteric. The fruit will freeze at about –3.2 to 2.4°C (Wright 1942).

Harvesting

Wardlaw (1937) indicated that they should be allowed to mature on the tree before harvesting. This is judged to be when the skin turns light green and they feel just soft to the touch, otherwise they will not ripen in storage. In work in Turkey, the quality of harvesting unripe figs of the cultivar Bursa Siyahi was poorer than that of ripe figs both before and after storage. Unripe figs (harvested about 2 days before ripening) had poor taste, low total soluble solids contents and high acidity, and pulp and skin colour were not fully developed. Unripe fruits had a higher respiration rate than ripe fruits when held at 20°C for 1 day after storage (Celikel and Karacali 1998). Dampness and cool conditions at harvest or ripening at too low a temperature may cause them to split (Pantastico 1975, Wardlaw 1937).

They are usually cut from the tree by hand with a 2–3 cm stalk. They bruise very easily and must be handled with great care (Wardlaw 1937).

Physiological disorders

Sunburn can cause the development of dark brown blotches or bands during subsequent storage (Pantastico 1975).

Disease control

Storage in atmospheres containing high CO_2 reduced mould growth without affecting the flavour of the fruit (Wardlaw 1937).

Coatings

Wardlaw (1937) found that coating with paraffin wax delayed deterioration during transport. However, this treatment would not be acceptable commercially since the whole of the fruit is eaten and once applied it would not be possible to get rid of the flavour of paraffin wax.

Precooling

In order to maximize their postharvest life, forced-air precooling was recommended by Turk (1989). The most appropriate method would be forced air at high humidity.

Storage

At room temperature of 20°C and 60% r.h. the fruit could be kept for 1–2 days (Mercantilia 1989). Refrigerated storage recommendations are as follows:

- 0°C for 44 days (Wardlaw 1937)
- 7.2°C for 1 month (Wardlaw 1937)
- 10°C and 85% r.h. for 21 days (Wardlaw 1937)
- 1.7°C and 4.4°C for 14 days (Australia) (Wardlaw 1937)
- –0.6 to 0°C and 85–90% r.h. for 7–10 days (Lutz and Hardenburg 1968)
- 0–1.7°C and 85–90% r.h. for 7 weeks with 11.5% weight loss (Pantastico 1975)
- 0°C and 90% r.h. for 1–2 weeks (Mercantilia 1989)
- 1°C and 85–95% r.h. for 6 weeks for the cultivar Bursa Siyahi if harvested at the firmness stage of 6 lb in^{-2} and precooled (Turk 1989)
- 7–10°C and 80–85% r.h. for 15–20 days (Hardenburg *et al.* 1990)
- 0°C and 90–95% r.h. for 1–2 weeks (Snowdon 1990)
- 0°C and 85–90% r.h. for 7–10 days (SeaLand 1991)
- 1°C and 85–90% r.h. for 4 weeks for fruits harvested when fully ripe (Celikel and Karacali 1998).

Controlled atmosphere storage

Storage in high-CO_2 atmospheres reduced mould growth without detrimentally affecting the flavour of the fruit Wardlaw (1937). Enriched CO_2 was suggested as a supplement to refrigeration by Hardenburg *et al.* (1990). Other storage recommendations were as follows:

- 0°C with 15% CO_2 and 5% O_2 (SeaLand 1991)
- 0–5°C with 15% CO_2 and 5% O_2 (Kader 1985)
- 0–5°C with 15–20% CO_2 and 5–10% O_2 (Kader 1989 and 1992).

Modified atmosphere packaging

Park and Jung (2000) found that storage in polyethylene bags retained fruit quality better. They also stored the cultivar Masudohin sealed in 0.04 mm polyethylene film bags with 0, 60, 70, 80 or 90% CO_2 at 0°C for up to 20 days. During the first 4 days of storage the atmosphere in the bags varied between 12.3 and 14.0% O_2 and 10.0 and 14.2% CO_2 depending on the CO_2O_2 treatment, and then subsequently stabilized at 13.2–14.0% O_2 and 1.1–1.2% CO_2. The ethylene content in polyethylene film bags increased in the first 2 days and was higher in the control compared with those treated with CO_2. The CO_2 treated fruit were firmer, had better visual quality and reduced the incidence of decay.

Gages, green gages

Prunus italica, Rosaceae.

Botany

The fruit is a drupe and since it resembles *P. domestica* it is probably also climacteric but no direct evidence could be found to confirm this.

Harvesting

They are mature when the skin has turned from dark green to a light green or a yellowish green. They also feel soft to the touch and have a distinct aroma at this time.

Storage

Refrigerated storage recommendation was 0°C for 3–6 weeks then 6–8°C for 6 weeks in Australia for Golden Gage (Fidler 1963).

Garlic

Allium sativum, Alliaceae.

Botany

The edible part is made up of a cluster of small bulbs that are called cloves, which are enclosed in white or pink papery skin. The cloves will freeze at about –4.1 to 3.5°C (Wright 1942) or –3.39 to 0.83°C (Weichmann 1987). It is thought to be a cultivated race of central Asian origin. It is used as flavouring for foods, which dates back to ancient Egypt.

Harvesting

Unlike other *Allium* spp., the bulbs are produced entirely underground. Harvest maturity is when the leaves die back and wither. They are lifted by pulling the tops and levering with a fork inserted in the soil just below the bulbs. Mechanical harvested that have been developed for bulb onions (*A. cepa*) can be used for garlic with only a slight adjustment to the cutting depth. In the cultivar Eusung, clipping the stems to 5 or 10 cm reduced sprouting, incidence of brown spotty disorder and weight loss during 7.5 months storage at 0°C compared with those stored without clipping (Park *et al.* 2000).

Sprouting

Maleic hydrazide at 15 l ha^{-1} applied 2 weeks before harvest or irradiation with 60 Gy using ^{60}Co γ-rays 30 days after harvest was used to control sprouting (Pellegrini *et al.* 2000). However, maleic hydrazide is banned in some countries and there is consumer resistance to irradiation.

Storage

At tropical ambient temperatures of 27–32°C and <70% r.h. they can be stored for up to 30 days (Tindall 1983). At a simulated room temperature of 20°C and 60% r.h. they could be stored for 3–4 weeks (Mercantilia 1989). In temperate countries such as the USA they are commonly stored in dry, well ventilated sheds over winter for up to 4 months (Lutz and Hardenburg 1968), and this practice still exists. Dried bulbs will tolerate temperatures as low as –6°C (McGuire 1929), but such a low temperature is not commonly used commercially. Refrigerated storage recommendations are as follows:

- 0°C and 85–90% r.h. for 1–3 months (Wardlaw 1937)
- 0°C and 70–75% r.h. for 6–8 months (Wardlaw 1937; Phillips and Armstrong 1967)
- –1.5 to 0°C and 70–75% r.h. for 6–8 months (Anon 1967)
- 0°C and 70–75% r.h. for 3–4 months (Lutz and Hardenburg 1968)
- 0°C and 65% r.h. for 28–36 weeks with 12.6% loss (Pantastico 1975)

- 0°C and <70% r.h. for up to 150 days (Tindall 1983)
- 0°C and 70% r.h. for 6–7 months (Mercantilia 1989)
- 0°C and 65–70% r.h. for 6–7 months (Hardenburg *et al.* 1990)
- 0°C and 65–70% r.h. (SeaLand 1991)
- 0°C and 70% r.h. for 6–8 months (Snowdon 1991).

Controlled atmosphere storage

Controlled atmosphere storage was recommended at 0°C with 0% CO_2 and 1–2% O_2 (SeaLand 1991).

Modified atmosphere packaging

Lee and Lee (1996) used low-density polyethylene film of 27 μm thickness for the preservation of a mixture of carrot, cucumber, garlic and green peppers. The steady-state atmosphere at 10°C inside the bags was 5.5–5.7% CO_2 and 2.0–2.1% O_2.

Genips, Spanish limes

Melicocca bijuga, Sapindaceae.

Botany

The fruit are borne in clusters and are ovoid in shape and usually about 3 cm long. They have a hard, green skin and inside is a large ovoid stone surrounded by a thin layer of sweet acid yellow pulp (Figure 12.58). It is a tree from tropical America and is widely grown in the West Indies. They are commonly sold at the roadside, particularly to children, and the skin is easily cracked with the teeth to gain access to the pulp.

Storage

No information on its postharvest life could be found.

Figure 12.58 Genip fruit *Melicocca bijuga*.

Giant taro

Alocasia macrorrhiza variety *macrorrhiza*, Araceae.

Botany

It is a giant, succulent, herbaceous perennial plant, reaching up to 4.5 m in height. It produces a basal corm surrounded by smaller cormels (Kay 1987). The fleshy aerial stems are also used as a vegetable. It is though to have originated in Sri Lanka.

Harvesting

They can be stored in the ground and therefore harvesting can be anything between 9 months and 4 years (Bradbury and Holloway 1988). They are dug from the ground by hand.

Storage

In the South Pacific they may be stored in barns. The recommended refrigerated storage temperature is 5°C but they can also be stored for about 1 month at 15°C (Bradbury and Holloway 1988).

Ginger

Zingiber officinale, Zingiberaceae.

Botany

The edible portion is the fleshy rhizome. It is a native of Asia and has been used there since ancient times and has been carried to Europe in its dried form for centuries. It is sold fresh, but also as preserved ginger, crystallized ginger and dried ginger. It also has medicinal properties.

Harvesting

It is harvested about 9–10 months after planting when the leaves begin to turn yellow and the stems fall down (Purseglove 1975). However, Okwuowulu and Nnodu (1988), from experiments, found that rhizomes harvested 6 months after planting were most suited to prolonged storage in moist sawdust. For making preserved ginger the rhizomes are harvested immature so that they have no fibres and have a milder flavour. In many countries they are harvested mechanically.

Disease control

Pythium ultimum, *Fusarium oxysporum* and *Verticillium chlamydosporium* were associated with storage rot of ginger. Prestorage treatment with 0.2% Topsin M (thiophanate-methyl) and 0.2% Bavistin (carbendazim) for 60 minutes reduced the incidence of storage rot, loss in rhizome weight, surface shrivelling and sprouting of ginger and increased the recovery of the rhizomes (Dohroo 2001). Postharvest applications of benomyl (750 p.p.m. a.i.) with gibberellic acid (150 p.p.m.) was effective in controlling rotting and reducing weight loss and sprouting (Okwuowulu and Nnodu 1988). However, the postharvest use of chemicals may not be permitted in some countries.

Storage

In some parts of India they are stored over winter in pits (Dohroo 2001). A high storage humidity gives rise to mould growth (Lutz and Hardenburg 1968). In spite of this, some sources recommend storage in humidity as high as 90% r.h. Refrigerated storage recommendations are as follows:

- 1.5–3.5°C and 85–90% r.h. for 15 weeks (Anon 1967)
- 7.2–10°C and 75% r.h. for 16–24 weeks with a loss of 18.9% (Pantastico 1975)
- 13°C and 65% r.h. for 6 months with a loss of 16% (Lutz and Hardenburg 1968; Hardenburg *et al.* 1990)
- 13.3°C and 85–90% r.h. for 90–100 days (SeaLand 1991)
- 13°C and 70% r.h. for 4–6 months (Snowdon 1991).

Transport

In shipments from Puerto Rico to the USA taking 9 days at 13°C, the rhizomes consistently arrived in good condition (Burton 1970).

Globe artichokes

Cynara scolymus, Compositae.

Botany

The edible portion is the immature flower, which is made up of the receptacle and the fleshy bases of involucral bracts and called 'chokes' (Figure 12.59). It is the fleshy bases that are eaten as a delicacy after cooking. They will freeze at about –1.9°C (Wright 1942) or –1.44 to 1.17°C (Weichmann 1987). It is produced on a low, thistle-like plant that probably originated in the Mediterranean region and was cultivated in Greek and Roman times.

Figure 12.59 Globe artichokes in a wholesale market.

Harvesting

They are harvested while still immature and tender, since otherwise they tend to be tough and stringy. Wardlaw (1937) recommended harvesting them when they reach their maximum size.

Storage

Storage in crates lined with perforated polyethylene film was recommended by Lutz and Hardenburg (1968). At room temperature of 20°C and 60% r.h. they could be kept for 2–3 days (Mercantilia 1989). Refrigerated storage recommendations are as follows:

- 0–1°C for 4 months (Wardlaw 1937)
- –0.5 to 0°C and 85–95% r.h. for 1–3 weeks (Anon 1967)
- 0°C and 95% r.h. for 15–20 days (Mercantilia 1989)
- 0°C and 90–95% r.h. for at least 1 month (Hardenburg *et al.* 1990)
- 0°C and 90–95% r.h. for 10–16 days (SeaLand 1991)
- 0–1°C and 95–100% r.h. for 3–7 weeks (Snowdon 1991).

Controlled atmosphere storage

Storage recommendations are:

- 3% CO_2 with 3% O_2 in cold storage for 1 month to reduce browning (Pantastico 1975)
- 0–5°C with 3–5% CO_2 and 2–3% O_2, which was said not to be used commercially (Kader 1985)

- 0–5°C with 2–3% CO_2 and 2–3% O_2, which had only a slight effect (Saltveit 1989; Kader 1992)
- 0°C and 90–95% r.h. with 3–5% CO_2 and 2–3% O_2 was recommended by SeaLand (1991).

Golden apple, otaheite apple

Spondias cytherea, Anacardiaceae.

Botany

The fruit is a globular to ovoid drupe; its shape is often irregular with indistinct ridges on the sides. It is up to 8 cm long with a tough, yellow skin when ripe and a large stone surrounded by juicy sub-acid yellow pulp.

Storage

No information on its postharvest life could be found.

Gooseberries

Ribes grossulariodes, Saxifragaceae.

Botany

Fruits are oval to round in shape and are borne singly or in clusters. They usually have a pale green skin, sometimes with darker green 'veins' running lengthways. They may develop a brownish red colour when mature. They have soft, fine hairs over the skin surface. The skin is crisp and juicy and contains a soft acid pulp with many small seeds. There are many cultivars. The cultivar London dates back to the early nineteenth century and produces very large fruit. Other desert fruit include Early Sulphur, Gage, Glencarse, Langley, Muscat, Warrington and Whitesmith. Cultivars more suitable for cooking include Careless, Lancer, Leveller and Whinham's Industry. The fruit will freeze at about –1.8 to 1.6°C (Wright 1942).

Harvesting

This is done by hand with care since the small bushes can be very thorny. Bushes should be pruned to keep them open, which not only facilities harvesting but also helps in the control of the American Gooseberry Mildew or Powdery Mildew (*Sphaerotheca mors-uvae*). Harvest maturity for the desert cultivars is when the fruit become a lighter green or develop a reddish brown colour. For cooking types they are harvested when they reach an acceptable size.

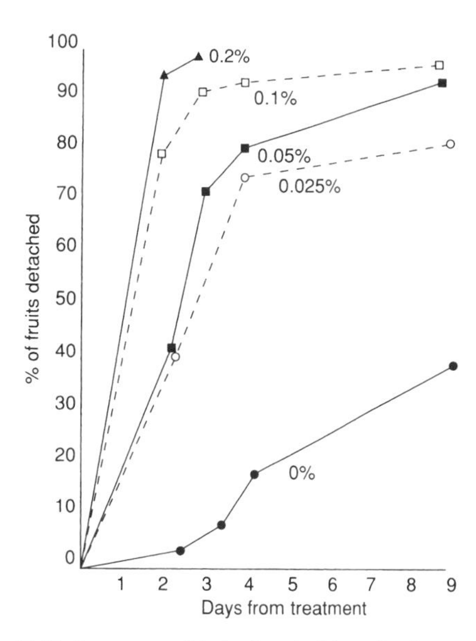

Figure 12.60 Percentage of fruits detached from Careless gooseberries 9 days after the application of various concentrations of Ethrel. Source: Hitchcock (1973).

Some work has been done on mechanical harvesters, but it was found that it was necessary to spray the bushes with a chemical to loosen the fruit before harvesting. Hitchcock (1973) showed the effectiveness of Ethrel treatment and the importance of concentration (Figure 12.60). In practical use the time between application of Ethrel and fruit fall was affected by fruit maturity and weather conditions. Ethrel can reduce the shelf-life of fruits and is not allowed by some supermarkets (John Love, personal communication 1994 quoted by Thompson 1996).

Precooling

Forced-air cooling with high humidity was the most acceptable method, air cooling in a conventional cold store was suitable and hydrocooling and vacuum cooling less suitable (MAFF undated).

Storage

Refrigerated storage recommendations are as follows:

- –0.5 to 0°C and 90–95% r.h. for 3–4 weeks with some losses mainly through collapsed berries, but they should be processed immediately after removal from storage (Hardenburg *et al.* 1990)

Table 12.32 Effects of temperature and reduced O_2 level on the respiration rate (CO_2 production in mg kg^{-1} h^{-1}) of Leveller gooseberries (Robinson *et al.* 1975)

	Temperature (°C)				
	0	5	10	15	20
In air	10	13	23	40	58
In 3% O_2	7		16		26

- −0.5 to 1°C and 90–95% r.h. for 2–4 weeks (Snowdon 1990)
- −0.5 and 90–95% r.h. for 14–28 days (SeaLand 1991).

Controlled atmosphere storage

No information could be found on research on controlled atmosphere storage, but it was shown to have a reduced respiration rate in reduced O_2 levels at temperatures between 0 and 20°C (Table 12.32).

Governor's plum

Flacourtia indica, Flacourtiaceae.

Botany

The subglobose fruits are up to 3 cm in diameter and are reddish purple to blackish. The pulp is reddish, juicy and contains 8–10 seeds. It is said to be a native of tropical Africa and Asia.

Storage

No information on its postharvest life could be found.

Granadillas, giant granadillas

Passiflora quadrangularis, Passifloraceae.

Botany

The edible portion is the soft pulp around the seeds of the fruit. The fruit, which are produced on a vine, can be up to 30 cm long and 15 cm in diameter. The skin is soft and below this the flesh is white and juicy and up to 4 cm thick. The fruit is green as it matures and turns pale yellow or yellowish green at maturity. The green ones can be boiled and used as a vegetable. It produces tuberous roots, which Purseglove (1968) states are usually considered poisonous, but are said to be eaten in Jamaica as a substitute for yams.

Harvesting

This is carried out by hand when the skin has lightened in colour or turned yellow.

Storage

Wardlaw (1937) quoted three different views on storage:

- 10°C for 17 days for green, turning and yellow fruits, immature fruit subsequently ripened normally at 21.1°C
- 7.2–10°C for 15 days for green and turning fruit with subsequent normal ripening at 21.1°C in 7–10 days, but fully yellow fruit were overripe after 7 days at 10°C
- mature fruits stored well over long periods at 8.3°C, but immature fruits were unpalatable after cold storage.

Other refrigerated storage recommendations are as follows:

- 7–10°C and 85–90% r.h. for 3–5 weeks (Snowdon 1990)
- 10°C and 85–90% r.h. for 21–28 days (SeaLand 1991).

Granadillos, sweet granadillas

Passiflora ligularis, Passifloraceae.

Botany

The edible portion is the orange–brown ovoid fruit that are up to 10 cm in diameter. They have a hard skin, inside which is a whitish aromatic pulp containing many seeds.

Harvesting

They are harvested by hand from the vines when the colour begins to change from green to orange.

Storage

Storage at 0 or 2.2°C for 2–3 months resulted in no deterioration in Hawaii (Wardlaw 1937). Fruits harvested at the green mature or slightly mature stage were stored at 5°C for 28 or 35 days. The juice pH was greater in slightly mature than green mature fruits. Total soluble solids contents were similar among treatments, but green mature fruits stored for 28 days had less colour development than those stored for 35 days (Parodi and Campbell 1995).

Chilling injury

Chilling injury was observed in fruits stored at 5°C after 35 days, but not after 28 days (Parodi and Campbell 1995).

Grapes

Vitus vinifera, Vitaceae.

Botany

The fruit, which is classified as non-climacteric, is a berry borne in what are called racemes or, more commonly, bunches. The fruit will freeze at about –5.3 to 2.9°C (Wright 1942). They originated from the part of Caucasia that lies south of the Black and Caspian seas in Asia Minor.

Harvesting

They must be harvested at peak maturity, which is judged on peel colour, pulp texture aroma, flavour and sweetness. The soluble solids content can be used to determine harvest maturity. This varies between cultivars and should be 18–20% for Thompson Seedless and 12–14% for Bangalore Blue (Pantastico 1975). If the berries separate easily from the fruit stalk they are fully mature, but this can be a problem when they are harvested fresh, since the market requires them to be attached to the bunch and many berries may fall off during handling. They are also more susceptible to decay at this stage (Anon 1968). Witbooi *et al.* (2000) found that an increase in harvest maturity was associated with increased susceptibility to *Aspergillus niger* and *Rhizopus stolonifer* infection during storage. In India, the cultivar Gulabi exhibited a double sigmoid pattern of growth and took 12 weeks from fruit set to attain harvest maturity (Rao *et al.* 1995).

Grapes for processing may be harvested by tractor-mounted machines which have combing fingers that are run up the stems pulling off the fruit bunches, as well as a high proportion of the leaves. The fruit are removed from the plants by small-amplitude, high frequency vibration by the rubber fingers being mounted on a vertical drum that rotates at the same velocity as the harvester moves forward. The fruit may subsequently be separated from the leaves by blowers and the juice extracted. In other cases the fruit is taken straight from the field and passed through a press and filter system or to be dried into currants or raisins.

Shattering

Shattering is where the individual berries become dislodged from the bunch and can occur during production, at harvest and during storage. There are two types of shattering: physiological shattering, which was due to the formation of an abscission layer at the *bourrelet*, and decay shattering, where a remnant of the pericarp was attached at the *bourrelet*. The *bourrelet* is the swelling at the base of the pedicel where it is attached to the berry. Several species of fungi were associated with shattering in the cultivars Kyoho, Muscat Bailey A and Delaware. Inoculation experiments indicated that the major fungi involved in the shattering were *Botrytis cinerea*, *Alternaria* sp. and *Rhizopus stolonifer*. The results indicate that fungal infection causes not only decay shattering, but also physiological shattering. Sprays of iprodione 65 and 9 days before harvest markedly decreased berry shattering during storage at 5°C for 2 months. Similar protection was observed when harvested clusters were wrapped in sulphur dioxide-generating paper sheets (Xu *et al.* 1999).

Disease control

Fumigation with sulphur dioxide is used for controlling postharvest diseases of *V. vinifera* fruit. The primary function of the treatment is to give control of *Botrytis cinerea*. One method is to place the boxes of fruit in a gas-tight room and introduce the gas from a cylinder to the appropriate concentration. A fumigation treatment that results in a residue of 5–18 μl litre^{-1} sulphur dioxide in the grapes was found to be sufficient to control decay (Pentzer and Asbury 1934). Its toxicity to *B. cinerea* spores was found to be proportional to temperature over the range 0–30°C where the toxicity of sulphur dioxide increased about 1.5-fold for every 10°C rise in temperature (Couey and Uota 1961). Treatment with 1% sulphur dioxide for 20 minutes was found to be effective (Ryall and Harvey 1959).

Sulphur dioxide can be corrosive, especially to metals, because it combines with atmospheric moisture to form sulphurous acid. If applied in too high a concentration if can remove the colour from black grapes. Special sodium metabisulphite–impregnated pads are available which can be packed into individual boxes of fruit and give a slow release of sulphur dioxide.

Acetaldehyde fumigation of Sultanina grapes at 5000 µl $litre^{-1}$ for 24 hours was shown to reduce subsequent decay by 92% compared with untreated fruit and left no residues or off-flavours in fruit (Avissar *et al.* 1989).

High levels of CO_2 (10–15%) can control *B. cinerea* for limited periods. Prolonged exposure to these levels can damage the fruit, but up to 2 weeks of exposure did not have a detrimental effect (Kader 1989). Witbooi *et al.* (2000) found that both *Aspergillus niger* and *Rhizopus stolonifer* developed on Thompson Seedless stored at 7.5°C for 14 days, but growth was inhibited at –0.5°C and 3°C.

Storage

Water loss is a major limiting factor in storage. Losses of only 1–2% can result in browning of the fruit stalk. This browning is often used by buyers as a good indicator of the fruits' previous storage. With losses of 3–5% the fruit lose their bright, turgid appearance. During storage where the temperature fluctuated from the average 6°C by ±5, ±3 or ±1°C every 12 or 6 hours, Ito and Nakamura. (1985) found that Muscat of Alexandria grapes stored in fluctuating temperatures were similar in condition to those stored at a constant temperature. The shelf-life at 20°C was about 1 week (Mercantilia 1989). Refrigerated storage recommendations are as follows:

- –2.2 to 0.6°C and 80–85% r.h. for 2 weeks to 6 months depending on variety (Wardlaw 1937).
- 1°C and 85–95% r.h. (Hulme 1971)
- –2.2 to 1.7°C (Anon 1968)
- 0°C and 90–95% r.h. (Mercantilia 1989)
- 2°C for 140 days (Mercantilia 1989)
- 4°C for 90 days (Mercantilia 1989)
- 8°C for 45 days (Mercantilia 1989)
- 12°C for 25 days (Mercantilia 1989)
- –1 to 0.5°C and 90–95% r.h. (Hardenburg *et al.* 1990)
- –1.1°C and 90–95% r.h. for 56–180 days (SeaLand 1991).

Recommendations for specific cultivars include the following.

Almera (Almeria)

- –1 to 0°C and 85–90% r.h. for 3–5 months (Anon 1967)
- –1 to 0.5°C and 90–95% r.h. for 3–6 months (Snowdon 1990)

Alphonse Lavalle

- –1 to 0.5°C and 90–95% r.h. for 3 months (Snowdon 1990).

Barlinka

- –1 to 0°C and 85–90% r.h. for 3–5 months (Anon 1967)
- –1 to 0.5°C and 90–95% r.h. for 4–6 months (Snowdon 1990).

Cardinal

- –1.1 to 0.6°C and 90–95% r.h. for 4–6 weeks (Anon 1968).

Chasselas

- –1 to 0°C and 85–90% r.h. for 2 months (Anon 1967).

Cornichon

- –1.1 to 0.6°C and 90–95% r.h. for 2–3 months (Anon 1968).

Emperor

- –1 to 0°C and 85–90% r.h. for 3–5 months (Anon 1967)
- –1.1 to 0.6°C and 90–95% r.h. for 3–5 months (Anon 1968).

Malaga

- –1.1 to 0.6°C and 90–95% r.h. for 2–3 months (Anon 1968).

Muscat

- –1 to 0°C and 85–90% r.h. for 2 months (Anon 1967)
- –1.1 to 0.6°C and 90–95% r.h. for 4–6 weeks (Anon 1968).

Red Emperor

- –1°C and 90–95% r.h. for 4–6 months (Snowdon 1990).

Red Malaga

- –1.1 to 0.6°C and 90–95% r.h. for 2–3 months (Anon 1968).

Ribier

- −1.1 to 0.6°C and 90–95% r.h. for 3–5 months (Anon 1968).

Ohanes (Ohanez)

- −1 to 0°C and 85–90% r.h. for 3–5 months (Anon 1967)
- −1.1 to 0.6°C and 90–95% r.h. for 3–5 months (Anon 1968).

Servant

- −1 to 0°C and 85–90% r.h. for 3–5 months (Anon 1967).

Sultanina

- −1–0°C and 85–90% r.h. for 2 months (Anon 1967).

Thompson seedless

- −1.1 to 0.6°C and 90–95% r.h. for 4–10 weeks (Anon 1968)
- −1°C and 90–95% r.h. for 2–3 months (Snowdon 1990).

Tokay

- −1.1 to 0.6°C and 90–95% r.h. for 4–10 weeks (Anon 1968).

Waltham Cross

- −1 to 0.5°C and 90–95% r.h. for 3 months (Snowdon 1990).

Controlled atmosphere storage

Controlled atmosphere storage recommendations were 1–3% CO_2 with 3–5% O_2 (SeaLand 1991). Kader (1989 and 1993) recommended 0–5°C with 1–3% CO_2 and 2–5% O_2, which was said to be incompatible with sulphur dioxide fumigation.

Hypobaric storage

Burg (1975) found that hypobaric storage extended their postharvest life from 14 days in cold storage to as much as 60–90 days. The effectiveness of short hypobaric treatments against postharvest rots was investigated by Romanazzi *et al.* (2001). On Italia table grape bunches, treatment with 0.25 atm applied for 24 hours significantly reduced the incidence of *B. cinerea.*

Ethylene

They showed no response to exposure to 100 μl $litre^{-1}$ of ethylene for 24 hours at any temperature over the range 5–35°C (Inaba *et al.* 1989). However, ethylene in storage was found to induce abscission layer formation by Xu *et al.* (1999) and thus increase shatter.

Transport

To reduce vibration injury, specially sprung lorries or bubble polyethylene film liners at the bottom of crates can be used for transport (John Love, personal communication, quoted by Thompson 1996).

Grapefruits

Citrus pardisi, Rutaceae.

Botany

The fruit is a berry or hesperidium, which is non-climacteric. The grapefruit is thought to have originated as a mutation of the pummello (*C. grandis*) in the West Indies. The fruit will freeze at about −2.2 to 1.7°C (Wright 1942). Respiration rate increases rapidly with increasing temperature, as would be expected (Table 12.33).

Harvesting

It is important, as with other citrus fruits, to harvest them at peak maturity since they do not ripen after harvest, only deteriorate (Anon 1968). The juice content increases as they mature on the tree. By taking representative samples of the fruit, extracting the juice in a standard and specified way and then relating the juice volume to the original mass of the fruit, it is possible to specify its maturity. In the USA the minimum value is 35%. The ratio between soluble solids and acidity can also be used with ranges of 5:1–7:1. They also turn from green to yellow as they mature and aromatic

Table 12.33 Effects of temperature on the respiration rate (CO_2 production in mg kg^{-1} h^{-1}) of grapefruit grown in Florida (Hallar *et al.* 1945)

Temperature (°C)	0	4.4	10.0	15.5	21.1	26.6	32.2	37.7
Respiration rate	2.6	4.5	6.9	13.2	16.0	19.0	27.5	51.4

compounds increase. Both of these should be taken into account when harvesting.

They should be clipped from the tree, not pulled, and not harvested when wet. Quailing for a short time before transport from the field has been recommended to reduce oleocellosis, toughen the skins and make them less prone to injury and infection (Wardlaw 1937). However, it is rarely practised today.

Curing

Sealing in plastic film bags and then exposing them to 34–36°C for 3 days resulted in the inhibition of *Penicillium digitatum* infection through lignification and an increase in antifungal chemicals in the peel of the fruit (Ben Yehoshua *et al.* 1989b). Sealing the fruit in plastic film before curing was said to be essential to reduce weight loss (Ben Yehoshua *et al.* 1989a).

Hot water brushing

Storage experiments using organically grown Star Ruby showed that hot water brushing at 56°C for 20 seconds reduced decay development by 45–55% and did not cause surface damage. Also, it did not influence fruit weight loss or internal quality parameters (Porat *et al.* 2000).

Physiological disorders

High water loss from fruits can cause areas of the skin to become brown and necrotic. This often starts from the stem end.

Degreening

Optimum temperatures for degreening range between 25 and 29°C and humidity as high as possible, generally between 90 and 95% r.h. with ethylene at between 1 and 10 μl $litre^{-1}$ for 24–72 hours. The actual choice of conditions will vary between different situations, particularly whether they have been grown in the tropics or subtropics. The cultivar and the stage of maturity at harvest should also be taken into account.

Storage

General refrigerated storage recommendations are:

- 7°C and 85–90% r.h. for 10 weeks (Anon 1967)
- 5.6–7.2°C and 85–90% r.h. for 8–12 weeks (Pantastico 1975).

Recommendations for different areas of production are as follows.

Australia

- 7.2–10°C (Wardlaw 1937).

Arizona

- 14.4–15.6°C and 85–90% r.h. (Anon 1968; Lutz and Hardenburg 1968)
- 14–15°C and 90% r.h. for 4–6 weeks (Snowdon 1990)
- 13.3°C and 85–90% r.h. for 28–42 days (SeaLand 1991).

California

- 14.4–15.6°C and 85–90% r.h. for 4–6 weeks (Anon 1968; Lutz and Hardenburg 1968)
- 14–16°C and 85–90% r.h. for 4–8 weeks (Mercantilia 1989)
- 14–15°C and 90% r.h. for 4–6 weeks (Snowdon 1990)
- 13.3°C and 85–90% r.h. for 28–42 days (SeaLand 1991).

Florida

- 10°C and 85–90% r.h. for 4–6 weeks (Mercantilia 1989; Hardenburg *et al.* 1990)
- 0°C for 3–4 weeks but must be marketed within 3–4 days (Hardenburg *et al.* 1990)
- 10°C and 90% r.h. for 4–6 weeks (Snowdon 1990)
- 13.3°C and 85–90% r.h. for 28–42 days (SeaLand 1991)
- 10°C for fruit harvested after January in Florida (Davies and Albrigo 1994)
- 15°C for fruit harvested before January in Florida (Davies and Albrigo 1994).

Israel

- 6–9°C and 85–90% r.h. for 3–4 months (Mercantilia 1989)
- 10–12°C and 90% r.h. for 10–16 weeks (Snowdon 1990).

Mexico

- 13.3°C and 85–90% r.h. for 28–42 days (SeaLand 1991).

South Africa

- 11°C and 90% r.h. for 12–14 weeks (Snowdon 1990).

Texas

- 10°C and 85–90% r.h. for 4–6 weeks (Hardenburg *et al.* 1990)
- 10°C and 90% r.h. for 4–6 weeks (Snowdon 1990)
- 13.3°C and 85–90% r.h. for 28–42 days (SeaLand 1991).

Trinidad

- 7.2 for 50–60 days (Wardlaw 1937).

USA

- 10–12.8°C and 88–92% r.h. for 60–90 days (Wardlaw 1937).

Chilling injury

Chilling injury can occur after prolonged exposure to low temperatures as skin pitting, which increases in size and eventually the fruit rots. Waxing reduced chilling injury of grapefruit, even though in other studies it increased such injury in limes. Certain differentially permeable transparent films, notably Pliofilm (rubber hydrochloride), established in-package atmospheres that reduced chilling injury and extended subsequent shelf-life, even though the seal had to be broken on leaving cold storage (Grierson 1971).

Marsh Seedless fruits that had been dipped in methyl jasmonate at 10 μM for 30 seconds and then stored for 4–10 weeks at 2°C had reduced the severity of chilling injury symptoms and the percentage of injured fruits compared with non-dipped fruit (Meir *et al.* 1996).

Controlled atmosphere storage

Controlled atmosphere storage experiments have given inconclusive results, with some indication that fruit stored in 10% CO_2 at 4.5°C for 3 weeks had less pitting than those stored in air. Pretreatment with 20–40% CO_2 for 3 or 7 days at 21°C reduced physiological disorders on fruit stored at 4.5°C for up to 12 weeks (Hardenburg *et al.* 1990). Other recommendations include:

- 5–10% CO_2 with 3–10% O_2 for grapefruit from California, Arizona, Florida, Texas and Mexico (SeaLand 1991)
- 10–15°C with 5–10% CO_2 and 3–10% O_2 had a fair effect on storage but controlled atmosphere storage was not used commercially (Kader 1985, 1993)
- 10–15°C and 85–90% r.h. with 6–8% O_2 and 3–10% CO_2 for 6–8 weeks (Burden 1997)
- In experiments with Star Ruby grown in Turkey, which compared 1, 3 and 5% CO_2, all at 3% O_2 and 10°C, the best conditions were 1% CO_2 and 3% O_2 giving 125 days of storage with little loss of quality (Erkan and Pekmezci 2000).

Hypobaric storage

Haard and Salunkhe (1975) mentioned two reports that suggested that low-pressure storage could extent the postharvest life compared with cold storage alone. In one report the increase was from 20 days at normal pressure to 3–4 months in hypobaric conditions and in the other from 30–40 days at normal pressure to 90–120 days in hypobaric storage. Hypobaric storage at 380 mmHg (the lower limit of the experimental equipment) and 4.5°C had no effect on the incidence of chilling injury (Grierson 1971).

Greater yam, Lisbon yam, white yam, water yam, Asiatic yam

Dioscorea alata, Dioscoreaceae.

Botany

The edible portion is the tuber that has a brown skin. There are many varieties, that vary in size shape and flavour. They may be cylindrical, globular, branched, lobed, flattened or fan-shaped. They commonly have white flesh, but can be pink or deep reddish purple. Tubers normally weigh 5–10 kg, but can be as heavy as 60 kg (Kay 1987). In Tunapuna in Trinidad and Tobago there is an annual yam festival and competition where enormous tubers are judged entirely on size; the biggest yam wins. Kay (1987) states that *D. alata* is not known in the wild, but was developed from species originating in Southeast Asia and now is cultivated throughout the tropics.

Harvesting

It normally matures 9–10 months after planting, but some varieties are harvested immature after about 6 months (Kay 1987). Harvesting is usually by digging the tubers by hand. Experimental mechanical harvesters were developed and tested in the University of the West Indies in Trinidad by Dr Theo Ferguson and

Dr Lewis Campbell. These were successfully used to harvest the tubers with minimum mechanical damage (Kay 1987).

Curing

For *D. alata*, 29–32°C and 90–95% r.h. for about 4 days directly after harvest was recommended by Gonzalez and de Rivera (1972). Ravindran and Wanasundera (1993) cured tubers by exposure them to sunlight for 3 days after harvest in mean ambient conditions of 29°C and 84% r.h.

Disease control

Ogali *et al.* (1991) treated tubers with either lime (calcium oxide) or local gin (fermented palm wine or sap from (*Elaeis guineensis*) and then stored them at 25–32°C with 62–90% r.h. for up to 32 weeks. After 24 weeks there was 25% and 5% rotting with gin and lime, respectively. After 32 weeks, the level of rot was comparable in all treatments. Four species were isolated: *Aspergillus niger*, *Fusarium oxysporum*, *Penicillium citrinum*, all of which were found to be mildly pathogenic, and *Fusarium moniliforme* [*Gibberella fujikuroi*], which did not produce any rot symptoms.

Sprouting

Tubers of the cultivars Sree Keerthi and Sree Roopa, stored at room temperature of 30–32°C and 80–85% r.h. in the dark, sprouted 70–80 days after harvest (Muthukumarasamy and Panneerselvam 2000). Ravindran and Wanasundera (1993) found that sprouting started after 60–90 days in 24–28°C and 70–90% r.h. A range of chemicals that had been shown successfully to prevent sprouting in potato tubers were studied on yams. They all proved ineffective (Cooke *et al.* 1988). However, Passam (1982) showed that postharvest treatment of *Dioscorea alata* and *D. esculenta* with gibberellins can delay sprouting.

Storage

Tubers were stored for 150 days as thin layers on the floor of a well ventilated room at room temperature of 24–28°C and 70–90% r.h. Crude protein, starch and ascorbic acid contents decreased significantly during the first 60 days of storage, but the decrease was gradual thereafter. Losses of some minerals and a significant reduction in levels of oxalates were also noted during storage (Ravindran and Wanasundera 1993). Yams could be stored in ventilated barns under West African ambient conditions of about 30°C and 60% r.h. for several months (Tindall 1983). In Nigeria, yams are stored in specially constructed barns (see Figures 7.6 and 7.7). The design of these structures varies but they usually consist of live stakes driven into the ground and the tubers are suspended from these by twine. The live stakes produce foliage that shades the tubers from the sun and provides a cool, moist environment. In a comparison between barn storage and pit storage, Ezeike (1985) recorded weight losses over a 5-month period of 60% for the former method but only 15–25% for the latter. Yams stored in slatted wooden trays in the coastal region of Cameroon had losses of 29–47% in only 2 months (Lyonga 1985). In the apical and basal regions of the tuber, the starch content decreased while the sugars and α-amylase activity increased during storage (Muthukumarasamy and Panneerselvam 2000). Refrigerated storage recommendations are as follows:

- 12.5°C for 8 weeks (Coursey 1961)
- For cured tubers, 15–17°C and 70% r.h. for 180 days with about 11% weight loss (Gonzales and de Rivera 1972)
- For non-cured tubers, 15–17°C and 70% r.h. for 150 days with about 12–25% weight loss (Gonzales and de Rivera 1972).

Chilling injury

Storage at either 3 or 12°C resulted in total physiological breakdown within 3–4 weeks (Czyhriniciw and Jaffe 1951). At 5°C they appeared to store well for 6 weeks but on removal to 25°C chilling injury symptoms were rapidly produced (Coursey 1961). Symptoms include flesh browning and a susceptibility to microorganism infection.

Green beans, kidney beans, snap beans, common beans

Phaseolus vulgaris, Leguminosae.

Botany

See also runner bean, *P. coccineus*. The edible portion is the pod or legume, which is long and slender, containing up to 12 seeds. It probably originated in the western parts of Guatemala and Mexico. The pods will freeze at about –1.3 to 1.1°C (Wright 1942) or

–0.89–0.72°C (Weichmann 1987). Inaba *et al.* (1989) found that there was increased respiration in response to ethylene at 100 μl $litre^{-1}$ for 24 hours at 20–35°C, but the effect was much reduced at lower temperatures.

Harvesting

This is carried out when they are still young, crisp and tender and before there is any obvious swelling to show the beans forming inside. It takes about 60–150 days from sowing to harvesting, depending on cultivar and growing conditions. If harvesting is delayed they will become tough and stringy. Harvesting is by hand every 3–4 days. In Malaysia they were harvested directly into bags worn around the neck that were carried to the field side when full and transferred into large baskets (Figure 12.61).

Figure 12.61 Hand harvesting of beans into a basket in Malaysia.

Discoloration

Mechanical harvesting of green beans for processing may lead to discoloration of the broken ends of the beans, especially if there is a delay in processing them (Sistrunk *et al.* 1989). Exposing them to sulphur dioxide at 0.7% for 30 seconds reduced broken end discoloration as a result of mechanical injury (Henderson and Buescher 1977). Discoloration of the cut end can also be prevented by exposure to 20–30% CO_2 for 24 hours (Hardenburg *et al.* 1990).

Moisture condensing on the pods may cause a dark discoloration on the skin surface. This is a particular problem when they are removed from cold storage to warmer temperatures for marketing.

Precooling

Forced air cooling with high humidity was the most acceptable method, hydrocooling was said to be suitable and air cooling in a conventional cold store and vacuum cooling less suitable (MAFF undated). However, beans are highly susceptible to bacterial diseases and condensation spotting and therefore any precooling that involves wetting the beans would normally be detrimental.

Storage

At room temperature of 20°C and 60% r.h. they could be kept for about 2 days (Mercantilia 1989). Refrigerated storage recommendations are as follows:

- 4.4°C and 85% r.h. (Platenius *et al.* 1934)
- 0–3.3°C and 85% r.h. for 3 weeks with 20% loss (Wardlaw 1937)
- 2°C and 90% r.h. for 2 weeks (Wardlaw 1937)
- 7.2°C and 80–85% r.h. to 22 days (Wardlaw 1937)
- 7°C for 5–8 days (Fidler 1963)
- 7.2°C and 85–90% r.h. for 8–10 days if cooled immediately after harvest (Anon 1967)
- 7.2°C and 90–95% r.h. for 1 week (Lutz and Hardenburg 1968)
- 4.4°C and 90–95% r.h. for 10 days if they are processed immediately after storage (Lutz and Hardenburg 1968)
- 7.2–10°C for 8–10 days, 3.3–5.6°C and 88% r.h. for 2–3 weeks with 18% loss (Pantastico 1975)
- 4–7°C and high humidity with some variation in storage life between cultivars, but rarely longer than 10 days (Tindall 1983)

- 10°C (Umiecka 1989)
- 5–6°C and 90–95% r.h. (Mercantilia 1989)
- 0°C for 2 days (Mercantilia 1989)
- 4 and 12°C for 4 days (Mercantilia 1989)
- 5°C for 7 days (Mercantilia 1989)
- 7.2°C and 90–95% r.h. for 10–14 days (SeaLand 1991)
- 4.4°C and 95% r.h. for 7–10 days for *haricot vert* (SeaLand 1991)
- 7–8°C and 95–100% r.h. for 1–2 weeks (Snowdon 1991).

Chilling injury

Chilling injury was reported to occur after 3–6 days at 4.4°C (Anon 1968) or 3 days at 4.4°C (Lutz and Hardenburg 1968). In both cases the symptoms did not occur until they were removed to a higher temperature.

Controlled atmosphere storage

Storage in 18% CO_2 at either 0 or 15°C resulted in off-flavours and injury (Pantastico 1975). The use of 5–10% CO_2 with 2–3% O_2 retarded yellowing (Hardenburg *et al.* 1990). SeaLand (1991) recommended the same conditions. Controlled atmosphere storage at 0–5°C with 5–10% CO_2 and 2–3% O_2 had a fair effect on storage but was of limited commercial use (Kader 1985). Saltveit (1989) recommended 5–10°C with 4–7% CO_2 and 2–3% O_2, which had only a slight effect, but for those destined for processing 5– 10°C with 20–30% CO_2 and 8–10% O_2 was recommended, but it had only a moderate effect.

Hypobaric storage

Pods could be kept in good condition in storage for 30 days at reduced pressure compared with 10–13 days in cold storage alone (Haard and Salunkhe 1975). McCaslan 42 pole beans and Sprite bush beans stored better at 76 and 152 mmHg and 7°C for 2 weeks than similar beans stored at 760 mmHg (Spalding 1980).

Guavas

Psidium guajava, Myrtaceae.

Botany

It was classified as a non-climacteric fruit by Biale and Barcus (1970). The fruit is a berry with a persistent calyx and is borne directly on the trunk or branches of the tree. They are usually pear shaped, but can be round and are up to 15 cm long (Figure 12.62, in the colour plates). Levels of polyphenols, particularly the high molecular weight polyphenols, in guava were shown to decrease as the fruit matures (Itoo *et al.* 1987). Young guava fruit contained 620 mg per 100 g fresh weight, of which 65% were condensed tannins (Itoo *et al.* 1987). Over 80 volatile aroma compounds have been identified in guava fruit (Askar *et al.* 1986). The production of volatile chemicals has been shown to change during maturation of guava. As the fruit matures the production of isobutanol, butanol and sesquiterpenes decreases and in mature and ripe fruit ethyl acetate, ethyl caproate, ethyl caprylate and *cis*-hexenyl acetate levels are high (Askar *et al.* 1986). β-Caryophyllene was identified as an important volatile associated with the aroma of guava (McLeod and de Troconis 1982).

Calcium treatment

In India, preharvest application of 1% calcium nitrate resulted in the lowest mean loss in weight (5.6%) and lowest respiration rate (64.9 mg CO_2 kg^{-1} h^{-1}) and maintained the firmness of fruits during storage longer than the untreated fruit or those sprayed with other calcium nitrate concentrations. This treatment also resulted in the highest mean total soluble solids and ascorbic acid contents (13.8% and 272.7 mg per 100 g pulp, respectively), and the lowest mean acidity value (0.64%) and lowest incidence of disease (caused by *Pestolotia psidii*) after 9 days of storage (Singh and Singh 1999). The storage life at 28–30°C was doubled to 12 days when fruits of the cultivar Baladi were dipped in a mixture of 2% calcium chloride and 2% corn flour (Amen 1987). Fruits immersed for 20 minutes in 0.5 or 1.0% calcium solution maintained marketable quality for up to 16 days at 10°C and 90% r.h. (Gonzaga-Neto *et al.* 1999).

Harvesting

Generally, harvest maturity is judged by the skin becoming smoother, loss of green colour and development of their distinct odour. Guavas have a strong aroma, which develops during ripening. The aroma volatiles may be evident at an early stage of ripeness; it is thought that these may be involved in attracting fruit flies that may attack the fruit as they begin to ripen. One way to reduce fruit fly infestation is to harvest at an early stage of maturity before the aroma volatiles have been synthesized. In India, the cultivar Sardar exhibited a sigmoid pattern of growth and took 18 weeks from fruit set to attain harvest maturity (Rao *et al.* 1995). A steady increase in total soluble solids of the cultivar Allahabad Safeda occurred up to 120 days from fruit set, while total pectic substances increased for 90 days and then declined (Majumder and Mazumdar 1998).

The trees are small so harvesting is by hand where mature fruits are easily selected. The meadow orchard system of production can be used to facilitate mechanical harvesting. This is where the whole of the tree is harvested just above the ground each year or every other year. The fruit are then separated from the stems in machines in the packhouse (Mohammed *et al.* 1984). The tree regrows from the stump left in the ground and the cycle continues.

Disease control

In Yemen, Kamal and Agbari (1985) recommended preharvest sprays with either metiram or propineb for control of fruit canker caused by *Pestalotiopsis psidii.* Snowdon (1990) also recommended orchard application of chemical fungicides in addition to postharvest fungicidal dips and low-temperature storage for *P. psidii* control. Preharvest fungicidal sprays can also control anthracnose caused by *Colletotrichum gloeosporiodes* (Pantastico 1975). Qadir and Hashinaga (2001) showed that postharvest fumigation with a mixture of 80% N_2O and 20% O_2 delayed the appearance of disease and reduced the lesion growth rate on inoculated fruit.

Storage

Storage at ambient temperatures is characterized by a relatively short shelf-life due to the rapid development of fungal rots. Wills *et al.* (1983) quoted a life of about 1 week at 20°C for the cultivars 1050 and GA 11–56. At 25–30°C Singh and Mathur (1954) showed that Safeda had a shelf-life of only 3 days. The cultivars Chittidar and Sardar had a storage life of 9 days, and Allahabad Safeda 6 days at 16–20°C and 80–85% r.h. (Singh *et al.* 1990). At 20°C and 60% r.h. the shelf-life was reported to be 7–10 days (Mercantilia 1989). Refrigerated storage recommendations are as follows:

- 8–10°C and 85–90% r.h. for 4 weeks for Safeda (Singh and Mathur 1954)
- 2–7°C for 1 week for fully ripe fruit with only minimal loss of ascorbic acid (Boyle *et al.* 1957)
- 7–10°C and 85–90% r.h. for about 3 weeks (Anon 1967)
- 8.3–10°C and 85–90% r.h. for 2–5 weeks with 14% weight loss (Pantastico 1975)
- 5°C (Wills *et al.* 1983)
- 10°C and 90% r.h. for 3 weeks (Mercantilia 1989)
- 5–10°C and 90% r.h. for 2–3 weeks (Hardenburg *et al.* 1990)
- 3.5–7°C for several weeks (Vazquez-Ochoa and Colinas-Leon 1990)
- 5–10°C and 90% r.h. for 2–3 weeks (Snowdon 1990)
- 10°C and 85–90% r.h. for 14–21 days (SeaLand 1991).

Chilling injury

This was reported to occur during storage at 0–2.2°C (Lutz and Hardenburg 1968). Wills *et al.* (1983) also showed that fruit stored at 0°C suffered from chilling injury, but had lower levels of rotting.

Modified atmosphere packaging

All fruits of the cultivar Kumagai stored well at 7°C and 80–90% r.h. for up to 37 days when sealed in low-density polyethylene, polypropylene or heat-shrinkable copolymer films. They were of good flavour and acceptable appearance compared with those stored unpacked. However, Yamashita and Benassi (1998) found that fruits sealed in polyester film developed off-flavour after 17 days. It was found that when fruits of the cultivar Paluma were harvested just before ripening (fruits reaching complete growth but still with green peel) they maintained marketable quality for up to 16 days when stored in polyethylene bags and stored at 10°C and 90% r.h. (Gonzaga-Neto *et al.* 1999). In experiments on various films, Jacomino *et al.* (2001) found that polyolefinic film with selective permeability showed the best results. In spite of affecting the normal ripening of the guavas, the fruits had a fresh appearance and uniform colour, normal flavour and absence of decay for up to 14 days of storage in 10°C and 85–90% r.h. followed by 3 days at 25°C and 70–80% r.h. During this same period, the control fruits became excessively mature and showed a high incidence of decay.

Hawthorne

Crataegus spp., Rosaceae.

Botany

There are several species of *Crataegus* that produce edible fruit. The common hawthorne, *C. monogyna*, grows wild through nothern Europe and the azarole, *C. azarolus*, fruit have an apple flavour and can be used for making jam. In Mexico, *C. mexicana* produces edible fruit that are round and slightly flattened, with high nutritive value (Martinez Damian *et al.* 1999).

Harvesting and storage

Martinez Damian *et al.* (1999) investigated *C. mexicana* fruits (selections San Pablo and Tetela) growing at Chapingo, Mexico. Fruit growth showed a simple sigmoidal pattern, reaching physiological maturity after 165–175 days, with a length:diameter ratio of 0.93, a fruit diameter of 3.15 cm, an ascorbic acid content of 97.5 mg per 100 g and a maximum respiration rate of 90 mg kg^{-1} h^{-1}. They found that they had a medium level of perishability.

Hog plum, yellow mombin

Spondias mombin, Anacardiaceae.

Botany

The fruit is oval in shape and up to 4 cm long. It contains a large seed surrounded by a thin layer of watery, juicy, aromatic, pleasantly sub-acid, pungent pulp.

Storage

No information could be found on its postharvest behaviour.

Horn of plenty

Craterellus (*Cratarellus*) *cornucopioides.*

Botany

It is a bracket fungus with a cap diameter of up to 8 cm and a height of up to 10 cm. The cap is irregular and funnel shaped with a wavy margin and is a grey–brown in colour and grey and slightly ribbed on the underside. The stalk is short and stout, tapering towards the base. In a comparison of eight species on edible mushrooms, Vetter (1993) found that *C. cornucopioides* was among the low-protein species that were tested.

Storage

No information could be found on its postharvest life, but from the literature it would appear that all types of edible mushrooms have similar storage requirements and the recommendations for the cultivated mushroom (*Agaricus bispsorus*) could be followed.

Horse mushroom

Agaricus arvensis.

Botany

It is stout with a cream to off-white cap that may become stained with yellow as it becomes over-mature. It is globular when it first appears, but as it develops the vale is broken, leaving a shallow dome and gills that are white, then change to pink and finally to brown. The stem is off-white and thick and tapers to the base and has two distinctive rings where the vale was attached.

Storage

No information could be found on its postharvest life, but from the literature it would appear that all types of edible mushrooms have similar storage requirements and the recommendations for the cultivated mushroom (*Agaricus bispsorus*) could be followed.

Horseradish

Armoracia rusticana, Cruciferae.

Botany

It is a herbaceous plant that produces long, cylindrical, fleshy roots whose white flesh has a very pungent flavour. The fleshy roots will freeze at about –3.8 to 2.4°C (Wright 1942). It is a very persistent plant: 'once a horseradish grower, always a horseradish grower.'

Harvesting

In temperate countries the roots are harvested when the tops have been killed by frost, and these store very well. Immature roots do not store well (Lutz and Hardenburg 1968; Kay 1987).

Storage

They may be stored over winter in cellars, pits or trenches. At room temperature of 20°C and 60% r.h. it was claimed that they could be kept for only 7–10 days (Mercantilia 1989). Refrigerated storage recommendations are:

- −1.1 to 0°C and 90–95% r.h. for 10–12 months (Lutz and Hardenburg 1968)
- −1 to 0°C and 95% r.h. for 10–12 months (Mercantilia 1989)
- 0°C and 95–100% r.h. for 300–350 days (SeaLand 1991).

Controlled atmosphere storage

Controlled atmosphere storage had a slight to no effect (SeaLand 1991) and is obviously not necessary since they store so well in refrigerated storage.

Huckleberries

Solanum nigrum guineese, Solanaceae.

Botany

The fruit is a berry and is similar in appearance to blueberry, but has a tougher skin, a sharper flavour and hard seeds. It is common in the southern states of the USA and Mark Twain used the name for his character Huckleberry Finn.

Storage

No information on its postharvest life could be found.

Hyssop

Hyssopus officinalis, Labiatae.

Botany

It is a herb that has been used from ancient times for salads and flavouring stews and meat. It can also be used medicinally as an expectorant. Its flavour is described as slightly bitter and minty and it is used in the liqueur Chartreuse. It was also used for ritual cleansing and was mentioned in the Bible no less than 12 times, for example in Numbers, chapter 19, verse 18, 'And a clean person shall take hyssop, and dip it in the water and sprinkle it upon the tent . . . .' The respiration rate increased from 484 W tonne^{-1} at 10°C to 1080 W tonne^{-1} at 20°C and 2041 W tonne^{-1} at 30°C (Bottcher and Gunther 2001).

Harvesting

It is usually cut just before or just after it begins to flower. Sometimes a second cut is taken.

Storage

Plants at approximately the 50% flowering stage were harvested and stored at 10°C with 98% r.h., 20°C with 95% r.h. or 30°C with 92–97% r.h. by Bottcher and Gunther (2001). The loss of dry matter was 1.6% at 10°C, 3.0% at 20°C and 6.5% at 30°C over a 24-hour period after harvest. The colour and external quality of the harvested herbs were best maintained at 10°C.

Intoxicating yam, Asiatic bitter yam

Dioscorea hispida, Dioscoreaceae.

Botany

Several globose to deeply lobed tubers are borne on each plant, which may be fused into a cluster weighing up to 35 kg (Kay 1987). They are pale skinned and covered with masses of fibrous roots. They are of southeast Asian origin and grow wild in India, Indonesia, the Philippines and Papua New Guinea, and thrive in tropical rain forest conditions. They are extremely poisonous and in the East it was said that the tubers are used for 'criminal and hunting purposes' and were a major famine food (Kay 1987).

Harvesting

They are mature about 12 months after planting and are harvested by digging the tubers by hand.

Detoxification

Tubers contain the toxin dioscorine, which is removed by slicing or grating, boiling them and finally placing them in running water or repeated changes of salt water (Kay 1987).

Storage

No information could be found on their postharvest life.

Jaboticaba

Myrciaria cauliflora, Myrtaceae.

Botany

The fruits are produced directly on the trunk and the older branches and are green as they develop and mature to a purplish black colour. The skin is tough and the pulp soft and juicy with a sub-acid and slightly aromatic flavour resembling the taste of grapes. There is usually just a single seed embedded in the pulp. The fruit are produced on small bushy trees. It is a native of Brazil.

Harvesting

Harvesting is by hand when the fruits have turned completely purplish black.

Storage

The only information on storage was 12.8°C and 90–95% r.h. for 2–3 days (SeaLand 1991).

Jackfruit, jaca

Artocarpus heterophyllus, A. integra, Moraceae.

Botany

Trees produce fruit that are a barrel-or pear-shaped syncarp, which can be up to 90 cm long and 50 cm in diameter and can weigh up to 40–50 kg (Figure 12.63). They have a distinctive yellow skin covered with short, hexagonal, fleshy spines. The flesh is aromatic, yellow, thick and waxy and since it is propagated by seed the fruit from some trees have a sweet and firm pulp whereas those from other trees can be soft and acid. The flesh contains numerous large seeds. The fruit arise from the trunk and main branches. Its respiratory pattern led to it being classified as an indeterminate fruit between climacteric and non-climacteric by Biale and Barcus (1970). However, during ripening, the fruit was found by Selvaraj and Pal (1989) to exhibit a typical climacteric pattern of respiration. They are a native of India and are used both for cooking and eating as fresh fruit.

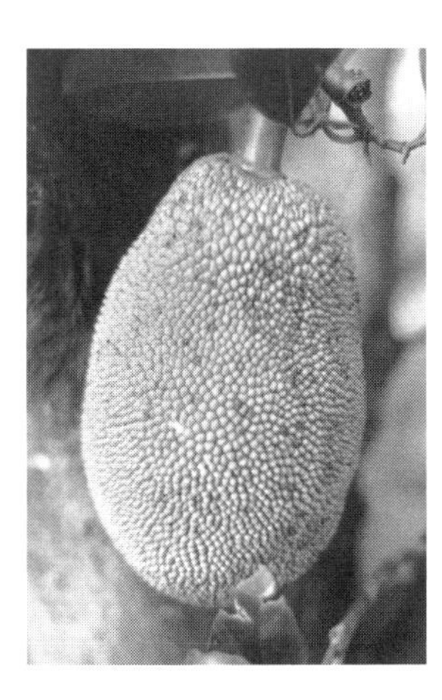

Figure 12.63 Jackfruit growing directly on the trunk of the tree in Malaysia.

Harvesting

Harvest maturity is judged to be when the last leaf of the fruit stalk turns yellow, the spines become spaced and yield to moderate pressure, it develops an aromatic odour and produces a hollow sound when tapped (Pantastico 1975). Similar results were obtained by Selvaraj and Pal (1989), who found that the fruit is judged to be ready for consumption when:

- it produces a strong ripe fruit flavour
- the skin colour, at least in part, changes to yellow
- there is a marked change in the pressure required to press the rind of the fruit with a finger.

Storage

At room temperature of 20°C and 60% r.h. the fruit could be kept for 3–5 days (Mercantilia 1989). Mathur *et al.* (1950) investigated refrigerated storage at seven temperatures ranging from 0–28°C and at various relative humidity conditions ranging from 55–90%. From their study, the optimum storage conditions were 11–13°C and 85–90% r.h. for about 6 weeks. Other refrigerated storage recommendations are as follows:

- 11.1–12.8°C and 85–90% r.h. for 6 weeks with 15.6% weight loss (Pantastico 1975)
- 13°C and 90% r.h. for 1–3 weeks (Mercantilia 1989)
- 13°C and 95% r.h. for 1–3 weeks (Snowdon 1990)
- 13.3°C and 85–90% r.h. for 14–45 days (SeaLand 1991).

Chilling injury

Mercantilia (1989) indicated that it was very sensitive to chilling injury, and Mathur *et al.* (1950) claimed that it occurred after exposure to 11°C.

Jerusalem artichoke

Helianthus tuberosus, Compositae.

Botany

The plant produces a number of tubers up to 8 cm in diameter and up to 20 cm long. They have large, pronounced eyes and a thin, white or sometimes red skin with white flesh. It will freeze at about –2.71 to 2.3°C (Wright 1942).

Harvesting

Harvesting can be carried out when the leaves begin to wither. In Britain they are commonly harvested in autumn, but harvesting can be delayed until spring since they keep well buried in the soil.

Storage

At room temperature of 20°C and 60% r.h. they could be kept for 1–2 weeks (Mercantilia 1989). At temperatures over 4°C they lost moisture rapidly and shrivelled while at 16°C in still air they developed mould growth in 2–3 weeks (Wardlaw 1937). Refrigerated storage recommendations are as follows:

- 0–1.7°C and 89–92% r.h. for 4 months (Wardlaw 1937)
- 0°C and 90–95% r.h. for 150 days with 20% loss due to decay and shrivelling (Tindall 1983)
- –0.5 to 0°C and 90–95% r.h. for 2–5 months (Anon 1967; Hardenburg *et al.* 1990)
- 0°C and 90% r.h. for 3–5 month (Mercantilia 1989)
- 0°C and 90–95% r.h. for 100–150 days (SeaLand 1991)
- –0.5 to 0°C and 90–95% r.h. for 2–5 months (Snowdon 1991).

Jew's ear

Hirneola auricula-judae.

Botany

It is a jelly fungus and has a rubbery texture, but is cultivated and eaten in China and is also used medicinally to cure sore throats and inflamed eyes. It is reddish brown and can be almost translucent. It is usually found growing on branches of elder (*Sambucus nigra*). As the common name implies, it resembles a human ear and the name is a corruption of Jude's ear since traditionally in the Christian religion Jude was supposed to have hanged himself on an elder tree.

Harvesting

It should be harvested while still young because as it matures it darkens in colour and becomes hard.

Storage

No information could be found on its postharvest life, but from the literature it would appear that all types of edible mushrooms have similar storage requirements and the recommendations for the cultivated mushroom (*Agaricus bispsorus*) could be followed.

Jujube, Chinese jujube

Zizyphus jujuba, Rhamnaceae.

Botany

This is a deciduous tree of temperate regions and has been cultivated in China for at least 4000 years (Purseglove 1968). The fruit is a drupe and its pulp has a crisp texture with a sweetish sub-acid flavour. Wu *et al.* (2001) stored the cultivars Zanhuang Dazao and Shenglizao at 0°C, which resulted in an initial accumulation of ascorbic acid, followed by a gradual decrease and the flesh firmness also gradually decreased. See also the Indian jujube, *Z. mauritiana.*

Prestorage treatment

Wu *et al.* (2001) showed that treatment with 1% calcium chloride or 1% calcium chloride + 15 mgl^{-1} 6-benzyladenine reduced the rate of ascorbic acid loss and of the decrease in flesh firmness and loss in fresh fruit. The effect of 15 mgl^{-1} 6-benzyladenine on respiration rate and fruit quality was lower than that of the calcium chloride treatment.

Harvesting

They are harvested from the tree by hand, usually when the skin has turned an orange–red colour with just a hint of brown and they are still firm.

Storage

Wu *et al.* (2001) used storage at 0°C, but no other information could be located on its postharvest life.

Ripening

They will ripen and become soft and sweet with a brown skin after 1–2 days at room temperature.

Jujube, Indian jujube, ber

Zizyphus mauritiana, Rhamnaceae.

Botany

It is a small evergreen tree and is widespread throughout the dryer parts of the tropics. The edible portion is the orange–brown-skinned fruit, which are round- to oblong-shaped drupes and 1–3 cm long. It has a single, hard central stone surrounded by acid pulp of variable thickness, which may be delicious or barely edible depending on the variety. Abbas (1997) indicated that the fruit is climacteric. He quotes figures for respiration rate at 20°C as 28 mg CO_2 kg^{-1} h^{-1} at harvest, progressively rising to 140 mg CO_2 kg^{-1} h^{-1} after 8 days then 113 mg CO_2 kg^{-1} h^{-1} after 9 days. Ethylene production peaked at 130 μl kg^{-1} h^{-1} after 6 days.

Harvesting

The cultivar Zaytoni takes about 120 days from flowering to maturity, but other cultivars can take up to 180 days (Abbas 1997). Maturity is when the first brown spots appear on the skin (Wardlaw 1937). Abbas (1997) mentioned that soluble solids and acidity are the most important criteria used in the Basrah region of Iraq. He also indicated that in India, specific gravity and a change in skin colour to golden yellow are the most suitable methods for judging maturity.

Diseases

Singh *et al.* (2000) found that 18 different fungal species caused postharvest decay of jujube fruits, but only *Aspergillus flavus* infection was of significance. Infection resulted in maximum loss of ascorbic acid from the fruit tissues and also induced aflatoxin production during pathogenesis. It was noted that approximately 85.7% of the *A. flavus* isolates associated with jujube decay were toxic.

Hot water treatment

Hot water treatment at 40°C was effective in increasing its storage life by slowing ripening and also reduced the rate of ethylene evolution (Anuradha *et al.* 1998).

Calcium treatment

Postharvest dipping of fruits in 0.17% calcium salts, as calcium chloride or calcium nitrate solutions, extended the shelf-life of mature green Umran and Kadaka fruits by 3 days (Panpatil *et al.* 2000).

Storage

The limiting factors in storage are cracking and wilting. Under ambient conditions of 26–30°C fruits showed a high degree of disease and loss in colour but could be stored for up to 9 days (Nath and Bhargava 1998). Abbas (1997) indicated that fruit picked at the proper stage of maturity have a storage life of 6–8 days at room temperature, depending on the cultivar. Optimum refrigerated storage was given as:

- 0°C for 45 days with 5 days subsequent shelf-life at room temperature (McGuire 1929)
- 4–10°C for 25–33 days for several cultivars (Abbas 1997)
- 6–8°C for up to 15 days (Nath and Bhargava 1998).

Modified atmosphere packaging

Modified atmosphere packaging alone had no significant effect on the storage life of fruits at room temperature, but packing fruits in sealed perforated plastic bags containing potassium permanganate-impregnated chalk sticks was the most effective and reduced the rate of ethylene evolution and slowed ripening (Anuradha *et al.* 1998). McGuire (1929) found that at 0°C in closed containers fruit could be stored successfully for 85 days.

Ripening

Fruits harvested while still green could be ripened to a brown colour with soft sweetish flesh and no off-flavours at 21.1°C in 1000 μl $litre^{-1}$ ethylene (Wardlaw 1937). Abbas (1997) found that dipping fruit in 500 mg l^{-1} Ethrel advanced ripening by 6–7 days compared with untreated fruit. In the latter work there was no indication of the temperature used; also, Ethrel is not permitted to be used commercially in many countries.

Kale

Brassica oleracea variety *acephala*, Cruciferae.

Kale is grown in Europe for animal feed, but there are varieties that are grown as a vegetable for human consumption; see collards (*Brassica oleracea* variety *acephala* f. *viridis*) and Chinese kale (*Brassica alboglabra*).

Kiwanos, horned melons, melano

Cucumis metuliferus, Cucurbitaceae.

Botany

It is an indehiscent fleshy berry, and is elongated to oval in shape. Its skin is thick, a bright golden orange and covered with soft spikes or bumps unevenly distributed over, the surface, giving it a striking and attractive appearance. The pulp is bright green and jelly-like, containing numerous elongated soft black seeds but otherwise similar to passionfruit. Its flavour is fairy bland and somewhat resembles a mixture of banana, lime and cucumber. In mature fruit, fibre content was about 4.0 g per 100 g fresh weight. Citric acid was the main organic acid with 0.9 g per 100 g fresh weight and potassium was the main mineral element with 266 mg per 100 g fresh weight, but sucrose was not found (Romeron Rodriguez *et al.* 1992). It originated in Africa, but is now grown commercially in the USA, New Zealand and Portugal.

Harvesting

Fruits required about 5 weeks to reach the mature green stage from flowering and another 2 weeks to reach full maturity (Benzioni *et al.* 1993).

Storage

It is reported to store very well and the shelf-life at 20–24°C was found by Benzioni *et al.* (1993) to be at least 3 months. Refrigerated storage recommendations are:

- 10°C and 90% r.h. (Snowdon 1990)
- 10°C and 90% r.h. for 180 days (SeaLand 1991).

Ripening

Ethylene treatment was beneficial and may enable growers to harvest the crop at the breaker stage and to store fruits for a longer period (Benzioni *et al.* 1993).

Kiwifruits, Chinese gooseberries, yang tao

Actinidia chinensis, *A. deliciosa*, Actinidiacea.

Botany

The fruit is classified as climacteric. In their review, Sfakiotakis *et al.* (1999) found that kiwifruits differ from other climacteric fruits in that they show no increase in ethylene production on the vine nor produce ethylene during cold storage. The fruits show a typical climacteric rise in respiration rate and ethylene production at temperatures of 17–34°C, whereas lower temperatures of <11–14.5°C they behave as non-climacteric fruits. The fruits show thermoregulation of ethylene production separate from the ripening process. Ethylene production was shown to be induced autocatalytically by exposure to exogenous ethylene or propylene, exposure to chilling temperatures, wounding and *Botrytis* infection.

Preharvest treatments

Vines of the cultivar Hayward were sprayed with liquid calcium (Nutrical) at 9.33 l ha^{-1} on three dates between fruit set and 10 days before harvest. Controlled atmosphere-stored fruits from calcium-treated vines were firmer and of better quality than fruits from other treatment combinations (Basiouny and Basiouny 2000).

Harvesting

In New Zealand, optimum harvest maturity of kiwifruit is about 23 weeks after flowering (Pratt and Reid 1974). Soluble solids could be used as a measure of maturity, where they should have a soluble solids content in the flesh of at least 8% (Pratt and Reid 1974). In subsequent work, the OECD (1992) stipulated that the minimum soluble solids content should be 6.2% based on the average of 10 sample fruits. A non-destructive technique for assessing the harvest maturity uses a ballistic collision technique (Harvey *et al.* 1995). Fruit are commonly harvested into picking bags by hand (Figure 12.64). Various handling practices were compared by Benge *et al.* (2000), who found that harvesting directly into trays, rather than harvesting into bags, transferring them to pallet boxes and packing in a packhouse, did not significantly affect fruit softening during storage.

Figure 12.64 Kiwifruit harvesting into bags in New Zealand.

Quality standards

The OECD, Paris, published a brochure of standards for kiwifruit.

Grading

Clarke (1996) described a machine where there are two cables along which the crop runs. They are made up from high-tensile, polyethylene-coated steel. When the crop reaches the prescribed point, ejector bars controlled by solenoid valves and operated by compressed air cylinders push them into a chute. The machine is very gentle to the crop and high drops are unnecessary, although padded chutes can be provided.

Curing

Curing is carried out in Italy to reduce rotting during subsequent storage, especially rots caused by *Botrytis cinerea* and *Phialophora* spp. It was possible to carry out curing directly in refrigerated rooms, by gradually reducing the temperature from 10–0°C over approximately 10 days. Postponing the establishment of controlled atmosphere storage conditions by 30–50 days avoided its negative impact of increasing *Botrytis* rots, without any adverse effects on fruit firmness (Tonini *et al.* 1999).

Precooling

Precooling is a common practice with kiwifruits, but the incidence of *B. cinerea* stem end rot, low-temperature breakdown and physiological pitting is likely to be higher in precooled fruits than non-precooled fruits. It was concluded by Lallu and Webb (1997) that it is preferable not to precool kiwifruits that are predisposed to stem end rots, low-temperature breakdown or physiological pitting.

Storage

At room temperature of 20°C and 60% r.h. the fruit could be kept for 7–10 days (Mercantilia 1989). Refrigerated storage recommendations are:

- 0°C and 90–95% r.h. for 2–3 months (Mercantilia 1989)
- –0.5 to 0.3°C and above 94% r.h. (Warrington and Weston 1990)
- –0.5 to 0°C for 2–3 months (Snowdon 1990)
- 0°C and 90–95% r.h. for 28–84 days (SeaLand 1991).

Chilling injury

Exposure of fruits to chilling temperatures for at least 12 days stimulated ethylene production upon rewarming. However, fruits removed from prolonged storage at 0°C in conventional or controlled atmosphere storage showed reduced capacity to produce ethylene, due to reduced ACC production and ACC synthase and ACC oxidase activities (Sfakiotakis *et al.* 1999). Low-temperature breakdown can affect fruit after several months of cold storage. One of its symptoms manifests as a grainy appearance of the outer pericarp, followed by water soaking associated with extreme softening of the fruit (Bauchot *et al.* 1999).

Controlled atmosphere storage

Pratella *et al.* (1985) found that CO_2 above 10% was toxic, resulting in the production of off-flavours (Harman and McDonald 1983). Also, 10% CO_2 concentrations in storage reduced the rate of propylene-induced ethylene production, and reduced O_2

concentrations (1–5%) inhibited propylene-induced ethylene production at the ACC synthesis level (Sfakiotakis *et al.* 1999). Ultra-low O_2 storage inhibited ethylene production by chilling, mainly by destroying the system that converts ACC to ethylene (Sfakiotakis *et al.* 1999). Storage recommendations are:

- 0°C and 90–95% r.h. with 5% CO_2 and 2% O_2 (SeaLand 1991)
- 3–5% CO_2 and 2% O_2 increase storage life by 30–40% (Pratella *et al.* 1985)
- 0–5°C and 5% CO_2 with 2% O_2 gave excellent results (Kader 1985)
- 0–5°C with 3–5% CO_2 and 1–2% O_2 (Kader 1989, 1992)
- 0°C and 90–95% r.h. in 3% CO_2 and 1–1.5% O_2 maintained fruit firmness during long-term storage, but 1% CO_2 with 0.5% O_2 gave the best storage conditions while maintaining an acceptable level of incidence of rotting (Brigati *et al.* 1989)
- –0.5 to 1°C at 92–95% r.h. and 4.5–5% CO_2, 2–2.5% O_2 and ethylene at 0.03 μl litre^{-1} or less delayed flesh softening but strongly increased the incidence of *Botrytis* stem end rot (Tonini *et al.* 1989)
- 0°C and either 5% O_2 with 5% CO_2 or 5% O_2 with 2 CO_2, but the shelf-life of these fruits was limited to 15 days after 6 months storage for Hayward in Turkey (Ozer *et al.* 1999)
- 0°C and 3% O_2 with 10% CO_2 (Basiouny and Basiouny 2000).

Intermittent storage

Storage at 0°C for 1 week in air and 1 week in air enriched with either 10% or 30% CO_2 reduced fruit softening during storage (Nicolas *et al.* 1989).

Modified atmosphere packaging

Plastic liners are commonly used in boxes of fruits for long-distance transport. Mineral powders are incorporated into some films for kiwifruit, particularly in Japan, where it was claimed that among other thing these films can remove or absorb ethylene, excess CO_2 and water (Industrial Material Research 1989, quoted by Abe 1990).

Ethylene

Very low levels of ethylene affect the texture of kiwifruit. Arpaia *et al.* (1985) found that kiwifruit softened more quickly at levels as low as 0.03 μl litre^{-1} ethylene compared with ethylene-free air at 0°C. Accumulation of ethylene from wounding or *Botrytis* infection can also cause excessive softening of the fruits in storage (Sfakiotakis *et al.* 1999).

Ethylene absorbents

The cultivar Bruno was stored at –1°C in sealed polyethylene bags of varying thickness, with and without an ethylene absorbent (Ethysorb). The ripening rate of the fruit and its keeping quality during 6 months of storage and 7 days of subsequent shelf-life at 20°C were compared with those of fruit stored in unsealed 0.02 mm polyethylene bags. Without ethylene removal from the storage atmosphere, there was no significant effect of the modified atmospheres in the sealed bags on the rate of fruit ripening and no improvement in its keeping quality. However, when ethylene was absorbed from the storage atmosphere, increasing the thickness of the polyethylene bag retarded fruit ripening and, as a result, the CO_2 level within it. The storage life was extended to 6 months by storing the fruit at –1°C in sealed 0.04–0.05 mm thick polyethylene bags containing the absorbent. The average composition of the atmosphere in these bags was 3–4% CO_2, 15–16% O_2 and <0.01 μl litre^{-1} C_2H_4 (Ben Arie and Sonego 1985).

Ripening

Starch was shown to be hydrolysed to glucose and fructose and to a lesser extent to sucrose during ripening (Okuse and Ryugo 1981). Recommended ripening temperatures were 18–21°C using ethylene at 10 μl litre^{-1} for 24 hours in 85–90% r.h. (Wills *et al.* 1989). At either 0°C or 14°C Wills *et al.* (2001) found that the time to ripen of the kiwifruit cultivar Hayward increased linearly with logarithmic decrease in ethylene concentration over the range <0.005, 0.01, 0.1 1.0 and 10 μl litre^{-1}.

Kohlrabi, turnip rooted cabbage

Brassica oleracea variety *gongylodes*, Cruciferae.

Botany

The edible portion is the spherical secondary thickened stem that is up to 10 cm in diameter. It bears leaves just like a normal stem and it commonly has green or purple skin with white flesh. It will freeze at about –1.3 to 1.0°C (Wright 1942).

Harvesting

If over-mature they can be tough and fibrous, so they should be harvested when they are of an acceptable size and the leaves still appearing fresh and healthy. Tindall (1983) found that they normally mature within 10 weeks from transplanting and should be harvested before they become fibrous and woody.

Coatings

Coating with an 8% emulsion of carnauba wax reduced weight loss during storage at 18–21°C (Weichmann 1987).

Precooling

They should be cooled to 4.4°C as soon as possible after harvest (Ryall and Lipton 1972).

Storage

Weichmann (1987) stated that it was highly perishable because it develops fibres during storage, making it unmarketable, and therefore it should not be kept longer than 2–4 days. In trials with the cultivar White Vienna, moderate irrigation, low nitrogen rates (50 kg ha^{-1}) and storage at 1.7°C gave the best keeping quality (Bhatnagar *et al.* 1986). Refrigerated storage recommendations are:

- 0–0.6°C (Wardlaw 1937)
- 0°C and 90–95% r.h. for 2–4 weeks (Phillips and Armstrong 1967; Lutz and Hardenburg 1968)
- 0°C and 95% r.h. for 1 month (Ryall and Lipton 1972)
- 0°C and 95% r.h. for up to 30 days (Tindall 1983)
- 0°C and 95–100% r.h. for 25–30 days (SeaLand 1991)
- 0–1°C and 95–100% r.h. for 2–4 weeks (Snowdon 1991).

Controlled atmosphere storage

Controlled atmosphere storage was reported to have only a slight to no effect (SeaLand 1991).

Modified atmosphere packaging

Topped kohlrabi can be stored in perforated film to retard moisture loss (Ryall and Lipton 1972).

Kumquats

Fortunella spp., Rutaceae.

Botany

There are two species grown for their fruit, the round kumquat (*F. japonica*) and the oval kumquat (*F. margarita*). The fruit is a berry up to 4 cm in diameter with a pulpy, mild, edible skin enclosing a sweet, acid pulp. They look like small oranges and their name in Cantonese, *kam kwat*, means golden orange, but they are not true citrus fruits. The whole fruit including the peel is eaten fresh (the peel is sweeter than the pulp) or made into delicious jam. They are widely cultivated at Japan and China and dwarf bushes are given as decorative presents in the Chinese New Year.

Harvesting

They are produced on small bushes and harvested by hand. They are mature when the fruit has turned from green to orange.

Chilling injury

Storage below about 8°C resulted in chilling injury and caused increased decay (Chalutz *et al.* 1989).

Storage

The fruit can be kept at 20°C and 60% r.h. for 1 week (Mercantilia 1989). Refrigerated storage recommendations are as follows:

- 10°C and 90% r.h. for 4 weeks (Mercantilia 1989)
- 8–11°C for several weeks (Chalutz *et al.* 1989)
- 4.4°C and 90–95% r.h. for 14–28 days (SeaLand 1991).

Langsat, lanzon, duku

Lansium domesticum, Meliaceae.

Botany

Langsat, lanzon, duku and longkong are considered separate types of fruit in Thailand because of their different appearance and flavour. Langsat was referred to as *Aglaia domestica* and longkong, duku nam and duku as *A. dookkoo* by Namsri *et al.* (1999), while longkong was referred to as *L. domesticum* by Rattanapong *et al.*

(1995). Earlier Prakash *et al.* (1977) referred to the duku and langsat as varieties of *L. domesticum.*

The edible portion of the tree is the fruit, which is a berry that is usually borne in clusters from older branches. They are round to oval in shape and vary in size depending on the type and can be up to 4 cm long. They have a thin, tough, milky skin and translucent, sweet, juicy pulp in which are embedded 0–5 seeds. They probably originated in the Malay Peninsula.

Preharvest treatment

Spraying clusters of fruit with 5% calcium chloride 10 or 11 weeks after fruit set reduced fruit drop and increased fruit firmness, total soluble solids content and the ratio of total soluble solids to titratable acidity (Rattanapong *et al.* 1995).

Harvesting

They are considered suitable for harvesting when they have lost their green colour and turned a dull yellow (Pantastico 1975). Over-mature fruits develop brown specks and lose their pubescent bloom, but are said to be of better flavour (Pantastico 1975). In duku, the time taken from floral initiation to anthesis was 7 weeks; ripe fruits were formed 15 weeks after flowering. Only 11% of the flowers normally set fruit, but on bagging the inflorescences at the floral-primordia stage, 21% flowers set fruit (Prakash *et al.* 1977).

Chilling injury

This was said to occur when fruits were exposed to temperatures below 10–12.8°C as brown specks with a water-soaked appearance (Wardlaw 1937; Pantastico 1975).

Storage

Fruits buried in boxes of sawdust had a shelf-life of 7 days in tropical ambient conditions compared with only 3 days for unprotected fruit (Wardlaw 1937), but wastage was high in both cases. Placing fruits in cold storage directly after harvest was said to extend their storage life (Wardlaw 1937). Refrigerated storage recommendations are as follows:

- 12.8–15.6°C for 13 days with 9.5% loss or 16 days with 50% loss (Wardlaw 1937)
- 11.1–14.4°C and 85–90% r.h. for 2 weeks with 24.3% weight loss (Pantastico 1975)
- 11.1°C and 85–90% r.h. for 14 days (SeaLand 1991).

Controlled atmosphere storage

Storage at 14.4°C with 0% CO_2 and 3% O_2 extended the storage life from 9 days in air to 16 days (Pantastico 1975).

Modified atmosphere packaging

Sealing fruits in polyethylene film bags extended their storage life but resulted in skin browning, which was due to CO_2 toxicity (Pantastico 1975).

Leeks

Allium ampeloprasum variety porrum, Alliaceae.

Botany

The edible portion is the leaves and the immature bulbs (Figure 12.65). They will freeze at about −1.9 to 1.2°C (Wright 1942). Respiration rate was given as 10–20 ml CO_2 kg^{-1} h^{-1} at 5°C (Weichmann 1987).

Harvesting

They are harvested by pulling the tops when they are judged to be thick enough for the market. On dry, heavy soils it may be necessary to prise the roots with a spade or fork. The roots are trimmed close to the base, taking care not to damage the base plate otherwise they can discolour and the outer leaves become detached.

Quality standards

They should be free from seed stalks and the bottom 2–3 cm white with the rest dark green.

Figure 12.65 Leeks being washed after preparation for the market.

Precooling

Forced-air cooling with high-humidity air and hydrocooling were the most suitable method, air cooling in a conventional cold store was also suitable and vacuum cooling was less suitable (MAFF undated).

Humidity

Humidity has a substantial effect on weight loss and therefore quality because when leeks lose a lot of weight they become flaccid and unmarketable. Storage losses due to decay and trimming at 0–1°C over a 14 week period were 45–75% in 90–95% r.h. and 25–55% in 98–100% r.h., while weight loss was 2.4–2.6% at 90–95% r.h. and 0.5–0.9% for 98–100% r.h. (Weichmann 1987).

Storage

At room temperature of 20°C and 60% r.h. the fruit could be kept for 1–2 days (Mercantilia 1989). Refrigerated storage recommendations are:

- 0°C and 85–90% r.h. for 1–3 months (Wardlaw 1937; Hardenburg *et al.* 1990)
- 0–1°C and 90–95% r.h. for 1–3 months (Anon 1967)
- 0°C and 90–95% r.h. for 1–3 months (Phillips and Armstrong 1967; Pantastico 1975)
- 0°C and 90–95% r.h. for 28–80 days (Tindall 1983)
- –1°C and 90–95% r.h. for 42 days (Mercantilia 1989)
- 1°C and 90–95% r.h. for 25 days (Mercantilia 1989)
- 4°C and 90–95% r.h. for 15 days (Mercantilia 1989)
- 8°C and 90–95% r.h. for 8 days (Mercantilia 1989)
- 0°C and 90–95% r.h. for 3–4 weeks; at –1°C the maximum storage period can be extended by a few days (Goffings and Herregods 1989)
- 0°C and 90–95% r.h. for 1 week (Hardenburg *et al.* 1990)
- 0°C and 95–100% r.h. (SeaLand 1991)
- 0–1°C and 95–100% r.h. for 1–3 months (Snowdon 1991).

Controlled atmosphere storage

Low O_2 levels were shown to reduce respiration rate at a variety of temperatures (Table 12.34).

CO_2 levels of 15–20% can cause injury (Lutz and Hardenburg 1968), but controlled atmosphere storage had variable effects as the following recommendations indicate:

Table 12.34 Effects of temperature and reduced O_2 level on the respiration rate (CO_2 production in mg kg^{-1} h^{-1}) of Musselburgh leeks (Robinson *et al.* 1975)

	Temperature (°C)				
	0	5	10	15	20
In air	20	28	50	75	110
In 3% O_2	10		30		57

- 1–3% O_2 with 5–10% CO_2 at 0°C for up to 4–5 months (Kurki 1979)
- 0–5°C with 3–5% CO_2 and 1–2% O_2, which had a good effect but was of no commercial use (Kader 1985, 1992)
- 0–5°C with 5–10% CO_2 and 1–6% O_2, which had only a slight effect (Saltveit 1989)
- 0°C and 94–96% r.h. with 2% CO_2, 2% O_2 and 5% CO for up to 8 weeks compared with 4 weeks at the same temperature in air (Goffings and Herregods (1989)
- 0°C with 3–5% CO_2 and 1–2% O_2 (SeaLand 1991).

Lemons

Citrus limon, Rutaceae.

Botany

The fruit is a berry or hesperidium, and is non-climacteric. The respiration rate increases rapidly with increasing temperature, as would be expected (Table 12.35). It will freeze at about –2.3 to 1.9°C (Wright 1942).

Harvesting

When the fruit matures it changes from green to yellow. Size is sometimes used to determine when to harvest, irrespective of the peel colour (Wardlaw 1937). A common size is 56–57 mm in diameter. The juice content increases as they mature on the tree. By taking representative samples of the fruit, extracting the juice in a standard and specified way and then relating the juice volume to the original mass of the fruit, it is possible to specify its maturity. In the USA the minimum value is 25%. The ratio between soluble solids and acidity, used with some other citrus fruits, is not important with lemons. They are usually clipped from the tree when the weather conditions are dry to avoid oleocellosis.

Table 12.35 Effect of temperature on the respiration rate (CO_2 production in mg kg^{-1} h^{-1}) of Eureka lemons (Hallar *et al.* 1945)

Temperature (°C)	0	4.4	10	15.5	21.1	26.6	32.2	37.7
Respiration rate	3.1	5.2	11.2	15.8	22.7	25.6	40.2	80.5

Handling

Tariq (1999), in a comparison of different types of damage, found that scuffing of the peel resulted in the greatest losses. Scuffing resulted in the peel turning brown and *Penicillium* spp. growing on the damaged area. Cuts and bruises were found to be less injurious to fruits than scuffing. Pekmezci (1983) showed that injured or mouldy lemons could generate relatively large amounts of ethylene.

Curing

Davies and Albrigo (1994) quoted 12.5–15°C with 95% r.h. for fruit harvested when they are light green. Sealing in plastic film bags and then exposing them to 34–36°C for 3 days resulted in the inhibition of *Penicillium digitatum* infection through lignification and an increase in antifungal chemicals in the peel of the fruit (Ben Yehoshua *et al.* 1989b). Sealing the fruit in plastic film before curing was said to be essential to reduce weight loss (Ben Yehoshua *et al.* 1989a). Tariq (1999) found that placing damaged fruits in sealed polyethylene film bags and exposing them to temperatures in the range 25–35°C for between 48 and 3 hours (the lower the temperature, the longer the exposure time required) completely eliminated the browning and fungal growth. The nature of curing appeared to be related only to temperature and humidity, since when polyethylene bags were used the thickness of the film (120 or 200 gauge) had no effect on the process, although it changed the O_2 and CO_2 content around the fruit. Curing could also be achieved without film if the surrounding atmosphere was maintained at 95–98% r.h. (Tariq 1999).

Prestorage treatments

The cultivar Verna, harvested at either colour break or uniform yellow, were vacuum-infiltrated with 100 μl $litre^{-1}$ gibberellic acid or heat treated at 45°C, and then stored at 15°C for 3 weeks. Both treatments increased fruit firmness during storage for both harvest maturities. Gibberellic acid treatment was the most effective in retarding the colour change during storage, especially in the colour break fruits; this was associated with the lowest levels of abscisic acid found in the fruit (Valero *et al.* 1998).

Disease control

Treating the cultivar Femminello Siracusano fruits by either immersion in *Penicillium digitatum* spores in 10% ethanol at 45°C for 3 minutes, followed by curing at 32°C for 1 day, or by the application of the yeast *Candida oleophila*, were as effective in controlling disease development as dipping fruits in 1 g l^{-1} a.i. imazalil (Lanza *et al.* 1998). Immersion of lemons in a 3% solution of sodium carbonate plus hot water treatment at 52°C were as effective as imazalil treatment against *P. digitatum* (Lanza *et al.* 1998).

Degreening

When harvested fruit arrive at the packhouse, they are graded into fully yellow and those still green or turning. The green or turning fruit are then degreened. Optimum temperatures for degreening ranged between 25 and 29°C and humidity as high as possible, generally between 90 and 95% r.h. with ethylene between 1 and 10 μl $litre^{-1}$ for 24–72 hours depending on the stage of maturity at harvest and the climate in which they were grown. Degreening of lemons can also be achieved by dipping fruit in 1000 μl $litre^{-1}$ Ethrel. This had the same effect as exposing the fruit to 50 μl $litre^{-1}$ ethylene gas for 24 hours (Amchem Information Sheet 46). Direct dipping in Ethrel of crops, which are to be eaten, may be governed by food legislation in some countries.

Storage

At a simulated room temperature of 20°C they could be kept for 1–3 weeks (Mercantilia 1989). Pekmezci (1983) showed that the Turkish cultivar Kutdiken could be stored for 8–9 months at 10°C and 85–90% r.h. Green lemons require slightly higher temperatures than yellow to prevent ethylene build-up. Air flushing to reduce ethylene accumulation in store does not need to be excessive unless a high level of fruit damage is suspected. Other refrigerated storage recommendations are as follows:

- 3–4°C and 65–70% r.h. for Italian lemons imported into Russia (Wardlaw 1937)
- 11–14.5°C and 85–90% r.h. for 1–4 months for green fruit (Anon 1967)
- 0–1.5°C and 85–90% r.h. for 3–6 weeks for coloured fruit (Anon 1967)
- 0–12.8°C and 85–90% r.h. up to 4 weeks for yellow fruit or 11.1–12.8°C for longer periods (Lutz and Hardenburg 1968)
- 15°C for up to 4 months after waxing (Anon 1968)
- 5.6–7.2°C and 85–90% r.h. for 6 weeks (Pantastico 1975)
- 14–15°C and 85–90% r.h. for 1–4 months for green fruit (Mercantilia 1989)
- 11–13°C and 85–90% r.h. for 3–6 weeks for yellow fruit (Mercantilia 1989)
- 8–12°C and 85–90% r.h. for 1–2 months (Hardenburg *et al.* 1990)
- 10–14°C and 90% r.h. for 2–5 months for green fruit (Snowdon 1990)
- 11°C and 90% r.h. for 3–6 months for yellow fruit (Snowdon 1990)
- 12.2°C and 85–90% r.h. for 30–180 days (SeaLand 1991)
- 10–12°C (Davies and Albrigo 1994).

Chilling injury

In spite of the recommendations given above, Cohen (1978) stated that yellow lemons should be kept above 10°C as lower temperatures encourage the earlier development of internal browning and peel pitting, especially where the fruit are harvested early. In earlier work Fidler (1963) also showed that green fruit were subject to chilling injury in storage below 12°C and another report indicated that it was 11.1°C (Anon 1968). Symptoms were given as pitting and spotting of the skin, staining and darkening of the membranes, dividing the pulp segments, water breakdown and off-flavours (Anon 1968).

Controlled atmosphere storage

The rate of colour change can be delayed with high CO_2 and low O_2 in the storage atmosphere but 10% CO_2 could impair their flavour (Pantastico 1975). Other recommendations are as follows:

- 10–15°C with 0–5% CO_2 and 5% O_2 10–15°C (Kader 1985) or with 0–10% CO_2 and 5–10% O_2 (Kader 1989) had a good effect but was not used commercially (Kader 1985, 1992)
- 12.2°C and 85–90% r.h. with 5–10% CO_2 with 5% O_2 (SeaLand 1991)
- 10–13°C and 85–90% r.h. with 5–10% O_2 and 0–10% CO_2 for 2–24 weeks (Burden 1997).

Ethylene

Ethylene has been known for a long time to increase the rate of respiration in lemons (Denny 1924).

Lemon balm

Melissa officinalis, Labiatae.

Botany

It is grown as a herb and the leaves are used for infusion to make tea. Its origin is the Mediterranean region and its generic name means honeybee in Greece, where it is much prized for its fragrant flavour. Bottcher *et al.* (2000) found that the harvested leaves have a high respiration rate at 10°C of 523 ± 109 W tonne^{-1}, at 20°C of 1190 ± 202 W tonne^{-1} and at 30°C of 2342 ± 232 W tonne^{-1}. In this study the comparatively high respiration levels were continued throughout a postharvest period of 80 hours.

Harvesting

It is normally harvested two or three times a year by cutting the plant close to the ground. The contents and composition of essential ingredients changed with cutting date and length of postharvest storage. In the more favourable second cut, the content of essential oils was twice as high as the first cut and contained higher quantities of citral A and B, geraniol, citronellal and geranyl acetate, but less linalool and β-caryophyllene (Bottcher *et al.* 2000).

Storage

As would be expected, postharvest quality was better after storage at 10°C than at 20 or 30°C (Bottcher *et al.* 2000).

Lesser yam, Asiatic yam, lesser Asiatic yam

Dioscorea esculenta.

Botany

Tubers are formed as swellings at the end of stolons that are borne in clusters, like potatoes, with up to 20 tubers per plant. They resemble long, narrow sweet potatoes, but can also be spindle shaped or branched. In the West Indies the tubers usually weigh up to 200 g and are up to 10 cm long and 5 cm in diameter, but in Papua New Guinea some varieties produce tubers weighing up to 3 kg (Kay 1987). Its origin was within the area of India, Vietnam, Papua New Guinea and the Philippines and it is now grown widely throughout the tropics (Kay 1987).

Curing

Ravindran and Wanasundera (1993) successfully cured tubers by exposure to sunlight for 3 days after harvest in mean ambient conditions of 29°C and 84% r.h.

Harvesting

This is usually carried out 6–10 months after planting when the vines begin to turn yellow. The dormancy period is very short so harvesting should not be delayed (Kay 1987). Commercial potato harvesting machines can be used for the lesser yam when they are planted on ridges (Kay 1987).

Sprouting

Tubers were stored for 150 days in thin layers on the floor of a well ventilated room at room temperature of 24–28°C and 70–90% r.h. Sprouting started after 120 days. Crude protein, starch and ascorbic acid contents decreased significantly during the first 60 days of storage, but the decrease was gradual thereafter. Losses of some minerals and a significant reduction in levels of oxalates were also noted during storage (Ravindran and Wanasundera 1993). Tubers of the cultivars Sree Priya and Sree Subrastored stored at room temperature of 30–32°C and 80–85% r.h. in the dark sprouted 70–80 days after harvest. In the apical and basal regions, the starch content decreased while the sugars and α-amylase activity increased (Muthukumarasamy and Panneerselvam 2000). Passam (1982) showed that postharvest treatment with gibberellins can delay sprouting, but no records were found of its commercial use and postharvest use of gibberellins is not permitted in some countries.

Storage

Ventilated storage at about 25°C for up to 60 days has been reported (Tindall 1983). Uninjured tubers can be stored for 4 months or more in ambient tropical temperatures in well ventilated stores (Kay 1987).

Lettuces

Lactuca sativa, Compositae.

Botany

The edible portion is the leaves. Some cultivars produce hearts where the leaves in the centre are closely compact, pale green or yellow and crisp and tender. They will freeze at about –0.6 to 0.3°C (Wright 1942) or –0.39 to 0.17°C (Weichmann 1987). Inaba *et al.* (1989) found that there was no response in respiration rate to ethylene at 100 μl litre^{-1} for 24 hours at any temperature between 5 and 35°C.

Harvesting

For 'hearting' cultivars, harvesting should be just when the heart is firm, but before it begins to bolt, that is, the flower stalk begins to grow. When they bolt they lose their crisp texture, become tough and may have a bitter taste. Firmness, or what is usually called 'solidity,' can be used for assessing harvest maturity. The harvester slightly presses the lettuce with the thumb and fingers to feel how much it yields to this pressure. Normally the back of the hand is used for testing the firmness of lettuce in order to avoid damage (John Love, personal communication (quoted by Thompson 1996). Some growers leave harvesting to late morning or early afternoon because when they are too turgid the leaves are soft and more susceptible to bruising (John Love, personal communication, quoted by Thompson 1996).

Diffuse reflectance measures the reflected light just below the surface of the crop and can be used for maturity measurement. In lettuce there was a shift in diffuse reflectance from 640–660 nm to 700–750 nm during maturation, which could be used to determine harvest maturity (Brach *et al.* 1982). A lettuce harvester was developed by Lenker and Adrian (1971) which used X-rays to determine which heads were sufficiently mature for harvesting. Garrett and Talley (1970) used γ-rays for the same purpose. The basis of these tests depends on the rate of transmission of the rays through the lettuce, since this depends on the density of the head, which increases as it matures.

Harvesting is by cutting them just below the soil surface or pulling them from the ground and trimming off the roots, outer leaves and any necrotic, diseased or damaged leaves. Harvesting should not be carried out when there is rain or dew, as they are then more easily damaged (Pantasico 1979).

Mechanical aids to harvesting involve cutting by hand and placing then on a conveyor on a mobile packing station, which is slowly conveyed across the field (Figure 12.66).

Figure 12.66 Gantry used to facilitate lettuce harvesting in California.

Packing

Perforated polyethylene film is used as head wraps or plastic film box liners to reduce weight loss. If the crop is to be vacuum cooled, the film must be perforated.

Precooling

Vacuum cooling was the most suitable, forced-air cooling with high humidity was also suitable and air cooling in a conventional cold store was less suitable and hydrocooling was generally unsuitable (MAFF undated). Vacuum cooling is used for most lettuce marketed in Europe and the USA, in spite of the high cost, mainly because the method is so quick. After 25–30 minutes of vacuum cooling at 4.0–4.6 mmHg with a condenser temperature of –1.7 to 0°C the lettuce temperature fell from 20–22°C to 1°C (Barger 1963). Top icing as soon as they are harvested is used when the lettuces are packed into crates or waxed cartons (Lutz and Hardenburg 1968). This may be done directly in the field or at the packhouse.

Storage

At room temperature of 20°C and 60% r.h. they could be kept for about 2 days (Mercantilia 1989). Refrigerated storage recommendations are as follows:

- 0°C for 30 days with 2.5% weight loss (Platenius *et al.* 1934)
- 4.4°C for 20 days (Platenius *et al.* 1934)
- 10°C for 6 days (Platenius *et al.* 1934)
- 0°C and 95% r.h. for 2–3 weeks (Wardlaw 1937; Phillips and Armstrong 1967; Anon 1968; Hardenburg *et al.* 1990)
- 0–1°C and 90–95% r.h. for 1–3 weeks or 4–6 weeks (Iceberg type) (Anon 1967)
- 0–1°C and 95% r.h. for 7 days (Shipway 1968)
- 0°C and 90–95% r.h. for 1 week (leaf) (Pantastico 1975)
- 0–3°C and 90–95% r.h. for up to 14 days with up to 15% weight loss (Tindall 1983)
- 0°C and 90–95% r.h. for 12 days for Butterhead (Mercantilia 1989)
- 0°C and 90–95% r.h. for 8–12 days for Butterhead (SeaLand 1991)
- 2°C for 8 days for Butterhead (Mercantilia 1989)
- 4°C for 6 days for Butterhead (Mercantilia 1989)
- 8°C for 4 days for Butterhead (Mercantilia 1989)
- 12°C for 2 days for Butterhead (Mercantilia 1989)

Table 12.36 Effects of temperature and reduced O_2 level on the respiration rate CO_2 production (mg kg^{-1} h^{-1}) of lettuce (Robinson *et al.* 1975)

	In air					In 3% O_2		
Cultivar	0°C	5°C	10°C	15°C	20°C	0°C	10°C	20°C
Unrivalled	18	22	26	50	85	15	20	55
Kordaat	9	11	17	26	37	7	12	25
Kloek	16	24	31	50	80	15	25	45

- 0°C and 90–95% r.h. for 14 days for Crisphead (Mercantilia 1989)
- 4°C for 9 days for Crisphead (Mercantilia 1989)
- 8°C for 6 days for Crisphead (Mercantilia 1989)
- 20°C for 2 days for Crisphead (Mercantilia 1989)
- 0°C and 90–95% r.h. for 9–14 days for Iceberg (SeaLand 1991)
- 0°C and 95–100% r.h. for 14–21 days for head lettuce (SeaLand 1991)
- 0–1°C and 95–100% r.h. for 1–4 weeks (Snowdon 1991).

Controlled atmosphere storage

Low O_2 levels in storage reduced the respiration rate of all three cultivars studied, but low O_2 was more effective at higher temperatures (Table 12.36).

CO_2 levels above 2.5% or O_2 levels below 1% can injure lettuce (Hardenburg *et al.* 1990). Storage in 1.5% CO_2 with 3% O_2 inhibited butt discoloration and pink rib at 0°C, but the effect did not persist during 5 days of subsequent storage at 10°C (Hardenburg *et al.* 1990). Physiological disorders associated with high CO_2 include brown stain on the midribs of leaves (Stewart and Uota 1971), which can be caused by storage in levels of CO_2 of 2% or higher, especially if combined with low O_2 (Haard and Salunkhe 1975). Storage for 42 hours in 30% CO_2, resulted in the lettuce producing the high levels of ethanol at 28°C but very little was produced by those stored at 8°C in the same CO_2 level (Yanez Lopez *et al.* 1998). Other storage recommendations are:

- 0°C with 0% CO_2 and 1–8% O_2 (Haard and Salunkhe 1975)
- 0–5°C with 0% CO_2 and 2–5% O_2 or 0–5°C with 0% CO_2 (Kader 1985)
- 0–5°C with 0% CO_2 and 1–3% O_2 for leaf, head and also cut or shredded lettuce; the effect on the former was moderate and on the cut or shredded it was high (Saltveit 1989)
- 1°C with 3% CO_2 and 1% O_2 for 21 days with less loss in ascorbic acid than for those stored in air (Adamicki 1989)
- 0°C with 2% CO_2 and 3% O_2 for 1 month (Hardenburg *et al.* 1990)
- 0°C with 0% CO_2 and 2–5% O_2 (SeaLand 1991)
- 1–3% O_2, which had a good effect and was of some commercial use when CO was added at the 2–3% level (Kader 1992).

Ozone toxicity

There is evidence that ozone is phytotoxic and lettuce is damaged by exposure to as little as 0.5 μl $litre^{-1}$ O_3.

Carbon monoxide

With levels of O_2 between 2 and 5%, CO can inhibit the discoloration of lettuce on the cut butts or from mechanical damage on the leaves. Concentrations of CO in the range 1–5% were effective, but the effect was lost when the lettuce was removed to normal air at 10°C. The respiration rate of lettuce was reduced during a 10-day storage period at 2.5°C when CO was added to the store. However, low concentrations of CO in the store atmosphere can increase the levels and intensity of the physiological disorder called russet spotting (Table 12.37).

Table 12.37 Effect of carbon monoxide on the development of russet spotting on harvested lettuce (source: Kader 1987)

CO concentration (μl l^{-1})	Russet spotting score[a]
0	0a
80	16c
400	22d
2000	24d
10 000	20d
50 000	6b

[a] Figures followed by the same letter were not significantly different ($p = 0.05$) by Duncan's Multiple Range Test.

Table 12.38 Effect of ethylene on the development of russet spotting on harvested lettuce (source: Kader 1987)

Ethylene concentration ($\mu l\ l^{-1}$)	Russet spotting score[a]
0	0a
0.01	4b
0.1	20d
1	36e
10	32e

[a] Figures followed by the same letter were not significantly different ($p = 0.05$) by Duncan's Multiple Range Test.

Ethylene

Discoloration of lettuce is associated with ethylene. This can take the form of russet spotting or a rusty brown discoloration on the leaves (Kader 1985). It is usually eliminated by storage at 0°C, but is increased by ethylene in the storage atmosphere (Table 12.38) (Anon 1968).

Hypobaric storage

It was claimed that hypobaric storage increased their storage life from 14 days in conventional cold stores to 40–50 days (Haard and Salunkhe 1975).

Lima beans, butter beans, Burma beans

Phaseolus lunatus, Leguminosae.

Botany

The edible portion is the seed, which is produced in a pod or legume. The pods are up to 12 cm long and 2.5 cm wide and contain up to six seeds. The seeds are variable in size and can be divided into the large-sized ones called *macrospermus* that are about 2.5 cm long, which are thought to have originated in Mexico, and the small-seeded type called *microspermus* or baby lima bean, which are about 1 cm long and probably originated in Peru. The seeds are usually white, but can be red, purple, brown, black or even mottled. It will freeze at about –1.2 to 0.9°C (Wright 1942) or –0.89 to 0.56°C (Weichmann 1987).

Harvesting

The small types usually mature between 90 and 130 days after sowing, but the large-seeded types take from 200–270 days depending on cultivar. Green lima beans are harvested by hand when the seeds have become fully developed, but are still green so that they are still soft and tender. Several pickings are made at 4–5 day intervals. Mature lima beans are usually harvested when about three-quarters of the pods are yellow and dry and the rest have begun to turn from green to yellow. In the USA, lima beans are harvested by machine and there are several types of harvester on the market.

Precooling

Hydrocooling was recommended for the shelled beans (Lutz and Hardenburg 1968).

Storage

Storage information in this section refers to green lima beans, which may be marketed in their pods or as shelled beans. In the latter case they become sticky and their colour fades if they are stored for too long (Phillips and Armstrong 1967). Refrigerated storage recommendations are as follows:

- 0°C for 15 days (Wardlaw 1937)
- 0–1°C and 85–90% r.h. for 2–3 weeks (Anon 1967)
- 0°C and 85–90% r.h. for 2 weeks or 4.4°C and 85–90% r.h. for 4 days (Phillips and Armstrong 1967)
- 0–4.4°C and 90% r.h. for beans in pods for 1 week (Lutz and Hardenburg 1968)
- 0°C for 10–14 days, 21.2°C for 8 days, 4.4°C for 4–7 days for shelled beans (Lutz and Hardenburg 1968)
- 4.4–7.2°C and 90–95% r.h. for 10–14 days with 12% loss for beans in pods (Pantastico 1975)
- 0–4.4°C and 90% r.h. for 2 weeks for shelled beans (Pantastico 1975)
- –2.2 to 0°C for 11 days (Purseglove 1968)
- 4–6°C with high humidity for 5–6 days (Tindall 1983)
- 10°C and high humidity for about 3 days (Tindall 1983)
- 0°C and 90–95% r.h. for 7–10 days (SeaLand 1991).

Chilling injury

After 3 or 4 weeks at 0°C, shelled beans became discoloured, which was attributed to desiccation and chilling injury by Wardlaw (1937).

Controlled atmosphere storage

Controlled atmosphere storage in CO_2 concentrations of 25–35% inhibited fungal and bacterial growth without adversely affecting the quality of the beans (Pantastico 1975).

Modified atmosphere packaging

Storage in perforated polyethylene bags was recommended by Lutz and Hardenburg (1968) or in sealed polyethylene bags by Pantastico (1975).

Limes

Citrus aurantifolia, *C. latifolia*, Rutaceae.

Botany

The fruit is a berry or hesperidium, which is non-climacteric. There are two main types of lime, the small diploid type, usually called the West Indian or Mexican lime, and the larger triploid seedless type, which is known as the Tahiti or Persian lime (Figure 12.67, in the colour plates). Thompson *et al.* (1974d) describes a local Sudanese variety, which is called Baladi, and is the same as the West Indian. It had a good, strong flavour but was generally smaller than the European market preferred, which is >42 mm in diameter. In studies it was found that the distribution of fruit sizes in a commercial harvest was as follows:

- 8% were >42 mm in diameter
- 25% were between 38 and 42 mm in diameter
- 67% were < 38 mm in diameter.

Fruit will freeze at about –1.8 to 1.4°C (Wright 1942). Oil is also extracted from the peel, which is a major by product of the lime juice extraction industry.

Harvesting

They are commonly harvested when fully grown, but still green. This is judged when the skin loses its irregular bumpy surface, becomes smooth and the peel may turn a lighter shade of green. There are no quality differences between this stage and when fruits turn fully yellow, it is just that many markets prefer limes to be green (Thompson *et al.* 1974d). In India, the cultivar Kagzi exhibited a sigmoid pattern of growth and took 26 weeks from fruit set to attain harvest maturity (Rao *et al.* 1995). As the fruit matures their chlorophyll levels are reduced and it is possible to separate limes into different classes on the basis of the chlorophyll levels of the peel, and therefore light transmission properties can be used to measure their maturity. With this method it was possible to grade fruits that were indistinguishable by visual examination. A photoelectric machine (ESM Model G) was developed by Jahn and Gaffney (1972) for colour sorting citrus fruit, including limes.

Disease control

Thompson *et al.* (1974d) found that rotting was generally low during storage and although levels were higher when fruit were stored in polyethylene bags it did not reach levels that merited the use of a fungicide. *Rhizopus nigricans* was the most common fungus, followed by *Penicillium* spp.

Physiological disorders

If the fruit are too turgid at harvest, due to recent rain, irrigation or even early morning dew, some of the oil cells in the peel may be ruptured, resulting in the disorder called oleocellosis.

Prestorage treatments

Limes coated with wax containing 0.1% of either thiabendazole or benomyl remained green and suitable for marketing after 3–4 weeks at 170 mmHg at 21.1°C (Spalding and Reeder 1976). However, the use of postharvest fungicides is being increasingly restricted. Dipping fruit in kinetin is commonly believed to retain their green colour during storage. Fioravanco *et al.* (1995) dipped Tahiti lime fruits in 25 or 50 μl l^{-1} kinetin for 5 min and found that it did not have any effect on chlorophyll-a or -b or total chlorophyll content of the peel, which declined throughout storage for 80 days at 7.5–8.5°C and 80–85% r.h. During 28 days of storage at 8 or 10°C, degreening was retarded compared with storage in ambient conditions. However, dipping the fruits in gibberellic acid at 250 mg before storage prevented degreening more than storage at low temperature, but the best treatment was gibberellic acid and storage at 10°C (Mizobutsi *et al.* 2000). Gibberellic acid is not a permitted postharvest treatment in many countries.

Storage

At room temperature of 20°C and 60% r.h. the fruit could be kept for about 2 weeks (Mercantilia 1989). Refrigerated storage recommendations are as follows:

- 7.2°C and 85–90% r.h. (Wardlaw 1937)
- 9°C and 85–90% r.h. (El Shiaty 1961)
- 8–10°C and 85–90% r.h. for up to 8 weeks (Anon 1967)

- 11.1–12.8°C and 85–90% r.h. for 8 weeks with 15% weight loss for yellow fruit or 7 weeks with 18% loss for green fruit (Pantastico 1975).
- 9–10°C and 85–90% r.h. for 6–8 weeks with some loss of green colour after 3–4 weeks for both Tahiti and Key (Hardenburg *et al.* 1990)
- 10°C and 90% r.h. for 6–8 weeks (Mercantilia 1989)
- 9–10°C and 90% r.h. for 1–2 months (Snowdon 1990)
- 12.2°C and 85–90% r.h. for 21–35 days for Persian and Tahiti (SeaLand 1991)
- 12.2°C and 85–90% r.h. for 42–56 days for Mexican and Key (SeaLand 1991).

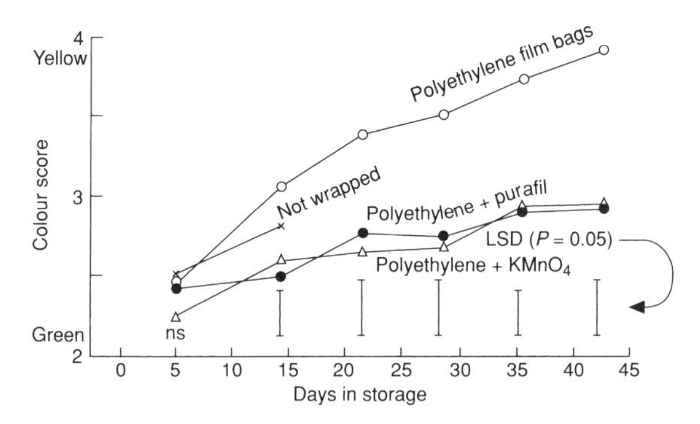

Figure 12.68 Effects of packaging on the colour of limes during storage at Sundanese ambient conditions of 31–34°C and 29–57% r.h. Source: Thompson et al. (1974).

Chilling injury

Limes stored for 4 weeks at 170 mmHg at 2.2°C and 98–100% r.h. developed as much chilling injury as comparable limes stored at normal atmospheric pressure. (Spalding and Reeder 1976), but storage in 7% O_2 reduced the symptoms of chilling injury compared with storage in air (Pantastico 1975).

Controlled atmosphere storage

Controlled atmosphere storage of Tahiti limes increased decay rind scald and reduced juice content (Pantastico 1975). Other recommendations are that Tahiti limes stored at 10°C for 6 weeks in 5% O_2 with 7% CO_2 maintained an acceptable green colour, but had low juice content, thick rinds and a high incidence of decay. Fruits stored in 21% O_2 with 7% CO_2 had acceptable juice content and were considerably greener than limes stored in air. Limes from all treatments had acceptable flavour (Spalding and Reeder 1974). Storage at 10–15°C with 0–10% CO_2 and 5% O_2 (Kader 1986) or with 0–10% CO_2 and 5% O_2 (Kader 1989 and 1992) had a good effect on storage but was not used commercially (Kader 1985). A temperature of 12.2°C with 0–10% CO_2 and 5% O_2 was recommended by SeaLand (1991).

Modified atmosphere packaging

A major source of deterioration of limes during marketing is their rapid weight loss. This can give their skin a hard texture and an unattractive appearance. In Sudanese, ambient conditions packing limes in sealed polyethylene film bags inside cartons resulted in a weight loss of only 1.3% in 5 days, but with all the fruits degreening more rapidly than to those which were packed just in cartons where the weight loss was about 14% (Thompson *et al.* 1974d). However, this degreening effect could be countered by including an ethylene absorbent in the bags (Figure 12.68). During storage for 80 days at 7.5–8.5°C and 80–85% r.h. and then for 5 days at 19–26°C and 55–75% r.h., waxing plus polyethylene bags gave better results than waxing or polyethylene bags alone (Fioravanco *et al.* 1995).

Hypobaric storage

Haard and Salunkhe (1975) stated that Tahiti limes can be stored for 14–35 days in cold storage, but this was extended to 60–90 days when it was combined with hypobaric storage. Spalding and Reeder (1974) stored Tahiti limes at 152 mmHg pressure for 6 weeks at 10°C and observed only small changes in colour, rind thickness, juice content, soluble solids, total acids and ascorbic acid. Decay averaged 7.8% compared with 8.3% for those stored in air. Those stored at 228 mmHg maintained an acceptable green colour, but were a slightly lighter green than limes stored at 152 mmHg. Those stored at 76 mmHg maintained an acceptable green colour, but had low juice content, thick rinds, and a high incidence of decay. Limes from all treatments had acceptable flavour. In subsequent work, Spalding and Reeder (1976) found that Tahiti limes retained green colour, juice content and flavour acceptable for marketing and had a low incidence of decay during storage at a pressure of 170 mmHg for up to 6 weeks at 10°C or 15.6°C with 98–100% r.h. Control fruits turned yellow within 3 weeks at normal atmospheric pressure.

Limequats

Citrus aurantifolia × *Fortunella* sp., Rutaceae.

Botany

Limequat is a hybrid between limes and kumquats. The fruit is a berry or hesperidium, and has a bright green or yellowish green, thin, edible skin, but they taste like limes and are too sour to be eaten raw. There are other crosses with kumquats including orangequats, a cross between kumquat and orange, and citrangequats, which is a tri-specific hybrid between lime, citron and orange. The purpose of these hybrids appears to be to introduce cold tolerance and dwarfing to the trees.

Storage

At a room temperature of 20°C and 60% r.h. limequats could be kept for about 1 week (Mercantilia 1989). Refrigerated storage recommendations are as follows:

- 10°C and 90% r.h. for 4 weeks (Mercantilia 1989)
- 10°C and 90–95% r.h. for 21–28 days (SeaLand 1991).

Litchi, lychee

Litchi chinensis, Sapindaceae.

Botany

Litchi is classified as a non-climacteric fruit that contains a single large, brown, shiny inedible stone or seed. They have a leathery scaly skin enclosing white, translucent, firm flesh that is sweet and fragrant (Figure 12.69). Several cultivars are grown in China; the best are large with small seeds, which means that they have a large edible portion. Two such cultivars, which are reported to have a good sweet flavour, are Nuo Mi Ci and Huai Zhi (Huang *et al.* 1990). Fruits were served as imperial tributes in China during the Qing dynasty between 1644 and 1911. The Bangkok Post of 2 July 2002 reported that a single litchi fruit from a 400-year-old tree that once produced fruit for the Chinese emperors had just sold in an auction in Guangdong province for 555 000 yuan (over US $60 000).

Figure 12.69 Litchi fruit ready for harvesting, also showing the internal structure. Source: Dr Wei Yuqing.

Harvesting

The cultivars Dehradun and Seedless took the same number of days (69–72) after full bloom to reach harvest maturity, with Calcuttia maturing 2 or 3 days later (Muthoo *et al.* 1999). Fruit size can be used for determining the harvest maturity of litchi. In South Africa it was shown that the size of the fruit at harvest could have a very large effect on its profitability during marketing (Table 12.39). However, the longer the fruits were left to mature on the tree, the higher were the postharvest losses, but even if 70 mm diameter fruit were harvested and had increased postharvest losses, it may still be economic (Blumenfeld 1993a).

Fumigation

Sulphur dioxide fumigation is used to prevent discoloration of the skins of litchis, which is caused by fungal infection. Small, simple but effective chambers can be used for this purpose (Figure 12.70). Fumigation with 1.2% sulphur dioxide for 10 minutes was shown to be effective in reducing skin discoloration in fresh litchi, especially if this was combined with a 2 minute dip in 1 M hydrochloric acid directly afterwards (Underhill *et al.* 1992). La-Ongsri *et al.* (1993) also showed that sulphur dioxide fumigation was an

Table 12.39 Effects of harvest maturity, as measured by fruit diameter, on weight, price and income from the fruit as a proportion of the base of 100 at 60 mm fruit diameter (source: Blumenfeld 1993)

	Fruit diameter (mm)		
	60	65	70
Weight	100	120	140
Price	100	115	130
Income	100	138	182

Figure 12.70 Chamber for on farm fumigation of litchis with sulphur dioxide in Thailand.

effective treatment, but in this case at 2%, followed by a dip in 1 M hydrochloric acid to stabilize the red colour and reduce skin browning in the cultivar Hong Huay in Thailand. Tongdee (1993) recommended for the cultivars Honghual and Emperor 5–125 ml sulphur dioxide kg^{-1} and for the cultivar Khom 125 ml sulphur dioxide kg^{-1}. In South Africa, burning 100–150 g of sulphur in a 5 m^3 sealed chamber prevented skin browning in litchis (Sauco and Menini 1989). This treatment was only effective for fruit stored for up to 2 weeks. Immediately after sulphur dioxide treatment, litchi fruit may appear a uniform yellow colour and then turn red again after 1–2 days (Jurd 1964). Milne (1993) described another sulphur treatment of litchi that retained a good postharvest colour. This involves dipping fruit in a sodium metabisulphite solution at 60 g $litre^{-1}$ for 15 minutes followed by 1 M hydrochloric acid treatment for 2 minutes.

Prestorage treatment

In experiments with the cultivar Huaizhi, a low-temperature hardening treatment of 5 days at 5°C prior to storage for 40 days at 1°C reduced the respiration rate during the latter part of storage. It also delayed increases in cell membrane permeability, browning and organic acid and ascorbic acid concentrations. Additionally it reduced the concentration of peroxides in the peel and extended the shelf-life of fruits after removal from cold storage (Zhang and Quantick 2000).

Storage

They can be stored at 20°C and 60% r.h. for 7–10 days (Mercantilia 1989). Storage at 1.1°C resulted in a loss of brightness (Hatton *et al.* 1966). Refrigerated storage recommendations are as follows:

- 1.7°C and 85–90% r.h. for 8–12 weeks (Pantastico 1975)
- 5°C and 90% r.h. for 4–6 weeks (Mercantilia 1989)
- 1.5°C and 90–95% r.h. for 3–5 weeks or 7.2°C for 2 weeks (Lutz and Hardenburg 1968; Hardenburg *et al.* 1990)
- 5°C for 30–35 days for the cultivars Nuo Mi Ci and Huai Zhi (Huang *et al.* 1990)
- 5–10°C and 90–95% r.h. for 4–6 weeks (Snowdon 1990)
- 1.7°C and 90–95% r.h. for 21–35 days (SeaLand 1991).

Chilling injury

Fruit stored at 1°C suffered from chilling injury (Huang *et al.* 1990). Storage at 0 or 2.5°C for 3 or 4 weeks also resulted in chilling injury symptoms of dark brown spots appearing on the inner side of the peel. Fruit stored at 5°C did not show these symptoms, but disease development occurred in fruit stored at 7.5°C after 4 weeks (La-Ongsri 1993).

Controlled atmosphere storage

Kader (1993) in his review recommended 3–5% CO_2 combined with 5% O_2 at 7°C, with a range of 5–12°C, and reported that the benefits of reduced O_2 were good and those of increased CO_2 were moderate. Vilasachandran *et al.* (1997) stored the cultivar Mauritius at 5°C in air (control) or 5, 10 and 15% CO_2 combined with either 3 or 4% O_2. After 22 days all fruit were removed to air at 20°C for 1 day. Fruits that had been stored in 15% CO_2 with 3% O_2 or 10% CO_2 with 3% O_2 were lighter in colour, retained total soluble solids better than the other treatments, but had the highest levels of off-flavours. All the controlled atmosphere-stored fruit had negligible levels of black spot and stem end rot compared with the controls. On the basis of the above, Vilasachandran *et al.* (1997) recommended 5% CO_2 with 3% O_2 or 5% CO_2 with 4% O_2.

Modified atmosphere packaging

Kader (1993) pointed out that modified atmosphere packaging was used to a limited extent. In some cases packing fruit in plastic film was actually detrimental during storage. Fruit were packed in punnets overwrapped with plastic film with 10% CO_2 or vacuum packed and stored for 28 days at 1°C followed by 3 days shelf-life at 20°C. It was found that skin browning was reduced in both of these plastic film packages compared with unwrapped fruit. However, their taste and flavour were unacceptable, whereas they remained acceptable for unwrapped fruits (Ahrens and Milne 1993). Fruits of the cultivar Brewster sealed in low-density polyethylene film had the lowest water loss throughout storage at 5°C, but there was condensation in the internal face of the film. On the other hand, poly(vinyl chloride) film effectively reduced fresh weight loss, without condensation. Storage at 5°C reduced pericarp browning, with best results in the poly(vinyl chloride) film, which delayed browning for 36 days after harvest. In storage at 27°C, perforated low-density polyethylene and poly(vinyl chloride) film reduced the rate of browning compared with the control, with 16% and 5% incidence of dark fruits 8 days after harvest (Fontes *et al.* 1999). Ragnoi (1989), working with the cultivar Hong Huai from Thailand, showed that fruits packed in sealed 150 gauge polyethylene film bags containing 2 kg of fruit and sulphur dioxide pads could be kept in good condition at 2°C for up to 2 weeks, whereas fruits that were stored without packaging were discoloured and unmarketable.

Loganberries

Rubus loganobaccus, Rosaceae.

Botany

The fruit is a collection of drupes and is classified as non-climacteric. They are dull red and more acid than blackberries (*R. ulmifolius*), and also the drupes remain tightly adhering to the receptacle even when fully mature. They originated in Santa Cruz in California, where they were found growing in the garden of Judge J.H. Logan in 1881. It is probably a true hybrid.

Harvesting

They are normally harvested by hand with the receptacle in place. When fully mature they have a soluble solids content of about 10% and an acid content of 1.02–3.12% w/w (Hulme 1971).

Storage

Refrigerated storage recommendations are:

- –0.6 to 0°C and 85–90% r.h. for 5–7 days (Phillips and Armstrong 1967)
- –0.6 to 0°C and 90–95% r.h. for 2–3 days (Lutz and Hardenburg 1968)
- –0.5 and 90–95% r.h. for 2–3 days (SeaLand 1991).

Longan

Euphoria longana, *Dimocarpus longan*, Sapindaceae.

Botany

The fruit is non-climacteric (Tongdee 1997) and is an ovoid-shaped berry, which can be up to 2 cm long. It is similar to litchi but is light brown or it may be red to orange–yellow with a smooth skin. It has a large, central, brownish black seed surrounded by a thin layer of sweet, translucent pulp. The fruit are borne in panicles or bunches (Figure 12.71). Tongdee (1997) found that their respiration rate was 10–16 ml CO_2 kg^{-1} h^{-1} at 22°C and 2–6 ml CO_2 kg^{-1} h^{-1} at 5°C, ethylene production from undetectable to trace levels. Its origin was probably in the area from Myanmar to South China.

Harvesting

It takes about 20–28 weeks from fruit set to fruit maturity. In Chiang Mai in Thailand the harvest season is

Figure 12.71 Longans growing in China. Source: Dr Wei Yuqing.

between late June and August and all the fruit on one tree are normally harvested at the same time (Tongdee 1997). Fruit size and weight were consistently shown to have a high correlation with eating quality (Onnap *et al.* 1993) and are used to determine harvest time. Tongdee (1997) indicated that soluble solids could be used to determine harvest maturity using 15.5–16 °Brix.

Harvesting is carried out by climbing the trees or using ladders. The bunches are collected in baskets which are then lowered to the ground (Figure 12.72).

Grading

The fruit are packed in the plantation in baskets or boxes still in bunches (Figure 12.73). Tongdee (1997) mentioned that fruit are graded by size for the market with the top grade 55–75 fruits kg^{-1} being the best grade and smaller fruit being considered to be of lower grade. She also indicated that the other factors related to quality are 'length of stalk, colour of rind, freedom of fruit from blemishes, rotting or apparent insect damage and percentage of loose fruit.'

Figure 12.72 Harvesting longans in Thailand.

Figure 12.73 Packing longans into plastic boxes in Thailand for export to Malaysia and Singapore.

Fumigation

For the cultivars Do and Biew Kew, Tongdee (1993) recommended fumigation with 200–300 ml sulphur dioxide kg^{-1} fruit as soon as possible after harvest. Excessive sulphur dioxide can damage fruit, causing irregular rusty brown circles or lines on the underside of the skin and eventually increased fungal growth (Tongdee 1997).

Storage

Refrigerated storage recommendations are as follows:

- 4°C and 90% r.h. (Snowdon 1990)
- 1.7°C and 90–95% r.h. for 21–35 days (SeaLand 1991)
- 1°C for the cultivar Shixia (Jiang 1999)
- 2.5°C for the cultivar Wuyuan (Jiang 1999)
- 2°C and 90% r.h. (Jiang and Li 2001).

Chilling injury

With the cultivars Shixia and Wuyuan, grown in Guangzhou in China, storage at 1°C resulted in chilling injury on the skin of Wuyuan, which increased rapidly after 20 days. Wuyuan appeared to be more chilling sensitive than Shixia. In contrast to the above, Tongdee (1997) in Thailand found that chilling injury occurred during storage at 5–7°C for 3–4 weeks as browning of the skin and increased fungal growth when removed to ambient temperatures.

Controlled atmosphere storage

Tongdee (1997) found that controlled atmospheres had no extra benefit on storage life compared with cold storage alone. However, in China storage of the cultivar Shixia at 1°C with 6–8% CO_2 and 4–6% O_2 gave the best disease control and fruit quality maintenance (Jiang 1999).

Longkong, longong

Aglaia dookkoo, Meliaceae

Botany

Longkong duku nam and duku are referred to as *A. dookkoo* by Namsri *et al.* (1999) and Chantaksinopas and Kosiyachinda (1987) also refer to longkong as *A. dookoo*, but Rattanapong *et al.* (1995) referred to longkong as *L. domesticum*. Chantaksinopas and Kosiyachinda (1987) found that over 95% of fruits had five carpels. The fruit is a berry borne in clusters and has a similar appearance to Langsat, but is larger in size and has a distinct flavour.

Harvesting

The growth curve from fruit set until maturity was single sigmoidal, and the growth period lasted 14 weeks. The fruit shape during the growth period up to 13 weeks was global and changed to spherical 1 week later. The rind thickness increased after fruit set until the nineth week, then decreased slightly until maturity. The percentage of soluble solids in fruit juice increased gradually until week 14, with a maximum of 17%. Acid quantity decreased until the fruit became mature, with 1% of titratable acids in week 15. The soluble solids:titratable acids ratio in week 15 was 18.1. Harvesting is recommended during weeks 14–16 after fruit set (Chantaksinopas and Kosiyachinda 1987). Bioassay-guided fractionation of stems and fruits of *A. elliptica* from Thailand showed that the compounds were very potent cytotoxic substances when evaluated against a panel of human cancer cell lines (Cui *et al.* 1997). Extracts from the fruit of another species, *A. roxburghiana*, were also shown to have medical uses.

Preharvest treatment

The effect of calcium chloride (0, 3 or 5%), applied 9, 10 or 11 weeks after fruit set, on the quality of *L. domesticum* fruits was evaluated. Treatment of clusters with calcium chloridet, 10 or 11 weeks after fruit set, reduced fruit drop and increased fruit firmness, total soluble solids content and the ratio of total soluble solids:titratable acidity Rattanapong *et al.* (1995).

Loquats, Japanese medlars

Eriobotrya japonica, Rosaceae

Botany

The fruit are ovoid or pear-shaped, up to 8 cm long with a pale orange or light yellow skin. The flesh is yellow to light orange and may contain up to 10 seeds. Hamauzu *et al.* (1997) found that the respiration rate decreased during maturation and Ding *et al.* (1998) confirmed that the cultivar Mogi was non-climacteric. They also found that the rate of respiration at 20°C was 62 ml CO_2 kg^{-1} h^{-1} on the first day of storage. Ethylene production increased simultaneously with the decrease in green colour and the appearance of reddish colour. Fructose, glucose and sucrose were the dominant sugars; sorbitol was a minor one in the cultivar Mogi.

Harvesting

Hamauzu *et al.* (1997) described eight stages of harvest maturity from stage 1 (green, small) through to stage 7 (yellowish orange) and stage 8 (slightly over-ripe) and indicated that stage 7 is the optimum harvest maturity. The sucrose content decreased from stage 5 to stage 8; fructose was the dominant sugar at maturity stage 8. Malic acid was the dominant organic acid; citric, fumaric and succinic acids were the minor ones. Malic acid content decreased through fruit maturation, whereas citric acid content remained nearly constant. Phenolic compounds and the ratio of *o*-diphenol to total phenolics increased during fruit maturation. β-Carotene was the dominant carotenoid at stage 2; cryptoxanthin became predominant at stage 7.

Storage

At a room temperature of 20°C and 60% r.h. Mercantilia (1989) indicated that the fruit could be kept for 3–5 days, whereas Ding *et al.* (1998) found that at 20°C they could be kept for up to 10 days. Refrigerated storage recommendations are as follows:

- 0°C and 90% r.h. for 2–3 weeks (Mercantilia 1989)
- 0°C and 90% r.h. for up to 3 weeks (Hardenburg *et al.* 1990)
- 0°C and 90–95% r.h. for 2–3 weeks (Snowdon 1990)
- 0°C and 90% r.h. for 14–21 days (SeaLand 1991)
- 1°C or 5°C for up to 30 days (Ding *et al.* 1998)
- 10°C for up to 15 days (Ding *et al.* 1998).

Modified atmosphere packaging

Perforated polyethylene film bags (0.15% perforation) were successfully used for the cultivar Mogi when stored at 1 or 5°C (Ding *et al.* 1998).

Lotus roots

Nymphaeas lotus, Nymphaeaceae.

Botany

This is the root of the tropical aquatic plant that bears the lotus flower, which is commonly used in Buddhist temples. The roots can be eaten fresh after boiling or dried and ground to make flour.

Modified atmosphere packaging

Mineral powders are incorporated into some films used for packing lotus root (Industrial Material Research 1989, quoted by Abe 1990).

Lovi lovi

Flacourtia inermis, Flacourtiaceae.

Botany

It is a small tree that produces edible fruit, which are subglobose, dark red when mature and contain many small seeds. The fruit is usually very astringent and must be cooked before eating, but there are low-astringent, sweet varieties.

Storage

No information on the postharvest life could be found.

Maitake

Grifola frondosa.

Botany

It is a bracket fungus and grows in oak and beech woods. It is cultivated in Japan where bag culture is the most frequently used system. Sawdust is mixed with rice and wheat bran. After adjustment of the moisture content, the mixture is packed in a polypropylene bag and is moulded into a square-shaped culture bed. Log and outdoor bed culture are also used.

Storage

Mizuno (2000) found that low-temperature storage was more effective in maintaining their quality and the contents of anti-tumour polysaccharides as health beneficial foods; these both deteriorated quickly during storage at 20°C.

Malay apple, pomerac, jambos, Malacca apple

Eugenia malaccensis, Myrtaceae.

Botany

The fruit is a berry with a persistent calyx and is borne directly of the main trunk or the branches of the tree. They are pear-shaped, bright red, pinkish or purple in colour with a thin skin and crisp, white flesh, and with a distinct rose scent. There is a central cavity containing a single, round seed. The fruit can be up to 10 cm long, but are usually considerably smaller. Its respiratory pattern led to it being classified as an indeterminate fruit by Biale and Barcus (1970). They originated in Malaysia and can be seen commonly in markets, especially in Southeast Asia. It is now grown throughout the tropical world, often as a garden tree. Basanta and Sankat (1995, 1997) and Sankat *et al.* (2000) all refer to pomerac as *Syzygium malaccense* and Le *et al.* (1998) refers to Malay apple also as *Syzygium malaccense*, while Purseglove (1968) indicates that *Syzygium malaccensis* (note the difference in spelling of the specific name) is synonymous with *Eugenia malaccensis.* Pommerosa (*E. jambos*) also has similar fruit characteristics and it seems more likely that its common name would be jambos.

Harvesting

Harvesting is by hand. Maturity can be judged from colour changes in the skin, but also the fruit becomes more rounded as it develops.

Storage

Fruits keep for 4–6 days under Trinidad ambient conditions of about 28°C, after which they deteriorate rapidly with increased fading of the fruit's bright, strikingly red skin colour (Sankat *et al.* 2000). Fruits stored at 10 or 15°C and 95% r.h. were shrivelled and showed decay and skin colour loss after 10–15 days, but fruits stored at 5°C had acceptable colour, firmness, taste and aroma, even after 20 days, although there were some signs of shrivelling (Basanta and Sankat 1995). In storage at 5°C for 30 days, those stored in the light had more fading of the red skin colour than those stored in the dark (Sankat *et al.* 2000).

Modified atmosphere packaging

Waxing reduced shrivelling at 5°C, but not as effectively as packing in polyethylene bags, which also preserved fruit colour and firmness (Basanta and Sankat 1995).

Controlled atmosphere storage

Sankat and Basanta (1997) stored fruits at 5°C for up to 30 days in:

- 5% CO_2 with 2% O_2
- 5% CO_2 with 4% O_2
- 5% CO_2 with 6% O_2
- 5% CO_2 with 8% O_2
- 5% CO_2 with 1% O_2
- 8% CO_2 with 1% O_2
- 11% CO_2 with 1% O_2
- 14% CO_2 with 1% O_2.

The optimum storage conditions were found to be 11% CO_2 with 1% O_2 and 14% CO_2 with 1% O_2 where fruits maintained their flavour and appearance for 25 days.

Transport

Wardlaw (1937) mentioned that a shipment of fruit had been made from Java to the Netherlands, but was said to have no commercial value on arrival. There was no mention of the shipping conditions.

Mamey, mamay apple, mammey apple

Mammea americana, Guttiferae.

Botany

The edible portion is the fruit that is produced on large evergreen trees. Fruits are up to 20 cm in diameter and have a rough brown skin, often with a beaked tip. The pulp is orange in colour and apricot in flavour and contains up to four large seeds. It is a native of South America and the West Indies.

Toxicity

The fruit is widely eaten raw, but Morris *et al.* (1952) found that all parts of the tree contained a principle that was toxic to guinea pigs. When cut, a bitter yellow resin is exuded from all parts of the tree except the fruit pulp. This yellow resin, which is also extracted from the seeds, can be used as an insecticide.

Storage

The only storage recommendation found was 12.8°C and 90–95% r.h. for 14–42 days (SeaLand 1991).

Mandarins

Citrus reticulata, *C. unshiu*, Rutaceae.

Botany

The fruit is a berry or hesperidium, which is non-climacteric. The term Mandarin is used throughout Japan, China, Italy and Spain and they are referred to as soft citrus in South Africa (Davies and Albrigo 1994). Inaba *et al.* (1989) found that there was increased respiration in response to ethylene at 100 μl litre^{-1} for 24 hours at 20–35°C, but the effect was much reduced at lower temperatures.

Harvesting

The fruit changes in colour from green to reddish orange as they mature on the tree and the skin becomes smoother. The juice content also increases. By taking representative samples of the fruit, extracting the juice in a standard and specified way and then relating the juice volume to the original mass of the fruit it is possible to specify its maturity. In the USA the minimum value is 33%. The ratio between soluble solids and acidity can also be used; it should be between 7:1 and 9:1.

Handling

Tariq (1999) found that scuffing of the peel was the worst damage, and resulted in the peel turning brown and *Penicillium* spp. growing on the damaged area. Cuts and bruises were found to be less injurious to fruits than scuffing.

Curing

Placing the damaged fruits in sealed polyethylene film bags and exposing them to temperatures in the range 25–35°C for between 3 and 48 hours completely eliminated the browning and fungal growth of damaged fruits. The lower the temperature, the longer was the exposure time required for the curing process. Curing could be achieved without film packaging at the same temperature if the surrounding atmosphere was maintained at the same temperature range at 95–98% r.h. (Tariq 1999).

Physiological disorders

Pantastico (1975) mentioned puffiness, where the peel becomes soft and very lose and may crack, which appears to be associated with water loss from the segments.

Degreening

Optimum temperatures range between 25 and 29°C and humidity as high as possible, generally between 90 and 95% r.h. with ethylene at between 1 and 10 μl litre^{-1} for 24–72 hours. The actual choice of conditions will vary between different situations, particularly whether they have been grown in the tropics or subtropics. The cultivar and the stage of maturity at harvest should also be taken into account.

Storage

Refrigerated storage recommendations are as follows:

- 5.6–7.2 and 85–90% r.h. for 8 weeks with 13% loss (Coorg main crop) or 6 weeks with 15.8% loss for Coorg in the rainy season crop (Pantastico 1975)
- 0–3°C and 85–90% r.h. for 8–10 weeks (Mercantilia 1989)
- 4–8°C and 90% r.h. for 3–8 weeks (Snowdon 1990)
- 7.2°C and 85–90% r.h. for 14–28 days (SeaLand 1991)
- 0–4°C for 2 months or more (Davies and Albrigo 1994)
- 4 or 8°C and 90% r.h. for 3 months for Malvasio Mandarins grown in Sardinia (Agabbio *et al.* 1999).

Mangoes

Mangifera indica, Anacardiaceae.

Botany

The fruit is a fleshy drupe that has a climacteric pattern of respiration (Biale and Barcus 1970). Vega-Pina *et al.* (2000) found that fruits of the cultivar Haden reached their climacteric peak of respiration after 6 days at 21°C. They will freeze at about –1.3 to 1.1°C (Wright 1942).

Preharvest disease control

The main disease problem is infection of the fruit during development on the tree. Preharvest fungal infections by *Glomerella cingulata*, and *Colletotrichum gloeosporiodes* causes anthracnose disease. The spores of this organism may infect fruit, germinate and form appressoria which remain quiescent on the skin of the fruit until the fruit begins to ripen. The fungus then invades the cells of the fruit by hyphae growing from the appressorium resulting in the anthracnose disease. Fruit, especially those that are to be exported, are harvested at a stage of maturity where the fungus has not penetrated the fruit but is firmly fixed to its surface at the resistant appressorial stage. Therefore, fruits that look perfectly healthy at harvest may develop the disease symptoms postharvest.

Thompson (1987) reviewed preharvest sprays of mangoes to control anthracnose. Recommendations are for field application of chemical fungicides often followed by postharvest hot water treatment, usually combined with a fungicide. In Florida (McMillan 1972), cupric hydroxide at 2.4 g litre^{-1} or tribasic copper sulphate at 3.6 g litre^{-1} plus the organic sticker Nu-Film 17 at 0.125% applied at monthly intervals at 57 l per tree from flowering to harvest gave good anthracnose control. Benomyl at 0.3 g litre^{-1} plus Triton B1956 at 0.15 ml litre^{-1} at 57 l per tree at monthly intervals from flowering to 30 days before harvesting was shown to be very effective against anthracnose on mangoes (McMillan 1973). Mancozeb, chlorothalonil and ferbam were shown to be equally effective as field sprays against mango anthracnose in Florida (Spalding 1982). Trials carried out on Keitt mangoes in South Africa showed that two pre-flowering applications of copper oxychloride then two applications of Bayfidan (triadimenol) during flowering followed by monthly applications of copper oxychloride from fruit set ensured effective control of anthracnose (Lonsdale 1992). Sprays were applied to run off at about 20 l per tree. In Australia mancozeb applied at 1.6 g litre^{-1} active ingredient as a weekly spray during flowering then as a monthly spray until just before harvesting gave good control of mango anthracnose (Grattidge 1980). In studies in the Philippines, field sprays with mancozeb or copper were effective in controlling mango anthracnose and superior to either captan or zineb (Quimio and Quimio 1974).

Soft brown rot of mangoes, caused by *Nattrassia mangiferae* in South Africa, is due to field infection that was shown to be controlled by applications of copper oxychloride in June (pre-flowering) and in January. Further applications after fruit set (between June and January) did not benefit the control of the disease (Lonsdale 1992).

Stem end rot of mangoes in Australia was shown to be caused by preharvest infections with *Dothiorella dominicana*, *D. mangiferae*, *Phomopsis mangiferae*, *Lasiodiplodia theobromae*, *Cytosphaeria mangiferae* or *Pestaloptiopsis* sp. (Johnson *et al.* 1992). They also showed that these fungi occur widely on mature, branches and, in certain conditions, colonize the inflorescence tissue as it matures, thus reaching the stem end of fruit through the pedicel. These infections remained latent until after the fruit has been harvested or until the fruit on the tree became senescent. However, Johnson *et al.* (1992) found a higher level of stem end rot caused by *D. dominicana* in a mango orchard that had been regularly sprayed with a copper fungicide than at another site that had not been sprayed. At the unsprayed site the mangoes were extensively diseased with anthracnose, which may have prevented *D. dominicana* infecting the fruit from the pedicel where it was detected.

In the area of St Thomas in Jamaica, mango trees gown close to the prevailing wind from the sea had no disease whatsoever. Those more inland protected from the prevailing wind had high levels of disease, especially anthracnose.

Harvesting

Harvest maturity is judged on skin colour, firmness of the flesh, size, shape and aroma. The cultivars Tommy Atkins and Kent were harvested at 15–17 lb pulp pressure by Lizana and Ochagavia (1997). The specific gravity of the fruits showed a decreasing trend up to 45 days after fruit set and a linear increase up to maturity of fruits in all the cultivars, whereas titratable acidity increased after fruit set and then slowly decreased towards maturity. Ascorbic acid and moisture content of the fruits decreased after fruit set to maturity and maximum values were observed in the cultivar Langra. However, total soluble solids and total sugar content of the fruits showed an increase after fruit set up to maturity and they were highest in Langra and Sunderja.

Mangoes also change shape during maturation on the tree. For some cultivars this change could be correlated in a systematic way with maturity. As the mango fruit matures, the relationship between the shoulders of the fruit and the point at which the fruit stalk is attached to the fruit may change. On very immature mango fruits the shoulders slope away from the fruit stalk; on more mature fruit the shoulders become level

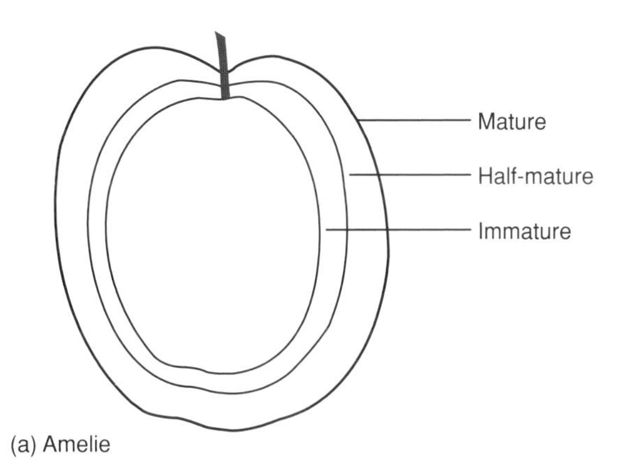

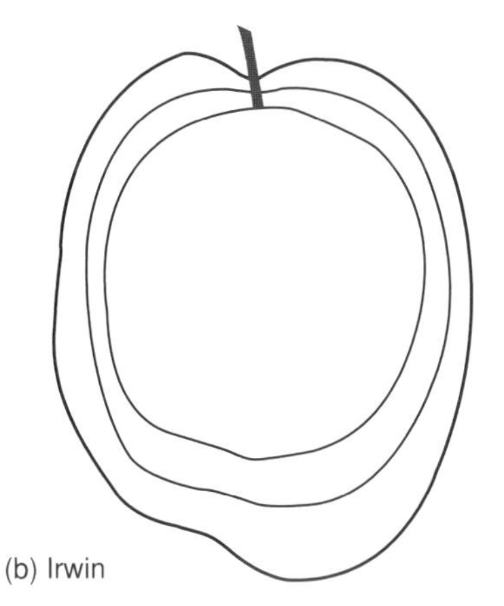

Figure 12.74 Mango harvest maturity for many varieties can be judged by the slope of the shoulders of the fruit to the pedicel insertion. Source: Wardlaw and Leonard (1936). Drawing Dr A. P. Medlicott.

with the point on attachment and on even more mature fruit the shoulders may be raised above the point of attachment (Figure 12.74). Wardlaw and Leonard (1936) gave the following definitions:

A. Fruits almost fully grown, green in colour and with shoulders level with the stem insertion.
B. Shoulders raised above the stem insertion due to further fruit growth, and the skin a lighter green in colour.
C. No further growth but fruits on the point of becoming soft, and colour present in the skin.

Using this method of determining mango fruit maturity, Thompson (1971) showed that the percentage of fruit still unripe after storage at 7°C for 28 days was 68, 57 and 41% for A, B and C maturity stages, respectively.

The time after flowering can also be used to determine harvest maturity. In India, Alphonso exhibited a sigmoid pattern of growth and took 16 weeks from fruit set to attain harvest maturity (Rao *et al.* 1995). Rajput *et al.* (1999) showed that the fruit growth of the cultivars Langra, Sunderja, Mallika and Amrapali also followed a sigmoid pattern and in general it was rapid between 30 and 90 days after fruit set.

Fruits of Karuthacolomban, Velleicolomban and Willard grown in Sri Lanka were harvested at 10, 11, 12 and 13 weeks after flowering. Peel colour development could be used to determine harvest maturity in Willard. Changes from dark green to light green with maturity could not be considered as a reliable index for the other two cultivars. Raising of shoulders with maturity is a better indicator of maturity than peel colour development, and it could be used as a reliable index to harvest all three cultivars. Although mature fruits of Willard and Velleicolomban passed the float test, Karuthacolomban did not respond consistently. The mean value of total soluble solids recorded from Karuthacolomban and Velleicolomban harvested 13 weeks after flowering was 18 °Brix, while titratable acidity was 0.3%. Total soluble solids and titratable acidity of Willard were similar to these values in the 12th week after flowering. The fruits harvested before the optimum stage of maturity contained significantly lower total soluble solids, higher titratable acidity and poorer sensory properties than mature fruits (Amarakoon *et al.* 1999).

As mentioned, specific gravity is used to grade crops into different maturities postharvest. To do this the fruit is placed in a tank of water and if they float they will be less mature that those which sink. To give greater flexibility to the test and make it more precise, a salt or sugar solution can be used in place of water in the tank. This changes the density of the liquid, resulting in fruits that would have sunk in water, floating in the salt or sugar solution. Lizada (1993) showed that 1% sodium chloride solution was suitable for grading Carabao mangoes in the Philippines.

Acoustic and vibration tests equipment puts vibration energy into a fruit, and measure its response to this input. Good correlations have been found using the second resonant frequency to determine mango maturity (Darko 1984; Punchoo 1988). Near-infrared reflectance has been studied in relation to measuring the internal qualities of fruit with good correlation with sugar content (Kouno *et al.* 1993a).

Hot water treatment for anthracnose control

This was developed mainly to control anthracnose disease (Figure 12.75). Postharvest treatment of the fruit with fungicide is generally reported to be ineffective in controlling anthracnose disease (Prakash and Pandey 2000), but immersing them in hot water, preferably containing an appropriate fungicide, can give good disease control. There are various recommendations for successful treatments (Table 12.40).

Figure 12.75 A batch mango hot water treatment facility erected on a farm in St James, Jamaica.

Table 12.40 Recommended conditions for hot water treatments to control anthracnose disease on mango fruits (source: Thompson 1987)

Water temperature (°C)	Dipping time (minutes)	Fungicide ($\mu g\ l^{-1}$)	Country	Reference
51–51.5	15	0	USA	Pennock and Montaldo (1962)
54.4–55.8	5	0	USA	Smoot and Segall (1963)
55	5	0	USA	Hatton and Reeder (1965)
54.4	5	0	USA	Spalding and Reeder (1972)
53–55	5	0	Mexico	Lakshminarayana *et al.* (1973)
53	10	0	Philippines	Quimio and Quimio (1974)
55	5	0	Australia	Muirhead (1976)
50	30	0	Brazil	Sampaio *et al.* (1980)
55	10	0	Brazil	Sampaio *et al.* (1980)
53	5	0	West Indies	Crucifix *et al.* (1989)
51–55	30	0	Unknown	Carlos (1990)
52	30	0	India	Prakash and Pandey 2000
51.5	5	500 benomyl	Australia	Muirhead (1976)
52	4	500 benomyl	Australia	Grattidge (1980)
53.5	4	500 benomyl	Australia	Muirhead (1983)
52	5	500 benomyl	Australia	Muirhead and Grattidge (1984)
55	5	1000 benomyl	South Africa	Jacobs *et al.* (1973)
54.4	5	1000 benomyl	USA	Spalding and Reeder (1972)
48.5	5	1000 benomyl	Australia	Muirhead (1976)
55	5	1000 benomyl	Brazil	Sampaio *et al.* (1980)
58–62	2	2000 benomyl	Zambia	Sufi *et al.* (1977)
54.4	5	1000 thiabendazole	USA	Spalding and Reeder (1972)
55	5	1000 thiabendazole	Brazil	Sampaio *et al.* (1980)
55	5	1350 captan	Brazil	Sampaio *et al.* (1980)
55	5	1000 iprodione	South Africa	Brodrick and Thord-Gray (1982)
55	5	2000–4000 kasugamycin	Brazil	Sampaio *et al.* (1983)
55	5	1000 benomyl plus 0.75 kGy irradiation	South Africa	Brodrick and Thord-Gray (1982)
52	15	0.1% carbendazim or 0.1% thiophanate-methyl	India	Prakash and Pandey (2000)

In some cases it has been shown that there is an interaction between water temperature and fungicide, where at lower temperatures there is a greater need to add fungicide (Muirhead 1976). Control of anthracnose was achieved when a 30 second dip in 1000 μg litre^{-1} flusilazole was preceded by a 5 minute dip in water at 55°C, but this treatment damaged the fruit. Good control of soft brown rot and anthracnose was achieved in Kent mangoes by dipping in water at 40°C for 5 minutes followed by a 30 second dip in 3000 μg litre^{-1} flusilazole plus 1000 μg litre^{-1} prochloraz. Good control of the two postharvest rots was obtained in Sensation when the hot water treatment was followed by an application of unheated penconazole and prochloraz, each at 1000 mg litre^{-1}, by dipping the fruit for 5 minutes in a mixture of 2000 mg litre^{-1} of penconazole and 500 mg litre^{-1} of prochloraz at temperature of 40–50°C or by a 30 second dip in unheated 6000–8000 mg litre^{-1} of penconazole plus 2000 mg litre^{-1} of prochloraz. However, the latter treatment resulted in poor colour development. Phytotoxicity resulted when Sensation mangoes were dipped for 5 minutes in 2500 mg litre^{-1} penconazole plus 1100 mg litre^{-1} prochloraz at temperatures between 40 and 55°C (Pelser and Lesar 1989).

Hot water treatment containing fungicide can also be effective in the control of stem end rot caused by the fungus *Dothiorella dominicana* (Johnson *et al.* 1990).

When preceded by a 5 minute dip in water at 40°C, flusilazole controlled soft brown rot caused by *Hendersonia creberrima in* Kent during simulated export by sea (3 weeks at 10°C followed by ripening at 20°C), but did not control anthracnose rot (Johnson *et al.* 1990).

The quarantine standard hot water treatment in Chile for fruit fly control was immersion at 46.5°C for 65 minutes (Muñoz *et al.* 1998). It was found that fruits exposed to 50% CO_2 with 2% O_2 for 5 days or 70–80% CO_2 with less than 0.1% O_2 for 4 days controlled fruit fly infestations and the fruit did not suffer adverse effects when they were subsequently ripened in air (Yahia *et al.* 1989).

Hot water treatment on quality and damage

In studies by Thompson (1987), hot water treatment had little or no effect on the quality or marketable life of the fruit. In other work, the skin colour was improved by both hot water and vapour heat treatment (McGuire 1991; Jacobi and Wong 1992; Coates *et al.* 1993). Jacobi (1993) observed little fruit injury in mangoes exposed to 46°C even for 75 minutes and in a comparison between hot water treatment, vapour heat treatment and untreated fruit, the time to fruit softening was 3.6, 3.3 and 4.1 days, respectively. Hot water treatments at 52°C for up to 30 minutes did not show any adverse effect on fruit ripening (Prakash and Pandey 2000).

In contrast, hot water treatment increased the rate of shrivelling and reduced the fruit firmness and acidity during subsequent storage (McIntyre 1993). Crucifix *et al.* (1989) and Crucifix (1990) also showed that exposure of Julie to 55°C for 5 minutes resulted in scorch.

Mangoes that were treated with hot water postharvest were found to be susceptible to damage as a result of the treatment if it was applied within 24 hours of harvesting if they were wet from rain (Coates *et al.* 1993; Cook and Johnson 1993). Hot water treatment did not damage fruit if there was a delay of 48 hours between harvest and treatment, but this delay could affect the efficiency of the hot water in disease control. Mature Kensington fruits were incubated at 22–42°C for 4–16 hours prior to a hot water treatment of 45°C for 30 minutes or 47°C for 15 minutes by Jacobi *et al.* (2000). They were then ripening at 22°C. Hot water injuries were reduced, and in some cases eliminated, by conditioning the fruits at 40°C for 8 hours. The conditioning temperature was more important than the duration of the hot water treatment in injury alleviation. Conditioning at 40°C prior to hot water treatment accelerated fruit ripening, increased weight loss, reduced fruit firmness, increased °Brix and lowered titratable acidity compared to untreated fruits and fruits receiving other heat treatments, but made them more resistant to postharvest diseases.

Vapour heat treatment

This treatment was developed to control infections of fruit flies. It involves stacking the boxes of freshly harvested fruit in a room that is heated and humidified by the injection of steam. The temperature and exposure time are adjusted to kill all stages of the insects (eggs, larvae, pupa and adult) but not damage the fruit. The most difficult stage to control by hot water or vapour heat treatment is the larval stage, because it tends to be further from the fruit surface and thus exposed to high temperatures for shorter periods (Jacobi *et al.* 1993). A recommended treatment is placing the fruit in an insulated room 43°C in saturated air for 8 hours, until the temperature at the centre of the fruit reaches 43°C, and then holding the temperature for a further 6 hours. Merino *et al.* (1985) and Esquerra and Lizada (1991) described a method of control of oriental fruit fly in Carabao mangoes. This involved exposing the fruit to 46°C and at least 95% r.h. The exposure time was judged by placing a temperature probe alongside the seed of a fruit and when it had reached 46°C they were kept under those conditions for 10 minutes. This treatment also resulted in significant reductions in both anthracnose and stem end rot diseases. A similar treatment was quoted by Coates *et al.* (1993, 1993a) against Queensland fruit fly for Kensington mangoes. However, McGuire (1991) found that hot water treatment at 46°C was more effective than hot air treatment at either 46 or 48°C with 58–90% r.h. in controlling both anthracnose and stem end rot. Coates *et al.* (1993) found that vapour heat treatment did not control stem end rot but reduced anthracnose. To achieve complete control of anthracnose they found that it was necessary to apply a combination of vapour heat treatment (95% r.h. and above with a core temperature of 46.5°C for 10 minutes) plus hot water plus benomyl (0.5 g l^{-1} active ingredient) at 52°C for 5 minutes or a prochloraz ambient dip (0.25 ml l^{-1} active ingredient) replacing the benomyl. Vapour heat treatment had little effect on stem end rot of mangoes caused by *Dothiorella dominicana* and *Lasiodiplodia theobromae* (Coates *et al.* 1993). Vapour heat treatment enhanced ripening of the cultivar Baneshan during 14 days of storage compared with the control, but it resulted in better marketability of fruits due to uniform peel colour development (Pal *et al.* 1999). Vapour heat treatment can cause injury to some fruits.

Physiological disorders

A physiological disorder of mangoes called 'jelly-seed' can develop during storage and marketing of the fruit. The disorder starts around the seed and progresses towards the outside of the fruit. It forms a characteristic water-soaked appearance in the flesh, giving the fruit an offensive flavour and aroma. The cause of the disorder may be partially genetic since brightly coloured cultivars such as Tommy Atkins seem particularly susceptible. However, Mayne *et al.* (1993), has shown that jelly-seed in Tommy Atkins is associated with flowering time. They showed that delaying flowering by removing all the inflorescences from the tree greatly reduced jelly-seed in fruit that developed from the subsequent flowering. These fruit were larger than those produced from trees where the inflorescences had not been removed but the number of fruit per tree was reduced.

In a study of the physiological disorder called lenticel damage on the cultivars Heidi, Sensation, Kent, Keitt, Tommy Atkins and Zill in South Africa, Oosthuyse (1998) found that cool, humid or wet conditions on the date of harvest strongly favour the postharvest occurrence of lenticel damage. Conversely, dry, hot conditions discouraged the postharvest occurrence of lenticel damage.

Quality standards

The Organisation de Coopération et de Développment Économiques, Paris, published a brochure of standards for mangoes in 1993. Medlicott *et al.* (1987) showed that Brazilian mangoes varied in quality for the same variety throughout the harvest season. The early maturing fruit tended to have better quality and postharvest characteristics than those that matured later.

Air and controlled atmosphere storage

General refrigerated storage recommendations are as follows:

- 12°C and 85% r.h. for 2–3 weeks (Mercantilia 1989)
- 20°C and 60% r.h. for 3–4 days (Mercantilia 1989)
- 13.3°C and 85–90% r.h. for 14–25 days (SeaLand 1991)
- 10–14°C and 85–90% r.h. with 2–5% O_2 and 5–10% CO_2 for 1–4 weeks (Burden 1997).

Riper fruits could be stored at lower temperatures as follows:

- 12°C and 3–4% O_2 with 25% CO_2 for 3 weeks for mature-green Tommy Atkins, Haden, Keitt and Kent (Bender *et al.* 2000)
- 8°C 3–4% O_2 plus 25% CO_2 tree-ripe Tommy Atkins, Haden, Keitt and Kent for 3 weeks at 8°C (Bender *et al.* 2000)
- 5°C 5% O_2 plus 10% CO_2 tree-ripe Tommy Atkins, Haden, Keitt and Kent for 3 weeks at with no evidence of chilling injury (Bender *et al.* 2000).

However, storing tree-ripe mangoes in controlled atmosphere storage at 12°C for even 2 weeks was not successful owing to zero shelf-life following storage (Bender *et al.* 2000). Storage at O_2 levels of 1% resulted in the production of 'off-flavours' and skin discoloration (Hatton and Reeder 1966). General controlled atmosphere storage recommendations were as follows:

- 12°C with 5% CO_2 and 5% O_2 for 20 days (Hatton and Reeder 1966)
- 13°C with 5% CO_2 and 5% O_2 (Pantastico 1975, SeaLand 1991)
- 10–15°C with 5% CO_2 and 5% O_2 (Kader 1986)
- 10–15°C with 5–10% CO_2 and 3–5% O_2 (Kader 1989 and 1992).

Storage recommendations for some varieties are as follows.

Alphonso

- 5.6–7.2°C and 85–90% r.h. (Mathur *et al.* 1953)
- 8–10°C and 85–90% r.h. for up to 4 weeks (Anon 1967).

Alphonso from India

- 7–9°C and 90% r.h. for 7 weeks (Snowdon 1990).

Alphonso from the USA

- 13°C and 90% r.h. for 2–3 weeks (Snowdon 1990).

Amelie

- 11°C with 5% O_2 and 5% CO_2 for 4 weeks (Medlicott and Jeger 1987).

Bangalore

- 7–9°C and 85–90% r.h. for 4–7 weeks (Anon 1967)
- 5.5–7°C and 90% r.h. for 7 weeks (Snowdon 1990).

Caraboa

- 7.2–10°C and 85–90% r.h. for 17–24 days with 5.1% loss (Pantastico 1975)

- 8°C with 10% CO_2 and 6% O_2 for 4 weeks for 6 weeks (Bleinroth *et al.* 1977).

Ceylon
- 10°C for 3 weeks (Wardlaw and Leonard 1936; Thompson 1971).

Haden
- 12.8°C and 85–90% r.h. for 2–3 weeks (Lutz and Hardenburg 1968)
- 8°C with 10% CO_2 and 6% O_2 for 4 weeks (Bleinroth *et al.* 1977)
- 10–11°C with under 2% O_2 and either 1 or 5% CO_2 for 6 weeks (Medlicott and Jeger 1987)
- 12–14°C and 90% r.h. for 2 weeks (Snowdon 1990).

Irwin
- 10°C and 85–90% r.h. for 3 weeks (Irwin,) (Lutz and Hardenburg 1968).

Jasmin
- 8°C with 10% CO_2 and 6% O_2 for 6 weeks (Bleinroth *et al.* 1977).

Julie
- 10°C for 3 weeks (Wardlaw and Leonard 1936, Thompson 1971)
- 11°C with 5% O_2 and 5% CO_2 for 4 weeks (Medlicott and Jeger 1987)
- 11–12°C and 90% r.h. for 2 weeks (Snowdon 1990).

Keitt
- 12.8°C and 85–90% r.h. for 2–3 weeks (Lutz and Hardenburg 1968)
- 12–14°C and 90% r.h. for 2 weeks (Snowdon 1990).

Kent
- 12°C with 10% CO_2 and 5% O_2 at 12°C from 21 days in air to 29 days (Lizana and Ochagavia 1997).

Khuddus
- 7–9°C and 85–90% r.h. for 4–7 weeks (Anon 1967).

Neelum
- 5.6–7.2°C and 85–90% r.h. (Mathur *et al.* 1953)
- 5.5–9°C and 90% r.h. for 5–6 weeks (Snowdon 1990).

Pedda
- 7–9°C and 85–90% r.h. for 4–7 weeks (Anon 1967).

Pico
- 7.2–10°C and 85–90% r.h. for 17 days with 6.2% loss (Pantastico 1975).

Raspuri
- 5.6–7.2°C and 85–90% r.h. (Mathur *et al.* 1953)
- 7–9°C and 85–90% r.h. for 4–7 weeks (Anon 1967)
- 8.3°C and 85–90% r.h. for 4 weeks with 6.8% loss (Pantastico 1975)
- 5.5–9°C and 90% r.h. for 5–6 weeks (Snowdon 1990).

Safeda
- 7–9°C and 85–90% r.h. for 4–7 weeks (Anon 1967)
- 5.5–7°C and 90% r.h. for 7 weeks (Snowdon 1990).

San Quirino
- 8°C with 10% CO_2 and 6% O_2 for 6 weeks (Bleinroth *et al.* 1977).

Seedling types
- 5.6–7.2°C and 85–90% r.h. (Mathur *et al.* 1953).

Tommy Atkins
- 12°C with 5% CO_2 and 5% O_2, for 31 days (Lizana and Ochagavia 1997).

Zill
- 10°C and 85–90% r.h. for 3 weeks (Lutz and Hardenburg 1968)
- 10°C and 90% r.h. for 3 weeks (Snowdon 1990).

Carbon dioxide and oxygen injury

Symptoms of CO_2 injury during storage of mangoes involved irreversible inhibition of ripening upon transfer to air at the ripening-conducive temperature of 20°C, which was characterized by abnormal, greyish, epidermal colour, inhibition of normal aroma development and appearance of off-flavours (Bender *et al.* 2000). Injurious levels of CO_2 irreversibly inhibited the production of ethylene by inhibiting the enzyme ACC oxidase. Lower levels of CO_2 acted at the same sites to inhibit ethylene production reversibly (Bender *et al.* 2000).

Low O_2 resulted in irreversible inhibition of ripening, especially chlorophyll degradation, which resulted in abnormal colour development and increased ethanol production. Injurious levels of O_2 inhibited the production of ethylene by the fruit in an irreversible manner by inhibiting the enzyme ACC synthase. Less extreme levels of O_2 acted at the same sites to inhibit ethylene production reversibly (Bender *et al.* 2000).

Chilling injury

Storage of mango cultivars Tommy Atkins and Keith at 12°C caused slight chilling injury symptoms on the fruit peel expressed as red spots around the lenticels (Pesis *et al.* 1999). Gonzalez-Aguilar *et al.* (2000) found that exposure of Tommy Atkins mangoes to methyl jasmonate vapour (10–4 M) for 24 hours at 25°C reduced chilling injury during subsequent storage for 21 days at 7°C and after 5 days of shelf-life at 20°C.

Intermittent warming

Storage at 6°C for 10 days, 13°C for the next 10 days and 6°C for the last 10 days were recommended by Muñoz *et al.* (1998). Fruits stored continuously at 6°C did not develop carotenoid pigments, while the maximum storage period at 13°C was less than 20 days.

Modified atmosphere packaging

Fruits stored in polyethylene film bags at 21°C had almost twice the storage life of fruits stored without wraps (Thompson 1971). After 3 weeks of storage of Tommy Atkins and Keith at 12°C plus 1 week at 20°C in either modified atmospheres or controlled atmospheres (5% CO_2 and 15% O_2) had less red spotting on the peel compared with controls. The most effective reduction in chilling injury symptoms, expressed as red spots, was found in fruits that were packed in the Xtend film. Another advantage of the Xtend film was a reduction in the level of fruit sap inside the package as compared with polyethylene film. This was attributed to the lower humidity in the Xtend film, less than 95% r.h., compared with 99% r.h. which was recorded in the polyethylene film (Pesis *et al.* 1999).

Ripening

A simple method of ripening was to place the fruit in baskets lined with banana leaves with calcium carbide, which gave fruit of uniform colour within 2–3 days at Malaysian ambient conditions, but with inferior flavour to fruits ripened without calcium carbide (Berwick 1940). Calcium carbide was also used to ripen fruit in South Africa, where they were placed in a room maintained at 21.1–26.7°C and 85–90% r.h. and calcium carbide was introduced at 1 oz per 72 ft^3 for 1–2 days with ventilation every 4 hours (Marloth 1947). A similar practice was used commercially on mangoes for airfreight export from Brazil, where the harvested fruit were kept under gas-tight tarpaulins for 2–3 days before export.

Ethylene is also used successfully for ripening. At 20°C Wills *et al.* (2001) found that the time to ripen of unripe mangoes increased linearly with logarithmic decrease in ethylene concentration over the range <0.005, 0.01, 0.1 1.0 and 10 μl $litre^{-1}$. Other recommendations are:

- 19–21°C gave better quality characteristics than those ripened at 28–30°C (Singh and Mathur 1952)
- 15.5–18.5°C was satisfactory but the fruit were a little tart and required up to 3 days of subsequent exposure to 21–24°C to develop a good flavour (Hatton *et al.* 1965)
- 21–24°C was recommended as the optimum ripening conditions (Hatton *et al.* 1965)
- 29–31°C using ethylene at 10 μl $litre^{-1}$ for 24 hours at 85–90% r.h. (Wills *et al.* 1989).

Mangosteen

Garcinia mangostana, Guttiferae.

Botany

The edible portion is the fruit, which is produced on a small, slow-growing tree and is subglobose in shape with a persistent calyx. It ripens to dark reddish purple colour often with brown scars and can be up to 8 cm in diameter. The skin is thick and tough and inside of which is white translucent flesh divided into 5–8 segments. Each segment may or may not contain a seed. It originated in Malaysia and is considered by some to be the most delicious of all fruits.

Harvesting

Harvest maturity is normally judged on the skin colour, which changes from green as it matures. The Malaysian Agricultural Research and Development Institute published a colour chart of changes in skin colour. Wardlaw (1937) mentioned that fruit that were 'just ripening' at harvest did not appear to ripen normally and had hard flesh and poor flavour. Those picked fully ripe were of good appearance and flavour. In Indonesia Daryono and Sosrodiharjo (1986) found that the best time for harvesting was when 25% of the fruit skin had developed a purple colour. Sugar content, acidity, sugar acid ratio and ascorbic acid content were 14.3%, 0.46%, 31.3 and 42.3 mg per 100 g, respectively, at harvest.

Physiological disorders

Translucent flesh is a common physiological disorder. Fruits exhibiting translucent flesh disorder had higher rind and flesh water contents than fruits with normal flesh. Specific gravity of translucent flesh fruit was >1 and that of normal flesh fruit was <1. Fruit specific gravity and natural transverse rind cracking were used to separate translucent-fleshed fruits from normal fruits. Translucent flesh may be associated with water infiltration during development, since it could be induced in normal fruits following vacuum infiltration with water (Pankasemsuk *et al.* 1996). Images obtained by either X-rays or nuclear magnetic resonance were suitable for detecting internal disorders and rotten pulp in fruits and NMR imaging was able to identify translucent flesh disorder (Yantarasri *et al.* 1998).

Storage

Weight loss and percentage of diseased fruit after 7 days of storage were 3.3% and 23.9%, respectively, at ambient temperature and 0 and 11% at 5°C. Cold storage caused no chilling injury and had no adverse effect on fruit quality (Daryono and Sosrodiharjo 1986). At a room temperature of 20°C and 60% r.h. the fruit could be kept for 2–3 weeks (Mercantilia 1989), but in other work at a room temperature of 27–30°C and 50–60% r.h., 30% of the fruit was spoiled after 10 days and 100% after 20 days (Raman *et al.* 1971). Refrigerated storage recommendations are as follows:

- 4.5–7.0°C and 80–90% r.h. for 30 days with 38% spoilage (Raman *et al.* 1971)
- 3.9–5.6°C and 85–90% for 7 weeks (Pantastico 1975)
- 4 or 8°C or for 42 days but fruit hardening occurred (Augustin and Azudin 1986)
- 5°C and 90% r.h. for 6–7 weeks (Mercantilia 1989)
- 10°C and 90% r.h. for 3–4 weeks (Snowdon 1990)
- 13.3°C and 85–90% r.h. for 14–25 days (SeaLand 1991).

Modified atmosphere packaging

Daryono and Sosrodiharjo (1986) found that storage in plastic bags slightly reduced the sugar:acid ratio in the fruit pulp but had no effect on the ascorbic acid content compared with storage without bags. However, the levels of weight loss and disease were greatly reduced when fruit were stored in polyethylene bags as follows:

Type of polyethylene bag	Weight losses (%)	Diseased fruit (%)
None	7.2	22.3
Perforated polyethylene bags	2.3	16.1
Closed polyethylene bags	0.3	13.0

Ripening

In Indonesia, fruits ripened normally in 1 day at ambient temperature when harvested when 25% of the skin was purple (Daryono and Sosrodiharjo 1986).

Transport

Wardlaw (1937) mentioned trial shipments from Burma to the UK at 3.9–5.6°C and 85–90% r.h. that took 30 days. Losses on arrival were measured at 40–55%, mainly due to *Diplodia* spp. rot and rind hardening, but the remainder were in good condition and had a shelf-life of 5 days at 12.8–15.6°C.

Marrow, squash

Cucurbita maxima, C. mixta, C. moschata, C. pepo, Cucurbitaceae.

Botany

These are highly polymorphic species and considerable confusion exists with their common names; see also the section on Pumpkin. The fruit that will be dealt with in this section are those with a fine-grained, mild-flavoured flesh, often pale green colour and with a hard skin or shell. They will freeze at about –1.7 to 0.4°C (Wright 1942) or –1.94 to 0.50°C (Weichmann 1987). The immature forms are dealt with in the section on Courgettes.

Harvesting

They are harvested when fully mature, which is usually judged from experience related to their size and gently pressing the skin to assess whether they are beginning to soften. Harvey *et al.* (1997) found that starch and dry matter did not accumulate significantly after 40 days from flowering. They also found that skin hardness and heat accumulation levels were the most effective means of estimating the optimum harvest date. Heat units should be between 240 and 300 growing degree days (base temperature 8°C) from flowering. When fruits of the cultivar Delica were left for longer on the vine, the skin hardened, the flesh became redder, dry matter increased then decreased, the soluble solids and sucrose content increased but sensory properties improved.

Penetrometers have been tested to measure firmness. Penetrometer readings increased as flesh colour and juice soluble solids increased. Harvey *et al.* (1997) showed that the earliest time to harvest fruits to ensure an acceptable level of sensory quality after simulated refrigerated shipment conditions was at a skin penetrometer measurement of 7 kg force. However, penetrometers will damage the fruit and make them unsuitable for marketing. An objective non-destructive technique for assessing the maturity of Buttercup squash used ballistic collision. This method was found to correlate well with destructive penetrometer measurements. However, the equipment was not robust enough for use in moving grading line processes (Harvey *et al.* 1995).

Curing

This was recommended by Phillips and Armstrong (1967) and Purseglove (1968) to harden the shell by exposure for 2 weeks at 23.9–29.4°C, but other work claimed that curing was either ineffective or detrimental (Wardlaw 1937; Lutz and Hardenburg 1968).

Chilling injury

This was found in Table Queen at both 0 and 4.4°C (Platenius *et al.* 1934).

Storage

Harvey *et al.* (1997) found that once harvested the flesh of the cultivar Delica continued to become redder, sucrose and soluble solids increased and starch and dry matter levels decreased. At 12.8–21.1°C Table Queen became yellow with stringy flesh within 5 weeks (Lutz and Hardenburg 1968). General refrigerated storage recommendations are as follows:

- 15.6–21.1°C and less than 60% r.h. for 5 months with 7% loss per month (Platenius *et al.* 1934)
- 4.4°C and 70% r.h. with 3% weight loss per month and a slow conversion of starch to sugar (Platenius *et al.* 1934)
- 12.8°C for 5 months with 20% loss or 23.3°C for 5 months with 63% loss (Wardlaw 1937)
- 12.8–15.6°C and 70–75% r.h. for 3–6 months (Wardlaw 1937)
- 12.8–15.6°C and 70–75% r.h. for 8–24 weeks with 4–15% loss (Pantastico 1975)
- 10–12.8°C and low humidity for 6 months (Purseglove 1968)
- 10–13°C and 60–70% r.h. for 2–6 months (Snowdon 1991).

Recommendations for specific cultivars and types are as follows.

Acorn

- 10°C and 70–75% r.h. for 5–8 weeks (Anon 1968, Lutz and Hardenburg 1968).

Buttercup

- 6.7–10°C and 70–75% r.h. for 3 months (Phillips and Armstrong 1967; Anon 1968).

Butternut

- 6.7–10°C and 70–75% r.h. for 2 months (Phillips and Armstrong 1967)
- 10°C and 70–75% r.h. for 2–3 months (Anon 1968; Lutz and Hardenburg 1968)
- 10–16°C and 60% r.h. for up to 50 days (Tindall 1983).

Calabaza

- 10°C and 50–70% r.h. for 60–90 days (SeaLand 1991).

Hard skinned winter squash

- 12.2°C with 70–75% r.h. for 84–150 days (SeaLand 1991).

Hubbard

- 6.7–10°C and 70–75% r.h. for 6 months (Phillips and Armstrong 1967; Anon 1968)
- 10–16°C and 60% r.h. for up to 180 days (Tindall 1983).

Table cream

- 6.7–10°C and 70–75% r.h. for 2 months (Phillips and Armstrong 1967; Anon 1968).

Table queen

- 10°C and 70–75% r.h. for 5–8 weeks (Anon 1968; Lutz and Hardenburg 1968).

Turban

- 6.7–10°C and 70–75% r.h. for 3 months (Phillips and Armstrong 1967; Anon 1968) (Figure 12.76, in the colour plates).

Controlled atmosphere storage

Controlled atmosphere storage was reported to have only a slight or no effect (SeaLand 1991).

Ripening

Fruits harvested at an immature or early stage required a postharvest ripening period to enhance sweetness and texture and to optimize other sensory quality (Harvey *et al.* 1997).

Matricaria

Matricaria recutita, *Chamomilla recutita*, Compositae.

Botany

It is an annual herb with a branched erect stem with solitary terminal flower heads. The flowers have white ray florets and yellow disc florets. The flowers are used medicinally and contain up to 1% of the essential oil azulene and also bisabolol, farnescene, flavonoid and coumarin glycosides. During storage of the cultivar Bodegold for 80 hours, Bottcher *et al.* (2001) showed that the harvested flowers had a high respiration rate of 999 ± 134 W tonne^{-1} at 10°C, 2438 ± 289 W tonne^{-1} at 20°C and 4552 ± 570 W tonne^{-1} at 30°C.

Harvesting

In Germany they are mechanically harvested from autumn- and spring-sown crops. The flowers need ventilating, cooling or drying directly after harvest (Bottcher *et al.* 2001).

Storage

In a comparison of three temperatures, 10, 20 and 30°C, Bottcher *et al.* (2001) found that losses were lowest at 10°C. Constituents of the flowers decreased only slightly during 80 hours storage. The valuable (–)-α-bisabolol and its A and B oxidized forms showed the greatest loss at 10°C, but responded differently in different trials with losses between 10 and 80% (Bottcher *et al.* 2001).

Medlar

Mespilus germanica, Rosaceae.

Botany

The fruit is five seeded and set in a receptacle with five conspicuous calyx lobes. Ripe fruit is reported to have a sugar content in the flesh of about 10.6% and a dry weight of 42% (Burton 1982). It probably originated in Asia Minor, Greece or Iran. There are several cultivars, with Dutch and Monstrous bearing fruits that can be 5 cm in diameter. Nottingham is said to have the best flavour with Royal with Stoneless being of poorer eating quality, but the latter is thought to store well.

Harvesting

In Britain the fruit does not become palatable until it is 'bletted,' which means it is soft and brown and half rotten. In the countries of origin it can be ripened on the tree.

Storage

No detailed information could be found on its postharvest life.

Melons, cantaloupes, musk melons

Cucumis melo, Cucurbitaceae.

Botany

C. melo is sometimes classified into two basic types: *reticulates* group, which are the rough-skinned netted or cantaloupe types, and *inodorus* group, which are smooth skinned types and include the Honeydew, Casaba, Crenshaw, Charentais, Hami, Persian, Spanish and Santa Claus types (Seymour and McGlasson 1993). It is an indehiscent fleshy berry, usually spherical with a hollow central cavity containing many seeds. Pratt (1971) showed that both Honeydew muskmelons and PMR-45 cantaloupes were climacteric and all melons were classified as being climacteric by Wills *et al.* (1989), but Biale (1960) classified all melons as non-climacteric. Ripe muskmelons will freeze at about –2.2 to 1.2°C (Wright 1942).

Harvesting

In melons when the leaf, in whose axis a fruit is borne, dies, then that fruit is judged to be ready for harvesting. In certain parts of the world total soluble solids is used to specify maturity, for example Honeydew must have a minimum soluble solids content of 10%. Shining a light on a fruit and measuring the amount transmitted can also be used to measure their soluble solids content. The transmission of near-infrared light has also been used as a non-destructive method for measuring the soluble solids content of Cantaloupe melons (Dull *et al.* 1989). A simpler method is to tap the fruit with the knuckle in the field to judge whether they are ready to be harvested. The sound changes as they mature. Consumers when purchasing melons sometimes use this method of testing maturity.

As part of the natural development of fruit, an abscission layer is formed in the pedicel, which is referred to as 'slip'. In cantaloupe melons, harvesting before this abscission layer is fully developed results in inferior-flavoured fruit compared with those left on the vine for the full period (Pratt 1971). However, fruit harvested at this full maturity will have only a short marketable life. The method is described as follows:

1. Non-slip, fruits fully grown, but still firmly attached.
2. Half slip abscission layer beginning to form with an obvious crack at the junction.
3. Full slip green crack quite distinct, circling the whole stalk and can be broken with gentle pressure.
4. Full slip yellow abscission complete and the fruit soft, fully yellow, with a distinct aroma.

Neri and Brigati (1996) found that skin colour was a good indicator of maturity, while for the cultivars Harper and Simphony, flesh colour was equally effective. For Soleado, fruit firmness was also a good indicator of maturity. For all cultivars total soluble solids was a suitable indicator of harvest maturity, but it is destructive. In experiments described by Moelich *et al.* (1996), Galia 5 fruits were harvested with <10% yellow colour break, 10–20% yellow colour break and 50–70% yellow colour break, and stored at 2°C for up to 26 days. Fruits with <10% yellow and 10–20% yellow fruits showed no senescent breakdown during storage, but failed to develop sufficient colour to be acceptable in the market place. Galia 5 with 50–70% yellow developed unacceptable levels of senescent breakdown. It was concluded that Galia 5 is not suitable for normal sea freight export from South Africa to Europe. Doral fruits harvested 31 and 35 days after anthesis were stored at 12°C for up to 33 days. The first evidence of senescent breakdown appeared after 21–25 days of storage. Fruits harvested 31 days after anthesis were of acceptable internal quality, but fruits harvested 35 days after anthesis showed unacceptable levels of senescent breakdown after storage for 4 weeks at 12°C.

Hot water treatment

Hot water treatment was not effective in controlling postharvest diseases. With the cultivar Galia F1, hot water at 55°C for 90 seconds resulted in high fungal infections after subsequent storage at 2°C and 85–90% r.h. *Alternaria* spp., *Penicillium* spp., *Cladosporium* sp., *Mucor* spp., *Botrytis cinerea* and *Aspergillus* spp. were found to be present in infected fruits. Fruit softening was also quicker in the hot water-treated fruits (Halloran *et al.* 1999).

Hot water brushing

Hot water brushing of melons was shown to removed soil, dust and fungal spores from the fruit surface, and partially or entirely sealed natural openings in the epidermis. It did this by smoothing the epicuticular waxes and thus covered and sealed stomata and cracks on the fruit surface, which could have served as potential pathogen invasion sites. The optimal hot water brushing treatment to reduce decay while maintaining fruit quality on Galia was 59°C for 15 seconds or 56°C for 20 seconds (Fallik *et al.* 2000). The hot water brushing treatment at 56°C did not cause surface damage, and did not influence fruit weight loss or internal quality parameters.

Precooling

This is not commonly necessary but they can be hydrocooled.

Storage

Refrigerated storage recommendations are as follows.

Cantaloupe type (reticulates group)

- 0°C and 85–90% r.h. for 2 weeks (Phillips and Armstrong 1967)
- 0–1°C and 85–90% r.h. for up to 7 weeks (Anon 1967)
- 2.2–4.4°C and 85–90% r.h. for 15 days (0.75 to full slip) (Anon 1968; Lutz and Hardenburg 1968)
- 0–1.1°C and 85–90% r.h. for 1 weeks (Lutz and Hardenburg 1968)
- 0–1.7°C and 85–90% r.h. for 5–14 days (full slip) (Anon 1968, Lutz and Hardenburg 1968)
- 1.7–3.3°C and 85–90% r.h. for 10 days with 7.2% loss (Pantastico 1975)
- 3–4°C and 85–90% r.h. for 6–12 days depending on cultivar (Tindall 1983)
- 3–5°C and 85–90% r.h. for 10–14 days (Mercantilia 1989)
- 4–5°C and 90% r.h. for 1–3 weeks (Snowdon 1990)
- 4.4°C and 85–90% r.h. for 10–14 days (SeaLand 1991).

Honeydew type (inodorus group)

- 7.2°C and 85–90% r.h. for 2–3 weeks (Honeydew) (Phillips and Armstrong 1967)
- 7.2–10°C and 85–90% r.h. for 2 weeks (Crenshaw, Persian), 3–4 weeks (Honeydew), 4–6 weeks (Casaba) (Anon 1968; Lutz and Hardenburg 1968)
- 7.2°C and 85% r.h. for 4–5 weeks (Honeydew) (Pantastico 1975)
- 10–14°C and 85–90% r.h. for 16–20 days (Honey) (Mercantilia 1989)
- 6–9°C and 85–90% r.h. for 10–14 days (Net) (Mercantilia 1989)
- 6–7°C and 85–90% r.h. for 10–14 days (Ogen) (Mercantilia 1989)
- 5–10°C and 90% r.h. for 1–4 weeks (melons) (Snowdon 1990)
- 6°C (Ogen, Galia) (Snowdon 1990)
- 10–15°C and 90% r.h. for 3 weeks for Honeydew from South Africa (Snowdon 1990)
- 5°C and 90% r.h. for 2–4 weeks for Honeydew from the USA (Snowdon 1990)
- 10°C and 85–90% r.h. for 21–28 days (SeaLand 1991)
- 10°C and 85–90% r.h. for 14–21 days (Persian) (SeaLand 1991).

Chilling injury

Symptoms are surface breakdown, softening, decay and the production of off-flavours especially when removed to higher temperatures. Susceptibility to chilling injury varies between cultivars and is a time–temperature factor. For example, Lutz and Hardenburg (1968) showed that periods of more than 15 days at 2.2–4.4°C were required to injure Cantaloupes. They also showed that the harvest maturity affects susceptibility to chilling injury, with the less mature fruit being more susceptible.

Controlled atmosphere storage

Recommendated conditions are:

- 10–15% CO_2 and 3–5% O_2 was recommended for Cantaloupe and 5–10% CO_2 with 3–5% O_2 for Honeydew and Casaba (SeaLand 1991)
- Kader (1985, 1992) recommended 3–7°C with 10–15% CO_2 and 3–5% O_2 for Cantaloupes which had a good effect but limited commercial use
- For Honeydew Kader (1985, 1992) recommended 10–12°C with 0% CO_2 and 3–5% O_2 which had a fair effect but was of no commercial use
- Saltveit (1989) recommended 5–10°C with 10–20% CO_2 and 3–5% O_2, which had only a slight effect.

Ripening

Exposing muskmelon cultivars to ethylene was said to be of no practical use since they are 'adequately self-ripening' (Pratt 1971). Recommended ripening temperatures for Honeydew melons were 18–21°C using ethylene at 10 μl $litre^{-1}$ for 24 hours at 85–90% r.h. and for Cantaloupe melons 18–21°C with no ethylene (Wills *et al.* 1989). Wardlaw (1937) recommended exposing fruit to 500 μl $litre^{-1}$ ethylene. Pantastico (1975) found that fruit ripened evenly at 21°C when exposed to ethylene at either 1000 or 5000 μl $litre^{-1}$ for 18–24 hours.

Ethylene

Exposing Honeydew melons to ethylene at 20°C for 24 hours reduced the incidence of chilling injury during subsequent storage at 2.5°C compared with melons which had not been exposed to ethylene before storage (Lipton *et al.* 1979). This is presumably a ripening effect, and the riper the melon the less susceptible it is to chilling injury.

Methi

Trigonella foenum-graecum, Papilionaceae.

Botany

Methi is an Indian vegetable grown for its leaves and is used like spinach.

Storage

It was reported by Waskar *et al.* (1998) that in India their storage life was increased from 1 day at room temperature of 17–30°C and 24–73% r.h. to 6.5 days in polyethylene bags of 200 gauge and 2% ventilation in storage at 8°C and 90–92% r.h.

Mint

Mentha spp, Labiatae.

Botany

There are many different species and therefore flavours of mint:

- *M. aquatica* var. *crispa* is the water mint or curly mint
- *M.* × *piperita* is the peppermint
- *M. spicata* is spearmint
- *M. rotundifolia* is apple mint or Bowles mint
- *M. citrata* is eau de Cologne mint
- *M. rotundifolia* variegata is pineapple mint
- *M. pulegium* is pennyroyal.

They are all perennial herbs that spread rapidly by underground stolons. The leaves have been used from ancient times for flavouring and they also have medicinal and digestive properties.

Storage

The only specific recommendation is 0–1°C and 95–100% r.h. for 2–4 weeks (Snowdon 1991).

Monstera

Monstera deliciosa, Araceae.

Botany

The fruit is a spadix covered by hexagonal plates with a green skin colour turning yellow when ripe. They are up to 25 cm long and contain a soft, white pulp.

Harvesting

When fully ripe the fruit are detached at the slightest touch. No information could be found on harvesting them immature and ripening them, but immature fruit contain calcium oxalate crystals that can irritate the mouth and throat.

Storage

No information could be found on its postharvest life.

Mora, Andes berry

Rubus glaucus, Rosaceae.

Botany

The fruit is a collection of drupes and is probably non-climacteric (Figure 12.77). They originate in the Andean region of South America and are very popular in Colombia, where they are eaten fresh and used in drinks, ice cream and a variety of puddings.

Harvesting

They are normally harvested, with their receptacle still in place, as they change colour from light to dark red. Harvesting is by hand.

Storage

Lutz and Hardenburg (1968) recommended 0°C with 85–90% r.h. for 1–2 weeks. No other specific

Figure 12.77 Mora fruit ready for harvesting in Colombia.

information could be found on their postharvest life, but from the author's observations in Colombia they appear to have a similar storage life to blackberries (*R. ulmifolius*).

Morles

Morchella esculenta.

Botany

It is a cup fungus with a variable-shaped head from rounded to irregular oval and up to 20 cm high. There are a honeycomb of ridges on the surface with deep pits that are brown to olive in colour. The head is usually hollow, giving them a sponge-like feel. The stem is stout, white, hollow and often swollen towards the base. They are delicious and much prized and can be found in Northern Europe growing wild in grassy woodlands and chalky soils in spring (Sterry 1995). Related species include *M. vulgaris*, *M. rotunda*, *M. deliciosa*, *M. elata* and *M. crassipes* and the black morel *M. angusticeps*. A histo-cytological study showed that hyphae of *M. rotunda* was associated with all parts of the root system of Norway spruce (*Pinus abies*) in a plantation near Strasbourg in France. A mycelial sheath surrounds the main suberized roots. Finer rootlets are enveloped in a cottony mycelium and their absorbing root tips are ectomycorrhizal. Morel hyphae also invaded older parts of a heterobasidiomycete ectomycorrhiza (Buscot and Kottke 1990).

Storage

No information could be found on its postharvest life, but from the literature it would appear that all types of edible mushrooms have similar storage requirements and the recommendations for the cultivated mushroom (*Agaricus bispsorus*) could be followed.

Mulberries

Morus nigra, Moraceae.

Botany

The fruit is the swollen calyces. They are green as they develop then turn red as they begin to mature and dark purple or black when fully mature. Burton (1982) gave its sugar content as 8.1% and the dry weight 54%. It probably originated in Western Asia, but was grown in Europe from ancient times and is referred to by both Greek and Roman writers. The White Mulberry (*M. alba*), whose fruit are white, pink or purple and taste sweet and a bit insipid, is grown mainly for its leaves, which are used as the food for the silk worm.

Harvesting

The fruits must be picked when they are fully ripe, but at this stage they are very juicy and easily squashed. Harvesting is actually best by collecting the fruit from the ground when they have fallen from the tree.

Storage

No information could be found on storage, but its postharvest life is very short and this has limited its commercial exploitation in spite of the fruit being delicious when fully mature.

Mume, Japanese apricot

Prunus mume, Rosaceae.

Botany

The fruit is a drupe that has a thin skin and juicy flesh with a single, hard seed in the middle. It is a climacteric fruit (Miyazak 1983) and the number of days from harvest to the peak of the climacteric, characterized by a rapid increase in CO_2 and ethylene production and yellowing of peel, decreased with increasing maturity at harvest (Goto *et al.* 1988). They are eaten fresh and used for liqueur production or pickling

Harvesting

In Japan they are usually harvested at the mature green stage in June (Miyazak 1983). Fruits of seven cultivars were harvested at 5-day intervals during maturation so that changes in the colour of the stone surface could be monitored. When embryos were immature, the stone surface was white–yellow. It then changed to light brown as the embryos matured and finally to brown. The colour did not differ significantly between cultivars. No significant yearly fluctuations in stone surface colour were observed in cultivar Nanko harvested at intervals in three successive years (Ishizawa *et al.* 1995). Goto *et al.* (1988) showed that the fruits harvested early were less sensitive to chilling than those harvested late. They also reported that the surface colour of the stone might be a useful criterion for assessing harvesting time.

A shaker-harvester was developed, which was mounted on a 2 horsepower engine and caused fruits to drop by vibration. It was shown to reduce working hours by 35% and costs by 41% compared with conventional methods (Anon 1993).

Storage

At an average temperature of 22°C their shelf-life is only 2 or 3 days (Miyazak 1983).

Chilling injury

Chilling injury occurred as surface pitting and/or peel browning during storage at 0 or 5°C. Sealing the fruit in polyethylene bags or storage in 5% CO_2 prevented or reduced injury, and the use of perforated bags delayed it (Iwata and Yoshida 1979).

Controlled atmosphere storage

Kaji *et al.* (1991) reported the following storage life at 20°C and 100% r.h.:

- 2–3% O_2 with 3% CO_2 for 15 days
- 2–3% O_2 with 8% CO_2 for 15 days
- 2–3% O_2 with 13% CO_2 for 19 days
- 2–3% O_2 with 18% CO_2 for 12 days.

Fruits held at O_2 concentrations of up to 1% developed browning injury and produced ethanol at a high rate; the percentage of fruits with water-core injury was high in conditions of low O_2 and no CO_2. Ethylene production rates were suppressed at high CO_2 concentrations or at O_2 levels of <5%. In 3% O_2, ethylene production rates peaked half or one day after the respiratory climacteric peak (Koyakumaru *et al.* 1995). Storage recommendations are:

- 10°C with 4–5% O_2 and 8–9% CO_2 for mature-green Hakuoukoume (Koyakumaru 1997)
- 25°C with 3–4% O_2 and 10% CO_2 delayed the respiratory climacteric rises slightly and increase ethanol production after the third day of storage (Koyakumaru *et al.* 1995)
- 25°C with 3–5% O_2 and 9–10% CO_2 combined with an ethylene scrubber was effective in preserving fruit quality during a 10-day storage period (Koyakumaru 1997).

Modified atmosphere packaging

Mature green Japanese apricots could be stored in sealed plastic film bags for at least 9 days at 20°C but showed increasing physiological damage when the CO_2 levels exceeded 20%. Putting an ethylene absorbent inside the bags greatly reduced levels of physiological damage (Osajima *et al.* 1987). Sealed packaging fruits in polyethylene, which reduced the O_2 level in the bags to 1–4%, and ethylene removal extended the shelf-life by 2 and 4 days, respectively, compared with no packaging. An O_2 level of <0.5% in the sealed bags caused metabolic abnormality in the fruits (Miyazak 1983). Sealed packaged fruits using linear low-density polyethylene film 30 μm thick with an ethylene scrubber had a shelf-life of 12 days at 20°C (Kaji *et al.* 1991). In the sealed permeable film pouches containing 1 kg of fruit, the atmosphere contained 3–4% O_2 and 15% CO_2, whereas in the sealed non-permeable film pouches, the O_2 concentration was around 2%, but the CO_2 concentration varied depending on the film (>20% for polyethylene and 10% for polymethylpentene). In sealed permeable film pouches, fruits ripened normally and few disorders were observed (Asami and Aoyagi 1997).

Ripening

Treating the fruits with ethylene advanced the onset of the climacteric and of ripening (Miyazak 1983).

Mushrooms, cultivated mushroom

Agaricus bispsorus, Agaricacea.

Botany

The edible part is a sporophore, which is a fungal fruiting body that will freeze at about –1.2 to 0.9°C (Wright 1942). The respiration rates of samples of harvested mushrooms of the strain X25 held in air at 10°C for 24 hours ranged from 1.0–2.25 mmol kg^{-1} h^{-1}, averaging 1.6 mmol kg^{-1} h^{-1}. Respiration rate remained constant in mushrooms stored in open punnets at 1°C with 95% r.h. for 4 days but increased linearly in mushrooms stored in open punnets at 10°C and 90% r.h. for 4 days, reaching ~3.1 mmol kg^{-1} h^{-1} (Varoquaux *et al.* 1999).

Preharvest factors

Addition of calcium chloride to irrigation water increased the calcium content and improved the quality of mushrooms independent of inherent calcium content (Varoquaux *et al.* 1999). Hartman *et al.* (2000) also showed that significant improvements in mushroom quality resulted from addition of 0.3% calcium chloride to the irrigation water. This practice resulted in about 2-fold increase in calcium content, improved initial whiteness at harvest and reduced postharvest browning compared with untreated controls. They were also more resistant to the negative effects of excessive handling or bruising, but the calcium chloride treatment slightly reduced crop yield. Sodium selenite added to the irrigation water increased mushroom resistance to postharvest browning and enhanced their nutritional value, but the treatment decreased solids content. When calcium chloride and sodium selenite were both added, the positive effects remained and all these negative trends were reversed.

Harvesting

It should be harvested when it reaches a size that is acceptable to the market, but while the velum is still unbroken. For certain markets the sporophore is allowed to open fully with the velum broken, the cap extended and the stalk elongated. Care must be taken not to damage or detach immature sporophores during harvesting.

The strain U1 was harvested 2 days and 1 day before the planned commercial harvesting date, on the commercial harvesting date and 1 day later by Braaksma *et al.* (1999). The earlier the mushrooms were harvested, the less cap opening occurred during subsequent storage for 3 days at 20°C and over 90% r.h. The smallest mushrooms, with a cap diameter of 15–20 mm, showed no cap opening during storage, irrespective of harvesting date. Mushrooms within one size class but harvested on different days were similar in size and appearance at the time of harvest but cap opening occurred in a larger proportion of the later harvested mushrooms and was completed earlier during storage.

A harvesting machine for mushrooms has been developed using robotics (Tillet and Reed 1990). This method uses a sucker end-effector that attaches to the cap of the mushroom and gently pulls it from the bed without damaging it (Figure 12.78). An image-processing algorithm that finds and selects the mushroom to be harvested on the basis of data with which it has

Figure 12.78 Robotic mushroom harvester developed at the Silsoe Research Institute. Source: Dr Robin Tillet.

been provided guides the robot (Tillet and Batchelor 1991). There are problems with this method, particularly where mushrooms are touching or growing out at acute angles to the bed surface. The development of neural networks with trainable algorithms can be used to overcome this and other image processing challenges.

Prestorage factors

Harvested mushrooms were coated with different gum-based coatings, including alginate and alginate–ergosterol, with or without emulsifier. Coated mushrooms had a better appearance, a better colour, and an advantage in reduced weight loss in comparison with non-coated mushrooms. The alginate–ergosterol–Tween coating combination was most suitable for maintaining the size and shape of coated mushrooms (Hershko and Nussinovitch 1998).

Storage

The main deteriorative factors are weight loss resulting in shrivelling, cap greying, stalk elongation, breaking of the velum, water soaking and microbial infections. At room temperature of 20°C and 60% r.h. it was claimed that they could be kept for only 1 day (Mercantilia 1989). However, Thompson *et al.* (1977) found that little loss in quality occurred, other than dry matter loss, when they were stored at room temperature of about 25°C for several days. Storage at 4°C led to the greatest inhibition of microbial contamination and washing in chlorinated water and incorporation of silica gel humidity absorber in the low-density polyethylene bags also decreased microbial contamination (Popa *et al.* 1999). Refrigerated storage recommendations are as follows:

- 0–1°C and 85–90% r.h. for 3–7 days (Anon 1967)
- 0°C and 85–90% r.h. for 5 days (Phillips and Armstrong 1967; Anon 1968; Hardenburg *et al.* 1990)
- 4.4°C for 2 days (Lutz and Hardenburg 1968)
- 10°C for 1 day (Lutz and Hardenburg 1968)
- 0°C and 95% r.h. for 10 days with 7.8% loss (Pantastico 1975)
- 0°C and 90–95% r.h. for 7 days (Mercantilia 1989)
- 2°C for 5 days (Mercantilia 1989)
- 4°C for 4 days (Mercantilia 1989)
- 8°C for 3 days (Mercantilia 1989)
- 12°C for 2 days (Mercantilia 1989)
- 0°C and 90–95% r.h. (SeaLand 1991).

Controlled atmosphere storage

High levels of CO_2 in storage can result in cap discoloration (Smith 1965). Storage recommendations were as follows:

- 4.4°C with 5–10% CO_2 (Lutz and Hardenburg 1968; Ryall and Lipton 1979)
- 0°C and 95% r.h. with 10–20% CO_2 inhibited mould growth and retarded cap and stalk development; O_2 levels of less than 1% were injurious (Pantastico 1975)
- 0–5°C with 10–15% CO_2 and 21% O_2 which had a fair effect but was of limited commercial use (Kader 1985, 1992)
- 0–5°C with 10–15% CO_2 and 21% O_2 which had a moderate level of effect (Saltveit 1989)
- 0°C and 90–95% r.h. with 10–15% CO_2 and 21% O_2 (SeaLand 1991).

Modified atmosphere packaging

Poly(vinyl chloride) over-wraps on consumer-sized punnets (about 400 g) greatly reduced weight loss, cap and stalk development and discoloration, especially when combined with refrigeration (Nichols 1971). Storage of mushrooms in modified atmosphere packaging at 10°C and 85% r.h. for 8 days delayed maturation and reduced weight loss compared with those stored without packaging (Lopez-Briones *et al.* 1993). Modified atmosphere packaging in microporus film delayed mushroom development, especially when combined with cool storage at 2°C (Burton and Twyning 1989). Storage in low-density polyethylene bags containing 3% O_2 with 5% CO_2, 5% O_2 with 5% CO_2, 1% O_2 with 5% CO_2, or 3% O_2 with 7% CO_2 and at 4, 18 and 26–28°C showed little effect of CO_2 level, but storage was best under the lowest O_2 regime (Popa *et al.* 1999). Thompson *et al.* (1977) found that there were no differences between mushrooms sealed polyethylene film of 0.03, 0.025 or 0.015 mm thickness during storage at either 0 or 20°C.

Decreases in dry matter of modified atmosphere packaged mushrooms matched the theoretical carbon consumption calculated from the constant respiration rate. Reducing O_2 from 20–1% did not affect the respiration rate at temperatures ranging from 5–20°C. Glucose and glycogen were rapidly catabolized, whereas mannitol consumption began 3 and 8 days after harvest at 20 and 10°C, respectively. It was concluded that no extension of mushroom shelf-life was attained through modified atmosphere packaging, but that controlling humidity within the package may be more effective (Varoquaux *et al.* 1999).

Ethylene

Ethylene can stimulate the stalk to elongate and the cap to expand. In button mushrooms, cap expansion may cause the veils, which join the cap to the stalk, to break and they cease to be button mushrooms.

Mustard, white mustard

Brassica alba, Cruciferae.

Botany

The plants are grown only to the seedling stage and the whole plant is eaten in salads often combined with cress. They are also grown for their mature seeds, which may be dried, ground and blended with other species to produce table mustard.

Harvesting

Mustard seedlings are usually harvested about 6 days after sowing (Purseglove 1968) before they become bitter and tough (Pantastico 1975).

Storage

No information could be located on postharvest factors.

Nameko

Pholiota nameko, Basidiomycetes.

Botany

It is a popular mushroom in Asia, especially Japan, where it is grown on logs or media such as rice bran, sawdust and wood chips.

Storage

Although Oda *et al.* (1975) referred to production conditions, the mycelium was not injured at –5°C, but the resistance of the mycelium to injurious micro-organisms was found to be weakened by freezing. This may well have implications for postharvest situations. The shelf-life was approximately 14–20 days at 1°C, 10 days at 6°C and 2–3 days at 20°C.

There was some browning of the pilei, gills and stipes and polyphenol oxidase activity was increased during storage at 20°C. The total free amino acid content was also greatly increased in storage at 20°C (Minamide *et al.* 1980).

Naranjilla, lulo, toronja

Solanum quitoense, Solanaceae.

Botany

Suarez and Duque (1992) referred to lulo as *S. vestissimum*. The fruit is a globular berry with a persistent calyx and up to 7 cm in diameter. They are green as they develop and turn yellow or bright orange as they mature. The prickly, short, brown hairs in the skin surface are usually rubbed off for marketing. The pulp is juicy, green and acid, somewhat resembles that of pineapple and guava in flavour and has numerous small seeds embedded in it. The fruit are borne on semi-perennial woody shrubs that grows up to 2 m high and have thorny stems, large leaves and white flowers (Santos and Nagai 1993). It originated from Ecuador and southern Colombia. The fruits are eaten raw or used for juices or preserves. It is a popular fruit in the Andean region, especially Colombia, and small quantities are exported to Europe and North America.

Harvesting

Harvesting is by hand and is based on the skin colour. They are usually harvested when the skins have begun to turn yellowish orange.

Storage

At room temperature in Colombia (24–26°C and 40% r.h.) they lost weight rapidly and became discoloured (Amezquita 1973). Refrigerated storage recommendations are as follows:

- 10°C and 90% r.h. for 1 week with no deterioration (Amezquita 1973)
- 7–10°C and 90% r.h. for 4–6 weeks (Snowdon 1990).

Ripening

Suarez and Duque (1992) showed that ripening was characterized by a significant increase in the amount of esters, mainly butyl acetate, methyl (*E*)-2-butenoate, 3-methylbutyl acetate, methyl (*E*)-2-methyl-2-butenoate, methyl hexanoate, (*Z*)-3-hexenyl acetate and methyl benzoate. A significant increase in linalool and α-terpineol concentrations was also observed, together with a moderate increase in geraniol, hotrienol and nerol concentrations. (*Z*)-3-Hexenol only increased significantly after the fourth ripening stage. In contrast, β-myrcene, limonene and terpinolene showed a slight decrease in their concentrations with increasing maturation. Detected aliphatic hydrocarbons showed irregular variations in their concentrations and aldehydes a slight increase.

Nectarines

Prunus persica variety nectarina, Rosaceae.

Botany

The fruit is a drupe and is classified as climacteric. It is the same species as the peach and shares a lot of its postharvest characteristics.

Harvesting

Determination and prediction of harvest maturity are based on parameters such as flesh firmness, background colour and soluble solids. The 'Twist Tester,' developed at Massey University in New Zealand, measures firmness as crush strength by radial movement of a small calibrated blade at any point within the non-peeled fruit. The system was devised by Griessel (1995), where the differences in rate of softening between the inner and outer mesocarp were used in the determination and prediction of harvest maturity for the cultivar May Glo. In Turkey, flesh firmness and soluble solids appeared to be the best practical methods for determining harvest maturity. Firmness values for minimum harvest maturity were 12–13 lb. for the cultivar Nectared-6 and 12 lb. for Independence. When combined with soluble solids, a firmness of 14–15 lb. with minimum soluble solids of 11% was suitable for determining minimum acceptable maturity for Nectared-6, and 15–16 lb. with a minimum of 10.5% soluble solids for Independence (Ozelkok *et al.* 1997).

Storage

Hardenburg *et al.* (1990) stated that storage recommendations are similar to those for peach for temperature and controlled atmosphere storage. At a room temperature of 20°C and 60% r.h. the fruit could be kept for 1–2 days (Mercantilia 1989). Specific refrigerated storage recommendations are as follows:

- –1 to 0°C and 85–90% r.h. for 3–7 weeks (Anon 1967)
- 0°C and 90–95% r.h. for 14 days (Mercantilia 1989)
- 2°C for 7 days (Mercantilia 1989)
- 4°C for 5 days (Mercantilia 1989)
- –1 to 0°C and 90–95% r.h. for 2–7 weeks (Snowdon 1990).

Chilling injury

Chilling injury occurred after storage at 0°C for 4 weeks in the cultivar Flavortop and resulted in the flesh having a woolly texture (Zhou *et al.* 2000). Both delayed storage, fruits held for 48 hours at 20°C before storage, and controlled atmosphere storage in 10% CO_2 with 3% O_2 alleviated or prevented chilling injury during storage at 0°C. Control fruits showed 80% chilling injury during ripening after 4 weeks at 0°C and 100% after 6 weeks. Delayed storage and controlled atmosphere storage were similar in their beneficial effect after 4 weeks but controlled atmosphere storage was better after 6 weeks. Delayed storage initiated ripening, which appeared to be the mechanism by which woolliness was prevented in that case (Zhou *et al.* 2000).

Controlled atmosphere storage

Recommendations include:

- 0–5°C with 5% CO_2 and 1–2% O_2 or 0–5°C with 3–5% CO_2 and 1–2% O_2, which had a good effect, but limited commercial use (Kader 1985)
- 5% CO_2 and 2.5% O_2 for 6 weeks (Hardenburg *et al.* 1990)
- –0.5°C with 90–95% r.h. with 5% CO_2 and 2% O_2 for 14–28 days (SeaLand 1991).

Oca

Oxalis tuberosa, Oxalidaceae

Botany

The edible portion is the tuber, which has a smooth skin bearing scales that cover long, deep eyes. They are small, up to 8 cm long, and are often turbinate in shape, although the shape and colour vary considerably between cultivars (Kay 1987; Purseglove 1968), but are commonly white, yellow or red. Cultivars are sometimes grouped into sweet and bitter. The bitter type contain calcium oxalate crystals, an anti-nutritional factor. High levels of oxalate were found by Hermann and Erazo (2001), ranging from 306–539 mg per 100 g in edible matter of freshly harvested tubers of the native Colombian cultivars ECU-1018, ECU-1025, CA-5054, MU-028 and HN-1146. The young shoots and leaves can be used in salads or boiled like spinach. It originated in the high Andes of South America.

Harvesting

In Colombia they mature about 8 months after planting and are dug by hand.

Curing

Oca are commonly left in the sun to cure for several days after harvesting. This is thought necessary for the bitter cultivars to reduce calcium oxalate levels (Kay 1987). In an experiment, Hermann and Erazo (2001) exposed harvested oca to direct sunlight for several days. This treatment increased the proportions of dry matter, soluble solids and sugars (mainly sucrose), reduced oxalate levels in dry matter by an average of 26% and decreased total acids, mainly owing to a large reduction in malate and glutarate.

Storage

In South America they are usually preserved by sun drying when they have a sort of dried fig flavour. At 21°C the fresh tubers can be kept for 12 weeks; the limiting factors are sprouting and deterioration in flavour (Kay 1987). Refrigerated storage recommendations were 4°C for 20 weeks (Kay 1987).

Okra, gumbo, lady's finger

Hibiscus esculentus, Malvaceae.

Botany

The fruit, which is the edible portion, is a beaked capsule with longitudinal furrows (Figure 12.79) that split, when fully mature, to release the seeds. There are many cultivars ranging from those where the skin is completely smooth to those which are hirsute or spiny.

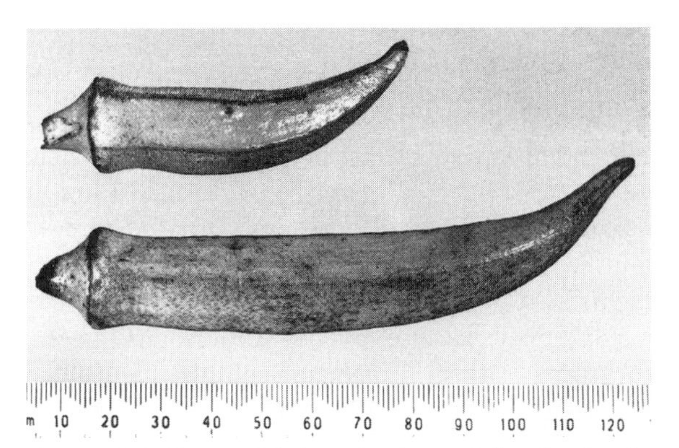

Figure 12.79 Clemson Spineless okra in the Sudan showing the maximum and minimum sizes commonly selected for harvesting.

Harvesting

They are harvested while still immature, tender, crisp and without fibre. In the tropics this is usually 4–5 days after flowering and harvesting should be every 1–2 days. They are commonly harvested by hand and placed in a suitable container (Figure 12.80). In the Sudan, polyethylene film, liners were used in field boxes to reduce the rate of dehydration. However, if they were kept too long inside plastic film condensation occurred on the surface causing water spotting. This is a black spotting on the skin that makes them unsightly (Thompson *et al.* 1973b).

Precooling

Rapid cooling helps to maintain their freshness, but top icing was reported to cause water spotting (Lutz and Hardenburg 1968). Forced-air precooling with high humidity air would cool them quickly without causing desiccation.

Figure 12.80 Harvesting okra into metal buckets in Mali. The fruit seem very large for most markets. Source: photo library of FAO, Rome.

Storage

Weight loss is a major cause of reduced storage life since it results in them losing their crispness. At a room temperature of 20°C and 60% r.h. they could be kept for 1–2 days (Mercantilia 1989). Refrigerated storage recommendations are as follows:

- 8.9°C and 90% r.h. for 2 weeks with 6.8% loss (Pantastico 1975)
- 7.2–10°C and 90–95% r.h. for 7–10 days (Lutz and Hardenburg 1968)
- 7–10°C and 95% r.h. for up to 10 days with up to 10% weight loss (Tindall 1983)
- 10°C and 90% r.h. for 7–10 days (Mercantilia 1989)
- 7–10°C and 95–100% r.h. for 1–2 weeks (Snowdon 1991)
- 10°C and 90–95% r.h. for 7–10 days (SeaLand 1991).

Chilling injury

Chilling injury can occur in storage at 7.2°C as pitting, surface discoloration and decay (Lutz and Hardenburg 1968).

Controlled atmosphere storage

Recommended conditions are:

- 7.2°C with 5–10% CO_2 kept them in good condition, but with levels of 10–12% CO_2 off-flavours may be produced (Pantastico 1975)
- 8–10°C with 0% CO_2 and 3–5% O_2, which had a fair effect but was of no commercial use, but CO_2 at 5–10% was beneficial at 5–8°C (Kader 1985)
- 7–12°C with 4–10% CO_2 and 21% O_2 which had only a slight effect (Saltveit 1989)
- 7–10°C with 5–10% CO_2 lengthened their shelf-life by about 1 week compared with air storage at the same temperature (Hardenburg *et al.* 1990)
- 10°C and 90–95% r.h. with 0% CO_2 with 3–5% O_2 (SeaLand 1991).

Olives

Olea europaea, Oleaceae.

Botany

The fruit, which are produced on small, slow-growing trees, has a single hard seed and is a small green drupe that may turn dark blue or purplish as it matures. They are classified as non-climacteric and are borne on wood of the previous year. It originated in the Mediterranean region and has been a source of food, medicine and oil since ancient times. It is now grown throughout the world in sub-tropical and warm-temperate regions. The fruit will freeze at about –2.4 to 1.4°C (Wright 1942).

Quality standards

To grade fruit into different maturities, specific gravity can be used by floating the fruit in a brine solution of the appropriate salt concentration (Hulme 1971). Magnetic resonance imaging has been used to obtain images of defective pits in olives (Chen *et al.* 1989).

Harvesting

Harvest maturity is based on fruit size and skin colour. The oil content, which increases throughout maturation, can be used as an indicator by analysing samples and when the level is high enough harvesting all the fruit from that orchard.

Storage

Refrigerated storage recommendations are as follows:

- 5–10°C and 85–90% r.h. for 4–6 weeks (Hardenburg *et al.* 1990)
- 7.2°C and 85–90% r.h. for 28–42 days (SeaLand 1991).

Chilling injury

Storage below 7.2°C resulted in the flesh turning brown at the stem end. Unusually, ripe fruit were more susceptible to chilling injury than less ripe fruit and it could occur even at 10°C within 1 month (Lutz and Hardenburg 1968).

Controlled atmosphere storage

Recommended storage conditions are:

- 8–10°C with 5–10% CO_2 and 2–5% O_2 had a fair effect, but was not being used commercially (Kader 1985)
- Kader (1989 and 1992) recommended 5–10°C with 0–1% CO_2 and 2–3% O_2
- Kader *et al.* (1989) found that green Manzanillo olives were damaged when exposed to CO_2 levels of 5% and above, but storage was extended in 2% O_2; they also found that in air their storage life was 8 weeks at 5°C, 6 weeks at 7.5°C and 4 weeks at 10°C, but in 2% O_2 the storage life was extended to 12 weeks at 5°C and 9 weeks at 7.5°C.

Onions, bulb onions

Allium cepa variety cepa, Alliaceae.

Botany

The edible part is the mature bulb. In temperate countries the bulb is produced in the first year of growth, it becomes dormant over winter, then regrows in spring when it flowers and produces seed. It will freeze at about –1.3 to 0.9°C (Wright 1942). Inaba *et al.* (1989) found that there was an increased respiration rate of bulbs in response to ethylene at 100 μl litre^{-1} for 24 hours at 20–35°C, but the effect was much reduced at lower temperatures.

Quality standards

The Organisation de Coopération et de Développment Économiques, Paris, published a brochure of standards for onions in 1984.

Harvesting

Bulb onions that are to be stored should be allowed to mature fully before harvest. This is judged to when the leaves bend just above the top of the bulb and fall over. In practice, the farmer calculates the percentage of bulbs that have fallen over in a field by doing sample counts, and when it reaches say 70% then the whole crop will be harvested. The tops may be cut off mechanically before harvesting. With some machines the crop may be left on the soil surface after harvesting; in others they may be lifted and loaded into a trailer for direct transport to the store or packhouse (Figure 12.81). Some harvesters have the facility to top and pack the crop directly in the field into wholesale or retail packs

Figure 12.81 Onion harvesting where the leaves have been cut off and the harvesting machine is lifting them and transferring them directly to a tractor and trailer alongside.

In a study by Maw *et al.* (1997) over three seasons, bulbs of the cultivar Granex 33 were harvested at three stages of maturity (early, optimum or late), cured for 24, 48 or 72 hours in heated air at 35–38°C, and stored at about 24°C in either high or low humidity. Of all treatments, harvest maturity had the greatest influence upon storability. Early-harvested bulbs exhibited the best storability. Storability was prolonged following 48 hours of curing, and was enhanced by low humidity. Storability was not enhanced by curing for >24 hours for late or >48 hours for optimally harvested bulbs. For early-harvested bulbs, curing for as long as 72 hours enhanced storability. Early-harvested bulbs had not reached their highest quality nor their maximum yield, so it was concluded that the optimal harvest maturity would have to be a compromise between yield, quality and storability. In New Zealand, bulbs lifted at 60–80% top down, field-cured and the foliage removed after curing was the simplest method and the best compromise to ensure postharvest quality and successful storage. Onions that were lifted 3 weeks after 50% top down and topped before curing had the greatest incidence of rots in store. Increasing harvest maturity increased their skin colour but also increased the number of loose skins (Wright *et al.* 2001).

Harvest maturity can affect physiological disorders. In trials carried out in Southeast Norway, late harvesting at about 100% leaf fall-over, high drying temperature (30°C) and a long drying duration produced the highest incidence of the physiological disorder translucent scales, the highest internal CO_2 levels and the lowest O_2 levels. Late harvesting and prolonged field curing gave the highest occurrence of leathery scales (Solberg and Dragland 1998).

Handling

If bulbs were dropped on to a hard surface they could be damaged even if the fall was as little as 30 cm (Thompson *et al.* 1972). Magnetic resonance imaging has been used to obtain images of onions in order to detect bruises during postharvest grading (Chen *et al.* 1989).

Packaging

Extruded plastic nets are commonly used for onions. Where two strands cross they are welded together. They are usually made in a long tube and cut to the required size and the bottom stitched to form a bag.

Curing and drying

See also the section on harvesting. Bulbs are normally dried before storage. This does not involve drying the crop to an even, low moisture content as in dehydration, but only drying the outer layers. The object of drying the outer layers of skin and the 'neck' of the onion is to provide a surface barrier to water loss and microbial infection. This process is sometimes referred to as curing, but since no cell regeneration or wound healing occurs it is clearer to refer to it as drying. Traditionally, drying of onions has been carried out in the field and called 'windrowing.' This involves pulling the bulbs from the ground when they are mature and laying them on their sides to dry for 1–2 weeks. In wet weather the bulbs can take a long time to dry and they may have increased levels of rot in subsequent storage. In hot climates the bulbs may be damaged by direct exposure to the sun. In such circumstances the bulbs should be windrowed in such a way that the drying leaves of one bulb covers the adjacent bulb to reduce the exposed surface. If the soil is wet then the side of the bulb in contact may develop brown stains, which can adversely affect the appearance and the value of the bulb.

Because of these problems, many bulbs are not now windrowed. Instead, they are taken straight from the field and dried in the store. This drying is achieved by constantly passing air across the surface of the bulbs at about 30°C and 70% r.h. If the humidity is too high, drying is delayed and levels of fungal disease can increase; if it is too low, the skins of the bulbs can split, which reduces their value and can increase subsequent losses. Drying may take 7–10 days and is judged to be complete when the necks of the bulbs have dried out and are tight and the skins rustle when held in the hand.

Sprout suppression

In India, Noor and Izhar (1998) found that sprouting started after 4 weeks of storage under ambient conditions of 35°C with 70% r.h. Some cultivars sprout later than this. Respiration rate and therefore sprouting of bulbs decreased with increasing temperature over the range 15–25°C (Karmarkar and Joshi 1941). Storage at 25–30°C was shown to reduce sprouting and root growth compared with storage at 10–20°C, but weight loss was high (Thompson *et al.* 1972).

Application of chemical sprout suppressants in storage does not work for onions because the meristematic region where sprouting occurs is deep inside the bulb and the chemicals do not normally penetrate so far into the tissue. It is therefore achieved by preharvest application of maleic hydrazide. Maleic hydrazide is applied to the leaves of the crop about 2 weeks before harvesting so that the chemical can be translocated into the middle of the bulb into the meristematic tissue (Thompson *et al.* 1972). Maleic hydrazide should be applied according to the manufacturer's instructions but 2500 μg litre^{-1} sprayed on the crop shortly before the leaves die down was recommended by Tindall (1983).

Noor and Izhar (1998) found that dipping the bulbs in 600, 800 or 1000 μg litre^{-1} salicylic acid before storage reduced sprouting and rotting. There also appears to be a residual effect of controlled atmosphere storage on sprouting after 2 weeks at 20° C. This could be highly beneficial in marketing (Gadalla 1997).

Sprout suppression can also be achieved by irradiating the bulbs with doses of 20–30 Gy. This level completely prevent sprouting if applied as soon as possible after harvest (Chachin and Ogata 1971), but irradiation is not permitted in all countries and where it is there is often consumer resistance to its use.

Simple storage structures

A windbreak was a traditional way of storing onions in Britain and farmers claimed that they could be stored in this way for up to 6 months. Windbreaks are constructed by driving wooden stakes into the ground in two parallel rows about 1 m apart. The stakes should be slightly sloping outwards and should be about 2 m high. A wooden platform is build between the stakes about 30 cm from the ground, often made from upturned wooden boxes. Chicken wire is fixed between the stakes and across the two ends of the windbreak. The bulbs are then loaded into the cage, covered with 15–20 cm on straw. On top of the straw a polyethylene film sheet of about 500 gauge is placed and weighted down with stones to protect it from the rain (see Figure 7.5). The windbreak should be sited with its longer axis at right-angles to the prevailing wind.

In Britain, cellars below houses were used for storing onions during the winter. The bulbs were usually spread out thinly on shelves to ensure good air circulation. In the Sudan, onions are stored in palm leaf barns called 'rakubas.' The bulbs are stacked on bamboo a platform inside the structure to a maximum depth of about 1 m. The nature of the structure allows good air circulation from the prevailing wind while protecting them from rain and direct sunlight. In many countries, bulbs harvested with their tops still attached can be plaited into bunches and suspended from hooks.

In a study of night-ventilated storage, Skultab and Thompson (1992) showed that onions could be maintained at an even temperature very close to the minimum night temperature. The design that was used is shown in Figure 7.11. A time switch can also be used which would be set, for example, to switch the fan on for 2 hours each day between 5.00 and 7.00 a.m. An onion store was built and tested in the Sudan which was similar to that described above. Losses were high but lower than those encountered using the traditional method of onion storage in the Sudan (Table 12.41).

Storage

At room temperature of 20°C and 75–85% r.h. bulbs could be kept for about 25 days (Mercantilia 1989). Relative humidity of 70–75% should be maintained during storage with regular ventilation. Bulk storage height should not exceed 3 m to prevent compression damage to the lower bulbs, or they should be stored in boxes. Where the downward force on the crop is above a threshold level, it can be bruised. This damage may also be a function of time, especially where the pressure is close to the threshold value. In boxes it may be as a result of overfilling them and then stacking the boxes so that the crop in the lower boxes supports the weight of the ones above. Refrigerated storage recommendations are as follows:

- 0°C and 70–75% r.h. (Wardlaw 1937)
- –3°C and 85–90% r.h. for 5–6 months (Wardlaw 1937)
- –0.6°C and 78–81% r.h. for 6–7 months (Wardlaw 1937)
- –3 to 0°C and 70–75% r.h. for 6 months (Anon 1967
- 0°C and 75–85% for 6 months (Shipway 1968)
- 0°C and 70–75% r.h. for 20–24 weeks with 16.3% loss for red onions (Pantastico 1975)
- 1.1°C and 70–75% r.h. for 16–20 weeks with 14.2% loss for white onions (Pantastico 1975)
- 0°C and 70–75% r.h. or 90–95% r.h. for up to 120 days (Tindall 1983)
- 30–35°C and 75–85% r.h. (Tindall 1983)
- 0°C and 65–70% r.h. for 6–8 months for Globe, 1–2 months for Bermuda (Hardenburg *et al.* 1990)
- –2°C and 75–85% r.h. for 300 days (Mercantilia 1989)
- 0°C for 230 days (Mercantilia 1989)
- 4°C for 170 days (Mercantilia 1989)
- 8°C for 120 days (Mercantilia 1989)
- 12°C for about 90 days (Mercantilia 1989)
- –1 to 0°C and 70–80% r.h. for 6–8 months (Snowdon 1991)
- 0°C and 65–75% r.h. (SeaLand 1991).

Controlled atmosphere storage

Low O_2 levels in storage reduced the respiration rate of bulbs at all temperatures studied (Table 12.42).

Storage recommendations are:

- 4.4° C with 10% CO_2 and 3% O_2 and the next best was 5% CO_2 and 5% O_2 Chawan and Pflug 1968)
- 1 or 5°C with 5% CO_2 and 3% CO_2 (Adamicki and Kepka 1974)

Table 12.41 Comparative storage losses of onion bulbs in a simple straw cottage (rakuba) and a night-ventilated store during 6 months of storage at an average of 36.1°C and 42% r.h. for the rakuba and 33.1°C and 60% r.h. for the night-ventilated store (source: Silvis and Thompson 1974)

Storage method	% sound by weight	% rotting	% sprouting
Rakuba	43	22	7
Night-ventilated	58	18	4

Table 12.42 Effects of temperature and reduced O_2 level on the respiration rate (CO_2 production in mg kg^{-1} h^{-1}) of Bedfordshire Champion onions (Robinson *et al.* 1975)

	Temperature (°C)				
	0	5	10	15	20
In air	3	5	7	7	8
In 3% O_2	2		4		4

- 0–5°C with 1–2% O_2, 0% CO_2 and 75% r.h. (Kader 1985)
- 0–5°C with 0–5% CO_2 and 1–2% O_2 (Kader 1989)
- 1°C with 5% CO_2 and 3% O_2 of Granex for 7 months resulted in 99% of the bulbs considered to be still marketable with 9% weight loss, and the bulbs kept well when removed from storage for marketing (Smittle 1989)
- 0–5°C with 0–5% CO_2 and 0–1% O_2, which had only a slight effect (Saltveit 1989)
- 1°C with 5% CO_2 and 3% O_2 (Adamicki 1989)
- 5% CO_2 with 3% O_2 were shown to reduce sprouting and root growth but controlled atmosphere storage conditions gave variable success and is not generally recommended (Hardenburg *et al.* 1990)
- 0°C with 0% CO_2 and 1–2% O_2 (SeaLand 1991)
- 0°C with 5 or 10% CO_2 and 1% O_2 (Mikitzel *et al.* (1993).

In contrast, Stoll (1974) reported that at 0, 2 and 4°C air storage gave better results than controlled atmosphere storage with 8% CO_2 and 1.5% O_2. Ogata and Inous (1957), Chawan and Pflug (1968), Adamicki and Kepka (1974), Adamicki *et al.* (1977), Smittle (1988) and Mikitzel *et al.* (1993) all found that higher concentrations of CO_2 (usually 10%) could result in internal spoilage. Internal spoilage or internal breakdown is a physiological disorder, which appears as breakdown in the fleshy scales which may become clear when bulbs are removed to room temperature (Smittle 1988).

In a detailed study of 15 different cultivars and controlled atmosphere storage gas combinations, Gadalla (1997) found that all the cultivars stored in 10% CO_2 for 6 months (but not for 3 months) or more had internal spoilage, with more spoilage when 10% CO_2 was combined with 3% or 5% O_2. All controlled atmosphere mixtures 1, 3 or 5% O_2 with 0 or 5% CO_2 reduced sprouting but those combinations that included 5% CO_2 were the most effective. Also, the residual effect of controlled atmosphere storage was still apparent after 2 weeks at 20° C. No external root growth was detected when bulbs were stored in 1% O_2 compared with all the bulbs in air. There was also a general trend to increased rooting with increased O_2 over the range 1–5%. Bulbs stored in 5% CO_2 had less rooting than those stored in 0% CO_2. Both of these latter effects varied between cultivars. Most of the cultivars stored in 1% O_2 with 5% CO_2 and some cultivars stored in 3% O_2 with 5% CO_2 were considered marketable after 9 months of storage.

Gadalla (1997) also found that controlled atmosphere storage was most effective when applied directly after curing, but a delay of 1 month was almost as good. It was therefore concluded that with a suitable cultivar and early application of controlled atmosphere conditions it is technically possible to store onions at 0°C for 9 months without chemical sprout suppressants.

Modified atmosphere packaging

Packing bulbs in plastic film is uncommon because the high humidity inside the bags can cause rotting and root growth (Table 12.43).

Transport

For international transport they are normally transported by sea in refrigerated holds or containers. However, ventilated containers can be used. These

Table 12.43 Effects of number and size of perforations in 1.36 kg 150 gauge polyethylene film bags of Yellow Globe onions on the relative humidity in the bags, rooting of the bulbs and weight loss after 14 days at 24°C (source: Hardenburg 1955)

No. of perforations	Perforation size (mm)	R.h. in bag (%)	Bulbs rooted (%)	Weight loss (%)
0	–	98	71	0.5
36	1.6	88	59	0.7
40	3.2	84	40	1.4
8	6.4	–	24	1.8
16	6.4	54	17	2.5
32	6.4	51	4	2.5
0[a]	–	54	0	3.4

[a] Kraft paper with film window.

are standard metal containers, usually 40 feet long by 8 feet wide and 8 feet high, which have no insulation or refrigeration unit but have some kind of ventilation system (see Chapter 11). They have been successfully used from Australia to transport onions to Britain.

Ethylene

Exposure of onions to ethylene gave them a milder flavour (Weichmann 1978).

Oranges

Citrus sinensis, Rutaceae.

Botany

The fruit is a berry or hesperidium (Figure 12.82), which is non-climacteric. It will freeze at about –2.3 to 2.0°C (Wright 1942). Oranges grown in the tropics tend to have a higher sugar and total solids content than those grown in the sub-tropics. However, tropical grown oranges tend to be less orange in colour and peel less easily. These two factors seem to be related more to the lower diurnal temperature variation that occurs in the tropics rather than the actual temperature difference between the tropics and subtropics. Respiration rate increases rapidly with increasing temperature, as would be expected (Table 12.44).

Figure 12.82 Oranges selected for export in China. Source: Dr Wei Yuqing.

Harvesting

Being non-climacteric, the fruit does not ripen after harvest and must therefore be picked at peak maturity. A visual method is to observe the size, smoothness and change in colour of the fruit from green to orange. The juice content increases as they mature on the tree. By taking representative samples of the fruit, extracting the juice in a standard and specified way and then relating the juice volume to the original mass of the fruit it is possible to specify its maturity. In the USA the minimum value for navel oranges is 30% and for other oranges 35%. The ratio between soluble solids and acidity can also be used with ranges between 7:1 and 9:1 for full maturity. Light transmission properties of fruit can be used to measure their maturity. With this method it was possible to grade fruits, that were indistinguishable by visual examination. A photoelectric machine was used for colour sorting citrus fruit (see Chapter 2) (Jahn and Gaffney 1972).

If oranges are too turgid at harvest, the oil glands in the skin can be ruptured, releasing phenolic compounds and causing damage to the peel, resulting in oleocellosis (Wardlaw 1938). In some cases fruit were kept in the basket in which they were harvested for a couple of hours before being transported to the packhouse. This allowed the fruit to lose a little moisture, a practice that is called 'quailing.'

Harvesting methods

Oranges harvested for juice extraction may be removed from the trees by powerful wind machines that are dragged through the orchard, followed by a device for collecting them from the ground. The grass beneath the trees is usually mown before harvesting and these collecting devices usually have nylon brushes that sweep up the fruit and transfer them to containers. Tree shakers can also be used (see Chapter 3) (Figure 12.83).

Grading

X-rays can be used to detect internal disorders of crops, such as hollow heart in potatoes (Finney and Norris 1973), split pit in peaches (Bowers *et al.* 1988) and

Table 12.44 Effects of temperature on the respiration rate (CO_2 production in mg kg^{-1} h^{-1}) of oranges grown in Florida (Hallar *et al.* 1945)

Temperature (°C)	0	4.4	10	15.5	21.1	26.6	32.2	37.7
Respiration	3.1	6.4	12.3	20.7	29.9	35.4	46.5	69.1

Figure 12.83 Tree shaker and catcher being used for apples in the UK.

granulation in oranges (Johnson 1985). Clarke (1996) described a machine where there are two cables along which the crop runs, made up from high-tensile, polyethylene-coated steel. When the crop reaches the prescribed point, ejector bars controlled by solenoid valves and operated by compressed air cylinders push it into a chute. The machine is very gentle to the crop and high drops are unnecessary, although padded chutes can be provided.

Waxing and washing

All citrus fruits can benefit from the application of waxes and subsequent polishing. This is not simply to enhance the appearance, which of course is important, but also to improve the storage quality of the product. When the crop is in the field or even during harvesting, transport, washing or grading, it can undergo some scratching or abrasion that not only removes the natural layer of wax but also scars the protective layer of skin around the fruit. Washing in particular is essential for the removal of bird droppings, insect marks, chemical residues and general field dirt, but can also remove much of the natural wax layer, especially if used with detergents. The natural layer of wax reduces moisture loss from the fruit. Citrus fruits may be improved in appearance by synthetic wax application and some producers may add colouring dyes to the wax in order to modify the colour in addition to providing an enhanced shine. Natural waxes are preferred, such as sugar cane wax, carnauba wax, shellac and various resins. These may be applied to the fruit in foam or liquid bath, as a liquid spray or by sponge or brush rollers. Their application rates may be governed by legislation in the country where the coating is being applied or in the country where the fruit and vegetables are to be exported. It is essential that these regulations and also the manufacturer's recommendations be strictly adhered to. Spray systems often apply an excess of fluid so the unused surplus flows through the conveyor back into the holding vessel ready for re-application. The crop must be dried after waxing to give good adhesion and retention of the wax. Drying is usually done in a drying tunnel and may also incorporate soft brushes. A blast of warm air is often directed at the crop as it passes along the conveyor under the fan or else the heaters are located under the crop and the fan draws the air through it by suction (Clarke 1996).

Storage experiments using organically grown Shamouti showed that hot water brushing at 56°C for 20 seconds reduced decay development by 45–55% and did not cause surface damage, and did not influence fruit weight loss or internal quality parameters (Porat *et al.* 2000).

Disease control

In Florida, orange trees that were sprayed with 300–500 μg litre^{-1} benomyl more than a month before harvest had less postharvest disease (*Penicillium* spp. and stem end rot) than unsprayed trees (Brown and Albrigo 1972). Exposing oranges to 30°C and 90–100% r.h. for several days after harvest was shown to reduce decay associated with *Penicillium* spp. infection (Hopkins and Cubbage 1948). Fumigating oranges with 2-AB was also recommended for postharvest disease control (Eckert 1969). With the restrictions on the postharvest use of chemicals, other methods of disease control are being used; see Chapters 6 and 8.

Physiological disorders

Oleocellosis is a browning and pitting of the skin. It is due to the oil vesicles being ruptured and releasing phenolic compounds that damage the cells of the skin.

The actual damage is the result of the tissue drying out, giving minuscule cavities, and during prolonged storage it may encroach on the albedo beneath (Wardlaw 1937).

Desiccation consists of wrinkled and brown areas of peel, spreading as it develops, more frequently at the stem end (Wardlaw 1937). Superficial rind pitting of Shamouti orange causes losses to growers. Some symptoms could be seen on the tree or in the packhouse; the majority only showed symptoms 3–5 weeks after harvest, during shipment and marketing. Ethylene increased the incidence of superficial rind pitting but storage at 5°C restricted its development (Tamim *et al.* 2000).

Degreening

Oranges are often degreened after harvest by exposing them to ethylene gas, but brown spotting on the skin may be due to excessive ethylene concentration during degreening. Concentrations in excess of 200 μl $litre^{-1}$ or too high a temperature can give this effect (Wardlaw 1937). Temperatures range between 25 and 29°C with humidity as high as possible, generally between 90 and 95% r.h. with ethylene at between 1 and 10 μl $litre^{-1}$ for 24–72 hours. Ladaniya and Singh (2001) found that the respiration rate increased from initial 35–80 mg CO_2 kg^{-1} h^{-1} in ethylene-exposed fruits and slowly declined after removal from the ethylene atmosphere. Fruit firmness, juice content, total soluble solids and ascorbic acid remained unchanged with ethylene treatment but titratable acidity declined and decay losses due to *Geotrichum candidum* were lower in degreened fruits. The actual choice of conditions will vary between different situations, particularly whether they have been grown in the tropics or subtropics. The cultivar and the stage of maturity at harvest should also be taken into account. Application of ethylene may be by an initial injection of the correct level or by continuously 'trickling' it into the room. The conditions for degreening are as follows:

- 27°C and 85–92% r.h. with ethylene gas trickled into the room to achieve 20–30 μl $litre^{-1}$ for 24–48 hours (Winston 1955)
- 20–25°C and 90% r.h. with 5–10 μl $litre^{-1}$ of ethylene at for up to 72 hours with air circulation of one room volume per minute and 1–2 fresh air changes per hour in California (Eaks 1977)
- 27.5–29.5°C and 90–95% r.h. with 1–5 μl $litre^{-1}$ ethylene at or up to 72 hours with good air circulation and one fresh air change per hour in Florida (McCornack and Wardowski 1977)
- 27–29°C and 90–95% r.h. with 5–10 μl $litre^{-1}$ ethylene for 48 hours with four air changes per hour and air circulation of 0.5–0.6 l s^{-1} kg^{-1} of fruit for the cultivar Mosambi in India (Ladaniya and Singh 2001).

Curing

Tariq (1999) showed that placing fruits in sealed polyethylene film bags and exposing them to temperatures in the range 25–35°C for between 48 and 3 hours (the lower the temperature, the longer the exposure time required) completely eliminated the browning and fungal growth on damaged fruits (see also Chapter 6).

Storage

Valencia oranges from Florida and the cultivar Pineapple can be stored for 15–20 weeks at 3°C and have a marketable life of 3–7 days after removal from storage. Temperatures of 0°C or less resulted in pitting and temperatures above 4°C led to excessive decay. The later into the season the fruit is harvested, the shorter is the storage life. Early-season cultivars (Pineapple and Hamlin) are more susceptible to chilling injury than fruits harvested later in the season (Valencia). An imbalance between CO_2 and O_2 in the fruit store may cause them to develop 'off-flavours.' Having good ventilation around the fruit can prevent this (Davies and Albrigo 1994). General refrigerated storage recommendations are as follows:

- 7.2°C (Trinidad) (Wardlaw 1937)
- 4.5°C (storage and shipment from South Africa) (Fidler 1963)
- 0–4°C and 85–90% r.h. for 1–4 months (Anon 1967)
- 0–4°C and 85–90% r.h. (Spain) (Anon 1967)
- 4–6°C and 85–90% r.h. for up to 6 months (Israel) (Anon 1967)
- 5.6–7.2°C and 85–90% r.h. for 18 weeks (India) (Anon 1967)
- 7.2°C and 85–90% r.h. for 21–56 days (California, Arizona) (SeaLand 1991)
- 2.2°C and 85–90% r.h. for 56–84 days (Florida, Texas) (SeaLand 1991)
- 0–4°C for 2 months or more (Davies and Albrigo 1994).

Specific storage recommendations for different cultivars and production areas are as follows:

Blood orange

- 4.4°C and 90–95% r.h. for 21–56 days (SeaLand 1991).

Camargo

- 2°C and 90% r.h. for 3 months from Brazil (Snowdon 1990)
- 4°C and 90% r.h. for 2–3 months from South Africa (Snowdon 1990).

Castellana

- 1–2°C and 90% r.h. for 2–3 months from Spain (Snowdon 1990).

Jaffa

- 8°C and 90% r.h. for 3 months from Israel (Snowdon 1990)
- 7.8°C and 85–90% r.h. for 56–84 days (SeaLand 1991).

Moroccan

- 2–3°C and 90% r.h. 2–3 months from Spain (Snowdon 1990).

Navel

- Navel oranges from California are sometimes stored from 2–6 weeks at 3–4°C during winter and spring in order to even out the supply (Stahl and Camp 1963)
- 3°C and 85–90% r.h. for 8–10 weeks from Spain (Mercantilia 1989)
- 2–7°C and 85–90% r.h. for 5–8 weeks from California (Mercantilia 1989)
- 4.5–5.5°C and 90% r.h. 1–2 months from Australia (Snowdon 1990).

Ponkan

- 4.4°C and 85–90% r.h. for 3–4 weeks with 7.5% loss (Pantastico 1975).

Salustiana

- 2°C and 90% r.h. for 3–4 months form Spain (Snowdon 1990).

Sathgudi

- 5.6–7.2°C and 85–90% r.h. for 16 weeks with 15% loss (Pantastico 1975).

Shamouti

- 4–5°C and 85–90% r.h. for 6–8 weeks from Israel (Mercantilia 1989)
- A combination of preharvest potassium spray with storage at 5°C gave good control of superficial rind pitting (Tamim *et al.* 2000).

Swikom

- 8.9–10°C and 85–90% r.h. for 4–5 weeks with 8% loss (Pantastico 1975).

Valencia

- 0°C and 85–90% r.h. for 8–12 weeks for fruit grown in Florida or Texas (Lutz and Hardenburg 1968)
- 4.4–6.7°C and 85–90% r.h. for fruit grown in California (Lutz and Hardenburg 1968)
- 8.9°C from Arizona harvested in March or 3.3°C from Arizona harvested in June (Lutz and Hardenburg 1968)
- 4.4–6.1°C and 88–92% r.h. for 5–6 weeks with 12% loss (Pantastico 1975)
- 0–1°C and 85–90% r.h. for 8–12 weeks from Florida (Mercantilia 1989)
- 2–7°C and 90% r.h. for 1–2 months from California (Snowdon 1990)
- 2–3°C and 90% r.h. for 2–3 months from Cyprus (Snowdon 1990)
- 0–1°C and 90% r.h. for 2–3 months from Florida or Texas (Snowdon 1990)
- 2°C and 90% r.h. for 2–3 months from Israel (Snowdon 1990)
- 2–3°C and 90% r.h. for 2 months from Morocco (Snowdon 1990)
- 4.5°C and 90% r.h. for 2–4 months from South Africa (Snowdon 1990)
- 2°C and 90% r.h. for 3–4 months from Spain (Snowdon 1990).

Verna

- 1–2°C and 90% r.h. for 3–4 months from Spain (Snowdon 1990).

Washington Navel

- 2–7°C and 90% r.h. for 1–2 months from the USA (Snowdon 1990).

Chilling injury

Symptoms begin as light brown discoloration surrounding the oil vesicles on the skin with a number of small irregular spots appearing at the same time, either localized or scattered over the surface. They increase in size, coalesce and the tissue in between the oil vesicles is depressed so that the vesicles themselves are prominent, giving a pinhead effect. The vesicle outlines usually appear darker than the centre. They continue to turn a darker brown and this may be accelerated when they are transferred to higher temperatures. The oil vesicles then collapse as the tissue dries out, giving minute cavities. During prolonged storage they may encroach into the white albedo beneath, but this varies with cultivar. Eventually the fruit suffer from watery breakdown (Wardlaw 1937). Hardenburg *et al.* (1990) showed that storage of Valencia grown in Florida at 1°C in 0 or 5% CO_2 and 15% O_2 followed by 1 week in air at 21°C had less skin pitting after 12 weeks than those stored in air.

Controlled atmosphere storage

It has been claimed (Anon 1968) that controlled atmosphere storage could have deleterious effects on fruit quality, particularly through increased rind injury and decay or on fruit flavour. In contrast, Hardenburg *et al.* (1990) showed that Valencia grown in Florida retained better flavour during storage for 12 weeks at 1°C in 0 or 5% CO_2 and 15% O_2 followed by 1 week in air at 21°C than those stored in air, but CO_2 levels of 2.5–5% especially when combined with 5 or 10% O_2, adversely affected flavour retention.
Other storage recommendations were as follows:

- 5–10°C with 5% CO_2 and 10% O_2 had a fair effect on storage but was not being used commercially (Kader 1985)
- 5–10°C with 0–5% CO_2 and 5–10% O_2 (Kader 1989, 1992)
- 7.2°C with 5% CO_2 and 10% O_2 (SeaLand (1991)
- 1–9°C and 85–90% r.h. with 5–10% O_2 and 0–10% CO_2 for 3–12 weeks (Burden 1997).

Hypobaric storage

The cultivar Fukuhara showed lower rates of ethylene production and respiration after low-pressure storage than fruits kept at normal pressure. Oranges stored at 190 mmHg and 98% r.h. decayed more rapidly after return to normal pressure than fruits stored at 190 mmHg and 75% r.h. or normal pressure and 86% r.h. (Kim and Oogaki 1986). However, the fruits stored at low humidity and low pressure would have been severely desiccated.

Ortanique, temple oranges, murcotts

Citrus reticulata × *C.* sinensis, Rutaceae.

Botany

The fruit is a berry or hesperidium, which is non-climacteric. They are bispecific hybrids, there are many different types and they have different morphological characteristics since they are different selections from the same type of bispecific cross. Generally they all have a deep orange-coloured peel and flesh and a delicious flavour.

Temple orange was found in Jamaica in the late 19th century, but it is now grown throughout the tropics and sub-tropics. They are easier to peel than oranges but more difficult than tangerines and are usually very seedy.

Murcott were first found in the south of the USA in the early part of the 20th century and are also seedy, with a semi-hollow central axis.

Ortanique is also called honey tangerine. They are sweet and delicious but seedy, and were also first found in Jamaica. The name ortanique is exclusive to those produced in Jamaica and in the 1970s other countries acquired them, including Cuba; Cuba was not permitted to export them as ortaniques and called them cubaniques.

Storage

Only one refrigerated storage recommendation could be found, 4–5°C and 90% r.h. for 3–5 weeks for ortaniques (Snowdon 1990).

Chilling injury

Temple oranges have been shown to be susceptible to chilling injury at 0°C, therefore 3.3°C for up to 4 weeks was recommended by Lutz and Hardenburg (1968).

Oyster mushrooms, hiratake mushroom

Pleurotus ostreatus.

Botany

The common name is derived from its appearance, which resembles an oyster shell. The cap surface is smooth brown or bluish grey when they emerge and turn a pale brown at maturity. The gills are whitish and the spore mass lilac. The stem is short and stout, arising almost directly from the host tree (Figure 12.84, in the colour plates).

Prestorage treatment

Treatment with chitosan prolonged storage life to 10 days, 4 days longer than those packed in antifogging film (Cho and Ha 1998). In India, investigations revealed that postharvest application of lactic acid bacteria such as *Lactobacillus* could inhibit the growth of moulds (Gogoi and Baruah 2000).

Storage

Storage life was approximately 14–20 days at 1°C, 10 days at 6°C and 2–3 days at 20°C. Browning of the pilei, gills and stipes and polyphenol oxidase activity were markedly increased during storage at 20°C, as was the total free amino acid content (Minamide *et al.* 1980).

An important parameter for extending the shelf-life of fresh oyster mushrooms was maintaining a suitable humidity (Martinez Soto *et al.* 1998). Bohling and Hansen (1989) recommended a temperature of 1 or 3.5°C for about 1 week. Storage at 8°C resulted in mushrooms that were still acceptable but had lost colour and firmness. After 7 days at 15°C, the mushrooms were spoiled (Bohling and Hansen 1989). During storage the hyphae can grow beyond the surface of the stem, cap and lamellae, giving the impression of mould infection (Bohling and Hansen 1989; Henze 1989).

Controlled atmosphere storage

Controlled atmosphere storage at 1°C and 94% r.h. with 30% CO_2 and 1% O_2 for about 10 days was recommended (Henze 1989). Controlled atmosphere storage was shown to have little effect on increasing storage life at either 1°C or 3.5°C but at 8°C with a combination of 10% CO_2 and 2% O_2 or 20 or 30% CO_2 and 21% O_2 (Bohling and Hansen 1989).

Modified atmosphere packaging

Packaging in low-density polyethylene films reduced transpiration and external browning and maintained a soft texture and general quality during storage at 5 or 10°C for 9 days (Martinez Soto *et al.* 1998). Rai *et al.* (2000) successfully stored them at 5 and 15°C, using highly permeable micro-perforated oriented polypropylene film and ordinary oriented polypropylene film packages. The gas concentrations inside the packages went as low as 0.4% O_2 and 12–13% CO_2. The effect of poly (vinylidene chloride)-coated oriented nylon, antifogging, wrap or vacuum packing in film before storage at 25 or 2°C was investigated by Chi *et al.* (1998). The best treatment for maintaining quality was packing in antifogging film, which increased storage life to 24 days at 2°C. At 0°C in sealed polyethylene bags their storage life was only 15 days (Chi *et al.* 1996). Popa *et al.* (1999) found that storage at 4°C led to the greatest inhibition of microbial contamination and washing in chlorinated water and incorporation of silica gel humidity absorber in the low-density polyethylene bags also decreased microbial contamination.

Papa criolla, criolla

Solanum stemonotum × *S. andigenum*, Solanaceae.

Botany

The edible portion is the stem tuber, which has deep eyes, a yellow skin and flesh that can vary from white to yellow. They resemble a small potato (*S. tuberosum*); the common name means local potato. They have a nice nutty texture and are popular in Colombia. They originated in the Andean region of South America.

Storage

The main limitation to storage is sprouting. This can occur within 1 week of harvesting at room temperature (24–26°C and 40% r.h.). Tubers also lost 25% in weight within 1 week at these temperatures and rapidly became discoloured and infected with fungi (Amezquita 1973). The optimum refrigerated storage temperature was given as 10–15°C with 85–90% r.h.

Papayas, pawpaws

Carica papaya, Caricaceae.

Botany

It is a climacteric fruit (Biale and Barcus 1970), with a fleshy berry with a large hollow internal cavity containing small, soft, round, black, shiny seeds that have a high oil content. The surface has grooves that run the length of the fruit, which are called funicles. There are three types of plant: male, female and hermaphrodite.

Fruit of the cultivar Solo from female plants are almost round or ovoid to oblong, whereas those from hermaphrodite plants are pear shaped or cylindrical. Fruit will freeze at about –1.2 to 0.9°C (Wright 1942).

Harvesting

The fruit turns from green to yellow during ripening. If the fruit are harvested when they are still green it may be possible to develop the fruit colour after harvest but not all the flavour characteristics (Thompson and Lee 1971). If the fruit is harvested just as the yellow colour begins to show in the funicles, it can eventually ripen to an acceptable flavour. In an experiment, fruits of the cultivar Pakchong 1 were hand harvested at the mature green stage, the 5% yellow and the 15% yellow stages were then ripened at 25°C and 80% r.h. by Kulwithit *et al.* (1998). When ripe, the 15% yellow fruits had the highest flesh firmness of 57.4 newton, soluble solids content of 13.1%, titratable acidity of 0.14% and ascorbic acid content of 179.0 mg 100 per ml of juice, but there were no significant differences between the mature green and the 5% yellow fruits. Delayed light emission was also used on papaya by Forbus *et al.* (1987) to grade fruit into different maturity groups after harvesting. Body transmittance spectroscopy was used for optical grading into ripening and non-ripening groups (Birth *et al.* 1984). With this method it was possible to grade fruits that were indistinguishable by visual examination.

Harvesting is carried out by hand. Fruits harvested by cutting the fruit stalk and packing them into export cartons in the field had a lower incidence of rotting during subsequent storage than fruit harvested by twisting and pulling them from the plant and transporting them to a packhouse (Table 12.45). Many supermarkets will not accept fruit that have been packed directly in the field, because they are concerned that the cartons will be contaminated by soil.

Disease and pest control

Vapour heat treatment to control pests and diseases is 43°C in saturated air for 8 hours and then holding the temperature for a further 6 hours was recommended by Jacobi *et al.* (1993). Vapour heat treatment can cause injury to some fruits. For control of fruit flies in Hawaii the fruit were first exposed to 43°C and 40% r.h. for 11 hours followed by 43°C and 100% r.h. for 8.75 hours (Pantastico 1975). The initial exposure of fruit to the lower humidity was said to increase the number of insects killed and the tolerance of the fruit to the treatment. Storage at 20°C with 0% CO_2 and <0.4% O_2, as a method of fruit fly control, resulted in increased incidence of decay, abnormal ripening and the development of off-flavour after 5 days of exposure (Yahia *et al.* 1989).

Maharaj and Sankat (1990) used hot water treatment at 48°C for 20 minutes or hot water plus benomyl (1.5 g $litre^{-1}$) at 52°C for 2 minutes to control anthracnose caused by *Colletotrichum gloeosporiodes*. In Yemen the control of anthracnose on papaya was achieved by preharvest sprays with Metiram 80% wettable powder at 200 g per 100 l of water, Propineb 70% at 200 g per 100 l of water or copper oxychloride 50% wettable powder at 400 per 100 l of water applied at 7–10-day intervals (Kamal and Agbari 1985). At 18°C an atmosphere containing 10% CO_2 reduced decay during storage (Hardenburg *et al.* 1990).

Storage

Kader *et al.* (1985) stated that the critical temperature above which injury will not occur could be as low as 6°C. Studies in Malaysia, however, indicated a preferred storage temperature of 20°C, since fruit kept for more than 7 days at 15°C and below failed to ripen (Nazeeb and Broughton 1978). At a room temperature of 20°C and 60% r.h. they could be kept for 2–3 days (Mercantilia 1989). Refrigerated storage recommendations are as follows:

Table 12.45 Effect of harvesting method on various parameters after storage of Solo papaya for 21 days at 7°C followed by 6 days at 25–28°C (source: Thompson and Lee 1971)

Harvesting method	Weight loss (%)[a]	Soluble solids (%)[b]	Flavour score (0–11)[c]	No of fungal lesions
Normal	5.4	9.2	5.6	13.6
Careful	4.1	10.5	6.7	4.2

[a] As a percentage of the weight at harvest.
[b] Where 0 = very poor and 11 = excellent.
[c] Mean number per fruit.

- 4–5.5°C and 85–90% r.h. for 5 weeks (Anon 1967)
- 7.2°C and 85–90% for 1–3 weeks (Lutz and Hardenburg 1968)
- 13°C for 21 days with subsequent normal ripening at 26°C (Thompson and Lee 1971)
- 10°C and 85–90% r.h. for 3–4 weeks with 5.8% loss for green fruit (Pantastico 1975)
- 8.3°C and 85–90% r.h. for 2–3 weeks for ripe fruit (Pantastico 1975)
- 10°C and 90% for 2–3 weeks (Mercantilia 1989)
- 7°C and 85–90% r.h. for 1–3 weeks (Hardenburg *et al.* 1990)
- 10°C and 90% r.h. for 3–4 weeks for green fruit (Snowdon 1990)
- 7°C and 90% r.h. for 2–3 weeks for turning fruits (Snowdon 1990
- 10°C and high humidity for 14 days subsequently ripened normally at 25°C in 4 days (Lam 1990)
- 12.2°C and 85–90% r.h. for 7–21 days (SeaLand 1991).

Chilling injury

They are sensitive to chilling injury. Symptoms include sensitivity to attack by *Alternaria* sp., failure to ripen normally and, sometimes, water-soaked tissues. Nazeeb and Broughton (1978) found that chilling injury could occur during storage at 15°C.

Controlled atmosphere storage

Storage recommendations are diverse and include the following:

- 13°C for 3 weeks in 5% CO_2 with 1% O_2 followed by ripening at 21°C gave fruit that were in a fair condition with little or no decay and were of good flavour (Hatton and Reeder 1969)
- 18°C with 10% CO_2 for 6 days for the cultivar Solo resulted in reduced decay compared with fruits stored in air, but they decayed rapidly when removed to ambient (Akamine 1969)
- 13°C in 1% O_2 for 6 days or 1.5% O_2 for 12 days ripened about 1 day slower than those stored in air (Akamine and Goo 1969)
- 10°C with 2% O_2 and 98% nitrogen was shown to extend the storage life in Hawaii compared with storage in air at the same temperature (Akamine and Goo 1969)
- 10°C and 85–90% r.h. with 5% CO_2 and 1% O_2 for 21 days (Pantastico 1975)
- 10–15°C with 5–10% CO_2 and 5% O_2 had a fair effect but was said to be 'not being used commercially' (Kader 1985, 1989, 1992)
- 1.5–5% O_2 delayed ripening of the cultivar Kapaho Solo compared with those stored in air and that there was no further benefits on storage by increasing the CO_2 level to either 2% or 10% (Chen and Paull 1986)
- 18°C with 10% CO_2 reduced decay (Hardenburg *et al.* 1990)
- 16°C with 5% CO_2 and 1.5–2% O_2 for 17 days for the cultivar Known You Number 1 compared with only 13 days in air (Sankat and Maharaj 1989)
- 16°C with 5% CO_2 and 1.5–2% O_2 for the cultivar Tainung Number 1 gave a maximum life of 29 days compared with 17 days stored in air (Sankat and Maharaj 1989)
- 12.2°C and 85–90% r.h. with 10% CO_2 and 5% O_2 (SeaLand 1991)
- 12°C, with a range of 10–15°C with 5–8% CO_2 and 2–5% O_2 (Kader 1993)
- 10–15°C with 5–8% CO_2 and 2–5% O_2 (Bishop 1996)
- 10°C with 8% CO_2 and 3% O_2 for 36 days for the cultivar Solo, still having an adequate shelf-life of 5 days at 25°C (Cenci *et al.* 1997)
- 7–13°C and 85–90% r.h. with 2–5% O_2 and 5–8% CO_2 for 1–3 weeks (Burden 1997).

Modified atmosphere packaging

Wills (1990) reported that modified atmosphere packaging extended their storage life. For maximum storage life in plastic films it is advantageous to include an ethylene absorbent (Nazeeb and Broughton 1978).

Ripening

Optimum ripening was achieved at 21.1–26.7°C (Wardlaw 1937). Wills *et al.* (1989) recommended ripening at 21–27°C with no ethylene. In South America, fruit are sometimes lightly scored on the skin, which is said to hasten ripening, probably because it stimulates the production of wound ethylene. Lam (1990) reported that fruits kept at 25°C for 11 days ripened normally.

Kulwithit *et al.* (1998) found that there was no significant difference in eating quality among ripened fruits of the various harvest maturities, although Thompson and Lee (1971) found that fruit harvested before they began to turn yellow were of inferior eating quality.

Ethylene

In Malaysia, Broughton *et al.* (1977) and Nazeeb and Broughton (1978) showed that scrubbing ethylene from the store containing the cultivar Solo Sunrise had no effects on their storage life, but they stated that 20°C with 5% CO_2 and ethylene removal was the optimum condition for 7–14 days. In later work it was shown that ethylene application could accelerate ripening by 25–50% (Wills 1990).

Papayuela, mountain papaya

Carica pubescens, *C. candamarcenis*, Caricaceae.

Botany

There is some confusion as to the species of this plant. Purseglove (1968) referred to mountain papaya as *C. candamarcenis* whereas Morales and Duque (1987) and Alarcon *et al.* (1998) referred to mountain papaya and papayuela as *C. pubescens*. The fruit is a fleshy, oval berry with acid flesh and a central core of seeds, and is up to about 10 cm long, with a green skin as it develops, turning yellow at maturity. It is similar to the papaya (*C. papaya*) but it is frost resistant and grows up to 2500 m above sea level in Colombia and Ecuador. It is rarely eaten fresh and is made into jam or purée, as it is fairly sour. Morales and Duque (1987) found at least 53 volatile components in the fruit, 45 of which were positively identified. Esters were the most abundant group of compounds (63% of the extract), followed by alcohols (30%). The major aroma compounds were ethyl butyrate, butanol, ethyl acetate, methyl butyrate and butyl acetate. Papaya is a climacteric fruit and therefore papayuela probably is also.

Harvesting

Maturity is when the skin of the fruit just begins to turn from green to yellow and it feels soft to the touch.

Storage

A refrigerated storage recommendation was 8–12°C and 85–90% r.h. for 1–2 weeks (Amezquita 1973).

Parasol mushrooms, parasol fungus

Macrolepiota procera, *Lepiota procera*, Agaricaceae.

Botany

It is a gill fungus with a cap up to 14 cm in diameter and 16 cm high. The cap is a creamy white to brown colour, covered in dark flaky scales. They are shaped like an organ stop as they emerge, eventually expanding to plate shaped with white flesh. The gills are white and closely packed with white spores and a whitish stalk covered with darker scales in patches and a thick double ring. It is a native of grassy areas and tastes delicious.

Storage

No information could be found on its postharvest life, but from the literature it would appear that all types of edible mushrooms have similar storage requirements and the recommendations for the cultivated mushroom (*Agaricus bispsorus*) could be followed.

Parsley

Petroselinum sativum, Umbelliferae.

Botany

The edible portion is the leaves and young stems. Flower shoots should be removed in the field as they appear because they will shorten the productive life of the crop. The taproot can also be eaten. The freezing point was given as –1.28 to 1.11°C (Weichmann 1987).

Precooling

Vacuum precooling was recommended by Aharoni *et al.* (1989).

Storage

Refrigerated storage recommendations for the foliage are as follows:

- 0°C and 90–95% r.h. for 1–2 months (Lutz and Hardenburg 1968)
- 0–1°C and 85–90% r.h. for 1–2 weeks (Amezquita 1973)
- 0–1°C and 95–100% r.h. for 1–2 months (Snowdon 1991).

The taproot can be stored for several months at 0°C and 90–95% r.h. (Lutz and Hardenburg 1968).

Controlled atmosphere storage

Aharoni *et al.* (1989) fond that chlorophyll degradation was delayed effectively when they were stored at 0°C in 5% CO_2 and 21% O_2.

Modified atmosphere packaging

In an experiment, green tops were bunched, with 25–30 g per bunch, packed in perforated or non-perforated polythene bags (20 × 25 cm) and stored at 2, 5, 10, 15 or 20°C. The control was kept unpacked. The tops stored better in non-perforated than in perforated bags and the longest satisfactory storage of 14–21 days was in non-perforated bags at 2°C. The non-packed bunches kept well for 1–2 days at 2°C but deteriorated rapidly at the higher temperatures (Umiecka 1973).

Hypobaric storage

It was claimed that hypobaric conditions extended their storage life from 5 weeks in cold storage to 8 weeks without appreciable losses of protein, ascorbic acid or chlorophyll contents (Bangerth 1974).

Ethylene

They showed no response to exposure to 100 μl $litre^{-1}$ of ethylene for 24 hours at any temperature over the range 5–35°C (Inaba *et al.* 1989).

Transport

Aharoni *et al.* (1989) studied freshly harvested parsley under simulated conditions of air transport from Israel to Europe, and also with an actual shipment, during which temperatures fluctuated between 4 and 15°C. Packaging in perforated polyethylene-lined cartons resulted in a marked retardation of both yellowing and decay.

Parsnips

Pastinaca sativa, Umbelliferae.

Botany

The edible portion is the swollen taproot. It has an off-white to light brown skin with concentric rings running around its circumference. The flesh is also off-white. They will freeze at about –2.0 to 1.5°C (Wright 1942) and are said not to be injured by slight freezing (Phillips and Armstrong 1967; Lutz and Hardenburg 1968). In contrast, quality and sweetness were shown to be improved by delaying harvesting so that they were exposed to frost for 2 months (Lutz and Hardenburg 1968).

Harvesting

This is carried out when they achieve a size acceptable for the market. In Northern Europe this is in autumn, but they may be harvested when immature and marketed as 'baby parsnips.' They can be harvested mechanically using modified carrot harvesters.

Precooling

Hydrocooling was the most suitable method, forced-air cooling with high humidity was also suitable, air cooling in a conventional cold store was less suitable and vacuum cooling was unsuitable (MAFF undated).

Storage

The main limiting factor in storage is root discoloration and shrivelling (Phillips and Armstrong 1967). At a room temperature of 20°C and 60% r.h. they could be kept for 4–6 days (Mercantilia 1989). An improvement in quality can be achieved by storage for 2–3 weeks at 0–1.1°C, which speeded the hydrolysis of starch to sucrose (Wardlaw 1937; Lutz and Hardenburg 1968). Refrigerated storage recommendations are as follows:

- 0°C and 90–95% r.h. for 2–4 months (Wardlaw 1937)
- 0°C and 95% r.h. for 2–4 months (Phillips and Armstrong 1967)
- 0°C and 90–95% r.h. for 2–6 months (Anon 1967; Lutz and Hardenburg 1968)
- 0°C and 90 r.h. for 2–6 months (Mercantilia 1989)
- 0°C and 98–100% r.h. for 4–6 months (Hardenburg *et al.* 1990)
- 0°C and 95–100% r.h. for 120–150 days (SeaLand 1991)
- 0–1°C and 95–100% r.h. for 2–6 months (Snowdon 1991).

Modified atmosphere packaging

Perforated polyethylene film liners in storage boxes reduced shrivelling (Lutz and Hardenburg 1968), as did packing in sphagnum moss (Phillips and Armstrong 1967), sand or clean soil (Lutz and Hardenburg 1968).

Ethylene

Ethylene was shown to increase respiration especially if it accumulates in stores when the parsnips were stored with climacteric fruit (Rhodes and Wooltorton 1971).

Passionfruits, maracuya

Passiflora edulis, Passifloraceae.

Botany

There are two distinct forms, the purple passionfruit (*P. edulis* forma *edulis*) and the yellow passionfruit (*P. edulis* forma *flavicarpa*). The purple passionfruit grows well at higher altitudes in the tropics and has the better flavour, whereas the yellow passionfruit grows better in the lowlands. The edible portion is the fruit, which are globular to oval in shape and have a dry, brittle skin, the gelatinous aromatic pulp containing numerous seeds. Their respiratory pattern led to it being classified as an indeterminate fruit by Biale and Barcus (1970). In contrast, Akamine *et al.* (1957) described how *P. edulis*, stored at 25°C, reached a climacteric peak of CO_2 production after 13–14 days.

Harvesting

The fully mature fruit have the best flavour, but a short shelf-life. Wardlaw (1937) therefore recommended that they should be harvested when they are about three-quarters yellow or purple. Immature fruits do not ripen properly after harvest. The purple passionfruit, forma *edulis*, may become reddish and shrivel if harvested too immature (Wardlaw 1937).

Waxing and pretreatments

Neither wax (Autocitrol), thiabendazole nor sodium hypochlorite application had a beneficial effect on fruit storability (Gama *et al.* 1991). In contrast, Gomez (2000) found that wax (Primafresh C) extended the shelf-life of fruits, reduced weight losses and maintained an adequate external appearance during storage at 12°C and 80% r.h. Purple passionfruit coated with 1% Semperfresh (an edible mixture of sucrose esters of fatty acids and carboxymethylcellulose) had lower storage weight losses, less shrivelling and were sweeter but had more fungal disease than untreated fruit (Ssemwanga 1990).

Storage

The major limitations to storage are fermentation of the pulp, shrivelling and fungal decay (Wardlaw 1937), although it was found that moisture was lost mainly from the skin rather than the pulp, and shrivelled samples had a similar flavour to those of smooth samples (Ssemwanga 1990). In storage trials on yellow passionfruit at 5, 10 and 15°C, the soluble solids percentage did not change over a 45-day period at 5 or 10°C but deceased linearly at 15°C and the external appearance of fruit deteriorated rapidly at 5 and 15°C (Arjona *et al.* 1992). Fruits used in the experiments were vine ripened and the percentage of pulp increased linearly during storage at 5°C, did not change at 10°C but at 15°C it increased up to 30 days then deceased from that point to the end of the experiment at 45 days (Arjona *et al.* 1992). Storage of vine-ripened yellow passionfruit at 29 ± 2°C retained their fresh, characteristic, ester-type aroma (ethyl butanoate and ethyl hexanoate) for only 3 days after harvest. There was a continuous increase in the concentration of ethyl acetate over the 15 days of storage. Hexyl butanoate levels were highest after 3 days of storage, decreasing thereafter. Benzaldehyde and hexanol levels decreased after 9 days of storage but 2-heptanone increased continuously during storage (Narain and Bora 1992.). At a room temperature of 20°C and 60% r.h. they could be kept for about 1 week (Mercantilia 1989). Other refrigerated storage recommendations are as follows:

- 5.5–7°C and 80–85% r.h. for 4–5 weeks (Anon 1967)
- 5.6–7.2°C and 85–90% r.h. for 3 weeks with 32% weight loss for purple passionfruits (Pantastico 1975)
- 8°C and 90% r.h. for 3–4 weeks (Mercantilia 1989)
- 7–10°C and 85–90% r.h. for 3–5 weeks (Snowdon 1990)
- 12.2°C and 90–95% r.h. for 14–21 days (SeaLand 1991).

Chilling injury

Storage below 10°C led to chilling injury in the form of blood-red discoloration of the skin, quickly followed by mould attack (Wardlaw 1937).

Modified atmosphere packaging

Storage in plastic bags can prolong their storage life. In Colombia the fruits of the cultivar Degener were considered unsaleable after only 5 days at a mean ambient temperature of 23°C and 76% r.h., but after storage for 14 days in non-perforated and perforated bags there were 99% and 80% of saleable fruits, respectively (Salazar and Torres 1977). Packing purple passionfruit fruits in polyethylene bags reduced weight loss and the fruits stored well at 6°C for 42 days. Ssemwanga (1990) also showed that storage of fruit in polyethylene bags

25 μm thick reduced moisture loss and shrivelling and extended their storage life to 28 days without adversely affecting the internal quality.

Ripening

Akamine *et al.* (1957) found that adding 500 μl litre^{-1} ethylene during storage at 25°C resulted in the climacteric peak being reached in only 2 days compared with 13–14 days in fruits without ethylene.

Pe-tsai, pak choi, pak choy, celery cabbage

Brassica chinensis, *B. chinensis* variety *pekinensis*, *Brassica rapa* variety *chinensis*, Cruciferae.

Botany

It is closely related to *B. pekinensis*, and tastes similar, but is said to have a more delicate flavour. The edible portion is the leaves that are dark green in colour with thick, white, fleshy leaf stalks and midribs (see Figure 12.10).

Harvesting

The whole plant is harvested when it has reached a marketable size, but while the leaves are still tender. The plants pulled from the soil and cut off just below the lowest leaf.

Storage

At room temperature of 20°C and 60% r.h. they could be kept for 2–3 days (Mercantilia 1989). Refrigerated storage recommendations are as follows:

- 0°C and 95% r.h. for 10–17 days with 15% loss (Pantastico 1975)
- 0°C and 90% r.h. for 30–40 days (Mercantilia 1989)
- 0°C and 90–95% r.h. for 30–40 days (SeaLand 1991).

Controlled atmosphere storage

10°C with 2% O_2 and 5% CO_2 increased shelf-life by 112% compared with cold storage alone (O'Hare *et al.* 2000).

Peaches

Prunus persica, Rosaceae.

Botany

The fruit is a drupe and is classified as climacteric. It will freeze at about –1.7 to 1.3°C (Wright 1942).

Harvesting

Bedford and Robertson (1955) showed that peaches harvested when too immature tended to develop a stale 'off-flavour' when ripe. Also, Phillips and Armstrong (1967) stated that immature fruits may be rubbery and astringent while over-mature fruit tended to be stringy. The assessment of harvest maturity by skin colour changes usually depends on the judgement of the harvester, but colour charts are used for some cultivars. A colour difference meter, Minolta Chroma CR-200 (see Figure 2.5), was successfully used to measure the surface colour of peaches and relate this to fruit pigments, soluble solids content and firmness (Kim *et al.* 1993). In India, the fruits of cultivar Flordasun were found to attain maturity 72–75 days after fruit set. Fruit firmness of around 6.5 lb in^{-2} was found to be a good index of harvest maturity together with total soluble solids, acidity and total sugars of about 10.9%, 0.50 and 7.90%, respectively (Badiyala 1998). In studies in Turkey, it was shown that the firmness of peaches changed significantly during the harvest season and these changes could be measured using a Magness and Taylor type pressure tester (Figure 12.85). A device which applied low-pressure air to opposite sides of fruit and then measured the surface deformation was described by Perry (1977) for measuring fruit firmness. Near-infrared reflectance (NIR) has been studied in relation to measuring the internal qualities of fruit, with good correlations with sugar content (Kouno *et al.* 1993a). Correlations between sugar content of peaches and NIR measurement have been shown (Kouno *et al.* 1993a). NIR measurement was achieved using a Nireco Model 6500 near-infrared spectrophotometer, by placing the fruit so that the light beam on the surface was are right-angles to the fruit surface and covered with a black cloth to avoid the influence of external light. Four places around the equator of the fruit were selected and the NIR beam was irradiated at 2 nm intervals from 400–2500 nm on to the fruit and the average absorbence was measured (Kouno *et al.* 1993b).

Grading

Peaches are extremely delicate fruit when ripe and can bruise very easily. Magnetic resonance imaging has been used to obtain images of bruises on peaches to facilitate grading out damaged fruits (Chen *et al.* 1989).

Physiological disorders

Peaches stored in 30% CO_2 at 28°C for 42 hours had high levels of ethanol compared with those stored at 8°C (Yanez Lopez *et al.* 1998). Split pit is an internal disorder of peaches probably caused by adverse weather and too much nitrogen fertilizer. It is not possible to tell from the external appearance whether a fruit is suffering from this disorder, but X-rays can be used to detect it (Bowers *et al.* 1988).

Precooling

Rapid cooling after harvest can give the fruit a stringy or woolly texture (Phillips and Armstrong 1967) and therefore is not recommended.

Storage

At room temperature of 20°C and 60% r.h. fruit could be kept for 1–2 days (Mercantilia 1989). Refrigerated storage recommendations are as follows:

- –0.5 to 1°C for 3 weeks then ripen at 18–25°C for Hales and Elberta (Fidler 1963)
- Storage at 18.3–21.1°C for several days until they are almost ripe then –0.6 to 0°C and 85–90% r.h. (Phillips and Armstrong 1967)
- –0.5 to 0°C and 85–90% r.h. for 2–3 weeks (Anon 1967)
- –1 to 1°C and 85–90% r.h. for 4–8 weeks for late cultivars (Anon 1967)
- –0.6 to 0°C and 90% r.h. for 2 weeks for freestone or 3–4 weeks for clingstone and some freestone (Lutz and Hardenburg 1968)
- 0°C and 90–95% r.h. for 14 days (Mercantilia 1989)
- 2°C for 7 days (Mercantilia 1989)
- 4°C for 5 days (Mercantilia 1989)
- –1 to 0°C and 90–95% r.h. for 2–6 weeks (Snowdon 1990)
- –0.5 to 0°C and 90–95% r.h. for 2–4 weeks (Hardenburg *et al.* 1990)
- –0.5°C with 90–95% r.h. for 14–28 days (SeaLand 1991).

Chilling injury

Prolonged low-temperature storage can result in internal browning or a woolly texture (Haard and Salunkhe 1975). The symptoms were described as flesh browning, mealy breakdown and abnormal peeling by Kajiura (1975).

Intermittent warming

Controlled atmosphere storage at –0.5 to 0°C in 5% CO_2 and 1% O_2 plus intermittent warming to 18°C in air for 2 days every 3–4 weeks was said to have shown promising results (Hardenburg *et al.* 1990).

Controlled atmosphere storage

J.E. Bernard in 1819 in France (quoted by Thompson 1998) showed that storage in zero O_2 gave fruit a life of 28 days. The cultivar Okubo was stored for 3 weeks at 1°C in air, 3% CO_2 with 3% O_2 or 0% CO_2 with 3% O_2 by Kajiura (1975). After removal from storage, fruits were ripened at 20°C. In cold-stored fruits, low-temperature injuries developed. In 3% O_2 with 0% CO_2, the ripening rate after storage was faster than that of fruits which had been placed in 20°C directly after harvest. They also exhibited chilling injury. Fruits that had been stored in 3% O_2 with 3% CO_2 had a ripening rate after storage that was the same as the directly ripened fruits, and chilling injury was almost completely eliminated.

Fruits of the cultivar Springcrest were harvested at two maturity stages (flesh firmness of 60 and 45 newton) and maintained at 20°C in air (control) or for 24 or 48 hours in streams of active ingredient containing either ultra-low oxygen (<1%) or high CO_2 (30%) concentration and then transferred to air for up to 8 days. The decline in flesh firmness was strongly reduced in ultra-low O_2 and 30% CO_2 in fruits from both harvests, although the effect was stronger in fruits picked earlier, in which ethylene biosynthesis remained at the basal level. Ethanol increased slightly during ripening in control fruits, but ultra-low O_2 and 30% CO_2 strongly stimulated ethanol production. When fruits were transferred to air, the ethanol concentration declined rapidly (Bonghi *et al.* 1999).

Other storage controlled atmosphere storage recommendations are as follows:

- –0.5 to 0°C with 5% CO_2 and 1% O_2 maintained quality and retarded internal breakdown for about twice as long as those stored in air (Hardenburg *et al.* 1990)
- 0.5°C with 90–95% r.h. with 5% CO_2 with 2% O_2 SeaLand (1991)
- 0–5°C with 5% CO_2 and 1–2% O_2 which had a good effect, but limited commercial use (Kader 1985)

- 0–5°C and 3–5% CO_2 and 1–2% O_2 for both freestone and clingstone peaches (Kader 1989 and 1992)
- 0 ± 5°C and 90 ± 5% r.h. with 5% CO_2 and 2% O_2 for Flavorcrest and Red Top, which reduced weight loss and postharvest decay (Eris and Akbudak 2001)

Ozone

Ozone has been shown to control fungal growth and rotting in fruit exposed to 1–2 μl litre^{-1}, but peaches were damaged by exposure to only 1 μ litre^{-1} O_3.

Hypobaric storage

Hypobaric storage prolonged the life of peaches from 7–27 days. It delayed chlorophyll and starch breakdown, carotenoid formation and the decrease in sugars and titratable acidity (Salunkhe and Wu 1973). Storage at 11–12°C in an aerated hypobaric system at either 190 or 76 mmHg retarded ripening compared with storage at the same temperature in air (Kondou *et al.* 1983). In the cultivar Okubo kept in hypobaric storage at 108 mmHg, the ripening rate after storage was slower than that of fruits that had been stored in air at 760 mmHg throughout. Mealy breakdown was reduced in hypobaric storage, but no effect was found on flesh browning or on abnormal peeling (Kajiura 1975).

Ripening

Watada *et al.* (1979) found a larger range and higher quantities of organic volatile compounds in tree-ripened fruit than in those harvested before they were ripe and ripened postharvest. Bedford and Robertson (1955) showed that fruits ripened at 24°C had a better flavour than those ripened at 29°C. At temperatures below 18°C rotting usually outstripped ripening. At 20°C Wills *et al.* (2001) found that the time to ripen of the cultivars Fragar, Coronet, Loring and Flordagold increased linearly with logarithmic decrease in ethylene concentration over the range <0.005, 0.01, 0.1 1.0 and 10 μl litre^{-1}. In South Africa, 24°C with 1% acetylene in the atmosphere for 24 hours was recommended (Fidler 1963).

In Australia, exposing the cultivar Victoria to 40°C and high humidity then storage at 24°C until ripe was recommended. To achieve this, air was passed over thermostatically controlled steam heaters. In the USA, immersing the fruit in water to bring its pulp temperature to 37°C for 3–3.5 minutes was recommended before ripening (Fidler 1963).

Pears

Pyrus communis, Rosaceae.

Botany

The fruit is a pome and is classified as climacteric. Ripe pears will freeze at about –2.7 to 2.1°C (Wright 1942). They are tall, vigorous trees and are grafted on to rootstocks to control their size for commercial production. There are many species of *Pyrus*, all of which are native to Europe and Asia. There are hybrids between *P. communis* and *P. sinensis* called Kieffer pears developed in the USA for disease resistance. Pears were cultivated for their fruit by the Phoenicians and Romans and were a delicacy for the ancient Persian kings. There may be as many as 5000 varieties and cultivars throughout the world. See also Chinese pears *P. bertschneideri* and Asian or Japanese pears *P. pyrifolia*.

Quality standards

The Organisation de Coopération et de Dévéloppment Économiques, Paris, published a brochure of standards for pears in 1983.

Physiological disorders

The effects of various factors that may influence the development of core breakdown, which is also called brownheart, during storage were studied by Lammertyn *et al.* (2000). They showed some difference in susceptibility between different cultivars. Conference was more likely to develop the disorder than Doyenne du Comice. Conference fruits were particularly susceptible when stored in high CO_2 and low O_2 levels. Large fruits and late-harvested fruits were also more likely to develop the disorder and it developed more during storage at 1°C than at –1°C.

Magnetic resonance imaging has been used to obtain images of bruises and insect damage in pears (Chen *et al.* 1989).

Precooling

They are not normally precooled. However, if it is thought necessary then air cooling in a conventional cold store was the most suitable, forced-air cooling with high humidity was also suitable and hydrocooling were less suitable and vacuum cooling was unsuitable (MAFF undated).

Storage and controlled atmosphere storage

The recommendations for storage vary considerably between cultivars, but there are general recommendations where the author does not specify the cultivar. Koelet (1992) recommended 0.5°C with 0.5–1.0% CO_2 and 2–2.2% O_2. SeaLand (1991) recommended −1.1°C and 90–95% r.h. with 0–1% CO_2 and 2–3% O_2. Lawton (1996) recommended 0°C and 93% r.h. with 0.5% CO_2 and 1.5% O_2. Storage at 0–5°C with 0–1% CO_2 and 2–3% O_2 was recommended (Kader 1985) or 0–5°C with 0–3% CO_2 and 1–3% O_2 (Kader 1992). Recommendations for specific cultivars are given in the following.

Anjou

- −1.1°C and 85–90% r.h. for 4.5–6 months (Phillips and Armstrong 1967)
- −0.5 to 0°C and 90–95% r.h. for 4–6 months (Mercantilia 1989)
- 0.8–1% CO_2 and 2–2.5% O_2 or 0.1% CO_2 or less with 1–1.5% O_2 (Hardenburg *et al.* 1990)
- Exposure of Anjou to 12% CO_2 for 2 weeks immediately after harvest had beneficial effects on the retention of ripening capacity (Hardenburg *et al.* 1990)
- −1.1°C and 90–95% r.h. for 120–180 days (SeaLand 1991).

Bartlett

- −1.1°C and 85–90% r.h. for 2–3 months (Phillips and Armstrong 1967)
- −1.1 to 0.6°C and 90–95% r.h. for 2–3 months (Anon 1968)
- A maximum of 5% CO_2 and a minimum of 2% O_2 (Fellows 1988)
- −0.5 to 0°C and 90–95% r.h. for 2.5–3 months (Mercantilia 1989)
- −1 to 0.5°C and 90–95% r.h. for 1–3 months (Snowdon 1990)
- −1.1°C and 90–95% r.h. for 60–90 days (SeaLand 1991).

Bosc

- −1.1 to 0.6°C and 90–95% r.h. 3–4 months (Anon 1968)
- −1.1°C and 90–95% r.h. for 60–90 days (SeaLand 1991).

Buerre Bosc

- −1.1°C and 85–90% r.h. for 3–3.5 months (Phillips and Armstrong 1967)
- −1°C and 90–95% r.h. for 3 months (Snowdon 1990).

Beurre d'Anjou

- After 6.5 months of storage, at −0.5°C the average percentage of fruits with symptoms of scald ranged from zero in fruits stored in 1% O_2 with 0% CO_2 or 1.5% O_2 with 0.5% CO_2 to 2% in fruits stored in 2.5% O_2 with 0.8% CO_2 and 100% in control fruits stored in air.

Buerre Hardy

- −1 to 0.5°C and 90–95% r.h. for 2–3 months (Snowdon 1990).

Clapps Favourite

- −1 to 0°C and 90–95% r.h. for 1–2 months (Snowdon 1990).

Comice (Table 12.46)

- −1.1°C and 85–90% r.h. for 2–3 months (Phillips and Armstrong 1967)
- −1.1 to 0.6°C and 90–95% r.h. for 2–3 months (Anon 1968)
- −1 to 0.5°C and 90–95% r.h. for 3–4 months (Snowdon 1990).

Concord

- −1 to 0.5°C with less than 1% CO_2 and 2% O_2 (Johnson 1994).

Conference *(Table 12.47)*

- 0°C with 2% CO_2 and 2% O_2 (Stoll 1972)
- 0.5–1°C with 5% CO_2 and 5% O_2 (Fidler and Mann 1972)

Table 12.46 Recommended controlled atmosphere storage conditions for Comice by country and region (Herregods 1993)

	Temperature (°C)	CO_2 (%)	O_2 (%)
Belgium	0–0.5	<0.8	2–2.2
France	0	5	5
Germany (Westphalia)	0	3	2
Italy	−1–0.5	3–5	3–4
Spain	−0.5	3	3
Switzerland	0	5	3
USA (Oregon)	−1	0.1	0.5

Table 12.47 Recommended controlled atmosphere storage conditions for Conference by country and region (Herregods 1993)

	Temperature (°C)	CO_2 (%)	O_2 (%)
Belgium	–0.5	<0.8	2–2.2
Denmark	–0.5–0	0.5	2–3
UK	–1–5	<1	2
Germany (Saxony)	1	1.1–1.3	1.3–1.5
Germany (Westphalia)	–1–0	1.5–2	1.5
Netherlands	–1–0.5	≤1	3
Italy	–1–0.5	2–4	2–3
Italy	–1–1.5	1–1.5	6–7
Slovenia	0	3	3
Spain	–1	2	3.5
Switzerland	0	2	2

- –1 to 0.5°C with less than 1% CO_2, and 2% O_2 (Sharples and Stow 1986; Johnson 1994)
- –1 to 0.5°C and 90–95% r.h. for 4–6 months (Snowdon 1990).

d'Anjou

- –1°C and 90–95% r.h. for 4–6 months (Snowdon 1990)
- Chen and Varga (1997) found that no controlled atmosphere storage regime containing 0.5–2% O_2 was feasible in controlling physiological disorders of d'Anjou pears during 6 or 8 months storage. They showed that storage in 0.5% O_2 with <0.1% CO_2 resulted in a high incidence of scald and black speck.

Doyenne Boussock

- –1.1°C and 85–90% r.h. for 2 months (Phillips and Armstrong 1967).

Doyenne du Comice

- 0.5–1°C with 5% CO_2 and 5% O_2 (Fidler and Mann 1972)
- 0°C with 2% CO_2 and 2% O_2 (Stoll 1972).

Flemish Beauty

- –1.1°C and 85–90% r.h. for 2 months (Phillips and Armstrong 1967).

Hardy

- –1.1°C and 85–90% r.h. for 2–3 months (Phillips and Armstrong 1967)
- –1.1 to 0.6°C and 90–95% r.h. for 2–3 months (Anon 1968).

Josephine

- 0°C and 90–95% r.h. for 5–6 months (Snowdon 1990).

Jules Guyot

- –1 to 0.5°C and 90–95% r.h. for 1–3 months (Snowdon 1990).

Kieffer

- –1.1°C and 85–90% r.h. for 2–3 months (Phillips and Armstrong 1967)
- –1.1 to 0.6°C and 90–95% r.h. for 2–3 months (Anon 1968).

Packham's Triumph (Table 12.48)

- –1°C and 90–95% r.h. for 3–6 months (Snowdon 1990).

Passecrassane

- –1 to 1°C and 90–95% r.h. for 5–6 months (Snowdon 1990).

Seckel

- –1.1 to 0.6°C and 90–95% r.h. 3–4 months (Anon 1968).

William's Bon Chretien

- 0.5–1°C with 6% CO_2 and about 15% O_2 (Fidler and Mann 1972)
- 0°C with 2% CO_2 and 2% O_2 (Stoll 1972).

Winter Cole

- –1°C and 90–95% r.h. for 4 months (Snowdon 1990).

Winter Nelis

- –1.1°C and 85–90% r.h. for 4.5–6 months (Phillips and Armstrong 1967)
- –1.1 to 0.6°C and 90–95% r.h. 4–6 months (Anon 1968)
- –1 to 0.5°C and 90–95% r.h. for 4–6 months (Snowdon 1990).

Controlled atmosphere storage conditions can affect the quality of the fruit. It was shown that con-

Table 12.48 Recommended controlled atmosphere storage conditions for Packham's Triumph by country and region (Herregods 1993)

	Temperature (°C)	CO_2 (%)	O_2 (%)
Australia (South)	–1	3	2
Australia (Victoria)	0	0.5	1
Australia (Victoria)	0	4.5	2.5
Germany (Westphalia)	–1–0	3	2
New Zealand	–0.5	2	2
Slovenia	0	3	3
Switzerland	0	2	2

Table 12.49 Effect of controlled atmosphere storage on the acidity and firmness of Spartlett pears stored at 0°C and their taste after 7 days of subsequent ripening at 20°C (source: Meheriuk 1989a)

Storage atmosphere		Storage time	Acidity	Taste	Firmness
CO_2 (%)	O_2 (%)	(days)	(mg per 100 ml)[a]	(9 = like, 1 = dislike)	(newton)[a]
0	21	60	450b	6.1	62a
5	3	60	513a	7.6	61ab
2	2	60	478ab	7.6	60b
0	21	120	382b	4.3	55b
5	3	120	488a	6.7	58a
2	2	120	489a	6.9	58a
0	21	150	342b	2.9	51b
5	3	150	451a	7.7	58a
2	2	150	455a	7.1	58a
0	21	180	318b	–	40b
5	3	180	475a	–	62a
2	2	180	468a	–	61a

[a] Figures followed by the same letter were not significantly different ($p = 0.05$) by the Duncan's Multiple Range Test.

trolled atmosphere storage resulted in the fruit being 'liked' more than those stored in air as well as better retaining their firmness and acidity (Table 12.49).

Hypobaric storage

Hypobaric conditions prolonged the storage life of pears by between 1.5 and 4.5 months compared with cold storage alone. Hyperbaric storage delayed chlorophyll and starch breakdown, carotenoid formation and the decrease in sugars and titratable acidity (Salunkhe and Wu 1973).

Ripening

Ripening conditions of 20°C (Phillips and Armstong 1967) or 20–22.5°C with high humidity (Anon 1968) were recommended. However, Wills *et al.* (1989) recommended 15–18°C using ethylene at 10 µl litre^{-1} for 24 hours at 85–90% r.h.

Peas, garden peas, mange tout, snow peas, sugar peas

Pisum sativum, Leguminosae.

Botany

In most cases the edible portion is the seeds, which are produced within the pod or legume. In some cases the pod containing the immature seeds is eaten and they are usually called mange-tout, snow pea or sugar pea. Some authorities term the garden pea, which are grown for their seeds, as *humile* and those grown for their pods as *macrocarpon*. Snow peas are also referred to as *P. sativum* var. *saccharatum*. Green peas will freeze at about –1 to 1.0°C (Wright 1942) or –1.17 to 0.61°C (Weichmann 1987).

Harvesting

As the peas develop they accumulate sugars and as they mature sugar is converted to starch and they become firmer and less tender. Harvest maturity for the seeds is when the pods are well filled, but they are still young, tender and sweet. Small-scale growers usually harvest the peas by hand as they mature every 5–10 days. For mange-tout, snow peas and sugar peas harvesting is carried out before the seeds have swollen in the pods.

For processing, sample peas are taken from the field and their texture is tested in a shear cell called a 'tenderometer' (Figure 12.85) (Knight 1991). The whole field of peas is harvested when a particular tenderometer value is reached. However, in many cases the farmer will simply sample the peas and squeeze them between the fingers to determine their firmness. The farmer will then taste them to test their sweetness. On the basis of experience the farmer will be able to tell whether or not to harvest that field of peas. Tenderometers are destructive tests that assume the sample taken is representative of the crop, but special processing cultivars have been bred where all the pea pods mature at about the same time. Some work was done on heat units in relation to harvest maturity. The number of degree hours above 4°C is calculated.

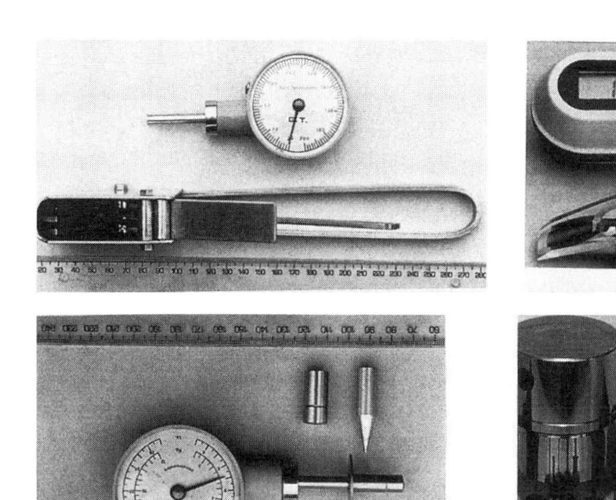

Figure 12.85 Texture measuring equipment, also with refractometers at the top right. Source: Dr James Ssemwanga.

This varies between cultivars and is usually within 30 000–35 000, but it has limited application. However, day degrees/heat units are used to compute harvest dates for vining peas (John Love, personal communication, quoted by Thompson 1996).

For processing they are harvested in combine machines called 'pea viners.' These machines harvest the whole plant; remove the pods from the vines and shell the pods, thus removing the peas. Physical damage during harvesting has been shown to have an adverse effect on the flavour of peas (Murray *et al.* 1976), but since they are frozen within an hour or so of harvesting this is not normally a problem in a well run industry.

Precooling

Forced-air cooling with high humidity was the most acceptable, hydrocooling was suitable and air cooling in a conventional cold store and vacuum cooling less suitable (MAFF undated). After 25–30 minutes of vacuum cooling at 4.0–4.6 mmHg with a condenser temperature of −1.7 to 0°C, the pea temperature fell to 6°C from 20–22.2°C (Barger 1963). Precooling with crushed ice may extend their storage life from about 1 week to 15–20 days (Tindall 1983).

Storage

For the fresh market, garden peas store better in the pods than when shelled. At a room temperature of 20°C and 60% r.h. peas could be kept for about 1 day and mange-tout for 1–2 days (Mercantilia 1989). Refrigerated storage recommendations are as follows:

- 0°C and 80–90% r.h. for 25 days with 5% loss or 32 days with 10% loss (Wardlaw 1937)
- 0.6°C and 85% r.h. for 4 weeks (Wardlaw 1937)
- −0.5 to 0°C and 85–90% r.h. for 1–3 weeks (Anon 1967)
- 0°C and 90–95% r.h. for 1–2 weeks (Lutz and Hardenburg 1968)
- 0°C and 85–90% r.h. for 1–2 weeks (Lutz and Hardenburg 1968)
- 0°C and 88–92% r.h. for 2–3 weeks with 8% loss (Pantastico 1975)
- 0°C and 95–100% r.h. for up to 10 days with a weight loss of up to 10% (Tindall 1983)
- 4°C and 95–100% r.h. for 6–7 days (Tindall 1983)
- 0°C and 90–95% r.h. for 7 days (Mercantilia 1989)
- 4°C for 4 days (Mercantilia 1989)
- 12°C for 2 days (Mercantilia 1989)
- 0°C and 90–95% r.h. for 1–2 weeks for mange-tout (Mercantilia 1989)
- 0°C and 95–98% r.h. for 1–2 weeks (Hardenburg *et al.* 1990)
- 0°C and 90–95% r.h. for 7–10 days for both peas and snow peas (SeaLand 1991)
- 0–1°C and 95–100% r.h. for 1–3 weeks (Snowdon 1991)
- 0–1°C and 95–100% r.h. for 1–2 weeks for mange-tout (Snowdon 1991).

Controlled atmosphere storage

Storage in either 2.5% O_2 with 5% CO_2 or 10% CO_2 with 5% O_2 concentrations resulted in the development slight off-flavours, but this effect was reversible since it was partially alleviated after ventilation (Pariasca *et al.* 2001). Other storage recommendations are:

- 0°C with 5–10% O_2 and 5–7% CO_2 for 20 days but at 0°C in air only 7–10 days (Pantastico 1975)
- 0°C with 5–7% CO_2 for 20 days (Hardenburg *et al.* 1990)
- 0–10°C with 2–3% CO_2 and 2–3% O_2 for sugar peas, but it had only a slight effect (Saltveit 1989).

Modified atmosphere packaging

Storage in polymethylpentene polymeric films of 25 and 35 μm thickness at 5°C modified the in-bag atmosphere concentration to approximately 5% O_2 and 5% CO_2. Storage in these conditions resulted in better maintenance of the pod appearance and colour, and also chlorophyll, ascorbic acid, and sugar contents, compared with those stored not wrapped. Sensory scores were also maintained for those packed in plastic films (Pariasca *et al.* 2001).

Ethylene

They showed no response to exposure to 100 μl litre^{-1} of ethylene for 24 hours at any temperature over the range 5–35°C (Inaba *et al.* 1989).

Pepinos, mishqui, tree melons

Solanum muricatum, Solanaceae.

Botany

The edible portion is the fruit. It has a smooth, yellow skin streaked with purple and is up to 15 cm long with pale yellow flesh. It can be eaten fresh and was said to have a sweet melon flavour with a hint of lemon and pineapple (Popenoe 1992). Citric acid and an approximately equal amount of malic and tartaric acid were present in all ripening stages (Huyskens-Keil *et al.* 2000). The seeds are also said to be edible. It probably originated in Peru.

Harvesting

Differences in maturity between clones of up to 20 days were found in the fruit growth period (Prohens and Nuez 2001), but generally harvest maturity is based on skin colour change from green to yellowish. In an experiment fruits were harvested at the mature-green, colour break and ripe stages by Huyskens-Keil *et al.* (2000). They found that fruits at the colour break stage showed a better postharvest behaviour and reduced quality losses, specifically at a storage temperature of 5°C.

Storage

Refrigerated storage recommendations are as follows:

- 7–10°C and 90% r.h. for 4–6 weeks (Snowdon 1990)
- 4.4°C and 85–90% r.h. for 30 days (SeaLand 1991)
- 5°C and 98% r.h. for up to 21 days (Waskar *et al.* 2000)
- 5°C for up to 21 days (Huyskens-Keil *et al.* 2000).

Controlled atmosphere storage

Controlled atmospheres resulted in higher quality fruit after storage compared with storage in air and short bursts of high CO_2 concentrations improved skin firmness and colour intensity (Waskar *et al.* 2000).

Ripening

Glucose and fructose contents declined during ripening at 18°C, whereas sucrose content increased significantly, leading to a higher ratio of dissacharides to monosaccharides in ripe fruits than in younger fruits. During ripening, the sugar:acid ratio increased until the fruits reached the colour break stage, but declined thereafter, with the ratio being 16:1 and 13:1 in colour break and ripe fruits, respectively (Huyskens-Keil *et al.* 2000). Prohens and Nuez (2001) studied different clones in response to Ethephon sprays (0, 500 or 2000 mg l^{-1}). Some clones did not respond to the sprays, whereas in others the ripening period was shortened by more than 60%. In general, the effect of Ethephon was greater in the clones with a longer ripening period. The effects of Ethephon on fruit quality characteristics were not significant in the majority of clones, although in five clones Ethephon resulted in skin degreening, a higher firmness and lower soluble solids content after storage compared with those not treated. On the other hand, Ethephon sprays did not affect either the postharvest behaviour or the sensitivity of fruit to bruising.

Persimmons, kaki, sharon fruit, date plums

Diospyros kaki, Ebenaceae.

Botany

The fruit is classified as climacteric and is green in colour as it develops, and in most cultivars it turns orange or red and becomes translucent as it reaches maturity. The fruit has a pale brown calyx and contains up to eight seeds, but many modern cultivars are seedless. They originated in Japan, where they are called 'the food of the gods.'

Harvesting

The soluble tannin content in the fruit varies between cultivars and in astringent cultivars it is particularly high in immature fruit (Ito 1971). It was shown that in Japan the time of harvest greatly affected tannin content and therefore astringency of the fruit (Table 12.50).

Grading

Diffuse reflectance measures the reflected light just below the surface of the crop. It was shown to be effective with persimmon with a diffuse reflectance of 680 nm suitable for automatic grading lines (Chuma *et al.* 1980a).

Astringency removal

Exposure to high levels of CO_2 can be used to reduce astringency as follows:

- Exposure to 90–95% CO_2 for a short duration at 20–25°C showed that astringency disappeared 3–4 days after removal from the CO_2 in the cultivar Hiratanenashi (Matsuo *et al.* 1976).
- Exposure to 60–90% CO_2 for 24 hours at 17–20°C removed astringency (Ben Arie and Guelfat-Reich 1976).
- Exposure to 80–90% CO_2 for 24 hours in a poly(vinyl chloride) film tent reduced astringency (Kitagawa and Glucina 1984).
- Storage at –1°C in 4% CO_2 for about 2 weeks before removal from storage followed by 6–18 hours in 90% CO_2 at 17°C removed astringency from fruits (Hardenburg *et al.* 1990).

Postharvest treatment with alcohol has been known for over 100 years in Japan to reduce their astringency. Spraying fruit with 35–40% ethanol at the rate of 150–200 ml per 15 kg of fruit was shown to have removed astringency 10 days later in ambient storage. This treatment took longer to remove the astringency than CO_2 treatment, but the fruit were considered of better quality (Kitagawa and Glucina 1984); see also Chapter 6.

Storage

At a room temperature of 20°C and 60% r.h. they could be kept for 5–6 days (Mercantilia 1989). Refrigerated storage recommendations are as follows:

- –1.1°C for 2 months with 8 days subsequent shelf-life (Wardlaw 1937)
- 0°C and 85% r.h. for 120 days in Australia (Wardlaw 1937)
- –1°C and 90% r.h. for 3–4 months (Lutz and Hardenburg 1968)
- 0–1.7°C and 85–90% r.h. for 7 weeks (Pantastico 1975)
- 0°C for about 2 months for Fuyu with an increase in colour intensity and some softening especially when ethylene is present (Kitagawa and Glucina 1984)
- 0°C and 90% r.h. for 2–3 months (Mercantilia 1989)
- –1 to 0°C and 90–95% r.h. for 2–4 months (Snowdon 1990)

Table 12.50 Variations in soluble tannin content (%) of four cultivars of persimmon harvested at different times (source: Ito 1971)

	Cultivar			
Harvest Date	Fuyu	Amahyakume	Aizumishirazu	Yokono
14 July	0.29	1.08	3.74	4.87
18 Aug.	0.00	0.42	2.13	3.89
14 Sept.	0.10	0.19	1.31	3.59
13 Oct.	0.00	0.16	1.35	2.49
7 Nov.	0.00	0.00	0.92	1.94

- 10°C and 90–95% r.h. for 35–84 days for Fuyu (SeaLand 1991)
- 5°C and 90–95% r.h. for 50–90 days for Hachiya (SeaLand 1991).

Controlled atmosphere storage

Recommendations are as follows:

- 1°C and 90–100% r.h. with 8% CO_2 and 3–5% O_2 for Fuyu (Hulme 1971)
- 0°C with 5–8% CO_2 and 2–3% O_2 for 5–6 months for Fuyu (Kitagawa and Glucina 1984)
- 0–5°C with 5–8% CO_2 and 3–5% O_2 was reported to have only a fair effect on storage and was not used commercially (Kader 1985, 1992)
- 5°C and 5–8% CO_2 with 3–5% O_2 SeaLand (1991)
- –0.5°C with 15% O_2 with 15% CO_2 maintained the highest flesh firmness of Fuyu after 100 days of storage and 4 days shelf-life at 14°C (Brackmann *et al.* 1999).

Modified atmosphere packaging

In Korea, fruits are stored commercially for 6 months at 2–3°C with minimal rotting or colour change, and storage in the same conditions with 10 fruits sealed in 60 μm thick polyethylene film bags was even better (A.K. Thompson, unpublished 1974). A problem of condensation on the inner surface of the bags was overcome by cooling the fruit before packaging. In an experiment in Brazil, fruits were stored in 40 μm micro-perforated polyethylene film, with or without ethylene absorption (sachets of potassium permanganate inside the packaging) and with or without postharvest dipping in iprodione (150 g active ingredient per 100 l). On average the levels of gases inside the packaging were 11.5% for O_2 and 6.7% for CO_2. Ethylene absorption reduced skin browning, but did not affect flesh firmness. Iprodione treatment reduced decay from 69–47% but also caused skin browning (Brackmann *et al.* 1999).

Ripening

Fruits held at 18.3°C in an atmosphere containing 1000 μl litre^{-1} ethylene for 50 hours or 500 μl litre^{-1} ethylene for 60 hours softened more quickly and had lower astringency levels than fruit ripened in air (Ito 1971). Ripening at 18–21°C using ethylene at 10 μl litre^{-1} for 24 hours at 85–90% r.h. was recommended by Wills *et al.* (1989).

Pineapples

Ananas comosus, Bromeliaceae.

Botany

It is classified as non-climacteric and is a parthenocarpic multiple fruit called a syncarp or sorosis, which is composed of some 100–200 berry-like fruitlets, sometimes called the eyes, attached to a central peduncle. The skin or peel is often referred to as the shell. Ripe pineapples will freeze at about –1.3 to 1.0°C (Wright 1942).

Harvesting

Harvest maturity is usually based on skin colour and the shape of the individual fruitlets. Colour is not an infallible means of measurement since Smooth Cayenne grown in Ghana can be fully mature with yellow sweet flesh while the skin is still green. This has presented a problem in marketing such fruits in Europe and in some cases they have been labelled to indicate they are ripe while the skin is green. Skin colour can vary due to season, rainfall, microclimate and field practices (Smith 1984). In other cases estimation of the skin colour can be effective. The eyes turn from green to yellow–orange from the bottom and maturity can be judged on the number of rows of eyes that have changed colour. A common harvest maturity is when two or three rows have changed colour. In Hawaii, skin colour is supplemented by a minimum reading of 12° Brix for fruit destined for fresh consumption (Bartholomew and Paull 1986).

Harvest maturity can also be estimated by relating it to the time after flowering or mid-flowering, but the number of days may vary from place to place. Bartholomew and Paull (1986) mentioned that in Madagascar the period from flower induction to ripening varied from 140–221 days and in South Africa from 234–283 days. For Smooth Cayenne in Hawaii, Dull (1971) reported that it took 110 days from the completion of flowering to harvest maturity, but Bartholomew and Paull (1986) found that it was between 70–120 days, that is, from mid-flowering to the onset of ripening. These variations were attributed to seasonal temperature variations.

In Queensland in winter, pineapples are picked at an immature stage to avoid black heart, and this stage is preferable for handling owing to high resistance to bruising (Smith 1984). Half ripe (50% of the shell yellowing) Smooth Cayenne fruit can be held for about 2 weeks at 7.5°C and still have about 1 week of shelf-life (Paull 1992), and at this stage the fruit was suitable for export (Saluke and Dasai 1984). Quarter yellow (25% yellowing) pineapples at harvest gain about one additional week's storage for every 6°C decrease in storage temperature (no temperature range was reported) and at 7°C the maximum storage life is 4 weeks (Dull 1971).

The sound of a fruit as it is tapped sharply with the knuckle of the finger can change during maturation and ripening. Consumers sometimes use the method of testing fruit. This method may only be used post-harvest to determine their maturity since the plants are so spiky that it would be very difficult to get at the fruit.

Preharvest ethylene treatment

Ethrel (Ethephon), has been used as a source of ethylene for decades and it is used to initiate flowering in pineapples. It has also been applied just before harvesting to accelerate degreening and therefore the development of the orange colour in the skin. In Queensland, Smooth Cayenne pineapples treated prior to harvest with Ethrel, at a concentration of 2.5 l in 1000 l of water, had superior eating quality, degreened more evenly, but had a shorter shelf-life due to accelerated skin senescence than untreated fruit 10 days after harvest. Treated fruit left on the plant for 23 days had inferior eating quality to the untreated fruit (Smith 1991). These effects are probably due to the effect of the ethylene speeding up the maturation of the fruit.

Physiological disorder

Internal browning, also known as blackheart, is a major storage disorder, often associated with chilling injury (see below). In Australia when the night-time temperature fell below 21°C, internal browning of the fruit could be detected postharvest (Smith and Glennie 1987). Selvarajah *et al.* (2001) found that treatment with 1-methylcyclopropene (1-MCP) at 0.1 ppm (4.5 nmol litre^{-1}) for 18 hours at 20°C effectively controlled internal browning during subsequent storage at 10°C for 4 weeks. These results suggest that ethylene may be involved in the internal browning. Self *et al.* (1993) have shown that there was a relationship between ultrasonic velocity and the internal browning in pineapples so that this technique may be used to detect its presence.

Disease control

Infection of the fruit is commonly through the cut fruit stalk. Dipping this area directly after it has been cut is normally sufficient to control postharvest disease. This allows for reduced levels of fungicide application that may therefore result in lower residue levels and lower the costs of the chemical used.

Vapour heat treatment

A recommended treatment was 43°C in saturated air for 8 hours and then holding the temperature for a further 6 hours (Jacobi *et al.* 1993).

Packaging

Pineapples will only fit into boxes in small numbers of around four or six packed vertically or horizontally with a crown of leaves cut to fit snugly into the box.

Storage

In a study in South Africa, the most suitable storage temperature for Queen (harvested at 50–80% yellow and less than 10% sugar content) was 14°C, with no chilling injury occurring in storage at about 12°C. However, some internal browning was noted after storage at all temperatures between 10–20°C (Swarts 1991). In subsequent work it was found that internal browning was associated with prestorage stress and was not a problem with initially sound fruit (Swarts 1991). Queen pineapples stored at 2 or 4°C developed a white, watery pulp while fruit stored at higher temperatures developed internal browning (van Lelyveld *et al.* 1991). Storage at 3 and 8°C for longer than 2 weeks resulted in the crown and shell appearance being unacceptable (Paull and Rohrbach 1985). At a room temperature of 20°C and 60% r.h. they could be kept for only about 3 days (Mercantilia 1989). Refrigerated storage recommendations are as follows:

- 10°C and 90% r.h. for 2–4 weeks for green fruit (Anon 1967)
- 4.5–7°C and 90% r.h. for 2–4 weeks for ripe fruit (Anon 1967)
- 8.6°C in South Africa (Hulme 1971)
- 8°C Smooth Cayenne could be stored for 1 week without chilling injury (Dull 1971)
- 7°C with a maximum storage period of 4 weeks (Akamine *et al.* 1975)
- 7°C for at least 7 days (Smith 1983)

- 8.3–10°C and 85–90% r.h. for 4–6 weeks with 4% loss for all green fruit (Pantastico 1975)
- 14.4–6.7°C and 85–90% r.h. for 1–2 weeks for 25% yellow fruit (Pantastico 1975)
- 10°C and 90–95% r.h. for 2–3 weeks for unripe fruit (Mercantilia 1989)
- 7–8°C and 90–95% r.h. for 5–7 days for ripe fruit (Mercantilia 1989)
- 10–13°C and 90% r.h. for 3–4 weeks for mature green fruit (Snowdon 1990)
- 7–10°C and 90% r.h. for 3–4 weeks for turning fruit (Snowdon 1990)
- 7°C and 90% r.h. for 2–4 weeks for ripe fruit (Snowdon 1990)
- 10°C and 85–90% r.h. for 14–36 days (SeaLand 1991)
- 8 or 12°C for <3 weeks (Haruenkit and Thompson 1993).

Chilling injury

Mature green pineapples are very susceptible to chilling injury when stored at temperatures below 10°C (Dull 1971). Rohrbach and Paull (1982) reported that immature fruit should not be shipped because they were prone to chilling injury. Chilling injury occurred in Smooth Cayenne below 7°C (Akamine *et al.* 1975; Bartholomew and Paull 1986). Dull (1971) reported that at 8°C Smooth Cayenne could be stored for 1 week without chilling injury. The characteristics of chilling injury in Smooth Cayenne as described by Dull (1971) are:

- failure of the green shell colour to turn to yellow
- the yellow shell fruit turning brown or dull colour
- shrivelling of the shell
- drying and wilting and discoloration of crown tip.

At low temperature the fruit's skin darkens in severe cases of injury (Paull 1992). The breaking up of internal tissue gives a watery appearance. The watery appearance is the advanced stage of the watery spots appearing at the base of fruitlets. Dull (1971) found that the symptoms developed when fruit was stored at temperatures within the range 18–30°C after storage below 7°C. The crown is more susceptible to low-temperature injury than the fruit itself (Paull and Rohrbach 1985) and for European markets a well developed crown is required. Crown leaves that turn brown at the edge or become limp detract from the appearance of the fruit and usually reduce its market value (King 1972). The effect of low-temperature storage on shell appearance and the crown was said to be due to a loss of water. The crown is the site of most of the fruit's initial weight loss, having stomata that appear to be open with no diurnal cycling (Paull 1992).

Controlled atmosphere storage

Storage recommendations are:

- 7.2°C with 2% O_2 for Smooth Cayenne (Akamine and Goo 1971)
- 10–15°C with 5% O_2 and 10% CO_2 at (Kader *et al.* 1985)
- 8°C with either 3% O_2 and 5% CO_2, or 3% O_2 and 0% CO_2 for Smooth Cayenne, but it did not suppress the internal browning symptoms (Paull and Rohrbach 1985)
- 22°C with 3% O_2 in the first week of storage followed by 8°C effectively reduced internal browning symptoms (Paull and Rohrbach 1985)
- 10–15°C with 10% CO_2 and 5% O_2 had a fair effect, but was not being used commercially (Kader 1985)
- 8–13°C with 5–10% CO_2 and 2–5% O_2 (Kader 1989, 1992)
- 10°C with 10% CO_2 with 5% O_2 (SeaLand 1991)
- 7–13°C and 85–90% r.h. with 2–5% O_2 and 5–10% CO_2 for 2–4 weeks (Burden 1997).

Hypobaric storage

Staby (1976) stated that storage of pineapples under hypobaric condition can extend the storage life for up to 30–40 days.

Transport

For transportation by sea freight a storage temperature of 7–8°C for ripe and 10°C for unripe fruit has been recommended (Anon 1989).

Pink-spored grisette, rose-gilled grisette

Volvariella speciosa.

Botany

It is a gill fungus with a cap diameter of up to 10 cm and a height of 10 cm. The cap surface is very sticky and pale grey–green to pale brown, darkening towards the centre with white flesh. The gills are dull orange or pink and closely packed together with pink spores. The stalk is whitish and tapers from the base, which is slightly bulbous and enclosed in a delicate cup-like volva that disintegrates easily when harvested. It is found growing in compost heaps, rotting straw bales and heavily manured pastures.

Harvesting

They should be picked while they are still young because they acquire an unpleasant smell as they develop.

Storage

No information could be found on its postharvest life, but from the literature it would appear that all types of edible mushrooms have similar storage requirements and the recommendations for the cultivated mushroom (*Agaricus bispsorus*) could be followed.

Pitahaya, dragon fruit

Cerus tringularis, Cactaceae.

Botany

The fruits are a deep yellow and up to 12 cm long, and are made up of many tightly packed carpels each terminated in a protuberance with a small central hole and a green tip. There are two distinct varieties, one with a yellow skin and the other with a fuchsia pink skin. The flesh of both is translucent white, dotted with a mass of small, black, edible seeds and very beautiful when cut in two. It is sweet and refreshing with the seeds giving it a kind of crunchiness. In one publication (Le *et al.* 1998) dragon fruit is referred to as *Hylocereus undatus*.

Harvesting

They should be harvested when the skin is fully coloured and they are ready to eat when they yield to gentle pressure from the thumb.

Storage

It is commonly seen in markets in Latin America and Asia, and occasionally in Europe, but no information on its postharvest life could be found.

Plantains

Musa AAB, Musaceae.

Botany

It is a climacteric fruit, which is a parthenocarpic berry. They are similar to bananas but have a higher starch content. Hulme (1971) gave this as about 30% for green fruit reducing to 5–10% when they are fully ripe. At 20°C Marriott *et al.* (1983) showed that starch degradation was slower than in Cavendish bananas and was not complete in over-ripe fruits. The total sugar content continued to increase after full maturity and was still increasing in over-ripe fruits. Sucrose comprised about 70% of total sugars when plantains became fully yellow but this proportion fell to about 50% when the fruits became over-ripe. Ripe plantains, fully yellow but showing no blackening, contained 3–12% starch and 18–28% total sugars on a fresh weight basis. Like most other fruits they are acid with a pulp pH below 4.5 (Von Loesecke 1949). They can contain high levels of phenolic compounds, especially in the peel (Von Loesecke 1949).

Harvesting

Maturity is usually assessed by the shape or 'angularity' of the fingers where the fruit in cross-section becomes progressively more rounded and less angular as it matures. The most common maturity is full or full three-quarters, where full is almost completely rounded in cross-section; see under bananas for an explanation. Increasing fruit maturity at harvest significantly increased the ripening rate of harvested fruits (Ferris *et al.* 1993). The optimum harvest time can vary with cultivar and the best harvesting stages in weeks before ripeness of the first finger on the bunch were given by Tchango *et al.* (1999) as:

- 1 week for French Clair with a flowering to harvesting interval of 79 days
- 1–2 weeks for Big Ebanga with a flowering to harvesting interval of 71–78 days
- 2–3 weeks for Batard with a flowering to harvesting interval of 68–75 days.

Handling

Damage can reduce the green life of fruits. The effects of different types of bruise on the ripening of plantains showed that the greatest effect was caused by abrasion. The ripening rate of French, True Horn and False Horn during storage in ambient conditions averaging 28°C and 82% r.h. was increased when fruit had been damaged by abrasion, more so in less mature fruits, and also by impact, but only in immature fruits (Ferris *et al.* 1993). For some local markets, damage and blackening of the peel do not adversely affect their marketability. A study carried out in Ghana showed that even fruit that had been crushed to a pulp were readily saleable and often had specialized markets (Ferris 1991).

Packaging

In experiments, it was shown that plantains stored in cartons without a packing material lost weight rapidly during ripening and became blackened and shrivelled. Those packed in dry coir dust lost less weight, and those in moist coir dust gained weight; no shrivelling was seen following the latter two treatments and there was less skin blackening (see Table 5.3). The fruits in coir dust remained green for long periods before ripening rapidly (Thompson *et al.* 1974b). However, they took up moisture and would eventually split if stored for too long.

Coating

Coating fruits in Semperfresh, a commercially available material based on sucrose esters of fatty acids and sodium carboxymethylcellulose, which had been shown to reduce the gaseous exchange between fruits and the outside air, did not impede initiation of ripening by exposure to ethylene. The major changes associated with ripening were generally slower for coated fruits than for fruits ripened without coating. There was some indication, especially where a thick layer of Semperfresh was applied, that the rate of degreening was slowed more than that of other ripening processes. The effects on fruit weight loss were generally related to the storage humidity, with the coating having a large inhibitory effect on weight loss in fruits ripened at lower humidity and little or no effect on weight loss in fruits ripened at higher humidity. Taste panels were unable to detect organoleptic differences between fruits ripened with and without coating (Al-Zaemey *et al.* 1989). Chukwu *et al.* (1995) found that the combination of vacuum infiltration with 1% calcium chloride for 10 minutes followed by immersion in a 1.5% solution of Semperfresh extended the green life of Agbagba and the hybrid TMP × 2796–5 without adverse effects. Demerutis *et al.* (2000) treated fruits of 11, 12 and 13 weeks of age with experimental waxes (from FMC Company) and commercial ethylene scrubbers. Laboratory and shipping trials from Venezuela to Europe indicated that SF 965 wax at 25% gave good results, but only when pathogens on the fruits were fully controlled. As this is difficult to attain, the use of ethylene scrubbers with bagged fruits is recommended.

Storage

Refrigerated storage recommendations are as follows:

- 10°C and 85–90% r.h. for green fruit for 5 weeks with 6% loss (Pantastico 1975)
- 7.2–10°C and 85–90% r.h. for ripe fruit for 10 days (Pantastico 1975)
- 10°C and 85–90% r.h. for 5 weeks (Snowdon 1990)
- 11–15.5°C and 85–90% r.h. for coloured fruit for 1–3 weeks (Snowdon 1990)
- 14.4°C and 85–95% r.h. for 10–35 days (SeaLand 1991)
- 12–14°C and 85–95% r.h. for 15 days for green mature fruits (Tchango *et al.* 1999).

Humidity

Exposing fruit to sufficient stress can initiate ripening. Low humidity in the storage atmosphere will result in stress and can initiate ripening and thus result in a reduced storage life (Thompson *et al.* 1972, 1974a). Maintaining high humidity around the fruit can help to keep it in the preclimacteric stage so that where fruit were stored in moist coir dust or individual fingers were sealed in polyethylene film they could remain green and preclimacteric for over 20 days in Jamaican ambient conditions (Thompson *et al.* 1972a, 1974). Such fruit ripened quickly and normally when removed. Fruits stored at 20°C and 96–100% r.h. ripened about 15 days later than those stored at 20°C and 55–65% r.h. (Ferris *et al.* 1993). Fruits kept in desiccators with air passed over them at 0% r.h. became visibly riper than those with air at 100% r.h. passing over them (Thompson *et al.* 1974).

Controlled atmosphere storage

The only controlled atmosphere storage recommendation was 14.4°C and 85–95% r.h. with 2–5% CO_2 and 2–5% O_2 (SeaLand 1991).

Modified atmosphere packaging

Plantains stored in modified atmospheres had a considerably longer postharvest life than those stored unwrapped. The degree of perforation in the bag has a large six fruits in a bag ripened in 14.6 days at 26–34°C compared with 18.5 days when fruits were packed individually (Thompson *et al.* 1972a).

Ripening

The pulp:peel weight ratio increased from 1.23 in green plantains to 1.60 when fully ripe. Starch contents for green and ripe fruits were 27.10 and 8.21%, respectively, total sugars contents were 1.13 and 19.91%, respectively, soluble solids 3.67 and 22.47%, respectively, pH 5.41 and 4.35, respectively and moisture content 70.63 and 71.87%, respectively (Fernandez *et al.* 1996). The pigments in the peel of plantains are chlorophylls and carotenoids. The change in colour of ripening fruits is associated with the breakdown of chlorophylls with carotenoid levels remaining relatively constant (Seymour 1985; Montenegro 1988). With plantains it was shown that complete chlorophyll destruction could occur even at 35°C (Seymour 1985; Seymour *et al.* 1987). As they ripen, the ratio of the mass of the pulp to the mass of the peel increases and the peel becomes progressively easier to detach from the pulp. The onset of the starch to sugar conversion has been shown to be influenced by harvest maturity, with more mature fruits responding earlier. In bananas the breakdown of starch is usually completed during ripening, but in plantains this breakdown is not complete even when the fruit is fully yellow and soft (George 1981). Tannins are perhaps the most important phenolic from the point of view of fruit utilization because they can give fruit an astringent taste. As fruit ripens their astringency becomes lower which seems to be associated with a change in structure of the tannins, rather than a reduction in their levels, in that they form polymers (Von Loesecke 1949; Palmer 1971).

At 30°C and 88% r.h., fruits of the cultivar Rose d'Ekona reached colour stage 7 (fully yellow with black spots) after only 6 days, compared with 10–12 days for French Clair, Big Ebanga, 2 Hands Planty (Horn plantain) and Batard. At stage 7, the weight loss of French Clair was 22.8%, compared with 12.6–16.8% for the other cultivars (Ngalani *et al.* 1998). 22–24°C was used by Fernandez *et al.* (1996) in Cuba. Optimum ripening condition were given as 22.2°C and 85–100% r.h. with 1000 μl litre^{-1} ethylene by Sanchez Nieva *et al.* (1970). In these conditions the fruit were fully yellow with soft pulp in 4–5 days.

Plums

Prunus domestica, Rosaceae.

Botany

The fruit is a drupe and is classified as climacteric. It will freeze at about –2.4 to 2.0°C (Wright 1942).

Harvesting

Harvest maturity is assessed by colour change from green to red or yellow, feeling soft to the touch and the development of an aroma. Destructive methods can be used on samples. These include soluble solids content, which should be at least 17–18% for Prune (Phillips and Armstrong 1967), 13.4% for Beauty (Hulme 1971) or more than 16% for Queen Ann (Hulme 1971). Pressure tests can be used (Hulme 1971), but it is usually easier to test texture by feel.

Precooling

Air cooling in a conventional cold store was the most suitable, forced-air cooling with high humidity was also suitable and hydrocooling were less suitable and vacuum cooling was unsuitable (MAFF undated).

Damage

Magnetic resonance imaging has been used to obtain images of bruises (Chen *et al.* 1989), which could be used for grading out damaged fruit in the packhouse.

Storage

At room temperature of 20°C and 60% r.h. they could be kept for only 2 days (Mercantilia 1989). General refrigerated storage recommendations are as follows:

- –0.5 to 1°C and 85–90% r.h. for 2–8 weeks (Anon 1967)
- 0°C and 90–95% r.h. for 20 days (Mercantilia 1989)
- 2°C for 15 days (Mercantilia 1989)
- 8°C for 8 days (Mercantilia 1989)
- –0.5°C with 90–95% r.h. for 14–28 days (SeaLand 1991).

Storage recommendations for specific cultivars are as follows.

Alu Bokharo

- 0–1.7°C and 85–90% r.h. for 2 weeks with 5.2–9.6% weight loss (Pantastico 1975).

Castleman

- –0.6 to 1°C and 90–95% r.h. for 1 month (Lutz and Hardenburg 1968)
- –0.5 to 0°C and 90–95% r.h. for 4–5 weeks for well matured fruit (Hardenburg *et al.* 1990).

Climax

- 0.5–0°C and 90–95% r.h. for 4–5 weeks for well matured fruit (Hardenburg *et al.* 1990).

Eldorado

- –0.6 to 1°C and 90–95% r.h. for 1 month (Lutz and Hardenburg 1968)
- –0.5 to 0°C and 90–95% r.h. for 4–5 weeks for well-matured fruit (Hardenburg *et al.* 1990).

Gaviota

- 0–1.7°C and 85–90% r.h. 3 weeks with 5.2–7.8% loss (Pantastico 1975)
- 0°C and 90–95% r.h. for 10 days followed by 7°C and 90–95% for 7 days (Snowdon 1990).

Giant Prune

- 0°C and 90–95% r.h. for 10 days followed by 10°C and 90–95% for 7 days (Snowdon 1990).

Hale

- 0–1.7°C and 85–90% r.h. for 4 weeks with 5.2–12.9% loss (Pantastico 1975).

Japanese types

- 4.4°C for early fruit (Phillips and Armstrong 1967).

Kelsey

- 0°C and 90–95% r.h. for 10 days followed by 10°C and 90–95% for 7 days in South Africa (Snowdon 1990).

Nubiana

- –0.6 to 1°C and 90–95% r.h. for 1 month (Lutz and Hardenburg 1968).

President

- 0°C for 3–6 weeks then 6–8°C for 6 weeks in Australia (Fidler 1963).

Queen Ann

- –0.6 to 1°C and 90–95% r.h. for 1 month (Lutz and Hardenburg 1968).

Rubio

- 0–1.7°C and 85–90% r.h. 3 weeks with 5.2–7.8% loss (Pantastico 1975).

Santa Rosa (Late)

- –0.6 to 1°C and 90–95% r.h. for 1 month (Lutz and Hardenburg 1968)
- –0.5 to 0°C and 90–95% r.h. for 4–5 weeks for well matured fruit (Hardenburg *et al.* 1990)
- 0°C and 90–95% r.h. for 3–4 weeks in California (Snowdon 1990)
- 0°C and 90–95% r.h. for 10 days followed by 7°C and 90–95% for 7 days in South Africa (Snowdon 1990).

Satsuma

- 0°C for 3–6 weeks then 6–8°C for 6 weeks in Australia (Fidler 1963).

Shiro

- 0–1.7°C and 85–90% r.h. for 4 weeks with 5.2–12.9% loss (Pantastico 1975).

Victoria

- 0–1°C and 90–95% r.h. for 3 weeks in the UK (Snowdon 1990).

Wickson

- 0°C for 3–6 weeks then 6–8°C for 6 weeks in Australia (Fidler 1963)
- –0.5 to 0°C and 90–95% r.h. for 4–5 weeks for well matured fruit (Hardenburg *et al.* 1990).

Chilling injury

Many cultivars are subject to chilling injury, but this is often not detected until they are removed from storage for marketing when they decay, shrivel and the flesh becomes brown (Lutz and Hardenburg 1968). It was also reported that the odour and colour may be abnormal (Anon 1986).

Intermittent warming

Fidler (1963) found that at –0.5°C Victoria could be stored for 3 weeks. However, if the fruit were warmed to 18°C for 2 days after 15 days in storage and then recooled, the storage life could be extended to 5 or 6 weeks in total. Hardenburg *et al.* (1990)

also showed that storage at –0.5°C and 1% O_2 with intermittent warming could increase the storage life of plums. Fidler (1963) stated that Japanese cultivars are subject to chilling injury and for export from South Africa to Europe they could be transported at alternating temperatures of –5°C and 7–10°C although he did not give the period at each temperature.

Controlled atmosphere storage

Storage recommendations are:

- In storage at –0.5°C chilling injury to Monarch was about 25% less in 2.5% CO_2 with 5% O_2 than in air (Anon 1968)
- 0–5°C with 0–5% CO_2 and 1–2% O_2 was reported to have a good effect on storage, but was not being used commercially (Kader 1985)
- At 1°C with 12% CO_2 and 2% O_2 the cultivar Buhler Fruhzwetsche had a good appearance, taste and firmness after 4 weeks storage, with no CO_2 injury detected during storage at concentrations below 16% (Steif 1989)
- –0.5°C with 0–5% CO_2 and 2% O_2 (SeaLand 1991).

Pomegranates

Punica granatum, Punicaceae.

Botany

Pomegranate fruit is a berry and is classified as a non-climacteric fruit with no detectable levels of ethylene produced during storage (Nanda *et al.* 2001). It has a tough, leathery skin, which is a brownish yellow colour with a red blush or it can be fully red. It has a persistent calyx. Inside are numerous seeds, each embedded in the sweet, pink juicy pulp called the aril, which is the edible portion and is easily dislodged. They were traditionally dislodged and eaten with a pin in Britain. It is a native of Iran and was known in the hanging gardens of Babylon and ancient Egypt. It has since spread throughout the world.

Harvesting

This should be carried out when they are mature, which is difficult to judge, but can be related to skin colour changes. If too mature they have a tendency to split. In Iran the optimum harvesting time was when the soluble solids content reached 17.5% (Sherafatian 1994). Wardlaw (1937) recommended clipping the fruit stalks, not pulling them.

Physiological disorders

Internal breakdown can occur and results in the pulp having a light streaky appearance and a 'flat' taste (Pantastico 1975). The cause of this disorder is not known.

Storage

At a room temperature of 20°C and 60% r.h. it was reported in Mercantilia (1989) that they could be kept for 1–2 weeks. The main losses in storage at 5°C were due to decay caused by *Penicillium* spp. (Artes *et al.* 2000). Refrigerated storage recommendations are as follows:

- 1.1–1.7°C and 80% r.h. (Wardlaw 1937)
- 0°C or 4.4°C for 7 months in India (Lutz and Hardenburg 1968)
- 0–1.7°C and 85–90% r.h. for 11 weeks for the cultivar Khandari (Pantastico 1975)
- 0°C and 90% r.h. for 2 months (Mercantilia 1989)
- 1, 5 and 10°C and 90–95% r.h. for 2 months for the cultivar Gok Bahce for 2 months with 4.9, 5.7 and 4.6% weight loss, respectively (Koksal 1989)
- 5–6°C and 90% r.h. for 3–4 months (Snowdon 1990)
- 0°C and 90% r.h. for 2–4 months (Hardenburg *et al.* 1990)
- 0°C and 90–95% r.h. for 1 month or 5°C and 90–95% r.h. for 2 months (Hardenburg *et al.* 1990)
- 3°C (Sherafatian 1994)
- 5°C and 90–95% r.h. for 28–56 days (SeaLand 1991)
- 1°C for the cultivars Malas Torsh, Saveh Hendes, Shirin Paeizeh and Sefid Torsh for 4.5 months (Askary and Shahedi 1994)
- 6°C and 85–90% r.h. for 5 months for the cultivar Hicaz (Kupper *et al.* 1995)
- 8 or 10°C and 85–90% r.h. for 50 days for the cultivar Hicaz (Kupper *et al.* 1995)
- 2°C and 95% r.h. (Artes *et al.* 2000).

Chilling injury

There was no decay during storage at 0°C but an increased risk of chilling injuries such as pitting and husk scald, while storage at 5°C reduced these injuries but fungal attacks were not inhibited. Intermittent warming alleviated chilling injury (pitting and husk scald) without any incidence of decay (Artes *et al.* 1998a); see below. In later work on the cultivar Mollar de Elche, Artes *et al.* (2000) found that 2°C was the optimum storage temperature.

Intermittent warming

Intermittent warming of 1 day at 20°C every 6 days during storage of the cultivar Mollar at 0°C and 95% r.h. for 80 days was the best treatment for maintaining the red skin colour. However, the red colour of the pulp was maintained better with intermittent warming at 5°C and 95% r.h. for 80 days than at 0°C (Artes *et al.* 1998). After shelf-life for 3 days at 15°C and 75% r.h., intermittent warming for 1 day every 6 days during storage at 2°C and 95% r.h. for 90 days was the only treatment that resulted in fruits with a flavour similar to that at harvest (Artes *et al.* 2000).

Controlled atmosphere storage

Controlled atmosphere storage had a slight to no effect (SeaLand 1991). Kupper *et al.* (1995) recommended 6°C and 85–90% r.h. with 6% CO_2 and 3% O_2 for 6 months for Hicaz.

Modified atmosphere packaging and coating

Individual shrink film-wrapped fruits using either BDF-2001 polyolefin film or D-955 polyolefin film could be stored for 12, 9 and 4 weeks at 8, 15 and 25°C, respectively, compared with 8, 6 and 2 weeks with Semperfresh coating and 7, 5 and 1 week for non-wrapped fruits under similar storage conditions (Nanda *et al.* 2001). Changes in acidity, sugars and ascorbic acid of the shrink-wrapped fruits were lower than those of non-wrapped fruits during 12 weeks of storage at 8°C. Fruit that were shrink wrapped also had a lower respiration rate than non wrapped fruit.

Potatoes

Solanum tuberosum, Solanaceae.

Botany

The edible portion is the swollen stem tubers that are borne on the ends of stolons below the soil level. Tubers will freeze at about −1.8 to 1.7°C (Wright 1942) or 1.44–0.83°C (Weichmann 1987). The respiration rate of tubers is low in main crop potatoes since they are natural storage organs and are dormant when harvested at full maturity (Table 12.51). Generally, low levels of O_2 and high levels of CO_2 reduce respiration rate (Thompson 1998), but in experiments carried out by Robinson *et al.* (1975) storage of main crop potatoes in 3% O_2 had no effect on their respiration rate, but it had a slight effect on immature tubers. However, Pal and Buescher (1993) described experiments in which Russet potatoes were held in air or in 10, 20 or 30% CO_2. Short-term exposure to 20 or 30% CO_2 resulted in an increased respiration rate. At 4°C Hartmans *et al.* (1990) showed that tuber respiration rate was lower in 3 or 1% O_2 compared with those in air and in 35% O_2.

Harvest maturity

If potatoes are to be stored, then the optimum harvest time is after the leaves and stems have died down. If they are harvested earlier, the skins are less resistant to harvesting and handling damage and are more prone to storage diseases. In certain cases where the leaves do not senesce naturally, the whole of the tops of the plant (haulm) may be cut off or removed chemically to produce tubers with firmer and stronger skins before they are harvested.

For processing, the maturity of tubers at the time of harvest determines the storage behaviour through the initial amount of the enzyme, or enzyme system, responsible for cold-induced sweetening and therefore the colour of crisps or chips (Hertog *et al.* 1997).

Table 12.51 Effects of temperature and reduced O_2 level on the respiration rate (CO_2 production in mg kg^{-1} h^{-1}) potatoes (Robinson *et al.* 1975)

	In air					In 3% O_2		
	0°C	5°C	10°C	15°C	20°C	0°C	10°C	20°C
Main crop (King Edward)	6	3	4	5	6	5	3	4
New (immature)	10	15	20	30	40	10	18	30

Harvesting methods

The harvesting method and equipment used can affect the amount of damage caused to the tubers, which in turn can affect their postharvest life (see the section on damage in Chapter 3). Mechanized harvesters remove the crop from the ground either by undercutting it or by inverting the soil ridges in which the crop is grown. Mouldboard ploughs, subsoilers and other tillage tools used in root crop harvesting may require a lot of power to remove them from the soil. Using vibrating digger blades can reduce the power required, which varies with the amplitude of the vibration (Kang *et al.* 1993). X-rays were used in prototype potato harvesters to differentiate between the tubers and extraneous matter such as clods of earth and stones (Whitney 1993). The equipment was mounted behind the chain lifter, but was found to be impractical because the height of fall of the tubers was too great.

With some machines, the crop may be left on the soil surface after harvesting; in others they may be lifted and loaded into a trailer for direct transport to the store or packhouse. Some harvesters have the facility to pack the crop directly in the field into wholesale or retail packs. Normally only early potatoes are lifted and bagged in the field (John Love 1994, quoted in Thompson 1996), while main crop potatoes are taken to the store or packhouse for grading, sometimes cleaning and packing.

Curing

The first work to be published on curing was by Priestley and Woffenden (1923) and subsequently by Artschwager (1927), who described the process in detail. Curing is carried out as soon as possible after harvest to heal any damaged areas on the tuber surface to reduce the chance of microorganism infection and reduce weight loss and desiccation. It involves suberin being deposited in the parenchymatous cells just below the damaged area of the tuber. Suberin refers to a group of fatty acids (a polymer containing phenolics and long chain decarboxylic, hydroxy and perhaps epoxy fatty acids) that provide initial protection to the tuber by forming a highly lipophilic coating within and on the surface of tissues. Burton (1989) suggested that its early response is probably due to enzymes induced or stimulated by an endogenous growth regulator. There are some indications that the fatty acids may also act as phytoalexins and the peroxides and epoxides may have toxic effects on infectious agents (Galliard 1975). Subsequently, below the suberized cells, a meristematic layer of cells is formed which is the periderm, but is also called the cork cambium. This produces new cells, which seal off the damaged area, and contain lignin.

Both these processes are temperature and humidity sensitive (Wigginton 1974; Burton 1989). The time required for suberization to begin was shown to be:

- 1 day at 21°C and above
- 2 days at 15°C
- 3 days at 10°C
- 5–8 days at 5°C
- >8 days at 2.5°C.

The time for formation of periderm after suberization depends on the cultivar but in general it takes about 1 day at 15°C and over, 3 days at 10°C and 5 days at 7°C. No periderm formation occurred within 10 days at 5°C (1974; Burton 1989). It has been stated that the curing process ceases at 4°C and therefore expressed the curing requirement in day degrees above 5°C (Anon 1985). That means that at 6°C theoretically it would require 150 days to complete suberization and periderm formation, that is, 150 divided by 1, or at 21°C it would be 150 divided by (21 – 5) = about 9.5 days. In practice this could be achieved as follows:

Curing temperature (°C)	Approximate curing time (days)
10	30
11	25
12	21.5
13	18.75
14	16.5
15	15
16	13.5
17	12.5
18	11.5
19	10.75
20	10
21	9.5

Another important factor that controls curing is the environmental humidity around the potato tuber. At 12°C and a vapour pressure deficit (VPD) of 3.8 mmHg, that is, about 60% r.h., no suberin or periderm was formed within 6 days. At the same temperature and 2.7 mmHg VPD (80% r.h.), 1–1.5 layers of suberized cells had formed, and at 0.6 mmHg VPD (94% r.h.), there were two layers of well suberized cells.

At the two higher VPDs no periderm was formed, but at the lower VPD a periderm was formed within 9 days. In another experiment which compared 2.2 mmHg VPD (68% r.h.) with 0.4 mmHg VPD at 6.5°C (94% r.h.) during 53 days of storage, suberization had occurred at both VPDs but periderm did not occur at the higher but was well developed at the lower VPD. Most chemical sprout suppressants inhibit periderm formation. High levels of CO_2 can inhibit curing (Butchbaker *et al.* 1967). Wounded and sound potatoes of the cultivar Superior were stored at 13, 15 or 18°C with 90% r.h by Kim *et al.* (1993). At 18°C, a good periderm layer was formed in 6 days. At 15 and 13°C periderm formation was completed after 10 and 12 days, respectively.

Damage

Types of damage are discussed in Chapter 3 and in potatoes susceptibility may be related to the strength of the skin. Muir and Bowen (1994) showed that no single skin characteristic was related to skin strength, except in the case of Record, where the strength of the skin was always related to its thickness. Only 25% of tubers showed signs of scuffing damage after pulling compared with 33.6% after chemical desiccation with diquat and 37.6% after flailing. Bowen *et al.* (1996) developed a standardized instrumental test of skin set, called a scuff meter. Using this meter 49 days after haulm destruction, skin adhesion strengths were similar in all six cultivars except cultivar Record, which had much greater skin adhesion strength. Differences in skin adhesion strength were poorly related to skin morphological characteristics such as skin thickness, cell size and suberin content. The method of haulm destruction affected skin adhesion strength, with skins adhering more rapidly and more strongly to the tuber following haulm pulling compared with either desiccation with diquat or mechanical flailing. The method of haulm destruction did not influence skin morphological characteristics.

Votoupal and Slavik (1986) quantified the level of storage losses of tubers due to mechanical damage incurred during harvesting. In severely damaged tubers losses due mainly to rotting were 40% for the very early cultivar Ostara, 61% for the early cultivar Karin, 50% for the semi-early cultivar Radka and 75% for the semi-late cultivar Nora. Changes in respiratory losses resulting from mechanical damage were small.

Physiological disorders

There are many storage diseases and disorders. These are dealt with in detail in Snowdon (1991).

Precooling

Potatoes are rarely, if ever, precooled, but forced-air cooling with high humidity was the most acceptable for new potatoes, air cooling in a conventional cold store was suitable and hydrocooling and vacuum cooling were said to be less suitable (MAFF undated), but hydrocooling could lead to bacterial rot and vacuum cooling would be so slow and expensive as to be irrelevant.

Sprouting

This is dealt with in Chapter 6. The limiting factor in storage is usually when the dormancy period ends and the tuber regrows. The length of the dormancy period and sprouting behaviour depend on the cultivar, the previous history of the tubers and the storage conditions. Even after the dormancy period has ended the tuber might not sprout if the conditions, especially temperature, are not conducive. Sprouting is normally suppressed at 5°C and below, but the natural dormancy period varies between cultivars. Sprouting can be suppressed to extend the storage life by applying growth-regulating chemicals to the crop (Burton 1989). The most commonly used is CIPC, which is commonly applied as a thermal fog into the stored crop. IPC (propham) is also used either by itself or in a formulation with CIPC. Maleic hydrazide is also used by spraying the plants before harvest so that a residue is translocated to the tubers. Other methods of sprout suppression include:

- irradiation of at least 35 Gy (Burton and Hannan 1957; Ranganna *et al.* 1997)
- controlled atmosphere storage 9.4% CO_2 with 3.6% O_2 at 5°C for 25 weeks (Khanbari and Thompson 1997)
- hot water treatment at 57.5°C for 20–30 minutes (Ranganna *et al.* 1998)

Storage

At a room temperature of 20°C and 60% r.h. it was reported in Mercantilia (1989) that tubers could be kept for about 3 weeks. General refrigerated storage recommendations are as follows:

- 4°C and 95–98% r.h. for 240 days (Mercantilia 1989)
- 8°C for 120 days (Mercantilia 1989).

Refrigerated storage recommendations for different types and uses are as follows:
Immature or early cultivars

- 3–4°C and 85–90% r.h. for a few weeks (Anon 1967)
- 10°C and 85–90% r.h. for a few days (Phillips and Armstrong 1967)
- 10°C and 90% r.h. for 3 months (Lutz and Hardenburg 1968)
- 4–5°C and 90–95% r.h. for 3–8 weeks (Snowdon 1991).

Late or ware cultivars

- 3–4°C and 85–90% r.h. for a few weeks (Anon 1967)
- 10°C and 85–90% r.h. for a few days (Phillips and Armstrong 1967)
- 10°C and 90% r.h. for 3 months (Lutz and Hardenburg 1968).

For processing

- 15.6–21.1°C and 85–90% r.h. for a few days for early cultivars (Phillips and Armstrong 1967)
- 10°C and 90–95% r.h. for 56–175 days (for processing) (SeaLand 1991)
- 7–8°C and 90–95% r.h. for 6–9 months (for processing into chips) (Snowdon 1991)
- 8–9°C and 90–95% r.h. for 6–9 months (for processing into crisps) (Snowdon 1991).

For seed

- 2.2–3.3°C and 85–90% r.h. (Phillips and Armstrong 1967)
- 2–7°C and 85–90% r.h. for 5–8 months (Anon 1967)
- 3.3–4.4°C and 85% r.h. for 34 weeks with 4.9% loss (Pantastico 1975)
- 4.4°C and 90–95% r.h. for 84–173 days (SeaLand 1991).

Gas levels in normal stores

CO_2 accumulates naturally in stores through tuber respiration and is reduced either by air leakage or ventilation. Mortimer and Bishop (1998) found CO_2 levels between 0.6 and 1.6% in commercial stores in the UK. When these levels were reproduced in simulated storage at 4, 9 or 25°C they found that there was a small increase in tuber sprouting. Mazza and Siemens (1990) studied the atmospheric CO_2 content in 18 commercial potato storage bins each containing 500–2500 tonnes of the cultivars Russet Burbank, Norchip, Kennebec and Monona in the USA. The CO_2 concentration in the storage bins tested ranged from 0.06–3.2%. Storage bins filled with Norchip had lower CO_2 levels than those filled with Russet Burbank, probably owing to differences in maturity and bruising at harvest. Values higher than 1% were found in about 50% of the storage bins in 1986–87 but in none in 1987–88. The highest CO_2 levels occurred primarily during the suberization period and during application of the sprout inhibitor CIPC.

Controlled atmosphere storage

The amounts of O_2 and CO_2 in the atmosphere of a potato store can affect the sprouting of the tubers, rotting, physiological disorders, respiration rate, sugar content and processing quality (Table 12.52). Fellows (1988) recommended 10% CO_2 and a minimum of 10% O_2.

Controlled atmosphere storage on sprouting

As early as 1918 (Anon 1919) work being carried out at the Low Temperature Research Station in Cambridge described 'Interesting results in stopping sprouting of potatoes have been obtained … proving the importance of the composition of the air.' Kidd was also working on the effects of CO_2 and O_2 on sprouting of potatoes (Kidd 1919). Kubo *et al.* (1989a) claimed that exposure of tubers to high CO_2 had little or no effect on sprouting. In contrast, Hartmans *et al.* (1990) found that sprouting was inhibited in 3% CO_2 or higher, but at 7% CO_2 partial deterioration occurred and there was an increase in sugar content. Khanbari and Thompson (1997) found that there was almost complete sprout inhibition, low weight loss and maintenance of a healthy skin for all cultivars stored in 9.4% CO_2 with 3.6% O_2 at 5°C for 25 weeks. When tubers from this treatment were stored for a further 20 weeks in air at 5°C, the skin remained healthy and they did not sprout while the tubers which had been previously stored in air or other controlled atmosphere combinations sprouted quickly. This residual effect of controlled atmosphere storage could have major implications in that it presents an opportunity to replace chemical treatments in controlling sprouting in stored potatoes. Schouten (1992) found that some inhibition of sprout growth occurred in 6% CO_2 and sprouting was

Table 12.52 Sugars (g per 100 g dry weight) in tubers of three potato cultivars stored for 25 weeks under different controlled atmospheres at 5 and 10°C and reconditioned for 2 weeks at 20°C (Khanbari and Thompson 1996)

	Gas combination		Storage temperature 5°C			Storage temperature 10°C		
Cultivar	CO_2 (%)	O_2 (%)	Sucrose	RS[a]	TS[b]	Sucrose	RS[c]	TS[d]
Record	9.4	3.6	0.76	0.22	0.97	0.91	0.49	1.40
	6.4	3.6	0.76	0.35	1.11	1.39	1.14	2.52
	3.6	3.6	0.62	0.53	1.16	1.60	0.75	2.35
	0.4	3.6	0.79	0.51	1.30	0.65	0.52	1.18
	0.5	21.0	0.32	0.73	1.05	1.00	0.63	1.63
Saturna	9.4	3.6	0.90	0.32	1.22	0.69	0.23	0.92
	6.4	3.6	0.33	0.61	0.99	0.64	0.24	0.88
	3.6	3.6	0.44	0.38	0.82	0.73	0.36	1.08
	0.4	3.6	0.29	0.22	0.51	0.80	0.12	0.92
	0.5	21.0	0.22	0.62	0.83	0.79	0.41	1.19
Hermes	9.4	3.7	0.37	0.48	0.85	0.26	0.22	0.48
	6.4	3.7	0.22	0.33	0.55	1.36	0.47	1.84
	3.6	3.7	0.49	0.74	1.23	0.88	0.27	1.15
	0.4	3.7	0.29	0.43	0.72	0.59	0.30	0.89
	0.5	21.0	0.70	0.93	1.63	0.62	0.51	1.13

[a] Reducing sugars.
[b] Total sugars.

strongly inhibited in 1% O_2 at 6°C. However, he found that patho-logical breakdown may develop at this O_2 level and concluded that controlled atmosphere storage at 6°C was not an alternative to chemical control of sprout growth for ware tubers. In the early phase of storage, Schouten (1994) showed that sprouting was stimulated in 3–6% CO_2 and inhibited in 9% CO_2. During the later phase all CO_2 concentrations inhibited sprouting, but reconditioning after storage stimulated sprouting (Schouten 1994).

Controlled atmosphere storage on processing quality

If potatoes are to be processed into crisps or chips, it is important that the levels of reducing sugars and amino acids are at a level to permit the Maillard reaction during frying. This gives them an attractive golden colour and their characteristic flavour. If levels of reducing sugars are too high the crisps or chips are too dark after frying and may be unacceptable to the processing industry (Schallenberger *et al.* 1959; Roe *et al.* 1990). Potatoes stored at low temperatures of around 4°C have higher levels of sugars than those stored at higher temperatures of 7–10°C (Khanbari and Thompson 1994), but the levels of CO_2 in the store can also influence their sugar levels (Table 12.53). With levels of 6% CO_2 and 2–15% O_2 in store during 178 days at 8 and 10°C, Reust *et al.* (1984) found that the fry colour was very dark.

Mazza and Siemans (1990), studying 1–3.2% CO_2, found that darkening of crisps occurred at the higher CO_2 levels in stores. At levels of CO_2 of 8–12% in stores the fry colour was darker than for tubers stored in air (Schmitz 1991). At O_2 levels of 2% reducing sugar levels were reduced or there was no accumulation at low temperature, but 5% O_2 was much less effective (Workman and Twomey 1969). The cultivar Bintje were stored at 6°C in atmospheres containing 0, 3, 6 or 9% CO_2 and 21, 18, 15 or 12% O_2. During the early phase, the unfavourable effect of high CO_2 concentration on chip colour increased to a maximum at 3 months. In the later phase, chip colour was also dark in tubers stored without CO_2 and CO_2 enrichment had less effect (Schouten 1994). Khanbari and Thompson (1994) stored Record tubers at 4°C under CO_2 at levels of 0.7–1.8% with 2.1–3.9% O_2 and found that this combination resulted in a significantly lighter crisp colour, low sprout growth and fewer rotted tubers compared with storage in air. Schouten (1994) described an experiment in which tubers of cultivar Bintje were stored at 6°C in atmospheres containing 0, 3, 6 or 9% CO_2 and 21, 18, 15 or 12% O_2 and chip colour was determined at intervals during early and late phases of storage. During the early phase, the unfavourable effect of high CO_2 concentration on chip

Table 12.53 Sugar content and fry colour (grey level) of tubers from three potato cultivars after being stored for 25 weeks in high CO_2 and another 20 weeks in air at 5°C and then reconditioned for 2 weeks at 20°C

		Sugar (g per 100 g dry weight)			
Tuber source	Cultivar	Fructose	Glucose	Sucrose	Grey level[a]
Processed directly after storage	Record	0.78	0.85	0.52	130
	Saturna	0.66	0.69	0.54	137
	Hermes	0.98	1.12	0.56	123
Processed after reconditioning	Record	0.63	0.56	0.90	133
	Saturna	0.38	0.34	0.49	152
	Hermes	1.03	1.13	1.13	124

[a] This is for the crisps processed from the potatoes above. Minimum acceptable grey level = 149 or above.

colour increased to a maximum at 3 months. In the later phase, chip colour deteriorated in tubers stored without CO_2 and CO_2 enrichment had relatively less effect. In Korea, tubers of cultivar Superior stored at a constant 15°C maintained adequate levels of reducing sugars acceptable for processing as chips (Choi *et al.* 1997). Storing tubers in anaerobic conditions of total nitrogen prevented the accumulation of sugars at low temperature but it had undesirable side effects on the tubers (Harkett 1971).

Reconditioning

Reconditioning of stored potato tubers involves exposing them to higher temperatures for a period after storage and before processing. Reconditioning was shown to be effective at:

- 20°C for 14 days (Khanbari and Thompson 1996)
- 25°C for 20 days (Jeong *et al.* 1996)
- 15°C for 14–28 days (Schouten 1994).

It has been shown to reduce their sugar content (Table 12.54) and improve their fry colour (Khanbari and Thompson 1996). Schouten (1994) showed that the treatment improved chip fry colour, but stimulated sprouting. Choi *et al.* (1997) also showed that reconditioning hastened tuber sprouting of the cultivars Atlantic and Superior grown in Korea when stored at 5°C for 90 days and reconditioned at 15°C for as long as 90 days.

Using a reducing sugar content of 0.25% as the limit of acceptability, reconditioning studies showed that storage at 10°C maintained contents below this level for up to 120 days of storage and at this level up to 150 days of storage. In contrast, Jeong *et al.* (1996) found that reconditioning after storage at 4°C could not maintain the reducing sugar content below 0.25% even at the early stage of storage.

Coatings

Use of Nature Seal 1020, a cellulose-based edible coating, as carrier of antioxidants, acidulants and preservatives prolonged the storage life of cut potato cultivar Russet by about 1 week when stored in overwrapped trays at 4°C. Similarly, the preservatives

Table 12.54 Sugar content of tubers from three potato cultivars after being stored for 25 weeks in 9.4% CO_2 and 3.7% O_2 then another 20 weeks in air, all at 5°C, with or without reconditioning for 2 weeks at 20°C after storage (Khanbari and Thompson 1996)

		Sugar (g per 100 g dry weight)				
Reconditioning	Cultivar	Fructose	Glucose	Sucrose	RS[a]	Grey level
No	Record	0.78	0.85	0.52	1.63	130
No	Saturna	0.66	0.69	0.54	1.35	137
No	Hermes	0.98	1.12	0.56	2.10	123
Yes	Record	0.63	0.56	0.90	1.19	133
Yes	Saturna	0.38	0.34	0.49	0.72	152
Yes	Hermes	1.03	1.13	1.13	2.16	124

[a] RS = reducing sugars.

sodium benzoate and potassium sorbate were more effective in controlling certain microbial populations when applied in Nature Seal than in aqueous solutions, but less effective for others. Adjustment of the coating pH to 2.5 gave optimal control of browning and microbial populations. Addition of soy protein to the original cellulose-based Nature Seal formulations reduced the coating permeability to O_2 and water vapour. Cellulose formulations with protein were effective in controlling weight loss, especially when the pH of the formulation was raised above that of the isoelectric point of the protein (Baldwin *et al.* 1996).

Modified atmosphere packaging

The cultivars Russet Burbank and Chieftain packaged in polyethylene bags were lowest in weight loss, ascorbic acid and nitrate nitrogen, and highest in discoloration, phenols and glycoalkaloids. Tubers packaged in mesh were lowest in discoloration and phenols, and highest in ascorbic acid. There were no significant differences in weight loss, glycoalkaloids or nitrate-N between potatoes packaged in mesh or paper bags (Gosselin and Mondy 1989).

Ethylene

Inaba *et al.* (1989) found that there was increased respiration in response to ethylene at 100 µl litre^{-1} for 24 hours at 20–35°C, but the effect was much reduced at lower temperatures.

Potato yams

Dioscorea bulbifera, Dioscoreaceae

Botany

The plant produces edible bulbils in the leaf axils above ground and tubers below ground. The tubers are hard, bitter and unpalatable (Purseglove 1972). The bulbils are eaten and can weigh up to 2 kg, but are more commonly around 500 g, and have a grey or brown colour with white or yellow mucilaginous flesh. Some types may require detoxification by soaking or boiling (Purseglove 1972; Kay 1987).

Harvesting

It may take 2 years from planting to harvesting (Kay 1987). Harvest maturity is when the vines begin to wither. At this point the bulbils may fall off naturally or they can easily be dislodged with a twist.

Curing

Both bulbils and tubers are resistant to fungal infections and harvest wounds heal quickly on leaving them in the shade in tropical ambient conditions (Kay 1987).

Storage

Kay (1987) indicated that both tubers and bulbils can be stored under dry, cool conditions away from sunlight.

Prickly pear, tuna, Indian fig, barberry fig, cactus fruit

Opuntia ficus-indica, *O. robusta*, *O. amyclaea*, Cactaceae

Botany

They are non-climacteric fruit, which have a thick peel that is purple, red, orange, yellow or green when mature, and are covered with sharp spines. It is ovoid and 5–7 cm long (Figure 12.86, in the colour plates). The pulp is sub-acid, juicy and contains numerous seeds.

Harvesting

Harvest maturity is usually judged on peel colour. In a study of the cultivar Gialla in Italy by Inglese *et al.* (1999), the number of days required to reach commercial harvest maturity changed with the time of bud burst, but the thermal time (40 × 103 growing degree hours) did not. Glochids (small spines) of cactus pear negatively affect harvest operations, quality and their acceptance, Corrales-Garcia and Gonzalez-Martinez (2001) investigated ways of removing them from *O. amyclaea*. To do this, gibberellic acid at 100 µg litre^{-1} was sprayed over flower buds in six consecutive applications every 7 days during their development, and also solutions of Ethephon (Ethrel) at 700 µg litre^{-1} active ingredient were sprayed over the buds immediately after fruit set, alone or in combination with gibberellic acid applications. After the mechanical effect of harvest, gibberellic acid + Ethephon caused up to 90% abscission of glochids compared with the controls (37%), whereas Ethephon alone caused 79% of glochids to abscise in the best case. None of the treatments negatively affected the quality of the fruit in terms of soluble solids concentration, acidity, colour, weight of pulp and weight of peel (Corrales-Garcia and Gonzalez-Martinez 2001).

Hot water treatment

Prestorage dipping of the fruit in water at 55°C for 5 minutes or conditioning at 38°C in saturated air for 24 hours improved their quality during marketing and could reduce the need for postharvest fungicides (Schirra *et al.* 1996). Hot water dipping at 52°C for 3 minutes and hot air treatment at 37°C for 24–72 hours sealed the microwounds and cracks normally present in untreated fruits and caused the disappearance of platelets, resulting in a relatively homogeneous skin surface, presumably due to melting of wax layers. Occlusion of possible entry points for wound pathogens by melting the surface may contribute to protecting the fruit against decay (D'hallewin *et al.* 1999).

Storage

At a room temperature of 20°C and 60% r.h., it was reported in Mercantilia (1989) that they could be kept for 5–7 days. Refrigerated storage recommendations are:

- 5°C and 90–95% r.h. for 3–4 weeks (Mercantilia 1989)
- 8–10°C and 90% r.h. for 2 weeks (Snowdon 1990)
- 5°C and 90–95% r.h. for 3–4 weeks for prickly pears (SeaLand 1991)
- 2.2°C and 90–95% r.h. for 3 weeks for cactus pears (SeaLand 1991).

Pummelos, pumelos, pomelo, shaddock

Citrus grandis, Rutaceae.

Botany

The fruit is a berry or hesperidium, which is non-climacteric. Superficially the fruit looks like a grapefruit (*C. pardisi*) and ranges in shape from ovoid to pyriform with a pronounced flattening of the stylar end in most cultivars. Its peel is very thick and easy to peel, and the flesh has very pronounced juice sacks with a rubbery texture and appearance (Figure 12.87). It has a sweet, mild flavour, not bitter like grapefruit. It probably originated in the region of Malaysia, the Indian archipelago and southern China. It was apparently first taken to the West Indies in a ship whose captain's name was Shaddock.

Figure 12.87 Pomelos in China showing the thick skin. In Thailand some varieties have much thinner skins and therefore a larger edible portion. Source: Dr Wei Yuqing.

Harvesting

Harvest maturity is judged on the skin texture. As the fruit develops it has a fairly rough skin that becomes progressively smoother as it matures. Harvesting is by hand by clipping the fruit stalk (Figure 12.88).

Curing

Curing at 34–35°C for 48–72 hours, not later than 48 hours after harvest, reduced decay and various

Figure 12.88 Pomelo ready for harvesting in China. Source: Dr Wei Yuqing.

blemishes without deleterious effects (Ben Yehoshua *et al.* 1989a). Sealing the fruit in plastic film before curing was said to be essential to reduce weight loss (Ben Yehoshua *et al.* 1989a).

Storage

Refrigerated storage recommendations are as follows:

- 7.2–8.9°C and 85–90% r.h. for 12 weeks (Pantastico 1975)
- 11°C and 90% r.h. for 8–12 weeks (Waks *et al.* 1988)
- 10–15°C and 90% r.h. (Snowdon 1990)
- 7.2°C and 85–90% r.h. for 84 days (SeaLand 1991).

Modified atmosphere packaging

Individual seal-packaging with plastic films reduced decay during long-term storage at 11°C (Waks *et al.* 1988).

Pumpkin

Cucurbita maxima, C. mixta, C. moschata, C. pepo, Cucurbitaceae.

Botany

Fruits are globular in shape and are differentiated from marrows by having a somewhat coarse, strong-flavoured yellow–orange flesh (Purseglove 1968). They will freeze at about –1.0 to 0.9°C (Wright 1942).

Harvesting

This is carried out when the fruits are fully mature and have reached an acceptable size, with golden yellow flesh and a good texture.

Storage

Pumpkins store very well and it was claimed that the cultivar Mammoth Cheese pumpkins have been kept for as long as 17 months (Wardlaw 1937). Refrigerated storage recommendations are as follows:

- 12.8–15.6°C and 70–75% r.h. for 2–3 months (Wardlaw 1937)
- 10–13°C and 70–75% r.h. for 2–6 months (Anon 1967)
- 10–12.8°C and 70–75% r.h. for 2–3 months for Connecticut Field and Cushaw (Anon 1968)
- 1.7–15.6°C and 70–75% r.h. for 24–36 weeks with 3.7% loss (Pantastico 1975)
- 12.2°C and 70–75% r.h. for 84–160 days (SeaLand 1991)
- 10–13°C and 60–70% r.h. for 2–5 months (Snowdon 1991).

Controlled atmosphere storage

SeaLand (1991) stated that controlled atmosphere storage had only slight to no effect.

Queensland arrowroot, edible canna

Canna edulis, Cannaceae.

Botany

It is a root crop that produces clusters of edible rhizomes up to 60 cm long with fleshy segments that superficially resemble corms and which are greyish white with violet scales.

Harvesting

Harvest maturity may be anything between 6 and 19 months after planting, depending on the climate of the country where they are grown and the use that will be made of the rhizomes. The rhizomes are produced just below the soil surface and are harvested by hand by pulling and levering, or mechanical harvesters are available (Kay 1987).

Storage

They can be stored for several weeks without deterioration in a cool dry place. In Japan they were often stored over winter in field pits 30 cm deep (Kay 1987).

Quinces

Cydonia oblonga, Rosaceae.

Botany

The fruit is a pome and is classified as climacteric. They are borne on trees that can be up to 6 m high. They can be either 'apple' or 'pear' shaped and are similar to those fruit in structure, but they have up to 20 carpels compared with only two in apples and pears. The flesh is greenish white and very acidic and are rarely eaten fresh. The fruit are commonly made into jam when the flesh turns a dull pink colour. It is probably a native of western Asia, where it grows wild and where it has been cultivated from ancient times. It is sometimes said that the golden apple given by Paris to Aphrodite was in fact a quince.

Harvesting

As the fruit develops the skin is downy and becomes smooth as it ripens and turns a golden yellow colour.

Storage

They store well at ambient temperature, but a refrigerated storage recommendation is –0.5°C and 90–95% r.h. for 60–90 days (SeaLand 1991).

Radishes, salad radishes

Raphanus sativus, Cruciferae.

Botany

The edible portion is the swollen taproot, which is small, round or oval in shape with a red skin and white pungent flesh and is used in salads. They will freeze at about –2.8 to 2.3°C (Wright 1942) or –1.22 to 0.72°C (Weichmann 1987).

Harvesting

This is carried out when they are young and tender and reach a size that is acceptable to the market. If they are over-mature they become fibrous, tough and hollow and eventually produce a flower stalk. They are pulled from the soil and either bunched with their leaves still attached or the tops are cut off for marketing.

Precooling

Hydrocooling was the most suitable method, but both cooling with high humidity and air cooling in a conventional cold store were also suitable and vacuum cooling was generally unsuitable (MAFF undated). Both top icing and hydrocooling were recommended to retain their crispness by Lutz and Hardenburg (1968) and Kay (1987) recommended hydrocooling to 4°C.

Storage

At a room temperature of 20°C and 60% r.h. it was reported in Mercantilia (1989) that they could be kept for 1 day with their tops still attached and 2 days if the tops are removed. Tindall (1983) also found that roots with leaves attached have approximately half the storage life of topped roots. General refrigerated storage recommendations are as follows:

- 0°C and 90–95% r.h. for 3–4 weeks (Anon 1967; Phillips and Armstrong 1967; Lutz and Hardenburg 1968)
- 0°C and high humidity for 28 days with about 10% weight loss (Tindall 1983)
- 7°C for less than 7 days (Tindall 1983)
- 0°C and 95–100% r.h. for 21–28 days (SeaLand 1991)
- 0°C and 95–100% r.h. for 120 days (Daikon) (SeaLand 1991)
- 0–1°C and 95–100% r.h. for 1–4 weeks (Snowdon 1991).

Without their leaves (topped)

- 0°C and 90% r.h. for 3–4 weeks (Wardlaw 1937)
- 7.2°C for at least a week (Lutz and Hardenburg 1968)
- 0°C and 88–92% r.h. for 3–5 weeks with 8% loss (topped) (Pantastico 1975)
- 0–1°C and 90–95% r.h. for 10–14 days (Mercantilia 1989)
- 2–5°C and 90–95% r.h. for 7–10 days (Mercantilia 1989).

With their leaves (bunched)

- 0°C and 90–95% r.h. for 1–2 weeks (Phillips and Armstrong 1967; Hardenburg *et al.* 1990)
- 0–1°C and 90–95% r.h. for 3–5 days (Mercantilia 1989)
- 2–5°C and 90–95% r.h. for 2–3 days (Mercantilia 1989).

Controlled atmosphere storage

Low-O_2 storage was shown to prolong their storage life (Haard and Salunkhe 1975). Condition 0.6°C with 5% O_2 or 5 or 10°C with 1–2% O_2 were recommended in Pantastico (1975). Saltveit (1989) recommended 0–5°C with 2–3% CO_2 and 1–2% O_2 for topped radishes, but it had only a slight effect. Controlled atmosphere storage was also said to have only a slight to no effect in SeaLand (1991).

Rambutan

Nephellium lappaceum, Sapindaceae.

Botany

The fruit is an ovoid-shaped berry that can be up to 6 cm long. It is red to orange–yellow when mature and covered in soft spines (Figure 12.89), which are referred to as spinterns. They contains one large seed surrounded by white, melting, subacid flesh (Purseglove 1968). In some varieties the pulp is attached to the seeds, in others it is free. They are non-climacteric (O'Hare *et al.* 1994), and are borne in clusters on handsomely spreading trees. It is a native of Malaysia.

Harvesting

The colours of the spines and the skin are used as guides to harvest maturity. The colour varies between different cultivars and fruit are considered overmature when the tips of the spines turn brown (Lam and Kosiyachinda 1987). Harvest maturity may also be judged on the time after full flowering. In Thailand this is 90–120 days, in Indonesia 90–100 days and in Malaysia 100–130 days (Kosiyachinda 1968).

Packaging

Muhammad (1972) showed that packing fruit in sawdust reduced fruit weight loss and kept them in good condition.

Disease control

Trichoderma harzianum (TrH 40) isolated in soil samples from rambutan orchards had an antagonistic effect against three postharvest pathogens, *Botryodiplodia theobromae*, *Colletotrichum gloeosporioides* and *Gliocephalotrichum microchlamydosporium*, causative fungi of stem end rot, anthracnose and brown spot, respectively. Treatment with *T. harzianum* had no detrimental effects on the overall quality and colour of the fruits (Sivakumar *et al.* 2000). Later, Sivakumar *et al.* (2002) found that the application of TrH 40 and potassium metabisulphite effectively controlled the incidence and severity of the three postharvest diseases and maintained the overall quality and colour of the fruit in storage at 13.5°C and 95% r.h. for 18 days.

Figure 12.89 Rambutans packed for the market in Sri Lanka.

Storage

At 20°C and 60% r.h. they had a storage life of only 3–5 days (Mercantilia 1989). Refrigerated storage recommendations are as follows:

- 10°C and 90–95% r.h. for 1–2.5 weeks with 6–12% weight loss (Pantastico 1975)
- 10°C and 90% r.h. for 1–2 weeks (Mercantilia 1989)
- 10–12°C and 95% r.h. for 1–2 weeks (Snowdon 1990)
- 7.5°C for 15 days for the cultivars R162, Jit Lee and R156 (O'Hare *et al.* 1994)
- 10°C for 11–13 days for the cultivars R162, Jit Lee and R156 (O'Hare *et al.* 1994)
- 13°C and 95–100% r.h. for 12 days (Kanlayanarat *et al.* 2000).

Chilling injury

Chilling injury was reported to occur at 7°C as browning of both the skin and spines (Mendoza *et al.* 1972; Somboon 1984).

Controlled atmosphere storage

Kader (1993), in a review, recommended 10°C, with a range of 8–15°C, combined with 7–12% CO_2 and 3–5% O_2. Kadar (1993) also claimed that they could tolerate CO_2 levels of >20% and that exposure of <1% O_2 resulted in an increase in decay. Storage of cultivar R162 at 7.5°C in 9–12% CO_2 retarded colour loss and extended shelf-life by 4–5 days, but storage in 3% O_2 did not significantly affect the rate of colour loss (O'Hare *et al.* 1994).

Modified atmosphere packaging

Fruit have been stored successfully in polyethylene film bags (Mendoza *et al.* 1988) with storage times of about 10 days in perforated bags and 12 days in non-perforated bags at 10°C. Muhammad (1972) recommended storage in sealed polyethylene film bags at 10°C for 12 days but at high storage temperatures (32–37°C) perforation of the bags was preferable. Dipping the fruit in 100 μg $litre^{-1}$ benomyl before packing in the trays could control storage decay. Kanlayanarat *et al.* (2000) showed that the cultivar Rong-Rien stored in 0.01 or 0.04 mm thick polyethylene bags at 13°C and 95–100% r.h. had the longest storage life of 18 days, compared with 16 days in fruits stored in 0.08 mm thick and 12 days in controls. All fruit in polyethylene bags had reduced moisture loss and delayed spine discoloration. The following atmospheres developed in the sealed bags:

- 0.01 mm polyethylene bags 15–16% O_2 and 2–3% CO_2
- 0.04 mm polyethylene bags 3–5% O_2 and 10–11% CO_2
- 0.08 mm polyethylene bags 1–2% O_2 and 15–16% CO_2.

Ethylene

Storage in 5 μl $litre^{-1}$ ethylene at 7.5°C or 10°C did not significantly affect the rate of colour loss (O'Hare *et al.* 1994).

Raspberries

Rubus idaeus, Rosaceae.

Botany

The fruit is a collection of drupes and is classified as a non-climacteric fruit. They will freeze at about –1.4 to 1.1°C (Wright 1942).

Harvesting

They are ready for harvest when they become a bright, even red and are easily detached from the receptacle. If harvesting is delayed the individual drupes may become detached from each other during harvesting and marketing. Mechanical harvesters have been developed for raspberries (Figure 12.90), but they still need hand grading to remove unsuitable fruit (Figure 12.91) and are really only suitable for fruit destined for processing.

Figure 12.90 Raspberry harvester in operation in the UK. Source: Professor H.D. Tindall.

Figure 12.91 Field grading of mechanically harvested raspberries in the UK. Source: Professor H.D. Tindall.

Precooling

Hydrocooling and vacuum cooling were generally unsuitable, with placing in a conventional cold storage acceptable, but forced-air cooling with high-humidity air was found to be the best method (MAFF undated).

Storage

At a room temperature of 20°C and 60% r.h. it was reported in Mercantilia (1989) that they could be kept for just 1 day. Refrigerated storage recommendations are as follows:

- –0.6 to 0°C and 85–90% r.h. for 5–7 days (Phillips and Armstrong 1967)
- 0°C and 85–90% r.h. for 3–5 days (Anon 1967)
- 0°C for 6 days then 18.3°C for 1 day for marketing or up to 11 days at 0°C (Wilkinson 1972)
- 0°C and 85–90% r.h. for 2–3 days (Shoemaker 1978)
- 0°C and 90–95% r.h. for 5 days (Mercantilia 1989)
- 8°C for 3–4 days (Mercantilia 1989)
- 14°C for 2 days (Mercantilia 1989)
- 0°C and 90–95% r.h. for 1–7 days (Snowdon 1990)
- –0.5 to 0°C and 90–95% r.h. for 2–3 days (Hardenburg *et al.* 1990)
- –0.5°C and 90–95% r.h. for 2–3 days (SeaLand 1991).

Controlled atmosphere storage

Controlled atmosphere storage with 20–25% CO_2 retarded the development of rot (Hardenburg *et al.* 1990). Kader (1989) recommended 0–5°C with 15–20% CO_2 and 5–10% O_2.

Transport

Fruit used to be transported in punnets from Scotland to the south on England in non-refrigerated trucks, but wastage was very high. Precooling and transporting in refrigerated vans reduced wastage on the journey (Hulme 1971).

Redcurrants, whitecurrants

Ribes sativum, Grossulariaceae.

Botany

They are closely related to the blackcurrant *R. nigrum* and originated in Europe. The fruits contain many small seeds but they are larger and fewer in number than in blackcurrant. *R. sativum* has about 970 seeds per 100 g of fruit weighing about 4.25 mg, whereas *R. nigrum* has about 4450 seeds per 100 g weighing about 1.03 g (Hulme 1971). The fruit are small and when ripe they turn shiny and either red or, in the case of whitecurrants, a beautiful translucent, silvery golden colour (Figure 12.92, in the colour plates). There is also a pink variety. The fruits are borne in clusters called 'strigs' and they mature in order along the strigs with the first one nearest the pedicel and the last being the terminal fruit. When mature, redcurrants have a total sugar content of about 4.4% and a total solids content of 16%, made up of 10.5% soluble solids and 5.5% insoluble solids, whereas whitecurrants have a total sugar content of 5.6% and a total solids content of 15.7% (Hulme 1971).

Harvesting

It is normal to harvest the whole cluster of fruit together by hand when almost all of them have fully changed colour, even though the cluster will contain some under-mature fruit (Hulme 1971). The currants can be stripped from the clusters by pulling them through the tines of a fork; the fully mature ones are easily removed.

Storage

No specific data could be found on *R. sativum*, but they are so closely related to blackcurrants that storage and handling conditions are probably the same.

Red whortleberries, cowberries, mountain cranberries

Vaccinium vitus-idaea, Vacciniaceae.

Botany

The fruit is a globose berry up to 8–10 mm in diameter, is red and is produced on evergreen shrubs 15–30 cm high. It is native to the British Isles and most of the cooler parts of the northern hemisphere. The blueberry (*V. corymbosum*) is also called the whortleberry.

Harvesting

Harvest maturity is judged on colour change of the skin and they are picked by hand.

Storage

At a room temperature of 20°C and 60% r.h. they could be kept for 5–7 days, and at 0°C and 90% r.h. for 3–4 weeks (Mercantilia 1989).

Rhubarb

Rheum rhaponticum, Polygonaceae.

Botany

The edible portion is the leaf stalks or petioles, which are normally red, but green ones are available. They are used as a stewed fruit or in pies. In Britain it can be grown by harvesting the rootstocks in autumn and growing them in the dark in a heated shed over winter. When grown this way it is called forced rhubarb and is earlier, more tender and less acid than those grown naturally in the open. Rhubarb will freeze at about –2.5 to 1.5°C (Wright 1942).

Harvesting

This is done by pulling the leaf stalks so that they come away cleanly from the root stock and trimming off the leaves. Harvest time is when they are large enough for the market, but before they become tough and bitter.

Precooling

Forced-air cooling with high humidity was the most suitable, vacuum cooling and air cooling in a conventional cold store were also suitable and hydrocooling was generally unsuitable (MAFF undated).

Storage

At room temperature of 20°C and 60% r.h., it was reported in Mercantilia (1989) that they could be kept for about 2 days. Refrigerated storage recommendations are as follows:

- 0°C and 90% r.h. for 2–3 weeks (Anon 1967)
- 0°C and 90–95% r.h. for 2–3 weeks (Phillips and Armstrong 1967)
- 0–1°C and 90–95% r.h. for up to 21 days (Tindall 1983)
- 10°C for about 7 days (Tindall 1983)
- 0°C and 90–95% r.h. for 20 days (Mercantilia 1989)
- 4°C for 12 days (Mercantilia 1989)
- 8°C for 8 days (Mercantilia 1989)
- 0°C and 90–95% r.h. for 2–4 weeks (Hardenburg *et al.* 1990)
- 0°C and 95–100% r.h. for 14–21 days (SeaLand 1991)
- 0–1°C and 95–100% r.h. for 2–4 weeks (Snowdon 1991).

Modified atmosphere packaging

Unsealed polyethylene film liners in crates reduced their weight loss (Lutz and Hardenburg 1968), but there was no evidence that they increased their postharvest life.

Rose apple, pommarosa

Eugenia jambos, Myrtaceae.

Botany

The edible portion is the fruit, which is a berry with a persistent calyx, oval or globular in shape and up to 5 cm long. The skin is white or yellow with a faint pinkish blush. The pulp is rose scented, crisp, juicy, sweet, and pale yellow with a central cavity containing up to three seeds. It is a native of the East Indies and is also used as an ornamental tree.

Storage

No information is available on storage, but it is related to the Malay apple (*E. malacensis*) and may well have similar storage characteristics.

Saffrom milk cap

Lactarius deliciosus.

Botany

It is a gill fungus up to 10 cm in diameter and 8 cm high. As it emerges, the cap is a flattened dome with a depressed centre and an inrolled margin, but as it expands it becomes more funnel shaped. It has a saffron yellow colour with concentric rings of darker orange with a slightly sticky surface. The stalk is an orange–buff colour, often with darker coloured blotches. The flesh is milky and orange–yellow in colour, turning green in places. The gills are orange–yellow with cream-coloured spores. It is found in coniferous woodland on chalky soils.

Storage

No information could be found on its postharvest life, but from the literature it would appear that all types of edible mushrooms have similar storage requirements and the recommendations for the cultivated mushroom (*Agaricus bispsorus*) could be followed.

Salak, snake fruit

Salacca edulis, Palmae.

Botany

The fruits are globose in shape up to 10 cm long and with a reddish brown skin with appressed imbricate snakeskin-like scales (Figure 12.93, in colour plates), which account for one of the common names. The creamy flesh is the edible part and is divided into four segments, each containing a large, brown, globular seed up to 3 cm in diameter. They have a sweet sour flavour with a slight taste of apple (Purseglove 1968). In Indonesia, they are said to give more mature women desirable characteristics.

Storage

At tropical ambient temperatures the skin will split within a few days (Purseglove 1968). However, Mahendra and James (1993) contradicted this and found that they could be stored at 29–32°C for 9 days and 22–24°C for 10 days. Refrigerated storage recommendations are as follows:

- 3–5°C for 25 days with symptoms of slight chilling injury beginning after 3 days (Mahendra and James 1993)
- 7–10°C for 23 days with symptoms of slight chilling injury beginning after 4 days (Mahendra and James 1993)
- 15°C for 13 days (Mahendra and James 1993).

Chilling injury

Symptoms of chilling injury were skin discoloration, surface pitting, necrotic areas and the flesh turning brown and soft (Mahendra and James 1993).

Figure 12.94 Salsify on the wholesale market in the UK.

Salsify

Tragopogen porrifolius, Compositae.

Botany

It is a native of Eurasia and the edible portion is the swollen taproot, which has a white skin and white flesh (Figure 12.94). It is also called the vegetable oyster because of its flavour. They will freeze at about −1.6 to 1.3°C (Wright 1942) or −1.1°C (Weichmann 1987). It is eaten boiled, baked or in soups. The tender young leaves can be used in salads.

Storage

The crop is not injured by light freezing but should be handled with care in this condition (Lutz and Hardenburg 1968). Refrigerated storage recommendations are:

- 0°C and 90–95% r.h. for 2–4 months (Wardlaw 1937; Hardenburg *et al.* 1990)
- 0°C and 95–100% r.h. for 60–100 days (SeaLand 1991).

Controlled atmosphere storage

Condition of 0°C with 3% CO_2 and 3% O_2 were recommended (Hardenburg *et al.* 1990), but controlled atmosphere storage was said to have only a slight effect or no effect (SeaLand 1991).

Sapodillas, sapota, zapota, chico chiko

Manilkara achras, synonymous with *Achras sapota*, Sapotaceae.

Botany

Fruits are globose to ovoid and up to 10 cm in diameter, with a thin, rusty brown to greyish skin. The

Figure 12.95 Sapodillas showing the correct harvest maturity and the internal structure.

pulp is yellowish brown and can have from 0–12 hard black seeds, which are easily separated from the pulp (Figure 12.95). Guadarrama *et al.* (2000) found fruit to be climacteric in at 22–28°C. In studies in Mexico fruits were also found to be climacteric and reached their respiratory peak of 27 ml CO_2 kg^{-1} h^{-1} at 20°C and 85% r.h. after 6 days (Baez *et al.* 1998). The ethylene peak preceded the rise in respiration by 1 day with maximum production of 1.7 μl kg^{-1} h^{-1}. Broughton and Wong (1979) also found fruits to be climacteric with the respiratory peak occurring at the same time as, or 1–2 days after, peak ethylene production. They also found that ascorbic acid and glucose levels increased with ripening but ascorbic acid decreased when the fruit became over-ripe.

Chicle, which is used in making chewing gum, is obtained from the latex of the tree. Trees are tapped, like rubber trees, every 2 or 3 years and the latex contains 20–40% chicle. Chicle was said to have been used by the Aztecs as chewing gum (Purseglove 1968). The tree is a tall forest tree up to 20 m high and is a native of Central America. In India, application of calcium to trees was found to improve their postharvest characteristics (Lakshmana and Reddy 1995).

Harvesting

In India, the cultivar Kalipatta exhibited a double sigmoid pattern of growth and took 42 weeks from fruit set to attain harvest maturity (Rao *et al.* 1995). The skin of the fruits changes colour from green to brown as they mature. Harvest maturity is usually determined on the colour of the skin (Table 12.55).

Baez *et al.* (1998) showed that at harvest, the fruits should have 13.2 °Brix, 0.24% titratable acidity and a firmness of 64.6 newton. Fruit weight loss after 6 days at 20°C reached 5.1%, while firmness dropped to 1.9 N. However, the soluble solids content increased to 25.8 °Brix and titratable acidity did not show any significant changes during storage. Pulp colour ranged from white–cream at harvest to light orange after 6 days at 20°C.

Astringency

Wardlaw (1937) reported that after harvesting, fully grown but unripe fruits could be placed in an atmosphere containing high levels of CO_2 to remove astringency.

Storage

Fruits were highly perishable if stored at room temperature and could not be kept beyond 8 days (Flores and Rivas 1975). The cultivar Kalipatti was stored at a room temperature of 31.7–36.9°C and 23–35% r.h. or in an evaporatively cooled chamber at 20.2–25.6°C and 91–95% r.h. in India by Waskar *et al.* (1999). The rate of change of physicochemical constituents was slower in fruits stored in a cool chamber than in fruits stored at room temperature. SeaLand (1991) differentiated between sapote, for which they recommended 12.2°C and 85–90% r.h., and sapodilla, for which they recommended 15.6°C and 85–90% r.h., both for 14–21 days. Other storage recommendations are as follows:

- 4.4–10°C in India (Wardlaw 1937)
- 7.2–10°C for 16 days then 2–3 days at 27–34°C in Jamaica, but when these conditions were tried in Trinidad the fruits failed to ripen normally (Wardlaw 1937)

Table 12.55 Effects of harvest maturity and storage temperature on the mean storage life, or the maximum time that they can be held and still ripen normally, of Philippines Sapodillas (Campo 1934)

	Storage temperature (°C)				
Harvest maturity	0	10	15	20	27.5
Green	5.6	8.4	18.0	12.0	6.6
Turning	6.4	11.0	16.4	10.4	5.4
Ripe	12.6	8.4	5.8	4.2	2.0

- 1.7–3.3°C for 8 weeks in India (Singh and Mathur 1954)
- 12°C for up to 24 days (Flores and Rivas 1975)
- 20°C, but short-term holding of the fruit at lower temperatures was also possible (Broughton and Wong 1979)
- 19.4–21.1 and 85–90% r.h. for 2.5 weeks with 12% loss for turning fruit (Pantastico 1975)
- 0–2.2°C and 85–90% r.h. for 2 weeks for ripe fruit (Pantastico 1975)
- 20°C and 90% r.h. for 2–3 weeks for turning fruit (Snowdon 1990)
- 0°C and 90% r.h. for 2 weeks for ripe fruit (Snowdon 1990)
- 10°C for 21 days for the variety Jantung (Abdul-Karim 1993)
- 15°C for 16 days for the variety Jantung (Abdul-Karim 1993)
- 20°C for 10 days for the variety Jantung (Abdul-Karim 1993).

Chilling injury

Storage below 1.7°C resulted in chilling injury, but not at 1.7–3.3°C (Singh and Mathur 1954). However, later workers have found chilling injury during storage at much higher temperatures. Fruits stored for 28 days at 10°C failed to ripen properly, indicating chilling injury (Abdul Karim 1993). Fruits stored at 5°C developed chilling injury and they failed to ripen properly even after 3 days at room temperature (Mohamed *et al.* 1996), but it was found that chilling injury was observed in non-wrapped fruits but not in fruits sealed in 0.05 mm low-density polyethylene film bags at 5°C.

Controlled atmosphere storage

There is little information in the literature on controlled atmosphere storage, but Broughton and Wong (1979) recommended 20°C with 5–10% CO_2 and complete removal of ethylene from the storage atmosphere. High concentrations of CO_2 or extreme humidities impaired the quality of stored fruits (Broughton and Wong 1979).

Modified atmosphere packaging

The fruits could be stored in polyethylene bags of 100 gauge and with 1.2% vents at room temperature for up to 9 days and in a cool chamber for 13–15 days (Waskar *et al.* 1999). Packaging in sealed 0.05 mm low-density polyethylene film bags, with two fruits per 10 × 15 cm bag, or sealed, double-layered 0.05 mm low-density polyethylene film bags, with two fruits per 10 × 15 cm bag, or vacuum-packed in 0.05 mm low-density polyethylene film or shrink wrapped in 0.025 mm poly(vinyl chloride) (under a hot-air tunnel at 150°C for 20 seconds) generally increased storage life to 4 weeks at 10°C and 3 weeks at 15°C (Mohamed *et al.* 1996). It was also found that fruit texture was maintained best in low-density polyethylene film packaging and these fruits had the highest sensory scores for taste, colour, texture and overall acceptability. Ascorbic acid content was highest in vacuum-packed fruits, followed by low-density polyethylene film-packaged fruits. The heating involved in shrink wrapping had deleterious effects on storage life. Non-wrapped fruits had the highest percentage of infected fruits and vacuum-packed fruits had the lowest; there was no difference in spoilage between the two low-density polyethylene films.

Ripening

Ripening was not affected by treatment with O_2, ethylene or indoleacetic acid (Broughton and Wong 1979). During ripening at 22–28°C and 60–70% r.h. of the cultivars Prolific and Martin grown in Venezuela, total chlorophyll and tannins decreased and total carotenoids, soluble solids and fresh weight were lost and softening increased, although the postharvest patterns were different in each cultivar (Guadarrama *et al.* 2000).

Sapote mamey, mamey zapote

Pouteria sapota, *Calocarpum sapota*, Sapotaceae.

Botany

The common name sapote is used for many tropical fruits. The fruit are ovoid or ellipsoid and can be up to 15 cm long. The skin has a roughened and scurfy surface and the flesh is somewhat granular and red to reddish brown, with a rich sweet flavour.

Hot water treatment

In hot water treatment at 60°C, the centre of a 500 g fruit took about 45 minutes to reach the water temperature. Once the centre of the fruit had reached the water temperature, they were held for an additional 60 minutes in the water. After immersion, fruits were allowed to dry and were held in storage at 25°C for 4–6 days. Hot water treatment had relatively minor effects on fruit quality and may have potential as an insect disinfestation treatment. Judges gave a higher score for the flesh colour of fruits treated at 60°C compared with those of untreated fruits as there was less flesh browning and they had firmer flesh (Diaz Perez *et al.* 2000).

Storage

Diaz Perez *et al.* (2000) found that temperature affected their speed of ripening as follows:

- 27°C ripened in 3.5 days after harvest
- 25°C ripened in 5 days after harvest
- 20°C ripened in 7 days after harvest
- 10°C fruits showed minor changes in colour and firmness and a slow rate of soluble solids increase.

Chilling injury

In Mexico, storage of fruits at 2.5°C showed abnormal ripening with partial browning of the flesh near to the skin, indicating chilling injury (Diaz Perez *et al.* 2000). Fruits stored at 10 or 15°C and then ripened at 20°C had portions of the flesh with a much higher firmness and poorer development of red colour compared with other parts of the fruit. This uneven ripening was probably a result of chilling injury. The number of fruits with injury was higher at 10°C than at 15°C, and increased with storage time. (Diaz Perez *et al.* 2000).

Ripening

Ripening was associated with flesh softening, an increase in soluble solids content and a change in flesh colour from yellow or pale pink to a dark pink or red. No changes in fruit skin colour or in flesh acidity were observed as ripening progressed. Ripe fruits had 30% or higher soluble solids content, orange or red flesh, acidity of 6–8 mM H^+, and flesh firmness (compression force) 50 newton. The flesh turned brown in overripe fruits (Perez 1999).

Satsumas

Citrus reticulata, *C. unshiu*, Rutaceae.

Botany

The fruit is a berry or hesperidium, which is non-climacteric. It is oblate to obovate in shape and large compared with other mandarin types (Figure 12.96).

Figure 12.96 Satsumas on sale at a wholesale market in Britain.

It has coarse peel, often with internal colour developing before peel colour, and they have a moderately hollow axis and are usually seedless. Fruits produced under cool conditions tend to have a deep orange peel colour. These are produced in the cool subtropical areas of Japan, Central China, Spain and South Africa (Davies and Albrigo 1994) and will freeze at about −2.3 to 1.8°C (Wright 1942).

Harvesting

Harvest maturity is usually judged on colour change of the skin to deep orange. Delayed light emission is a technique that has been used on satsumas, to grade by maturity based on chlorophyll content (Chuma *et al.* 1977). The fruit is subjected to a bright light and then placed in darkness, where a sensor measures the amount of light emitted. A portable, easy-to-use instrument was developed from this work although at this time it does not appear to be commercially available. Harvesting is by hand, usually clipping the fruit stalk.

Physiological disorders

Pantastico (1975) referred to zebra skin, which forms dark, sunken areas that follow the divisions between the fruit segments. Incidence appears to be related to rough handling, heavy irrigation during production, prolonged degreening, high ethylene levels during degreening or in fully coloured fruit mixed with green ones during degreening.

Degreening

Optimum temperatures range between 25 and 29°C and humidity as high as possible, generally between 90 and 95% r.h., with ethylene at 1–10 μl litre^{-1} for 24–72 hours. The cultivar and the stage of maturity at harvest should be taken into account.

Storage

Refrigerated storage recommendations are as follows:

- 4.4°C (Wardlaw 1937)
- 0°C for 3–6 weeks then 6–8°C for 6 weeks in Australia for Wickson, Satsuma, Golden Gage and President (Fidler 1963)
- 4°C and 85–90% r.h. for 8–12 weeks (Mercantilia 1989)
- 3.9°C and 85–90% r.h. for 56–84 days (SeaLand 1991).

Savory, summer savory

Satureja hortensis, Labiateae.

Botany

It is an annual herb closely related to winter savory (*S. montana*), which is a perennial. The summer savory is a bushy, tender plant up to 30 cm tall with dark green leaves that are used for flavouring. Bottcher *et al.* (2000a) showed that freshly harvested leaves of the cultivar Aromatawas have a high respiration rate of 696 W tonne^{-1} at 10°C, 1578 W tonne^{-1} at 20°C and 2536 W tonne^{-1} at 30°C. The values declined only by 49, 46 and 41% of the initial rate for each temperature range over an 80 hour storage period. It originated in the Eastern Mediterranean and has been used in Britain from Saxon times. It is used for seasoning foods but principally beans and is known in Germany as bohnenkraut, the bean herb (Loewenfeld 1964).

Storage

The high respiration rate resulted in high storage losses. At 10°C they lost 1.65% of fresh weight in 24 hours. It was found that their colour and exterior quality were best maintained by dehydration and storage in relative cool conditions (Bottcher *et al.* 2000a).

Scarlet runner beans, runner beans

Phaseolus coccineus, Leguminosae.

Botany

The edible portion is the pod of the fruit, which is called a legume. Pods are normally up to 30 cm long and contain up to 10 seeds. It is a non-climacteric fruit. It originated in Mexico, but is now widely distributed throughout the world. In some cases the seed is allowed to develop and they are harvested as dry beans.

Harvesting

Harvest maturity is when the seeds are just beginning to form within the pod. If harvesting is delayed the pod becomes stringy. In Britain they are usually harvested on a 4–5-day cycle.

Storage

At low temperatures the pods may develop brown discoloration (Fidler 1963); however, refrigerated storage recommendations are as follows:

- 4.5°C for up to 7 days (Fidler 1963)
- 0–6°C and 85–90% r.h. for 1–3 weeks (Anon 1967).

Scorzonere

Scorzonera hispanica, Compositae.

Botany

It is a root crop and the edible part is the long, slender, dark brown taproot about 30 cm long. Its white flesh does not contain starch but it does contain inulin, which is a carbohydrate composed of units of fructose and can be eaten by diabetics. Inulin is an uncommon storage carbohydrate, but it is found particularly in some genera of Compositae. Scorzonere are thought to have originated in Spain, but they are cultivated throughout Europe. They are eaten boiled, baked or in soups in a similar way to salsify and are sometimes called black salsify, but they have a sweetish flavour. Their highest freezing point is –1.1°C (Weichmann 1987). The tender young leaves can also be used in salads.

Harvesting

In Europe they are usually harvested in autumn. They can withstand heavy frost periods before harvesting

(Weichmann 1987), hence harvesting can be delayed and they can be stored in the soil. In Belgium it was reported that the dry matter content was lower in later harvests, but in Poland there was no effect of harvest date (Weichmann 1987).

Damage

When the roots are damaged they exude a milky sap that darkens when in contact with air and they are therefore usually cooked by boiling with their skins intact. Weichmann (1987) reported that the broken roots could be stored with minimal losses since the milky fluid exuded from the damaged areas may facilitate callus production.

Storage

At a room temperature of 20°C and 60% r.h. it was reported in Mercantilia (1989) that they could be kept for 5–7 days. They are susceptible to shrivelling and therefore humidity above 96% r.h. was recommended, since during 4 months of storage at 0°C the weight loss was 7.1% in 93% r.h. and only 4.7% in 97% r.h. (Weichmann 1987). Refrigerated storage recommendations are:

- 0°C and 90% r.h. for 4 months (Mercantilia 1989)
- 0°C and 95–98% r.h. for 56–84 days (SeaLand 1991).

Modified atmosphere packaging

Weichmann (1987) recommended the use of perforated plastic films or the 'Marcellin–Teinturier' system to maintain high storage humidity, but there was no mention of any modified atmosphere effect.

Seville oranges, bitter oranges

Citrus aurantium, Rutaceae.

Botany

The fruit is a berry or hesperidium, and is non-climacteric. They have dark orange peel that is thick and fleshy with large pores. The fruit has a distinct, pleasant, aromatic flavour and the peel is used for the production of liqueur and the whole fruit for marmalade. It originated in southern Asia, probably India, and was probably brought to Spain during Moorish times.

Harvesting

They are picked by hand when the peel has turned from green to fully orange. Care must be taken in harvesting since the mature trees are high, up to 15 m, and the stems thorny.

Storage

At a room temperature of 20°C and 60% r.h. it was reported in Mercantilia (1989) that they could be kept for 2–3 weeks. Refrigerated storage recommendations are:

- 10°C and 90% r.h. for 3 months (Mercantilia 1989)
- 10°C and 85–90% r.h. for 90 days (SeaLand 1991).

Shaggy ink cap, lawyer's wig

Coprinus comatus, Basidiomycetes.

Botany

It is a basidiomycete gill fungus with a cap height of up to 15 cm. The cap is cylindrical in shape as it emerges and is covered with white shaggy fibres and scales. It tends to become brown at the centre of the cap. The cap continues to expand and becomes more bell-shaped and darkens especially around the margins. The gills are initially pale pink but turn black and then liquefy. The spores are blackish and the stem is slender and white with a delicate ring. Hybrid forms showed considerable variation in cap shape and colour and stalk length (Loon and Fritsch 1988). It is related to *C. psychromorbidus*, which can cause postharvest rots in apples and pears.

Harvesting

They must be harvested when young before the gills blacken and begin to liquefy.

Storage

Loon and Fritsch (1988) showed that the shelf-life varied very widely between strains and the wild accession Com.2 and one of its crosses with Com.7 (c2/12 × c7/5) exhibited much improved storage quality compared with Com.7 and LM62.

Shallots

Allium cepa variety *aggregatum*, Alliaceae.

Botany

For many years they were considered to be a separate species to onions and given the specific name *A. ascalonicum*, but they are now considered to be onions and belong to the *aggregatum* group. This group differs from onion in that they multiply while they are still growing and produce many lateral bulbs. They are mainly used for pickling.

Harvesting

In Britain there is a proverb that says 'plant on the shortest day and harvest on the longest day'. They are harvested by undercutting the bulbs and pulling them from the ground, either by hand or an onion harvesting machine.

Storage

High nitrogen content in bulbs, early harvesting and poor postharvest handling were associated with short keeping quality in Thailand (Ruaysoongnern and Midmore 1994). Snowdon (1991) recommended 0–1°C and 95–100% r.h. for 1–2 weeks, but since the bulbs are closely related to the bulb onion the humidity seems high, and should probably be about 70% r.h., and the storage period seems too short. During storage where the temperature fluctuated from the average 6°C by ±5, ±3 or ±1°C every 12 or 6 hours, Ito and Nakamura (1985) found that fluctuating temperature had a particularly deleterious effect on the quality of shallots.

Shiitake

Lentinus edodes, *Lentinula edodes*, Basidiomycetes

Botany

This is an Agaricales s.l., Basidiomycetes fungus whose cap is a greyish brown with a slender stalk and white gills. The respiration rate at 20°C was found to be 264–653 μl O_2 g^{-1} h^{-1} (Sun *et al.* 1999).

Harvesting

They are grown on sticks and branches and are harvested by cutting the mushrooms from the sticks by hand. In China, Liang *et al.* (2000) described a shiitake stalk-cutting machine, but they found problems in operation and it had not been developed to the stage at which it could be used commercially.

Calcium infiltration

Vacuum infiltration with calcium chloride solution at concentrations of 0.005, 0.01, 0.03, 0.06 and 0.09 μg l^{-1} markedly inhibited respiration and ethylene production, decreased the loss of soluble solids, protein, reducing sugars, starch, organic acids, ascorbic acid and fibre, significantly restricted the increase of cell membrane permeability, and thus enhanced the quality of shiitake in storage. The best concentrations were 0.01, 0.03 and 0.06 μg l^{-1} (Li *et al.* 2000).

Storage

They are often preserved by drying, but fresh ones were said to be preferred to dehydrated ones (Lui *et al.* 1989). The shelf-life was approximately 14–20 days at 1°C and 10 days at 6°C and 2–3 days at 20°C. Browning of the pilei, gills and stipes and polyphenol oxidase activity were markedly increased during storage at 20°C. The total free amino acid content was also greatly increased at 20°C (Minamide *et al.* 1980). The storage life in ambient conditions was 1–3 days (quoted by Pujantoro *et al.* 1993). Storage at 5°C was for 7 days (quoted by Pujantoro *et al.* 1993). During storage where the temperature fluctuated from the average 6°C by ±5, ±3 or ±1°C every 12 or 6 hours, Ito and Nakamura (1985) found that fluctuating temperature quickly reduced their quality. Low-temperature storage is more effective in maintaining the quality and the contents of anti-tumour polysaccharides as health-beneficial foods (Mizuno 2000).

Controlled atmosphere storage

Browning was low in samples stored in 10% CO_2 with 1% O_2 with ethanol vapour. From these results, it is suggested that the inhibition of browning results from endogenous ethanol accumulation under high-CO_2 and low-O_2 conditions (Gong *et al.* 1993). Other controlled atmosphere storage studies were as follows:

- 0°C with 5, 10, 15 and 20% CO_2 and 1, 5 and 10% O_2 plus an air control was studied by Pujantoro *et al.* (1993), who found that the lower O_2 levels resulted in poor storage
- 40% CO_2 and 1–2% O_2 at an unspecified temperature extended their storage life four times longer than those stored in air (Minamide 1981, quoted by Bautista 1990)
- 20°C with 2% O_2 resulted in a marked reduction in gill browning and polyphenol oxidase activity and a rise in free amino acid content (Minamide *et al.* 1980b)
- At CO_2 levels of 5–80%, polyphenol oxidase activity was reduced and amino acids increased but fermentation also occurred, resulting in off-flavours after 2 days. Some volatiles, especially formaldehyde and ethanol, increased in 100% CO_2 on exposure to air, and disappeared after 6 hours. Keeping quality was best at 40% CO_2 and 1–2% O_2, with a shelf-life up to four times longer than storage in air (Minamide *et al.* 1980b).

Modified atmosphere packaging

The shelf-life in non-sealed polyethylene bags (20 × 26 cm, 30 μm thickness) was about 18 days at 1°C, 14 days at 6°C, 7 days at 15°C and 4 days at 20°C. For the best results they should be held at about 1°C for a short period immediately after harvest and then at a constant storage temperature. The shelf-life at 20°C was longest following precooling at 1°C for 24 h in hermetically sealed bags 80 μm thick and containing CO_2 instead of air (Minamide *et al.* 1980a). Storage at 5°C within a composite paper/plastic bag with the inclusion of a natural substance to reduce the 'peculiar smell,' prevent browning, reduce weight loss and loss of flavour for about 15 days was described by Lui *et al.* (1989).

The concentration of ethanol in tissues of shiitake in polyethylene film bags increased during storage at 20°C to 15 mM after 6 days of storage and there was a negative correlation between ethanol concentration and browning index (Gong *et al.* 1993).

Ethylene

Shiitake showed no response to exposure to 100 μl litre^{-1} ethylene for 24 hours at any temperature over the range 5–35°C (Inaba *et al.* 1989).

Sloes

Prunus spinosa, Rosaceae.

Botany

Sloes are the fruit of the blackthorn shrub. They are drupes and are probably climacteric and resemble a tiny plum with a bluish black skin covered with a slight bloom. The flesh is green and acidic. They are seen growing wild, particularly in hedgerows, throughout Europe and are astringent even when ripe and are used for jam, liqueurs, wine and to flavour gin.

Harvesting

They are harvested in autumn when they change colour from green to blue and are thought to be better if they are harvested after frost. Care must be taken in harvesting since they are borne on spiny shrubs.

Storage

It is probably not grown commercially and no information on its postharvest life could be found.

Snake gourds

Trichosanthes cucumerina, Cucurbitaceae.

Botany

The edible portion is the fruit, which is a pale green to white pepo, up to 150 cm long, very slender in shape, sculptured with flesh containing many brownish seeds.

Harvesting

They are harvested when they reach an acceptable size, but before they are over-mature otherwise the flesh becomes bitter and fibrous.

Storage

Refrigerated storage recommendations are as follows:

- 18.3–21.1°C and 85–90% r.h. for 2 weeks (Pantastico 1975)
- 16–17°C and 85–90% r.h. for 10–14 days (Tindall 1983).

Sorrel, Jamaican sorrel, roselle

Hibiscus sabdariffa variety *sabdariffa*, Malvaceae.

Botany

It is bushy shrub, up to 2 m tall, usually with red stems and red, inflatable, edible calyces surrounding the seed pod (Figure 12.97, in the colour plates). These calyces are infused with boiling water to make a delicious acid drink (mainly citric acid) that is a beautiful red colour similar to cranberry juice. They are also used for making jam. It is probably a native of West Africa, but is grown throughout the tropical world. Purseglove (1968) claimed that it was taken by the slave trade to Brazil in the 17th century and was first recorded in Jamaica in 1707. It is the national drink of St Vincent and is a good mixer with rum. A dehydrated powder is produced commercially in the Sudan, which is marketed as Instant Kirkady.

Harvesting

The calyces should be harvested about 15–20 days after flowering while they are still tender and fleshy.

Storage

The calyces are usually used directly after harvest and no work appears to have been done on their storage. They can be kept in good condition for considerable periods of time in a domestic refrigerator.

Sorrel, French sorrel

Rumex acetosa, *R. scutatus*, Polygonacae.

Botany

It is a perennial plant with a deep taproot. The leaves are harvested when young and used in salads.

Precooling

Vacuum precooling was recommended by Aharoni *et al.* (1989).

Modified atmosphere packaging

Packaging in sealed polyethylene-lined cartons was recommended by Aharoni *et al.* (1989). They also found that chlorophyll degradation was delayed effectively when exposed to 5% CO_2 in air in a flow-through system.

Transport

Aharoni *et al.* (1989) studied freshly harvested sorrel under simulated conditions of air transport from Israel to Europe, and also with an actual shipment, during which temperatures fluctuated between 4 and 15°C. Packaging in sealed polyethylene-lined cartons resulted in a marked retardation of both yellowing and decay. However, sealed film packaging was applicable only if the temperature during transit and storage was well controlled, otherwise perforated polyethylene was better.

Sour cherries

Prunus cerasus, Rosaceae.

Botany

The fruit is a drupe and is classified as non-climacteric. It will freeze at about –2.4 to 1.9°C (Wright 1942).

Storage

In experiments with the cultivar Shattenmorelle it was found that after 9 days at 0°C in air and subsequent storage for 3 days at 20°C the percentage of rotten fruits was 4% (Nabialek *et al.* 1999). Blue mould (*Penicillium expansum*) was identified as the main cause of rotting. Refrigerated storage recommendations are as follows:

- 0°C for only a few days as they are said to be unsuitable for storage (Lutz and Hardenburg 1968)
- 0–1°C and 90–95% r.h. for 1–2 weeks (Snowdon 1990)
- –0.5°C and 90–95% r.h. for 3–7 days (SeaLand 1991).

Controlled atmosphere storage

Storage at –1.1°C in 10–12% CO_2 with 3–10% O_2 was recommended by SeaLand (1991). Nabialek *et al.*

(1999) found that after 24 days at 2°C in 10% CO_2 with 3% O_2 and subsequent storage for 3 days at 20°C the percentage of rotten fruits was 3%. Elimination of CO_2 from the atmosphere resulted in the proportion of rotten fruit being increased to 21%.

Soursop, graviola, guanabana

Annona muricata, Annonaceae.

Botany

It is a tree that produces a dark green, spiny aggregate fruit made up of berries fused together with associated flower parts. The fruits are irregular in shape but generally ovoid and up to 25 cm long with white, milky flesh full of hard, black seeds (Figure 12.98). It is classified as a multiple climacteric fruit owing to the berries that make up a single fruit being of different maturities and thus ripening at different times (Biale and Barcus 1970).

Figure 12.98 Soursop (Annona muricata) growing in Brasil where it is called graviola.

Storage

It has been reported that it cannot withstand low-temperature storage at any stage of maturity (Wardlaw 1937) and 10–15°C and 90% r.h. for 1–2 weeks was recommended by Snowdon (1990).

Spanish plum, Jamaican plum, red mombin

Spondias purpurea, Anacardiaceae.

Botany

The fruit is oval in shape and up to 4 cm long. It contains a large seed surrounded by a thin layer of watery, pleasantly sub-acid, pungent pulp.

Storage

No information could be found on its postharvest behaviour.

Spinach

Spinaca oleraceace, Chenopodiaceae.

Botany

The edible portion is the leaves. They will freeze at about –1.0 to 0.8°C (Wright 1942) or –0.50 to 0.28°C (Weichmann 1987).

Harvesting

The leaves are harvested when they are young and tender, but before the flower stalks begin to develop. They are usually cut from the taproot and yellowing, dead and diseased leaves are removed before bunching for the market.

Precooling

They are cooled in fresh, clean water and crushed ice, used as top icing, is also recommended (Wardlaw 1937).

Storage

At a room temperature of 20°C and 60% r.h. it was reported in Mercantilia (1989) that they could be kept for only about 1 day. Refrigerated storage recommendations are as follows:

- –0.5 to –1°C and 90–95% r.h. for 1–2 weeks (Wardlaw 1937; Anon 1967, 1968; Phillips and Armstrong 1967; Lutz and Hardenburg 1968)
- 0°C and 90–95% r.h. for 8 days (Mercantilia 1989)
- 2°C for 6 days (Mercantilia 1989)
- 8°C for 3 days (Mercantilia 1989)
- 0°C and 95–100% r.h. for 10–14 days (SeaLand 1991)
- 0–1°C and 95–100% r.h. for 1–2 weeks (Snowdon 1991).

Controlled atmosphere storage

In New Zealand, storage in high concentrations of CO_2 could damage the tips of leaves, but later recommendations contradict this observation (Wardlaw 1937). Physiological disorders associated with high CO_2 include the production of off-flavours (McGill *et al.* 1966). Reduced O_2 levels in the store atmosphere had very little effect on respiration rate and there was some inexplicable indication that it might increase it (Table 12.56).

Spinach leaves were stored at 10°C in 10% CO_2 with 0.8% O_2 and 89.2% nitrogen by Hodges and Forney (2000). The rate and severity of senescence were similar between the leaves stored in air or controlled atmosphere until day 35, at which point the ambient air-stored leaves exhibited a sharp increase in antioxidants, which is a measure of senescence. Other storage recommendations were as follows:

- 5°C with 10% CO_2 retarded leaf yellowing (Pantastico 1975)
- 0–5°C with 10–20% CO_2 and 21% O_2, which had a fair effect but was of no commercial use (Kader 1985)
- 0–5°C with 5–10% CO_2 and 7–10% O_2, which had only a slight effect (Saltveit 1989)
- 10–40% CO_2 and 10% O_2 retarded yellowing (Hardenburg *et al.* 1990)
- 10–20% CO_2 and 21% O_2 (SeaLand 1991).

Table 12.56 Effects of temperature and reduced O_2 level on the respiration rate (CO_2 production in mg kg^{-1} h^{-1}) of Prickly True spinach (Robinson *et al.* 1975)

	Temperature (°C)				
	0	5	10	15	20
In air	50	70	80	120	150
In 3% O_2	51		87		137

Ethylene

Hodges and Forney (2000) found that spinach stored at 10°C in air plus 10 µl $litre^{-1}$ ethylene showed a rapid decrease in levels of almost all the antioxidants assessed, which indicated rapid senescence. Inaba *et al.* (1989) found that there was no response in respiration rate to ethylene at 100 µl $litre^{-1}$ for 24 hours at any temperature between 5 and 35°C.

Spring onions, green onions, escallion, scallions

Allium cepa, Alliaceae.

Botany

The edible portion is the young leaves and immature bulbs. They are immature and therefore have a high respiration rate and deteriorate quickly after harvest.

Harvesting

This is normally done by hand when the bulbs are as thick as or slightly thicker than the neck, which is the bottom of the leaves just above the bulb. Roots should be cut off close to the base but taking care not to damage the stem plate.

Precooling

They should be cooled within 3–4 hours of harvesting. Hydrocooling was the most suitable, vacuum cooling and forced-air cooling with high humidity were also suitable and air cooling in a conventional cold store was less suitable (MAFF undated). Top icing is also effective. Barger (1963) found that after 25–30 minutes of vacuum cooling at 4.0–4.6 mmHg with a condenser temperature of –1.7 to 0°C, the temperatured green onions fell to 1°C from 20–22.2°C, but this is an expensive method.

Storage

Thompson (1972a) found that in Jamaica scallions lost 14% in weight during the first day of marketing at ambient temperature of about 28°C and 71% r.h. If kept at too high a humidity they become discoloured and mouldy. Refrigerated storage recommendations are as follows:

- 0°C and 90–95% r.h. for 2 weeks (Anon 1967)
- 0°C and 95–98% r.h. (Shipway 1968)
- 0°C and 90–95% r.h. for a few days (Anon 1968; Hardenburg *et al.* 1990)
- 0°C and 95–100% r.h. (SeaLand 1991)
- 0–1°C and 95–100% r.h. for 1–3 weeks (Snowdon 1991)
- 0–1°C and 95–100% r.h. for 1–3 weeks (Snowdon 1991).

Controlled atmosphere storage

The following were recommended:

- 0–5°C with 10–20% CO_2 and 1–2% O_2 for green onions, but it had only a fair effect and was of limited commercial use (Kader 1985)
- 0–5°C with 0–5% CO_2 and 2–3% O_2, which had only a slight effect (Saltveit 1989)
- 5% CO_2 with 1% O_2 0°C for 6–8 weeks (Hardenburg *et al.* 1990)
- 10–20% CO_2 with 2–4% O_2 (SeaLand 1991).

Modified atmosphere packaging

Thompson (1972a) found that bundles of scallions packed in polyethylene film bags, with or without holes, lost little weight during marketing at 28°C, but soon became discoloured and had mould growth and were therefore unsuitable.

Strawberries

Fragaria × *ananassa*, Rosaceae.

Botany

The edible part of the fruit is the fleshy receptacle, which is bright red, juicy and covered with small achenes. In non-climacterics such as strawberry there was little change in ethylene production during maturation (Manning 1993). However, it was shown that auxin produced by the achenes could affect the colour (anthocyanin production) of the maturing fruit. Removal of the achenes from developing strawberry fruit resulted in quicker colouring of the area from which they were removed. They will freeze at about –1.4 to 1.1°C (Wright 1942).

Harvesting

Maturity is judged to be when they reach an even red colour over the whole surface. Being non-climacteric, the fruit does not ripen after harvest. Strawberries are very susceptilbe to bruising and harvesting directly into retail punnets can reduce damage (Figure 12.99). However, mechanical harvesters have been developed (Figure 12.100), but the fruit are damaged and are only used for processing shortly after harvest.

Figure 12.99 Harvesting strawberries in Paraguay directly into retail packs.

Figure 12.100 A mechanical harvester being used for strawberries for processing. Source: Professor H.D. Tindall.

Precooling

Forced-air cooling with high humidity was the most acceptable, air cooling in a conventional cold store was suitable and hydrocooling and vacuum cooling less suitable (MAFF undated). Certainly hydrocooling would result in high levels of decay and vacuum cooling would be ineffective.

Storage

Most postharvest rotting is due to infections by *Botrytis cinerea*, which causes grey mould, and is usually the limiting factor in storage. At a room temperature of 20°C and 60% r.h. it was reported in Mercantilia (1989) that they could be kept for only about 1 day. Refrigerated storage recommendations are as follows:

- 0°C and 85–90% r.h. for 1–5 days (Anon 1967)
- –0.6 to 0°C and 85–90% r.h. for up to 10 days (Phillips and Armstrong 1967)
- 0°C and 90–95% r.h. for 5 days (Mercantilia 1989)
- 6°C for 3–4 days (Mercantilia 1989)
- 12°C for 2–3 days (Mercantilia 1989)
- 0°C and 90–95% r.h. for 5–7 days (Hardenburg *et al.* 1990)
- 0°C and 90–95% r.h. for 5–10 days (SeaLand 1991)
- 0°C and 90–95% r.h. for 2–7 days (Snowdon 1990).

Controlled atmosphere storage

CO_2 levels of <20% during storage at 0–2°C can reduce *B. cinerea*. This was even effective at 5 or 15°C (Harris and Harvey 1973). Woodward and Topping (1972) also showed that high levels of CO_2 in the storage atmosphere at 1.7°C reduced rotting caused by infection with *B. cinerea*, but levels as high as 20% CO_2 were injurious to the fruit. High CO_2 can result in the production of off-flavours (Couey and Wells 1970) and Hardenburg *et al.* (1990) also found that 30% CO_2 or <2% O_2 may cause off-flavours to develop. In contrast, firmness of the cultivar Pajaro was enhanced during storage in air at 0°C by up to 3 days after CO_2 treatment at 40% at the beginning of storage (Harker *et al.* 2000). Low O_2 in the store atmosphere was shown to reduce their respiration rate particularly at the higher temperature (Table 12.57).

Other storage recommendations are as follows:

- 0°C and 90–95% r.h. with 15–20% CO_2 and 10% O_2. (SeaLand 1991)

Table 12.57 Effects of temperature and reduced O_2 level on the respiration rate (CO_2 production in mg kg^{-1} h^{-1}) of Cambridge Favourite strawberries (Robinson *et al.* 1975).

	Temperature (°C)				
	0	5	10	15	20
In air	15	28	52	83	127
In 3% O_2	12		45		86

- 3°C with 20% CO_2 for 10 days, but after 15–20 days there was a distinct loss of flavour (Woodward and Topping 1972)
- 0–5°C with 15–20% CO_2 and 10% O_2, or 5–10% O_2 (Kader 1985, 1989)
- 20°C with either 20 and 25% CO_2 for 7 days resulted in the lowest decay, which was 14 and 11%, respectively for the cultivar Tangi (Brackmann *et al.* 1999).

Modified atmosphere packaging

Fruits of the cultivar Camarosa were packed in 500 g capacity polypropylene punnets and wrapped automatically with 25 μm thick polypropylene film with micro-perforations. Fruits in these packages had lower losses during storage and the degree of 'fruit ripeness and nutritional value' was said to be better preserved. However, this modified atmosphere packaging seemed to reduce fruit colour, with lower anthocyanin levels, and led to off-flavour development (Sanz *et al.* 1999). Wrapping in PVC film reduced ascorbic acid loss by 5-fold at 1 and 10°C and 2-fold at 20°C in the cultivars Chandler, Oso Grande and Sweet Charlie stored for 8 days. A combination of wrapping in PVC film and storage at 1 or 10°C reduced ascorbic acid loss 7–5 fold compared with unwrapped fruits stored at 20°C (Nunes *et al.* 1998). The cultivar Tangi packed in different polyethylene films had high decay levels but the thicker gauge plastic was better than the thinner gauge (Table 12.58).

Hypobaric storage

It was reported by Haard and Salunkhe (1975) that the cultivars Tioga and Florida 90 could be stored for 21 days under reduced pressure compared with only 5–7 days in cold storage. The effectiveness of short hypobaric treatments against postharvest rots was investigated by Romanazzi *et al.* (2001). On Pajaro , the greatest reductions of *B. cinerea* and *Rhizopus stolonifer* rot were observed on fruits treated for 4 hours at 0.25 and 0.50 atm, respectively.

Ethylene

They showed no response to exposure to 100 μl $litre^{-1}$ of ethylene for 24 hours at any temperature over the range 5–35°C (Inaba *et al.* 1989).

Strawberry guava, cattley guava

Psidium cattleianum, Myrtaceae.

Botany

The edible portion is the round, purplish red fruit up to 3 or 4 cm in diameter with white pulp. They have a sweet, acid flavour somewhat resembling the strawberry. *P. cattleianum* var. *lucidum* has yellow fruits. Strawberry Guava was said to have a promising future because of the pleasant taste of the delicate fruits (Paniandy *et al.* 1999). It is a native of Brazil.

Harvesting

The timing was not clearly defined, but Paniandy *et al.* (1999) suggested just before optimal maturity.

Table 12.58 Effects of thickness of polyethylene film on percentage decay of strawberries (Brackmann *et al.* 1999)

	After 4 days		After 7 days	
Thickness (μm)	0°C	20°C	0°C	20°C
15	2	18	–	79
30	1	7	28	–
60.5	0	8	27	–

Storage

In Reunion it has a very short shelf-life in ambient conditions, which does not exceed 2 days, but fruits could be stored at 12–14°C for up to 12 days (Paniandy *et al.* 1999).

Straw mushrooms, paddy straw mushrooms

Volvariella volvacea.

Botany

It is a gill fungus. Other edible species include *V. esculenta, V. bombycina* and *V. speciosa.*

Harvesting

The content of volatile flavour compounds accumulated steadily with maturation with more flavour compounds, including both aroma and taste compounds, in sporaphores harvested when the stipe had elongated and the cap opened. Mau *et al.* (1997) defined this as stages 4 and 5.

Disease control

In India, investigations revealed that postharvest application of lactic acid bacteria such as *Lactobacillus* could inhibit the growth of moulds on straw mushrooms (Gogoi and Baruah 2000).

Storage

Yen (1992) recommended storage at 4°C. In Nigeria it deteriorated quickly after harvest, but storage at 12°C preserved their texture and nutrient composition for 1 day but longer storage periods resulted in discoloration and stickiness and greatly altered nutrient composition (Fasidi 1996).

Sugar cane, noble cane

Saccharum officinarum, Gramineae.

Botany

They produce thick stems and have been grown for chewing from ancient times in the Pacific Islands and Southeast Asia. This practice has spread throughout the tropics wherever sugar cane is grown and an export trade had developed to European and North American countries. There are various species of *Saccharum* grown commercially for sugar extraction, but the noble cane is preferred for chewing since it is relatively soft and has both high juice and sugar contents (Purseglove 1972).

Harvesting

This should be carried out when the sugar content has reached its peak and is usually 14–18 months after planting and 12 months for ratoons. Retarding the growth rate by withholding irrigation and fertilizers is used to hasten maturation (Purseglove 1972).

Peeling and cutting

The stems are normally cut into short lengths before being sold and may often be peeled. Peeling and cutting stimulated an increase in respiration rate of sugarcane more than 8-fold. The shorter the sugarcane section, the more the respiration rate increased (Mao and Liu 2000).

Storage

During storage of the cultivar Badila in China, the activities of sucrose invertases and polyphenol oxidase increased gradually, titratable acid, alcohol and reducing sugar contents increased, and sucrose and soluble sugar contents decreased. Peeled sugarcane, combined with vacuum packaging, could be stored for 20 days at 0°C (Mao and Liu 2000).

Summer white button mushroom

Agaricus bitorquis, Agaricaceae.

Botany

The sporophore is a fungal fruiting body. It will freeze at about –1.2 to 0.9°C (Wright 1942). The cap and stalk are greyish white and resemble the common mushroom. In Turkey, Yalcin and Tekinsen (1992) found that the average nutritional values on a dry matter basis of *Agaricus bisporus*, *A. campestris* and *A. bitorquis* were as follows: crude protein, fat and available carbohydrate were 6.31–8.87,

2.54–3.70, 0.18–0.39 and 0.38–1.87%, respectively. The energy value was 9.59–17.85 kcal per 100 g. Levels of calcium, phosphorus, copper and zinc were 1.64–1.86, 113.78–156.38, 0.52–1.01 and 0.54–1.56 mg per 100 g, respectively. They concluded that these mushrooms are low in fat, available carbohydrate and available energy but rich in protein and mineral content.

Harvesting

Harvesting is carried out when they reach a size that is acceptable to the market, but normally while the velum is still unbroken. For certain markets the sporophore is allowed to open fully with the velum broken, the cap extended and the stalk elongated.

Pretreatments

Pretreatment with 0.25% potassium metabisulphite solution dip had no positive effects and actually resulted in faster loss of quality than in untreated controls (Dhar 1992).

Precooling

Precooling for 6 hours in a cold room at 4–5°C increased the shelf-life (Gupta and Thakore 1998).

Storage

Storage recommendations included 0–1°C (Maaker 1971) and 10°C (Gupta and Thakore 1998). Fresh weight loss during storage at 18 ± 0.5°C for 5 days was about 10% per day (Smith *et al.* 1993). Stork and Schouten (1978) described three cultivars, Horst K26, Horst K32 and Les Miz K46, which remained white during storage, whereas Les Miz K48 deteriorated in colour and were also the most open.

Controlled atmosphere storage

In trials in which *A. edulis* and *A. bisporus* sporophores were held at 18°C for 5 days in air streams enriched to 5, 10 or 15% CO_2, stalk elongation was markedly less in 15% CO_2. When held in nitrogen streams enriched with 0.25–2.0% O_2, sporophores developed normally in O_2 at 1% or more, but O_2 levels below 0.5% arrested development (Nichols and Hammond 1976).

Modified atmosphere packaging

Packing in punnets covered with cellophane reduced weight loss without any detrimental effects (Maaker 1971). Storage of the sporophores in non-perforated bags showed better keeping quality than those stored in perforated bags. Of the various packing materials tried, polyethylene bags were best for storage of this mushroom (Gupta and Thakore 1998). In experiments by Nichols and Hammond (1976) at 18°C for 5 days, non-wrapped mushrooms lost 41% of their original weight and became wrinkled, but the caps remained relatively creamy white and the stalks elongated only slightly. In contrast, weight loss in over-wrapped packs with PVC films (VF 70, VF 71 or RMT 68 H) was no more than 6%, but many of the stalks elongated, split and distorted. Whether over-wrapped or not, storage at 2°C for 6 days resulted in little change in whiteness, and stalk and cap development was retarded compared with storage at 18°C. Dhar (1992) also found that mushrooms exhibited better shelf-life in non-perforated bags, maintaining quality for up to 6 days with little change in colour at either 15 or 20°C. They also found that sporophores stored in perforated bags began to rot within 4 days. Intense browning, extensive veil opening, increased cap diameter and heavy water condensation were observed in perforated bags at 20 and 25°C.

Swamp taro, giant swamp taro, gallan

Cyrtosperma chamissonis, Araceae.

Botany

It is a giant succulent herbaceous perennial plant with normally 6–8 enormous leaves up to 4 m high arising from a short subterranean stem. The base of the stem is thickened to form a corm, which can be conical, cylindrical or spherical in shape and weigh 15–25 kg, although it can be as much as 100 kg or more on 10-year-old plants (Kay 1987). Cormels are produced around the main corm, which develop into suckers about 3 years after planting. Both the corms and the cormels are eaten after cooking by boiling, steaming or roasting. It is though to have originated in Indonesia.

Harvesting

They can be stored in the ground and therefore harvesting can be carried out anything between 9 months and 4 years (Bradbury and Holloway 1988), although Kay (1987) stated that they require 2–3 years of growth before harvesting. They are dug from the soil by hand.

Storage

In the South Pacific they may be stored in barns, submerged in water or buried in wet sand (Bradbury and Holloway 1988). Kay (1987) stated that they are sometimes buried in a damp place and kept there for up to 6 months. They can also be stored for 2–3 months in lined pits covered with stones or soil for about 1 month at 15°C (Bradbury and Holloway 1988).

Swedes, rutabagas

Brassica napus variety *napobrassica*, *Brassica napobrassica*, Cruciferae.

Botany

It produces a swollen taproot that is joined to the tuberized stem base by the epicotyls, forming a neck. This distinguished it from the turnip (*B. rapa*). They will freeze at about –2.0 to 1.2°C (Wright 1942) or –1.50 to 1.06°C (Weichmann 1987).

Coatings

Coating with paraffin wax reduced weight loss, but if this coating was too thick the reduced O_2 supply could result in internal breakdown (Lutz and Hardenburg 1968).

Storage

At a room temperature of 20°C and 60% r.h. it was reported in Mercantilia (1989) that they could be kept for about 2 weeks. Refrigerated storage recommendations are as follows:

- 0°C and 90–95% r.h. for 6 months (Phillips and Armstrong 1967)
- 4°C for 4 months (Mercantilia 1989)
- 12°C for about 7 weeks (Mercantilia 1989)
- 0°C and 90–95% r.h. for 6 months (Mercantilia 1989)
- 0.5°C and 97–98% r.h. in an ice bank store where they remained turgid, firm and with no regrowth for 34–38 weeks (Geeson *et al.* 1989)
- 2–2.5°C and 90–95% r.h. in a conventional store but they were flaccid with root and leaf growth after 25–38 weeks (Geeson *et al.* 1989)
- 0°C and 90–95% r.h. for 2–4 months (Hardenburg *et al.* 1990)
- 0°C and 95–100% r.h. for 60–120 days (SeaLand 1991)
- 0–1°C and 95–100% r.h. for 4–6 months (Snowdon 1991).

Sprouting

Storage at 0°C prevented sprouting but at higher storage temperatures sprouting could occur but could be prevented by treatment with maleic hydrazide (Phillips and Armstrong 1967), where this is permitted.

Controlled atmosphere storage

Controlled atmosphere storage had a slight to no effect on keeping quality (SeaLand 1991).

Ethylene

Ethylene was shown to increase the respiration rate of stored swedes especially if it accumulated in stores because they are being stored with a climacteric fruit (Rhodes and Wooltorton 1971).

Sweetcorn, babycorn

Zea mays variety *saccharata*, Gramineae.

Botany

The edible portion is the caryopsis, commonly called the kernel or grain, which is borne on a modified spike that together form the ear or cob. They are formed in a leaf axil. Their freezing point was given as –0.94 to 0.61°C (Weichmann 1987).

Harvesting

Partially mature cobs are marketed as sweetcorn while even more immature, and thus smaller, cobs are

Figure 12.101 Baby corn in Thailand showing the ranges of size and quality for export.

marketed as babycorn (Figure 12.101). For several cultivars studied in the USA, the highest level of soluble solids occurred about 15 days after pollination (Splitter and Shipe 1972). There are several methods of determining harvest maturity, including kernel size, shear strength, percentage pericarp, succulometer level, moisture content, sugar content and organoleptic acceptability (Khalil 1971). All these methods are destructive and in practice the experienced harvester easily assesses maturity by the degree of browning of the silks, which are green on emergence, then turn red and finally brown. This is usually coupled with the feel of the pod and if still in doubt the outer scale is drawn back to observe the kernel colour, which becomes a deeper yellow as maturity proceeds.

Quality

Sweetness is the most important factor in judging acceptability, followed by tenderness, juiciness and colour (Splitter and Shipe 1972).

Precooling

Sugar is converted to starch and the kernels become tougher after harvest. The rate at which these processes take place depends on temperature. Forced-air cooling with high humidity was the most suitable, vacuum cooling was also suitable and air cooling in a conventional cold store and hydrocooling were less suitable (MAFF undated). Sweetcorn is a low-acid vegetable and once wet it would be difficult to dry and might lead to bacterial growth that could cause rotting, so hydrocooling should not be used. However, cobs were cooled with ice fragments in an insulated box immediately after harvest and stored in a warehouse at temperatures of 0–20°C for 15 days by Kim *et al.* (1997). The total sugar loss during storage was lowest in cobs when they were cooled, compared with non-cooled cobs stored at room temperature (25–30°C) or in the low-temperature warehouse. No mention was made of increased rotting.

Storage

Cobs from three Supersweet cultivars incorporating the *sh2* gene were given consistently higher panel test ratings for sweetness, flavour and overall quality than normal sugary or sugary enhanced cultivars directly after harvest and throughout 24 days storage in an ice bank (0.5°C and 97% r.h.) (Geeson *et al.* 1991). Textural ratings showed no obvious differences between cultivars and showed relatively little change during storage (Geeson *et al.* 1991). At a room temperature of 20°C and 60% r.h. it was reported in Mercantilia (1989) that both sweetcorn and babycorn could be kept for only about 1–2 days. Refrigerated storage recommendations are as follows:

- –0.6 to 0°C and 90–95% r.h. for 1–4 weeks with considerable loss of sugar (Wardlaw 1937)
- 0°C and 90–95% r.h. for 8 days (Phillips and Armstrong 1967)
- –0.5 to 0°C and 85–90% r.h. for 4–8 days (Anon 1967)
- 0°C and 90–95% r.h. for a few days (Lutz and Hardenburg 1968)
- 0.6–1.7°C and 90–95% r.h. for 1 week with 3.2% loss (Pantastico 1975)
- 1°C and 90–95% r.h. for up to 8 days (Tindall 1983)
- 4°C for up to 5 days (Tindall 1983)
- 10°C for about 2 days (Tindall 1983)
- 0°C and 90% r.h. for 4–8 days for both sweetcorn and babycorn (Mercantilia 1989)
- 0°C and 90–95% r.h. for 4–6 days (SeaLand 1991)
- 0°C and 90–95% r.h. for 4–8 days for babycorn (SeaLand 1991)
- 0–1°C and 95–100% r.h. for 4–8 days (Snowdon 1991).

Controlled atmosphere storage

Low O_2 in the store atmosphere was shown to reduce their respiration rate, particularly at higher temperatures (Table 12.62).

Controlled atmosphere storage has shown that cobs may be injured with >20% CO_2 or <2% O_2, but in an atmosphere of 2% O_2 the sucrose content remained higher (Hardenburg *et al.* 1990). Other work recommended 10–20% CO_2 with 2–4% O_2 (SeaLand 1991). Storage in high CO_2 retarded the conversion of starch to sugar (Wardlaw 1937). Kader (1985) recommended 0–5°C with 10–20% CO_2 and 2–4% O_2, which had a good effect but was of no commercial use. Saltveit (1989) recommended 0–5°C with 5–10% CO_2 and 2–4% O_2, which had only a slight effect.

Modified atmosphere packaging

The best packaging proved to be non-perforated film or film with the minimum perforation, that is one 0.4 mm hole per square inch (6.5 cm^2). Cobs packed in polypropylene film with no perforations showed a reduced rate of toughening, weight loss and loss of sweetness, but they tended to develop off-odours. Packing cobs with the sheath still attached but partly cut away to reveal the seeds below was the best treatment from both the storage and presentational aspects (Othieno and Thompson 1993). Retail packages of sweetcorn (film-wrapped trays containing a pair of trimmed cobs) were stored at 2°C within additional plastic liners by Rodov *et al.* (2000). The atmosphere generated in these nested packages was within the recommended range of 5–10% CO_2. Opening the liner after transfer to non-refrigerated conditions compensated for the respiration rate rise caused by elevated temperature, maintained the desirable modified atmosphere range and prevented fermentation and off-flavour development. The produce kept for 2 weeks at 2°C within nested packages, and for an additional 4 days at 20°C, which combined relatively low microbial spoilage with acceptable organoleptic quality.

Mineral powders are incorporated into some films that can be used for sweetcorn (Industrial Material Research 1989, quoted by Abe 1990).

Table 12.62 Effects of temperature and reduced O_2 level on the respiration rate (CO_2 production in mg kg^{-1} h^{-1}) of sweetcorn (Robinson *et al.* 1975)

	Temperature (°C)				
	0	5	10	15	20
In air	31	55	90	142	210
In 3% O_2	27		60		120

Hypobaric storage

Haard and Salunkhe (1975) found that storage at sub-atmospheric pressures could increase their storage life to 21 days from 4–8 days in a conventional cold storage.

Sweet passionfruit

Passiflora alata, Passifloraceae.

Botany

In Brazil, they differentiate between two similar looking fruit: sweet passionfruits (Maracujas Doce, *P. alata*) and acid passionfruits (Maracujas Acido, *P. edulis*). Their flowers open early in the morning and last about 12 hours, and emit a sweet aroma.

Harvesting

They are normally harvested after they turn from green to yellow, when they are considered mature (Veras *et al.* 2000). It was found that they could be harvested up to 4 days before fruit abscission without affecting quality.

Storage

Silva *et al.* (1999) found that pretreatment with StaFresh wax (1:2) helped to maintain their quality. Treatment with 1% calcium chloride at –25 kPa vacuum for 1 minute and +25 kPa pressure for 1 minute gave the lowest percentages of weight loss and delayed changes in peel colour during storage at 9°C and 85–90% r.h. for 30 days (Silva *et al.* 2000). During storage at 9°C for 28 days, fruits that had been treated with benzyladenine maintained ascorbic acid and soluble solids levels better than untreated fruits (Silva *et al.* 1999a). Silva *et al.* (1999) successfully stored the cultivar Dryander for 30 days at 9°C and 85–90% r.h.

Sweet potatoes

Ipomoea batatas, Convulvulaceae.

Botany

The edible portion is the root tuber, which is fusiform to globular in shape with a white, brown, red or purple skin. The flesh is white or yellow, but it may also be reddish. It will freeze at about –2.2 to 1.8°C (Wright 1942) or –1.89 to 1.06°C (Weichmann 1987).

Harvesting

There are simple tests that can be used by farmers to determine when they are ready for harvest. Mature tubers have a high starch content and this can be empirically judged by cutting samples in to pieces. If they dry out quickly when exposed to air they are mature (Pantastico 1975). Another test is that the sap from immature tubers rapidly turns black on exposure to air (Kay 1987). In the tropics, harvest maturity is often assessed by when the leaves turn yellow and begin to be shed. Time after planting has also been used, but this varies between cultivars and weather and climatic conditions.

In North America, harvesting is carried out before the first frosts or just after the first light frosts. They are left in the ground for as long as possible to maximize yield, but if the tubers are affected by frost they will quickly rot during storage.

The tubers can be harvested by hand by digging with a fork or spade, being careful not to cut or damage the tubers, then carefully pulling at the vines. Mechanical harvesters are also used, similar to those used for potatoes.

Curing

The curing process for sweet potatoes was first described by Artschwager and Starrett (1931). The processes of suberization and periderm formation are similar to those for potatoes but the conditions recommended are:

- 29–32°C and 85–90% r.h. for 7–15 days (Lutz 1952)
- 33°C and 95–97% r.h. (Cadiz and Bautista 1967)
- 29.4°C and 85–90% r.h. for 4–7 days (Kushman and Wright 1969)
- 32–36°C and 100% r.h. for 4 days (Thompson 1972a).

Cured sweet potatoes had a lower weight loss, less disease and maintained their quality during subsequent storage better than those which were not cured. The interior flesh tissues accumulate antifungal compounds under curing conditions of 30°C and 90–95% r.h. for 24 hours. The presence of antifungal compounds, including 3,5-dicaffeoylquinic acid, in the external tissues of sweet potatoes helps explain why shallow injuries can be resistant to infection. (Stange *et al.* 2001). Adequate ventilation should be supplied to sweet potato tubers during curing to prevent a buildup of CO_2 that can adversely affect the process. In the tropics, a simple method of curing is to keep the tubers in plastic-lined boxes at ambient temperatures of 27–29°C.

Disease control

Several species of fungi attack stored tubers and a range of chemical treatments have been shown to be successful in controlling these organisms. Careful handling and curing soon after harvesting are also very effective in disease control. Increased resistance to black rot disease was shown to occur in the presence of ethylene (Stahmann *et al.* 1966). Hot water treatment has been applied to sweet potatoes. Scriven *et al.* (1988) showed that there was a significant delay in disease development on roots that had been exposed to 90°C for 2 seconds, 80°C for 2, 4 or 10 seconds, 70°C for 10 seconds and 40°C for 2 minutes.

Storage

The cultivar Georgia Jet was stored for 5 months with of disinfection of the roots with iprodione (fogging) in conjunction with the curing. At the end of the storage period, 14% of the roots had decayed following this combined treatment compared with 61% with curing alone, 60% with iprodione alone and 100% in the control group (Afek *et al.* 1998). At a room temperature of 20°C and 60% r.h. it was reported in Mercantilia (1989) that they could be kept for 2–3 weeks. Refrigerated storage recommendations are as follows:

- 10–12.8°C and 80–90% r.h. for 13–20 weeks with 8.5% weight loss for 4–6 months (Wardlaw 1937)
- 11–13°C and 85–90% r.h. for 13 weeks (Anon 1967)
- 12.8–15.6°C and 85–90% r.h. for 4–6 months with 2% weight loss per month (Lutz and Hardenburg 1968)
- 10–12.8°C and 80–90% r.h. for 13–20 weeks with 8.5% weight loss (Pantastico 1975)
- 13–16°C and 85–90% r.h. for up to 140 days with up to 10% losses (Tindall 1983)
- 14°C and 90% r.h. for 3–6 months (Mercantilia 1989)
- 13.3°C and 85–90% r.h. for 90–120 days for sweet potato (SeaLand 1991)
- 12.8°C and 85–90% r.h. for 120–150 days for boniato (SeaLand 1991)
- 13–16°C and 85–90% r.h. for 4–7 months (Snowdon 1991).

Chilling injury

Chilling injury was reported to occur when the crop was exposed to 10°C for a few days and for shorter periods at lower temperatures (Lutz and Hardenburg 1968). Tubers may develop a hard texture after they have been cooked. This disorder is called 'hardcore' and is associated with chilling injury. However, the effect may be increased by exposure to ethylene. Timbie and Haard (1977) showed that sweet potatoes stored at 2°C for 3 days had more severe hardcore when exposed to 92 µl litre^{-1} ethylene than to ethylene-free air.

Controlled atmosphere storage

Controlled atmosphere storage in 2–3% CO_2 with 7% O_2 kept roots in better condition than those stored in air, but CO_2 levels >10% or O_2 levels <7% could give alcoholic or off-flavours to the roots (Hardenburg *et al.* 1990). SeaLand (1991) indicated that controlled atmosphere storage had a slight or no effect.

Ethylene

Undesirable flavours were reported in sweet potatoes exposed to ethylene in storage (Buescher *et al.* 1967). Exposure of tubers to ethylene increased their respiration rate from 18–22 µl O_2 g^{-1} h^{-1} (Solomos and Biale 1975) and increased the production of phenolic compounds. Inaba *et al.* (1989) also found that there an increased respiration rate in response to ethylene at 100 µl litre^{-1} for 24 hours at 20–35°C, but the effect was much reduced at lower temperatures. Ethylene may also be implicated in the disorder called hardcore; see above on chilling injury.

Sweetsops, sugar apples, custard apples

Annona squamosa, Annonaceae.

Botany

It is an aggregate fruit, which is made up of loosely coherent small berries or carpals, subglobose or ovoid in shape, with a yellowish green skin covered with a bluish white bloom (Figure 12.102). The

Figure 12.102 Sweetsop (Annona squamosa) growing in Malaysia.

flesh is white and custard-like with a sweet flavour and a granular texture containing numerous small brown shiny seeds. When ripe the carpels pull away easily and they may split on the tree if left to mature fully. During storage at 25 or 20°C fruit had a clear climacteric peak, whereas those stored at 15 and 10°C did not show any distinct rise in respiration rate. The ethylene peak was 2.40 μl kg^{-1} h^{-1} and it coincided with the respiratory climacteric at 25°C, which also corresponded with the peaks in total soluble solids, sugars, ascorbic acid and acidity (Prasanna *et al.* 2000). Chen and Zhang (2000) found that the climacteric peak was after 2–3 days during storage at 27–33°C and 85–90% r.h. and coincided with fruit softening and splitting. Total soluble solids, total sugar and reducing sugar contents increased during storage, while titratable acid and ascorbic acid contents decreased (Bhadra and Sen 1999). The name custard apple is used for *A. squamosa* in some Asian countries, but it more commonly refers to *A. reticulata*. It is a native of Central America and the Caribbean region.

Harvesting

Harvest maturity can be assessed on colour change of the skin. In practice, this is usually combined with feeling the fruit to judge texture, smelling the odour and looking at the bloom on the surface. In China, colour change has been developed to measure percentage maturity (Chen and Chen 1998). Fruits harvested at 90% maturity exhibited the best quality compared with those harvested at an earlier stage. However, they had the shortest storage life of 2 days compared with 4 and 6 days for 80 and 70% mature fruits, respectively, at a room temperature of 25–33°C. There was no significant difference in quality between fruits picked at 90% maturity and those picked at 80% maturity. Peduncle browning was the lowest in fruits picked at 80% maturity.

Storage

Refrigerated storage recommendations are as follows:

- 7.2°C and 85–90% r.h. for 4 weeks (turning) (Pantastico 1975)
- 1.1°C and 85–90% r.h. for 2 weeks (ripe) (Pantastico 1975)
- 10–15°C and 90% r.h. for 1–2 weeks (Snowdon 1990)
- 7.2°C and 85–90% r.h. for 28 days (SeaLand 1991)
- 10°C and 85–90% r.h. (Bhadra and Sen 1999)
- 15–20°C for 4–6 days (Prasanna *et al.* 2000).

Chilling injury

Fruit stored at 11°C were found to turn brown or black (Smith 1936).

Controlled atmosphere storage

Controlled atmosphere storage of the cultivar Tsulin at 20°C with 5% CO_2 and 5% O_2 resulted in reduced ethylene production levels and delayed ripening (Tsay and Wu 1989).

Modified atmosphere packaging

Placing fruits in sealed polyethylene bags with potassium permanganate ethylene absorbent reduced fruit spoilage, retained their natural colour and increased their storage life. Storage in perforated polyethylene also retarded ripening (Bhadra and Sen 1999).

Ripening

The major changes during ripening were a continuous decrease in fruit firmness and starch content and a continuous increase in total soluble solids and sugars (Prasanna *et al.* 2000). Fruits were ripe after 4 days at 25°C, 6 days at 20°C and 9 days at 15°C (Prasanna *et al.* 2000). They also found that the colour of the pulp, texture, taste and flavour were superior when ripened at 25 and 20°C, followed by fruits stored at 15°C, but fruits ripened at 10°C became hard with surface blackening, 'messy' pulp and less sweetness. Dipping in Ethrel at 1.5 or 2.5 ml $litre^{-1}$ enhanced fruit ripening at 28–32°C and 70–75% r.h. (Bhadra and Sen 1999).

Tamarillos, tree tomatoes

Cyphomandra betacea, Solanaceae.

Botany

The fruit are oval in shape with a wine red skin and reddish yellow pulp that is divided into two lobes

Figure 12.103 Tree tomatoes showing the range of harvest maturities for the export market from Colombia.

containing many black seeds (Figure 12.103). They are classified as non-climacteric.

Harvesting

This is done by hand when the skin has changed colour from green to red and they feel slightly soft to the touch (Figure 12.104, in the colour plates).

Storage

At a room temperature of 20°C and 60% r.h. it was reported in Mercantilia (1989) that they could be kept for 3–4 days. Refrigerated storage recommendations are as follows:

- 4°C and 90% r.h. for 3 weeks (Mercantilia 1989)
- 3–4°C and 90% r.h. for 6–10 weeks (Snowdon 1990)
- 3.9°C and 85–90% r.h. for 21–70 days for tree tomato (SeaLand 1991)
- 0°C and 90–95% r.h. for 28–42 days for tamarillo (SeaLand 1991).

It is not clear what the difference is between tree tomato and tamarillo referred to in SeaLand (1991).

Tamarind

Tamarindus indica, Leguminosae.

Botany

The fruit is a brown pod up to 15 cm long and about 2 cm wide containing several large seeds embedded in a sweet acid, brown pulp with a distinctive flavour. The pulp can have up to 10–14% acid, mainly tartaric acid, and up to 41% sugar. They may be eaten fresh by breaking the pod and sucking the pulp or the pulp is extracted and used for curries, sauces and chutney and it is made into a refreshing beverage. The fruit are borne on a large but slow-growing tree, which probably originated in Africa despite its specific name. The seeds can also be eaten after roasting or boiling and the flowers and young leaves can also be eaten. The pulp is also dried and sold as flavouring.

Harvesting

They are harvested by hand when the pod turns from green to dark brown and it gives off a distinctive odour.

Storage

The only recommendation that could be found was 7.2°C and 90–95% r.h. for 21–28 days by SeaLand (1991).

Tangerines

Citrus reticulata, *C. unshiu*, Rutaceae.

Botany

The fruit is a berry or hesperidium, which is non-climacteric. The term tangerine refers to most mandarin types in most of the USA and to the more deeply pigmented mandarins in Australia and China (Davies and Albrigo 1994). They will freeze at about –1.8 to 1.4°C (Wright 1942).

Degreening

Optimum temperatures range between 25 and 29°C and humidity as high as possible, generally between 90 and 95% r.h. with ethylene at between 1 and 10 μl litre^{-1} for 24–72 hours.

Storage

The only recommendation that could be found was 7.2°C and 85–90% r.h. for 14–28 days by SeaLand (1991).

Tannia, new cocoyam

Xanthosoma spp., *X. sagittifolium*, Araceae.

Botany

It is an herbaceous plant that produces a main corm with 10 or more lateral cormels up to 25 cm long (Kay 1987). *Xanthosoma* is an edible aroid and is sometimes confused with the polymorphic species *Colocasia esculenta*, which is also an edible aroid, some forms of which are called old cocoyams in West Africa.

Harvesting

In Trinidad, harvesting is usually carried out about 9 months after planting when the leaves begin to turn yellow to die (Purseglove 1972). It is carried out by carefully removing the soil from around the plant and harvesting the cormels while leaving the main corm in the ground to produce a further crop. Alternatively, the whole plant may be dug up, the cormels removed and the main corm replanted (Campbell and Gooding 1962).

Postharvest treatment

Waxing, fungicidal treatment and chlorine washes were all shown to reduce storage losses of the cormels (Burton 1970).

Storage

In common well ventilated stores (an average of 26°C and 76% r.h.) they had a weight loss of about 1% per week but sprouted after 6 weeks and were still considered edible after 9 weeks (Kay 1987). Traditional pit storage is generally more satisfactory than ventilated room or barn storage (Tindall 1983). Recommendations for refrigerated storage include:

- 7.2°C and 80% r.h. for 18 weeks (Gooding and Campbell 1961)
- 7°C and 80% r.h. for 120–130 days (Tindall 1983)
- 7–10°C and 80% r.h. for 4–5 months (Snowdon 1991).

Transport

Export shipments at 13–14°C from Puerto Rico to Chicago in the USA taking 9 days were generally in poor condition on arrival. In subsequent storage at 15°C and 65% r.h. for 30 days, 35% of the corms had decayed (Burton 1970).

Taro, dasheen, eddoe, cocoyam, malanga, old cocoyam

Colocasia esculenta, Araceae.

Botany

There are over 1000 recognised cultivars, but these all fall into two varieties: *antiquorum*, which has a small corm surrounded by large cormels, and *esculenta*, which has a large central corm (Figure 12.105) with a few small cormels (Kay 1987).

Figure 12.105 The main corm of dasheen (Colocasia esculenta).

Harvesting

Harvesting is usually carried out about 9 months after planting when the leaves turn yellow and begins to die (Kay 1987).

Disease control

Rotting of corms is a major factor limiting their storage life. Quevedo *et al.* (1991) studied various ways of disease control. They found that field sanitation and the application of benomyl to the crop during the growing period consistently decreased the severity of decay of corms in storage. Combinations of pruning, mulching and fungicide application at 15 days prior to harvest delayed decay development. Dipping corms in hot water at 50°C with 200 ppm benomyl suspension for 5 minutes was also effective, whilst immersion for longer periods damaged the corm tissue. Packing the corms in plastic bags reduced decay during 2 weeks in storage and was more effective when corms were pretreated with benomyl.

Storage

This is a traditional staple crop in many countries and simple storage techniques have been developed over the centuries. Corms can be stored in shaded pits where they may remain in good condition for up to 120 days (Tindall 1983). However, in the Solomon Islands they were said to be considered unfit for human consumption after only 1 or 2 weeks in ambient conditions, mainly owing to rotting caused by fungal infections (Gollifer and Booth 1973). In the Cameroons corms stored in trays for 6 months in ambient conditions had a loss of 95%. At a room temperature of 20°C and 60% r.h. it was reported in Mercantilia (1989) that they could be kept for 2–4 weeks. Recommendations for cold storage include:

- 6.1–7.2°C and 80% r.h. (Wardlaw 1937)
- 7–10°C and 85–90% r.h. for 4–5 months (Lutz and Hardenburg 1968)
- 11.1–12.8°C and 85–90% r.h. for 21 weeks (Pantastico 1975)
- 11–13°C and 85–90% r.h. for 150 days for taro and cocoyam (Tindall 1983)
- 10°C for up to 180 days (dasheen) (Tindall 1983)
- 10°C for up to 6 months (Kay 1987)
- 4.4°C for 3.5 months (Kay 1987)
- 12°C and 90% r.h. for 5 months (Mercantilia 1989)
- 7.2°C and 85–90% r.h. for 120–150 days for taro root (SeaLand 1991)
- 13.3°C and 85–90% r.h. for 42–140 days for dasheen or taro (SeaLand 1991)
- 7.2°C and 70–80% r.h. for 90 days (Malanga) (SeaLand 1991)
- 11–13°C and 85–90% r.h. for 5 months (Snowdon 1991).

Transport

Burton (1970) monitored export shipments from Puerto Rico to the USA. They were shipped at 13–14°C and took an average of 9 days. Most cormels were in poor condition on arrival and during subsequent storage at 15°C and 65% r.h. for 30 days 35% of the cormels had decayed. This was reduced to 13% if the cormels were washed in 2500 μg litre^{-1} chlorine then waxed or 18% if they were dipped in 1000 μg litre^{-1} SOPP plus 900 μg litre^{-1} of Dichloran then waxed. Since that time there has been more restriction on the use of postharvest chemicals.

Tayberries

Rubus sp., Rosaceae.

Botany

It is a hybrid between the tetraploid raspberry (*R. idaeus*) and the American blackberry (*R. ulmifolius*) and was bred in Scotland, hence its common name. Related types are sunberries, tummelberries and wineberries. The fruit is a collection of drupes and is classified as a non-climacteric fruit. They are bright red when mature and elongated in shape (Figure 12.106, in the colour plates). They have an aromatic and slightly acidic flavour and are borne on long spiny canes.

Storage

No specific information on storage could be found but they can probably stored in similar conditions to raspberries at 0°C and 85–90% r.h. for about 2–3 days.

Tomatoes

Lycopersicon esculentum, Solanaceae.

Botany

The fruit is a berry and is usually red when ripe, but yellow cultivars and other shades are grown. Mature green fruits will freeze at about –1.0 to 0.8°C and ripe fruits at –1.0 to 0.7°C (Wright 1942) or –1.28 to 0.50°C (Weichmann 1987). Genetic engineering has produced fruit that do not soften normally. These can be harvested green and ripened in an atmosphere containing ethylene. One genetically engineered cultivar of tomato called Flavr Savr was marketed in the USA in 1993, but owing to restrictions on marketing products of biotechnology in the European Union and other countries and consumer resistance, the market was restricted. More than 400 substances have been shown to contribute to the odour of tomatoes, but no single compound or simple combination of these compounds has the typical smell of ripe fruit (Hobson and Grierson 1993).

Harvesting

Harvest maturity is determined by skin colour and the Organisation de Coopération et de Développment Économiques, Paris, has produced a colour chart that shows photographs of 12 fruits ranging in colour from green to red (Figure 12.107, in the colour plates). Tomatoes that are fully red at harvest may be soft and susceptible to handling damage or become over-ripe and unacceptable in the marketing chain. In practice, fruit are normally harvested at the stage when they just turning from green to red. At this stage they are still firm and bruise less during handling, grading and packing. Titratable acidity was lower in fruits harvested at the firm red stage and decreased during storage (Chiesa *et al.* 1998) and the best flavoured fruit were those that ripened fully on the plant.

Harvesting is by hand when each individual fruit reaches the acceptable colour. They are twisted so that the pedicel breaks at the abscission point so that they are harvested with the calyx still attached. The stem scar is the major pathway for internal and external gas exchange (Yang *et al.* 1999) and leaving the pedicel attached may allow for internal depletion of O_2 and increase in CO_2, slowing the rate of ripening. Some cultivars are available where all the fruit in a truss ripen at the same time and the whole truss can be harvested by clipping it from the plant. In some cases they are harvested fully ripe and packed and marketed on the truss at a premium price.

Quality

The Organisation de Coopération et de Développment Économiques, Paris, published a brochure of standards for tomatoes in 1988 which also included a colour chart to facilitate harvest and grading standards. Much research work has been done on objective methods of measuring the maturity or ripeness of tomatoes. Much of this work is really for quality grading since it must be done postharvest. Delayed luminescence can be used as a non-destructive indicator of soluble solids and dry matter of cherry tomato fruit (Triglia *et al.* 1998). Acoustic and vibration tests equipment puts energy into a fruit, and measures its response to this input. Good correlations have been found using the second resonant frequency to determine tomato maturity (Darko 1984; Punchoo

1988). Delayed light emission has also been used on tomatoes (Forbus *et al.* 1985) objectively to grade fruit into different maturity groups postharvest. This is based on the chlorophyll content of the fruit, which reduces during maturation. The fruit is exposed to a bright light and then switched off so the fruit is in the dark. A sensor measures the amount of light emitted from the fruit, which is proportional to its chlorophyll content and thus its maturity. The wavelength of peak transmittance of light through fruits was shown to have a good correlation with their maturity (Birth and Norris 1958). This was tested on tomatoes and a portable, easy-to-use instrument was developed which was non-destructive. Mechanical resonance techniques have been applied to tomatoes (de Baerdemaeker 1989). The stiffness factor in tomatoes was shown to decrease during storage and was used to develop a small device to measure fruit maturity objectively.

Prestorage treatments

Postharvest application of calcium has been shown to inhibit ripening of tomatoes (Wills and Tirmazi 1979). Treatment of fruits prior to storage in atmospheres of total nitrogen was also shown to retard subsequent ripening of tomatoes (Kelly and Saltveit 1988).

Yang *et al.* (1999) found that sealing of the stem scar of the cultivar Floradade greatly reduced the ripening rate and extended the storage life. After 70 days at room temperature, the sealed fruits were still in good condition with some uneven red colour on the shoulder, while the unsealed fruits were fully ripe within 20 days. After sealing the stem scar, the O_2 level in sealed fruits decreased rapidly and dropped below 3% within 2 days, while the internal CO_2 level increased to about 6% in the following few days.

Precooling

Crops such as tomatoes, which have a relatively thick wax cuticle, are not suitable for vacuum cooling, but they can be successfully hydrocooled or cooled in a forced-air precooler.

Storage

Low-temperature storage can inhibit the decline in acids during ripening and thus affect their flavour (Hulme 1971). Refrigerated storage recommendations are as follows:

Mature green fruits

- 1.5–3°C and 85–90% r.h. for 6 weeks (Anon 1967)
- 12.8–15.6°C and 85–90% r.h. for 2–6 weeks (Phillips and Armstrong 1967)
- 12.8–21.1°C and 85–90% r.h. for up to 2 weeks (Anon 1968; Lutz and Hardenburg 1968)
- 8.9–10°C and 85–90% r.h. for 4–5 weeks with 5.2% loss (VC lines) (Pantastico 1975)
- 1.7–3.3°C and 85–90% r.h. for 6 weeks with 4.8% loss (Oxheart, Hybrid 6, Marathi) (Shipway 1968)
- 7.5–8°C and 85% r.h. for 10 days (Moneymaker which are one-quarter ripe) (Shipway 1968)
- 13.3°C 90–95% r.h. (SeaLand 1991)
- 12–15°C and 90% r.h. for 1–2 weeks for mature green (Snowdon 1991)
- 10–12°C and 90% r.h. for 1–2 weeks for turning (Snowdon 1991).

Red fruits

- 0°C and 85–90% r.h. for 1–3 weeks (Anon 1967)
- 10°C and 85–90% r.h. (Phillips and Armstrong 1967)
- 0–3.3°C (Phillips and Armstrong 1967)
- 2–4°C and 85–90% r.h. for 2–4 weeks (Anon 1967)
- 7.2–10°C and 85–90% r.h. for several days (Lutz and Hardenburg 1968)
- 2–4°C (Shipway 1968)
- 1–4°C and 90% r.h. for 19 days (Manalucie) if marketed quickly after removal from storage (Thompson 1972a)
- 7.2°C and 90% r.h. for 1 week (VC lines) (Pantastico 1975)
- 0–1.7°C and 85–90% r.h. for 2 weeks (Sioux) (Pantastico 1975)
- 2°C and 85–90% r.h. for 40 days (Tindall 1983)
- 10°C and 85–90% r.h. for 35 days (Tindall 1983)
- 21°C and 85–90% r.h. for 21 days (Tindall 1983)
- 11–14°C and 80–85% r.h. for firm ripe for 2 weeks (Mercantilia 1989)
- 8–10°C and 80–85% r.h. (Mercantilia 1989)
- 8–10°C and 90% r.h. for 1 week (Snowdon 1991)
- 7–14°C and 90–95% r.h. for breaker to light pink stage (SeaLand 1991).

Chilling injury

Chilling injury in storage is related to the stage of ripeness of fruits. Thompson (1972a) showed that tomatoes at the green and turning stages suffered from chilling injury below 13°C, but fully ripe fruits could be stored at 0°C. The presence of ethylene in the storage atmosphere had no effect on chilling sensitivity. This was shown for green tomatoes where storage at 5°C with or without 50 μl $litre^{-1}$ ethylene had the same chilling injury effects (Manzano-Mendez *et al.* 1984).

Intermittent warming

In an experiment, the cultivar Daniela F1 at the breaker stage was treated with iprodione (1 g $litre^{-1}$) and stored for 14 or 21 days at 6, 9 or 12°C with 90–95% r.h. Some fruits stored at 6 and 9°C were exposed to intermittent warming (20°C for 1 day) at 7-day intervals. After storage, fruits were ripened at 20°C and 75–80% r.h. for 3 days. Compared with cold storage, intermittent warming enhanced the colour and promoted ripening. After ripening, no significant differences were observed in soluble solids content, but intermittent warming did reduce their titratable acidity (Artes *et al.* 1998).

Controlled atmosphere storage

Storage in low concentrations of O_2 had less effect on tomatoes that were subsequently ripened than the stage of maturity at harvest (Kader *et al.* 1978). Increases in ethylene evolution and respiration, red colour development and loss of fruit firmness were delayed in mature green tomato fruits held in 3% O_2 compared with control fruits held in air, which showed normal ripening-related changes at 20°C (Kim *et al.* 1999). Low O_2 in the store atmosphere was shown to reduce their respiration rate with the effect being greater at the higher temperatures (Table 12.59).

Table 12.59 Effects of temperature and reduced O_2 level on the respiration rate (CO_2 production in mg kg^{-1} h^{-1}) of Eurocross BB tomatoes (Robinson *et al.* 1975)

	Temperature (°C)				
	0	5	10	15	20
In air	6	9	15	23	30
In 3% O_2	4		6		12

High CO_2 can result in uneven ripening (Morris and Kadar 1977) and surface blemishes (Tomkins 1965). Pantastico (1975) also found that storage in atmospheres containing >4% CO_2 or <4% O_2 gave uneven ripening. Batu (1997) stored the cultivar Criterium in controlled atmosphere conditions for 60 or 70 days at 15°C followed by air for another 10 days at 20°C for ripening. All tomatoes stored either in 1–3% O_2 or 3–9% CO_2 kept their green colour for 40–50 days. Acidity levels were similar in all treatments, except that the fruits stored in 1% O_2 or 9% CO_2 were significantly lower. Soluble solids of fruit stored in air or in 1% O_2, particularly after 40 days of storage, were significantly lower than in other treatments. While the greatest deterioration occurred in fruits stored in air or in 1% O_2 with 5% CO_2, the least occurred in 3% CO_2 with 5.5% O_2.

Storage recommendations were:

- 0% CO_2 with 3–5% oxygen for breaker or light pink, 0–3% CO_2 with 3–5% O_2 for mature green (SeaLand 1991)
- 12.8°C and 0% CO_2 with 3% O_2 for 6 weeks (Pantastico 1975)
- 12.8°C and 1–2% CO_2 with 4–8% O_2 at for breaker or pink fruit (Pantastico 1975)
- 12°C with 5% CO_2 with 5% O_2 for fruits harvested at the yellow or tinted stage (Wardlaw 1937)
- 12–20°C with 0% CO_2 and 3–5% O_2 for mature green fruits, which had a good effect but was of limited commercial use (Kader 1985)
- 8–12°C with 0% CO_2 and 3–5% O_2 for partially ripe fruit, which had a good effect but was of limited commercial use (Kader 1985)
- 12–20°C with 2–3% CO_2 and 3–5% O_2, which had only a slight effect (Saltveit 1989)
- 12–13°C with either 0% CO_2 and 2% O_2, 5% CO_2 and 3% O_2 or 5% CO_2 and 5% O_2 for 6–10 weeks for mature green tomatoes (Adamicki 1989).

Carbon monoxide

The addition of carbon monoxide to the storage atmosphere can have beneficial effects on tomatoes. A combination of 4% O_2, 2% CO_2 and 5% carbon monoxide was shown to be optimum in delaying ripening and maintaining good quality of mature green tomatoes stored at 12.8°C after being subsequently ripened at 20°C (Morris *et al.* 1981a). The level of chilling injury symptoms was reduced, but not eliminated, when carbon monoxide was added to the store. Carbon monoxide was also found to stimulate respiration rate, but this effect could be minimized by reducing the level of O_2 in the store to below 5% (Kader 1987). Decay in stored mature green tomatoes stored at 12.8°C was reduced when 5% CO was included in the storage atmosphere (Morris *et al.* 1981a).

Modified atmosphere packaging

Partially ripe tomatoes, packed in suitably permeable plastic film giving 4–6% CO_2 and 4–6% O_2 with about 90% r.h., can be expected to have 7 days longer shelf-life at ambient temperature than those stored without wrapping (Geeson 1989). Under these conditions the eating quality should not be impaired but film packaging that results in higher CO_2 and lower O_2 levels may prevent ripening and result in tainted fruit when the fruit are ripened after removal from the packs. Geeson (1989) also showed that film packages with lower water vapour transmission properties can encourage rotting of the tomatoes. Mineral powders incorporated into films can remove or absorb ethylene, excess CO_2 and water and can be used for tomatoes (Industrial Material Research 1989, quoted by Abe 1990).

Ripening physiology

At the green stage of maturity tomatoes contain a toxic alkaloid called solanine, which decreases during ripening. Chlorophyll levels are progressively broken down into phytol. It was observed that in cherry tomatoes the total chlorophyll level was reduced from 5490 μg per 100 g fresh weight in green fruit to 119 μg per 100 g fresh weight in dark red fruit (Laval-Martin *et al.* 1975). Concurrent with this degradation process lycopene, carotenes and xanthophylls are synthesised, giving the fruit its characteristic colour, usually red (Grierson and Kader 1986). Total caroteniods in cherry tomatoes increased from 3297 μg per 100 g fresh weight in green fruit to 11 694 μg per 100 g fresh weight in dark red fruit (Laval-Martin *et al.* 1975). This colour development occurs in both the pulp and the flesh. The optimum temperature for colour development is 24°C; at 30°C and above lycopene is not formed (Laval-Martin *et al.* 1975).

Exposing tomatoes to ethylene can result in the rapid breakdown of chlorophyll (Lougheed and Franklin 1970). Application of ethylene to stored tomatoes was shown to increase their carotenoid and lycopene contents (Boe and Salunkhe 1967). This appears to be additional to the increase that is normally associated with ripening. Green tomatoes exposed to high levels of ethylene (8000 ppm) for 1 day and then ripened had 16% more ascorbic acid when ripe than untreated fruit (Watada *et al.* 1976). It has been shown that ethylene can reduce the growth of *Fusarium oxysporum* and rotting in tomatoes infected with the fungus (Lockhart *et al.* 1969).

Non-ripening mutant cultivars have been produced. Crosses between these and wild types produce F_1 hybrids that will over-ripen very slowly (Hobson 1989). Evenness of ripening, greater depth of colour and some disease resistance have also been introduced into these transgenic tomatoes (Hobson 1989; Hobson and Grierson 1993).

Ripening

The best quality in terms of colour and flavour is achieved by allowing the fruit to ripen fully before harvest (Hobson 1989). Harvesting fruit before they have fully ripened and then ripening them under controlled conditions resulted in fruit of inferior flavour and aroma than those allowed to ripen fully on the plant (Hayase *et al.* 1984). Under-ripe fruit that are stored under refrigeration and not allowed to recover before consumption have eroded flavour (Hobson and Grierson 1993). In spite of this, storage of mature green fruit at 10°C for 10 days then ripening at 21°C for 2–6 days followed by storage at 10°C for a further 8–10 days was recommended for the production of high-quality fruit (Hulme 1971). Storage at 30°C inhibited fruit ripening in that they became orange or yellow rather than red, but coloration may be improved by reducing the temperature to 18–24°C (Tindall 1983). At 13–18°C and 85–90% r.h. fruits are likely to ripen in 14–16 days (Tindall 1983).

Ripening at 18.3–20°C could be shortened by 2 days if the atmosphere was enriched with 200 μl $litre^{-1}$ of ethylene (Lutz and Hardenburg 1968). Ripening at 13–22°C with 10 μl $litre^{-1}$ of ethylene continuously was recommended by Wills *et al.* (1989). Fruit exposed to 100 μl $litre^{-1}$ of ethylene for 48 hours had a higher rate of ripening at 20°C, resulting in a higher reduction in ascorbic acid content when fruits were table ripe than fruits ripened without exogenous ethylene (Kader *et al.* 1978). However, there was no significant difference in flavour between tomatoes ripened with or without ethylene. At 20°C Wills *et al.* (2001) found that the time to ripen increased linearly with logarithmic decrease in ethylene concentration over the range <0.005, 0.01, 0.1 1.0 and 10 μl $litre^{-1}$.

Topee tambo

Calathea allouia, Marantaceae.

Botany

The plant produces clusters of round or ovoid tubers up to 6 cm long. They usually have a white or light brown, thin skin and white or yellow crisp flesh, which has a pleasant nutty flavour. It is common in Colombia and is thought to be a native of northern parts of South America and some of the Caribbean islands (Kay 1987).

Harvesting

Harvest maturity is about 9–12 months after planting. Harvesting is usually done by hand using a garden fork (Kay 1987).

Storage

No information could be found on storage.

Truffles

Tuber spp.

Botany

They grow wild underground in woods, especially beech or oak woods, and produced a globular edible fruiting body that can be up to 10 cm in diameter. There are a number of species including the red truffle, *T. rufum*, and the white truffle, *T. aestivum*. The major species of truffles cultivated in Italy are *T. magnatum*, grown principally in medium clay soils, and *T. melanosporum*, grown in drier soils and called the French Périgord truffle. *T. melanosporum* is used to make paté de foie gras. They are noted for their delicious flavour and the rarer species command very high prices.

Harvesting

In the wild, specially trained dogs or pigs are used to hunt them and dig them up.

Storage

Conditions of 0 or 5°C for up to 40 days were recommended by Mencarelli *et al.* (1997). In *T. melanosporum* two aldehydes (2- and 3-methylbutanal) and two alcohols (2- and 3-methylbutanol) were shown to have a major effect on flavour; during storage at 25°C all these compounds are lost by oxidation or evaporation, but at 0°C a strong amylic fermentation occurred, resulting in an increase in their concentrations (Bellesia *et al.* 1998).

Controlled atmosphere storage

When truffles were stored in either low O_2 (1%) or high CO_2 (60%) at 5 and 10°C ethylene production and weight loss were reduced by the high CO_2 with little effect with low O_2 (Mencarelli *et al.* 1997). Truffles stored in high CO_2 at 5°C retained firmness, gumminess and chewiness similar to those of fresh samples. The CO_2 treatment at 5°C kept their strong, typical odour better than samples kept in low O_2 (Mencarelli *et al.* 1997).

Turnips

Brassica rapa variety *rapa*, Cruciferae.

Botany

The edible portion is the swollen spherical taproot, commonly purplish in colour with white flesh. The cultivar White Globe will freeze at about –1.0 to 0.5°C (Wright 1942) or, for unspecified cultivars, –1.39 to 1.06°C (Weichmann 1987).

Prestorage treatment

Lutz and Hardenburg (1968) recommended coating with paraffin wax prior to marketing.

Storage

At a room temperature of 20°C and 60% r.h. it was reported in Mercantilia (1989) that they could be kept for about 2–3 days. Refrigerated storage recommendations are as follows:

- 0°C and 90–95% r.h. for up to 6 months (Phillips and Armstrong 1967)
- –0.6 to 0°C (Wardlaw 1937)
- 0.6°C and 85% r.h. (Wardlaw 1937)
- 1.1–2.8°C (Wardlaw 1937)
- 0°C and 90–95% r.h. for 4–5 months (Anon 1968; Lutz and Hardenburg 1968)
- 1–2°C and 95% r.h. for 80–90 days (Tindall 1983)
- 0°C and 95–100% r.h. for 60–120 days (SeaLand 1991)
- 0–1°C and 90–95% r.h. for 4–5 months (Snowdon 1991)
- 0°C and 95–100% r.h. for 10–14 days (turnip greens) (SeaLand 1991).

Controlled atmosphere storage

Controlled atmosphere storage had a slight to no effect (SeaLand 1991). However, low O_2 in the store atmosphere was shown to reduce their respiration rate with the effect being similar at all three temperatures (Table 12.60).

Turnip greens

Brassica rapa variety *rapa*, Cruciferae.

Botany

These are the leaves of the turnip, which may be used like spinach and are called turnip greens.

Precooling

Top icing is recommended to keep them fresh (Lutz and Hardenburg 1968).

Table 12.60 Effects of temperature and reduced O_2 level on the respiration rate (CO_2 production in mg kg^{-1} h^{-1}) of bunching turnips with leaves (Robinson *et al.* 1975)

	Temperature (°C)				
	0	5	10	15	20
In air	15	17	30	43	52
In 3% O_2	10		19		39

Storage

Refrigerated storage recommendations are as follows:

- 0–1.1°C (Wardlaw 1937)
- 0°C and 90–95% r.h. (Lutz and Hardenburg 1968)
- 0°C and 95–100% r.h. for 10–14 days (SeaLand 1991).

Turnip rooted parsley, hamburgh parsley

Petroselinum crispum variety *tuberosum*, Umbelliferae.

Botany

The edible portion is the off-white tap root, which resembles a small parsnip but has a flavour similar to that of celeriac. The leaves resemble flat-leafed parley and can be used as a herb.

Storage

At a room temperature of 20°C and 60% r.h. it was reported in Mercantilia (1989) that they could be kept for about 1 week. Refrigerated storage recommendations are as follows:

- 0°C and 90–95% r.h. for 1–2 months (Lutz and Hardenburg 1968)
- 0–1°C and 85–90% r.h. for 1–2 weeks (Amezquita 1973)
- 0°C and 95–100% r.h. for 2–2.5 months (Hardenburg *et al.* 1990)
- 0°C and 95% r.h. for 4–6 months (Mercantilia 1989)
- 0°C and 95–100% r.h. for 30–60 days (SeaLand 1991).

Controlled atmosphere storage

Controlled atmosphere storage with 11% CO_2 and 10% O_2 helped to retain the green colour of the leaves of *P. crispum* during storage (Hardenburg *et al.* 1990).

Uglifruits, mineolas, minneolas, tangelos

Citrus reticulata × *C. paradisi*, Rutaceae.

Botany

Uglifruits, mineolas and tangelos have different morphological characteristics since they are different hybrid selections. The fruit is a berry or hesperidium, which is non-climacteric.

Hot water brushing

Storage experiments using various organically grown minneolas showed that hot water brushing at 56°C for 20 seconds reduced decay development by 45–55% and did not cause surface damage. Also, it did not increase weight loss or internal quality parameters (Porat *et al.* 2000).

Storage

At a room temperature of 20°C and 60% r.h. it was reported in Mercantilia (1989) that they could be kept for 7–10 days. Refrigerated storage recommendations are as follows:

- 3.3°C and 85–90% r.h. for to 4 weeks for Orlando Tangelo (Lutz and Hardenburg 1968)
- 4°C and 90% r.h. for 3–5 weeks (Mercantilia 1989)
- 4–5°C and 90% r.h. for 3–5 weeks (Snowdon 1990)
- 4.4°C and 90–95% r.h. for 14–21 days for Uglifruit (SeaLand 1991)
- 3.3°C and 90–95% r.h. for 21–35 days for Mineola (SeaLand 1991).

Ullocu, ulloco

Ullucus tuberosus, Basellaceae.

Botany

They are an important food of the high Andes of South America. They produce tubers at the end of adventitious roots, which are spherical or cylindrical in shape and usually with a yellow skin, but can also be purple, brown, red, pink or white. The flesh is yellow and mucilaginous. It is a potato-like tuber that is rich in ascorbic acid and has a silky texture and nutty taste (Popenoe 1992).

Storage

They are usually consumed directly after harvest and are rarely stored (Kay 1987).

Velvet shank

Flammulina velutipes, Basidiomycetes.

Botany

It is a basidiomycete gill fungus and at first the cap is conical, but it flattens with age. It is bright yellow or orange, often darkening towards the centre when wet and the surface is shiny and sticky and can become undulated with age. The stem is tough, orange–brown, darkening and velvety towards the base and usually curving upwards. The flesh is yellowish.

Storage

No information could be found on its postharvest life, but from the literature it would appear that all types of edible mushrooms have similar storage requirements and the recommendations for the cultivated mushroom (*Agaricus bispsorus*) could be followed.

Watermelon

Citrullus lanatus, Cucurbitaceae.

Botany

The fruit is classified as climacteric and is a large fleshy berry, which is oblong to globular in shape. The skin is usually mottled green or striped, but dark green and brown fruit are grown. The flesh is juicy with many black or brown seeds and usually dark red, but can be pink, white or yellow. The application of potassium to watermelons during growth was shown to decrease the respiration rate of the fruit after harvest (Cirulli and Ciccarese 1981). The flesh will freeze at about –1.7 to 1.4°C (Wright 1942).

Harvesting

If fruit are too immature at harvest they will not ripen (Wardlaw 1937). The tendril accompanying the fruit becomes shrivelled when the fruit is mature and can be used to judge harvest maturity. Also, the portion resting on the ground changes colour from white to a creamy yellow and it produces a dull, hollow sound when tapped. The measurement of electrical properties can be used to determine fruit ripeness. In watermelons the specific electrical resistance appeared to decrease with increasing sugar content (Nagai *et al.* 1975). In Japan, the measurement of acoustic impulse responses has been developed and a machine installed in commercial packing facilities (Kouno *et al.* 1993). It is being used for watermelons to detect both the ripeness and any hollowness in the fruit. Nine machines were installed and they are as accurate in grading melons as skilled inspectors using 'the traditional slapping method.'

Figure 12.108 Watermelons after being transported to the market stocked in the back of a lorry in Peshawar, Pakistan.

Losses

In a study in Mexico, fruits were transported loose in lorries and on arrival at the market were unloaded by hand and stacked in piles. Most defects were superficial, resulting from mechanical injuries during production and postharvest handling. Severe mechanical damage (Figure 12.108), aggravated by over-maturity of some fruits, accounted for 5.7% loss during transport and an estimated 2% during subsequent marketing (Noon 1979).

Disease control

A range of fungi can attack the fruit postharvest, but careful handling and application of copper sulphate to the end of the fruit stalk directly after harvest gave good control (Wardlaw 1937). Snowdon (1990) reviewed other diseases and their control.

Physiological disorders

White heart has been associated with adverse weather conditions during fruit growth (Wardlaw 1937). Imbalance of fertilizers can result in the physiological disorder of watermelon called blossom end rot (Cirulli and Ciccarese 1981).

Storage

In Jamaica, it is claimed that the cultivar Charleston Grey can occasionally be kept at room temperature for several months without decay, while at other times it spoils within a few days of harvest. In the Kordofan area of the Sudan, a round, brown skin type can be kept in good condition for 1 year in very hot arid desert conditions and is used as a source of drinking water. At a room temperature of 20°C and 60% r.h. it was reported in Mercantilia (1989) that they could be kept for only 8–12 days. Refrigerated storage recommendations are as follows:

- 7.2–15.6°C and 80–90% r.h. for 2 weeks with 2.2–4.5°C and 85–90% r.h. for 2–3 weeks (Anon 1967)
- 2.2–4.4°C and 85–90% r.h. for 2–3 weeks (Phillips and Armstrong 1967)
- 4.4–10°C and 80–85% r.h. for 2–3 weeks (Lutz and Hardenburg 1968)
- 2.8–3.9 for 2–3 months (Russia) (Lutz and Hardenburg 1968)
- 8–15°C and 80–85% r.h., since chilling injury can occur in storage below 7°C and loss of skin colour may occur during storage below 10°C (Tindall 1983)
- 5–6°C and 80–85% r.h. for 16–20 days (Mercantilia 1989)
- 10–15°C and 90% r.h. for 2–3 weeks (Snowdon 1990)
- 10–12°C and 90% r.h. for 2–3 weeks (Snowdon 1991)
- 10°C and 85–90% r.h. for 14–21 days (SeaLand 1991).

Chilling injury

Fruits held for 1 week at 0°C showed surface pitting and had objectionable flavours (Phillips and Armstrong 1967; Lutz and Hardenburg 1968) and those stored at 1.7°C for 2 months appeared to be sound externally but were inedible when cut open (Wardlaw 1937). Chilling injury can occur in storage below 7°C and loss of skin colour may occur during storage below 10°C (Tindall 1983).

Controlled atmosphere storage

Controlled atmosphere storage had only a slight to no effect (SeaLand 1991).

Ripening

Flesh red colour was intensified during storage at 21.1°C (Lutz and Hardenburg 1968).

Water spinach, tong cai, kang kong

Ipomea reptans, *I. aquatica*, Convolvulaceae.

Botany

As the name implies, it is an aquatic herbaceous plant cultivaed throughout Asia. The young terminal shoots and leaves are used as spinach. It is also used as an animal feed (Purseglove 1968).

Storage

The shoots are washed after harvest and tied in bundles of 8–10 shoots. They can be packed in banana leaves or polyethylene-lined crates to prevent wilting (Tindall 1983).

West Indian gooseberries, otahiete gooseberries

Phyllanthus acidus, Euphorbiaceae.

Botany

It produces pale green or yellowish fruit with six ribs, each containing a flat, brown seed, and are up to 3 cm in diameter. They resemble the gooseberry, *Ribes grossularia*, in flavour, but are even more acidic. When stewed with sugar they turn bright red. They originated from India and Madagascar.

Storage

No information is available on storage.

White radishes, Japanese radishes, mooli, daikon

Raphanus sativus variety *acanthiformis*, Cruciferae.

Botany

The edible portion is the white taproot, which is cylindrical and tapering towards the bottom and can be as much as 50 cm long and weigh up to 2 kg. It has a strong, fiery taste due to the presence of mustard oils (Mercantilia 1989).

Harvesting

In Australia the preferred market weight (1 kg) was reached in 60 days in the spring–summer period and 82–85 days in the autumn–winter period (Harris *et al.* 2000).

Physiological disorder

The physiological disorder pithiness is characterized by the formation of air spaces in the root that will eventually leave it spongy, dry and hollow. It developed both in the field and in storage (Harris *et al.* 2000). The development of pithiness was related to cultivar, but not to senescence, since it developed in the field in Long White from an early stage of growth.

Storage

Harris *et al.* (2000) found that Tomas was the best cultivar overall, and market quality was maintained during a simulated export period of 4 weeks, provided that it was harvested at the preferred market weight of 1 kg. All leaf material should be trimmed from the radishes before storage. Postharvest life was limited mainly by the physiological disorder pithiness (Harris *et al.* 2000). At a room temperature of 20°C and 60% r.h. it was reported in Mercantilia (1989) that they could be kept for about 1 week. Refrigerated storage recommendations are as follows:

- 0°C and 90–95% r.h. for 4 months (Mercantilia 1989)
- 4°C for 2 months (Mercantilia 1989)
- 8°C for 1 month (Mercantilia 1989)
- 0°C and 90–95% r.h. for 120 days (SeaLand 1991).

White sapote, zapote

Casimiroa edulis, Rutaceae.

Botany

The fruit is a berry that is yellowish green in colour, ovoid in shape and up to 8 cm in diameter. It has a thin skin and a custard-like pulp, which is creamy yellow and contains up to five large seeds. Its respiratory pattern led to it being classified as an indeterminate fruit by Biale and Barcus (1970).

Harvesting

Harvest maturity is when the skin colour changes from green to yellowish, but the fruit is still firm.

Storage

Mature fruit will soften at tropical ambient conditions in 2–3 days. A refrigerated storage recommendation was 19.4°C and 85–90% r.h. for 14–21 days (SeaLand 1991).

White yams, negro yams, guinea yams

Dioscorea rotundata, Dioscoreaceae.

Botany

The edible portion is the tuber, which is normally cylindrical, but can be palmate. They have brown skin and white flesh and normally weigh around 2–5 kg, but can be a much as 25 kg.

Harvesting

This is usually carried out about 7–9 months after planting when the vines die down (Purseglove 1972; Kay 1987). Harvesting is by hand, but some success has been achieved with mechanical harvesting for plants grown on high ridges (Kay 1987).

Pest and disease control

Yam tubers are traditionally dusted with wood ash or sometimes lime (calcium hydroxide) directly after harvesting to reduce subsequent storage rot, by small-scale farmers in Nigeria (Coursey 1961) and Jamaica (Thompson 1972a). Tubers were treated with either lime (calcium oxide) or local gin (fermented palm wine or sap from *Elaeis guineensis*) by Ogali *et al.* (1991) and stored at 25–32°C with 62–90% r.h. for 32 weeks. After 24 weeks, the tubers that had been treated with gin showed no rotting, whereas 22.5% of the lime-treated yams were rotten. After 32 weeks, the level of rotting was comparable in all treatments. Four species were isolated: *Aspergillus niger*, *Fusarium oxysporum*, *Penicillium citrinum*, all of which were found to be mildly pathogenic, and *Fusarium moniliforme* [*Gibberella fujikuroi*], which did not produce any rot symptoms. Postharvest dipping in a benzimidazole fungicide effectively controlled storage rot (Thompson 1972a), but its use is now restricted in many countries. Tubers of the cultivar Iyawo inoculated by spraying with a conidiospore suspension of *Trichoderma viride* in potato dextrose broth showed a drastic reduction in the range and number of mycoflora, including the pathogens *Rhizopus* sp., *Rhizoctonia* sp., *Aspergillus niger*, *A. flavus*, *Penicillium oxalicum*, *Botryodiplodia theobromae*, *Fusarium oxysporum*, *F. solani*, *Neurospora* sp. and *Choanephora* sp., on the tuber surface during 5 months of storage in a traditional yam barn (Okigbo and Ikediugwu 2001).

Infection with parasitic nematodes in the field can cause tissue just below the skin to die. This necrosis can increase during ambient storage, because the nematodes increase in numbers. Storage at 13°C resulted in a rapid decrease in nematode numbers in the tubers and thus prevented spoilage (Thompson *et al.* 1973a).

Curing

Exposure of the tubers to 35–40°C and 95–100% r.h. for 24 hours was found to initiate the curing process (Thompson 1972). The effect of the curing process was to form a suberized layer and a cork cambium in a similar way to that reported for potatoes (Figure 12.109) (Been *et al.* 1976). Cured tubers had a lower weight loss throughout storage than to those which had not been cured (Figure 12.110). There were problems with bacterial soft rots developing during subsequent storage, but these were overcome by loading the tubers into the curing room, heating them until the tubers reached 40°C and then injecting the steam to increase the humidity. Tubers cured in this way had reduced weight loss, fungal infection and necrotic tissue after storage compared with tubers which had not been cured. However, cured tubers sprouted earlier by about 2 weeks

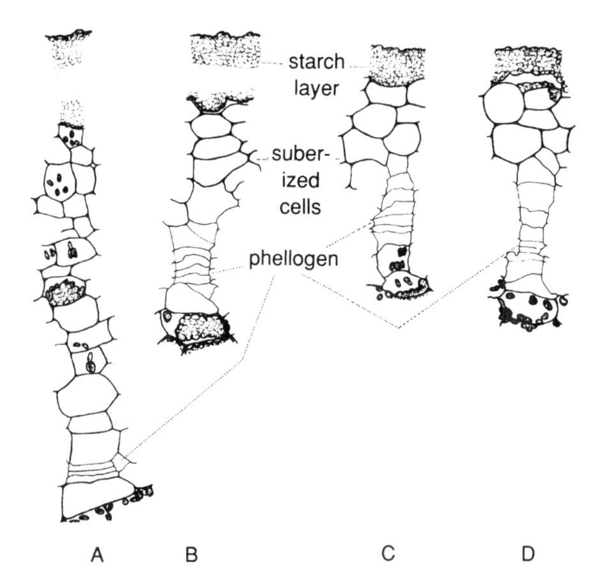

Figure 12.109 Histological sections through the cut surfaces of yam tubers (*Dioscorea cayenensis*) after 7 days of curing and 8 days of subsequent storage. Curing conditions were (A) direct sunlight; (B) 26°C and 66% r.h.; (C) 30°C and 91% r.h.; (D) 40°C and 98% r.h. Source: Been *et al.* (1976).

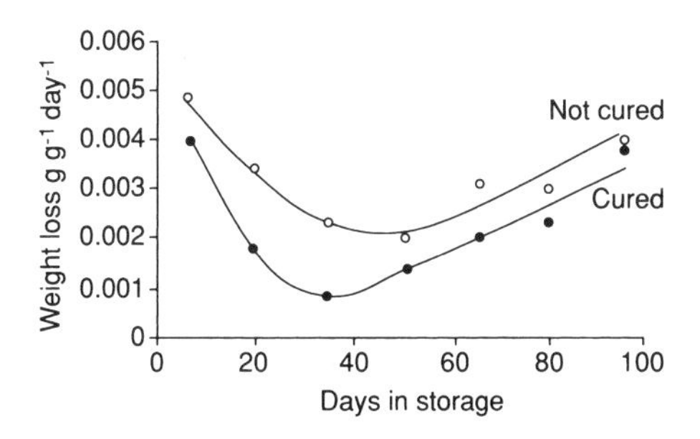

Figure 12.110 Effect of curing at 35–40°C and 95–100% r.h. compared with tubers which were not cured on the storage loss rate of yams (*Dioscorea rotundata*) during storage in Jamaican ambient conditions of 25–34°C and 64–92% r.h. Source: Thompson *et al.* (1972a) and unpublished data.

when stored at tropical ambient temperatures, probably owing to the increased cambial activity associated with curing.

Storage

In tropical ambient conditions, Kay (1987) indicated that sound tubers can be stored for up to 4 months, but desiccation and sprouting could be problems. In Nigeria, yams are stored in specially constructed barns (see Figures 7.6 and 7.7). In a comparison between barn storage and pit storage for yams in Nigeria, Ezeike (1985) recorded weight losses over a 5-month period of 60% for the former method but only 15–25% for the latter. Yams stored in slatted wooden trays in the coastal region of Cameroon had losses of 29–47% in only 2 months (Lyonga 1985). Refrigerated storage recommendations include 16°C and 80% r.h. for several months (Tindall 1983).

Sprouting

Tubers of the cultivar Sree Latha stored at room temperature of 30–32°C and 80–85% r.h. in the dark sprouted 70–80 days after harvest. In the apical and basal regions, the starch content decreased while the sugars and α-amylase activity increased (Muthukumarasamy and Panneerselvam 2000). In Nigerian ambient conditions, sprouting occurred within 8 weeks (Coursey 1961), whereas in ambient conditions in Jamaica sprouting could occur after only 1 month (Thompson 1972).

Chilling injury

At 1.1°C, total physiological breakdown occurred within 10 days and chilling injury was observed on tubers stored at 12.5°C (Coursey 1968). In later work, tubers stored at 11.7°C kept in better condition than those stored in tropical ambient conditions. However, when the storage period exceeded 1-week, the tubers lost weight rapidly and became soft, necrotic and infected with fungi shortly after removal to higher temperatures. This confirms Coursey's findings on chilling injury (Thompson 1972a; Thompson *et al.* 1977a).

Modified atmosphere packaging

Cured tubers were dipped in 500 µl litre–1 thiabendazole, dried and individually sealed in 0.04 mm thick polyethylene film. During storage for 5 months at 12–13°C, tubers lost only 3% in weight, did not sprout, had no internal necrosis and no surface lenticel proliferation (Thompson 1972a). This latter effect has previously been shown to occur on plastic-wrapped yam tubers at ambient temperatures and can be seen as small white dots on the tuber surface. All the tubers removed from storage to ambient conditions sprouted within 3 weeks.

Wild cucumbers, pepinos

Cyclanthera pedata variety *edulis*, Cucurbitaceae.

Botany

The fruit is a fleshy berry-like structures called a pepo and is yellowish or green, slightly spiny, ovoid in shape, about 5 cm long and hollow with many seeds. It is a native of Mexico and has a strong smell.

Storage

The only recommendation that could be found is 7–10°C and 85–90% r.h. for 1–2 weeks (Hardenburg *et al.* 1990).

Winged beans

Psophocarpus tetragonolobus, Leguminosae.

Botany

It is a climbing tropical perennial plant, which produces four-sided pods with characteristic serrated wings running down the four corners and are up to 35 cm long (Kay 1979). They contain up to 20 seeds that can be white, yellow, brown or black or even mottled. The edible portion is both the immature pods and the mature seeds. They are thought to have originated in Africa.

Harvesting

The pods are harvested while still immature and tender, about 2 weeks after fertilization. If harvesting is delayed they will become fibrous. When grown for seed they are left on the plant until the swollen seeds are clearly visible within the pods. Harvesting is done by hand.

Seed storage

Seeds with moisture contents of 12.5, 14.5, 16.5 or 18.5% were stored for 24 weeks. All seeds stored at 8°C retained good colour and odour qualities and high percentage germination. Seeds with a 14.5% moisture content stored above 15°C and seeds with higher moisture contents stored above 8°C lost their aesthetic qualities; this loss was associated with the incidence of invasion by storage fungi (*Aspergillus glaucus*, *A. flavus* and *A. candidus*) (Onesirosan 1986). Germination was reduced by 50% when fresh seed of 20–21% moisture content was stored for 2 weeks and there was no germination after 2 months of storage (Hew and Lee 1981).

Pod storage

Refrigerated storage recommendations are as follows:

- 10°C and 90% r.h. for up to 21 days with up to 21% loss (Tindall 1983)
- 10°C and 90% r.h. for 28 days (SeaLand 1991).

Wood blewits

Lepista nuda

Botany

It is a gill fungus up to 10 cm in diameter and 10 cm high. The cap is domed and conical as it emerges and gradually flattens and may even become depressed in the centre. It is brown in colour, sometimes tinged reddish when young, with a smooth surface with a waxy feel and lilac-coloured flesh. The gills are an intense bluish purple, fading to buff as they develop. They have pink spores. The stalk has a bulbous base and tapers slightly and is bluish white, sometimes streaked and lined. Its natural habitat is deciduous woods and hedgerows and it tastes delicious. A related species, *L. nebularis*, was found by Vetter (1993) to have a protein content of 39% of its dry weight.

Storage

No information could be found on its postharvest life, but from the literature it would appear that all types of edible mushrooms have similar storage requirements and the recommendations for the cultivated mushroom (*Agaricus bispsorus*) could be followed.

Wood mushroom

Agaricus silvaticus.

Botany

It is closely related to the cultivated mushroom *A. bisporus* and has a brown, scaly cap with a stalk that tends to be bulbous towards the base and has a distinct ring. It has a mild flavour, but is good to eat and has a sour smell.

Storage

No information could be found on its postharvest life, but from the literature it would appear that all types of edible mushrooms have similar storage requirements and the recommendations for the cultivated mushroom (*Agaricus bispsorus*) could be followed.

Yacon, jiquima, aricuma

Polymnia sonchifolia, Compositae.

Botany

This is an ancient staple crop of the high Andes. The edible portion is the fusiform tuberous root, which is up to 20 cm long and 10 cm in diameter and can weigh as much as 2 kg, although 100–500 g is more usual. They mainly have a soft, purplish, bark-like skin and transparent watery flesh (Kay 1987).

Harvesting

The tubers are lifted by hand, usually by levering with a spade from below and pulling the stems.

Storage

The tubers can be stored for several months in a cool, dark place (Kay 1987).

Yam bean, jicama

Pachsyrrhizus erosus, Leguminosae.

Botany

It is a native of Mexico and the edible portion is the white-fleshed tuberous root, which is generally turbinate in shape, but this can vary in different growing conditions (Figure 12.111). Immature seed pods can be eaten, but mature seeds are toxic since they may contain the insecticide rotenone (Purseglove 1968). Edem *et al.* (1990) referred to yam bean as *Sphenostylis stenocarpa.*

Harvesting

Schroeder (1967) indicated that harvest maturity was when the roots had reached 10–15 cm in diameter. Before this the tuberous roots were said to have insufficient flavour and older ones tended to be woody due to lignification.

Storage

Delaying harvesting for 2–3 months can be used as a storage method and is achieved by cutting off the tops and not irrigating. Just before harvesting the crop is irrigated to rehydrate the tuberous roots (Schroeder 1967). In Mexico the roots may be harvested, washed and stored in baskets in a cool environment. Moisture loss was reported to be the most important storage problem especially with immature tuberous roots

Figure 12.111 Yam bean (*Pachsyrrhizus erosus*) harvested into baskets in Malaysia.

(Schroeder 1967). Refrigerated storage recommendations are as follows:

- 0°C for 2 months (Kay 1987)
- 9–10°C arrests the tendency of the skin to change colour from creamy white to purplish brown (Kay 1987)
- 12.8°C and 65–70% r.h. for 30–60 days (SeaLand 1991).

Yams

Dioscorea spp., Dioscoreaceae.

Botany

This is a general summary of the yam. There are several species of *Dioscorea* that are used for food, many of them are major staple crops in tropical countries with a few in temperate and subtropical countries. The edible portion is the tuber, in some cases also the bulbils, most of which are borne singly, although several small tubers are borne in some species. They have a dark brown skin and usually white flesh, although some have other colours including yellow. Specific information is also found under the headings for the different types of yam. These include:

- *Dioscorea alata*, Asiatic yam, Lisbon yam, White yam
- *Dioscorea bulbifera*, Potato yam
- *Dioscorea cayenensis*, Yellow yam
- *Dioscorea dumetorum*, Bitter yam
- *Dioscorea esculentum*, Asiatic yam
- *Dioscorea hispida*, Asiatic bitter yam
- *Dioscorea opposita*, Chinese yam
- *Dioscorea rotundata*, White yam
- *Dioscorea trifida*, Yampie, Cush cush.

Amorphophallus spp. are also referred to as elephant yams.

Storage

Storage recommendations, which did not define which species was being referred to, are as follows:

- 16°C and 65% r.h. for 4 months (Mercantilia 1989)
- 13.3°C 85–90% r.h. for 50–115 days (SeaLand 1991).

Yampies, cush cush, elephant yams, Indian yams

Dioscorea trifida, *D. brasiliensis*, Dioscoreaceae.

Botany

They produce several small tubers on each plant at the end of short stolons. These have a brown skin and white flesh and are up to 20 cm long and club shaped or spherical. It is believed to have originated in Guyana.

Harvesting

They are normally harvested about 10 or 11 months after planting by digging by hand.

Handling

In a study of handling of yampies (Thompson 1972a), it was shown that subjecting the tubers to impacts commonly experienced during handling resulted in greatly increased losses during subsequent storage (Table 12.61).

Sprouting

Generally, Kay (1987) indicated that sprouting in tropical conditions takes between 1 and 8 weeks after harvesting. In Jamaica, tubers began to sprout within 3 weeks in storage at 20–29°C and 46–62% r.h. (Thompson 1973).

Table 12.61 Effects of dropping tubers of *Dioscorea trifida* from a height of 2 m on to a concrete floor on their subsequent losses during storage at about 28°C for 64 days (source: Thompson 1972a)

	Loss in weight(%)	Fungal score[a]	Internal necrosis (%)	Sprouting (%)
Dropped	47.5	1.5	61	64
Not dropped	23.6	0.2	5	97

[a] See Table 12.66.

Storage

Tubers can be stored in a cool, dry, well ventilated location, possibly for up to 1 year (Kay 1987), or in barns at 16–18°C and 60–65% r.h. (Tindall 1983). Refrigerated storage recommendations are as follows:

- 3°C for 1 month (Kay 1987)
- 10°C for several months (Tindall 1983).

Chilling injury

In trials in Jamaica, Thompson (1972) found that tubers stored well at 12°C with about 84% r.h., but on removal to ambient temperature they lost weight rapidly, became soft, developed internal necrosis and had copious surface mould growth, which appeared to be due to chilling injury.

Controlled atmosphere storage

Controlled atmosphere storage had a slight to no effect (SeaLand 1991).

Modified atmosphere packaging

Sealing tubers in polyethylene film bags reduced losses (Table 12.62) (Thompson 1972a).

Yanagimatsutake mushrooms

Agrocybe cylindracea, Basidiomycetes

Botany

It is a basidiomycete gill fungus that grows on clumps of poplar and willow and has a cap of up to 10 cm diameter, pale buff in colour, darkening towards the centre. The gills are brown and the stem has a ring and is pale buff.

Storage

No information could be found on its postharvest life, but from the literature it would appear that all types of edible mushrooms have similar storage requirements and the recommendations for the cultivated mushroom (*Agaricus bispsorus*) could be followed.

Yellow yams, twelve months yams

Dioscorea cayenensis, Dioscoreaceae.

Botany

Some authorities consider yellow yams to be a variety of D. rotundata. The edible portion is the tuber, which is borne singly and can weigh up to 10 kg, but more usually are within the range 500–2 kg. They have brown skin and yellow flesh.

Harvesting

There is no clear dormancy period, but the vines die back about 12 months after planting and then

Table 12.62 Effects of packaging material on the quality of *D. trifida* after 64 days at 20–29°C and 46–62% r.h.

Packaging type	Weight loss (%)	Fungal score (0–5)[a]	Necrotic tissue (%)[b]
Paper bags	24	0.2	5
Polyethylene bags with 0.15% of the area as holes	16	0.2	7
Sealed 0.03 mm thick polyethylene bags	5	0.4	4

[a] Fungal score was 0 = no surface fungal growth to 5 = tuber surface entirely covered with fungi.
[b] Necrotic tissue was estimated by cutting the tuber into two lengthways and measuring the area of necrosis and expressing it as a perentage of the total cut surface.

regrowth. At this time the vines may be cut off and the tubers may be harvested and dug from the ground. Jamaica is a major producer and there they are harvested by digging up the tuber while the vine is still attached. The top of the tuber, called the head end, is cut off with the vine attached and replanted in the same place. This means that there is a large cut surface on the harvested tuber that is easily infected by microorganisms.

Pest and disease control

A common way of controlling postharvest diseases is to dip the tubers in a fungicide. However, dichloran treatment was found actually to increase rotting (Thompson *et al.* 1977a). In Jamaica, benzimidazole fungicides were used for many years. However, during commercial application, a rot caused by infection with *Penicillium sclerotigenum* was frequently observed on the treated tubers (Plumbley *et al.* 1984). In *in vitro* tests this organism was found to be tolerant to benomyl. This tolerance was confirmed in *in vivo* tests, but the organism was highly susceptible to the fungicide imazalil (see Table 8.2). However, the use of benomyl or imazalil is now not permitted in many importing countries and an alternative method of disease control is needed (Thompson *et al.* 1977a).

Field infestation of yellow yam tubers with parasitic nematodes was shown to increase when they were stored in tropical ambient conditions, resulting in areas of necrotic tissues. However, when they were stored at 13°C there was no increase in nematode population in the tubers and no increase in necrosis (Thompson *et al.* 1973a).

Curing

Exposure of the tubers to 35–40°C and 95–100% r.h. for 24 hours initiates the curing process and controls storage rots. This treatment could well replace chemical fungicides for postharvest disease control. See the section on *D. rotundata* for details of the process.

Storage

In ventilated storage in ambient conditions of 24–31°C and 52–68% r.h., tubers lost 41% in weight after 4 months (Thompson *et al.* 1973). The storage period was too long and should probably be confined to a maximum of 1 month. Refrigerated storage recommendations are as follows:

- 13°C and 95% r.h. for less than 4 months with 29% weight loss and internal necrosis (Thompson *et al.* 1973)
- 16°C and 80% r.h. for 60 days (Tindall 1983).

Sprouting

At tropical temperatures tubers sprouted about 4 weeks after harvesting (Coursey 1961). The most popular variety of yellow yam in Jamaica is called Roundleaf and that began to sprout about 90 days after harvest compared with the variety Common that sprouted after about 50 days. This study was at ambient temperatures of 25–34°C and 64–92% r.h. (Thompson 1972a). A proprietary potato sprout suppressant containing CIPC and IPC in a dust formulation had no effect on sprouting. Low-temperature storage can control sprouting. No sprouting occurred on any tubers at 13°C during 5 months of storage, but there was chilling injury at temperatures below 15°C when stored for over 1 month (Thompson *et al.* 1973), so this method was not appropriate.

Youngberries

Rubus sp., *R. ursinus derivative*, Rosaceae

Botany

Some authorities indicate that it is a dewberry (*R. caesius*) × loganberry (*R. loganobaccus*) hybrid, others that it is a loganberry derivative. Berries are classified as non-climacteric fruit. The fruit are a collection of drupes and closely resemble elongated blackberries, and are sometimes erroneously referred to as a blackberry variety. They taste similar to loganberries, but are sweeter. They originated in the USA, and were introduced into New Zealand, where they are grown commercially (Wood *et al.* 1999).

Harvesting

They are normally harvested by hand with the receptacle in place.

Storage

Refrigerated storage recommendations are –0.6 to 0°C and 85–90% r.h. for 2–4 days (Phillips and Armstrong 1967).

Zapotes chupa chupa

Matisia cordata, Steruliaceae.

Botany

The edible portion is the fruit, which is oval in shape with an apical projection and a persistent calyx. The skin is grey–brown with a felt-like texture. The flesh is fibrous and contains five seeds.

Storage

The only reference to postharvest behaviour is storage at 6–9°C and 90–95% r.h. (Amezquita 1973).

Appendix: glossary of terms

Abbreviations

These conform to British Standard 5775, Parts 0–15, and are the same for singular and plural. SI units are Le Système International d'Unités and are accepted internationally as standards.

Carbon dioxide and oxygen in controlled atmosphere stores

Increasingly in scientific journals, the carbon dioxide (CO_2) and oxygen (O_2) levels in controlled atmosphere stores is expressed as kPa (kilopascals), which is a unit of pressure. Normal atmospheric pressure is 1013.25 millibars (mbar) and 1 m is equivalent to 0.1 kPa. Hence one atmosphere (1atm) is 101.325 kPa, and kPa is a pressure of approximately 1% of one atmosphere. If 1 atm was exactly 1000 mbar, then 1 kPa would have been exactly 1% of 1 atm. In terms of percentages, levels of CO_2 and O_2 in store are referred to % by volume rather than by mass. For a constituent of a total atmosphere (in this case CO_2 and O_2 as a constituent of atmospheric air) the volume of constituent divided by the volume of total = partial pressure of constituent divided by the pressure of the total. Hence the volume of a constituent is divided by volume of the total. For 5% CO_2, it would be 5 divided by 101.35 = 4.9% CO_2.

In this book, in order to simplify the information and provide comparisons between the work of different authors, all the levels of gases are expressed as % on the basis that 1% ≈ approximately 1 kPa.

Concentration of chemicals and ethylene in a store

Traditionally they have been referred to as parts per million (p.p.m.), which is based on mass for mass or volume for volume. It has become more conventional to express these terms as microlitres per litre and abbreviated as μl^{-1}. They are the same so that 100 μl $litre^{-1}$ ethylene is the same as 100 p.p.m of ethylene.

Film thickness

Plastic films are used in postharvest technology of fruits and vegetables, but their thickness is expressed in different units. The approximate relationships between the units used to describe the thickness of plastic films are given as follows:

Gauge		mm		Microns (μm)		Inches
4	=	0.001	=	1	=	0.00004
48	=	0.012	=	12	=	0.00048
50	=	0.0125	=	12.5	=	0.0005
75	=	0.01875	=	18.75	=	0.00075
100	=	0.025	=	25	=	0.001
125	=	0.03125	=	31.25	=	0.00125
150	=	0.0375	=	37.5	=	0.0015
160	=	0.04	=	40	=	0.0016
180	=	0.045	=	45	=	0.0018
200	=	0.05	=	50	=	0.002
300	=	0.075	=	75	=	0.003
400	=	0.1	=	100	=	0.004
500	=	0.125	=	125	=	0.005

Humidity

The amount of atmospheric moisture in a store is also referred to as the humidity and is expressed at relative humidity or % r.h. It may be defined as the ratio of actual water vapour pressure in air to the maximum possible water vapour pressure at the existing temperature. That is the amount of water vapour relative to saturation (100% r.h.) at that temperature. It can also be expressed as vapour pressure deficit (VPD). VPD is the difference between the saturation vapour pressure and the actual vapour pressure at the existing temperature. VPD can be expressed in millibars (mbar) or millimetres of mercury (mmHg).

Temperature (°C)	Vapour pressure deficit (mmtlg)	Approximate relative humidity (%)
12	3.8	60
12	2.7	80
12	0.6	94
6.5	0.4	94

Hypobaric storage

Hypobaric storage or low-pressure storage has been used for several decades as a way of affecting the partial pressure of oxygen in fruit and vegetable stores. Pressure is referred to as millimetres of mercury (mmHg) where atmospheric pressure at sea level is 760 mmHg. This can also be referred to 1 atm and reduced pressures as a fraction of normal pressure.

References

Aalbersberg, W.G.L. and Limalevu, L. 1991. Cyanide content in fresh and processed Fijian cassava (*Manihot esculenta*) cultivars. *Tropical Science* **31**, 249–256.

Abbas, M.F. 1997. Jujube. In Mitra, S. (Editor), *Postharvest Physiology* and *Storage of Tropical and Subtropical Fruits.* CAB International, Oxford, 405–416.

Abdul Karim, M.N.B., Nor, L.M. and Hassan, A. 1993. The storage of sapadilla (*Manikara achras L.*) at 10, 15, and 20°C. *Postharvest Handling of Tropical Fruit. Proceedings of an International Conference held in Chiang Mai, Thailand, 19–23 July 1993. Australian Centre for International Agricultural Research Proceedings* **50**, 443.

Abdul Raouf, U.M., Beuchat, L.R. and Ammar, M.S. 1993. Survival and growth of *Escherichia coli* O157:H7 on salad vegetables. *Applied and Environmental Microbiology,* **59**, 1999–2006.

Abdullah, H. and Pantastico, E.B. 1990. *Bananas.* Association of Southeast Asian Nations, COFAF, Jakarta, 147 p.p.

Abdullah, H. and Tirtosoekotjo, S. 1989. *Association of Southeast Asian Nations Horticulture Produce Data Sheets.* Association of Southeast Asian Nations Food Handling Bureau, Kuala Lumpur.

Abe, Y. 1990. Active packaging: a Japanese perspective. In Day, B.P.F. (Editor), *International Conference on Modified Atmosphere Packaging Part 1.* Campden Food and Drinks Research Association, Chipping Campden.

Abu Baker, F. and Abdul Karim, M.N.B. 1994. Chemical treatments for microbial control on sapota. *Association of Southeast Asian Nations Food Journal* **9**, 42–43.

Acedo, A.L., Jr. 1997. Storage life of vegetables in simple evaporative coolers.

Adamicki, F. 1989. Przechowywanie warzyw w kontrolowanej atmosferze. *Biuletyn Warzywniczy, Supplement* **I**, 107–113.

Adamicki, F. and Elkner, K. 1985. Effect of temperature and a controlled atmosphere on cauliflower storage and quality. Wplyw temperatury i kontrolowanej atmosfery na przechowywanie i jakosc kalafiorow. *Biuletyn Warzywniczy* **28**, 197–224.

Adamicki, F. and Kepka, A.K. 1974. Storage of onions in controlled atmospheres. *Acta Horticulturae* **38**, 53–73.

Adamicki, F., Dyki, B. and Malewski, W. 1977. Effect of CO_2 on the physiological disorders observed in onion bulbs during CA storage. *Quality of Plant Foods and Human Nutrition* **XXVII**, 239–248.

Adesina, A.A. and Aina, J.O. 1990. Preservation of African star apple fruits (*Chrysophyllum albidum*) by osmotic dehydration. *Tropical Science* **30**, 249–253.

Adesuyi, N.O. 1973. Advances in yam storage research in Nigeria. *Proceedings of the Third International Symposium on Tropical Root Crops, Ibadan, Nigeria,* 428–433.

Afek, U., Orenstein, J. and Nuriel, E. 1998. Increased quality and prolonged storage of sweet potatoes in Israel. *Phytoparasitica* **26**, 307–312.

Agabbio, M.D, Aquino, S., Piga, A. and Molinu, M.G. 1999. Agronomic behaviour and postharvest response to cold storage of Malvasio mandarin fruits. *Fruits Paris* **54**, 103–114.

Agar, I.T., Cetiner, S., Garcia, J.M. and Streif, J. 1994. A method of chlorophyll extraction from fruit peduncles: application to redcurrants. *Turkish Journal of Agriculture and Forestry* **18**, 209–212.

Agar, T., Bangerth, F. and Streif, J. 1994a. Effect of high CO_2 and controlled atmosphere concentrations on the ascorbic acid, dehydroascorbic acid and total vitamin C content of berry fruits. *Acta Horticulturae* **398**, 93–100.

Agar, T., Streif, J. and Bangerth, F 1991. Changes in some quality characteristics of red and black currants stored under CA and high CO_2 conditions. *Gartenbauwissenschaft* **56**, 141–148.

Aharoni, N. and Ben Yehoshua, S. 1973. Delaying deterioration of romaine lettuce by vacuum cooling and modified atmosphere produced in polyethylene packages. *Journal of the American Society for Horticultural Science* **98**, 464–468.

Aharoni, Y. and Houck, L.G. 1980. Improvement of internal color of oranges stored in O_2-enriched atmospheres. *Scientia Horticulturae* **13**, 331–338.

Aharoni, Y. and Houck, L.G. 1982. Change in rind, flesh, and juice color of blood oranges stored in air supplemented with ethylene or in O_2-enriched atmospheres. *Journal of Food Science* **47**, 2091–2092.

Aharoni, N., Reuveni, A. and Dvir, O. 1989. Modified atmospheres in film packages delay senescence and decay of fresh herbs. *Acta Horticulturae* **258**, 255–262.

Aharoni, Y., Nadel Shiffman, M. and Zauberman, G. 1968. Effects of gradually decreasing temperatures and polyethylene wraps on the ripening and respiration of avocado fruits. *Israel Journal of Agricultural Research* **18**, 77–82.

Ahmadi, H., Biasi, W.V.and Mitcham, E.J. 1999. Control of brown rot decay of nectarines with 15% carbon dioxide atmospheres. *Journal of the American Society for Horticultural Science* **124**, 708–712.

Ahmed, A., Clarke, B. and Thompson, A.K. 2001. Banana harvest maturity and fruit position on the quality of ripe fruit. *Annals of Applied Biology* **139**, 329–335.

Ahrens, F.H. and Milne, D.L. 1993. Alternatiewe verpakkingsmetodes vir see uitvoer van lietsjies om SO_2-behandeling te vervang. *South African Litchi Growers' Association Yearbook* **5**, 29–30.

Akamine, E.K. 1969. Controlled atmosphere storage of papayas. *Proceedings of the 5th Annual Meeting, Hawaii Papaya Industry Association, September 12–20, 1969. Hawaii Cooperative Extention Service Miscellaneous Publication* **764**, 23 pp.

Akamine, E.K. and Goo, T. 1969. Effects of controlled atmosphere storage on fresh papayas *Carica papaya* L. variety Solo with reference to shelf-life extention of fumigated fruits. *Hawaii Agricultural Experimental Station, Honolulu, Bulletin* **144**, 22 pp.

Akamine, E.K. and Goo, T. 1973. Respiration and ethylene production during ontogeny of fruit. *Journal of the American Society for Horticultural Science* **98**, 381–383.

Akamine, E.K., Beaumont, J.H., Bowers, F.A.I., Hamilton, R.A., Nishida, T., Sherman, G.D., Shoji, K. and Storey, W.B. 1957. Passionfruit cultivars in Hawaii. *University of Hawaii Extension Circular*, 35 pp.

Ake, M., Eba, B., Malan, A.K. and Atindehou, E. 2001. Liquid chromatographic determination of patulin in fruit juices sold in Cote d'Ivoire. Détermination de la patuline dans les jus de fruits commercialisés en Cote d'Ivoire. *Sciences des Aliments* **21**, 199–206.

Akhimienho, H.D. 1999. *The control of vascular streaking in fresh cassava.* MPhil thesis, Cranfield University.

Akinaga, T. and Kohda, Y. 1993. Environmental condition during air shipment of horticultural products from Okinawa to Tokyo. *Proceedings of ICAMPE '93 October 19–22 KOEX,* Korean Society for Agricultural Machinery, Seoul, Korea, 413–422.

Al Zaemey, A.B.S. Falana, I.B. and Thompson, A.K. 1989. Effects of permeable fruit coatings on the storage life of plantain and bananas. *Aspects of Applied Biology* **20**, 73–80.

Al Zaemey, A.B.S., Magan, N. and Thompson, A.K. 1993. Studies on the effect of fruit coating polymers and organic acids on growth of *Colletotrichium musae in vitro* and on postharvest control of anthracnose of bananas. *Mycological Research* **97**, 1463–1468.

Al Zaemey, A.B.S., Magan, N. and Thompson, A.K. 1994. *In vitro* studies of the effect of environmental conditions on the anthracnose pathogen of bananas, *Colletotrichum musae. International Biodeterioration and Biodegradation* 369–381.

Alarcon, L., Benavides, L. and Parodi, G. 1998. Effect of temperature and storage period on papayuela *Carica pubescens* Lenne et Koch germination. *Proceedings of the Interamerican Society for Tropical Horticulture* **41**, 242–245.

Ali Azizan 1988, quoted by Abdullah and Pantastico (1990).

Ali Niazee, M.T., Richardson, D.G., Kosittrakun, M. and Mohammad, A.B. 1991. Non insecticidal quarantine treatments for apple maggot control in harvested fruits. *Proceedings of the Fifth International Controlled Atmosphere Research Conference, Wenatchee, Washington, USA, 14–16 June*, 1989, **1**, 193–205.

Alique, R. and Oliveira, G.S. 1994. Changes in sugars and organic acids in cherimoya *Annona cherimola* Mill. fruit under controlled atmosphere storage. *Journal of Agricultural and Food Chemistry.* **42**, 799–803.

Allen, F.W. 1953. The influence of growth regulator sprays on the growth, respiration and ripening of Bartlett pears. *Proceedings of the American Society of Horticultural Science* **62**, 279–298.

Allen, F.W. and McKinnon, L.R. 1935. Storage of Yellow Newtown apples in chambers supplied with artificial atmospheres. *Proceedings of the American Society for Horticultural Science* **32**, 146.

Allen, F.W. and Smock, R.M. 1938. CO_2 storage of apples, pears, plums and peaches. *Proceedings of the American Society for Horticultural Science* **35**, 193–199.

Allsop, D.J. 1991. *The non destructive testing of apples for determination of harvest maturity.* MPhil thesis, Silsoe College, Cranfield Institute of Technology. 104 pp.

Allwood, M.E. and Cutting, J.G.M. 1994 Progress report: gas treatment of 'Fuerte' avocados to reduce cold damage and increase storage life. *Yearbook South African Avocado Growers' Association* **17**, 22–26.

Alper, Y., Erez, A. and Ben Arie, R. 1980. A new approach to mechanical harvesting of fresh market peaches grown in a meadow orchard. *Transactions of the American Society of Agricultural Engineers* **23**, 1084–1088.

Alvindia, D.G., Kobayashi, T., Yaguchi, Y. and Natsuaki, K.T. 2000. Evaluation of cultural and postharvest practices in relation to fruit quality problems in Philippine non chemical bananas. *Japanese Journal of Tropical Agriculture* **44**, 178–185.

Amarakoon, R., Sarananda, K.H. and Illeperuma, D.C.K. 1999. Quality of mangoes as affected by stage of maturity. *Tropical Agricultural Research* **11**, 74–85.

Amen, K.I.A. 1987 Effect of post harvest calcium treatments on the storage life of guava fruits. *Assiut Journal of Agricultural Sciences* **18**, 127–137.

Ameny, M.A. 1990. Traditional post-harvest technology of cassava in Uganda. *Tropical Science* **30**, 41–50.

Amezquita Garcia, A. 1973. *Almacenamiento refrigerado de frutas y hortalizas.* Departamento de Mercadeo, Prodesarrollo, Federacion National de Cafeteros de Colombia, Bogota, 36 pp.

Amos, N.D., Cleland, D.J., Cleland, A.C. and Banks, N.H. 1993. The effect of coolstore design and operation on air relative humidity. *Proceedings of ICAMPE '93, October 19–22 KOEX,* Korean Society for Agricultural Machinery, Seoul, Korea, 433–442.

Anandaswamy, B., and Iyengar, N.V.R. 1961. Pre packaging of fresh snap beans *Phaseolus vulgaris Food. Science* **10**, 279.

Anandaswamy, B. *et al.*, 1963. Pre packaging of fresh produce. IV. Okra *Hibiscus esculentus. Food Science* **12**, 332.

Anderson, R.E., Hardenburg, R.E. and Vaught, H.C. 1963. Controlled atmosphere storage studies with cranberries. *Proceedings of the American Society of Horticultural Science* **83**, 416–422.

Andre, P., Blanc, R., Buret, M., Chambroy,Y., Flanzy, C., Foury, C., Martin, F. and Pelisse, C. 1980. Globe artichoke storage trials combining the use of vacuum pre refrigeration, controlled atmospheres and cold. Essais de conservation d'artichauts par l'utilisation combinée de la preréfrigeration par le vide, d'atmosphères modifiées et du froid. *Revue Horticole* **211**, 33–40.

Andre, P., Buret, M., Chambroy, Y., Dauple, P., Flanzy, C. and Pelisse, C. 1980a. Conservation trials of asparagus spears by means of vacuum prerefrigeration, associated with controlled atmospheres and cold storage. Essais de conservation de turions d'aspèrge au moyen de la preréfrigeration par le vide, associée à des atmosphères modifiées et au froid. *Revue Horticole* **205**, 19–25.

Andrich, G. and Fiorentini, R. 1986. Effects of controlled atmosphere on the storage of new apricot cultivars. *Journal of the Science of Food and Agriculture* **37**, 1203–1208.

Anelli, G., Mencarelli, F. and Massantini, R. 1989. One time harvest and storage of tomato fruit: technical and economical evaluation. *Acta Horticulturae* **287**, 411–415.

Anon. 1919 Food Investigation Board. *Department of Scientific and Industrial Research Report for the Year 1918,* 21–23.

Anon. 1920 Food Investigation Board. *Department of Scientific and Industrial Research Report for the Year 1920,* 16–25.

Anon. 1941. Annual Report for 1940. *Massachusetts Agricultural Experiment Station,* Bulletin 378.

Anon. 1958. Food Investigation Board. *Department of Scientific and Industrial Research Report for the Year 1957,* 35–36.

Anon. 1964. State Storage Regulations for Certified Controlled Atmosphere of Apples. *US Department of Agriculture, Federal Extension Service,* December, 3 pp.

Anon. 1967. *Recommended Conditions for Cold Storage of Perishable Produce.* International Institute of Refrigeration, Paris,, Basic Publication, 100 pp.

Anon. 1968. *Fruit and Vegetables.* American Society of Heating, Refrigeration and Air conditioned Engineering, *Guide and Data Book* 361.

Anon. 1974, *Atmosphere Control in Fruit Stores.* Ministry of Agriculture, Fisheries and Food, Agricultural Development and Advisory Service, Short Term Leaflet 35.

Anon. 1978. Report of the Steering Committee for Study on Postharvest Food Losses in Developing Countries. National Research Council, National Science Foundation, Washington, DC, 206 pp.

Anon. 1978a. Banana CA storage. *Bulletin of the International Institute of Refrigeration* **18**, 312.

Anon. 1980. Peaches, small is beautiful. *Economist* May 24, 113.

Anon. 1987. Freshtainer makes freshness mobile. *International Fruit World* **45**, 225–231.

Anon. 1987a. *The Role of the Potato Marketing Board in Great Britain.* Potato Marketing Board Leaflet, June 1987, 24 pp.

Anon. 1988. *Shipping Guide for Perishables.* SeaLand Service Incorporated, Iselin, NJ, 26 pp.

Anon. 1988a. *Standards for Fresh Fruits and Vegetables.* United Nations, New York, 398 pp.

Anon. 1988b. *Sanyo Technical Review, Japan* **20**, 1116.

Anon. 1990. *Solar Driven Cold Store Project. Final Report. Stork Product Engineering, Amsterdam,* 44 pp.

Anon. 1993. *Future of Potato Marketing Scheme.* Press Release by the British Ministry of Agriculture Fisheries and Food, 420/93, 30 November 1993.

Anon. 1993a. *The Fresh Produce Desk Book.* Lockwood Press.

Anon. 1993b. Onion in store without MH. H.D.C. Progress Report. *Grower* April 12–13.

Anon. 1993c. Development of ume shaker harvester. *RNAM Newsletter* **47**, 13.

Anthony, R.W.V. 1991. *Electrostatic application of low volume sprays for postharvest treatment of bananas.* MPhil thesis, Silsoe College, Cranfield Institute of Technology.

Anuradha, G., and Saleem, S. 1998. Certain biochemical changes and the rate of ethylene evolution in ber fruits *Zizyphus mauritiana* Lamk. cultivar Umran as affected by the various post harvest treatments during storage. *International Journal of Tropical Agriculture* **16**, 233–238.

Apeland, J. 1971. Factors affecting respiration and colour during storage of parsley. *Acta Horticulturae* **20**, 43–53.

Apeland, J. 1985. Storage of Chinese cabbage *Brassica campestris* L. *pekinensis* Lour Olsson in controlled atmospheres, *Acta Horticulturae* **157**, 185–191.

Apelbaum, A., Aharoni, Y and Temkin Gorodeiski, N. 1977. Effects of subatmospheric pressure on the ripening processes of banana fruits. *Tropical Agriculture Trinidad* **54**, 39–46.

Aracena, J.J., Sargent, S.A., Brecht, J.K., Campbell, C.A. and Saltveit, M.E. 1993. Environmental factors affecting vascular streaking, a postharvest physiological disorder of cassava root *Manihot esculenta* Crantz. *Acta Horticulturae* **343**, 297–299.

Archbold, D.D., Hamilton Kemp, T.R. and Fallik, E. 2000. Aroma volatiles as modulators of postharvest mold development on fruit: *in vivo* role and fumigation tools. *Acta Horticulturae* **518**, 87–92.

Arjona, H.E., Matta, F.B. and Garner J.O., Jr. 1992. Temperature and storage time affect quality of yellow passionfruit. *HortScience* **27**, 809–810.

Arpaia, M.L., Mitchell, F.G., Kader A.A. and Mayer, G. 1985. Effects of 2% O_2 and various concentrations of CO_2 with or without ethylene on the storage performance of kiwifruit. *Journal of the American Society for Horticultural Science* **110**, 200–203.

Arras, G. and Arru, S. 1999. Inhibitory activity of antagonist yeasts against citrus pathogens in packinghouse trials. *Agricoltura Mediterranea* **129**, 249–255.

Arras, G. and Usai, M. 2001. Fungitoxic activity of 12 essential oils against four postharvest citrus pathogens: chemical analysis of *Thymus capitatus* oil and its effect in subatmospheric pressure conditions. *Journal of Food Protection* **64**, 1025–1029.

Artes, F. and Martinez, J.A. 1999. Quality of cauliflower as influenced by film wrapping during shipment. *European Food Research and Technology* **209**, 330–334.

Artes, F., Sanchez, E. and Tijskens, L.M.M. 1998. Quality and shelf life of tomatoes improved by intermittent warming. *Lebensmittel Wissenschaft und Technologie* **31**, 427–431.

Artes, F., Tudela, J.A. and Gil, M.I. 1998a. Improving the keeping quality of pomegranate fruit by intermittent warming. *Zeitschrift für Lebensmittel-Untersuchung und-Forschung. A, Food Research and Technology* **207**, 316–321.

Artes, F., Tudela, J.A. and Villaescusa, R. 2000. Thermal postharvest treatments for improving pomegranate quality and shelf life. *Postharvest Biology and Technology* **18**, 245–251.

Artes Calero, F., Escriche, A., Guzman, G. and Marin, J.G. 1981. Violeta globe artichoke storage trials in a controlled atmosphere. Essais de conservation d'artichaut 'Violeta' en atmosphere controlée. *30 Congresso Internazionale di Studi sul Carciofo, 1979,* 1073–1085.

Artschwager, E.F. and Starrett, R.C. 1931. Suberization and wound periderm formation in sweetpotato and gladiolus, as affected by temperature and humidity. *Journal of Agricultural Research* **43**, 353–364.

Asami, I. and Aoyagi, M. 1997. Studies on the shipping of high quality mature green Japanese apricot mume, *Prunus mume* Sieb. et Zucc. fruits. III. Effects of permeable plastic film pouch packaging on the freshness of fruits. *Research Bulletin of the Aichi ken Agricultural Research Center* **29**, 231–237.

Asen, S., Parsons, C.S. and Stuart, N.W. 1964. Experiments aimed at prolonging Narcissus display life. *Florists Review* **134**, 1–2.

Askar, A., El Nemr, S.E. and Bassionny, S.S. 1986. Aroma constituents in white and pink guava fruits. *Alimenta* **25**, 162–167.

Askary, M.A. and Shahedi, M. 1994. The effect of storage temperature on some qualitative and quantitative characters of the fruits of pomegranate cultivars. *Iranian Journal of Agricultural Sciences* **25**, 53–64.

Atkinson, D., Jackson, J.E., Sharples, R.O. and Waller, W.M. 1980. *Mineral Nutrition of Fruit Trees.* Butterworths, London. 435 pp.

Augustin, M.A. and Azudin, M.N. 1986. Storage of mangosteen *Garcinia mangostana* L. *Association of Southeast Asian Nations Food Journal* **2**, 78–80.

Avissar, I., Marinansky, R. and Pesis, E. 1989. Postharvest decay control of grape by acetaldehyde vapors. *Acta Horticulturae* **258**, 655–660.

Awad, M.A.G., Jager, A., Roelofs, F.P.M.M., Scholtens, A. and De Jager, A. 1993. Superficial scald in Jonagold as affected by harvest date and storage conditions. *Acta Horticulturae* **326**, 245–249.

Azizan 1988, quoted by Abdullah and Pantastico (1990).

Baab, G. 1987. Improvement of strawberry keeping quality. Verbessern der Erdbeer Haltbarkeit. *Obstbau,* **12**, 265–268.

Babarinsa, F.A., Williams, J.O. and Osanu, F.C. 1997. Storage of carrots in a brick wall cooler under semi-arid conditions. *Tropical Science* **37**, 21–27.

Badiyala, S.D. 1998. Studies on maturity standards in peach *Prunus persica* L. Batsch cultivar Flordasun. *Himachal Journal of Agricultural Research* **23**, 77–80.

Badran, A.M. 1969. *Controlled atmosphere storage of green bananas.* US Patent 3 450 542.

Baez, M.A., Siller, J.H., Heredia, J.B., Portillo, T., Araiza, E., Garcia, R.S. and Muy, M.D. 1998. Postharvest physiology of sapodilla *Achras sapota* L. fruits during marketing conditions. Fisiologia poscosecha de frutos de chicozapote *Achras sapota* L. durante condiciones de mercadeo. *Proceedings of the Interamerican Society for Tropical Horticulture* **41**, 209–214.

Banana Ripening Guide 1972. Why are bananas ripened commercially? CSIRO *Division of Food Research Circular* 8, Food Research Laboratory, North Ryde, NSW, 1670 Australia, 3–13.

Bancroft, R.D. 1989. *The effect of surface coating on the development of postharvest fungal rots of pome fruit with special reference to 'Conference' pears.* PhD thesis, University of Cambridge, 517 pp.

Bandyopadhyay, C. and Tewari, G.M.J. 1976. Lachrymatory factor in spring onion. *Journal of the Science of Food and Agriculture* **27**, 733–735.

Bandyopadhyay, C., Gholap, A.S. and Mamdapur, V.R. 1985. Characterisation of alkenyl resorcinol in mango *Mangifers indica* latex. *Journal of Agriculture and Food Chemistry* **33**, 377–379.

Bangarth, F. 1974. Hypobaric storage of vegetables. *Acta Horticulturae* **38**, 23–32.

Banks, N.H. 1984. Some effects of Tal prolong coating on ripening of bananas. *Journal of Experimental Botany* **35**, 127–137.

Banks, N.H. 1990. Factors affecting the severity of deflowering latex stain on banana bunches in the Winward Islands. *Tropical Agriculture Trinidad* **67**, 111–115.

Barger, W.R. 1949. Further tests with vacuum precooling on fruits and vegetables. *United States of America Department of Agriculture Market Research Report* 244.

Barger, W.R. 1961. Factors affecting temperature reduction and weight loss in vacuum cooled lettuce. *United States of America Department of Agriculture Market Research Report* **469**, 20 pp.

Barger, W.R. 1963. Vacuum precooling. A comparison of cooling of different vegetable. *United States of America Department of Agriculture Market Research Report* **600**, 12 pp.

Barkai Golan, R. and Phillips, D.J. 1991. Postharvest heat treatment of fresh fruits and vegetables for decay control. *Plant Disease* **75**, 1085–1089.

Barmore, C.R. 1987. Packing technology for fresh and minimally processed fruits and vegetables. *Journal of Food Quality* **10**, 207–217.

Barnell, H. R. 1943. Studies in tropical fruits. Carbohydrate metabolism of the banana fruit during storage at 53 °F. *Annals of Botany, New Series* **9**, 1–22.

Barnes, H. 1943. A wireway for bananas. *Queensland Agricultural Journal* **57**, 207–211.

Basanta, A.L. and Sankat, C.K. 1995. Storage of the pomerac *Eugenia malaccensis. Technologias de Cosecha y Postcosecha de Frutas y Hortalizas. Proceedings of a Conference held in Guanajuato, Mexico, 20–24 February 1995*, 567–574.

Basiouny, F.M. and Basiouny, A. 2000. Effects of liquid calcium and control atmosphere on storability and quality of kiwifruit *Actinidia chinensis* Planch cultivar 'Hayward.' *Acta Horticulturae* **518**, 213–221.

Bastrash, S. Makhlouf, J., Castaigne, F. and Willemot, C. 1993. Optimal controlled atmosphere conditions for storage of broccoli florets, *Journal of Food Science* **58**, 338–341.

Batten, D.J. 1990. Effect of temperature on ripening and post harvest life of fruit of atemoya *Annona cherimola* Mill. × *A. squamosa* L. cultivar 'African Pride'. *Scientia Horticulturae* **45**, 129–136.

Batu, A. 1998. Effects of long term controlled atmosphere storage on postharvest ripening qualities of tomatoes. Kontrollu atmosfer kosullarinda uzun sureli depolamanin domatesin hasat sonrasi olgunlasma kalitesine etkileri. *Bahce* **26**, 37–47.

Batu, A. and Thompson, A.K. 1993. Effects of crosshead speed and probe diameter on instumental measurement of tomato firmness. *Proceeding of ICAMPE '93, October 19–22, KOEX*, Korean Society for Agricultural Machinery, Seoul, Korea, 1340–1345.

Batu, A. and Thompson, A.K. 1994. The effects of harvest maturity, temperature and thickness of modified atmosphere packaging films on the shelf life of tomatoes. *Commissions C2, D1, D2/3 of the International Institute of Refrigeration International Symposium, June 8–10 Istambul Turkey.*

Batu, A. and Thompson, A.K. 1995. Effects of controlled atmosphere storage on the extension of postharvest qualities and storage life of tomatoes. *Workshop of the Belgium Institute for Automatic Control, Ostend, June 1995*, 263–268.

Bauchot, A.D., Hallett, I.C., Redgwell, R.J. and Lallu, N. 1999. Cell wall properties of kiwifruit affected by low temperature breakdown. *Postharvest Biology and Technology* **16**, 245–255.

Baumann, H. 1989. Adsorption of ethylene and CO_2 by activated carbon scrubbers. *Acta Horticulturae* **258**, 125–129.

Bautista, O.K. 1990. *Postharvest Technology for Southeast Asian Perishable Crops.* Technology and Livelihood Resource Center, 302 pp.

Bautista Banos, S., Long, P.G. and Ganesh, S. 2000. The role of relative humidity during early storage on *Botrytis cinerea* incidence of kiwifruit. *Revista Mexicana de Fitopatologia* **18**, 92–96.

Beaudry, R., Schwallier, P. and Lennington, M. 1993. Apple maturity prediction: an extension tool to aid fruit storage decisions. *HortTechnology* **3**, 233–239.

Beaudry, R.M. and Gran, C.D. 1993. Using a modified atmosphere packaging approach to answer some post harvest questions: factors influencing the lower O_2 limit. *Acta Horticulturae* **326**, 203–212.

Been, B. O., Perkins, C. and Thompson, A. K. 1976. Yam curing for storage, *Acta Horticulturae* **62**, 311–316.

Been, B.O., Marriott, J. and Perkins, C. 1975. Wound periderm formation in dasheen and its affects on storage. *Proceedings of the Caribbean Food Crop Society, Trinidad* **13**.

Been, B.O., Thompson, A.K. and Perkins, C. 1974. Effects of curing on stored yams. *Proceedings of the Caribbean Food Crops Society, Trinided* **12**, 38–42.

Behrsing, J., Winkler, S., Franz, P. and Premier R. 2000. Efficacy of chlorine for inactivation of *Escherichia coli* on vegetables. *Postharvest Biology and Technology* **19**, 187–192.

Bellesia, F., Pinetti, A., Bianchi, A. and Tirillini, B. 1998. The volatile organic compounds of black truffle *Tuber melanosporum*. Vitt. from middle Italy. *Flavour and Fragrance Journal* **13**, 56–58.

Ben Arie, R. 1996. Fresher via boat than airplane. *Peri News* **11**, Spring.

Ben Arie, R. and Guelfat Reich, S. 1976. Softening effects of CO_2 treatment for the removal of astringency from stored persimmon fruits. *Journal of the American Society for Horticultural Science* **101**, 179–181.

Ben Arie, R. and Or, E. 1986. The development and control of husk scald on 'Wonderful' pomegranate fruit during storage. *Journal of the American Society for Horticultural Science* **111**, 395–399.

Ben Arie, R. and Sonego, L. 1985. Modified atmosphere storage of kiwifruit *Actinidia chinensis* Planch with ethylene removal. *Scientia Horticulturae* **27**, 263–273.

Ben Arie, R., Levine, A., Sonego, L. and Zutkhi,Y. 1993. Differential effects of CO_2 at low and high O_2 on the storage quality of two apple cultivars. *Acta Horticulturae* **326**, 165–174.

Ben Arie, R., Roisman, Y., Zuthi, Y. and Blumenfeld, A. 1989. Gibberellic acid reduces sensitivity of persimmon fruits to ethylene. *Advances in Agricultural Biotechnology* **26**, 165–171.

Ben Yehoshua, S., Fang, DeQiu, Rodov, V., Fishman, S. and Fang, D.Q. 1995. New developments in modified atmosphere packaging. Part II. *Plasticulture* **107**, 33–40.

Ben Yehoshua, S., Kim, J.J. and Shapiro, B. 1989a. Curing of citrus fruit, applications and mode of action. *International Controlled Atmosphere Conference, Fifth, Proceedings, June 14–16, 1989, Wenatchee, Washington. Volume 2. Other Commodities and Storage Recommendations*, 179–196.

Ben Yehoshua, S., Kim, J.J. and Shapiro, B. 1989b. Elicitation of resistance to the development of decay in sealed citrus fruit by curing. *Acta Horticulturae* **258**, 623–630.

Bender, R.J., Brecht, J.K. and Campbell, C.A. 1994. Responses of 'Kent' and 'Tommy Atkins' mangoes to reduced O_2 and elevated CO_2. *Proceedings of the Florida State Horticultural Society* **107**, 274–277.

Bender, R.J., Brecht, J.K., Sargent, S.A. and Huber, D.J. 2000. Low temperature controlled atmosphere storage for tree ripe mangoes *Mangifera indica* L. *Acta Horticulturae* **509**, 447–458.

Benge, J.R., Banks, N.H., Tillman, R and Nihal de Silva, H.N. 2000. Pairwise comparison of the storage potential of kiwifruit from organic and conventional production systems. *New Zealand Journal of Crop and Horticultural Science* **28**, 147–152.

Benzioni, A., Mendlinger, S. and Ventura, M. 1993. Germination, fruit development, yield and postharvest characteristics of *Cucumis metuliferus*. In Huyskens, S., Janick, J. and Simon, J.E. (editors), *Proceedings of the Second National Symposium: New Crops, Exploration, Research and Commercialization, Indianapolis, Indiana, October 6–9, 1991*. Wiley, New York, 553–557.

Berard, J.E. 1821. Memoire sur la maturation des fruits. *Annales de Chimie et de Physique* **16**, 152–183.

Berard, L.S. 1985. Effects of CA on several storage disorders of winter cabbage. Controlled atmospheres for storage and transport of perishable agricultural commodities. *Fourth National Controlled Atmosphere Research Conference, July 1985*. North Carolina State University, Raleigh, NC, 150–159.

Berard, L.S., Vigier, B, Crete, R. and Chiang, M. 1985. Cultivar susceptibility and storage control of grey speck disease and vein streaking, two disorders of winter cabbage. *Canadian Journal of Plant Pathology* **7**, 67–73.

Berard, S.L., Vigier, B. and Dubuc Lebreux, M.A. 1986. Effects of cultivar and controlled atmosphere storage on the incidence of black midrib and necrotic spot in winter cabbage. *Phytoprotection* **67**, 63–73.

Berger, H., Galletti, Y.L., Marin, J., Fichet, T. and Lizana, L.A 1993. Efecto de atmosfera controlada y el encerado en la vida postcosecha de cherimoya *Annona cheimola* Mill. cv. Bronceada. *Proceedings of the Interamerican Society for Tropical Horticulture* **37**, 121–130.

Berlage, A.G. and Langmo, R.D. 1979. Shake harvesting tests with fresh market apples. *Transactions of the American Society of Agricultural Engineers* **22**, 733–738, 745.

Bernardin, J., Sankat, C.K. and Willemot, C. 1994. Breadfruit shelf-life extension by CaCl2 treatment. *Inter-American Institute for Cooperation on Agriculture, Tropical Fruits Newsletter* **11**, 6.

Berrang, M.E., Brackett, R.E. and Beuchat, L.R. 1989. Growth of *Listeria monocytogenes* on fresh vegetables stored under controlled atmosphere. *Journal of Food Protection* **52**, 702–705.

Berrang, M.E., Brackett, R.E. and Beuchat, L.R. 1989a. Growth of *Aeromonas hydrophila* on fresh vegetables stored under a controlled atmosphere. *Applied and Environmental Microbiology* **55**, 2167–2171.

Berrang, M.E., Brackett, R.E. and Beuchat, L.R. 1990. Microbial, color and textural qualities of fresh asparagus, broccoli, and cauliflower stored under controlled atmosphere. *Journal of Food Protection* **53**, 391–395.

Bertolini, P., Pratella, G.C., Tonini, G. and Gallerani, G. 1991. Physiological disorders of 'Abbe Fetel' pears as affected by low O_2 and regular controlled atmosphere storage. Paper presented at the Conference *Technical Innovations in Freezing and Refrigeration of Fruits and Vegetables, held in Davis, California, 9–12 July, 1989*, 61–66.

Berwick, E.J.H. 1940. Mangoes in Krian. *Malayan Agricultural Journal* **28**, 517.

Best, R.A. 1993. *The development of branding in the United Kingdom fresh fruit markets*. MSc thesis, Silsoe College, Cranfield University.

Betts, G.D. (Editor). 1996. *A Code of Practice for the Manufacture of Vacuum and Modified Atmosphere Packaged Chilled Foods*. Campden and Chorleywood Food Research Association Guideline 11.

Bhadra, S. and Sen, S.K. 1999. Post harvest storage of custard apple *Annona squamosa* L. fruit var. Local Green under various chemical and wrapping treatments. *Environment and Ecology* **17**, 710–713.

Bhatnagar, D.K. and Sharma, N.K. 1994. Maturity studies in bottlegourd. *Research and Development Reporter* **11**, 34–37.

Bhatnagar, D.K., Singh, B.P. and Mangal, J.L. 1986. Keeping quality of knol khol as influenced by different irrigation intensities and nitrogen doses. *Haryana Journal of Horticultural Sciences* **15**, 233–238.

Biale, J.B. 1950. Postharvest physiology and biochemistry of fruits. *Annual Review of Plant Physiology* **1**, 183–206.

Biale, J.B. 1960. Respiration of fruits. In Ruhland, W. (Editor), *Handbuch der Pflanzenphysiologie. Encyclopedia of Plant Physiology*, Vol. **12**, 536–592.

Biale, J.B. and Barcus, D.E. 1970. Respiratory patterns in tropical fruits of the Amazon basin. *Tropical Science* **12**, 93–104.

Biale, J.B. and Young, R.E. 1962. The biochemistry of fruit maturation. *Endeavour* **21**, 164–174.

Birth, G.A. 1983. Optical Radiation. In Mitchell, B.W. (Editor), *Instumentation and Measurement for Environmental Sciences.* American Society of Agricultural Engineers Special Publication, 13–82.

Birth, G.S. and Norris, K.H. 1958. An instrument using light transmittance for nondestructive measurement of fruit maturity. *Food Technology* **12**, 592–595.

Birth, G.S., Dull, G.G., Magee, J.B., Chan, H.T. and Cavaletto, C.G. 1984. An optical method for estimating papaya maturity. *Journal of the American Society for Horticultural Science* **109**, 62–69.

Bishop, C.F.H. 1992. *Energy efficiency of cooling systems for potato storage.* PhD thesis, Silsoe College, Cranfield University.

Bishop, D. 1996. Controlled atmosphere storage. In (Editor), Dellino, C.J.V. *Cold and Chilled Storage Technology.* Blackie, London.

Blackbourn, H.D., Jeger, M.J., John, P., and Thompson, A.K. 1990. Inhibition of degreening in the peel of bananas ripened at tropical temperatures, III. Changes in plastid ultrastructure and chlorophyll protein complexes accompanying ripening in bananas and plantains. *Annals of Applied Biology* **117**, 147–161.

Blackbourn, H.D., John, P. and Jeger, M. 1989. The effect of high temperature on degreening in ripening bananas. *Acta Horticulturae* **258**, 271–278.

Blankenship, S.M. and Sisler E.C. 1991. Comparison of ethylene gassing methods for tomatoes. *Postharvest Biology and Technology* **1**, 59–65.

Blanpied, G.D. 1960. Guides used in determining maturity as an aid to picking, and the relative merits of each method. *105th Annual Meeting of the New York State Horticultural Society* 177–184.

Blednykh, A.A., Akimov, Yu.A., Zhebentyaeva, T.N, and Untilova, A.E. 1989. Market and flavour qualities of the fruit in sweet cherry following storage in a controlled atmosphere. *Byulleten' Gosudarstvennogo Nikitskogo Botanicheskogo Sada* **69**, 98–102.

Bleinroth, E.W., Garcia, J.L.M. and Yokomizo 1977. Conseracao de quatro variedades de manga pelo frio e em atmosfera controlada. *Coletanea de Instituto de Tecnologia de Alimentos* **8**, 217–243.

Bliss, M.L. and Pratt, H.K. 1979. Effects of ethylene, maturity and attachment to the parent plant on production of volatile compounds by muskmelons. *Journal of the American Society for Horticultural Science* **104**, 273–277.

Blumenfeld, A. 1993. Maturation and harvesting criteria for avocado. *Postharvest Handling of Tropical Fruit. Proceedings of an International Conference held in Chiang Mai, Thailand 19–23 July 1993. Australian Centre for International Agricultural Research Proceedings* **50**, 343.

Blumenfeld, A. 1993a. When to harvest maturity standards versus harvesting indices. *Postharvest Handling of Tropical Fruit. Proceedings of an International Conference held in Chiang Mai, Thailand, 19–23 July, 1993. Australian Centre for International Agricultural Research Proceedings* **50**, 104.

Blythman, J. 1996. The Food We Eat. Michael Joseph, London.

Bonghi, C., Ramina, A., Ruperti, B., Vidrih, R. and Tonutti, P. 1999. Peach fruit ripening and quality in relation to picking time, and hypoxic and high CO_2 short term postharvest treatments. *Postharvest Biology and Technology* **16**, 213–222.

Boon-Long, P. Achariyaviriya, S. and Johnson, G.I. 1994. Mathematical modelling of modified atmosphere conditions. *Development of Postharvest Handling Technology for Tropical Tree Fruits: a Workshop Held in Bangkok, Thailand, 16–18 July 1992. Australian Centre for International Agricultural Research Proceedings* **58**, 63–67.

Booth, R.H. 1975. *Cassava Storage.* International Potato Centre, Lima, Series EE 16.

Booth, R.H. 1977. Storage of fresh cassava *Manihot esculenta.* II. Simple storage techniques. *Experimental Agriculture* 13, 119–128.

Botondi, R., Crisa, A., Massantini, R. and Mencarelli, F. 2000. Effects of low oxygen short term exposure at 15°C on postharvest physiology and quality of apricots harvested at two ripening stages. *Journal of Horticultural Science and Biotechnology* 75, 202–208.

Bottcher, H. and Gunther, I. 2001. Physiological postharvest response of hyssop (*Hyssopus officinalis* L.). Physiologisches Nacherntevehalten von Ysop (*Hyssopus officinalis* L.). *Zeitschrift für Arznei und Gewurzpflanzen* 6, 73–78.

Bottcher, H., Gunther, I. and Franke, R. 2000. Physiological postharvest response of lemon balm (*Melissa officinalis* L.). Physiologisches Nacherntevehalten von Zitronenmelisse (*Melissa officinalis* L.). *Zeitschrift für Arznei und Gewurzpflanzen* 5, 145–153.

Bottcher, H., Gunther, I. and Warnstorff, K. 2000a. Physiological postharvest response of savory (*Satureja hortensis* L.). Physiologisches Nacherntevehalten von Bohnenkraut (*Satureja hortensis* L.). *Gartenbauwissenschaft* 65, 22–29.

Bottcher, H., Gunther, I. Franke, R. and Warnstorff, K. 2001. Physiological postharvest responses of matricaria (*Matricaria recutita* L.) flowers. *Postharvest Biology and Technology* 22, 39–51.

Bouman, H. 1987. CA storage of head cabbage has a future with long storage. CA bewaring van sluitkool heeft toekomst bij lange bewaring. *Groenten en Fruit* 42, 156 157.

Bowen, S.A., Muir, A.Y. and Dewar, C.T. 1996. Investigations into skin strength in potatoes: factors affecting skin adhesion strength. *Potato Research* 39, 313–321.

Bower, J.P., Cutting, J.G.M. and Truter, A.B. 1990. Container atmosphere, as influencing some physiological browning mechanisms in stored Fuerte avocados. *Acta Horticulturae* 269, 315–321.

Bowers, S.V., Dodd, R.B. and Han, Y.J. 1988. Nondestructive testing to determine internal quality of fruit. *American Society of Agricultural Engineers*, Paper 88–6569.

Boyle, F.P., Seagrave Smith, H., Sakata, S. and Sherman, G.D. 1957. *Hawaii Agricultural Experimental Station Bulletin* 111.

Braaksma, A., Schaap, D.J. and Schipper, C.M.A. 1999. Time of harvest determines the postharvest quality of the common mushroom *Agaricus bisporus*. *Postharvest Biology and Technology* 16, 195–198.

Brach, E.J., Phan, C.T., Poushinsky, B., Jasmin, J.J. and Aube, C.B. 1982. Lettuce maturity detection in the visible 380–720 nm, far red 680–750 nm and near infrared 800–1850 nm wavelength band, *Lactuca sativa. Agronomique Science Prod. Vég. et de l'Environ.* 2, 685–694.

Brackman, A. 1989. Effect of different CA conditions and ethylene levels on the aroma production of apples. *Acta Horticulturae* 258, 207–214.

Brackmann, A. and Balz, G. 2000. Effects of precooling and postharvest treatments on the quality of Gala apples after CA storage. Efeito do pre-resfriamento e tratamentos quimicos pos-colheita sobre a qualidade da maca 'Gala' apos o armazenamento em atmosfera controlada. *Revista Cientifica Rural* 5, 12–20.

Brackmann, A. and Waclawovsky, A.J. 2000. Storage of apple (*Malus domestica Borkh.*) cv. Braeburn. Conservacao da maca (*Malus domestica* Borkh.) cv. Braeburn. *Ciencia Rural* 30, 229–234.

Brackmann, A., Hunsche, M. and Mazaro, S.M. 1999. Effect of high CO_2 and modified atmosphere storage on the postharvest quality of strawberries *Fragaria ananassa* L. cultivar Tangi. Efeito do alto CO_2 e do armazenamento em atmosfera modificada sobre a qualidade pos colheita de morangos *Fragaria ananassa* L. cultivar Tangi. *Revista Cientifica Rural* 4, 58–63.

Brackmann, A., Steffens, C.A. and Mazaro, S.M. 1999a. Storage of kaki *Diospyros kaki* L., cultivar Fuyu, in modified and controlled atmosphere storage. Armazenamento de caqui *Diospyros kaki* L., cultivar Fuyu, em condicoes de atmosfera modificada e controlada. *Revista Brasileira de Armazenamento* 24, 42–46.

Brackmann, A., Streif, J. and Bangerth, F. 1993. Relationship between a reduced aroma production and lipid metabolism of apples after long term controlled atmosphere storage. *Journal of the American Society for Horticultural Science* 118, 243–247.

Brackmann, A., Streif, J. and Bangerth, F. 1995. Influence of CA and ULO storage conditions on quality parameters and ripening of preclimacteric and climacteric harvested apple fruits. II. Effect on ethylene, CO_2, aroma and fatty acid production. Einfluss von CA bzw. ULO Lagerbedingungen auf Fruchtqualitat und Reife bei praklimakterisch und klimakterisch geernteten Apfeln. II. Auswirkung auf Ethylen, CO_2, Aroma und Fettsaureproduktion. *Gartenbauwissenschaft* 60, 23.

Bramlage, W.J., Bareford, P.H., Blanpied, G.D., Dewey, D.H., Taylor, S., Porritt, S.W., Lougheed, E.C., Smith, W.H. and McNicholas, F.S. 1977. CO_2 treatments for 'McIntosh' apples before CA storage. *Journal of the American Society for Horticultural Science* 102, 658–662.

Brash, D.W., Corrigan, V.K. and Hurst, P.L. 1992. Controlled atmosphere storage of 'Honey 'n' Pearl' sweet corn. *Proceedings of Annual Conference, Agronomy Society of New Zealand* 22, 35–40.

Bredmose, N. 1980. Effects of low pressure on storage life and subsequent keeping quality of cut roses. *Acta Horticulturae* **113**, 73–79.

Brewster, J.L. 1987. The effect of temperature on the rate of sprout growth and development within stored onion bulbs. *Annuals of Applied Biology* **111**, 463–464.

Brigati, S., Pratella, G.C. and Bassi, R. 1989. CA and low O_2 storage of kiwifruit: effects on ripening and disease. *International Controlled Atmosphere Conference, Fifth, Proceedings, June 14–16, 1989, Wenatchee, Washington, United States of America. Volume 2. Other Commodities and Storage Recommendations*, 41–48.

Brodrick, H.T. and Thord Gray, R. 1982. *Research Report of the South African Mango Growers Association* **2**, 223–226.

Brooks, C. 1932. Effect of solid and gaseous CO_2 upon transit disease of certain fruits and vegetables. *US Department of Agriculture, Technical Bulletin* 318, September, 6.

Brooks, C. 1940. Modified atmospheres for fruits and vegetables in storage and in transit. *Refrigerating Engineering* **40**, 233–236.

Brooks, C. and Cooley, J. S. 1917. Effect of temperature, aeration and humidity on Jonathan spot and scald of apples in storage. *Journal of Agricultural Research* **12**, 306–307.

Brooks, C., and McColloch, L.P. 1938. Stickiness and spotting of shelled green lima beans. *US Department of Agriculture, Technical Bulletin* 625.

Broughton, W.J. and Wong, H.C. 1979. Storage conditions and ripening of chiku fruits *Achras sapota* L. *Scientia Horticulturae* **10**, 377–385.

Broughton, W.J., Chan, B.E. and Kho, H.L. 1978. Maturation of Malaysian fruits. II. Storage conditions and ripening of banana *Musa sapientum* var. 'Pisang Emas.' *Malaysian Agricultural Research and Development Institute, Research Bulletin* 7, 28–37.

Brown, B.I. and Scott, K.J. 1985. Cool storage conditions for custard apple fruit *Annona atemoya* Hort. *Singapore Journal of Primary Industries* **13**, 23–31.

Brown, G.E. and Albrigo, L.G. 1972. Grove application of benomyl and its persistence in orange fruit. *Phytopathology* **62**, 1434–1438.

Brown, G.E., Davis, C. and Chambers, M. 2000. Control of citrus green mold with Aspire is impacted by the type of injury. *Postharvest Biology and Technology* **18**, 57–65.

Buescher, R.W. 1977. Hardcore in sweetpotato roots as influenced by cultivar, curing and ethylene. *HortScience* **12**, 326.

Buescher, R.W., Sistrunk, W.A. and Brady, P.L. 1975. Effects of ethylene on metabolic and quality attributes in sweet potato roots. *Journal of Food Science* **40**, 1018–1020.

Bull, C.T., Wadsworth, M.L., Pogge, T.D., Le, T.T., Wallace, S.K. and Smilanick, J.L. 1998. Molecular investigations into mechanisms in the biological control of postharvest diseases of citrus. *Molecular Approaches in Biological Control, Delemont, Switzerland, 15–18 September, 1997. Bulletin OILB SROP* **21**, 1–6.

Burden, J.N. 1997. Postharvest handling for export. In Mitra, S. (Editor), *Postharvest Physiology and Storage of Tropical and Subtropical Fruits.* CAB International, Oxford, 1–20.

Burden, O.J. 1969. Control of postharvest diseases in bananas. *Queensland Agricultural Journal* **95**, 621–624.

Burden, O.J. and Griffee, P.J. 1974. A simple machine for application of fungicide to harvested green bananas. *PANS* **20**, 358–364.

Burden, O.J. and Wills, R.B.H. 1989. *Prevention of Postharvest Food Losses: Fruits, Vegetables and Root Crops.* Food and Agriculture Organization of the United Nations, Rome, Training Series 17/2, 157 pp.

Burg, S.P. 1973. Hypobaric storage of cut flowers. *HortScience* **8**, 202–205.

Burg, S.P. 1975. Hypobaric storage and transportation of fresh fruits and vegetables. In Haard, N. F. and Salunkhe, D. K. (Editors). *Postharvest Biology and Handling of Fruits and Vegetables.* AVI Publishing, Westport, 172–188.

Burg, S.P. 1993. Current status of hypobaric storage. *Acta Horticulturae* **326**, 259–266.

Burg, S.P and Burg, E.A. 1962. The role of ethylene in fruit ripening. *Plant Physiology* **37**, 179–189.

Burg, S.P and Burg, E.A. 1967. Molecular requirements for the biological activity of ethylene. *Plant Physiology* **42**, 144–152.

Burton, C.L. 1970. Diseases of tropical vegetables on the Chicago market. *Tropical Agriculture Trinidad* **47**, 303–313.

Burton, C.L., Tennes, B.R. and Levin, J.H. 1974. Controlling market diseases of blueberries with a semi commercial hot water treatment system. *Proceedings of the Third North American Blueberry Research Workers Conference, Michigan State University, East Lansing*, 189–202.

Burton, K.S. and Twyning, R.V. 1989. Extending mushroom storage life by combining modified atmosphere packaging and cooling. *Acta Horticulturae* **258**, 565–571.

Burton, W.G. 1952. Studies on the dormancy and sprouting of potatoes. III. The effect upon sprouting of volatile metabolic products other than carbon dioxide. *New Phytologist* **51**, 154–161.

Burton, W.G. 1957 The dormancy and sprouting of potatoes. *Food Science Abstracts* **29**, 1–12.

Burton, W.G. 1958. The effect of the concentration of CO_2 and O_2 in the storage atmosphere upon the sprouting of potatoes at 10°C. *European Potato Journal* **1**, 47–57.

Burton, W.G. 1961. The physiology of the potato: problems and present status. *Proceedings of the First Triennial Conference of the European Association for Potato Research, Braunschweig, 1960,* 17–41.

Burton, W.G. 1963. Concepts and mechanism of dormancy. In Ivins J.D. and Milthorpe, F.L. (Editors). *The Growth of the Potato.* Butterworth, London, 17–41.

Burton, W.G. 1965. The sugar balance in some British potato varieties during storage. 1: Preliminary observations. *European Potato Journal* **8**, 80–91.

Burton, W.G. 1968 The effect of oxygen concentration upon sprout growth on the potato tuber. *European Potato Journal* **11**, 249–265.

Burton, W.G. 1972. The response of the potato plant and tuber to temperature. In A. R. Rees, A.R., Cockshull K.E., Hand, D.W. and Hurd, R.G. (Editors). *Crop Processes in Controlled Environments.* Academic Press, London, 217–233.

Burton, W.G. 1973 Physiological and biochamical changes in the tubers as affected by storage conditions. *Proceedings of the Fifth Triennial Conference of the European Association for Potato Research, Norwich, England,* 1972, 63–81.

Burton, W.G. 1974. Some biophysical principles underlying the controlled atmosphere storage of plant material. *Annals of Applied Biology* **78**, 149–168.

Burton, W.G. 1974a. The O_2 uptake, in air and in 5% O_2, and the CO_2 output of stored potato tubers. *Potato Research* **17**, 113–37.

Burton, W.G. 1982. *Postharvest Physiology of Food Crops.* Longmans., London, 339 pp.

Burton, W.G. 1989. *The Potato*, 3rd edition. Longman Scientific and Technical, Horlow, 742 pp.

Burton, W.G. and Hannan, R.S. 1957. Use of gamma radiation for preventing sprouting of potatoes. *Journal of the Science of Food and Agriculture* **12**, 707–715.

Butchbaker, A.F., Nelson, D.C. and Shaw, R. 1967. Controlled atmosphere storage of potatoes. *Transactions of the American Society of Agricultural Engineers* **10**, 534.

Caceres, M.I., Martinez, E. and Torres, F. 1998. Effect of tiabendazole and extracts of grapefruit seeds to control diseases of *Passiflora mollissima* before and after harvest in Nuevo Colon Boyaca. Efecto de tiabendazol y extractos de semilla de toronja en el control de enfermedades de curuba en pre y poscosecha en Nuevo Colon Boyaca. *Fitopatologia Colombiana* **22**, 35–38.

Cadiz, T.G. and Bautista, O.K. 1967. Sweetpotato. In Knott, J.E. and Deanon, J.R. (Editors). *Vegetable Production in South East Asia.* University of the Phillippines Press.

Calara, E.S. 1969. *The effects of varying CO_2 levels in the storage of 'Bungulan' Bananas.* BS thesis, University of the Phillipines, Los Baños, Laguna.

Callesen, O. and Holm, B.M. 1989. Storage results with red raspberry. *Acta Horticulturae* **262**, 247–254.

Callister, S.M. and Agger, W.A. 1987. Enumeration and characterization of *Aeromonas hydrophila* and *Aeromonas caviae* isolated from grocery store produce. *Applied Environmental Microbiology* **53**, 249–253.

Cameron, A.C., Boylan Pett, W. and Lee, J. 1989. Design of modified atmosphere packaging systems: modeling O_2 concentrations within sealed packages of tomato fruits. *Journal of Food Science* **54**, 1413–1421.

Campbell, C.A. 1989. Storage and handling of Florida carambola. *Proceedings of the Interamerican Society for Tropical Horticulture* **33**, 79–82.

Campbell, J.S. and Gooding, H.J. 1962. Recent developments in the production of food crops in Trinidad. *Tropical Agriculture Trinidad* **39**, 261.

Campo, J.H. 1934. Studies on the storage requirements of chico. *Phillipine Agriculture* **23**, 14.

Cantwell, M.I. 1995. Post harvest management of fruits and vegetable stems. *FAO Plant Production and Protection Paper 132,* 120–136.

Cantwell, M.I., Reid, M.S., Carpenter, A., Nie, X. and Kushwaha, L. 1995. Short term and long term high CO_2 treatments for insect disinfestation of flowers and leafy vegetables. *Technologias de Cosecha y Postcosecha de Frutas y Hortalizas. Proceedings of a Conference held in Guanajuato, Mexico, 20–24 February 1995,* 287–292.

Carlos, J. 1990. Hot water treatment for mango black spot. *Spore* **30**, 11.

Carpenter, A. 1993. Controlled atmosphere disinfestation of fresh supersweet sweet corn for export. *Proceedings of the Forty Sixth New Zealand Plant Protection Conference, 10–12 August 1993,* 57–58.

Carrouche, P. 1991. *The problem of temperature in palletised bananas.* MSc thesis, Silsoe College, Cranfield Institute of Technology.

Castoria, R., de Curtis, F., Lima, G., Caputo, L., Pacifico, S. and de Cicco, V. 2001. *Aureobasidium pullulans* (LS–30) an antagonist of postharvest pathogens of fruits: study on its modes of action. *Postharvest Biology and Technology* **22**, 7–17.

Cayley, G.R., Hide, G.A., Lewthwaite, R.J., Pye, B.J. and Vojvodic, P.J. 1987. Methods of applying fungicide sprays to potato tubers and description and use of a prototype electrostatic sprayer. *Potato Research* **30**, 301–317.

Celikel, F.G. and Karacali, I. 1998. Effects of harvest maturity and precooling on fruit quality and longevity of 'Bursa Siyahi' figs *Ficus carica* L. *Acta Horticulturae* **480**, 283–288.

Chachin, K. and Ogata, K. 1971. Effects of delay between harvest and irradiation and of storage temperatures on the sprout inhibition of onions by gamma radiation. *Journal of Food Science and Technology Japan* **18**, 378.

Chachin, K., Imahori, Y.and Ueda, Y. 1999. Factors affecting the postharvest quality of MA packaged broccoli. *Acta Horticulturae* **483**, 255–264.

Chalutz, E. and Stahman, M.A. 1969. Induction of pisatin by ethylene. *Phytopathology* **59**, 1972–1973.

Chalutz, E., DeVay, J.E. and Maxie, E.C. 1969. Ethylene induced isocoumarin formation in carrot root tissue. *Plant Physiology* **44**, 235–241.

Chalutz, E., Lomaniec, E. and Waks, J. 1989. Physiological and pathological observations on the post-harvest behaviour of kumquat fruit. *Tropical Science* **29**, 199–206.

Chantaksinopas, S. and Kosiyachinda, S. 1987. Fruit growth and development of longkong (*Aglaia dookoo* Griff.). *Kasetsart Journal, Natural Sciences* **21**, 142–150.

Chambroy, V., Souty, M., Reich, M., Breuils, L., Jacquemin, G. and Audergon, J.M. 1991. Effects of different CO_2 treatments on post harvest changes of apricot fruit. *Acta Horticulturae* **293**, 675–684.

Chapon, J.F. and Trillot, M. 1992. Apples. Long term storage in northern Italy. Pomme. L'entreposage longue durée en Italie du Nord. *Infos Paris* **78**, 42–46.

Charoenpong, C. and Peng, A.C. 1990. Changes in beta carotene and lipid composition of sweet potatoes during storage. *Ohio Agricultural Research and Development Center Special Circular* **121**, 15–20.

Chase, W.G. 1969. Controlled atmosphere storage of Florida citrus. *Proceedings of the First International Citrus Symposium* **3**, 1365–1373.

Chawan, T. and Pflug, I.J. 1968. Controlled atmosphere storage of onion. *Michigan Agricultural Station Quarterly Bulletin* **50**, 449–475.

Cheah, L.H., Irving, D.E., Hunt, A.W. and Popay, A.J. 1994. Effect of high CO_2 and temperature on Botrytis storage rot and quality of kiwifruit. *Proceedings of the Forty Seventh New Zealand Plant Protection Conference, Waitangi Hotel, New Zealand, 9–11 August 1994,* 299–303.

Chen, P. and Sun, Z. 1991. A review of non destructive methods for quality evaluation and sorting of agricultural products. *Journal of Agricultural Engineering Research* **49**, 85–98.

Chen, P., McCarthy, M.J. and Kauten, R. 1989a. NMR for internal quality evaluation of fruits and vegetables. *Transactions of the American Society of Agricultural Engineers* **32**, 1747–1753.

Chen, P., McCarthy, M.J., Kauten, R., Sarig, Y. and Han, S. 1993. Maturity evaluation of avocados by NMR methods. *Journal of Agricultural Engineering Research* **55**, 177–187.

Chen, P., Sun Z. and Huang, 1992. Factors affecting acoustic responses of apples. *Transactions of the American Society of Agricultural Engineers* **35**, 1915–1920.

Chen, R.Y., Chang, T.C., Liu, M.S. and Tsai, M.J. 1989. Postharvest handling and storage of bamboo shoots *Bambuso oldhamii* Munro. *Acta Horticulturae* **258**, 309–316.

Chen, W.H. 1998. Effects of different harvest maturity on storage quality and life of *Annona squamosa* fruits. *Advances in Horticulture* **2**, 270–273.

Chen, W.H. and Zhang, F.P. 2000. Physiological changes of sugarapple fruit during post harvest storage. *Plant Physiology Communications* **36**, 114–116.

Chen, X.H., Grant, L.A. and Caruso, F. 2000. Effect of BioSave R and carnauba wax on decay of cranberry. *Proceedings of the Florida State Horticultural Society* **112**, 116–117.

Chi, J.H., Ha, T.M. and Kim, Y.H. 1998. Effects of packing materials for the keeping freshness of *Pleurotus ostreatus* and *Flammulina velutipes*. *RDA Journal of Industrial Crop Science* **40**, 58–64.

Chien, C.Y and Kramer, G.F. 1989. Effect of low oxygen storage on polyamine levels and senescence in Chinese cabbage, zucchini squash and McIntosh apples. *International Controlled Atmosphere Conference, Fifth, Proceedings, June 14–16 1989, Wenatchee, Washington, United States of America. Volume 2. Other Commodities and Storage Recommendations,* 19–28.

Chien, Y.W. and Zuo, L.J. 1988. Abscisic acid and A.C.C. content of Chinese cabbage during low oxygen storage. *Journal of the American Society for Horticultural Science* **113**, 881–883.

Chien, Y.W. and Zuo, L.J. 1989. Effect of low oxygen storage on chilling injury and polyamides in Zucchinin squash. *Scientia Horticulturae* **39**, 1–7.

Chiesa, A., Moccia, S., Frezza, D. and Filippini de Delfino, S. 1998. Influence of potassic fertilization on the postharvest quality of tomato fruits. *Agricultura Tropica et Subtropica* **31**, 71–81.

Cho, S.I. and Krutz, G.W. 1989. Fruit ripeness detection by using NMR. *American Society of Agricultural Engineers Paper* 89–6620.

Cho, S.S. and Ha, T.M. 1998. The effect of supplementary package materials for keeping freshness of fresh mushroom at ambient temperature. *RDA Journal of Industrial Crop Science* **40**, 52–57.

Choo, C.G. and Choon, S.C. 1972. Comparative evaluation of some quality aspects of banana *Musa acuminata* Colla. II. Juiciness and texture. *Malayasian Agricultural Research* **1**, 118.

Choon, S.C. and Choo, C.G. 1972. Comparative evaluation of some quality aspects of banana *Musa acuminata* Colla in flavour, sourness and sweetness. *Malayasian Agricultural Research* **1**, 54–59.

Choudhury, M.M., de Lima, M.A.C., Soares, J.M. and Faria, C.M.B. 1999. Effect of nitrogen sources and calcium application on postharvest quality of cv. Italia grapes. Influencia de fontes de nitrogenio e aplicacao de calcio na qualidade pos colheita da uva cv. Italia. *Revista Brasileira de Fruticultura* **21**, 322–326.

Chu, C.L., Liu, W.T., Zhou, T. and Tsao, R. 1999. Control of postharvest gray mold rot of modified atmosphere packaged sweet cherries by fumigation with thymol and acetic acid. *Canadian Journal of Plant Science* **79**, 685–689.

Chukwu, E.U., Olorunda, A.O. and Ferris, R.S.B. 1995. Extension of ripening period of Musa fruit using calcium chloride infiltration, Semperfresh, and Brilloshine 1 coating. *MusAfrica* **8**, 14–15.

Chuma, Y., Nakaji, K. and Okura, M. 1980. Maturity evaluation of bananas by delayed light emission. *Transactions of the American Society of Agricultural Engineers* **23**, 1043.

Chuma, Y., Sein, K., Kawano, S. and Nakaji, K. 1977. Delayed light emission as a means of automatic selection of Satsuma oranges. *Transactions of the American Society of Agricultural Engineers* **20**, 996.

Chuma, Y., Shiga, T. and Morita, K. 1980a. Evaluation of surface color of Japanese persimmon fruits by light reflectance mechanized grading systems. *Journal of the Society of Agricultural Machilary* **42**, 115–120.

Chung, D.S. and Son, Y.K. 1994. Studies on CA storage of persimmon *Diospyros kaki* T. and plum *Prunus salicina* L. *RDA Journal of Agricultural Science, Farm Management, Agricultural Engineering, Sericulture, and Farm Products Utilization* **36**, 692–698.

Chung, H.L. and Ripperton, J.C. 1929. Utilization and composition of oriental vegetables. *Hawaii Agricultural Experimental Station, Honolulu, Bulletin* **60**.

Church, I.J. and Parsons, A.L. 1995. Modified atmosphere packaging technology: a review. *Journal of the Science of Food and Agriculture* **67**, 143–152.

CIP. 1981. *Principles of Potato Storage.* Annual Report of the International Potato Centre, Lima, Peru.

CIP. 1992. *Postharvest Management.* Annual Report of the International Potato Centre, Lima, Peru, 125–150.

Cirulli, M. and Ciccarese, F. 1981. Effect of mineral fertilizers on the incidence of blossom end rot of watermelon. *Phytopathology* **71**, 50–53.

Clark, C.J. and Burmeister, D.M. 1999. Magnetic resonance imaging of browning development in 'Braeburn' apple during controlled atmosphere storage under high CO_2. *HortScience* **34**, 915–919.

Clarke, B. 1996. Packhouse operations for fruit and vegetables. In Thompson, A.K. (Editor), *Posthavest Technology of Fruits and Vegetables.* Blackwell Science, Oxford. 189–217.

Claypool, L. 1959. Maturity standards with freestone peach varieties in 1958 and 1959. *Department of California Fresh Peach Advisory Board Report.*

Claypool, L.L. and Allen, F.W. 1951. The influence of temperature and O_2 level on the respiration and ripening of Wickson plums. *Hilgardia* **21**, 129–160.

Claypool, L.L. and Ozbek, S. 1952. Some influences of temperature and CO_2 on the respiration and storage life of the Mission fig. *Proceedings of the American Society of Horticultural Science* **60**, 226–230.

Claypool, L., Maxie, E.C. and Esau, P. 1955. Effect of aeration rate on the respiratory activity of some deciduous fruits. *Proceedings of the American Society of Horticultural Science* **66**, 125–134.

Coates, L.M. and Gowanlock, D. 1993. Infection processes of *Colletotrichum* species in subtropical and tropical fruit. *Postharvest Handling of Tropical Fruit. Proceedings of an International Conference held in Chiang Mai, Thailand, 19–23 July 1993. Australian Centre for*

International Agricultural Research Proceedings **50**, 162–166.

Coates, L.M., Johnson, G.I. and Cooke, A.W. 1993a. Postharvest disease control in mangoes using high humidity hot air and fungicide treatments. *Annals of Applied Biology* **123**, 441–448.

Cockburn, J.T. and Sharples, R.O. 1979. A practical guide for assessing starch in Conference pears. *Report of East Malling Research Station for 1978*, 215–216.

Codex Alimentarius. 1985. *Recommended International Code of Practice. General Principles of Food Hygiene*, Vol. A, 2nd Revision. Food and Agricultural Organization of the United Nations, World Health Organization, Rome, 1988.

Coffee, R.A. 1980. *Electrodynamic Spraying. Spraying Systems for the 1980s*. British Crop Protection Council, 95–107.

Cohen, E. and Schiffmann Nadel, M. 1978. Storage capability at different temperatures of lemons grown in Israel. *Scientia Horticulturae* **9**, 251–257.

Cohen, E., Lurie, S. and Shalom, Y. 1988. Prevention of Red Blotch in degreened lemon fruit. *HortScience* **23**, 864–865.

Colelli, G. and Martelli, S. 1995. Beneficial effects on the application of CO_2-enriched atmospheres on fresh strawberries *Fragaria × ananassa* Duch. *Advances in Horticultural Science* **9**, 55–60.

Colelli, G., Mitchell, F.G. and Kader, A.A. 1991. Extension of postharvest life of 'Mission' figs by CO_2 enriched atmospheres. *HortScience* **26**, 1193–1195.

Coles, G.D., Lammerink, J.P. and Wallace, A.R. 1993. Estimating potato crisp colour variability using image analysis and a quick visual method. *Potato Research* **36**, 127–134.

Collazos, E.O., Bautista, G.A., Millan, M.B. and Mapura, M.B. 1984. Effect of polyethylene bags on passion fruit *Passiflora edulis* var. *flavicarpa* Degener, tacso P. *mollissima* HBK Bailey and tomato *Lycopersicum esculentum* Miller storage. Efecto de bolsas de polietileno en la conservacion de maracuya *Passiflora edulis* var. *flavicarpa* Degener, curuba *P. mollissima* HBK Bailey y tomate *Lycopersicum esculentum* Miller. *Acta Agronomica* **34**, 53–59.

Condit, 1919. *Report of the California Experimental Station.* Cited by Wardlaw, C.W. (1937).

Conway, W.S., Leverentz, B., Saftner, R.A., Janisiewicz, W.J., Sams, C.E. and Leblanc, E. 2000. Survival and growth of *Listeria monocytogenes* on fresh cut apple slices and its interaction with *Glomerella cingulata* and *Penicillium expansum. Plant Disease* **84**, 177–181.

Cooke, J.R. 1970. A theoretical analysis of the resonance of intact apples. *American Society of Agricultural Engineers* 70–345, St. Joseph, MI, 26 pp.

Cooke, R.D., Rickard, J.E. and Thompson, A.K. 1985. Nutritional aspects of cassava storage and processing. *Proceedings of the Sixth Symposium of the International Society for Tropical Root Crops.*

Cooke, R.D., Rickard, J.E. and Thompson, A.K. 1988. The storage of tropical root and tuber crops – cassava, yam and edible aroids. *Experimental Agriculture* **24**, 457–470.

Coquinot, J.P. and Richard, L. 1991. Methods of controlling scald in the apple Granny Smith without chemicals. Méthodes de contrôle de l'échaudure de la pomme Granny Smith sans adjuvants chimiques. *Neuvieme Colloque sur les Recherches fruitières, 'La Maitrise de la Qualité des Fruits Frais,' Avignon*, 4–6 Decembre 1990, 373–380.

Corrales Garcia, J. 1997. Physiological and biochemical responses of 'Hass' avocado fruits to cold storage in controlled atmospheres. *Seventh International Controlled Atmosphere Research Conference. CA '97. Proceedings, Volume 3: Fruits Other than Apples and Pears, Davis, California, USA, 13–18 July, 1997. Postharvest Horticulture Series, Department of Pomology, University of California* **17**, 69–74.

Corrales-Garcia, J. and Gonzalez-Martinez, P. 2001. Effect of gibberellic acid and (2-chloroethane) phosphonic acid on glochid abscission in cactus pear fruit (*Opuntia amyclaea* Tenore). *Postharvest Biology and Technology* **22**, 151–158.

Costa, M.A.C., Brecht, J.K., Sargent, S.A. and Huber, D.J. 1994. Tolerance of snap beans to elevated CO_2 levels. *Proceedings of the Florida State Horticultural Society* **107**, 271–273.

Couey, H.M. and Uota, M. 1961. Effects of concentration, exposure time, temperature and relative humidity in the toxicity of sulfur dioxide to the spores of *Botrytis cinerea. Phytopathology* **51**, 815–819.

Couey, H.M. and Wells, J.M. 1970. Low oxygen and high carbon dioxide atmospheres to control postharvest decay of strawberries. *Phytopathology* **60**, 47–49.

Coultate, T.P. 1989. Food: *the Chemistry of its Components*, 2nd edn. Royal Society of Chemistry, Cambridge.

Council Directive 93/43/EEC 1993. Council Directive 93/43/EEC of 14 June 1993 on the hygiene of foofstuffs. *Official Journal of the European Communities* L175/1.

Coursey, D.G. 1961. The magnitude and origins of storage losses in Nigerian yams. *Journal of the Science of Food and Agriculture* **12**, 574–580.

Coursey, D.G. 1967. *Yams.* Longmans, London.

Cox, M.A., Zhang, M.I.N. and Willison, J.H.M. 1993. Apple bruise assessment through electrical impedance measurements. *Journal of Horticultural Science* **68**, 393–398.

Cox, S.W.R. 1988. Robotics in agriculture. In Reznicek, R. (Editor), *Physical Properties of Agricultural Materials and Products.* Hemisphere, Washington, DC, 47–52.

Crank, J. 1975. The Mathematics of Diffusion, 2nd edn., Claredon Press, Oxford.

Crawford, I. and Selassie, H. 1993. Market information systems for fruit and vegetable marketing. *International Workshop of Agricultural Produce Wholesale Marketing, Ministry of Agriculture Chengdu, Sichuan Province, China, 25–30 November 1993.*

Crucifix, D.C. 1990. Quality assessment of horticultural produce in the United Kingdom, August 1990, from member countries of the OECS. *Natural Resorces Institute, United Kingdom Report.*

Crucifix, D.C., Richards, F., McIntyre, A.A. and Bellot, C. 1989. Use of postharvest fungicide treatment to control market diseases of Julie mango in Dominica. *Final Report of a Two Year Assignment, Division of Agriculture Botanical Gardens, Roseau, Dominica.*

Cui, L., Chai, H.B., Santisuk, T., Reutrakul, V., Farnsworth, N.R., Cordell, G.A., Pezzuto, J.M. and Kinghorn, A.D. 1997. Novel cytotoxic 1*H*-cyclopenta[*b*]benzofuran lignans from *Aglaia elliptica. Tetrahedron* **53**, 17625–17632.

Curd, L. 1988. *The design and testing using strawberry fruits of an ethylene dilution system.* MSc thesis, Silsoe College, Cranfield Institute of Technology.

Currah, L. and Proctor, F.J. 1990. Onions in tropical regions. *Natural Resources Institute Bulletin* 35.

Czabaffy, A. 1984. Attempts to elaborate a non destructive optical method of measuring cherry ripeness. *Acta Alimentaria* **13**, 83–95.

Czabaffy, A. 1985. Attempts to elaborate a non destructive optical method for measuring the ripeness of Magyar Kajszi aprocots. *Acta Alimentaria* **14**, 125–138.

Czyhriniciw, N. and Jaffe, W. 1951. Modificaciones quemicas durante la conservacion de raices y tuberculos. *Archivos Venexolanos de Nutricion* **2**, 49–67.

Dalrymple, D.G. 1954. World's most modern apple storage. *American Fruit Grower,* November 20, 6–7.

Dalrymple, D.G. 1967. *The Development of Controlled Atmosphere Storage of Fruit.* Division of Marketing and Utilization Sciences, Federal Extension Service, US Department of Agriculture 56 pp.

Damen, P.M.M. 1985. Verlengen afzetperiode vollegrondsgroenten. *Groenten en Fruit* **40**, 82 83.

Daniels, J.A., Krishnamurthi, R. and Rizvi, S.S. 1985. A review of effects of CO_2 on microbial growth and food quality. *Journal of Food Protection* **48**, 532–537.

Darko, J. 1984. *An investigation of methods for evaluating the potential storage life of perishable food crops.* PhD thesis, Silsoe College, Cranfield Institute of Technology.

Daryono, M. and Sosrodiharjo, S. 1986. Practical methods for harvesting mangosteen fruits and their storage characteristics. Cara praktis penentuan saat pemanenan buah manggis dan sifat sifatnya selama penyimpanan. *Buletin Penelitian Hortikultura* **14**, 38–44.

Das, G.P., Lagnaoui, A., Salah, H.B. and Souibgui, M. 1998. The control of the potato tuber moth in storage in Tunisia. *Tropical Science* **38**, 78–80.

Davies, D.H., Elson, C.M. and Hayes, E.R. 1988. *N,O*-Carboxymethyl chitosan, a new water soluble chitin derivative. *Fourth Intenational Conference on Chitin and Chitosan, 22–24 August 1988, Trondheim, Norway, 6.*

Davies, F.S. and Albrigo, L.G. 1994. *Citrus.* CAB International, Oxford.

Davies, G. 1992. Two ways in which retailers can be brands. *International Journal of Retailing and Distribution Management* **20**, 24–34.

Davis, H., Taylor, J.P., Perdue, J.N., Selma G.N., Jr, Humphreys J.M., Jr, Rountree R, III and Greene, K.D. 1988. A shigellosis outbreak traced to commercially distributed shredded lettuce. *American Journal of Epidemiology* **128**, 1312–1321.

Day, B.P.F. 1994. Modified atmosphere packaging and active packaging of fruits and vegetables. *Minimal Processing of Foods, 14–15 April 1994, Kirkkonummi, Finland. VTT Symposium* **142**, 173–207.

Day, B.P.F. 1996. High O_2 modified atmosphere packaging for fresh prepared produce. *Postharvest News and Information* **7**, 31N–34N.

de Baerdemaeker, J. 1989. The use of mechanical resonance measurements to determine fruit texture. *Acta Horticulturae* **258**, 331–339.

de Buckle, T.S., Castelblanco, H., Zapata, L.E., Bocanegra, M.F., Rodriguez, L.E. and Rocha, D. 1973. Preservation of fresh cassava by the method of waxing. *Review de Instituto Investigaciones de Tecnologia,* **15**, 33–47.

de Chernatorny, D. and McDonald, M. 1992. *Creating Powerful Brands. The Stategic Route to Success in Commercial Brands in Consumer, Industrial, and Service Markets*, 1st edn. Butterworth Heinmann, Oxford.

de Costa, D.M. and Subasinghe, S.S.N.S. 1998. Antagonistic bacteria associated with the fruit skin of banana in controlling its postharvest diseases. *Tropical Science* **38**, 206–212.

de Ruiter, M. 1991. *Effect of gases on the ripening of bananas packed in banavac.* MSc thesis, Silsoe College, Cranfield Instute of Technology.

de Wild, H. 2001. 1-MCP can make a big breakthrough for storage. 1-MCP kan voor grote doorbraak in bewaring zorgen. *Fruitteelt Den Haag* **91**, 12–13.

Deck, S.H., Varghese, Z., Morrow C.T., Heinemann, P., Sommer, H.J. and Tao, Y. 1991. Neural networks vs. traditional classifiers for machine vision inspection. *American Society of Agricultural Engineers* 913502.

DeEll, J.R. and Prange, R.K 1993. Postharvest physiological disorders, diseases and mineral concentrations of organically and conventionally grown McIntosh and Cortland apples. *Canadian Journal of Plant Science* **73**, 223–230.

DeEll, J.R., Prange, R.K. and Murr, D.P. 1995. Chlorophyll fluorescence as a potential indicator of controlled atmosphere disorders in 'Marshall' McIntosh apples. *HortScience* **30**, 1084–1085.

Dekazos, E.D, and Birth, G.S. 1970. A maturity index for blueberries using light transmittance. *Journal of the American Society for Horticultural Science* **95**, 610–614.

Delate, K.M. and Brecht, J.K. 1989. Quality of tropical sweet potatoes exposed to controlled atmosphere treatments for postharvest insect control. *Journal of the American Society for Horticultural Science* **114**, 963–968.

Delate, K.M., Brecht, J.K. and Coffelt, J.A. 1990. Controlled atmosphere treatments for control of sweet potato weevil *Coleoptera: Curculionidae* in stored tropical sweet potatoes. *Journal of Economic Entomology* **82**, 461–465.

DeLong, J.M., Prange, R.K., Harrison, P.A. and McRae, K.B. 2000. Comparison of a new apple firmness penetrometer with three standard instruments. *Postharvest Biology and Technology* **19**, 201–209.

Demerutis Pena, C.F., Villarreal Gonzalez, E.A. and Crane, J.H. 2000. Prolonged storage of plantain *Musa* AAB, ABB in an immature state for extending shelf life. Conservacion prolongada del platano *Musa* AAB, ABB en estado inmaduro para aumentar su vida de anaquel. *Proceedings of the Interamerican Society for Tropical Horticulture* **42**, 246–251.

Dennis, C. 1975. Effect of pre harvest fungicides on the spoilage of soft fruit after harvest. *Annals of Applied Biology* **81**, 227–234.

Dennison, R.A. and Ahmed, E.M. 1975. Irradiation treatment of fruits and vegetables. In *Symposium: Postharvest Biology and Handling of Fruits and Vegetables.* AVI Publishing, Westport, CT, 118–129.

Denny, F.E. 1924. Hastening the coloration of lemons. *Journal of Agricultural Research* **27**, 757–771.

Denny, F.E. and Thornton, N.C. 1941. CO_2 prevents the rapid increase in the reducing sugar content of potato tubers stored at low temperatures. *Contributions of the Boyce Thompson Institute of Plant Research* **12**, 79–84.

Desai, B.B. and Deshpande, P.B. 1975. Chemical transformations in three varieties of banana Musa paradisica Linn. fruits stored at 20°C. *Mysore Journal of Agricultural Science* **9**, 634–643.

Deschene, A., Paliyath, G., Lougheed, E.C., Dumbroff, E.B. and Thompson, J.E. 1991. Membrane deterioration during postharvest senescence of broccoli florets: modulation by temperature and controlled atmosphere storage. *Postharvest Biology and Technology* **1**, 19–31.

Devkota, L.N. and Ghale, M.S. 1992. In Bhatti, M.H., Hafeez, Ch. A., Jaggar, A. and Farooq, Ch. M. (Editors). *Postharvest Losses of Vegetables. A Report on a Workshop Held Between 17 and 22 October 1992 at the Pakistan Agricultural Research Council, Islamabad, Pakistan.* Food and Agriculture Organization of the United Nations Regional Cooperation for Vegetable Research and Development RAS/89/41.

Dewey, D.H., Ballinger, W.E. and Pflug, I.J. 1957. Progress report on the controlled atmosphere storage of Jonathan apples. *Quarterly Bulletin Michigan Agricultural Experiment Station* **39**, 692.

D'hallewin, G., Schirra, M. and Manueddu, E. 1999. Effect of heat on epicuticular wax of cactus pear fruit. *Tropical Science* **39**, 244–247.

Dhillon, W.S., Bindra, A.S, Cheema, S.S. and Singh, S. 1992. Effects of graded doses of nitrogen on vine growth, fruit yield and quality of Perlette' grapes. *Acta Horticulturae* **321**, 667–671.

Dhua, R.S., Ghosh, S.K. and Sen, S. 1992. Standardisation of harvest maturity of banana cultivar Giant Governor in West Bengal. *Advances in Horticulture and Forestry* **2**, 24–35.

Diaz Perez, J.C., Bautista, S. and Villanueva, R. 2000. Quality changes in sapote mamey fruit during ripening and storage. *Postharvest Biology and Technology* **18**, 67–73.

Diaz-Perez, J.C., Mejia, A., Bautista, S., Zavaleta, R., Villanueva, R. and Lopez-Gomez, R. 2001. Response of sapote mamey [*Pouteria sapota* (Jacq.) H.E. Moore and Stearn] fruit to hot water treatments. *Postharvest Biology and Technology* **22**, 159–167.

Dick, E. and Marcellin, P. 1985. Effect of high temperatures on banana development after harvest. Prophylactic tests. Effects des temperatures elevées sur l'évolution des bananes après récolte. Tests prophylactiques. *Fruits Paris* **40**, 781–784.

Diener, R.G. and Fridley R.B. 1983. *Collection and catching.* In O'Brien, M., Cargill, B.F. and Fridley, R.B. (Editors), *Harvesting and Handling Fruits and Nuts.* AVI Publishing, Westport, CT, 245–303.

Dilley, D.R. 1990. Historical aspects and perspectives of controlled atmosphere storage. In colderon, M. and Barkai Golan (Editors), *Food Preservation by Modified Atmospheres.* CRC Press, Boca Raton, FL, 187–196.

Dillon, M., Hodgson, F.J.A., Quantick, P.C. and Taylor, D.J. 1989. Use of the APIZYM testing system to assess the state of ripeness of banana fruit. *Journal of Food Science,* **54**, 1379–1380.

Dimick, P.S. and Hoskins, J.C. 1983. Review of apple flavour. State of the Art. *CRC Critical Review, Food Science and Nutrition* **18**, 387–409.

Ding, C.K., Chachin, K., Hamauzu, Y., Ueda, Y. and Imahori, Y. 1998. Effects of storage temperatures on physiology and quality of loquat fruit. *Postharvest Biology and Technology* **14**, 309–315.

Dingman, D.W. 2000. Growth of *Escherichia coli* O157:H7 in bruised apple (*Malus domestica*) tissue as influenced by cultivar, date of harvest, and source. *Applied and Environmental Microbiology* **66**, 1077–1083.

Dixie, G. 1989. Horticultural marketing. *Food and Agriculture Organization of the United Nations, Agricultural Services Bulletin* **76**, 118 pp.

Doesburg, J.J. 1961. Some notes on development and maturation of fruits in relation to their suitability for storage. *Bulletin of the International Institute for Refrigeration Annex* 1961–1, 29–34.

Dohroo, N.P. 2001. Etiology and management of storage rot of ginger in Himachal Pradesh. *Indian Phytopathology* **54**, 49–54.

Dourtoglou, V.G., Yannovits, N.G., Tychopoulos, V.G. and Vamvakias, M.M. 1994. Effect of storage under CO_2 atmosphere on the volatile, amino acid, and pigment constituents in red grape *Vitis vinifera* L. var. Agiogitiko. *Journal of Agricultural and Food Chemistry* **42**, 338–344.

Dover, C.J. 1989. The principles of effective low ethylene storage. *Acta Horticulturae* **258**, 25–36.

Dover, C.J. and Bubb, M. undated. Report on Ethysorb as an absorbant of ethylene. *Report of East Malling Research Station, Maidstone, Kent,* 6 pp.

Dover, C.J. and Stow J.R. 1993. The effects of ethylene removal rate and period in a low ethylene atmosphere on ethylene production and softening of Cox apples. *Sixth Annual Controlled Atmosphere Research Conference, Cornell University, Ithaca, New York.*

Dowker, B.D., Fennell, J.F.M., Horobin, J.F., Crowther, T.C., Morgan, Sandra, J. and Carter, Philippa, J. 1980. Onions. *Report of the National Vegetable Reseach Station for 1979,* 55.

Drake, S.R. 1993. Short term controlled atmosphere storage improved quality of several apple cultivars. *Journal of the American Society for Horticultural Science* **118**, 486–489.

Drake, S.R. and Eisele, T.A. 1994. Influence of harvest date and controlled atmosphere storage delay on the color and quality of 'Delicious' apples stored in a purge type controlled atmosphere environment. *HortTechnology* **4**, 260–263.

Drake, S.R. and Elfving, D.C. 1999. Quality of 'Fuji' apples after regular and controlled atmosphere storage. *Fruit Varieties Journal* **53**, 193–198.

Drake, S.H. and Spayd, S.E. 1983. Influence of calcium treatment on 'Golden Delicious' apple quality. *Journal of Food Science* **48**, 403–405.

Draughon,F.A. and Mundt, J.O. 1988. Botulinal toxin production in fresh tomatoes infected with the molds *Fusarium, Alternaria, and Rhizoctonia. Tennessee Farm and Home Science* **147**, 12–15.

Droby, S., Porat, R., Cohen, L., Weiss, B., Shapiro, B., Philosoph Hadas, S. and Meir, S. 1999. Suppressing green mold decay in grapefruit with postharvest jasmonate application. *Journal of the American Society for Horticultural Science* **124**, 184–188.

Dubodel, N.P. and Tikhomirova, N.T. 1985. Controlled atmosphere storage of mandarins. *Sadovodstvo* **6**, 18.

Dubodel, N.P., Panyushkin, Y.A., Burchuladze, A.S. and Buklyakova, N.N. 1984. Changes in sugars of mandarin fruits in controlled atmosphere storage. *Subtropicheskie Kul'tury* **1**, 83–86.

Dull, G.G. 1986. Non-destructive evaluation of stored fruits and vegetables. *Food Technology* **40**, 106–110.

Dull, G.G., Birth, G.S., Smittle, D.A. and Leffler, R.G. 1989. Near infra red analysis of soluble solids in intact cantaloupe. *Journal of Food Science* **54**, 393–395.

Duque, P. and Arrabaca, J.D. 1999. Respiratory metabolism during cold storage of apple fruit. II. Alternative oxidase is induced at the climacteric. *Physiologia Plantarum* **107**, 24–31.

Duque, P. and Arrabaca, J.D. 1999a. *Journal of the American Society for Horticultural Science* **107**, 262–265.

Eaks, I.L. 1977. Physiology of degreening summary and discussion of related topics. *Proceedings of the International Society for Citriculture* **1**, 223–226.

Eaks, I.L. 1990. Change in the fatty acid composition of avocado fruit during otogeny, cold storage and ripening. *Acta Horticulturae* **269**, 141–152.

Eaks, I.L. and Morris, L. 1956. Respiration of cucumber fruits associated with physiological injury at chilling temperatures. *Plant Physiology* **31**, 308–314.

Eastman Kodak 1983. *Ergonomic Design for People at Work,* Vol 1. Lifetime Learning Publications, Belmont, CA.

Eaves, C.A. 1935. The present status of gas storage research with particular reference to studies conducted in Great Britain and preliminary trials undertaken at the Central Experiment Farm, Canada. *Scientific Agriculture* **15**, 548–554.

Eaves, C.A. and Forsyth, F.R. 1968. The influence of light modified atmospheres and benzimidazole on Brussel sprouts. *Journal of Horticultural Science* **43**, 317.

Eckert, J.W. 1969. Chemical treatments for control of postharvest diseases. *World Review of Pest Control* **8**, 116–137.

Eckert, J.W., Rubio, P.P., Mattoo, A.A. and Thompson, A.K. 1975. Diseases of tropical crops and their control. In Pantastico, Er.B. (Editor), *Postharvest Physiology, Handling and Utilization of Tropical and Sub Tropical Fruits and Vegetables.* AVI Publishing, Westport, CT., 415–443.

Edem, D.O., Amugo, C.I. and Eka, O.U. 1990. Chemical composition of yam beans (*Sphenostylis stenocarpa*). *Tropical Science* **30**, 59–63.

Edney, K.L., Burchill, R.T., and Chambers, D.A. 1977. The control of Gloeosporium storage rot in apple by orchard spray programme. *Annals of Applied Biology* **87**, 51–56.

Eeden Van, S.J., Combrink, J.C., Vries, P.J. and Calitz, F.J. 1992. Effect of maturity, diphenylamine concentration and method of cold storage on the incidence of superficial scald in apples. *Deciduous Fruit Grower* **42**, 25–28.

El Ghaouth, A., Arul, J., Ponnampalam, R. and Boulet, M. 1991. Use of chitosan coating to reduce water loss and maintain quality of cucumber and cell pepper fruits. *Journal of Food Process Preservation* **15**, 359.

El Shiaty, M.A., Esawi, M.T. and Atwa, A.A. 1961. Studies on the storage of Egyptian lime fruit. *Agricultural Research Review, United Arab Republic* **46**, 3.

Elansari, A.M., Hussein, A.M. and Bishop, C.F.H. 2000. Performance of wet deck (ice bank) pre-cooling systems with export produce from Egypt. *Landwards* **55**, 20–25.

Elgar, H.J., Burmeister, D.M. and Watkins, C.B. 1998. Storage and handling effects on a CO related internal browning disorder of 'Braeburn' apples. *HortScience* **33**, 719–722.

Elgayyar, M., Draughon, F.A., Golden, D.A. and Mount, J.R. 2001. Antimicrobial activity of essential oils from plants against selected pathogenic and saprophytic microorganisms. *Journal of Food Protection* **64**, 1019–1024.

Ellis, G. 1995 Potential for all year round berries. *The Fruit Grower* December 17–18.

Endt, D.J.W. 1981. The babaco – a new fruit in New Zealand to reach commercial production. *Orchardist of New Zealand* **54**, 58–59.

Engel, K.H., Heidlas, J. and Tressl, R. 1990. In Merton, I.D. and Macleod, A.J. (Editors), *Flavour of Tropical Fruits in Food Flavours.* Elsevier, Amsterdam, 195–220.

Eris, A. and Akbudak, B. 2001. Changes in some quality criteria during controlled atmosphere (CA) storage of peaches. *International Journal of Horticultural Science* **7**, 58–61.

Erkan, M. and Pekmezci, M. 2000. Investigations on controlled atmosphere CA storage of Star Ruby grapefruit grown in Antalya conditions. Antalya kosullarinda uretilen 'Star Ruby' altintopunun kontrollu atmosferde KA muhafazasi uzerinde arastirmalar. *Bahce* **28**, 87–93.

Ertan, U., Ozelkok, S., Celikel, F. and Kepenek, K. 1990. The effects of pre cooling and increased atmospheric concentrations of CO_2 on fruit quality and postharvest life of strawberries. Cilekte on sogutma ve yuksek karbondioksit uygulamalarinin meyve kalitesi ve pazarlama suresi uzerine etkileri. *Bahce*, **19**, 59–76.

Ertan, U., Ozelkok, S., Kaynas, K. and Oz, F. 1992. Comparative studies on the effect of normal and controlled atmosphere storage conditions on some important apple cultivars I. Flow system. Bazi onemli elma cesitlerinin normal ve kontrollu atmosferde depolanmalari uzerinde karsilastirmali arastirmalar I. Akici sistem. *Bahce* **21**, 77–90.

Escriche, I., Serra, J.A., Gomez, M. and Galotto, M.J. 2001. Effect of ozone treatment and storage temperature on physicochemical properties of mushrooms (*Agaricus bisporus*). *Food Science and Technology International* 7, 251–258.

Esquerra, E.B. and Lizada, M.C.C. 1990. The postharvest behaviour and quality of 'Carabao' mangoes subjected to vapor heat treatment. *Association of Southeast Asian Nations Food Journal* 5, 6–12.

Esquerra, E.B., Brown, E.O., Briones, J.P. and Lizada, M.C.C. 1984. Trial shipments of 'SABA' bananas from South Cotabato to Manila. *Postharvest Research Notes* 1, 110–112.

Esquerra, E.B., Kawada, K. and Kitagawa, H. 1992. Removal of astringency in 'Amas' banana Musa AA group with postharvest ethanol treatment. *Acta Horticulturae* **321**, 811–820.

Eun, J.B., Kim, J.D., Park, C.Y. and Gorny, J.R. 1997. Storage effect of LDPE film embedded with silver coated ceramic in enoki mushroom. *Seventh International Controlled Atmosphere Research Conference. CA '97. Proceedings, Volume 5: Fresh Cut Fruits and Vegetables and MAP, Davis, California, USA, 13–18 July, 1997. Postharvest Horticulture Series Department of Pomology, University of California* **19**, 158–163.

Evelo, R.G. 1995. Modelling modified atmosphere systems. COST 94. The post harvest treatment of fruit and vegetables: systems and operations for post harvest quality. *Proceedings of a Workshop, Milan, Italy, 14–15 September,* 1993, 147–153.

Everaarts, A.P. 2000. Nitrogen and post-harvest yellowing of Brussels sprouts. Acta Horticulturae **533**, 393–396.

Ezeh, N.O.A. 1992. Economic appraisal of an improved yam storage barn. *Tropical Science* **32**, 27–32.

Ezeike, G.O.I. 1985. Experimental analysis of yam Dioscorea spp. tuber stability in tropical stoage. *Transactions of the American Society of Agricultural Engineers* **28**, 1641–1645.

Ezeike, G.O.I., Ghaly, A.E., Ngadi, M.O. and Okafor, O.C. 1989. Engineering features of a controlled underground yam storage structure. *American Society of Agricultural Engineers Paper* 89 5020, 23 pp.

Fan, X., Argenta, L. and Mattheis, J.P. 2000. Inhibition of ethylene action by 1-methylcyclopropene prolongs storage life of apricots. *Postharvest Biology and Technology* **20**, 135–142.

Faragher, J.D., Borochov, A. and Halevy, A.H. 1983. Effects of low temperature storage on the physiology of carnation flowers. *Acta Horticulturae* **138**, 269–272.

Farber, J.M. 1991. Microbiological aspects of modified atmosphere packaging technology: a review. *Journal of Food Protection* **54**, 58–70.

Fellers, P.J. and Pflug, I.J. 1967. Storage of pickling cucumbers. *Food Technology* **21**, 74.

Fellows, P.J. 1988. *Food Processing Technology.* Ellis Horwood, Chichester.

Feng, S.Q., Chen,Y.X., Wu, H.Z. and Zhou, S.T 1991. The methods for delaying ripening and controlling postharvest diseases of mango. *Acta Agriculturae Universitatis Pekinensis* **17**, 61–65.

Ferguson, I., Volz, R., Woolf, A. and Cavalieri, R.P. 1999. *Preharvest factors affecting physiological disorders of fruit. Postharvest Biology and Technology* **15**, 255–262.

Fernandez, M., Vicente, I. and Garcia, M. 1996. Characterization of green and ripe Burro CEMSA plantains and their peels, harvested in Cuba. Caracterizacion del platano Burro CEMSA y de las cascaras, verde y maduro, cosechado en Cuba. *Alimentaria* **34**, 115–117.

Ferris, R.S.B. 1991. *Effects of damage and storage environment on the ripening of cooking bananas with implications for postharvest loss.* PhD thesis, Silsoe College, Cranfield Institute of Technology, 139 pp.

Ferris, R.S.B., Hotsonyame, G.K. Wainwright, H. and Thompson, A.K. 1993. The effects of genotype, damage, maturity, and environmental conditions on the postharvest life of plantain. *Tropical Agriculture* **70**, 45–50.

Fideghelli, C., Cappellini, P. and Monastra, F. 1967. *Progressive Agricultura, Bologna* **13**, 405.

Fidler, J.C. 1963. Refrigerated storage of fruits and vegetables in the U.K., the British Commonwealth, the United States of America and South Africa. *Ditton Laboratory Memoir* **93**, 22 pp.

Fidler, J.C. 1968. Low temperature injury of fruit and vegetables. *Recent Advances in Food Science* **4**, 271–283.

Fidler, J.C. and Mann, G. 1972. *Refrigerated storage of apples and pears: a practical guide.* Commonwealth Agricultural Bureau, Oxford.

Fidler, J.C., Wilkinson, B.G., Edney, K.L. and Sharples R.O. 1973. The biology of apple and pear storage. *Commonwealth Agricultural Bureaux Research Review,* **3**, 235 pp.

Finney, E.E. 1970. Mechanical resonance within red delicious apples and its relation to fruit texture. *Transactions of the American Society of Agricultural Engineers* **13**, 177–180.

Finney, E.E. and Norris, K.H. 1973. X ray images of hollow heart potatoes in water. *American Potato Journal* **50**, 1–8.

Fioravanco, J.C., Manica, I. and Paiva, M.C. 1995. Use of cytokinin and fruit coverings on Tahiti limes in cold storage. Uso de citocinina e recobrimentos em limao

'Tahiti' armazenado em temperatura controlada. *Pesquisa Agropecuaria Brasileira* **30**, 81–87.

Firth, J. 1958. Controlled Atmosphere Regulations. *Proceedings of the One Hundred and Third Meeting of the New York State Horticultural Society* 192–193.

Fischer, G., Buitrago, M. and Ludders, P. 1990. *Physalis peruviana* L. cultivation and research in Colombia. *Erwerbsobstbau* **32**, 229–232.

Fisher, D.V. 1939. Storage of delicious apples in artificial atmospheres. *Proceedings of the American Society for Horticultural Science* **37**, 459–462.

Flagge, H.H. 1942. Controlled atmosphere storage for Jonathan apples as affected by restricted ventilation'. *Refrigerating Engineering,* **43**, 215–220.

Flores, G.A. and Rivas, D. 1975. Studies on the ripening and frozen storage of sapodilla *Achras sapota L.* Estudios de maduracion y almacenamiento refrigerado de nispero *Achras sapota L. Fitotecnia Latinoamericana* **11**, 43–51.

Flynn, J.E. 1979. The IES approach to recommendations regarding levels of illumination. *Lighting Designs and Applications* **9**, 74–77

Folchi, A., Pratella, G.C., Tian, S.P. and Bertolini, P. 1995. Effect of low O_2 stress in apricot at different temperatures. *Italian Journal of Food Science* **7**, 245–253.

Fontes, V.L., de Moura, M.A., Vieira, G. and Finger, F.L. 1999. Influence of plastic films and temperature on postharvest pericarp browning in litchi *Litchi chinensis.* Efeito de filmes plasticos e temperatura de armazenamento na manutencao da cor do pericarpo de lichia *Litchi chinensis. Revista Brasileira de Armazenamento* **241**, 56–59.

Forbus, W.R., Senter, S.D. and Chan, H.T. 1987. Measurement of papaya maturity by delayed light emission. *Journal of Food Science* **52**, 356–360.

Forbus, W.R., Senter, S.D. and Wilson, R.L. 1985. Measurement of tomato maturity by delayed light emission. *Journal of Food Science* **50**, 750.

Forney, C.F., Rij, R.E. and Ross, S.R. 1989. Measurement of broccoli respiration rate in film wrapped packages. *HortScience* **24**, 111–113.

Francile, A.S. 1992. Controlled atmosphere storage of tomato. Conservacion de tomate en atmosfera controlada. *Rivista di Agricoltura Subtropicale e Tropicale* **86**, 411–416.

Francile, A.S. and Battaglia, M. 1992. Control of superficial scald in Beurre d'Anjou pears and Granny Smith apples. Control de escaldadura superficial en peras Beurre d'Anjou y en manzanas Granny Smith. *Rivista di Agricoltura Subtropicale e Tropicale* **86**, 397–410.

Frenkel, C. 1975. Oxidative turnover of auxins in relation to the onset of ripening in Bartlett pear. *Plant Physiology* **55**, 480–484.

Frenkel, C. and Patterson, M.E. 1974. Effect of CO_2 on ultrastructure of 'Bartlett pears'. *Horticultural Science* **9**, 338–340.

Fuchs, Y. and Temkin Gorodeiski, N. 1971. The course of ripening of banana fruits stored in sealed polyethylene bags. *Journal of the American Society for Horticultural Science* **96**, 401–403.

Fuglie, K.O., Khatana, V.S., Ilangantileke, S.G., Scott, G.J., Singh, J. and Kumar, D. 2000. Economics of potato storage in northern India. *Quarterly Journal of International Agriculture* **39**, 131–148.

Fulton, S.H. 1907. The cold storage of small fruits. *US Department of Agriculture, Bureau of Plant Industry Bulletin* **108**, September 17.

Gallerani, G., Pratella, G.C. and Cazzola, P.P. 1994. Superficial scald control via low O_2 treatment timed to peroxide threshold value. In Eccher Zerbini, P., Woolfe M.L., Bertolini P., Haffner K., Hribar J., Hohn, E. and Somogyi, Z. (editors), *COST 94. The Post Harvest Treatment of Fruit and Vegetables: Controlled Atmosphere Storage of Fruit and Vegetables. Proceedings of a Workshop, Milan, Italy, 22–23 April, 1993,* 51–60.

Gama, F.S.N.da, Manica, I., Kist, H.G.K. and Accorsi, M.R. 1991. Additives and polyethylene bags in the preservation of passion fruits stored under refrigeration. Aditivos e embalagens de polietileno na conservacao do maracuja amarelo armazenado em condicoes de refrigeracao. *Pesquisa Agropecuaria Brasileira* **26**, 305–310.

Gane, R. 1934. Production of ethylene by some ripening fruits. *Nature* **134**, 1008.

Gane, R. 1936. A study of the respiration of bananas. *New Phytologist* **35**, 383–402.

Garibaldi, E.A. 1983. Postharvest physiology of Mediterranian carnations: partial characterisation of a bacterial metabolite inducing wilt in flowers. *Acta Horticulturae* **138**, 255–260.

Gariepy, Y., Raghavan, G.S.V., Plasse, R., Theriault, R. and Phan, C.T. 1985. Long term storage of cabbage, celery, and leeks under controlled atmosphere. *Acta Horticulturae* **157**, 193–201.

Gariepy, Y., Raghavan, G.S.V. and Theriault, R. 1984. Use of the membrane system for long term CA storage of cabbage. *Canadian Agricultural Engineering* **26**, 105–109.

Gariepy, Y., Raghavan, G.S.V. and Theriault, R. 1987. Cooling characteristics of cabbage. *Canadian Agricultural Engineering* **29**, 45–50.

Gariepy, Y., Raghavan, G.S.V., Theriault, R. and Munroe, J.A. 1988. Design procedure for the silicone membrane system used for controlled atmosphere storage of leeks and celery. *Canadian Agricultural Engineering* **30**, 231–236.

Garrett, M. 1992. Applications of controlled atmosphere containers. *BEHR'S Seminare, Hamburg, 16–17 November 1992, Munich Germany.*

Garrett, R.E. and Furry, R.D. 1974. Velocity of sonic pulses in apples. *American Society of Agricultural Engineers Paper* 71–331.

Garrett, R.E. and Talley, W.K. 1970. Use of gamma ray transmission in selecting lettuce for harvest. *Transactions of the American Society of Agricultural Engineers* **13**, 820–823.

Geeson, J. 1984. Improved long term storage of winter white cabbage and carrots. Fruit, *Vegetables and Science* **19**, 21.

Geeson, J.D. 1989. Modified atmosphere packaging of fruits and vegetables. *Acta Horticulturae* **258**, 143–150.

Geeson, J.D., Browne, K.M. and Dennis, C, 1983. Carrots. Cashing in on ice banks for summer supplies. *Grower* **100**, 49–52.

Geeson, J.D., Browne, K.M. and Everson, H.P. 1988. Storage diseases of carrots in East Anglia 1978–82, and the effects of some pre and post harvest factors. *Annals of Applied Biology* **112**, 503–514.

Geeson, J.D., Browne, K.M. and Everson, H.P. 1989. Long term refrigerated storage of swedes. *Journal of Horticultural Science* **64**, 479–483.

Geeson, J.D., Browne, K.M. and Griffiths, N.M. 1991. Quality changes in sweetcorn cobs of several cultivars during short term ice bank storage. *Journal of Horticultural Science* **66**, 409–414.

Genay, J P. 1991. *Efficiency of the microstat sprayer in the control of the crown rot of harvested green bananas.* MSc thesis, Silsoe College, Cranfield Institute of Technology.

George, J.B. 1981. *Storage and ripening of plantains.* PhD Thesis, University of London, 143 pp.

Gerhardt, F., English, H. and Smith, E. 1942. Respiration, internal atmosphere and moisture studies of sweet cherries during storage. *Proceedings of the American Society of Horticultural Science* **41**, 119–123.

Ghafir, S.A.M. and Thompson, A.K. 1994. Destructive and non destructive apple maturity assessment. *Agricultural Engineer* **49**, 40–43.

Gil, M.I., Ferreres, F. and Tomas Barberan, F.A. 1998. Effect of modified atmosphere packaging on the flavonoids and vitamin C content of minimally processed Swiss chard *Beta vulgaris* subspecies *cycla. Journal of Agricultural and Food Chemistry* **46**, 2007–2012.

Girard, B. and Lau, O.L. 1995. Effect of maturity and storage on quality and volatile production of 'Jonagold' apples. *Food Research International* **28**, 465–471.

Goffings, G. and Herregods, M. 1989. Storage of leeks under controlled atmospheres. *Acta Horticulturae* **258**, 481–484.

Goffings, G., Herregods, M. and Sass, P. 1994. The influence of the storage conditions on some quality parameters of Jonagold apples. *Acta Horticulturae* **368**, 37–42.

Gogoi, P. and Baruah, P. 2000. Post harvest fungal contaminants of *Pleurotus sajor-caju* and *Volvariella volvaceae* and its biological management. *Science and Cultivation of Edible Fungi. Proceedings of the 15th International Congress on the Science and Cultivation of Edible Fungi, Maastricht, Netherlands, 15–19 May, 2000*, 741–744.

Golias, J. 1987. Methods of ethylene removal from vegetable storage chambers. Zpusoby odstraneni etylenu z atmosfery komory se skladovanou zeleninou. *Bulletin, Vyzkumny a Slechtitelsky Ustav Zelinarsky Olomouc* **31**, 51–60.

Gollifer, D.E. and Booth, R.H. 1973. Storage losses of taro corms in the British Solomon Islands Protectorate. *Annals of Applied Biology* **73**, 349–356.

Gomez, E., Labarca, J., Guerrero, M., Marin, M. and Bracho, B. 1999. Postharvest performance of coriander *Coriandrum sativum* l. under refrigeration. Comportamiento poscosecha de cilantro *Coriandrum sativum* l. bajo refrigeracion. *Revista de la Facultad de Agronomia, Universidad del Zulia* **16**, 146–150.

Gomez, P.K. 2000. Effect of storage temperatures and use of wax on the respiratory activity and internal composition of passion fruit *Passiflora edulis* f. *flavicarpa* Degener cultivar 'Maracuya'. Efecto de la temperatura de almacenamiento y uso de cera sobre la actividad respiratoria y algunos atributos de calidad de frutos de parchita *Passiflora edulis* f. *flavicarpa* Degener cultivar 'Maracuya'. *Revista de la Facultad de Agronomia, Universidad del Zulia* **17**, 1–9.

Gong, Y. Abe, K. and Chachin, K. 1993. Relationship between endogenous ethyl alcohol and browning in shiitake *Lentinus edodes* Sing. mushroom during storage in polyethylene film bags. *Journal of the Japanese Society for Food Science and Technology* **40**, 708–712.

Gonzaga Neto, L., Cristo, A.S. and Choudhury, M.M. 1999. Postharvest conservation of guava cultivar Paluma fruits. Conservacao pos colheita de frutos de goiabeira, variedade Paluma. *Pesquisa Agropecuaria Brasileira* **34**, 1–6.

Gonzales M.A. and de Rivera A.C. 1972. Storage of fresh yams *Dioscorea alata* L. under controlled conditions. *Journal of Agriculture of the University of Puerto Rico* **56**, 46–56.

Gonzalez Aguilar, G., Vasquez, C., Felix, L., Baez, R. and Siller, J. 1995. Low O_2 treatment before storage in normal or modified atmosphere packaging of mango. In Ait Oubahou A. and El Otmani M. (Editors), *Postharvest Physiology, Pathology and Technologies for Horticultural Commodities: Recent Advances. Proceedings of an International Symposium Held at Agadir, Morocco, 16–21 January 1994*, 185–189.

Gonzalez Aguilar, G.A., Fortiz, J., Cruz, R., Baez, R. and Wang, C.Y. 2000. Methyl jasmonate reduces chilling injury and maintains postharvest quality of mango fruit. *Journal of Agricultural and Food Chemistry* **48**, 515–519.

Gonzalez Aguilar, G.A., Wang, C.Y., Buta, J.G. and Krizek, D.T. 2001. Use of UV C irradiation to prevent decay and maintain postharvest quality of ripe 'Tommy Atkins' mangoes. *International Journal of Food Science and Technology* **36**, 767–773.

Goodenough, P.W. and Thomas, T.H. 1980. Comparitive physiology of field grown tomatoes during ripening on the plant or retarded ripening in controlled atmosphere. *Annals of Applied Biology*, **94**, 445.

Goodenough, P.W. and Thomas, T.H. 1981. Biochemical changes in tomatoes stored in modified gas atmospheres. I. Sugars and acids. *Annals of Applied Biology* **98**, 507.

Gooding, H.J. and Campbell, J.S. 1961. Preliminary trials of West African *Xanthosoma* cultivars. *Tropical Agriculture Trinidad* **38**, 145.

Goodrich Tanrikulu, M., Mahoney, N.E. and Rodriguez, S.B. 1995. The plant growth regulator methyl jasmonate inhibits aflatoxin production by *Aspergillus flavus. Microbiology Reading* **141**, 2831–2837.

Gore, H.C. 1911. Studies on fruit respiration. *United States Department of Agriculture, Chemistry Bulletin* **142**, 40 pp.

Gorini, F. 1988. Storage and postharvest treatment of brassicas. I: Broccoli. Conservazione e trattamenti post raccolta delle brassicacee. Nota I: Il broccolo. *Annali dell'Istituto Sperimentale per la Valorizzazione Tecnologica dei Prodotti Agricoli, Milano* **19**, 279–294.

Gorini, F., Rizzolo, A. and Polesello, A. 1989. Risultati delle ricerche sulle tecniche di depurazione dell'etilene. *Revista di Frutticoltura e di Ortofloricoltura* **51**, 73–81.

Gosselin, B. and Mondy, N.I. 1989. Effect of packaging materials on the chemical composition of potatoes. *Journal of Food Science* **54**, 629–631.

Goszczynsha, D. and Rudnicki, R. M. 1988. Storage of cut flowers. *Horticultural Review* **10**, 35–62.

Goto, M., Minamide, T. and Iwata, T. 1988. The change in chilling sensitivity in fruits of mume Japanese apricot, *Prunus mume* Sieb. et Zucc. depending on maturity at harvest and its relationship to phospholipid composition and membrane permeability. *Journal of the Japanese Society for Horticultural Science* **56**, 479–485.

Goulart, B.L., Evensen, K.B., Hammer, P. and Braun, H.L. 1990. Maintaining raspberry shelf life: Part 1. The influence of controlled atmospheric gases on raspberry postharvest longevity. *Pennsylvania Fruit News* **70**, 12–15.

Goulart, B.L., Hammer, P.E., Evensen, K.B., Janisiewicz, B. and Takeda F. 1992. Pyrrolnitrin, captan with benomyl, and high CO_2 enhance raspberry shelf life at 0 or 18°C. *Journal of the American Society for Horticultural Science* **117**, 265–270.

Goyette, B. and Vigneault, C. 1997. Development of a system for precooling using continuous liquid ice operation. *Canadian Society for Engineering in Agricultural Annual Conference, Sherbrooke, Quebec, Canada, 28–30 May, 1997*, **312**, 1–6.

Graell, J. and Recasens, I. 1992. Effects of ethylene removal on 'Starking Delicious' apple quality in controlled atmosphere storage. *Postharvest Biology and Technology* **2**, 101–108.

Graell, J., Recasens, I., Salas, J. and Vendrell, M. 1993. Variability of internal ethylene concentration as a parameter of maturity in apples. *Acta Horticulturae* **326**, 277–284.

Graham, D. 1990. Chilling injury in plants and fruits: some possible causes with means of amelioration by manipulation of postharvest storage conditions. *Proceedings of the International Congress of Plant Physiology, New Delhi, India, 15–20 February 1988, Vol. 2*, 1373–1384.

Graham, D.C. and Hamilton, G.A. 1970. Control of potato gangrene and skin spot diseases by fumigation of tubers with sec-butylamine. *Nature* **227**, 297–298.

Grattidge, R. 1980. Mango anthracnose control. *Queensland Department of Primary Industries, Australia Farm Note AGDEX 234/633 F18/Mar 80.*

Grierson, D. 1993. Chairman's remarks. *Postharvest Biology and Handling of Fruit, Vegetables and Flowers. Meeting of the Association of Applied Biologists, London, 8th December 1993.*

Grierson, D. and Kader, A.A. 1986. Fruit ripening and quality. In Atherton, J.G. and Rudich, J. (Editors), *The Tomato Crop.* Chapman and Hall, London, 680 pp.

Grierson, W. 1971. Chilling injury in tropical and subtropical fruits: IV. The role of packaging and waxing in minimizing chilling injury of grapefruit. *Proceedings of the Tropical Region, American Society for Horticultural Science* **15**, 76–88.

Griessel, H.M. 1995. The use of internal and positional variations in flesh firmness of 'May Glo' nectarines as a maturity determinant. *Journal of the Southern African Society for Horticultural Sciences* **5**, 81–84.

Griffee, P.J. and Burden, O.J. 1976. Fungi associated with crown rot of boxed bananas in the Windward Islands. *Phytopathologische Zeitschrift* **85**, 149–158.

Griffee, P.J. and Pinegar, J.A. 1974. Fungicides for the control of the banana crown rot complex; *in vivo* and *in vitro* studies. *Tropical Science* **16**, 107–120.

Guadarrama, A., Ortiz, V. and Ogier, J.P. 2000. Postharvest comparative study of two cultivars of sapodilla (*Manilkara zapote* L.) fruits. *Acta Horticulturae* **536**, 363–367.

Guan, J.F., Shu, H.R. and Zhang, L.C. 1998. Effect of calcium infiltration on H_2O_2 content and enzyme activities of postharvest apple fruits. *Acta Horticulturae Sinica* **25**, 391–392.

Guerzoni, M.E., Gianotti, A., Corbo, M.R. and Sinigaglia, M. 1996. Shelf life modelling for fresh cut vegetables. *Postharvest Biology and Technology* **9**, 195–207.

Gull, D.D. 1981. Ripening tomatoes with ethylene. *Vegetable Crops Fact Sheet* VC 29, Vegetable Crops Department, University of Florida, Gainsville, FL.

Gunasekaran, S., Paulsen, M.R. and Shove, G.C. 1985. Optical methods for nondestructive quality evaluation of agricultural and and biological materials. *Journal of Agricultural and Engineering Research* **32**, 209–241.

Haard, N.F. and Hultin, H.O. 1969. Abnormalities in ripening and mitochondrial succinoxidase resulting from storage of preclimacteric banana fruit at low relative humidity. *Phytochemistry* **8**, 2149.

Haard, N.F. and Salunkhe, D.K. 1975. *Symposium: Postharvest Biology and Handling of Fruits and Vegetables.* AVI Publishing, Westport, CT. 193 pp.

Haard, N.F., Sharma, S.C. Wolfe, R. and Frenkel, C. 1974. Ethylene induced isoperoxidase changes during fibre formation in postharvest asparagus. *Journal of Food Science* **39**, 452.

Haffner, K.E. 1993. Storage trials of 'Aroma' apples at the Agricultural University of Norway. *Acta Horticulturae* **326**, 305–313.

Hall, E.G. 1972. Precooling and container shipping of citrus fruits. *CSIRO Food Research Quarterly* **32**, 1–10.

Haller, M.H. and Harding, P.L. 1939. Effect of storage temperature on peaches. *United States Department of Agriculture, Technical Bulletin* **680**, 32 p.

Haller, M.H., Harding, P.L., Lutz, J.M. and Rose, D.H. 1932. The respiration of some fruits in relation to temperature. *Proceedings of the American Society of Horticultural Science* **28**, 583–589.

Haller, M.H., Rose, D.H. and Harding P.L. 1941. Studies on the respiration of strawberry and raspberry fruits. *United States Department of Agriculture, Circular* **613**, 13pp.

Haller, M.H., Rose, D.H., Lutz, J.M. and Harding, P.L. 1945. Respiration of citrus fruits after harvest. *Journal of Agricultural Research* **71**, 327–359.

Hallman, G.J. 1995. Cold storage and hot water immersion as quarantine treatments for canistel infested with Caribbean fruit fly. *HortScience* **30**, 570–572.

Halloran, N., Kasim, M.U., Cagiran, R. and Karakaya, A. 1999. The effect of postharvest treatments on storage duration of cantaloupes. *Acta Horticulturae* **492**, 207–212.

Halos, P.M. and Divinagracia, G.G. 1970. Histopathology of mango fruits infected by *Diplodia natalensis. Philippine Phytopathology* **6**, 16–28.

Hamauzu, Y., Chachin, K., Ding, C.K. and Kurooka, H. 1997. Differences in surface color, flesh firmness, physiological activity, and some components of loquat fruits picked at various stages of maturity. *Journal of the Japanese Society for Horticultural Science* **65**, 859–865.

Hamilton, A.J., Lycett, G.W. and Grierson, D. 1990. Antisense gene that inhibits synthesis of the hormone ethylene in transgenic plants. *Nature* **346**, 284–287.

Han, T., Li, L.P. and Ge, X. 2000. Effect of exogenous salicylic acid on postharvest physiology of peach fruit. *Acta Horticulturae Sinica* **27**, 367–368.

Hancock, C.T. and Epperson, J.E. 1990. Temporal cost analysis of a new development in controlled atmosphere storage: the case of Vidalia onions. *Journal of Food Distribution Research* **21**, 65–72.

Hansen, H. 1975. Storage of Jonagold apples preliminary results of storage trials. Die Lagerung von Apfeln der Sorte Jonagold erste Ergebnisse von Lagerungsversuchen. *Erwerbsobstbau* **17**, 122–123.

Hansen, H. 1977. Storage trials with less common apple varieties. Lagerungsversuche mit neu im Anbau aufgenommenen Apfelsorten. *Obstbau Weinbau* **14**, 223–226.

Hansen, H. 1986. Use of high CO_2 concentrations in the transport and storage of soft fruit. Der Einsatz hoher CO_2 Konzentrationen bei Transport und Lagerung von Weichobst. *Obstbau* **11**, 268–271.

Hansen, H. and Bohling, H. 1989. Studies on the metabolic activity of oyster mushrooms *Pleurotus ostreatus* Jacq. *Acta Horticulturae* **258**, 573–578.

Hansen, K., Poll, L., Olsen, C.E. and Lewis, M.J. 1992. The influence of O_2 concentration in storage atmospheres on the post storage volatile ester production of 'Jonagold' apples. *Lebensmittel Wissenschaft und Technologie* **25**, 457–461.

Hansen, M., Olsen, C.E., Poll, L. and Cantwell, M.I. 1993. Volatile constituents and sensory quality of cooked broccoli florets after aerobic and anaerobic storage. *Acta Horticulturae* **343**, 105 111.

Harb, J., Streif, J., Bangerth, F. and Sass, P. 1994. Synthesis of aroma compounds by controlled atmosphere stored apples supplied with aroma precursors: alcohols, acids and esters. *Acta Horticulturae* **368**, 142–149.

Hardenburg, R.E. 1955. Ventilation of packaged produce. Onions are typical of items requiring effective perforation of film bags. *Modern Packaging* **28**, 140, 199, 200.

Hardenburg, R.E. and Anderson, R.E. 1962. *Chemical Control of Scald on Apples Grown in Eastern United States.* United States Department of Agriculture, Agricultural Research Service, 51–54.

Hardenburg, R.E., Anderson, R.E. and Finney, E.E, Jr. 1977. Quality and condition of 'Delicious' apples after storage at 0 deg C and display at warmer temperatures. *Journal of the America Society for Horticultural Science* **102**, 210–214.

Hardenburg, R.E., Watada, A.E. and Wang C.Y. 1990. The commercial storage of fruits, vegetables and florist and nursery stocks. *United States Department of Agriculture, Agricultural Research Service, Agriculture Handbook* **66**, 130 pp.

Harjadi, S.S. and Tahitoe, D. 1992. The effects of plastic film bags at low temperature storage on prolonging the shelf life of rambutan *Nephelium lappacem* cv Lebak Bulus. *Acta Horticulturae* **321**, 778–785.

Harker, F.R., Elgar, H.J., Watkins, C.B., Jackson, P.J. and Hallett, I.C. 2000. Physical and mechanical changes in strawberry fruit after high carbon dioxide treatments. *Postharvest Biology and Technology* **19**, 139–146.

Harkett, P.J. 1971. The effect of O_2 concentration on the sugar content of potato tubers stored at low temperature. *Potato Research* **14**, 305–311.

Harman, J.E. 1983. Preliminary studies on the postharvest physiology of babaco fruit *Carica* × *heilbornii* Badillo nm. *pentagona* Heilborn Badillo. *New Zealand Journal of Agricultural Research* **26**, 237–243.

Harman, J.E. and McDonald, B. 1983. Controlled atmosphere storage of kiwifruit: effect on storage life and fruit quality. *Acta Horticulturae* **138**, 195–201.

Harris, C.M. and Harvey, J.M. 1973. Quality and decay of California strawberries stored in CO_2 enriched atmospheres. *Plant Disease Reporter* **57**, 44–46.

Harris, D.R., Nguyen, V.Q., Seberry, J.A., Haigh, A.M. and McGlasson, W.B. 2000. Growth and postharvest performance of white radish (*Raphanus sativus* L.). *Australian Journal of Experimental Agriculture* **40**, 879–888.

Harris, S. and McDonald, B. 1975. Physical data for kiwifruit *Actinidia chinensis. New Zealand Journal of Science* **18**, 307–312.

Hartman, S.C., Beelman, R.B., Simons, S. and van Griensven, L.J.L.D. 2000. Calcium and selenium enrichment during cultivation improves the quality and shelf life of *Agaricus* mushrooms. *Proceedings of the 15th International Congress on the Science and Cultivation of Edible Fungi, Maastricht, Netherlands, 15–19 May, 2000,* 499–505.

Hartmann, C. 1957. Quelques aspects du metabolisme des cerises et des abricots au cours de la maturation et de la senescence. *Fruits Paris* **12**, 45–49.

Hartmans, K.J., Es van, A. and Schouten, S. 1990. Influence of controlled atmosphere CA storage on respiration, sprout growth and sugar content of cv. Bintje during extented storage at 4°C. *11th Triennial Conference of the European Association for Potato Research, Edinburgh, UK, 8–13 July, 1990,* 159–160.

Haruenkit, R. and Thompson, A.K. 1993. Storage of fresh pineapples. *Postharvest Handling of Tropical Fruit. Proceedings of an International Conference held in Chiang Mai, Thailand, 19–23 July 1993. Australian Centre for International Agricultural Research Proceedings* **50**, 422–426.

Haruenkit, R. and Thompson, A.K. 1996. Effect of O_2 and CO_2 levels on internal browning and composition of pineapples Smooth Cayenne. *Proceedings of the International Conference on Tropical Fruits, Kuala Lumpur, Malaysia, 23–26 July 1996*, 343–350.

Harvey, J.M. 1978. Reduction of losses in fresh marketing fruits. *Annual Review of Phytopathology* **16**, 321–341.

Harvey, R.B. 1928. Artificial ripening of fruits and vegetables. *Minnesota Agricultural Experimental Station Bulletin* **247**, 36 pp.

Harvey, W., Lush, A. and Stuart, C. 1995. Too hard? Too soft? *Commercial Grower* **50**, 17–18.

Harvey, W.J., Grant, D.G. and Lammerink, J.P. 1997. Physical and sensory changes during the development and storage of buttercup squash. *New Zealand Journal of Crop and Horticultural Science* **25**, 341–351.

Hashem, M.Y. 2000. Suggested procedures for applying carbon dioxide CO_2 to control stored medicinal plant products from insect pests. *Zeitschrift fur Pflanzenkrankheiten und Pflanzenschutz* **107**, 212–217.

Hassan, A. and Atan, R.M. 1983. The development of black heart disease in Mauritius pineapple *Ananas comosus* cv. Mauritius during storage at low temperatures. *Malaysian Agricultural Research and Developments Institute Research Bulletin* **11**, 309–319.

Hassan, A. and Pantasico, E.B. 1990. *Bananas.* Postharvest Horticultural Training and Research Center, University of the Philippines at Los Banos, College of Agriculture, Laguna Philippines, 147 pp.

Hassan, A., Atan R. M. and Zain, Z.M. 1985. Effect of modified atmosphere on black heart development and ascorbic acid content in 'Mauritius' pineapple *Ananas comosus* cv. *Mauritius* during storage at low temperature. *Association of Southeast Asian Nations Food Journal* **1**, 15–18.

Hasselbrink, H. 1927. Carbohydrate transformation in carrots during storage. *Plant Physiology* **2**, 3.

Hatfield, S.G.S. 1975. Influence of post storage temperature on the aroma production by apples after controlled atmosphere storage. *Journal of the Science of Food and Agriculture* **26**, 1611–1612.

Hatfield, S.G.S. and Patterson, B.D. 1974. Abnormal volatile production by apples during ripening after controlled atmosphere storage. In *Facteurs and Regulation de la Maturation des Fruits. Colloeques Internationaux.* CNRS, Paris, 57–64.

Hatton, T.T. and Cubbedge, R.H. 1977. Effects of prestorage CO_2 treatment and delayed storage on stem end rind breakdown of 'Marsh' grapefruit. *HortScience* **12**, 120–121.

Hatton, T.T. and Reeder, W.F. 1966. Controlled atmosphere storage of 'Keitt' mangoes. *Proceedings of the Caribbean region of the American Society for Hoticultural Science* **10**, 114–119.

Hatton, T.T. and Reeder, W.F. 1972. Quality of Lula avocados stored in controlled atmospheres with and without ethylene. *Journal of the American Society of Horticultural Science* **97**, 339–341.

Hatton, T.T., Reeder, W.F. and Campbell, C.W. 1965. Ripening and storage of Florida mangoes. *Market Reseach Report* 752. United States Department of Agriculture, Washington, DC, 9 pp.

Hatton, T.T., Reeder, W.F. and Campbell, C.W. 1965a. Ripening and storage of Florida avocados. *Market Research Report* 697. United States Department of Agriculture, Washington, DC, 697, 13 pp.

Hatton, T.T., Reeder, W.F. and Kaufman, J. 1966. Maintaining market quality of fresh lychees during storage and transit. *Market Reseach Report* 770. United States Department of Agriculture, Washinng ton, DC.

Hauschild, A.H.W. 1989. *Clostridium botulinum.* In Doyle, M.P. (Editor), *Foodborne Bacterial Pathogens.* Marcel Dekker, New York, 111–189.

Hayase, J., Chung, T Y. and Kato, H. 1984. Changes in volatile components of tomato fruits during ripening. *Food Chemistry* **14**, 113–124.

Hayes, E.R. 1986. *N,O-Carboxymethyl chitosan and preparative method thereof.* US Patent 4 619 995.

Hayward, A.W. and Walker, H.M. 1961. The effects of preharvest foliar sprays of M.H. on the storage of yams. *Annual Report of the West African Stored Products Research Unit* **19**, 107.

Heap, R.D. 1989. Design and performance of insulated and refrigerated ISO intermodal containers. *International Journal of Refrigeration* **12**, 137–145.

Heather, N. 1993. Disinfestation: non-chemical options. *Postharvest Handling of Tropical Fruit. Proceedings of an International Conference held in Chiang Mai, Thailand, 19–23 July 1993. Australian Centre for International Agricultural Research Proceedings* **50**, 272–279.

Heather, N.W. 1985. Alternatives to EDB fumigation as post harvest treatment of fruit and vegetables. *Queensland Journal of Agriculture,* Nov./Dec., 321–326.

Heheriuk, M. and Herregods, M. 1993. Apple storage conditions. *Sixth Annual Controlled Atmosphere Research Conference, Cornell University, Ithaca, New York.*

Heisick, J.E., Wagner, D.E., Nierman, M.L. and Peeler, J.T. 1989. *Listeria* spp. found on fresh market product. *Applied Environmental Microbiology* **55**, 1925–1927.

Hellickson, M.L., Adre, N., Staples, J. and Butte, J. 1995. Computer controlled evaporator operation during fruit cool down. *Technologias de Cosecha y Postcosecha de Frutas y Hortalizas. Proceedings of a Conference held in Guanajuato, Mexico, 20–24 February 1995,* 546–553.

Henderson, D.W. 1993. The case for wholesale markets. *International Workshop of Agricultural Produce Wholesale Marketing, Ministry of Agriculture Chengdu, Sichuan Province, China, 25–30 November 1993.*

Henderson, J.R. and Buescher, R.W. 1977. Effect of sulfur dioxide and controlled atmospheres on broken end discoloration and processed quality atributes in snap beans. *Journal of the American Society for Horticultural Science* **102**, 768–770.

Henze, J. 1989. Storage and transport of Pleurotus mushrooms in atmospheres with high CO_2 concentrations. *Acta Horticulturae* **258**, 579–584.

Hermann, M. and Erazo, C. 2001. Compositional changes of oca tubers following postharvest exposure to sunlight. *Scientist and Farmer: Partners in Research for the 21st Century. Program Report 1999–2000. Centro Internacional de la Papa, Lima, Peru,* 391–396.

Herregods, M. 1993. Personal communication.

Herregods, M., Champ, B.R. and Highley, E. 1995. Current research on the postharvest handling of fruits and vegetables. *Australian Centre for International Agricultural Research Proceedings* **60**.

Hershko, V. and Nussinovitch, A. 1998. Relationships between hydrocolloid coating and mushroom structure. *Journal of Agricultural and Food Chemistry* **46**, 2988–2997.

Hesselman, C.W. and Freebairn, H.T. 1969. Rate of ripening of initiated bananas as influenced by oxygen and ethylene. *Journal of the American Society for Horticultural Science,* **94**, 635.

Hew, C.S. and Lee, Y.H. 1981. Germination and short term storage of winged bean seeds. *Winged Bean Flyer* **3**, 15.

Heydendorff, R. and Dobreanu, M. 1972. Factors influencing the commercial keeping quality of eggplant and green peppers. *Lucrari Stiintifice R.* **3**, 133.

Hguyen-The, C. 1991. Microbiological quality of ready-to-use vegetables. Qualité microbiologique des végétaux prêts a l'emploi. *Comptes Rendus de l'Académie d'Agriculture de France* **77**, 7–13.

Hicks, J.R. and Ludford, P.M. 1981. Effects of low ethylene levels on storage of cabbage. *Acta Horticulturae* **116**, 65–73.

Hill, G.R., Jr. 1913. Respiration of fruits and growing plant tissue in certain cases, with reference to ventilation and fruit storage. *Cornell University, Agricultural Experiment Station, Bulletin* **330**, 26 pp.

Hill, R.W. and Selassie, H. 1993. Observations on wholesale marketing systems for fruit, vegetables and fish in developed and developing countries. *International Workshop of Agricultural Produce Wholesale Marketing, Ministry of Agriculture Chengdu, Sichuan Province, China, 25–30 November 1993.*

Hironaka, K., Ishibashi, K. and Hakamada, K. 2001. Effect of static loading on sugar contents and activities of invertase, UDP-glucose pyrophosphorylase and sucrose 6-phosphate synthase in potatoes during storage. *Potato Research* **44**, 33–39.

Hitchcock, D. 1973. *Design of a gooseberry harvester.* MSc thesis, Silsoe College, Cranfield Institute of Technology.

Hobson, G.E. 1989. Manipulating the ripening of tomato fruit low and high technology. *Acta Horticulturae* **258**, 593–600.

Hobson, G.E. and Grierson, D. 1993. Tomato. In Seymour, G.B., Taylor, J.E. and Tucker, G.A. (Editors), *Biochemistry of Fruit Ripening.* Chapman and Hall, London, 405–442.

Hodges, D.M. and Forney, C.F. 2000. The effects of ethylene, depressed oxygen and elevated carbon dioxide on antioxidant profiles of senescing spinach leaves. *Journal of Experimental Botany* **51**, 645–655.

Hofman, P.J. and Smith, L.G. 1993. Preharvest effect on postharvest quality of subtropical and tropical fruits. *Postharvest Handling of Tropical Fruit. Proceedings of an International Conference held in Chiang Mai, Thailand, 19–23 July 1993. Australian Centre for International Agricultural Research Proceedings* **50**, 261–268.

Holder, G.D. and Gumbs, F.A. 1983. Agronomic assessment of the relative suitability of the banana cultivars 'Robusta' and 'Giant Cavendish' Williams Hybrid to irrigation. *Tropical Agriculture Trinidad* **60**, 17–24.

Holt, J.B. and Sharp, J.R. 1989. A simple inclined cup conveyor for vegetable harvesting. *Divisional Note, Agricultural and Food Research Council, Institute of Engineering Research,* DN 1499, 9 pp.

Hong, Q.Z., Sheng, H.Y., Chen, Y.F. and Yang, S.J. 1983. Effects of CA storage with a silicone window on satsuma oranges. *Journal of Fujian Agricultural College* **12** 53–60.

Hope Mason, D. 1984. Bananas the life blood of the Windward Islands. *Fruit Trades Journal* March 23rd, 22–29.

Hopkins, E.F. and Cubbage, R.H. 1948. A curing procedure for the reduction of mold decay in citrus fruits. *Bulletin of the Florida Agricultural Experimental Station* **450**.

Hosaka Y. 1987. Evaluation of taste of rice by means of near infrared. *Japanese Society of Agricultural Machinery, Proceedings of the International Symposium on Agricultural Mechanization and International Cooperation in High Technology Era, April 3 1987,* 357–360.

Hotchkiss, J.H. and Banco, M.J. 1992. Influence of new packaging technologies on the growth of microorganisms in produce. *Journal of Food Protection* **55**, 815–820.

Houck, L.G., Aharoni, Y. and Fouse, D.C. 1978. Colour changes in orange fruit stored in high concentrations of O_2 and in ethylene. *Proceedings of the Florida State Horticultural Society 1978,* **91**: 136–139.

Howarth, M.S., Brandon, J.R., Searsy, S.W. and Kehtarnavaz, N. 1992. Estimation of tip shape for carrot classification by machine vision. *Journal of Agricultural Engineering Research* **53**, 123–134.

Howell G.S., Jr, Stergios, B.G., Stackhouse, S.S, Bittenbender, H.C., and Burton, C.L. 1976. Ethephon as mechanical harvesting aid for highbush blueberries. *Journal of the American Society for Horticultural Science* **101**, 111–115.

Hribar, J., Plestenjak, A., Vidrih, R., Simcic, M. and Sass, P. 1993. Influence of CO_2 shock treatment and ULO storage on apple quality. *Acta Horticulturae* **368**, 634–640.

Hruschka, H.W. 1966. Storage and shelf life of packaged rhubarb. *Market Research Report* 771. United States Department of Agriculture, Washington, DC, 17 pp.

Hruschka, H.W. 1971. Storage and shelf life of packaged kale, *Market Research Report* 923. United States Department of Agriculture, Washingtin, DC., 19 pp.

Hruschka, H.W. 1974. Storage and shelf life of packaged green onions. *Market Research Report* 1015. United States Department of Agriculture, Washington DC, 21 pp.

Hruschka, H.W. 1978. Storage and shelf life of packaged leeks. *Market Research Report* 1984. United States Department of Agriculture, Wsshington DC, 19 pp.

Hruschka, H.W. and Wang, C.Y. 1979. Storage and shelf life of packaged watercress, parsley and mint. *Market Research Report* 1102, United States Department of Agriculture, Washington, D.C, 19 pp.

HSE 1991. *Confined Spaces.* Health and Safety Executive UK, Information Sheet 15.

Hu, X.-Q., Yu, X. and Chen, L.-G. 2001. Studies of Chinese bayberry fruits on some physiological characters during the storage. *Journal of Zhejiang University Agriculture and Life Sciences* **27**, 179–182.

Huang, X.-Y, Kang, D.–M. and Ji Z.–L. 1990. The optimum storage temperature for litchi fruits and chilling injury of them. *Journal of the South China Agricultural University* **11**, 13–18.

Hubbard, N., Pharr, D.M. and Huber, S.C. 1990. The role of sucrose phosphate synthase in sucrose biosynthesis in ripening bananas and its relationship to the respiratory climacteric. *Plant Physiology* **94**, 201–208.

Huber, D.J. 1983. Polyuronide degredation and hermicellulose modifications in ripening tomato fruits. *Journal of the American Society for Horticultural Science* **108**, 405–409.

Huelin, F.E. 1933. Effects of ethylene and of apple vapours on the sprouting of potatoes. *Report of the Food Investigation Board, London for 1932,* 51–53.

Hughes, P.A., Thompson, A.K., Plumbley, R.A. and Seymour, G.B. 1981. Storage of capsicums *Capsicum annum* [L.] Sendt. under controlled atmosphere, modified atmosphere and hypobaric conditions. *Journal of Horticultural Science* **56**, 261–265.

Hui, C.K.P. Vigneault, C. and Goyette, B. 1999. Optimization of openings in plastic containers for hydrocooling of fruits and vegetables. *American Society of Agricultural Engineers Paper* 996031.

Hulme, A.C. 1956. CO_2 injury and the presence of succinic acid in apples. *Nature* **178**, 218.

Hulme, A.C. 1970. *The Biochemistry of Fruits and Their Products,* Vol. 1. Academic Press, London.

Hulme, A.C. 1971. *The Biochemistry of Fruits and their Products,* Vol. 2. Academic Press, London.

Huyskens-Keil, S., Widayat, H.P., Ludders, P., Schreiner, M. and Peters, P. 2000. Physiological changes of pepino (*Solanum muricatum* Ait.) during maturation and ripening. *Acta Horticulturae* **531**, 251–256.

Ikeda, Y. 1986. On the nondestructive and realtime measurement of mechanical properties of agricultural products. *Memoirs of the College of Agriculture, Kyoto University* **129**, 47–54.

Ilangantileke, S. and Salokhe, V. 1989. Low pressure atmosphere storage of Thai mango. *Proceedings of the Fifth International Controlled Atmosphere Research Conference, Wenatchee, Washington, USA, 14–16 June. Volume 2. Other Conmodities and Storage Recommandatious,* 103–117.

Ilinski, A.S., Goudkovski, V.A. and Mitrokhin, M.A. 2000. Fruit quality after CA storage utilizing initial O_2 reduction by respiration of non pre cooled fruit. *Acta Horticulturae* **518**, 29–36.

Imahori, Y., Kota, M. Ueda, Y. and Chachin, K. 1998. Effects of low oxygen atmospheres on quality and ethanol and acetaldehyde formation of ethylene treated bananas. *Journal of the Japanese Society for Food Science and Technology* **45**, 572–576.

Imakawa, S. 1967. Studies on the browning of Chinese yam. *Memoir of the Faculty of Agriculture, Hokkaido University, Japan* **6**, 181.

Inaba, A., Kiyasu, P. and Nakamura, R. 1989. Effects of high CO_2 plus low O_2 on respiration in several fruits and vegetables. *Scientific Reports of the Faculty of Agriculture, Okayama University* **73**, 27 33.

Inglese, P., Barbera, G. and la Mantia, T. 1999. Seasonal reproductive and vegetative growth patterns and resource allocation during cactus pear fruit growth. *HortScience* **34**, 69–72.

Ingram, J.S. and Humphries, J.R.O. 1972. Cassava storage: a review. *Tropical Science* **14**, 131.

International Institute of Refrigeration. 1979. *Recommendations for Chilled Storage of Perishable Produce.* International Institute of Refrigeration, Paris, 148 pp.

Isenburg, F.M.R. and Sayle, R.M. 1969. Modified atmosphere storage of Danish cabbage. *Proceedings of the American Society of Horticultural Science* **94**, 447–449.

Isenberg, F.M.R. 1979. Controlled atmosphere storage of vegetables. *Horticultural Review,* **1**, 337–394.

Isenberg, F.M.R., Thomas, T.H., Abed Rahaman, M., Pendergrass, A., Carroll, J.C. and Howell, L. 1974. The role of natural growth regulators in rest, dormancy and regrowth of vegetables during winter storage. *Acta Horticulturae* **38**, 95–125.

Isenberg, F.M.R., Thomas, T.H. and Pendergrass, M. 1977. Hormone and histological differences between normal and malic hydrazide treated onions stored over winter. *Acta Horticulturae,* **62**, 95–122.

Ishii, K. and Okubo, M. 1984. The keeping quality of Chinese chives *Allium tuberosum* Rottler by low temperature and seal packaging with polyethylene bags. *Journal of the Japanese Society for Horticultural Science* **53**, 87–95.

Ishizawa, Y., Kyoutani, H., Nishimura, K., Yamaguchi, M. and Kakiuchi, N. 1995. Color charts for the evaluation of harvest time of mume fruits. *Bulletin of the Fruit Tree Research Station* **28**, 15–24.

Itai, A., Kawata, T., Tanabe, K., Tamura, F., Uchiyama, M., Tomomitsu, M. and Shiraiwa, N. 1999. Identification of 1-aminocyclopropane–1-carboxylic acid synthase genes controlling the ethylene level of ripening fruit in Japanese pear (*Pyrus pyrifolia* Nakai). *Molecular and General Genetics* **261**, 42–49.

ITC, 1988. *Manual on the Packaging of Fresh Fruits and Vegetables.* International Trade Centre, UNCTAD/GATT, Geneva. ITC/035/C1/88-XII, 241 pp.

ITC, 1992. *International Fibreboard Case Code, 7th edn.* Export Packaging Note. International Trade Centre, UNCTAD/GATT, Geneva, 41pp.

Ito, S. 1971. The persimmon. In Hulme, A.C. (Editor). *The Biochemistry of Fruits and their Products,* Vol. 2. Academic Press, London, 281–302.

Ito, S., Kakiuchi, N., Izumi, Y. and Iba, Y. 1974. Studies on the controlled atmosphere storage of satsuma mandarin. *Bulletin of the Fruit Tree Research Station, B. Okitsu* **1**, 39–58.

Ito, T. and Nakamura, R. 1985. Tolerance of fresh fruits and vegetables to fluctuating temperature. *Journal of the Japanese Society for Horticultural Science* **54**, 257–264.

Itoo, S., Matsuo, T. Ibushi, Y. and Tamari, N. 1987. Seasonal changes in the levels of polyphenols in guava fruit and leaves and some of their properties. *Journal of the Japanese Society of Horticultural Science* **56**, 107–113.

Izumi, H. Watada, A.E. and Douglas, W. 1996. Optimum O_2 or CO_2 atmosphere for storing broccoli florets at various temperatures. *Journal of the American Society for Horticultural Science* **121**, 127–131.

Jabati, M. 1988. *The market for mangoes in the United Kingdom.* MSc thesis, Silsoe College, Cranfield Institute of Technology.

Jacobi, K.K. Coates, L. and Wong, L. 1993. Quality of 'Kensington' mango (*Mangifera indica*) following hot water and vapour heat treatement. *Postharvest Biology and Technology* **1**, 349–359.

Jacobi, K.K., MacRae, E.A. and Hetherington, S.E. 2000. Effects of hot air conditioning of 'Kensington' mango fruit on the response to hot water treatment. *Postharvest Biology and Technology* **21**, 39–49.

Jacobs, C.T., Brodrick, H.T., Swarts, H.D., Mulder, N.J. 1973. *Plant Disease Reporter* **57**, 173–176.

Jacomino, A.P., Sigrist, J.M.M., de Sarantopoulos, C.I.G., Minami, K. and Kluge, R.A. 2001. Packagings for cold storage of guavas. Embalagens para conservacao refrigerada de goiabas. *Revista Brasileira de Fruticultura* **23**, 50–54.

Jaffee, S. 1993. Kenya's horticultural export marketing: a transaction cost perspective. In *Agricultural and Food Marketing in Developing Countries: Selected Readings.* CAB International, Technical Centre for Agricultural and Rural Cooperation, 388–403.

Jahn, O.L. and Gaffney, J.J. 1972. Photelectric color sorting of citrus fruits. *United States Department of Agriculture, Technical Bulletin* **1448**, 56 pp.

Jankovic, M. and Drobnjak, S. 1994. The influence of cold room atmosphere composition on apple quality changes. Part 2. Changes in firmness, mass loss and physiological injuries. *Review of Research Work at the Faculty of Agriculture, Belgrade* **39**, 73–78.

Jarimopas, B. and Therdwongworakul, A. 1994. Postharvest mechanical drenching and disease control of mango. *Kasetsart Journal, Natural Sciences* **28**, 616–625.

Jeffery, D., Smith, C., Goodenough, P.W., Prosser, T. and Grierson, D. 1984. Ethylene independent and ethylene dependent biochemical changes in ripening tomatoes. *Plant Physiology*, **74**, 32.

Ji, Z.L. and Wang, S.Y. 1988. Reduction of abscisic acid content and induction of sprouting in potato *Solanum tuberosum* by thidiazuron. *Journal of Plant Growth Regulation* **7**, 37–44.

Jiang, Y.M. 1999. Low temperature and controlled atmosphere storage of fruit of longan (*Dimocarpus longan Lour.*). *Tropical Science* **39**, 98–101.

Jiang, Y.M. and Li, Y.B. 2001. Effects of chitosan coating on postharvest life and quality of longan fruit. *Food Chemistry* **73**, 139–143.

Jiang, Y.-M, Joyce, D.C. and MacNish, A.J. 1999. Extension of the shelf life of banana fruit by 1-methylcyclopropene in combination with polyethylene bags. *Postharvest Biology and Technology* **16**, 187–193.

Jobling, J., McGlasson, W.B., Miller, P. and Hourigan, J. 1993. Harvest maturity and quality of new apple cultivars. *Acta Horticulturae* **343**, 53–55.

Johanson, A., Proctor, F.J., Cox, J.R. and Jeger, M.J. 1989. Control of crown rot in the Windward Islands. *Aspects of Applied Biology* **20**, 89–90.

Johnson, D.S. 1994. Storage conditions for apples and pears. *East Malling Research Association Review 1994–1995.*

Johnson, D.S. and Ertan, U 1983. Interaction of temperature and O_2 level on the respiration rate and storage quality of Idared apples. *Journal of Horticultural Science* **58**, 527–533.

Johnson, D.S., Dover, C.J. and Pearson, K. 1993. *Acta Horticulturae* **326**, 175–182.

Johnson, G., Boag, T.S., Cooke, A.W., Izard, M., Panitz, M. and Sangchote, S. 1990. Interaction of post harvest disease control treatments and gamma irradiation on mangoes. *Annals of Applied Biology* **116**, 245–257.

Johnson, G.I., Mead, A.J., Cooke, A.W., and Wells, I.A. 1992. Stem end rot disease of tropical fruit mode of infection in mango, and prospects for control. *Acta Horticulturae* **321**, 882–890.

Johnson, G.I., Sangchote, S. and Cooke, A.W. 1990a. Control of stem end rot *Dothiorella dominicana* and other postharvest diseases of mangoes cultivar Kensington Pride during short and long term storage. *Tropical Agriculture* **67**, 183–187.

Johnson, M. 1985. Automation in citrus sorting and packing. *Proceedings of Agri Mation Conference and Expo, Chicago, USA*, 63–68.

Johnston, J.W., Hewett, E.W., Banks, N.H., Harker, F.R. and Hertog, M.L.A.T.M. 2001. Physical change in apple texture with fruit temperature: effects of cultivar and time in storage. *Postharvest Biology and Technology* **23**, 13–21.

Jones, W.W. 1942. Respiration and chemical changes of papaya fruit in relation to temperature. *Plant Physiology* **17**, 481–486.

Jooste, M.M. and Taylor, M.A. 1999. Effect of harvest maturity and cold storage period on the overall quality of Bebeco apricots, with special reference to gel breakdown and variates used to accurately establish harvest maturity. *Deciduous Fruit Grower* **49**, S1–S10.

Jordan, V.W.L., and Richmond, D.V. 1974. The effect of benomyl on sensitive and tolerant isolates of *Botrytis cinerea* infecting strawberries. *Plant Pathology* **23**, 81–83.

Ju, Z.Q., Duan, Y.S. and Ju, Z.G. 2000. Plant oil emulsion modifies internal atmosphere, delays fruit ripening, and inhibits internal browning in Chinese pears. *Postharvest Biology and Technology* **20**, 243–250.

Jurd, J.R. 1964. Reactions involved in sulfite bleaching of anthocyanins. *Journal of Food Science* **29**, 16–19.

Kader, A.A. 1983. Physiological and biochmical effects of carbon monoxide added to controlled atmospheres of fruit. *Acta Horticulturae* **138**, 221–226.

Kader, A.A. 1985. Ethylene induced senescence and physiological disorders in harvested horticultural crops. *HortScience* **20**, 54.

Kader, A.A. 1985a. Modified atmosphere and low pressure systems during transport and storage. In Kader, A.A., Kasmire, R.F., Mitchell, F.G., Reid, M.S., Sommer, N.F. and Thompson, J.F. (Editors). *Postharvest Technology of Horticultural Crops*. Cooperative Extension, University of California, Division of Agriculture and Natural Resorces, 59–60.

Kader, A.A. 1986. Biochemical and physiological basis for effects on controlled and modified atmospheres on fruits and vegetables. *Food Technology* **40**, 99–104.

Kader, A.A. 1987. Respiration and gas exchange in vegetables. In Weichmann, J. (Editor), *Postharvest Physiology of Vegetables*. Marcel Dekker, New York, 25–44.

Kader, A.A. 1988. Comparison between 'Semperfresh' and 'Nutri Save' coatings on 'Granny Smith' apples. *Cooperative Extension, University of California, Perishables Handling, Postharvest Technology of Fresh Horticultural Crops* **63**, 4–5.

Kader, A.A. 1989. A summary of CA requirements and recommendations for fruit other than pome fruits. *International Controlled Atmosphere Conference Fifth, Proceedings, June 14–16, 1989, Wenatchee, Washington. Volume 2. Other Commodities and Storage Recommendations*, 303–328.

Kader, A.A. 1993. Modified and controlled atmosphere storage of tropical fruits. *Postharvest Handling of Tropical Fruit. Proceedings of an International Conference held in Chiang Mai, Thailand, 19–23 July 1993. Australian Centre for International Agricultural Research Proceedings* **50**, 239–249.

Kader, A.A., Brecht, P.E., Woodruff, R. and Morris L.L. 1973. Influence of carbon monoxide, CO_2 and O_2 levels on brown stain, respiration rate and visual quality of lettuce. *Journal of the American Society for Horticultural Science* **98**, 485–488.

Kader, A.A., Chordas, A. and Elyatem, S. 1984. Responses of pomegranates to ethylene treatment and storage temperature. *California Agriculture* **38**, 14–15.

Kader, A.A., Lyons, J.M. and Morris L.L. 1974. Post harvest responses of vegetables to pre harvest field treatment. *HortScience* **9**, 523–527.

Kader, A.A., Morris,L., Stevens, M.A. and Albright Holton, M. 1978. Composition and flavour quality of fresh market tomatoes as influenced by some postharvest handling procedures. *Journal of the American Society for Horticultural Science* **103**, 6–13.

Kader, A.A., Nanos, G.D. and Kerbel, E.L. 1989. Responses of 'Manzanillo' olives to controlled atmosphere storage. *International Controlled Atmosphere Conference, Fifth, Proceedings, June 14–16, 1989, Wenatchee, Washington. Volume 2. Other Commodities and Storage Recommendations*, 119–125.

Kader, A.A., Zagory, D. and Kerbel, E.L. 1989a. Modified atmosphere packaging of fruits and vegetables. *Critical Reviews in Food Science and Nutrition* **28**, 1–30.

Kagy, V. 1989. *The development of a partially destructive electrical technique to assess melon quality*. MSc thesis, Silsoe College, Cranfield Institute of Technology.

Kaji, H., Ikebe, T. and Osajima, Y. 1991. Effects of environmental gases on the shelf life of Japanese apricot. *Journal of the Japanese Society for Food Science and Technology* **38**, 797–803.

Kajiura, I. 1972. Effects of gas concentrations on fruit. V. Effects of CO_2 concentrations on natsudaidai fruit. *Journal of the Japanese Society for Horticultural Science* **41**, 215–222.

Kajiura, I. 1973. The effects of gas concentrations on fruits. VII. A comparison of the effects of CO_2 at different relative humidities, and of low O_2 with and without CO_2 in the CA storage of natsudaidai. Journal of the *Japanese Society for Horticultural Science* **42**, 49–55.

Kajiura, I 1975. CA storage and hypobaric storage of white peach 'Okubo.' *Scientia Horticulturae* **3**, 179–187.

Kajiura, I. and Iwata, M. 1972. Effects of gas concentrations on fruit. IV. Effects of O_2 concentration on natsudaidai fruits. *Journal of the Japanese Society for Horticultural Science* **41**, 98–106.

Kale, P.N., Warade, S.D. and Jagtap, K. 1991. Effect of different cultural practices on storage life of onion bulbs. *Onion News Letter for the Tropics* **3**, 25–26.

Kamal, M. and Agbari, A.A.A. 1985. *Manual of Plant Diseases in the Yemen Arab Republic. Precision Press*, London,, 144 pp.

Kamath, O.C., Kushad, M.M. and Barden, J.A. 1992. Postharvest quality of 'Virginia Gold' apple fruit. *Fruit Varieties Journal* **46**, 87–89.

Kanellis, A.K. and Kalaitzis, P. 1992. Cellulase occurs in multiple active forms in ripe avocado fruit mesocarp. *Plant Physiology* **98**, 530–534.

Kang, W.S., Kim, S.H. and Lee, G.H. 1993. Design and performance parameters of vibrating potato diggers. *Proceeding of ICAMPE '93, October 19–22 KOEX*, Korean Society for Agricultural Machinery, Seoul, Korea, 734–743.

Kanlayanarat, S., Wong Aree, C. and Maneerat, C. 2000. Use of film thickness for modified atmosphere packaging to prolong storage life of rambutan cultivar 'Rong Rien.' *Acta Horticulturae* **518**, 107–113.

Kapse, B.M. and Katrodia, J.S. 1997. Studies on hydrocooling in Kesar mango (*Mangifera indica* L.). *Acta Horticulturae* **455**, 707–717.

Karaoulanis, G. 1968. The effect of storage under controlled atmosphere conditions on the aldehyde and alcohol contents of oranges and grapes. *Annual Report of the Ditton Laboratory*, 1967–68.

Karmarker, D.V. and Joshi, B.M., 1941. Investigations on the storage of onions. *Indian Journal of Agricultural Science* **11**, 82–94.

Karup, C. 1990. Initial weight loss, packaging and conservation of asparagus. *Acta Horticulturae* **271**, 477–482.

Kawagoe, Y., Morishima, H., Seo, Y. and Imou, K. 1991. Development of a controlled atmosphere storage system with a gas separation membrane. Part 1. Apparatus and its performance. *Journal of the Japanese Society of Agricultural Machinery* **53**, 87–94.

Kay, D.E. 1973. *Root Crops. Crop and Product Digest.* Tropical Products Institute, London, 245 pp.

Kay, D.E. 1987. *Crop and Product Digest No. 2 - Root Crops*, 2nd edn, revised by Gooding, E.G.B. Tropical Development and Research Institute, London, 380 pp.

Kays, S.J. 1991. *Postharvest Physiology of Perishable Plant Products.* AVI Publishing, New York.

Ke, D. and Kader, A.A. 1989. Tolerance and responses of fresh fruits to O_2 levels at or below 1%. *International Controlled Atmosphere Conference, Fifth, Proceedings, June 14–16, 1989, Wenatchee, Washington. Volume 2. Other Commodities and Storage Recommendations*, 209–216.

Ke, D. and Kader, A.A. 1992. Potential of controlled atmospheres for postharvest insect disinfestation of fruits and vegetables. *Postharvest News and Information* **3**, 31N–37N.

Ke, D. and Kader, A.A. 1992a. External and internal factors influence fruit tolerance to low O_2 atmospheres. *Journal of the American Society for Horticultural Science* **117**, 913–918.

Ke, D., El Wazir, F., Cole, B., Mateos, M., and Kader, A.A. 1994. Tolerance of peach and nectarine fruits to insecticidal controlled atmospheres as influenced by cultivar, maturity, and size. *Postharvest Biology and Technology* **4**, 135–146.

Ke, D., Zhou, L. and Kader, A.A, 1994a. Mode of O_2 and CO_2 action on strawberry ester biosynthesis. *Journal of the American Society for Horticultural Science* **119**, 971–975.

Kelly, M.O. and Saltveit, M.E. 1988. Effect of endogenously synthesised and exogenously applied ethanol to tomato fruit ripening. *Plant Physiology* **88**, 143–147.

Kemp, D.C. and Matthews, M.D.P. 1977. Banana conveyor. *Overseas Development Technical Bulletin 7.* National Institute of Agricultural Engineering UK, 15 pp.

Kepczynski, J. and Kepczynski, E. 1977. Effect of ethylene on germination of fungal spores causing fruit rot. *Fruit Science Research Reports 4.* Research Institut of Pomology, Skiernewice, Poland, 31.

Kerbel, E.and Ke, D. 1991. Tolerance of 'Fantasia' nectarines to low O_2 and high CO_2 atmospheres. *Technical Innovations in Freezing and Refrigeration of Fruits and Vegetables.* Paper presented at the conference *held in Davis, California, USA, 9–12 July, 1989*, 325–331.

Kerbel, E.L., Mitchell, F.G., Kader, A.A. and Mayer, G. 1989. Effects of 'Semperfresh' coating on postharvest life, internal amosphere modification and quality maintainance of 'Granny Smith' apples. *International Controlled Atmosphere Conference, Fifth, Proceedings, June 14–16, 1989, Wenatchee, Washington. Volume 1. Pome Fruits*, 247–254.

Kester, J.J. and Fennema, O.R. 1986. Edible films and coatings – a review. *Food Technology* **40**, 46–57.

Ketsa, S. and Klaewkasetkorn, O. 1992. Postharvest quality and losses of 'Rongrein' rambutan fruits in wholsale markets. *Acta Horticulturae* **321**, 771–777.

Ketsa, S. and Leelawtana, K. 1992. Effect of pre and poststorage acid dipping on browning of lychee fruits. *Acta Horticulturae* **321**, 726–731.

Ketsa, S. and Pangkool, S. 1995. Ripening behaviour of durians at different temperatures. *Tropical Agriculture* **72**, 141–145.

Ketsa, S. and Pangkool, S. 1996. Effect of maturity stages and ethylene treatment on ripening of durian fruits. *Proceedings of the Australasian Postharvest Horticulture Conference 'Science and Technology for the Fresh Food Revolution,' Melbourne, Australia, 18–22 September, 1995*, 67–72.

Ketsa, S. and Prabhasavat, T. 1992. Effects of skin coating on shelf life and quality of 'Nang Klangwan' mangoes. *Acta Horticulturae* **321**, 764–770.

Ketsa, S. and Raksritong, T. 1992b. Effects of PVC film wrapping and temperature on storage life and quality of 'Nam Dok Mai' mango fruits on ripening. *Acta Horticulturae* **321**, 756–763.

Khalil, T.S. 1971. Histological and histochemical studies of sweet corn pericarp as influenced by maturity and processing. *Dissertation Abstracts* **11**, 6678.

Khanbari, O.S. and Thompson, A.K. 1993. Effects of amino acids and glucose on the fry colour of potato crisps. *Potato Research* **36**, 359–364.

Khanbari, O.S. and Thompson A.K. 1993a. Potato News and Developments. Silsoe College Research. *Snack Food International* **11**, 3–5.

Khanbari, O.S. and Thompson, A.K 1994. The effect of controlled atmosphere storage at 4°C on crisp colour and on sprout growth, rotting and weight loss of potato tubers. *Potato Research*, **37**, 291–300.

Khanbari, O.S. and Thompson, A.K. 1996. Effect of controlled atmosphere, temperature and cultivar on sprouting and processing quality of stored potatoes. *Potato Research* **39**, 523–531.

Khitron Ya, I. and Lyublinskaya, N.A 1991. Increasing the effectiveness of storing table grapes. *Sadovodstvo i Vinogradarstvo* **7**, 19–21.

Kidd, F. 1916. The controlling influence of CO_2: Part III. The retarding effect of CO_2 on respiration. *Proceedings of the Royal Society, London* **89B**, 136–156.

Kidd, F. and West, C. 1917. The controlling influence of carbon dioxide. IV. On the production of secondary dormancy in seeds of *Brassica alba* following a treatment with carbon dioxide, and the relation of this phenomenon to the question of stimuli in growth processes. *Annals of Botany* **34**, 439–446.

Kidd, F. 1919. Laboratory experiments on the sprouting of potatoes in various gas mixtures nitrogen, O_2 and CO_2. *New Phytologist* **18**, 248–52.

Kidd, F. and West, C. 1920. The role of the seed coat in relation to the germination of immature seed. *Annals of Botany* **34**, 439–446.

Kidd, F. and West, C. 1923. Brown heart – a functional disease of apples and pears. *Special Report of the Food Investigation Board, Department of Scientific and Industrial Research* **12**, 54 pp.

Kidd, F. and West, C. 1925. The course of respiratory activity throughout the life of an apple. *Report of the Food Investigation Board London for 1924,* 27–34.

Kidd, F. and West, C. 1927. Gas storage of fruit. *Special Report of the Food Investigation Board, Department of Scientific and Industrial Research,* 30.

Kidd, F. and West, C. 1927a. A relation between the respiratory activity and the keeping quality of apples. *Report of the Food Investigation Board London for 1925 and 1926,* 37–41.

Kidd, F. and West, C. 1927b. A relation between the concentration of O_2 and CO_2 in the atmosphere, rate of respiration, and the length of storage of apples. *Report of the Food Investigation Board London for 1925,* 41–42.

Kidd, F. and West, C. 1928. Forecasting the life of an apple. *Report of the Food Investigation Board London for 1927,* 23–27.

Kidd, F. and West, C. 1930. The gas storage of fruit. II. Optimum temperatures and atmospheres. *Journal of Pomology and Horticultural Science* **8**, 67–77.

Kidd, F. and West, C. 1934. Injurious effects of pure O_2 upon apples and pears at low temperatures. *Report of the Food Investigation Board London for 1933,* 74–77.

Kidd, F. and West, C. 1935. Gas storage of apples. *Report of the Food Investigation Board London for 1934,* 103–109.

Kidd, F. and West, C. 1935a. The Refrigerated Gas Storage of Apples. *Department of Scientific and Industrial Research, Food Investigation Leaflet* 6.

Kidd, F. and West, C. 1936. The cold storage of English grown Conference and Doyenne de Comice pears, *Report of the Food Investigation Board London for 1935,* 85–96.

Kidd, F. and West, C. 1937. Recent advances in the work on refrigerated gas storage. *Journal of Pomology and Horticultural Science* **14**, 304–305.

Kidd, F. and West, C. 1938. The action of CO_2 on the respiratory activity of apples. *Report of the Food Investigation Board London for 1937,* 101–102.

Kidd, F. and West, C. 1939. The gas storage of Cox's Orange Pippin apples on a commercial scale, *Report of the Food Investigation Board London for 1938,* 153–106.

Kidd, F. and West, C. 1940. Resistance of the skin of the apple fruit to gaseous exchange, *Report of the Food Investigation Board London for 1939,* 64–68.

Kidd, F. and West, C. 1945. Respiratory activity and duration of life of apples gathered at different stages of development and subsequently maintained at a constant temperature, *Plant Physiology Lancaster* **20**, 467–504.

Kim, Y.H. and Kim, Y.C. 1992. Food and Agriculture Organization of the United Nations regional cooperation for vegetable research and development RAS/89/41. In Bhatti, M.H., Hafeez, A., Jaggar, and Farooq, M. (Editors). Postharvest losses of vegetables. *A Report on a Workshop Held Between 17 and 22 October 1992 at the Pakistan Agricultural Research Council, Islamabad, Pakistan.*

Kim, J.K., Solomos, T. and Gross, K.C. 1999. Changes in cell wall galactosyl and soluble galactose content in tomato fruit stored in low oxygen atmospheres. *Postharvest Biology and Technology* **17**, 33–38.

Kim, M. and Oogaki, C. 1986. Characteristics of respiration and ethylene production in fruits transferred from low pressure storage to ambient atmosphere. *Journal of the Japanese Society for Horticultural Science* **55**, 339–347.

Kim, S.L., Hwang, J.J., Son, Y.K. and Kim, S.Y. 1997. Effects of ice cooling storage on chemical components in vegetable corn. *Korean Journal of Crop Science* **42**, 95–103.

Kim, Y.H., Kim, C.S. and Lee, C.H. 1993. Correlation between non destructive quality evaluation perameter and special reflectance of peaches. *Proceedings of ICAMPE '93, October 19–22, KOEX*, Korean Society for Agricultural Machinery, Seoul, Korea, 334–343.

Kinay, P., Yildiz, M., Droby, S., Yildiz, F., Cohen, L. and Weiss, B. 1998. Evaluation of antagonistic activity of epiphytic yeasts against rot pathogens of mandarin orange and grapefruit. *Bulletin OILB SROP* **21**, 291–296.

King, S. 1984. *Developing New Brands*. J. Walter Thompson, London.

Kitagawa, H. 1993. The market for tropical fruits in Japan. *Australian Centre for International Agricultural Research Proceedings* **50**, 90–93.

Kitagawa, H. and Glucina, P.G. 1984. Persimmon culture in New Zealand. *New Zealand Department of Scientific and Industrial Research, Information Series* **159**, 74 pp.

Kitinoja, L and Kader, A.A. 1993. Small scale postharvest handling practices. Department of Pomology, University of California, Davis, California, *Postharvest Horticulture Series* **8**, 188 pp.

Klein, J.D. and Lurie, S. 1992. Prestorage heating of apple fruit for enhanced postharvest quality: interaction of time and temperature. *HortScience* **27**, 326–328.

Klein, J.D. and Thorpe, T.G. 1987. Feijoas: postharvest handling and storage of fruit. *New Zealand Journal of Experimental Agriculture* **15**, 217–221.

Kleinhenz, V., Gosbee, M., Elsmore, S., Lyall, T.W., Blackburn, K., Harrower, K. and Midmore, D.J. 2000. Storage methods for extending shelf life of fresh, edible bamboo shoots [*Bambusa oldhamii* (Munro)]. *Postharvest Biology and Technology* **19**, 253–264.

Kleinkopf, G.E., Brandt, T.L., Frazier, M.J. and Moller, G. 1997. CIPC residues on stored Russet Burbank potatoes: 1. Maximum label application. *American Potato Journal* **74**, 107–117.

Klieber, A. and Wills, R.B.H. 1991. Optimisation of storage conditions for 'Shogun' broccoli. *Scientia Horticulturae* **47**, 201–208.

Kluge, K. and Meier, G. 1979 Flavour development of some apple cultivars during storage. Geschmacksentwicklung einiger Apfelsorten wahrend der Lagerung. *Gartenbau* **26**, 278–279.

Knavel, D.E. and Kemp, R.T. 1973. Ethephon and CPTA on color development in bell pepper fruits. *HortScience* **97**, 192.

Knee, M. 1973. Effects of controlled atmosphere storage on respiratory metabolism of apple fruit tissue. *Journal of the Science of Food and Agriculture* **24**, 289–298.

Knee, M. 1976. Influence of ethylene on the ripening of stored apples. *Journal of the Science of Food and Agriculture* **27**, 383–392.

Knee, M. and Bubb, M. 1975. Storage of Bramley's's Seedling apples. II. Effects of source of fruit, picking date and storage conditions on the incidence of storage disorders. *Journal of Horticultural Science* **50**, 121–128.

Knee, M. and Looney, N.E. 1990. Effect of orchard and postharvest application of daminozide on ethylene synthesis by apple fruit. *Journal of Plant Growth Regulation* **9**, 175–179.

Knee, M., Hatfield, S.G.S. and Bramlage, W.J. 1987. Responses of developing apple fruits to ethylene treatement. *Journal of Experimental Botany* **38**, 972–979.

Knee, M., Proctor, F.J. and Dover, C.J. 1985. The technology of ethylene control: use and removal in postharvest handling of horticultural commodities. *Annals of Applied Biology* **107**, 581–595.

Knight, C. 1991. Crop production harvesting and storage. In Arthey, D. and Dennis, C. (Editors). *Vegetable Processing*. VCH Publishers, New York, 12–41.

Knight, C., Cutts, D.C. and Colhoun, J. 1977. The role of *Fusarium semitectum* in causing crown rot of bananas. *Phytopathologische Zeitschrft* **89**, 170–176.

Knowles, L., Trimble, M.R. and Knowles, N.R. 2001. Phosphorus status affects postharvest respiration, membrane permeability and lipid chemistry of European seedless cucumber fruit (*Cucumis sativus* L.). *Postharvest Biology and Technology* **21**, 179–188.

Koelet, P.C. 1992. *Industrial Refrigeration*. Macmillan, London.

Kohne, J.S. *et al.* 1992. *Yearbook of the South African Avocado Growers Association* **15**.

Koksal, A.I. 1989. Research on the storage of pomegranate c.v. Gok Bahce under different conditions. *Acta Horticulturae* **258**, 295–302.

Kollas, D.A. 1964. Preliminary investigation of the influence of controlled atmosphere storage on the organic acids of apples. *Nature* **204**, 758–759.

Kondou, S., Oogaki, C. and Mim, K. 1983. Effects of low pressure storage on fruit quality. *Journal of the Japanese Society for Horticultural Science* **52**, 180–188.

Korsten, L., de Villers, E.E. de Jager, E.S., van Harmelen and Heitmann, A. 1993a. Biological control of litchi fruit diseases. *South African Litchi Growers' Association Yearbook* **5**, 36–40.

Korsten, L., de Villers, E.E., Wehner, F.C. and Kotze, J.M. 1993. A review of biological control of postharvest diseases of subtropical fruit. *Postharvest Handling of Tropical Fruit. Proceedings of an International Conference held in Chiang Mai, Thailand, 19–23 July 1993. Australian Centre for International Agricultural Research Proceedings* **50**, 172–185.

Kosiyachinda, P. 1968. Postharvest technology of mangosteen, durian and rambutan, Part 2. *Keha Karnkaset* **10**, 37–41.

Kosiyachinda, S. and Young, R. E. 1975. Ethylene production in relation to the initiation of respiratory climacteric in fruit. *Plant and Cell Physiology* **16**, 595–602.

Kotler, P. 1991. *Marketing Management: Analysis, Planning, Implementation and Control*, 7th edn. Prentice Hall, Englewood Cliffs, NJ.

Koto, K. 1987. Non destructive measurements of fruit quality by electrical impedance. *Research Report on Agricultural Machinery*. 17, Kyoto University, Kyoto.

Kouno, Y., Mizuno, T. and Maeda, H 1993a. Feasibility study into NIR techniques for measurement of internal qualities of some tropical fruits. *Proceedings of ICAMPE '93, October 19–22, KOEX*, Korean Society for Agricultural Machinery, Seoul, Korea, 326–333.

Kouno, Y., Mizuno, T. and Maeda, H. 1993b. The development of a device to measure the ripeness and internal quality of watermelons. *Proceedings of ICAMPE '93, October 19–22, KOEX*, Korean Society for Agricultural Machinery, Seoul, Korea, 1346–1353.

Kouno, Y., Mizuno, T., Maeda, H., Akinaga, T., Tanaba, T. and Kohda, Y. 1993c. Feasibility study into NIR technique for measurement of internal quality of some tropical fruits. *Postharvest Handling of Tropical Fruit. Proceedings of an International Conference held in Chiang Mai, Thailand, 19–23 July 1993. Australian Centre for International Agricultural Research Proceedings* **50**.

Koyakumaru, T. 1997. Effects of temperature and ethylene removing agents on respiration of mature green mume *Prunus mume* Sieb. et Zucc. fruit held under air and controlled atmospheres. *Journal of the Japanese Society for Horticultural Science* **66**, 409–418.

Koyakumaru, T., Adachi, K., Sakoda, K., Sakota, N. and Oda, Y. 1994. Physiology and quality changes of mature green mume *Prunus mume* Sieb. et Zucc. fruits stored under several controlled atmosphere conditions at ambient temperature. *Journal of the Japanese Society for Horticultural Science* **62**, 877–887.

Koyakumaru, T., Sakoda, K., Ono. Y. and Sakota, N. 1995. Respiratory physiology of mature green mume *Prunus mume* Sieb. et Zucc. fruits of four cultivars held under various controlled atmospheres at ambient temperature. *Journal of the Japanese Society for Horticultural Science* **64**, 639–648.

Ku, V.V.V., Wills, R.B.H. and Ben Yehoshua, S. 1999. 1-Methylcyclopropene can differentially affect the postharvest life of strawberries exposed to ethylene. *HortScience* **34**, 119–120.

Kubo, Y., Inaba, A. and Nakamura, R. 1989. Effects of high CO_2 on respiration in various horticultural crops. Journal of the *Japanese Society for Horticultural Science* **58**, 731–736.

Kulwithit, K. and Kosittrakun, M. 1998. Quality of ripened 'Pakchong 1' papaya fruits as influenced by harvest maturity. *Thai Journal of Agricultural Science* **31**, 379–384.

Kumar, D.A., Indira, P. and Nambisan, B. 1998. Effect of light and growth regulators on sprouting of *Amorphophallus* tubers. *Tropical Science* **38**, 187–189.

Kupper, W., Pekmezci, M. and Henze, J. 1994. Studies on CA storage of pomegranate *Punica granatum* L., cv. Hicaz. *Acta Horticulturae* **398**, 101–108.

Kurki, L. 1979. Leek quality changes during CA storage. *Acta Horticulturae* **93**, 85–90.

Kushman, J. and Wright, F.S. 1969. Sweetpotato storage. *US Department of Agriculture Handbook* **358**, 35 pp.

La Ongsri, S., Gomolmanee, S. and Onnop, W.A. 1993. Alleviating chilling injury in lychees by sulfur dioxide fumigation. *Postharvest Handling of Tropical Fruit. Proceedings of an International Conference held in Chiang Mai, Thailand, 19–23 July 1993. Australian Centre for International Agricultural Research Proceedings* **50**.

Lacroix, C.R. and Carmentran, M. 2001. Fertilizers and the strawberry plant: yield and fruit quality. Fertilisation du fraisier: rendement et qualité des fruits. *Infos Ctifl* **170**, 41–44.

Ladaniya, M.S. and Singh, S. 2001. Use of ethylene gas for degreening of sweet orange (*Citrus sinensis* Osbeck) cv. Mosambi. *Journal of Scientific and Industrial Research* **60**, 662–667.

Lafuente, M.T., Cantwell, M., Yang, S.F. and Rubatzky, U. 1989. Isocoumarin content of carrots as influenced by eyhylene concentration, storage temperature and stress conditions. *Acta Horticulturae* **258**, 523–534.

Lakshmana, and Reddy, T.V. 1995. Postharvest quality of sapota fruits as influenced by soil application of calcium. *Current Research, University of Agricultural Sciences Bangalore* **24**, 36–38.

Lakshminarayana, S., Valasco, C.J.J. and Sarmiento, L. 1973. *CONAFRUIT Serie Investigaciones Fisiologicas* **2**, 19 pp.

Lallu, N. and Webb, D.J. 1997. Physiological and economic analysis of precooling kiwifruit. *Acta Horticulturae* **444**, 691–697.

Lallu, N., Yearsley, C.W. and Elgar, H.J. 2000. Effects of cooling treatments and physical damage on tip rot and postharvest quality of asparagus spears. *New Zealand Journal of Crop and Horticultural Science* **28**, 27–36.

Lam, P.F. 1990. Respiration rate, ethylene production and skin colour change of papaya at different temperatures. *Acta Horticulturae* **269**, 257–266.

Lam, P.F. and Kosiyachinda, S. 1987. *Rambutan, Fruit Development, Postharvest Physiology and Marketing in ASEAN*. Association of Southeast Asian Nations Food Handling Bureau, Kuala Lumpur, Malaysia, 82 pp.

Lammertyn, J., Verlinden, B., Nicolai, B. and de Baerdemaeker, J. 2000. Relation between core breakdown disorder and storage conditions of *Pyrus communis*. *Acta Horticulturae* **518**, 115–120.

Landfald, R. 1966. Temperature effects on apples during storage. *Bulletin of the International Institute of Refrigeration, Annexe* **1966–1**, 453–460.

Lange, E., Nowacki, J. and Saniewski, M. 1993. The effect of methyl jasmonate on the ethylene producing system in preclimacteric apples stored in low oxygen and high carbon dioxide atmospheres. *Journal of Fruit and Ornamental Plant Research* **1**, 9–14.

Lanza, G., Aleppo, E. di M. and Strano, M.C. 1998. Alternative means to synthetic fungicides on green mold control in citrus fruit. Interventi alternativi ai fungicidi di sintesi nel controllo del marciume verde dei frutti di agrume. *Italus Hortus* **5**, 61–66.

Larson, J., Vender, R. and Camuto, P. 1994. Cholestatic jaundice due to akee fruit poisoning. *American Journal of Gastroenterology* **89**, 1577–1578.

Laszlo, J.C. 1985. The effect of controlled atmosphere on the quality of stored table grapes. *Deciduous Fruit Grower* **35**, 436–438.

Lau, O.L. 1983. Effects of storage proceedures and low O_2 and CO_2 atmospheres on storage quality of 'Spartan' apples. *Journal of the American Society for Horticultural Science* **108**, 953–957.

Lau, O.L. 1998. Effect of growing season, harvest maturity, waxing, low O_2 and elevated CO_2 on flesh browning disorders in 'Braeburn' apples. *Postharvest Biology and Technology* **14**, 131–141.

Lau, O.L. 1999. Factors affecting 'Braeburn browning disorder'. *Tree Fruit Postharvest Journal* **10**, 6–9.

Lau, O.L. and Yastremski, R. 1993. The use of 0.7% storage O_2 to attenuate scald symptoms in 'Delicious' apples: effect of apple strain and harvest maturity. *Acta Horticulturae* **326**, 183–189.

Laval Martin, D., Quennemet, J. and Moneger, R. 1975. Remarques sur l'évolution lipochromique et ultrastructurales des plastes durant la maturation du fruit de tomate 'cerise'. Facteurs et regulation da la Maturation des Fruits. *Colloques Internationaux du C.N.R.S.* 374.

Lavrik, I.P., Pomortseva, T.I., Zagoskina, V.I., Karaseva, L.G., Pivovarova, A.A. and Kulakova, M.S. 2000. Using Sorbilen for the storage of fruit and vegetable and floral production. *Izvestiya Timiryazevskoi Sel'skokhozyaistvennoi Akademii* **2**, 137–145.

Law, S.E. 1982. Spatial distribution of electrostatically deposited sprays on living plants. *Journal of Economic Entomology* **75**, 542–544.

Lawand, B.T., Patil, V.K. and Patil, P.V. 1992. Effects of different water regimes on fruit quality of pomegranate *Punica granatum*. *Acta Horticulturae* **321**, 677–683.

Lawton, A.R. 1996. *Cargo Care*. Cambridge Refrigeration Technology, Cambridge, UK (these recommendations were given as general guidelines and CRT accept no responsibility for their use).

Lazan, H., Ali, Z.M. and Sani, H.A. 1990. Effects of Vapor Gard on polygalacturonase, malic enzyme and ripening of *Harumanis mango*. *Acta Horticulturae* **269**, 359–366.

Le, H.T., Hancock, J.F. and Ton That Trinh. 1998. The fruit crops of Vietnam: introduced species and their native relatives. *Fruit Varieties Journal* **52**, 158–168.

Leberman, K.W., Nelson, A.L. and Steinberg, M.P. 1968. Postharvest changes of broccoli stored in modified atmospheres. I. Respiration of shoots and colour of flower heads. *Food Technology* 22, 143.

Lee, B.Y., Kim, Y.B. and Han, P.J. 1983. Studies on controlled atmosphere storage of Korean chestnut, *Castanea crenata* var. *dulcis Nakai. Research Reports, Office of Rural Development, South Korea. Soil Fertilizer, Crop Protection, Mycology and Farm Products Utilization* **25**, 170–181.

Lee, D.S., Hagger, P.E., Lee, J. and Yam, K.L. 1991. Model for fresh produce repiration in modified atmospheres based on the principles of enzyme kinetics. *Journal of Food Science* **56**, 1580–1585.

Lee, G.R., Proctor, F.J. and Thompson, A.K. 1973. Transport of papaya fruits from Trinidad to Britain. *Tropical Agriculture Trinidad* **50**, 303–306.

Lee, K.S. and Lee, D.S. 1996. Modified atmosphere packaging of a mixed prepared vegetable salad dish. *International Journal of Food Science and Technology*, **31**, 7–13.

Lee, M.J., Kang, H.M. and Park, K.W. 2000. Effects of selenium on growth, storage life, and internal quality of coriander *Coriandrum sativum* L. during storage. *Journal of the Korean Society for Horticultural Science* **41**, 490–494.

Lee, S.K. and Young, R.E. 1984. Temperature sensitivity of avocado fruit in relation to ethylene. *Journal of the American Society for Horticultural Science* **109**, 689–692.

Lee, S.K., Shin, I.S. and Park, Y.M. 1993. Factors involved in skin browning of non astringent 'Fuju' persimmon. *Acta Horticulturae* **343**, 300–303.

Leeuwen van, G. and Van de Waart, A. 1991. Delaying red currants is worthwhile. Verlating rode bes is de moeite waard. *Fruitteelt Den Haag* **81**(33), 14–15.

LeFlufy, M.J. 1983. Apple harvesting by combing technique. *Transactions of the American Society of Agricultural Engineers* **26**, 661–664.

Lenker, D.H. and Adrian, P.A. 1971. Use of X rays for selecting mature lettuce heads. *Transactions of the American Society of Agricultural Engineers* **4**, 894–898.

Lentza-Rizos, C. and Balokas, A. 2001. Residue levels of chlorpropham in individual tubers and composite samples of postharvest-treated potatoes. *Journal of Agricultural and Food Chemistry* **49**, 710–714.

Levin, A., Sonego, L., Zutkhi, Y., Ben Arie, R., and Pech, J.C. 1992. Effects of CO_2 on ethylene production by apples at low and high O_2 concentrations. Cellular and molecular aspects of the plant hormone ethylene. *Current Plant Science and Biotechnology in Agriculture* **16**, 150–151.

Levitt, T. 1986. *The Marketing Imagination.* Free Press, London.

Lewis, D.A. and Morris, L. 1956. Effect of chilling storage on respiration and deterioration of several sweet potato varieties. *Proceedings of the American Society of Horticultural Science* **68**, 421–428.

Li, H.Y. and Yu, T. 2001. Effect of chitosan on incidence of brown rot, quality and physiological attributes of postharvest peach fruit. *Journal of the Science of Food and Agriculture* **81**, 269–274.

Li, J.Y., Huang, W.N., Cai, L.X. and Hu, W.J. 2000. Effects of calcium treatment on physiological and biochemical changes in shiitake *Lentinus edodes* during the post harvest period. *Fujian Journal of Agricultural Sciences* **15**, 43–47.

Li, W. and Huang, C. 1992. Food and Agriculture Organization of the United Nations regional cooperation for vegetable research and development, RAS/89/41. In Bhatti, M.H., Hafeez, A., Jaggar A. and Farooq, M. (Editors). *Postharvest Losses of Vegetables. A Report on a Workshop Held Between 17 and 22 October 1992 at the Pakistan Agricultural Research Council, Islamabad, Pakistan.*

Lidster, P.D., Lawrence, R.A., Blanpied, G.D. and McRae, K.B. 1985. Laboratory evaluation of potassium permanganate for ethylene removal from controlled atmosphere apple storages. *Transactions of the American Society of Agricultural Engineers* **28**, 331–334.

Lieten, F. and Marcelle, R.D. 1993. Relationships between fruit mineral content and the 'albinism' disorder in strawberry. *Annals of Applied Biology* **123**, 433–439.

Lill, R.E. and Borst, W.M. 2001. Spear height at harvest influences postharvest quality of asparagus (*Asparagus officinalis*). *New Zealand Journal of Crop and Horticultural Science* **29**, 187–194.

Lill, R.E. and Corrigan, V.K. 1996. Asparagus responds to controlled atmospheres in warm conditions. *International Journal of Food Science and Technology* **31**, 117–121.

Lima, G., de Curtis, F., Castoria, R. and de Cicco, V. 1998. Activity of the yeasts *Cryptococcus laurentii and Rhodotorula glutinis* against post harvest rots on different fruits. *Biocontrol Science and Technology* **8**, 257–267.

Lindsey, R.T. and Neale, M.A. 1977. *Proceedings of the Vegetable Cooling and Storage Conference,* National Agricultural Centre, Stoneleigh.

Liner, H.L. 1971. A case study of blueberry mechanization economics in North Carolina. *Proceedings of the 1971 Highbush Blueberry Mechanization Symposium,* North Carolina State University, Raleigh, NC, 57–65.

Link, H. 1980. Effects of nitrogen supply on some components of fruit quality in apples. In Atkinson, D., Jackson, J.E., Sharples R.O. and Waller, W.M. (Editors), *Mineral Nutrition of Fruit Trees.* Butterworths, London, 285.

Lipton, W.J. 1957. *Physiological changes in harvested asparagus Asparagus officinalis as related to temperature.* PhD thesis, University California, Davis, CA, 116 pp.

Lipton, W.J. 1967. Some effects of low O_2 atmospheres on potato tubers. *American Potato Journal* **44**, 292.

Lipton, W.J. 1968. Market quality of asparagus: effects of maturity at harvest and of high CO_2 atmospheres during simulated transit. *USDA Markketing. Research Report* 817.

Lipton, W.J. 1972. Market quality of radishes stored in low O_2 atmospheres. *Journal of the American Society for Horticultural Science* **97**, 164.

Lipton, W.J. and Harris, C.M. 1974. Controlled atmosphere effects on the market quality of stored broccoli *Brassica oleracea* L. Italica group. *Journal of the American Society for Horticultural Science* **99**, 200–205.

Lipton, W.J. and Mackey, B.E. 1987. Physiological and quality responses of Brussels sprouts to storage in controlled atmospheres. *Journal of the American Society for Horticultural Science* **112**, 491–496.

Lipton, W.J., Aharoni, Y. and Elliston, E. 1979. Rates of carbon dioxide and ethylene production and of ripening of 'Honey Dew' muskmelons at a chilling temperature after pretreatment with ethylene. *Journal of the American Society for Horticultural Science* **104**, 846–848.

Lipton, W.J., Asai, W.K. and Fouse, D.C. 1981. Deterioration and CO_2 and ethylene production of stored mung bean sprouts. *Journal of the American Society for Horticultural Science* **106**, 817–820.

Littmann, M.D. 1972. Effect of water loss on the ripening of climacteric fruits. *Queensland Journal of Agriculture and Animal Science* **29**, 103.

Liu, F.W. 1976. Banana response to low concentrations of ethylene. *Journal of the American Society for Horticultural Science* **101**, 222–224.

Liu, F.W. 1976a. Ethylene inhibition of senescent spots on ripe bananas. *Journal of the American Society for Horticultural Science* **101**, 684–686.

Liu, F.W. 1978. Ripening bananas with ethephon in three polymeric film packages. *HortScience* **13**, 688–690.

Lizada, M.C.C. 1993. Fruit handling systems for developing countries. *Postharvest Handling of Tropical Fruit. Proceedings of an International Conference held in Chiang Mai, Thailand, 19–23 July 1993. Australian Centre for International Agricultural Research Proceedings* **50**, 109–115.

Lizada, M.C.C. and Novenario, V. 1983. The effect of prolong on patterns of physico chemical and physiological changes in the ripening banana. *Postharvest Horticultural Training and Research Center, University of the Philippines at Los Baños, College of Agriculture, Laguna, Annual Report.*

Lizada, M.C.C. Pantastico, E.B., Abdullah Shukor, A.R. and Sabari, S.D. 1990. Ripening of banana. In (Abdulla, H. and Pantastico, E.B. (Editors), *Banana.* Association of Southeast Asian Nations Food Handling Bureau, 65–84.

Lizana, L.A. 1975. The influence of water stress and elevated temperatures in banana ripening. *Proceedings of the Tropical Region, American Society for Horticultural Science* **19**, 137–145.

Lizana, L.A. and Espina, S. 1991. Effect of storage temperature on the postharvest behaviour of Cape gooseberry *Physalis peruviana* L. fruits. Efecto de la temperatura de almacenaje sobre el comportamiento en postcosecha de frutos de fisalis *Physalis peruviana* L. *Proceedings of the Interamerican Society for Tropical Horticulture* **35**, 278–284.

Lizana, L.A. and Figueroa, J. 1997. Effect of different CA on postharvest life of Hass avocado. *Seventh International Controlled Atmosphere Research Conference CA '97. Proceedings, volume 3: Fruits Other than Apples and Pears, Davis, CA, 13–18 July, 1997. Postharvest Horticulture Series Department of Pomology, University of California* **17**, 219–224.

Lizana, L.A. and Ochagavia, A. 1997. Controlled atmosphere storage of mango fruits (*Mangifera indica* L.) cvs Tommy Atkins and Kent. *Acta Horticulturae* **455**, 732–737.

Lizana, L.A., Fichet, T. Videla, G., Berger, H. and Galletti, Y.L. 1993. Almacenamiento de aguacates pultas cv. Gwen en atmosfera controlada. *Proceedings of the Interamerican Society for Tropical Horticulture* **37**, 79–84.

Lockhart, C.L., Eaves, C.A. and Chitman, E.W. 1969. Suppression of rots of four varieties of mature green tomatoes in controlled atmosphere storage. *Canadian Journal of Plant Sciences* **49**, 265–269.

Loewenfeld, C. 1964. *Herb Gardening.* Faber and Faber, London, 256 pp.

Lonsdale, J.H. 1992. Pre harvest fungicide sprays for the control of postharvest diseases of mangoes. *Suid Afrikaanse Mangowekersvereniging Jaarboek* **12**, 28–31.

Lopez Briones, G., Varoquaux, P., Bareau, G. and Pascat, B. 1993. Modified atmosphere packaging of common mushroom. *International Journal of Food Science and Technology* **28**, 57–68.

Lopez Briones, G., Varoquaux, P., Chambroy, Y., Bouqant, J., Bareau, G. and Pascat, B. 1992. Storage of common mushrooms under controlled atmospheres. *International Journal of Food Science and Technology* **27**, 493–505.

Lopez, M.L., Lavilla, M.T., Recasens, I., Graell, J. and Vendrell, M. 2000. Changes in aroma quality of 'Golden Delicious' apples after storage at different oxygen and carbon dioxide concentrations. *Journal of the Science of Food and Agriculture* **80**, 311–324.

Lougheed, E.C. 1987. Interactions of O_2, CO_2, temperature, and ethylene that may induce injuries in vegetables. *HortScience* **22**, 791–794.

Lougheed, E.C. and Franklin, E.W. 1970. Ethylene evolution from 2-chloroethane phosphonic acid under nitrogen atmospheres. *Canadian Journal of Plant Science* **50**, 586.

Lougheed, E.C. and Lee, R. 1991. Ripening, CO_2 and C_2H_4 production, and quality of tomato fruits held in atmospheres containing nitrogen and argon. *Proceedings of the Fifth International Controlled Atmosphere Research Conference, Wenatchee, Washington, 14–16 June, 1989. Volume 2. Other Commodities and Storage Recommendations*, 141–150.

Lui, R.X., Lui, M.Y., Shen, P., Wang, Z.L., Gong, X.R. and Wang, F.M. 1989. Studies on the techniques of keeping shiitake mushroom fresh. *Zhongguo Shiyongjun* **3**, 3–5.

Lukezic, F.L., Kaiser, W.J. and Martinez, M. 1967. The incidence of crown rot of boxed bananas in relation to microbial populations of the crown tissue. *Canadian Journal of Botany* **45**, 413–421.

Luo, Y. and Mikitzel, L.J. 1996. Extension of postharvest life of bell peppers with low O_2. *Journal of the Science of Food and Agriculture* **70**, 115–119.

Lurie, S., Aharoni, N. and Ben Yehoshua, S. 1997. Modified atmosphere storage of stone fruits. *14th International Congress on Plastics in Agriculture, Tel Aviv, Israel, March 1997*, 536–541.

Lurie, S., Pesis, E. and Ben Arie, R. 1991. Darkening of sunscald on apples in storage is a non enzymatic and non oxidative process. *Postharvest Biology and Technology* **1**, 119–125.

Lutz, J.M. 1938. Factors influencing the quality of American grapes in storage. *United States Department of Agriculture, Technical Bulletin* 606, 27 pp.

Lutz, J.M. 1952. Influence of temperature and length of curing period on keeping quality of 'Puerto Rico' sweetpotatoes. *Proceedings of the American Society for Horticultural Science* **59**, 421.

Lutz, J.M. and Culpepper, C.W. 1937. Certain chemical and physical changes produced in Kieffer pears during ripening and storage. *United States Department of Agriculture, Technical Bulletin* **590**, 37 pp.

Lutz, J.M. and Hardenburg, R.E. 1968. The commercial storage of fruits, vegetables and florist and nursery stocks. *United States Department of Agriculture, Agriculture Handbook* **66**, 94 pp.

Lyonga, S.N. 1985. Studies on the fertilization of yams and on yam tuber storage in Cameroon. *Proceedings of the Sixth International Symposium on Tropical Root Crops, Lima, Peru.*

Lyons, J.M. and Rappaport, L. 1959. Effect of temperature and respiration and quality of Brussels sprouts during storage *Proceedings of the American Society of Horticultural Science* 73, 361–366.

Lyons, J.M. and Rappaport, L. 1962. Effect of controlled atmospheres on storage quality of Brussel sprouts. *Proceedings of the American Society of Horticultural Science* **81**, 324.

Maas, J.L. and Smith, W.L., Jr. 1972. Preharvest fungicide treatments for reduction of Penicillium decay of strawberry. *Plant Disease Reporter* **56**, 881–887.

Mack, W.B. 1927. The action of ethylene in accelerating the blanching of celery. *Plant Physiology* **2**, 103.

Macku, C. and Jennings, W.G. 1987. Production of volatiles by ripening bananas. *Journal of Agricultural and Food Chemistry* **35**, 845–848.

Madamba, S.P., Baes, A.U. and Mendoza, D.B., Jr. 1977. Effect of maturity on some biochemical changes during ripening of banana *Musa sepientum* cv. Lakatan. *Food Chemistry* **2**, 177–83.

Madrid, M. and Lopez Lee, F. 1998. Differences in ripening characteristics of controlled atmosphere or air stored bananas. *Acta Horticulturae* **464**, 357–362.

Maekawa, T. 1990. On the mango CA storage and transportation from subtropical to temperate regions in Japan. *Acta Horticulturae* **269**, 367–374.

MAFF. Undated. *Refrigerated Storage of Fruit and Vegetables.* Ministry of Agriculture, Fisheries and Food, Bulletin RB 324. Her Majesty's Stationery Office, London.

Magness, J.R. and Ballard, W.S. 1926. The respiration of Bartlett pears. *Journal of Agricultural Research* **32**, 801–832.

Magness, J.R. and Diehl, H.C. 1924. Physiological studies on apples in storage. *Journal of Agricultural Research* **27**, 33–34.

Magness, J.R. and Taylor, G.F. 1925. An improved type of pressure tester for the determination of fruit maturity. *United States Department of Agriculture Circular* 350.

Magomedov, M.G. 1987. Technology of grape storage in regulated gas atmosphere. *Vinodelie i Vinogradarstvo SSSR* **2**, 17–19.

Maharaj, R. and Sankat, C.K. 1990. Storability of papayas under refrigerated and controlled atmosphere. *Acta Horticulturae* **269**, 375–385.

Maharaj, R. and Sankat, C.K. 1990a. The shelf life of breadfruit stored under ambient and refrigerated conditions. *Acta Horticulturae* **269**, 411–424.

Mainland, C.M. 1971. Fruit recovery and bush damage with mechanical harvesting. *Proceedings of the 1971 Highbush Blueberry Mechanization Symposium, North Carolina State University, Raleigh, NC*, 34–37.

Mainland, C.M., Kushman,J. and Ballinger, W.E. 1975. The effect of mechanical harvesting on yield, quality of fruit and bush damage of highbush blueberry. *Journal of the American Society for Horticultural Science* **100**, 129–134.

Majumder, K. and Mazumdar, B.C. 1998. Changing pattern of pectic substances in developing fruits of *Psidium guajava* L. *Indian Journal of Plant Physiology* **3**, 42–45.

Makhlouf, J., Castaigne, F., Arul, J., Willemot, C. and Gosselin, A. 1989. Long term storage of broccoli under controlled atmosphere. *HortScience* **24**, 637–639.

Makhlouf, J., Willemot, C., Arul, J., Castaigne, F. and Emond, J.P. 1989a. Regulation of ethylene biosynthesis in broccoli flower buds in controlled atmospheres. *Journal of the American Society for Horticultural Science* **114**, 955–958.

Malaysian Standard Export Specification for Fresh 'Mas' Bananas, MS 1075: 1987, UDC 582.58.004.1. Standards and Industrial Research Institute of Malaysia Established by Act of Parliament.

Malcolm G.L. and Gerdts, D.R. 1995. Review and prospects for use of controlled atmosphere technology in Mexican agribusiness. *Technologias de Cosecha y Postcosecha de Frutas y Hortalizas. Proceedings of a Conference held in Guanajuato, Mexico, 20–24 February 1995*, 530–537.

Mann, K. and Lewis, D.A. 1956. Rest and dormancy in garlic. *Hilgardia* **26**, 161–189.

Mannheim, C. 1980. Development and trends in the preservation of perishable produce. *Eighth Ami Shachori Memorial Lecture, 23 September, London.*

Manning, K. 1993. Regulation by auxin of ripening genes from strawberries, a non climacteric fruit. *Postharvest Biology and Handling of Fruit, Vegetables and Flowers. Meeting of the Association of Applied Biologists, London, 8 December 1993.*

Manzano Mendez, J., Hicks, J.R. and Mastes, J.F. 1984. Influence of storage temperature and ethylene on firmness, acids and sugars of chilling sensitive and chilling tolerent tomato. *Journal of the American Society for Horticultural Science* **109**, 273–277.

Mao, L.C. and Liu, W.X. 2000. Study on postharvest physiological changes and storage techniques of sugarcane. *Scientia Agricultura Sinica* **33**, 41–45.

Marcellin, P. 1973. Controlled atmosphere storage of vegetables in polyethylene bags with silicone rubber windows. Conservation de légumes en atmosphère controlée dans des sacs en polyethylene avec fenêtre d'élastomère de silicone. *Acta Horticulturae* **38**, 33–45.

Marcellin, P. and Chaves, A. 1983. Effects of intermittent high CO_2 treatment on storage life of avocado fruits in relation to respiration and ethylene production. *Acta Horticulturae* **138**, 155–163.

Marcellin, P. and LeTeinturer, J. 1966. Etude d'une installation de conservation de pommes en atmosphère controleé. *International Instiution of Refrigeration Bulletin, Annex 1966–1*, **141–152**.

Marcellin, P., Pouliquen, J. and Guclu, S. 1979. Refrigerated storage of Passe Crassane and Comice pears in an atmosphere periodically enriched in CO_2 preliminary tests. *Bulletin de l'Institut International du Froid* **59**, 1152.

Marchal, J. and Nolin, J. 1990. Fruit quality. Pre and post harvest physiology. *Fruits Paris* Special Issue, 119–122.

Marin-Thiele, F., Saenz-Murillo, M.V. and Crane, J.H. 2000. Diagnosis of postharvest handling of chayote (*Sechium edule* Sw) in Costa Rica: II. Evaluation of packed fruits. Diagnostico de manejo poscosecha de chayote (*Sechium edule* Sw) en Costa Rica: II. Evaluacion de fruto empacado. *Proceedings of the Interamerican Society for Tropical Horticulture, Barquisimeto, Venezuela, 27 September–2 October, 1998*, **42**, 443–450.

Marloth, R.H. 1947. *Bulletin of the Department of Agriculture of South Africa* 282.

Marriott, J. 1980. Bananas: physiology and biochemistry of storage and ripening for optimum quality. *Critical Review of Food Science NUTB* **13**, 41–87.

Marriott, J., and New, S. 1975. Storage physiology of bananas from new tetrapolid clones. *Tropical Science* **17**, 155–163.

Marriott, J., New, S., Dixon, E.A. and Martin, K.J. 1979. Factors affecting the preclimacteric period of banana fruit bunches. *Annals of Applied Biology* **93**, 91–100.

Marriott, J., Robinson, M. and Karikari, S.K. 1983. Changes in composition during the ripening of three plantain cultivars. *Fruits Paris* **38**, 343–347.

Martinez, J.A. and Artes, F. 1999. Effect of packaging treatments and vacuum cooling on quality of winter harvested iceberg lettuce. *Food Research International* **32**, 621–627.

Martinez Cayuela, M., Plata, M.C., Sanchez de Medina, L., Gil, A. and Faus, M.J. 1986. Changes in various enzyme activities during ripening of cherimoya in controlled atmosphere. Evolucion de diversas actividades enzimaticas durante la maduracion del chirimoyo en atmosfera controlada. *ARS Pharmaceutica* **27**, 371–380.

Martinez Damian, M.T., Velez Ortega, E. and Saucedo Veloz, C. 1999. Physiological and biochemical changes of two hawthorn (*Crataegus mexicana* H.B.K.) selections during fruit growth and postharvest. Cambios fisiologicos y bioquimicos en dos selecciones de tejocote (*Crataegus mexicana* H.B.K.) en pre y postcosecha. *Revista Chapingo, Serie Horticultura* **5**, 41–48.

Martinez Soto, G., Paredes Lopez, O., Ocana Camacho, R. and Bautista Justo, M. 1998. Oyster mushroom *Pleurotus ostreatus* quality as affected by modified atmosphere packaging. *Micologia Neotropical Aplicada* **11**, 53–67.

Massignan, L., Lovino, R. and Traversi, D. 1999. Postharvest treatment and cold storage of organic dessert grapes. Trattamenti post raccolta e frigoconservazione di uva da tavola 'biologica'. *Informatore Agrario, Supplemento* **55**, 46–48.

Mateos, M., Ke, D., Cantwell, M. and Kader, A.A. 1993. Phenolic metabolism and ethanolic fermentation of intact and cut lettuce exposed to CO_2-enriched atmospheres. *Postharvest Biology and Technology* **3**, 225–233.

Mathur, P.B., Singh, K.K. and Kapur, N.S. 1953. Cold storage of mangoes. *Indian Journal of Agricultural Sciences* **23**, 65.

Matsuo, T., Agravante, J.U. and Kitagawa, H. 1990. An improved method for starch determination in banana fruits. *Technical Bulletin of the Faculty of Agriculture, Kagawa University* **42**, 181–184.

Matsuo, T., Shinohara, J. and Itoo, S. 1976. An improvement on removing astringency in persimmon fruits by carbon dioxide gas. *Agricultural and Biological Chemistry* **40**, 215- 217.

Mattheis, J.P., Buchanan, D.A. and Fellman, J.K. 1991. Change in apple fruit volatiles after storage in atmospheres inducing anaerobic metabolism. *Journal of Agricultural and Food Chemistry* **39**, 1602–1605.

Mattus, G.E. 1963. Regular and automatic CA storage. *Virginia Fruit* June, 41.

Maw, B.W., Smittle, D.A. and Mullinix, B.G. 1997. The influence of harvest maturity, curing and storage conditions upon the storability of sweet onions. *Applied Engineering in Agriculture* **13**, 511–515.

Maxie, E.C., Catlin, P.B. and Hartmann, H.T. 1960. Respiration and ripening of olive fruits. *Proceedings of the American Society of Horticultural Science* **75**, 275–291.

Maxie, E.C., Mitchell, F.G. and Greathead, A.S. 1959. Studies on strawberry quality. *Californian Agriculture* **13**, 11–16.

Maxie, E.C., Sommer, N.F. and Mitchell, F.G. 1971. Infeasibility of irradiating fresh fruits and vegetables. *HortScience* **6**, 202–204.

Mayne, D., Vithanage, V. and Aylward, J.H. 1993. Management of 'jelly seed' in mango *Mangifera indica* cv. Tommy Atkins. *Postharvest Handling of Tropical Fruit. Proceedings of an International Conference held in Chiang Mai, Thailand, 19–23 July 1993. Australian Centre for International Agricultural Research Proceedings* **50**, 470.

Mazza, G. and Siemens, A. J. 1990. CO_2 concentration in commercial potato storages and its effect on quality of tubers for processing. *American Potato Journal* **67**, 121–132.

McCarthy, A.I., Palmer, J.K., Shaw, C.P. and Anderson, E.E. 1963. Correlation of gas chromatographic data for flavor profiles of fresh banana fruits. *Journal of Food Science* **28**, 379–384.

McCarthy, M.J., Chen, P., Kauten, R. and Sarig, Y. 1989. Maturity evaluation of avocados by NMR methods. *The American Society of Agricultural Engineers Paper* 89–3548.

McCornack, A.A. and Wardowski, W.F. 1977. Degreening Florida citrus fruit: procedures and physiology. *Proceedings of the International Society for Citriculture* **1**, 211–215.

McDonald, J.E. 1985. Storage of broccoli. *Annual Report, Research Station, Kentville, Nova Scotia,* 114.

McGarry, A. 1993. Mechanical properties of carrots. *Postharvest Biology and Handling of Fruit, Vegetables and Flowers. Meeting of the Association of Applied Biologists, London, 8 December 1993.*

McGill. J.N., Nelson, A.I. and Steinberg, M.P. 1966. Effect of modified storage atmosphere on ascorbic acid and other quality characteristics of spinach. *Journal of Food Science* **31**, 510.

McGlasson, W.B. and Wills, R.B.H. 1972. Effects of O_2 and CO_2 on respiration, storage life and organic acids of green bananas. *Australian Journal of Biological Sciences* **25**: 1, 35–42.

McGuire, P. 1929. Fruit and vegetable storage. *Tropical Agriculture Trinidad* **6**, 279.

McGuire, R.G. 1991. Concomitant decay reductions when mangoes are treated with heat to control infestations of Caribbean fruit flies. *Plant Disease* **75**, 946–949.

McGuire, R.G. 1993. Application of *Candida guilliermondii* in commercial citrus waxes for biocontrol of Penicillium on grapefruit. *Postharvest Handling of Tropical Fruit. Proceedings of an International Conference held in Chiang Mai, Thailand, 19–23 July 1993. Australian Centre for International Agricultural Research Proceedings* **50**, 464–468.

McGuire, R.G. and Hagenmaier, R.D. 2001. Shellac formulations to reduce epiphytic survival of coliform bacteria on citrus fruit postharvest. *Journal of Food Protection* 1756–1760.

McIntyre, A.A. 1993. *Optimising hot water treatments for the control of fruit fly and anthracnose and the maintenance of quality in mango cultivars Julie and Long.* MPhil thesis, University of the West Indies, 155 pp.

McLendon, B.D. and Brown, R.H. 1971. Dielectric properties of peaches as a maturity index. *American Society of Agricultural Engineers Paper* 71–332.

McLeod, A.J. and de Troconis, N.G. 1982. Volatile flavour components of mango fruit. *Phytochemistry* **21**, 2523–2526.

McMillan, R.T. 1972. Enhancement of anthracnose control on mangos by combining copper with Nu Film 17. *Proceedings of the Florida State Horticultural Society* **85**, 268–270.

McMillan, R.T. 1973. Control of anthracnose and powdery mildew of mango with systemic and non systemic fungicides. *Tropical Agriculture Trinidad* **50**, 245–248.

McRae, D.C. 1985. A review of developments in potato handling and grading. *Journal of Agricultural Engineering Research* **31**, 115–138.

Medlicott, A.P. and Jeger, M.J. 1987. The development and application of postharvest treatments to manipulate ripening of mangoes. In Prinsley, R.A. and Tucker, G. (Editors), Mangoes. *A Review.* Commonwealth Science Council, 56–77.

Medlicott, A.P. and Thompson, A.K. 1985. Analysis of sugars and organic acids in ripening mango fruits *Mangifera indica* var. Keitt by high performance liquid chromatography. *Journal of the Science of Food Agriculture* **36**, 561–566.

Medlicott, A.P., Bhogol, M. and Reynolds, S.B. 1986. Changes in peel pigmentation during ripening of mango fruit *Mangifera indica* var. Tommy Atkins. *Annals of Applied Biology* **109**, 651–656.

Medlicott, A.P., Reynolds, S.B. and Thompson, A.K. 1987. Effects of temperature on the ripening of mango fruit. *Journal of the Science of Food Agriculture* **37**, 469–474.

Medlicott, A.P., Reynolds, S.B., New, S.W. and Thompson, A.K. 1987a. Harvest maturity effects on mango fruit ripening. *Tropical Agriculture Trinidad* **65**, 153–157.

Medlicott, A.P., Semple, A.J., Thompson, A.J., Blackbourne, H.R. and Thompson, A.K. 1992. Measurement of colour changes in ripening bananas and mangoes by instrumental, chemical and visual assessments. *Tropical Agriculture Trinidad* **69**, 161–166.

Medlicott, A.P., Sigrist, J.M.M., Reynolds, S.B. and Thompson, A.K. 1987b. Effects of ethylene and acetylene on mango fruit ripening. *Annals of Applied Biology* **111**, 439–444.

Meheriuk, M. 1989. CA storage of apples. *International Controlled Atmosphere Conference, Fifth, Proceedings, June 14–16, 1989, Wenatchee, Washington. Volume 2. Other Commodities and Storage Recommendations,* 257–284.

Meheriuk, M. 1989a. Storage chacteristics of Spartlett pear. *Acta Horticulturae* **258**, 215–219.

Meheriuk, M. 1993. CA storage conditions for apples, pears and nashi. *Proceedings of the Sixth International Controlled Atmosphere Conference, June 15–17, 1983, Cornell University,* 819–839.

Mehlschau, J.J., Chen, P., Claypool, L. and Fridley, R.B. 1981. A deformer for nondestructive maturity detection of pears. *Transactions of the American Society of Agricultural Engineers* **24**, 1368–1371, 1375.

Meinl, G., Nuske, D. and Bleiss, W. 1988. Influence of ethylene on cabbage quality under long term storage conditions. Einfluss von Ethylen auf die Qualität von Kopfkohl bei der Langzeitlagerung. *Gartenbau* **35**, 265.

Meir, S., Akerman, M., Fuchs, Y. and Zauberman, G. 1993. Prolonged storage of Hass avocado fruits using controlled atmosphere. *Alon Hanotea* **47**, 274–281.

Meir, S., Naiman, D., Akerman, M., Hyman, J.Y., Zuuberman, G. and Fuchs, Y. 1997. Prolonged storage of 'Hass' avocado fruit using modified atmosphere packaging. *Postharvest Biology and Technology* **12**, 51–60.

Meir, S., Philosoph Hadas, S., Gloter, P. and Aharoni, N. 1992. Nondestructive assessment of chlorophyll content in watercess leaves by a tristimulus reflectance colorimeter. *Postharvest Biology and Technology* **2**, 117–124.

Meir, S., Philosoph Hadas, S., Lurie, S., Droby, S., Akerman, M., Zauberman, G., Shapiro, B., Cohen, E. and Fuchs, Y. 1996. Reduction of chilling injury in stored avocado, grapefruit, and bell pepper by methyl jasmonate. *Canadian Journal of Botany* **74**, 870–874.

Mencarelli, F. 1987. Effect of high CO_2 atmospheres on stored zucchini squash. *Journal of the American Society for Horticultural Science* **112**, 985–988.

Mencarelli, F. 1987a. The storage of globe artichokes and possible industrial uses. La conservazione del carciofo e possibili utilizzazioni industriali. *Informatore Agrario* **43**, 79–81.

Mencarelli, F., Fontana, F. and Massantini, R. 1989. Postharvest practices to reduce chilling injury CI on eggplants. *Proceedings of the Fifth International Controlled Atmosphere Research Conference, Wenatchee, Washington, 14–16 June, 1989. Volume 2. Other Commodities and Storage Recommendations*, 49–55.

Mencarelli, F., Fontana, F. and Massantini, R. 1990. Postharvest behaviour and chilling injury in babaco fruit. *Acta Horticulturae* **269**, 223–231.

Mencarelli, F., Lipton, W.J. and Peterson, S.J. 1983. Responses of 'zucchini' squash to storage in low O_2 atmospheres at chilling and nonchilling temperatures. *Journal of the American Society for Horticultural Science* **108**, 884–890.

Mencarelli, F., Massantini, R. and Botondi, R. 1997. Physiological and textural response of truffles during low temperature storage. *Journal of Horticultural Science* **72**, 407–414.

Mendoza D.B., Jr. 1968. Respiration of banana fruits. *Philippine Agriculture* **51**, 747–756.

Mendoza, D.B. 1978. Postharvest handling of major fruits in the Philippines. *Aspects of Postharvest Horticulture in Association of Southeast Asian Nations* 23–30.

Menezes, J.B., de Castro, E.B., Praca, E.F., Grangeiro, L.C. and Costa, L.B.A. 1998. Effect of postharvest exposure to solar radiation on the quality of melon fruits. Efeito do tempo de insolacao pos colheita sobre a qualidade do melao amarelo. *Horticultura Brasileira* **16**, 80–81.

Mercantilia. 1989. *Guide to Food Transport Fruit and Vegetables*. Mercantilia Publishers, Copenhagen, 247 pp.

Meredith, D.S. 1963. Latent infections in *Pyricularia grisea* causing pitting disease of banana fruits in Costa Rica. *Plant Disease Reporter* **47**, 766–768.

Merino, S.R., Eugenio, M.M., Ramas, A.U. and Hernandez, S.U. 1985. Fruitfly disinfestation of mangoes *Mangifera indica* var. `Manila Super' by vapor heat treatment. *Ministry of Agriculture and Food, Bureau of Plant Industry, Manila, Philippines.*

Merodio, C. and De la Plaza, J.L. 1989. Interaction between ethylene and CO_2 on controlled atmosphere storage of 'Blanca de Aranjuez jears'. *Acta Horticulturae* **258**, 81–88.

Mertens, H. 1985. Storage conditions important for Chinese cabbage. Bewaarcondities belangrijk bij Chinese kool. *Groenten en Fruit* **41**, 62–63.

Mertens, H. and Tranggono 1989. Ethylene and respiratory metabolism of cauliflower *Brassica olereacea* convar. *botrytis* in controlled atmosphere storage. *Acta Horticulturae* **258**, 493–501.

Meyers, J.B. and Prussia, S.E. 1987. Optimizing dynamic visual inspection performance using the theory of signal detection. *American Society of Engineers Paper* 87–1104.

Miccolis, V. and Saltveit, M.E., Jr. 1988. Influence of temperature and controlled atmosphere on storage of 'Green Globe' artichoke buds. *HortScience* **23**, 736–741.

Micke, W.C., Mitchell, F.G. and Maxie, E.C. 1965. Handling sweet cherries for fresh shipment. *Californian Agriculture* **19**, 12–13.

Miem, D.D.L. 1980. *Development and regulation of senescent spotting of 'Bungulan' bananas.* Student project, University of the Phillipines, Los Baños, Laguna.

Miller B.K. and Delwiche, M.J. 1989. Peach defect detection with machine vision. *American Society of Agricultural Engineers Paper* 89–6019.

Miller, E.V. and Brooks, C. 1932. Effect of CO_2 content of storage atmosphere on carbohydrate transformation in certain fruits and vegetables. *Journal of Agricultural Research* **45**, 449–459.

Miller, E.V. and Dowd, O.J. Effect of CO_2 on the carbohydrates and acidity of fruits and vegetables in storage. *Journal of Agricultural Research* **53**, 1–7.

Miller, W.F., Ruhkugler, G.E., Pellerin, R.A., Throop, J.A. and Bradley, R.B. 1973. Tree fruit harvester with insertable multilevel catching system. *Transactions of the American Society of Agricultural Engineers* **16**, 844–850.

Miller, W.R. and McDonald, R.E. 1998. Reducing irradiation damage to 'Aarkin' carambola by plastic packaging or storage temperature. *HortScience* **33**, 1038–1041.

Miller, W.R. and McDonald, R.E. 2000. Carambola quality after heat treatment, cooling and storage. *Journal of Food Quality* **23**, 283–291.

Milne, D.L. 1993. Postharvest handling of avocado, mango and litchi for export from South Africa. *Postharvest Handling of Tropical Fruit. Proceedings of an International Conference held in Chiang Mai, Thailand, 19–23 July 1993. Australian Centre for International Agricultural Research Proceedings* **50**, 73–89.

Minamide, T., Habu, T. and Ogata, K. 1980. Effect of storage temperature on keeping freshness of mushrooms after harvest. *Journal of Japanese Society of Food Science and Technology* **27**, 281–287.

Minamide, T., Nishikawa, T. and Ogata, K. 1980b. The effects of CO_2 and O_2 on the shelf life of shiitake Lentinus edodes after harvest. *Journal of the Japanese Society of Food Science and Technology* **27**, 505–510.

Minamide, T., Tsurata, M. and Ogata, K. 1980a. Studies on shelf life of shiitake *Lentinus edodes* after harvest. *Journal of the Japanese Society of Food Science and Technology* **27**, 498–504.

Mitchell, F.G. 1992. Preparation for fresh market. I. fruits. In Kader, A.A., Kasmire, R.F., Mitchell, F.G., Reid, M.S., Sommer, N.F. and Thompson, J.F. (Editors), *Postharvest Technology of Horticultural Crops. Cooperative Extension, University of California, Division of Agriculture and Natural Resorces,* 31–38.

Miyazaki, T. 1983. Effects of seal packaging and ethylene removal from sealed bags on the shelf life of mature green Japanese apricot *Prunus mume* Sieb. – Zucc. fruits. *Journal of the Japanese Society for Horticultural Science* **52**, 85–92.

Mizobutsi, G.P., Borges, C.A.M. and de Siqueira, D.L. 2000. Post harvest conservation of Tahiti limes *Citrus latifolia* Tanaka treated with gibberellic acid and stored at three temperatures. Conservacao pos colheita da lima acida 'Tahiti' *Citrus latifolia* Tanaka, tratada com acido giberelico e armazenada em tres temperaturas. *Revista Brasileira de Fruticultura* **22**, Special issue, 42–47.

Mizuno, M. 2000. Anti-tumor polysaccharides from mushrooms during storage. **12**, 275–281.

Moelich, D.H., Dodd, M.C. and Huysamer, M. 1996. Fruit maturity studies in 'Galia 5' and 'Doral' melons. II. The effect of harvest maturity on the postharvest quality after simulated shipping. *Journal of the Southern African Society for Horticultural Sciences* **6**, 64–68.

Mohamed, S., Khin, K.M.M. and Idris, A.Z. 1992. Effects of various surface treatments palm oil, liquid paraffin, Semperfresh or or starch surface coatings and LDPE wrappings on the storage life of guava *Psidium guajava* at 10°C. *Acta Horticulturae* **321**, 786–794.

Mohamed, S., Taufik, B. and Karim, M.N.A. 1996. Effects of modified atmospheric packaging on the physicochemical characteristics of ciku *Achras sapota* L. at various storage temperatures. *Journal of the Science of Food and Agriculture* **70**, 231–240.

Mohammed, M. 1993. Storage of passionfruit in polymeric films. *Proceedings of the Interamerican Society for Tropical Horticulture* **37**, 85–88.

Mohammed, S., Wilson, L.A. and Prendergast, N. 1984. Guava meadow orchard: effect of ultra high density planting and growth regulators on growth, flowering and fruiting. *Tropical Agriculture Trinidad* **61**, 297–301.

Montenegro, E.H. 1988. *Postharvest behaviour of banana harvested at different stages of maturity.* BS student project, University of the Phillipines, Los Baños, Laguna.

Montoya, J., Marriott, J., Quimi V. H. and Caygill. J. C. 1984. Age control of banana harvesting' under Ecuadorian conditions. *Fruits Paris* **39**, 293–296.

Moon, B.W., Lim, S.T., Choi, J.S. and Suh, Y.K. 2000. Effects of pre or post harvest application of liquid calcium fertilizer manufactured from oyster shell on the calcium concentration and quality in stored 'Niitaka' pear fruits. *Journal of the Korean Society for Horticultural Science,* **41**, 61–64.

Morales, A.L. and Duque, C. 1987. Aroma constituents of the fruit of the mountain papaya *Carica pubescens* from Colombia. *Journal of Agricultural and Food Chemistry* **35**, 538–540.

Moretti, C.L., Sargent, S.A., Balaban, M.O. and Puschmann, R. 2000. Electronic nose: a non-destructive technology to screen tomato fruit with internal bruising. Nariz eletronico: tecnologia nao-destrutiva para a deteccao de desordem fisiologica causada por impacto em frutos de tomate. *Horticultura Brasileira* **18**, 20–23.

Morris, J.R., Spayd, S.E., Brooks, J.G. and Cawthon, D.L. 1981. Influence of postharvest holding on raw and processed quality of machine harvested blackberries. *Journal of the American Society for Horticultural Science* **106**, 769–775.

Morris, L. and Kader, A.A. 1977. Physiological disorders of ceratin vegetables in relation to modified atmosphere. *Second National Controlled Atmosphere Research Conference. Proceedings, Michigan State University Horticultural Report* **28**, 266–267.

Morris, L., Yang, S.F. and Mansfield, D. 1981a. Postharvest physiology studies. *Californian Fresh Market Tomato Advisory Board Annual Report* 1980–1981, 85–105.

Morris, M.P., Pagan, C. and Garcia, J. 1952. Es el Mamey una fruta venenosa? *Revista de Agricultura de Puerto Rica, Supplemento Alimentos Nutritional* **43**, 288a–288b.

Morris, L., Pratt, H.K. and Tucker, C.L. 1955. Lettuce handling and quality. *Western Grower and Shipper* **26**, 14–18.

Morton, J.F. 1983. Why not select and grow superior types of canistel? *Proceedings of the Tropical Region, American Society for Horticultural Science* **27A**, 43–52.

Muir, A.Y. and Bowen, S.A. 1994. Measurements of the adhesion strengths of potato skin related to haulm treatments. *International Agrophysics* **8**, 531–536.

Muirhead, I.F. 1976. Postharvest control of mango anthracnose with benomyl and hot water. *Australian Journal of Experimental Agriculture and Animal Husbandry* **16**, 600–603.

Muirhead, I.F. and Grattidge, R. 1984. Postharvest diseases of mangoes, the Queensland experience. *Proceedings of the CSIRO First Australian Mango Research Workshop*, 248–252.

Munasque, V.S. and Mendoza, D.B., Jr. 1990. Developmental physiology and ripening behaviour of 'Senorita' banana Musa sp. L. fruits. *Association of Southeast Asian Nations Food Journal* **5**, 152–157.

Muñoz, V.R.S., Lizana, M.L.A., Galletti, G.L. and Luchsinger, L. 1998. Effects of postharvest temperature fluctuation on carotenoids and sugar in mangoes *Mangifera indica* L. cultivar Piqueno. Efectos de la fluctuacion termica en post cosecha sobre carotenos y azucares de frutos de mangos *Mangifera indica* L. cultivar Piqueno. *IDESIA* **15**, 7–13.

Murphy, J.M. 1990. *Brand Strategy.* Director Books, Cambridge.

Murray, K.E., Shipton, J., Whitfield, F.B. and Last, J.H. 1976. *Journal of the Science of Food and Agriculture* **27**, 1093.

Musa, S.K. and Thompson, A.K. 1976. Estimation of vegetable retailer's losses and suggested size grades. *Sudan Journal Food Science Technology* **8**, 23–31.

Muthoo, A.K., Raina, B.L. and Nathu, B.L. 1999. Fruit cracking studies in some commerical cultivars of litchi *Litchi chinensis* Sonn. *Advances in Plant Sciences* **12**, 543–547.

Muthukumarasamy, M. and Panneerselvam, R. 2000. Carbohydrate metabolism in yam tubers during storage and sprouting. *Tropical Science* **40**, 63–66.

Mziray, R.S., Imungi, J.K. and Karuri, E.G. 2000. Changes in ascorbic acid, beta carotene and sensory properties in sundried and stored *Amaranthus hybridus* vegetables. *Ecology of Food and Nutrition* **39**, 459–469.

Nabawanuka, J. 1993. *Osmotic dehyration of bananas.* MSc thesis, Silsoe College, Cranfield University.

Nabialek, A., Ben, J.M., Cisek, B. and Michalczuk, L. 1999. Effect of various temperatures and CO_2 and O_2 concentrations on the control of blue mould rot development in 'Shattenmorelle' sour cherry. *Acta Horticulturae* **485**, 281–285.

Nagai, M. 1975. Physical properties of fruit and vegetables. Measurement of specific electrical resistance of watermelons. *Bulletin of the Faculty of Agriculture, Mie University, Japan* **49**, 301–308.

Nair, H. and Tung, H.F. 1988. Postharvest physiology and storage of Pisang Mas. *Proceedings of the UKM Simposium Biologi Kebangsaan Ketiga, Kuala Lumpur, November 1988*, 22–24.

Namsri, U., Saelim, M., Naunsri, C. and Oates, C.G. 1999. Viability and morphology of pollen of Longkong, Duku (*Aglaia dookkoo* Griff.) and Langsat (*Aglaia domestica* Pelleg.). *37th Kasetsart University Annual Conference, 3–5 February, 1999*, 149–155.

Nanda, S., Rao, D.V.S. and Krishnamurthy, S. 2001. Effects of shrink film wrapping and storage temperature on the shelf life and quality of pomegranate fruits cv. Ganesh. *Postharvest Biology and Technology* **22**, 61–69.

Narain, N. and Bora, P.S. 1992. Post harvest changes in some volatile flavour constituents of yellow passion fruit *Passiflora edulis* f. *flavicarpa*. *Journal of the Science of Food and Agriculture* **60**, 529–530.

Narayan, C. and Ghosh, S.N. 1993. Extension of storage life of cashew apple. *Cashew* **7**, 12–14.

Nath, V. and Bhargava, R. 1998. Shelf life of ber (*Ziziphus mauritiana* Lamk.) as affected by postharvest treatments and storage environment. *Progressive Horticulture* **30**, 158–163.

Neale, M.A., Lindsay, R.T. and Messer, H.J.M. 1981. An experimental cold store for vegetables. *Journal of Agricultural Engineering Research* **26**, 529–540.

Nelson, J.W. 1966. Mechanical harvesting of highbush blueberries. *Proceedings of the Second North American Blueberry Workers Conference. Main Agricultural Experimental Station Miscellaneous Report* **118**, 58–59.

Nelson, S.O. 1979. Microwave dielectric properties of fresh fruits and vegetables. *American Society of Agricultural Engineers Paper* **79**, 3546.

Nelson, S.O. 1983. Dielectric properties of some fresh fruits and vegetables at frequencies of 2.45–22 GHz. *Transactions of the American Society of Agricultural Engineers* **26**, 613–616.

Neri, F. and Brigati, S. 1996. Qualitative indexes for melon harvesting. Gli indici qualitativi per la raccolta del melone. *Colture Protette* **25**, 45–48.

Neuwirth, G.R. 1988. Respiration and formation of volatile flavour substances in controlled atmosphere stored apples after periods of ventilation at different times. Respiration und Bildung fluchtiger Aromastoffe controlled atmosphere gelagerter Apfel nach zeitlich gestaffelten Zwischenbeluftungen. *Archiv für Gartenbau* **36**, 417–422.

New, S., Baldry, J. Marriott, J. and Dixon, E.A. 1976. Fruit quality factors affecting selection of banana clones. *Acta Horticulturae* **57**, 205–212.

Ng, J.K.H. and Mendoza, J.B. 1980. *An introductory study on spot development spotting of 'Bungulan' bananas.* Student project, University of the Phillipines, Los Baños, Laguna.

Ngalani, J.A., Tchango, J.T., Nkeng, M.N., Noupadja, P. and Tomekpe, K. 1998. Physicochemical changes during ripening in some plantain cultivars grown in Cameroon. *Tropical Science* **38**, 42–47.

Nicholas, J., Rothan, C. and Duprat, F. 1989. Softening of kiwifruit in storage. Effects of intermittent high CO_2 treatments. *Acta Horticulturae* **258**, 185–192.

Nichols, R. 1971. A review of the factors affecting the deterioration of harvested mushrooms. *Glasshouse Crops Research Institute, Littlehampton, United Kingdom,* Report 174.

Nicolas, J., Rothan, C. and Duprat, F. 1989. Softening of kiwifruit in storage. Effects of intermittent high CO_2 treatments. *Acta Horticulturae* **258**, 185–192.

Niedzielski, Z. 1984. Selection of the optimum gas mixture for prolonging the storage of green vegetables. Brussels sprouts and spinach. *Industries Alimentaires et Agricoles* **101**, 115 118.

Nielsen, L.W. 1968. Accumulation of respiratory CO_2 around potato tubers in relation to bacterial soft rot. *American Potato Journal* **45**, 174.

Nigro, F., Ippolito, A., Lattanzio, V., Venere, D.D. and Salerno, M. 2000. Effect of ultraviolet-C light on postharvest decay of strawberry. *Journal of Plant Pathology* **82**, 29–37.

Noomhorm, A. and Tiasuwan 1988. Effect of controlled atmosphere storage for mango. *N. Paper, American Society of Agricultural Engineers* 886589, 8 pp.

Noon, R.A. 1979. Report on an assignment as plant pathologist to CONAFRUT, Mexico City, March 1977–September 1979. *Tropical Products Institute Report* R923, 34 pp.

Noor,. and Izhar, A. 1998. Postharvest application of salicylic acid to onion bulbs for sprout inhibition. *Mati ur Rehman Sarhad Journal of Agriculture* **14**, 541–547.

Norwood, C.T. 1930. CO_2 storage of fruits, vegetables and flowers. *Industrial and Engineering Chemistry* **22**, 1186–1189.

Nowak, J. and Rudnicki, R.M. 1990. *Postharvest Handling and Storage of Cut Flowers, Florist Greens and Potted Plants.* Chapman and Hall, London, 210 pp.

Nunes, M.C.N., Brecht, J.K., Morais, A.M.M.B. and Sargent, S.A. 1998. Controlling temperature and water loss to maintain ascorbic acid levels in strawberries during postharvest handling. *Journal of Food Science* **63**, 1033–1036.

Nuske, D. and Muller, H. 1984. Preliminary results on the industrial type storage of headed cabbage under CA storage conditions. Erste Ergebnisse bei der industriemassigen Lagerung von Kopfkohl unter CA Lagerungsbedingungen. *Nachrichtenblatt fur den Pflanzenschutz in der DDR* **38**, 185–187.

Nussinovitch, A., Kopelman, I.J. and Mizrahi, S. 1990. Mechanical criteria of banana ripening. *Journal of the Science of Food and Agriculture* **53**, 63–71.

Nwaiwu, M.Y., Efiuvwevwere, B.J.O. and Princewill, P.J.T. 1995. Biodeterioration of African breadfruit (*Treculia africana*) in south-eastern Nigeria and its control. *Tropical Science* **35**, 22–29.

Nyanjage, M.O., Wainwright, H., Bishop, C.F.H. and Cullum, F.J. 2000. A comparative study on the ripening and mineral content of organically and conventionally grown Cavendish bananas. *Biological Agriculture and Horticulture* **18**, 221–234.

Oda, K., Takizawa, N. and Shida, H. 1976. Growth of three edible fungi, *Pholiota nameko, Pleurotus ostreatus* and *P. cornucopiae*, inoculated on the sawdust box medium in cold winter. *Journal of the Hokkaido Forest Products Research Institute* 2, 6–9.

Odigboh, E.U. 1991. Single row model II cassava harvester. *Agricultural Mechanization in Asia, Africa and Latin America* 22, 63–70.

OECD. 1992. *Kiwifruit. International Standardisation of Fruit and Vegetables.* Organization for Economic Cooperation and Development, Paris, 69 pp.

Ogali, E.L., Opadokun, J.S. and Okobi, A.O. 1991. Effect of lime and local gin on post-harvest rot of yams. *Tropical Science* 31, 365–370.

Ogata, K. and Inous, T. 1957. Studies on the storage of onions. *Proceedings of the XIX International Horticultual Congress, Warsaw, September 1974.*

Ogata, K., Yamauchi , N. and Minamide, T. 1975. Physiological and chemical studies on ascorbic acid in fruits and vegetables. 1. Changes in ascorbic acid content during maturation and storage of okra. *Journal of the Japanese Society for Horticultural Science* **44**, 192–195.

Ogundana, S.K., Naqui, S.H.Z. and Ekundago, J.A. 1970. Fungi associated with soft rot of yams stored in Nigeria. *Transactions of the British Mycological Society* **54**, 445.

O'Hare, T.J. 1993. Postharvest physiology and storage of carambola starfruit: a review. *Postharvest Biology and Technology* 2, 257–267.

O'Hare, T.J. and Johnson, G.I. 1992. Postharvest physiology and storage of rambutan: a review. *Development of Postharvest Handling Technology for Tropical Tree Fruits: a Workshop Held in Bangkok, Thailand, 16–18 July. Australian Centre for International Agricultural Research Proceedings* **58**, 15–20.

O'Hare, T.J. and Prasad, A. 1993. The effect of temperature and CO_2 on chilling symptoms in mango. *Acta Horticulturae* **343**, 244–250.

O' Hare, T.J., Prasad, A. and Cooke, A.W. 1994. Low temperature and controlled atmosphere storage of rambutan. *Postharvest Biology and Technology* **4**, 147–157.

O'Hare, T.J., Wong, L.S., Prasad, A., Able, A.J. and King, G.J. 2000. Atmosphere modification extends the postharvest shelflife of fresh-cut leafy Asian brassicas. *Acta Horticulturae* **539**, 103–107.

Okigbo, R.N. and Ikediugwu, F.E.O. 2001. Biological control of tuber surface mycoflora of yams (*Dioscorea rotundata*). *Tropical Science* **41**, 85–89.

Okuse, I. and Ryugo, K. 1981. Compositional changes in the developing Hayward kiwifruit in California. *Journal of the American Society for Horticultural Science* **106**, 73–76.

Okwuowulu, P.A. and Nnodu, E.C. 1988. Some effects of prestorage chemical treatments and age at harvesting on the storability of fresh ginger rhizomes (*Zingiber officinale* Roscoe). *Tropical Science* **28**, 123–125.

Olney, A.J. 1926. Temperature and respiration of ripening bananas. *Botanical Gazette* **82**, 415–416.

Onesirosan, P.T. 1986. Effect of moisture content, temperature and storage duration on the level of fungal invasion and germination of winged bean *Psophocarpus tetragonolobus* L. D.C. *Seed Science and Technology* **14**, 355–359.

Onnop, W.A., Sroymano, D., Gomolmanee, S. and La Ongsri, S. 1993. Development of maturity indices for longan. *Postharvest Handling of Tropical Fruit. Proceedings of an International Conference held in Chiang Mai, Thailand, 19–23 July 1993. Australian Centre for International Agricultural Research Proceedings* **50**, 341–342.

Oosthuyse, S.A. 1998. Effect of environmental conditions at harvest on the incidence of lenticel damage in mango. *Yearbook, South African Mango Growers' Association* **18**, 15–17.

Osajima, Y., Wada, K. and Ito, H. 1987. The effects of ethylene + acetaldehyde removing agents and seal packaging with plastic film on the keeping quality of Japanese apricot *Prunus mume* Sieb. et Zucc. and Kabosu *Citrus sphaerocarpa* hort. ex. Tanaka fruits. *Journal of the Japanese Society for Horticultural Science* **55**, 524–530.

Osman, A. and Mustaffa, R. 1996. The storage quality of carambola variety B10 as affected by different precooling conditions. *Association of Southeast Asian Nations Food Journal* **11**, 85–94.

Othieno, J.K. and Thompson, A.K. 1993. Modified atmosphere packaging of sweetcorn. *Post Harvest Treatment of Fruit and Vegetables. COST'94 Workshop, September 14–15, 1993, Leuven, Belgium.*

Othieno, J.K., Thompson, A.K. and Stroop, I.F. 1993. Modified atmosphere packaging of vegetables. *Post Harvest Treatment of Fruit and Vegetables. COST'94 Workshop, September 14–15, 1993, Leuven, Belgium* 247–253.

Otma, E.C. 1989. Controlled atmosphere storage and film wrapping of red bell peppers *Capsicum annuum* L. *Acta Horticulturae* **258**, 515–522.

Oudit, D.D. 1976. Polythene bags keep cassava tubers fresh for several weeks at ambient temperature. *Journal of the Agricultural Society of Trinidad and Tobago* **76**, 297–298.

Overholser, E.L. 1928. Some limitations of gas storage of fruits. *Ice and Refrigeration* 74, 551–552

Ozcan, M. 1999. The importance of standardization and packaging in fruit and vegetable marketing. Meyve sebze pazarlanmasinda standardizasyonun ve ambalajin onemi. *Karadeniz Bolgesinde Tarimsal Uretim ve Pazarlama Sempozyumu, 15–16 Ekim '99, Buyuk Otel Samsun, Turkey,* 102–107.

Ozelkok, S., Ertan, U. and Kaynas, K. 1997. Maturity and ripening concepts on nectarines. A case study on 'Nectared 6' and 'Independence.' *Acta Horticulturae* **441**, 287–294.

Ozer, M.H., Eris, A., Turk, R., Sivritepe, N. and Michalczuk, L. 1999. A research on controlled atmosphere storage of kiwifruit. *Acta Horticulturae* **485**, 293–300.

Pal, R.K. and Buescher, R.W. 1993. Respiration and ethylene evolution of certain fruits and vegetables in response to CO_2 in controlled atmosphere storage. *Journal of Food Science and Technology Mysore* **30**, 29–32.

Pal, R.K., Thomas, R.J., Lal, B., Singh, N.M. and Gupta, S. 1999. Influence of postharvest treatment with vapour heat and hydrogen peroxide based chemical on the quality of mango cv. Baneshan. *Journal of Applied Horticulture Lucknow* **1**, 108–111.

Pala, M., Damarli, E. and Alikasifoglu, K. 1994. A study of quality parameters in green pepper packaged in polymeric films. *Commissions C2, D1, D2/3 of the International Institute of Refrigeration International Symposium, June 8–10, Istanbul Turkey,* 305–316

Palma, T., Stanley, D.W., Aguilera, J.M. and Zoffoli, J.P. 1993. Respiratory behavior of cherimoya *Annona cherimola* Mill. under controlled atmospheres. *HortScience* **28**, 647–649.

Palmer, J.K. 1971. The Banana. In Hulme, A.C. (Editor), *The Biochemistry of Fruits and Their Products.* Vol. 2. Academic Press, London.

Palou, L., Smilanick, J.L., Crisosto, C.H. and Mansour, M. 2001. Effect of gaseous ozone exposure on the development of green and blue molds on cold stored citrus fruit. *Plant Disease* **85**, 632–638.

Paniandy, J.C., Normand, F. and Reynes, M. 1999. Factors affecting the conservation of fresh strawberry guavas produced on Reunion Island. Facteurs intervenant sur la conservation en frais de la goyave fraise a l'ile de la Réunion. *Fruits Paris* **54**, 49–56.

Pankasemsuk, T., Garner, J.O., Jr, Matta, F.B. and Silva, J.L. 1996. Translucent flesh disorder of mangosteen fruit *Garcinia mangostana* L. *HortScience* **31**, 112–113.

Panpatil, N.V., Munjal, S.V., Patil, S.R. and Damame, S.V. 2000. Effect of post-harvest application of calcium salts on physico-chemical parameters and shelf-life of ber (Zizyphus mauritiana Lamk) fruits at ambient storage. *Orissa Journal of Horticulture* **28**, 25–30.

Pantastico, Er.B. (Editor). 1975. *Postharvest Physiology, Handling and Utilisation of Tropical and Sub Tropical Fruits and Vegetables.* AVI Publishing, Westport, CT.

Pantastico, Er.B. and Mendoza, D.B., Jr. 1970. Note. Production of ethylene and acetylene during ripening and charring, *Philippine Agriculture* **53**, 477–483.

Pantastico, Er.B., Mendoza, D.B., Jr and Abilay, R.M. 1969. Some chemical and physiolgical changes during storage of lanzones *Lansium domesticum* Correa. *Philippine Agriculture* **52**, 505.

Papadopoulou Mourkidou, E. 1991. Postharvest applied agrochemicals and their residues in fresh fruits and vegetables. *Journal of the Association of Official Analytical Chemists* **74**, 745–765.

Pariasca, J.A.T., Miyazaki, T., Hisaka, H., Nakagawa, H. and Sato, T. 2001. Effect of modified atmosphere packaging (MAP) and controlled atmosphere (CA) storage on the quality of snow pea pods (*Pisum sativum* L. var. *saccharatum*). *Postharvest Biology and Technology* **21**, 213–223.

Park, J.C., Park, S.M., Yoo, K.C. and Jeong, C.S. 2001. Changes in postharvest physiology and quality of hot pepper fruits by harvest maturity and storage temperature. *Journal of the Korean Society for Horticultural Science* **42**, 289–294.

Park, N.P., Choi, E.H. and Lee, O.H. 1970. Studies on pear storage. II. Effects of polyethylene film packaging and CO_2 shock on the storage of pears, cv. Changsyprang. *Korean Journal of Horticultural Science* **7**, 21–25.

Park, Y.M. and Kwon, K.Y. 1999. Prevention of the incidence of skin blackening by postharvest curing procedures and related anatomical changes in 'Niitaka' pears. *Journal of the Korean Society for Horticulural Science* **40**, 65–69.

Park, Y.M., Hwang, J.M. and Ha, H.T. 2000. Storability of garlic bulbs as influenced by postharvest clipping treatments and storage temperature. *Journal of the Korean Society for Horticultural Science* **41**, 315–318.

Park, Y.S. and Jung, S.T. 2000. Effects of CO_2 treatments within polyethylene film bags on fruit quality of fig fruits during storage. *Journal of the Korean Society for Horticultural Science* **41**, 618–622.

Parkin, K.L. and Schwobe, M.A. 1990. Effects of low temperature and modified atmosphere on sugar accumulation and chip colour in potatoes *Solanum tuberosum*. *Journal of Food Science* **55**, 1341–1344.

Parodi, M.G. and Campbell, R.J. 1995. Study of the behaviour of granadilla Passiflora ligularis Juss. subjected to low storage temperature for different durations.Estudio del comportamiento de la granadilla *Passiflora ligularis* Juss. sometida a una baja temperatura de almacenamiento por diferentes periodos. *Proceedings of the Interamerican Society for Tropical Horticulture* **39**, 151–155.

Parsons, C.S. 1960. Effects of temperature, packaging and sprinkling on the quality of stored celery. *Procedings of the American Society for Horticultural Science* **75**, 403.

Parsons, C.S., Anderson, R.E. and Penney, R.W. 1974. Storage of mature green tomatoes in controlled atmospheres. *Journal of the American Society for Horticultural Science* **95**, 791–794.

Parsons, C.S., Gates, J.E. and Spalding D.H. 1964. Quality of some fruits and vegetables after holding in nitrogen atmospheres. *American Society for Horticultural Science* **84**, 549–556.

Passam, H.C. 1982. Gibberellin dip slows yam decay. *International Agriculture Development* October, 10–11.

Passam, H.C., Maharaj, D.S. and Passam, S. 1981. A note on freezing as a method of storage of breadfruit slices. *Tropical Science* **23**, 67–74.

Paull RE 1996. Postharvest atemoya fruit splitting during ripening. *Postharvest Biology and Technology* **8**, 329–334.

Paull, R.E. and Chen, N.J. 1999. Heat treatment prevents postharvest geotropic curvature of asparagus spears *Asparagus officinalis* L. *Postharvest Biology and Technology* **16**, 37–41.

Paull, R.E. and Rohrbach, K.G. 1985. Symptom development of chilling injury in pineapple fruit. *Journal of the American Society for Horticultural Science* **110**, 100–105.

Peacock, B.C. 1975. 'Mixed ripe' a problem fot the banana industry. *Queensland Agricultural Journal* **101**, 201–204.

Peacock B.C. 1987. Simulated commercial export of mangoes using controlled atmosphere container technology. *Australian Centre for International Agricultural Research Proceedings* **23**, 40–44.

Peirs, A., Lammertyn, J., Nicolai, B. and de Baerdemaeker, J. 2000. Non-destructive quality measurements of apples by means of NIR-spectroscopy. *Acta Horticulturae* **517**, 435–440.

Pekmezci, M. 1983. Studies on the storage of lemon fruits cv Kutdiken. *Acta Horticulturae* **138**, 203–214.

Peleg, K. 1985. *Produce Handling, Packaging and Distribution.* AVI Publishing Westport, CT.

Peleg, K. 1993. Comparison of non-destructive and destructive measurement of apple firmness. *Journal of Agricultural Engineering Research* **55**, 227–238.

Peleg, K., Ben Hanan, U. and Hinga, S. 1990. Classification of avocado by firmness and maturity. *Journal of Texture Studies* **21**, 123–139.

Pelleboer, H. 1983. A new method of storing Brussels sprouts shows promise. Nieuwe methode van spruitenbewaring biedt perspectief. *Bedrijfsontwikkeling* **14**, 828–831.

Pelleboer, H. 1984. A future for CA storage of open grown vegetables. Toekomst voor CA bewaring vollegrondsgroenten. *Groenten en Fruit* **39**, 62–63.

Pelleboer, H. and Schouten, S.P. 1984. New method for storing Chinese cabbage is a success. Nieuwe methode voor bewaren Chinese kool succes. *Groenten en Fruit* **40**, 51.

Pellegrini, C.N., Orioli, G.A. and Croci, C.A. 2000. Identification of the method used to inhibit sprouting in garlic. *Acta Horticulturae* **518**, 55–61.

Pelser, P.T. du and Lesar, K. 1989. Decay control in South African mangoes by flusilazol, penconazole and prochloraz during simulated shipment to Europe. *Aspects of Applied Biology* **20**, 41–48.

Pendergrass, A., Isenburg, F.M.R., Howell, L. and Caroll, J.E. 1976. Ethylene induced changes in appearance and hormone content of Floida grown cabbage. *Canadian Journal of Plant Sciences* **56**, 319

Pennock, W. and Maldonaldo, G. 1962. Hot water treatments of mango fruit to reduce anthracnose decay. *Journal of Agriculture, University of Puerto Rico* **46**, 272–283.

Pentzer, W.T. and Asbury, W.R. 1934. Sulphur dioxide as an aid in the preservation of grapes in transit and storage. *Blue Anchor* **11**, 2–4 23.

Pentzer, W.T., Asbury, C.E. and Hamner, K.C. 1933. The effect of sulfur dioxide fumigation on the respiration of Emperor grapes. *Proceedings of the American Society of Horticultural Science* **30**, 258–260.

Perera, O.D.A.N. and Karunaratne, A.M. 2001. Response of bananas to postharvest acid treatments. *Journal of Horticultural Science and Biotechnology* **76**, 70–76.

Perez Tello, G.O., Martinez Tellez, M.A., Briceno, B.O., Vargas Arispuro, I. and Diaz Perez, J.C. 1999. Effect of three temperatures of storage on the activities of polyphenoloxidase and peroxidase in mamey sapote fruits *Pouteria sapota*. *In* Hagg, M., Ahvenainenm R., Evers, A.M., Tiilikkala, K. (Editors), *Agri Food Quality II: Quality Management of Fruits and Vegetables From Field to Table, Turku, Finland, Eds.. 22–25 April, 1998,* 174–176.

Perry, J.S. 1977. A nondestructive firmness NDF testing unit for fruit. *Transactions of the American Society of Agricultural Engineers* **20**, 762–767.

Pertot, I. and Perin, L. 1999. Influence of N fertilization on rot caused by Botrytis cinerea on kiwifruit in cold store. Influenza della concimazione azotata sui marciumi causati da *Botrytis cinere*a su kiwi in frigoconservazione. *Notiziario ERSA* **12**, 39–41.

Pesis, E. and Marinansky, R. 1992. Influence of fruit coating on papaaya quality. *Acta Horticulturae* **321**, 659–666.

Pesis, E., Aharoni, D., Aharon, Z., Arie, R.B. and Aharoni, N. 1999. Storage in modified atmosphere to reduce chilling injury in mango fruit. *Alon Hanotea* **53**, 259–265.

Pesis, E., Aharoni, D., Aharon, Z., Arie, R.B., Aharoni, N. and Fuchs, Y. 2000. Modified atmosphere and modified humidity packaging alleviates chilling injury symptoms in mango fruit. *Postharvest Biology and Technology* **19**, 93–101.

Pesis, E., Copel, A., Ben Arie, R., Feygenberg, O. and Aharoni, Y. 2001. Low-oxygen treatment for inhibition of decay and ripening in organic bananas. *Journal of Horticultural Science and Biotechnology* **76**, 648–652.

Pesis, E., Fuchs, Y. and Zauberman, G. 1976. Cellulase activity and fruit softening in avocado. *Plant Physiology* **61**, 416–419.

Pesis, E., Levi, A., Sonego, L. and Ben Arie, R. 1986. The effect of different atmospheres in polyethylene bags on deastringency of persimmon fruits. *Alon Hanotea* **40**, 1149–1156.

Pesis, E., Marinansky, R., Zuaberman, G. and Fuchs, Y. 1993. Reduction of chilling injury symptoms of stored avocado fruit by pre storage treatment with high nitrogen atmosphere. *Acta Horticulturae* **343**, 251–255.

Peters, P. and Seidel, P. 1988. Gentle harvesting of Brussels sprouts and recently developed cold storage methods for preservation of quality. Gutschonende Rosenkohlernte verbunden mit neuen Methoden der qualitätserhaltenden Kuhllagerung. *Tagungsbericht, Akademie der Landwirtschaftswissenschaften der DDR.* **262**, 301 309.

Peters, P., Jeglorz, J. and Kastner, B. 1986. Investigations over several years on conventional and cold storage of Chinese cabbage. Mehrjahrige Untersuchungen zur Normal und Kuhllagerung von Chinasalat. *Gartenbau* **33**, 298–301.

Pflug, I.J. and Dewey, D.H. 1960. Unloading soft fleshed fruit from bulk boxes. *Michigan Agricultural Experimental Station Quarterly Bulletin* **43**, 132–141.

Phan, C.T. 1987. Temperature: effects on metabolism. In Weichmann, J. (Editor), *Postharvest Physiology of Vegetables.* Marcel Dekker, New York, 173–178.

Phillips, W.R. and Armstrong, J.G. 1967. *Handbook on the Storage of Fruits and Vegetables for Farm and Commercial Use.* Canada Department of Agriculture, Publication 1260, 50 pp.

Pieh, K. 1965. Cooling of fruits and vegetables. *Deutsch. Gartenbau* **13**, 136–140.

Pirov, T.T. 2001. Yield and keeping ability of onions under different irrigation regimes. *Kartofel' I Ovoshchi* **2**, 42.

Platenius, H. 1942. Effect of temperatureon the respiration rate and respiratory quotient on some vegetables. *Plant Physiology, Lancaster* **17**, 179–197.

Platenius, H., Platenius, H., Jamieson, F.S. and Thompson, H.C. 1934. Studies on cold storage of vegetables. *Cornell University Agricultural Experimental Station Bulletin* **602**, 24 pp.

Plumbley, R.A. and Rickard, J.E. 1991. Post harvest deterioration of cassava. *Tropical Science* **31**, 295–303.

Plumbley, R.A., Cox, J., Kilminster, K., Thompson, A.K. and Donegan, L. 1985. The effect of imazalil in the control of decay in yellow yam caused by *Penicillium sclerotigenum. Annals of Applied Biology* **106**, 277–284.

Plumbley, R.A., Hernandez Montes, A. and Thompson, A.K. 1984. Benomyl tolerance in a strain of *Penicillium sclerotigenum* infecting yams and the use of imazalil as a means of control. *Tropical Agriculture Trinidad* **61**, 182–185.

Poenicke, E.F., Kays, S.J., Smittle, D.A. and Williams R.E. 1977. Ethylene in relation to postharvest quality deterioration in processing cucumbers. *Journal of the American Society of Horticultural Science* **102**, 303–306.

Polderdijk, J.J., Boerrigter, H.A.M., Wilkinson, E.C., Meijer, J.G. and Janssens, M.F.M. 1993. The effects of controlled atmosphere storage at varying levels of relative humidity on weight loss, softening and decay of red bell peppers. *Scientia Horticulturae* **55**, 315–321.

Poma Treccarri, C. and Anoni, A. 1969. Controlled atmosphere packaging of polyethylene and defoliation of the stalks in the cold storage of artichokes. *Rivista della Ontoflorofrutticoltura Italiana* **53**, 203.

Popenoe, H.L. 1992. New crops for South America's farmers. *Acta Horticulturae* **318**, 209–210.

Povolny, P. 1995. Influence of exposure to light and cultivation system on resistance to *Phoma foveata and Fusarium solani* var. *coeruleum* in potato tubers: a pilot study. *Swedish Journal of Agricultural Research* **25**, 47–50.

Prakash, O. and Pandey, B.K. 2000. Control of mango anthracnose by hot water and fungicide treatment. *Indian Phytopathology* **53**, 92–94.

Prakash, N., Lim, A.L. and Manurung, R. 1977. Embryology of Duku and Langsat varieties of *Lansium domesticum. Phytomorphology* **27**, 50–59.

Rattanapong, J. Lim, M. and Sadoodee, S. 1995. The effect of calcium chloride on the quality of longkong (*Lansium domesticum* Corr.) fruit. *Khon Kaen Agriculture Journal* **23**, 67–73.

Prange, R.K. and Lidster, P.D. 1991. Controlled atmosphere and lighting effects on storage of winter cabbage. *Canadian Journal of Plant Science* **71**, 263–268.

Praquin, J.Y. and Miche, J.C. 1971. Essai de conservation de taros et macabos au Cameroun. *Inst. Rech. Agron. Trop. Report* **1**, 21 pp.

Prasad, B.K., Sahoo, D.R., Kumar, M. and Narayan, N. 2000. Decay of chilli fruits in India during storage. *Indian Phytopathology* **53**, 42–44.

Prasanna, K.N.V., Rao, D.V.S. and Krishnamurthy, S. 2000. Effect of storage temperature on ripening and quality of custard apple *Annona squamosa* L. fruits. *Journal of Horticultural Science and Biotechnology* **75**, 546–550.

Pratt, H.K. 1971. Melons. In Hulme, A.C. (Editor), *The Biochemistry of Fruits and their Products*, Vol. 2. Academic Press, London, 303–324.

Pratt, H.K. and Mendoza, D.B., Jr. 1980. Influence of nut removal on growth and ripening of the cashew apple. *Journal of the American Society for Horticultural Science* **105**, 540–542.

Pratt, H.K. and Morris, L. 1958. Some physiological aspects of vegetable and fruit handling. *Food Technology in Australia* **10**, 407–417.

Pratt, H.K. and Reid, M.S. 1974. Chinese gooseberry: seasonal patterns of fruit growth and maturation, ripening, respiration and the role of ethylene. *Journal of the Science of Food and Agriculture* **25**, 747–757.

Priestley, J.H. and Woffenden, M. 1923. The healing of wounds in potato tubers and their propagation by cut sets. *Annals of Applied Biology* **10**, 96–115.

Prohens, J. and Nuez, F. 2001. Improvement of mishqui (*Solanum muricatum*) earliness by selection and ethephon applications. *Scientia Horticulturae* **87**, 247–259.

Prussia, S.E. and Woodroof, J.G. 1986. Harvesting handling and holding fruit. In Woodroof, J.G. and Luh, B.S. (Editors), *Commercial Fruit Processing*, 2nd edn. AVI Publishing, Westport, CT.

Pujantoro, Tohru, S. and Kenmoku, A. 1993. The changes in quality of fresh shiitake Lentinus edodes in storage under controlled atmosphere conditions. *Proceedings of ICAMPE '93, October 19–22, KOEX*, Korean Society for Agricultural Machinery, Seoul, Korea, 423–432.

Punchoo, R. 1988. *Viscoelastic properties as a measure of fruit storability*. MSc thesis, Silsoe College, Cranfield Institute of Technology.

Purseglove, J.W. 1968. *Tropical Crops, Dicotyledons*. Longmans, London, 719 pp.

Purseglove, J.W. 1972. *Tropical Crops, Monocotyledons*. Longmans, London, 606 pp.

Qadir, A. and Hashinaga, F, 2001. Inhibition of postharvest decay of fruits by nitrous oxide. *Postharvest Biology and Technology* **22**, 279–283.

Quazi, H.H. and Freebairn, H.T. 1970. The influence of ethylene, oxygen and carbon dioxide on the ripening of bananas. *Botanical Gazette* **131**, 5.

Quevedo, M.A., Sanico, R.T. and Baliad, M.E. 1991. The control of post-harvest diseases of taro corms. *Tropical Science* **31**, 359–364.

Quimio, A.J., and Quimio, T.H. 1974. Postharvest control of Philippine mango anthracnose by benomyl. *Philippine Agriculturist* **58**, 147–155.

Rabus, C. and Streif, J. 2000. Effect of various preharvest treatments on the development of internal browning disorders in 'Braeburn' apples. *Acta Horticulturae* **518**, 151–157.

Raghavan, G.S.V., Gariepy, Y., Theriault, R., Phan, C.T. and Lanson, A. 1984. System for controlled atmosphere long term cabbage storage. *International Journal of Refrigeration* **7**, 66–71.

Raghavan, G.S.V., Tessier, S., Chayet, M., Norris, E.G. and Phan, C.T. 1982. Storage of vegetables in a membrane system. *Transactions of the American Society of Agricultural Engineers* **25**, 433–436.

Ragni, L. 1997. Analysis of the mechanical damage of products in lines for grading and packaging apples. Analisi del danneggiamento meccanico del prodotto in linee di cernita e confezionamento delle mele. *Rivista di Frutticoltura e di Ortofloricoltura* **59**, 51–55.

Ragni, L., Berardinelli, A., Barchi, G.L. and Baraldi, G. 2001. Damage to peaches during postharvest treatment and transport. Danneggiamenti delle pesche nelle lavorazioni post raccolta e nel trasporto. *Informatore Agrario* **57**, 55–59.

Ragnoi, S. 1989. *Development of the market for Thai lychee in selected European countries.* MSc thesis, Silsoe College, Cranfield Institute of Technology.

Rahman, A.S.A., Huber, D. and Brecht, J.K. 1993a. Physiological basis of low O_2 induced residual respiratory effect in bell pepper fruit. *Acta Horticulturae* **343**, 112–116.

Rahman, A.S.A., Huber, D. and Brecht, J.K. 1993b. Respiratory activity and mitochondrial oxidative capacity of bell pepper fruit following storage under low O_2 atmosphere. *Journal of the American Society for Horticultural Science* **118**, 470–475.

Rahman, A.S.A., Huber, D.J. and Brecht, J.K. 1995. Low O_2-induced poststorage suppression of bell pepper fruit respiration and mitochondrial oxidative activity. *Journal of the American Society for Horticultural Science* **120**, 1045–1049.

Rahman, N.A., Wei, Y. and Thompson, A.K. 1995a. Temperature and modified atmosphere packaging effects on the ripening of banana. *Technologias de Cosecha y Postcosecha de Frutas y Hortalizas. Proceedings of a Conference held in Guanajuato, Mexico, 20–24 February 1995*, 313–321.

Rai, D.R., Masuda, R. and Saito, M. 2000. Effect of modified atmosphere packaging on low molecular weight carbohydrates of oyster mushrooms. *Journal of Food Science and Technology Mysore* **37**, 384–387.

Rajput, S.S., Pandey, S.D. and Sharma, H.G. 1999. A study on physico-chemical changes associated with growth and development of mango *Mangifera indica* L. fruits. *Orissa Journal of Horticulture* **27**, 17–22.

Raman, K.R., Raman, N.V. and Sadasivam, R. 1971. A note on the storage behaviour of mangosteen *Garcinia mangostana* L. *South Indian Horticulture* **19**, 85–86.

Ramsey, A.M. 1985. Mechanical harvesting of raspberries: development of a system for Scottish conditions. *Scottish Institute of Agricultural Engineering Technical Report* 7, 58 pp.

Ramsey, M.D., Daniells J.W and Anderson, V.J. 1990. Effects of Sigatoka leaf spot Mycosphaerella *musicola Leach* on fruit yields, field ripening and greenlife of bananas in North Queensland. *Scientia Horticulturae* **41**, 305–313.

Ramsey, M.D., Vawdrey, L.L., and Schipke, L.G. 1987. Evaluation of systemic and protectant fungicides for the control of Sigatoka leaf spot *Micosphaerella musicola* Leach of bananas in North Queensland. *Australian Journal of Experimental Agriculture* **27**, 919–923.

Ranganna, B., Kushalappa, A.C. and Raghavan, G.S.V. 1997. Ultraviolet irradiance to control dry rot and soft rot of potato in storage. *Canadian Journal of Plant Pathology* **19**, 30–35.

Ranganna, B., Raghavan, G.S.V. and Kushalappa, A.C. 1998. Hot water dipping to enhance storability of potatoes. *Postharvest Biology and Technology* **13**, 215–223.

Rao, M.M., Rokhade, A.K., Shankaranarayana, H.N., Prakash, N.A., Sulladmath, V.V. and Hittalmani, S.V. 1995. Developmental patterns and duration of growth and development of some tropical and sub tropical fruits under mild tropical rainy climate. *Mysore Journal of Agricultural Sciences* **29**, 149–154.

Rappaport, L. and Watada, A.E. 1958. Effect of temperature on artichoke quality. *Proceedings of the Conference on Transport of Perishables. University of California, Davis, February 3–5*, 142–146.

Rappel, M., Cooke, A.W., Jacobi, K.K. and Wells, I.A. 1991. Heat treatments for postharvest disease control in mangoes. *Acta Horticulturae* **291**, 362–371.

Rattanapong, J., Lim, M. and Sadoodee, S. 1995. The effect of calcium chloride on the quality of longkong *Lansium domesticum* Corr. fruit. *Khon Kaen Agriculture Journal* **23**, 67–73.

Ravindran, G. and Wanasundera, J.P.D. 1993. Chemical changes in yam tubers (*Dioscorea alata* and *D. esculenta*) during storage. *Tropical Science* **33**, 57–62.

Recasens, D.I., Graell, J., Pinol, J. and Serra, J. 1989. Assessing fruit maturity of ten apple cultivars by internal ethylene concentrations. *Acta Horticulturae* **258**, 437–443.

Reichel, M. 1974. The behaviour of Golden Delicious during storage as influenced by different harvest dates. Das Lagerverhalten von 'Golden Delicious' unter dem Einfluss unterschiedlicher Erntetermine. *Gartenbau* **21**, 268–270.

Reid, M. 2001. Advances in shipping and handling of ornamentals. *Acta Horticulturae* **543**, 277–284.

Reid, M.S. and Pratt, H.K. 1970. Ethylene and the respiration climacteric. *Nature* **226**, 976.

Reilly, K., Han, Y.H., Tohme, J. and Beeching, J.R. 2001. Isolation and characterisation of a cassava catalase expressed during post-harvest physiological deterioration. *Biochimica et Biophysica Acta, Gene Structure and Expression* **1518**, 317–323.

Renault, P., Houal, L., Jacquemin, G. and Chambroy, Y. 1994. Gas exchange in modified atmosphere packaging. 2. Experimental results with strawberries. *International Journal of Food Science and Technology* **29**, 379–394.

Renel, L. and Thompson, A.K. 1994. Carambola in controlled atmosphere. *Inter American Institute for Cooperation on Agriculture, Tropical Fruits Newsletter* **11**, 7.

Rennie, T.J., Raghavan, G.S.V., Vigneault, C. and Gariepy, Y. 2001. Vacuum cooling of lettuce with various rates of pressure reduction. *Transactions of the American Society of Agricultural Engineers* **44**, 89–93.

Resnizky, D. and Sive, A. 1991. Storage of different varieties of apples and pears cv. Spadona in 'ultra ultra' low O_2 conditions. *Alon Hanotea* **45**, 861–871.

Reust, W., Schwarz, A. and Aerny, J. 1984. Essai de conservation des pommes de terre en atomsphère controlée. *Potato Research* **27**, 75–87

Reyes, A.A. 1988. Suppression of Sclerotinia sclerotiorum and watery soft rot of celery by controlled atmosphere storage. *Plant Disease* **72**, 790–792.

Reyes, A.A. 1989. An overview of the effects of controlled atmosphere on celery diseases in storage. *International Controlled Atmosphere Conference, Fifth, Proceedings, June 14–16, 1989, Wenatchee, Washington. Volume 2. Other Commodities and Storage Recommendations*, 57–60.

Reyes, A.A. and Smith, R.B. 1987. Effect of O_2, CO_2, and carbon monoxide on celery in storage. *HortScience* **22**, 270–271.

Reynaldo, I.M. 1999. The influence of rootstock on the postharvest behaviour of Ruby Red grapefruit. *Cultivos Tropicales* **20**, 37–40.

Rhodes, M.J.C. and Wooltorton, S.C. 1971. The effect of ethylene on the respiration and on the activity of phenylamine ammonia lyase in swede and parsnip root tissue. *Phytochemistry* **10**, 1989.

Richard, P C. and Wickens, R. 1977. The effect of the time of harvesting of spring sown dry bulb onions on their yield keeping ability and skin quality. *Horticulture* **29**, 45–51.

Richardson, D.G. and Meheriuk, M. 1982. Controlled atmospheres for storage and transport of perishable agricultural commodities. *Proceedings of the Third International Controlled Atmosphere Research Conference.* Timber Press, Beaverton, OR.

Richardson, D.G. and Meheriuk, M. 1989. CA recommendations for pears including Asian pears. *International Controlled Atmosphere Conference, Fifth, Proceedings, June 14–16, 1989, Wenatchee, Washington. Volume 2. Other Commodities and Storage Recommendations*, 285–302.

Rickard, J.E. and Coursey, D.G. 1981. Cassava storage. Part 1: storage of fresh cassava roots. *Tropical Science* **23**, 1–32.

Rickard, J.E., Burden, O.J. and Coursey, D.G. 1978. Studies on the insolation of tropical horticultural produce. *Acta Horticulturae* **84**, 115–122.

Righetti, T.L. and Curtis, D. 1989. Sampling approaches for fruit testing program on 'Anjou' pear. *International Controlled Atmosphere Conference, Fifth, Proceedings, June 14–16, 1989, Wenatchee, Washington. Volume 1 Pome Fruits*, 63–74.

Risse, L.A. 1989. Individual film wrapping of Florida fresh fruit and vegetables. *Acta Horticulturae* **258**, 263–270.

Robbins, J.A. and Fellman, J.K. 1993. Postharvest physiology, storage and handling of red raspberry. *Postharvest News and Information* **4**.

Roberts, J.A. and Tucker, G.A. 1985. *Ethylene and Plant Development.* Butterworths, London.

Roberts, R. 1990. An overview of packaging materials for MAP. *International Conference on Modified Atmosphere Packaging. Part 1.* Campden Food and Drinks Research Association, Chipping Campden.

Robinson, J. 1990. *Free convective airflow in fruit stores.* PhD thesis, Silsoe College, Cranfield Institute of Technology.

Robinson, J. 1993. *Remote Temperature Logger.* Green PC, Ramornie, Aydon Road, Corbridge, Northumberland NE45 5DT, United Kingdom.

Robinson, J.E., Brown, K.M. and Burton, W.G. 1975. Storage characteristics of some vegetables and soft fruits. *Annals of Applied Biology* **81**, 339–408.

Rodov, V., Copel, A., Aharoni, N., Aharoni, Y., Wiseblum, A., Horev, B. and Vinokur, Y. 2000. Nested modified atmosphere packages maintain quality of trimmed sweet corn during cold storage and the shelf life period. *Postharvest Biology and Technology* **18**, 259–266.

Roe, M.A., Faulks, R.M. and Belsten, J. 1990. Role of reducing sugars and amino acids in fry colour of chips from potato grown under different nitrogen regimes. *Journal of the Science of Food and Agriculture* **52**, 207–214.

Roelofs, F. 1992. Supplying red currants until Christmas. Rode bes aanvoeren tot kerst. *Fruitteelt Den Haag* **82**, 11–13.

Roelofs, F. 1993. CA storage of plums. Storage for longer than three weeks gives too much wastage. CA bewaring van pruimen: langer dan drie weken bewaren geeft te veel uitval. *Fruitteelt Den Haag* **83**, 18–19.

Roelofs, F. 1993a. Research results of red currant storage trials 1992: CO_2 has the greatest influence on the storage result. Onderzoeksresultalten bewaarproeven rode bes 1992: koolzuurgas heeft grootste invloed op het bewaarresultaat. *Fruitteelt Den Haag,* **83**, 22–23.

Roelofs, F. 1994. Experience with storage of red currants in 1993: unexpected quality problems come to the surface. Ervaringen bewaren rode bes 1993: onverwachte kwaliteitsproblemen steken de kop op. *Fruitteelt Den Haag,* **84** 16–17.

Roelofs, F. and Breugem, A. 1994. Storage of plums. Choose for flavour, choose for CA. Pruimen bewaren. Kies voor smaak, kies voor CA. *Fruitteelt Den Haag* **84**, 12–13.

Roelofs, F. and Van de Waart, A.J.P. 1993. Long term storage of red currants under controlled atmosphere conditions. *Acta Horticulturae* **352**, 217–222.

Rogiers, S.Y. and Knowles, N.R. 2000. Efficacy of low O_2 and high CO_2 atmospheres in maintaining the postharvest quality of saskatoon fruit (*Amelanchier alnifolia* Nutt.). *Canadian Journal of Plant Science* **80**, 623–630.

Romanazzi, G., Nigro, F., Ippolito, A. and Salerno, M. 2001. Effect of short hypobaric treatments on postharvest rots of sweet cherries, strawberries and table grapes. *Postharvest Biology and Technology* **22**, 1–6.

Romero Rodriguez, M.A., Vazquez Oderiz, M.L., Lopez Hernandez, J. and Simal Lozano, J. 1992. Physical and analytical characteristics of the kiwano. *Journal of Food Composition and Analysis* **5**, 319–322.

Romo Parada, L., Willemot, C., Castaigne, F., Gosselin, C. and Arul, J. 1989. Effect of controlled atmospheres low O_2, high CO_2 on storage of cauliflower *Brassica oleracea* L., *Botrytis* group. *Journal of Food Science* **54**, 122–124.

Romphophak, T. and Palakul, S. 1990. Quality of 'Chanee' durian *Durio zibethinus* stored at 5°C. *Acta Horticulturae* **269**, 213–216.

Romphophak, T., Kunprom, J., Siriphanich, J., Romphophak, T., Kunprom, J. Siriphanich, J. 1997. Storage of durians on a semi commercial scale. *Kasetsart Journal, Natural Sciences* **31**, 141–154.

Rood, P. 1957. Development and evaluation of objective maturity indices for Californian freestone peaches. *Proceedings of the American Society for Horticultural Science* **70**, 104–112.

Rosen, J.C. and Kader, A.A. 1989. Postharvest physiology and quality maintenance of sliced pear and strawberry fruits. *Journal of Food Science* **54**, 656–659.

Rosenfeld, H.J., Meberg, K.R., Haffner, K. and Sundell, H.A. 1999. MAP of highbush blueberries: sensory quality in relation to storage temperature, film type and initial high oxygen atmosphere. *Postharvest Biology and Technology* **16**, 27–36.

Roser, B. and Colaco, C. 1993. A sweeter way to fresher food. *New Scientist* 15 May, 25–28.

Ross, G. 1993. *Plant Health and the Single Market.* Plant Health Newsletter 93/5. Ministry of Agriculture Fisheries and Food, London, 5 pp.

Rouhani, I. 1978. Effect of silver ion application on the ripening and physiology of stored green bananas. *Journal of Horticultural Science* **53**, 317–321.

Roura, S.I., Davidovich, L.A. and del Valle, C.E. 2000. Postharvest changes in fresh Swiss chard Beta vulgaris, type cycla under different storage conditions. *Journal of Food Quality* **23**, 137–147.

Rowe, R.W. 1980. Future analytical requirements in the fruit industry. In Atkinson, D., Jackson, J.E., Sharples R.O. and Waller, W.M (Editors), *Mineral Nutrition of Fruit Trees. Butterworths, London,* 399–406.

Roy, S., Anantheswaran, R.C. and Beelman, R.B. 1995. Fresh mushroom quality as affected by modified atmosphere packaging. *Journal of Food Science,* **60**, 334–340.

Rukavishnikov, A.M., Strel'tsov, B.N., Stakhovskii, A.M. and Vainshtein, I.I. 1984. Commercial fruit and vegetable storage under polymer covers with gas selective membranes. *Khimiya v Sel'skom Khozyaistve* **22**, 26–28.

Rutherford, P.P. and Whittle, R. 1982. The carbohydrate composition of onions during long term cold storage. *Journal of Horticultural Science* **57**, 349–356.

Rutledge, P. 1991. Preparation procedures. In Arthey, D. and Dennis, C. (Editors), *Vegetable Processing* Blackie, Glasgow 42–68.

Ryall, A.L. 1963. *Proceedings of the Seventeenth National Conference on Handling Perishables, Perdue, USA, 11–14 March.*

Ryall, A.L. and Harvey, J.M. 1959. The cold storage of vinifera grapes. *United States Department of Agriculture Handbook* 159.

Ryall, A.L. and Lipton, W.J. 1972. *Handling, Transportation and Storage of Fruits and Vegetables,* Vol. 1. AVI Publishing, Westport CT, 473 pp.

Ryall, A.L. and Pentzer, W.T. 1974. *Handling Transportation and Storage of Fruits and Vegetables* Vol. 2. AVI Publishing, Westport CT, 545 pp.

Rylski, I., Rappaport, L. and Pratt, H.K. 1974. Duel effects of ethylene on potato dormancy and sprout growth. *Plant Physiology* **53**, 658.

Sacalis, J.N. 1978. Ethylene evolution by petioles of sleeved poinsettia plants. *HortScience* **13**, 594–596.

Safley, C.D. 1985. Harvesting, sorting and packaging expenses. *Proceedings of the Nineteenth Open House Southeastern Blueberry Council, North Carolina State University, Raleigh*, 89.

Saijo, R. 1990. Post harvest quality maintenance of vegetables. *Tropical Agriculture Research Series* **23**, 257–269.

Salazar, C.R. and Torres, M.R. 1977. Storage of *Passiflora edulis* var. *flavicarpa* fruits in polyethylene bags. Almacenamiento de frutos de maracuya *Passiflora edulis* var. *flavicarpa*, Degener en bolsas de polietileno. *Revista Instituto Colombiano Agropecuario* **12**, 1–11.

Salih, H.J.M. and Ruhni, C.M. 1993. Design and preformance of a prototype motorised mango harvester. *Proceedings of ICAMPE '93, October 19–22, KOEX*, Korean Society for Agricultural Machinery, Seoul, Korea, 744–751.

Salih, O.M. and Thompson, A.K. 1975. Storage of oranges in the Sudan. *Sudan Journal Food Science Technology* **7**, 41–46.

Salleh, M.M., Zulkifly, S. and Osman, M. 1993. Fruit fly problem and disinfestation research in Malaysia. *Postharvest Handling of Tropical Fruit. Proceedings of an International Conference held in Chiang Mai, Thailand, 19–23 July 1993. Australian Centre for International Agricultural Research Proceedings* **50**, 379.

Saltveit, M.E. 1989. A summary of requirements and recommendations for the controlled and modified atmosphere storage of harvested vegetables. *International Controlled Atmosphere Conference, Fifth Proceedings, June 14–16, 1989, Wenatchee, Washington. Volume 2. Other Commodities and Storage Recommendations*, 329–352.

Salunkhe, D.K. and Desai, B.B. 1984. *Postharvest Biotechnology of Fruits*, Vol. II. CRC Press, Boca Raton, FL, 147 pp.

Salunkhe, D.K. and Wu, M.T. 1973. Effects of subatmospheric pressure storage on ripening and associated chemical changes of certain deciduous fruits. *Journal of the American Society for Horticultural Science* **98**, 113–116.

Salunkhe, D.K. and Wu, M.T. 1974. Subatmospheric storage of fruits and vegetables. *Lebensmittel Wissenschaft und Technologie* **7**, 261–267.

Salunkhe, D.K. and Wu, M.T. 1975. Subatmospheric storage of fruits and vegetables. In Haard, N.F. and Salunkhe, D.K. (Editors), *Postharvest Biology and Handling of Fruits and Vegetables*. AVI Publishing, Westport, CT, 153–171.

Salunkhe, D.K., Kadam, S.S. and Jadhav, S.J. 1991. *Potato Production, Processing and Products*. CRC Press, Boca Rotan, FL, 292 pp.

Samisch, R.M. 1937. Observations on the effect of gas storage upon Valencia oranges. *Proceedings of the American Society for Horticultural Science* **34**, 103–106.

Sampaio, V.R., Barbin, D. and Guirado, N. 1983. *Solo* **75**, 31–35.

Samsoondar, J., Maharaj, V. and Sankat, C.K. 2000. Inhibition of browning of the fresh breadfruit through shrink wrapping. *Acta Horticulturae* **518**, 131–136.

Sanchez Nieva, F., Hernadez, I. and Bueso de Viñas, G. 1970. Studies in the ripening of plantains under controlled conditions. *Journal of the University of Puerto Rico* **54**, 517.

Sankat, C.K. and Balkissoon, F. 1990. Refrigerated storage of the carambola. *American Society of Agricultural Engineers Paper* 90–6513, 15 pp.

Sankat, C.K. and Maharaj, R. 1989. Controlled atmosphere storage of papayas. *International Controlled Atmosphere Conference, Fifth, Proceedings, June 14–16, 1989, Wenatchee, Washington. Volume 2. Other Commodities and Storage Recommendations*, 161–170.

Sankat, C.K., Basanta, A. and Maharaj, V. 2000. Light mediated red colour degradation of the pomerac *Syzygium malaccense* in refrigerated storage. *Postharvest Biology and Technology* **18**, 253–257.

Santos, A.R.L. dos, Reinhardt, D.H., Silveira, W.R., Oliveira, J.R.P. and Caldas, R.C. 1999. Postharvest quality of acerola for processing as influenced by ripening stage and storage environment. Qualidade pos colheita de acerola para processamento, em funcao de estadios de maturacao e condicoes de armazenamento. *Revista Brasileira de Fruticultura* **21**, 365–371.

Santos, R.H.S., da Silva, F., Casali, V.W.D. and Conde, A.R. 2001. Postharvest storage of lettuce cultivated with organic compost. Conservacao pos-colheita de alface cultivada com composto organico. *Pesquisa Agropecuaria Brasileira* **36**, 521–525.

Santos, R.R. dos and Nagai, H. 1993. What is naranjilla? O que e naranjilla? *Agronomico* **45**, 11–13.

Sanz, C., Perez, A.G., Olias, R. and Olias, J.M. 1999. Quality of strawberries packed with perforated polypropylene. *Journal of Food Science* **64**, 748–752.

Saray, T. 1988. Storage studies with Hungarian paprika *Capsicum annuum* L. var. *annuum* and cauliflower *Brassica cretica* convar. *botrytis* used for preservation. *Acta Horticulturae* **220**, 503–509.

Sarkar, T.K., Pradhan, U. and Chattopadhyay, T.K. 1993. Storability and quality changes of capegooseberry fruit as influenced by packaging and stage of maturity. *Annals of Agricultural Research* **14**, 396–399.

Sarker, S.K. and Phan, C.T. 1979. Naturally occurring and ethylene induced compounds in the carrot root. *Journal of Food Production* **42**, 526–534.

Satyan, S., Scott, K.J. and Graham, D. 1992. Storage of banana bunches in sealed polyethylene tubes. *Journal of Horticultural Science* **67**, 283–287.

Saucedo Veloz, C., Aceves Vega, E. and Mena Nevarez, G. 1991. Prolonging the duration of cold storage and marketing of Hass avocado fruits by treatment with high concentrations of CO_2. Prolongacion del tiempo de frigoconservacion y comercializacion de frutos de aguacate 'Hass' mediante tratamientos con altas concentraciones de CO_2. *III Symposium on Management, Quality and Postharvest Physiology of Fruits, Vina del Mar, Chile, 7 12 Oct, 1991*, Proceedings of the Interamerican Society for Tropical Horticulture **35**, 297–303.

Sauco, V.G. and Menini, U.G. 1989. Litchi cultivation. *Food and Agriculture Organization of the United Nations Plant Production and Protection Paper* 83, 136 pp.

Sayed, A.M., Nakano, K. and Maezawa, S. 2000. Measuring system of the respiration rate of fresh produce under continuous changes in CO_2 or O_2 concentration. *Research Bulletin of the Faculty of Agriculture, Gifu University* **65**, 49–57.

Schaik, A. van. 1994. CA storage of Elstar. Elstar can be stored longer with retention of quality. CA bewaring van Elstar. Elstar is langer te bewaren met behoud van kwaliteit. Van. *Fruitteelt Den Haag* **84**, 10–11.

Schaik, A. van. 1994a. Influence of a combined scrubber/separator on fruit quality: percentage of storage disorders ap67 to be reduced slightly. *Fruitteelt Den Haag* **84**, 14–15.

Schaik, A. van and Bevers, N. 1994. O_2 content for storage of Conference cannot be lowered further. Possibilities for storage of Jonathan in 0.9% O_2. *Fruitteelt Den Haag* **84**, 18–19.

Schallenberger, R.S., Smith, O. and Treadaway, R.H. 1959. Role of sugars in the browning reaction in potato chips. *Journal of Agricultural and Food Chemistry* **7**, 274.

Scheer, A., Molema, G.J. and Lokhorst, C. 2000. New storing and handling methods improve produce quality in the potato handling chain. *Proceedings of the Fourth International Conference, Wageningen, 25–26 May 2000*, 637–644.

Schirra, M., Barbera, G., D'Aquino, S., la Mantia, T. and McDonald, R.E. 1996. Hot dips and high-temperature conditioning to improve shelf quality of late-crop cactus pear fruit. *Tropical Science* **36**, 159–165.

Schlimme, D.V. and Rooney, M.L. 1994. Packaging of minimally processed fruits and vegetables. In Waley, R.C (Editor), *Minimally Processed Refrigerated Fruits and Vegetables*. Chapman and Hall, New York, 135–182.

Schmitz, S.M. 1991. *Investigation on alternative methods of sprout suppression in temperate potato stores*. MSc thesis, Silsoe College, Cranfield Institute of Technology.

Scholz, E.W., Johnson, H.B. and Buford, W.R. 1963. Heat evolution rates of some Texas grown fruits and vegetables. *Journal of the Rio Grande Valley Horticultural Society* **17**, 170–175.

Schomer, H.A. and Sainsbury, G.F. 1957. Controlled atmosphere storage of Starking Delicious apples in the Pacific Northwest. *US Department of Agriculture, Agricultural Marketing Service* 178, March.

Schouten, S.P. 1985. New light on the storage of Chinese cabbage. Nieuw licht op het bewaren van Chinese kool. *Groenten en Fruit* **40**, 60–61.

Schouten, S.P. 1992. Possibilities for controlled atmosphere storage of ware potatoes. *Aspects of Applied Biology* **33**, 181–188.

Schouten, S.P. 1994. Increased CO_2 concentration in the store is disadvantageous for the quality of culinary potatoes. Erhohter CO_2 Gehalt im Lager nachteilig für die Qualität von Konsumkartoffeln. *Kartoffelbau* **45**, 372–374.

Schouten, S.P. and Stork, H.W. 1977. Ethylene damage in eggplants. Do not keep with tomatoes. *Groenten en Fruit* **32**, 1688–1689.

Schroeder, C.A. 1967. The Jicama, a root crop from Mexico. *Proceedings of the Tropical Region of the American Society for Horticultural Science* **11**, 65.

Schulz, F.A. 1974. The occurrence of apple storage rots under controlled conditions. Uber das Auftreten von Apfellagerfaulen unter kontrollierten Bedingungen. *Zeitschrift für Pflanzenkrankheiten und Pflanzenschutz* **81**, 550–558.

Scott, K.J. and Chaplin, G.R. 1978. Reduction of chilling injury in avocados stored in sealed polyethylene bags. *Tropical Agriculture* **55**, 87–90.

Scott, K.J. and Wills, R.B.H. 1973. Atmospheric pollutants destroyed in an ultra violet scrubber. *Laboratory Practice* **22**, 103–106.

Scott, K.J. and Wills, R.B.H. 1974. Reduction of brown heart in pears by absorption of ethylene from the storage atmosphere. *Australian Journal of Experimental Agriculture and Animal Husbandry* **14**, 266–268.

Scott, K.J., Blake, J.R., Strachan, G., Tugwell, B.L. and McGlasson, W.B. 1971. Transport of bananas at ambient temperatures using polyethylene bags. *Tropical Agriculture Trinidad* **48**, 245–253.

Scott, K.J., Brown, B.T., Chaplin, G.R., Wilcox, M.E., and Bain, J.M. 1982. The control of rotting and browning of litchi fruit by hot benomyl and plastic film. *Scientia Horticulturae* **16**, 253–262.

Scott, K.J., McGlasson, W.B. and Roberts E.A. 1970. Potassium permanganate as an ethylene absorbent in polyethylene bags to delay ripening of bananas during storage. *Australian Journal of Experimental Agriculture and Animal Husbandry* **10**, 237–240.

Scriven, F.M., Gek, C.O. and Wills, R.B.H. 1989. Sensory differences between bananas ripened with and without ethylene. *HortScience* **24**, 983–984.

Scriven, F.M., Ndungguru, G.T. and Wills, R.B.H. 1988. Hot water dips for the control of pathological decay in sweet potatoes. *Scientific Horticulture* **35**, 1–5.

SeaLand 1991. *Shipping Guide to Perishables.* SeaLand Services, Iselim, NJ.

Sealy, T., Hart, H. and Black, C.V. 1984. *Jamaica's Banana Industry.* Jamaica Banana Producers Association, Kingston, Jamaica, 114 pp.

Seberry, J.A., Harris, D.R. and Galan Sauco, V. 1998. Postharvest evaluation of FHIA 01 and other new banana varieties for subtropical Australia. *Acta Horticulturae* **490**, 537–546.

Self, G.K., Ordozgoiti, E., Povey, M.J.W. and Wainwright, H. 1992. Ultrasonic evaluation of avocado fruit ripening. *Postharvest Biology and Technology* **4**, 111–116.

Self, G.K., Wainwright, H. and Povey, M.J.W. 1993. Non destructive quality determination for fruit and vegetables. *Postharvest News and Information* **4**, 18N–19N.

Selvaraj, Y. and Pal D.K. 1989. Biochemical changes during the ripening of jackfruit (*Artocarpus heterophyllus* Lam.). *Journal of Food Science and Technology* **26**, 304–307.

Selvarajah, S., Bauchot, A.D. and John, P. 2001. Internal browning in cold-stored pineapples is suppressed by a postharvest application of 1-methylcyclopropene. *Postharvest Biology and Technology* **23**, 167–170.

Semple, A.J. and Thompson, A.K. 1988. Influence of the ripening environment on the development of finger drop in bananas. *Journal of the Science of Food Agriculture* **46**, 139–146.

Senter, S.D., Bailey, J.S. and Cox, N.A. 1987. Aerobic microflora of commercially harvested, transported and cryogenically processed collards *Brassica oleracea.* *Journal of Food Science* **52**, 1020–1021.

Seymour, G.B. 1985. *The effects of gases and temperature on banana ripening.* PhD thesis, University of Reading, 191 pp.

Seymour, G.B. and McGlasson, W.B. 1993. Melon. In Seymour, G.B., Taylor, J.E. and Tucker, G.A. (Editors), *Biochemistry of Fruit Ripening.* Chapman and Hall, London, 273–290.

Seymour, G.B. and Tucker, G.A. 1993. Avocado. In Seymour, G.B., Taylor, J.E. and Tucker, G.A. (Editors), *Biochemistry of Fruit Ripening.* Chapman and Hall, London, 53–81.

Seymour, G.B., John, P. and Thompson, A.K. 1987. Inhibition of degreening in the peel of bananas ripened at tropical temperatures. 2. Role of ethylene, oxygen and carbon dioxide. *Annals of Applied Biology* **110**, 153–161.

Seymour, G.B., Thompson, A.K. and John, P. 1987a. Inhibition of degreening in the peel of bananas ripened at tropical temperatures. I. Effects of high temperature changes in the pulp and peel during ripening. *Annals of Applied Biology* **110**, 145–151.

Seymour, G.B., Thompson, A.K., Hughes, P.A. and Plumbley, R.A. 1981. The influence of hydrocooling and plastic box liners on the market quality of capsicums. *Acta Horticulturae* **116** 191–196.

Sfakiotakis, E., Antunes, M.D., Stavroulakis, G. and Niklis, N. 1999. Relationship between ethylene production and ripening in cultivar Hayward fruits at harvest and during storage. Rapporti fra produzione di etilene e maturazione dei frutti della cultivar Hayward nelle fasi di raccolta e conservazione. *Rivista di Frutticoltura e di Ortofloricoltura* **61**, 59–65.

Sfakiotakis, E., Niklis, N., Stavroulakis, G. and Vassiliadis, T. 1993. Efficacy of controlled atmosphere and ultra low O_2, low ethylene storage on keeping quality and scald control of 'Starking Delicious' apples. *Acta Horticulturae* **326**, 191–202.

Shan-Tao, Z. and Liang, Y. 1989. The application of carbon molecular sieve generator in CA storage of apple and tomato. *International Controlled Atmosphere Conference, Fifth, Proceedings, June 14–16, 1989, Wenatchee, Washington. Volume 2. Other Commodities and Storage Recommendations*, 241–248.

Sharp, J.L. 1993. Microwaves as a quarantine treatment to disinfest commodities of pests. *Postharvest Handling of Tropical Fruit. Proceedings of an International Conference held in Chiang Mai, Thailand, 19–23 July 1993. Australian Centre for International Agricultural Research Proceedings* **50**, 362–364.

Sharples, R.O. 1967. A note on the effect of N-dimethylaminosuccinamic acid on the maturity and storage quality of apples. *Annual Report of the East Malling Research Station for 1966*, 198–201.

Sharples, R.O. 1980. The influence of orchard nutrition on the storage quality of apples and pears grown in the United Kingdom. In Atkinson, D., Jackson, J.E., Sharples, R.O. and Waller, W.M. (Editors), *Mineral Nutrition of Fruit Trees.* Butterworths, London, 17–28.

Sharples, R.O. 1984. The influence of preharvest conditions on the quality of stored fruit. *Acta Horticulturae* **157**, 93–104.

Sharples, R.O. 1986. Obituary: Cyril West. *Journal of Horticultural Science,* **61**, 555.

Sharples, R.O. 1989. Storage of perishables. In Cox, S.W.R. (Editor), *Engineering Advances for Agriculture and Food.* Institution of Agricultural Engineers Jubilee Conference, 1988, Cambridge. Butterworths, London, 251–260.

Sharples, R.O. and Johnson, D.S. 1987. Influence of agronomic and climatic factors on the response of apple fruit to controlled atmosphere storage. *HortScience* **22**, 763.

Sharples, R.O. and Stow, J.R. 1986. Recommended conditions for the storage of apples and pears. *Report of the East Malling Research Station for 1985*, 165–170.

Sharples, R.O., Reid, M.S. and Turner, N.A. 1979. The effects of postharvest mineral element and lecithin treatments on the storage disorders of apple. *Journal of Horticultural Science* **54**, 299–304.

Sharples, R.O, Kidd, F. and West, C. 1989. In Janick, J. (Editor), *Classical Papers in Horticultural Science.* Prentice Hall, Englewood Cliffs, NJ, 213–219.

Shear, C.B. and Faust, M. 1980. In Janick, J. (Editor), *Nutritional Ranges in Deciduous Tree Fruits and Nuts. Horticultural Reviews 2.* AVI Publishing, Westport, CT, 142–163.

Shearer, A.E.H., Strapp, C.M. and Joerger, R.D. 2001. Evaluation of a polymerase chain reaction based system for detection of *Salmonella enteritidis, Escherichia coli* O157:H7, *Listeria* spp., and *Listeria monocytogenes* on fresh fruits and vegetables. *Journal of Food Protection* **64**, 788–795.

Shearer, S.A. and Payne, F.A. 1990. Machine vision sorting of peppers. *Proceedings of the 1990 Conference on Food Processing Automation, Lexington, Kentucky,* 289–300.

Sherafatian, D. 1994. The effect of harvesting date and temperature during storage on keeping quality of pomegranate. *Seed and Plant* **10**, 25–31.

Sherman, M. and Ewing, E.E. 1983. Effects of temperature and low oxygen atmospheres on respiration, chip colour, sugars and malate of stored potatoes. *Journal of the American Society of Horticultural Science* **108**, 129.

Shewfelt, R.L. 1986. Flavor and color of fruits as affected by processing. In Woodroof, J.G. and Luh, B.S. (Editors), *Commercial Fruit Processing.* AVI Publishing, Westport, CT.

Shewfelt, R.L. and Prussia, S.E. 1993. *Postharvest Handling.* Academic Press, San Diego, 356 pp.

Shibairo, S.I., Upadhyaya, M.K. and Toivonen, P.M.A. 1998. Influence of preharvest water stress on postharvest moisture loss of carrots (*Daucus carota* L.). *Journal of Horticultural Science and Biotechnology* **73**, 347–352.

Shipway, M.R. 1968. The refrigerated storage of vegetables and fruits. *Ministry of Agriculture Fisheries and Food UK* 324, 148 pp.

Shorter, A.J,. Scott, K.J.and Graham, D. 1987. Controlled atmosphere storage of banana in bunches at ambient temperatures. *CSIRO Food Research Quarterly* **47**, 61–63.

Silva, A.P. da and Vieites, R.L. 2000. Effect of post harvest treatment with calcium chloride applied by infiltration on the physical characteristics of sweet passion fruit. Tratamento pos colheita com cloreto de calcio aplicado por infiltracao, nas caracteristicas fisicas do maracuja doce. *Revista Brasileira de Fruticultura* **22**, Special Issue, 73–76.

Silva, A.P. da, Domingues, M.C.S., Vieites, R.L. and Rodrigues, J.D. 1999a. Role of plant growth regulators in the post harvest conservation of sweet passion fruit *Passiflora alata* Dryander stored under refrigeration. Fitorreguladores na conservacao pos colheita do maracuja doce *Passiflora alata* Dryander armazenado sob refrigeracao. *Ciencia e Agrotecnologia* **23**, 643–649.

Silva, A.P. da, Vieites, R.L. and Cereda, E. 1999. Conservation of sweet passion fruit using wax and cold shock. Conservacao de maracuja doce pelo uso de cera e choque a frio. *Scientia Agricola* **56**, 797–802.

Silvis, H. and Thompson, A.K. 1974. Onion storage in the Sudan. *Food and Agricultural Organization of the United Nations, Report Sud 70/543,* 13, 6 pp.

Silvis, H. and Thompson, A.K. 1974a. Preliminary investigations on improving the quality of bananas for the local market in the Sudan. *Sudan Journal of Food Science and Technology* **6**, 7–17.

Silvis, H., Thompson, A.K., Musa, Sulafa K., Salih, O.M. and Abdulla, Y.M. 1976. Reduction of wastage during postharvest handling in the Sudan. *Tropical Agriculture Trinidad* **53**, 89–94.

Simmonds, J.H. 1941. Latent infection in tropical fruit discussed in relation to the part played by species of *Gloeosporium* and *Colletotrichum*. *Proceedings of the Royal Society of Queensland* **52**, 92–120.

Simmonds, N.W. 1966. *Bananas*, 2nd edn. Longmans, London, 468 pp.

Sinclair, J. 1988. *Refrigerated Transportation.* Container Marketing Limited, London, 137 pp.

Sinco, 1985. Envirotainers: perishables shipping has come a long way. *The Packer June 29,* **18B**.

Singh, A.K., Kashyap, M.M., Gupta, A.K. and Bhumbla, V.K. 1993. Vitamin C during controlled atmosphere storage of tomatoes. *Journal of Research, Punjab Agricultural University* **30**, 199–203.

Singh, J.P. and Singh, S.P. 1999. Effect of pre-harvest spray of calcium nitrate on shelf-life of guava (*Psidium guajava* L.) fruits cv. Allahabad Safeda. *Journal of Applied Biology* **9**, 149–152.

Singh, J.V., Kumar, A. and Singh, C. 1998. Studies on the storage of onion (*Allium cepa* L.) as affected by different levels of phosphorus. *Indian Journal of Agricultural Research* **32**, 51–56.

Singh, K.K. and Mathur, P.B. 1952. Ripening of mangoes. *Current Science* **21**, 295.

Singh, K.K. and Mathur, P.B. 1953. Studies in cold storage of cashew apples. *Indian Journal of Horticulture* **10**, 115–121.

Singh, K.K. and Mathur, P.B. 1953a. Cold storage of tapioca root. *Bulletin of the Central Food Technology Research Institute, Mysore, India* **2**, 181.

Singh, K.K. and Mathur, P.B. 1954. A note on the cold storage of sapotas. *Indian Journal of Agricultural Science* **24**, 149–150.

Singh, K.K. and Mathur, P.B. 1954a. Cold storage of guavas. *Indian Journal of Horticulture* **11**, 1–5.

Singh, Y.P. and Sumbali, G. 2000. Ascorbic acid status and aflatoxin production in ripe fruits of jujube infected with *Aspergillus flavus. Indian Phytopathology* **53**, 38–41.

Siriphan Sophonsathian 1982. Effect of temperature and packing methods on the storage life of Chinese kale *Brassica alboglagra* Bailey. Phon khong unhaphum lae withi banchu hipho to kan kep raksa khana chin. *Faculty of Agriculture. Department of Horticulture, Kasetsart University, Bangkok, Thailand,* 11 pp.

Sirisomboon, P., Tanaka, M., Akinaga, T. and Kojima, T. 2000. Evaluation of the textural properties of Japanese pears. *Journal of Textural Studies* **31**, 665–677.

Sistrunk, W.A., Gonzales, A.R. and Moore, K.J. 1989. In Eskin, N.A.M. (Editor), *Quality and Preservation of Vegetables.* CRC Press Boca Raton, FL, 185.

Sites, J.W. and Reitz, H.J. 1949. The variation of individual 'Valencia' oranges from different locations on the tree as a guide to sampling methods and spot picking for quality. I. Soluble solids in the juice. *Proceedings of the American Society for Horticultural Science* **54**, 1.

Sites, J.W. and Reitz, H.J. 1950a. The variation of individual 'Valencia' oranges from different locations on the tree as a guide to sampling methods and spot picking for quality. II. Titratable acidity and the soluble solids titratable acidity ration of the juice. *Proceedings of the American Society for Horticultural Science* **55**, 73.

Sites, J.W. and Reitz, H.J. 1950b. The variation of individual 'Valencia' oranges from different locations on the tree as a guide to sampling methods and spot picking for quality. III. Vitamin C and juice content of the fruit. *Proceedings of the American Society for Horticultural Science* **56**, 103.

Sivakumar, D., Wijeratnam, R.S.W., Wijesundera, R.L.C., Marikar, F.M.T. and Abeyesekere, M. 2000. Antagonistic effect of *Trichoderma harzianum* on postharvest pathogens of rambutan (*Nephelium lappaceum*). *Phytoparasitica* **28**, 240–247.

Sive, A. and Resnizky, D. 1985. Experiments on the CA storage of a number of mango cultivars in 1984. *Alon Hanotea* **39**, 845–855.

Sivakumar, D., Wijeratnam, R.S.W., Wijesundera, R.L.C. and Abeyesekere, M. 2002. Combined effect of Generally Regarded As Safe (GRAS) compounds and *Trichoderma harzianum* on the control of postharvest diseases of rambutan. *Phytoparasitica* **30**, 43–51.

Sive, A. and Resnizky, D. 1989. Thermal fogging with DPA and ethoxyquin. *International Controlled Atmosphere Conference, Fifth, Proceedings, June 14–16, 1989, Wenatchee, Washington. Volume 1. Pome Fruits,* 465–468.

Skrzynski, J. 1990. Black currant fruit storability in controlled atmospheres. I. Vitamin C content and control of mould development. *Folia Horticulturae* **2**, 115–124.

Skultab, K. 1987. *Design, construction and testing of a night ventilated store for onions in the tropics.* MSc thesis, Silsoe College, Cranfield Institute of Technology.

Skultab, K. and Thompson, A.K. 1992. Design of a night ventilated onion store for the tropics. *Agricultural Mechanization in Asia, Africa and Latin America* **23**, 51–55.

Smith, F.E.V. 1936. *Jamaica Department of Science and Agriculture, Annual Report* 70.

Smith, G. 1991. Effects of ethephon on ripening and quality of fresh market pineapples. *Australian Journal of Experimental Agriculture* **31**, 123–127.

Smith, G. and Glennie, J.D. 1987. Blackheart development in growing pineapples. *Tropical Agriculture Trinidad* **64**, 7–11.

Smith, N.J.S. 1989. *Textural and biochemical changes during ripening of bananas.* PhD thesis, University of Nottingham, 131 pp.

Smith, N.J.S. and Seymour, G.B. 1990. Cell wall changes in bananas and plantains. *Acta Horticulturae* **269**, 283–289.

Smith, N.J.S. and Thompson, A.K. 1987. The effects of temperature, concentration and exposure time to acetylene on initiation of banana ripening. *Journal of the Science of Food Agriculture* **40**, 43–50.

Smith, N.J.S., Seymour, G.B. and Thompson, A.K. 1986. Effects of high temperatures on ripening responses of bananas to acetylene. *Annals of Applied Biology* **108**, 667–672.

Smith, R.B. 1992. Controlled atmosphere storage of 'Redcoat' strawberry fruit. *Journal of the American Society for Horticultural Science* **117**, 260–264.

Smith, R.B. and Reyes, A.A. 1988. Controlled atmosphere storage of Ontario grown celery. *Journal of the American Society for Horticultural Science* **113**, 390–394.

Smith, R.B. and Skog, L.J. 1992. Postharvest CO_2 treatment enhances firmness of several cultivars of strawberry. *HortScience* **27**, 420–421.

Smith, R.B., Skog, L.J., Maas, J.L. and Galletta, G.J. 1993. Enhancement and loss of firmness in strawberries stored in atmospheres enriched with CO_2. *Acta Horticulturae* **348**, 328–333.

Smith, W.H. 1952. The commercial storage of vegetables. *Department of Scientific and industrial Research London UK, Food Investigation Leaflet* 15, 8 pp.

Smith, W.H. 1957. Storage of black currants. *Nature* **179**, 876.

Smith, W.H. 1957a. The production of carbon dioxide and metabolic heat by horticultural produce. *Modern Refrigeration* **60**, 493–496.

Smith, W.H. 1964. The storage of mushrooms. *Agricultural Research Council, Ditton and Covent Garden Laboratories Annual Report for 1963–64*, 18.

Smith, W.H. 1967. The storage of gooseberries. *Agricultural Research Council, Ditton and Covent Garden Laboratories Annual Report for 1965–66*, 13–14.

Smittle, D.A. 1988. Evaluation of storage methods for 'Granex' onions. *Journal of the American Society for Horticultural Science* **113**, 877–880.

Smittle, D.A. 1989. Controlled atmosphere storage of Vidalia onions. *International Controlled Atmosphere Conference, Fifth, Proceedings, June 14–16, 1989, Wenatchee, Washington. Volume 2. Other Commodities and Storage Recommendations*, 171–177.

Smock, R.M. 1938. The possibilities of gas storage in the United States. *Refrigeration Engineering* **36**, 366–368.

Smock, R.M. 1956. Marketing controlled atmosphere apples. *Cornell University, Department of Agricultural Economics, A.E.* 1028, 4–7.

Smock, R.M. and Gross, C.R. 1950. Studies on respiration of apples. *New York Cornell Agricultural Experimental Station Memorandum* 197, 47 pp.

Smock, R.M. and Van Doren, A. 1938. Preliminary studies on the gas storage of McIntosh and Northwestern Greening. *Ice and Refrigeration* **95**, 127–128.

Smock, R.M. and Van Doren, A. 1939. Studies with modified atmosphere storage of apples. *Refrigerating Engineering* **38**, 163–166.

Smock, R.M., Mendoza, D.B., Jr, and Abilay, R.M. 1967. Handling bananas. *Philippines Farms and Gardens* **4**, 12–17.

Smoot, J.J. 1969. Decay of Florida citrus fruits stored in controlled atmospheres and in air. *Proceedings of the First International Citrus Symposium* **3**, 1285–1293.

Smoot, J.J. and Segall, R.H. 1963. Hot water as a postharvest control of mango anthracnose. *Plant Disease Reporter* **47**, 739–742.

Snowdon, A.L. 1990. *A Colour Atlas of Postharvest Diseases and Disorders of Fruits and Vegetables. Volume 1. General Introduction and Fruits.* Wolfe Scientific, 302 pp.

Snowdon, A.L. 1992. *A Colour Atlas of Postharvest Diseases and Disorders of Fruits and Vegetables. Volume 2. Vegetables.* Wolfe Scientific, 416 pp.

Sohonen, I. 1969. On the storage of cabbage in refrigerated stores. *Acta Agriculturae Scandanavica* **19**, 18.

Solberg, S.O. and Dragland, S. 1998. Effects of harvesting and drying methods on internal atmosphere, outer scale appearance, and storage of bulb onions *Allium cepa* L. *Journal of Vegetable Crop Production* **4**, 23–35.

Solomos, T. and Biale, J.B. 1975. Respiration in fruit ripening. *Colloque Internationaux du Center National de la Recherche Scientifique* **238**, 221–228.

Son, Y.K., Yoon, I.W., Han, P.J and Chung, D.S. 1983. Studies on storage of pears in sealed polyethylene bags. *Research Reports, Office of Rural Development, South Korea, Soil Fertilizer, Crop Protection, Mycology and Farm Products Utilization* **25**, 182–187.

Sornsrivichai, J., Uthaibutra, J. and Yantarasri, T. 2000. Controlling of peel and flesh color development of mango by perforation of modified atmosphere package at different temperatures. *Acta Horticulturae* **509**, 387–394.

Soucedo Veloz, C., Aceves Vega, E. and Mene Nevarez, G. 1991. Prolongacion del tiempo de frigoconservacion y comercialacion de frutos de aguacate 'Hass' mediante tratamientos con altas concentraciones de CO_2. *Proceedings of the Interamerican Society for Tropical Horticulture* **35**, 297–303.

Spalding, D.H. 1980. Low pressure hypobaric storage of several fruits and vegetables. *Proceedings of the Florida State Horticultural Society* **92**, 201–203.

Spalding, D.H. 1982. Resistance of mango pathogens to fungicides used to control postharvest diseases. *Plant Disease* **66**, 1185–1186.

Spalding, D.H. and Reeder, W.F. 1972. Postharvest disorders of mangoes as affected by fungicides and heat treatments. *Plant Disease Reporter* **56**, 751–753.

Spalding, D.H. and Reeder, W.F. 1974. Quality of 'Tahiti' limes stored in a controlled atmosphere or under low pressure. *Proceedings of the Tropical Region, American Society for Horticultural Science* **18**, 128–135.

Spalding, D.H. and Reeder, W.F. 1975. Low oxygen, high carbon dioxide controlled atmosphere storage for the control of anthracnose and chilling injury of avocados. *Phytopathology* **65**, 458–460.

Spalding, D.H. and Reeder, W.F. 1976. Low pressure hypobaric storage of limes. *Journal of the American Society for Horticultural Science* **101**, 367–370.

Sparrow, A.H. and Christensen, E. 1954. Improved storage quality of potato tubers after exposure to Co60 gammas. *Nucleonics* **12**, 16–17.

Spencer, D. 1957. Proposed legislation to regulate modified atmosphere storage. *Proceedings of the 102nd Meeting of the New York State Horticultural Society*,.218–219.

Splitter, J.L. and Shipe, W.F. 1972. Changes in quality during maturation of selected sweet corn cultivars. *Journal of Food Science* **37**, 257.

Srikul, S. and Turner, D.W. 1995. High N supply and soil water deficits change the rate of fruit growth of bananas (cv. Williams) and promote tendency to ripen. *Scientia Horticulturae* **62**, 165–174.

Srivastava, V.K. and Sachan, S.C.P. 1969. Grow ash gourd the efficient way. *Indian Horticulture* **14**, 13–15.

Sriyook, S., Siriatiwat, S. and Siriphanich, J. 1994. Durian fruit dehiscence water status and ethylene. *HortScience* **29**, 1195–1198.

Ssemwanga, J.R.K. 1990. *Effects of storage conditions, fruit coatings and packaging on the marketable life of passionfruit.* MSc thesis, Cranfield Insitute of Technology.

Staby, G. L. 1976. Hypobaric storage an overview. *Combined Proceedings of the International Plant Propagation Society* **26**, 211–215.

Stahl, A.L. and Cain, J.C. 1937. Cold storage studies of Florida citrus fruit. III. The relation of storage atmosphere to the keeping quality of citrus fruit in cold storage. *Florida Agricultural Experiment Station, Bulletin* 316, October, 44 pp.

Stahl, A.L. and Cain, J.C. 1940. Storage and preservation of micellaneous fruits and vegetables. *Florida Agricultural Experiment Station, Annual Report* 88.

Stahmann, M.A., Clare, B.G. and Woodbury, W. 1966. Increased disease resistance and enzyme activity induced by ethylene and ethylene formation by black rot infected sweet potato tissue. *Plant Physiology* **41**, 1505.

Stamford, D.C., Burden, O.J., Thompson, A.K. and Been, B.O. 1971. Plan of a trial for investigation into the effect of transport conditions within Jamaica on damage sustained by bananas transported both as stems and in fibreboard boxes. *Jamaican Banana Board Report*, 13 pp.

Stange R.R., Jr, Midland, S.L., Holmes, G.J., Sims, J.J. and Mayer, R.T. 2001. Constituents from the periderm and outer cortex of *Ipomoea batatas* with antifungal activity against *Rhizopus stolonifer*. *Postharvest Biology and Technology* **23**, 85–92.

Steiner, I., Werner, D. and Washuttl, J. 1999. Patulin in fruit juices I. Analysis and content in Austrian apple and pear juices. Patulin in Obstsaften. I. Analytik und Gehalte in osterreichischen Apfel und Birnensaften. *Ernahrung* **23**, 202–208.

Stenning, B.C. and Thompson, A.K. 1991. Cold storage of commodities and expectations for the future. *Gida Teknolojisi Arastirma Enstitusu, Bursa, Turkey* 28–34.

Stenvers, N. 1977. Hypobaric storage of horticultural products. Hypobarische bewaring van tuinbouwprodukten. *Bedrijfsontwikkeling* **8**, 175–177.

Sterry, P. 1995. *Country Guides to Fungi of Britain and Northern Europe.* Chancellor Press, London, 160 pp.

Stevens, C., Khan, V.A., Lu, J.Y., Wilson, C.L., Pusey, P.L., Igwegbe, E.C.K., Kabwe, K., Mafolo, Y., Liu, J., Chalutz, E. and Droby, S. 1997. Integration of ultraviolet UV C light with yeast treatment for control of postharvest storage rots of fruits and vegetables. *Biological Control* **10**, 98–103.

Stewart, J.K. and Uota, M. 1971. CO_2 injury and market quality of lettuce held under controlled atmosphere. *Journal of the American Society for Horticultural Science* **96**, 27–31.

Stoll, K. 1974. Storage of vegetables in modified atmospheres. *Acta Horticulturae* 13–23

Stoll, K. 1976. Storage of the pear cultivar Louise Bonne. Lagerung der Birnensorte 'Gute Luise.' *Schweizerische Zeitschrift für Obst und Weinbau* **112**, 304–309.

Stover, R.H. 1972. *Banana, Plantain and Abacca Diseases.* Commonwealth Agricultural Bureaux, UK.

Stover, R.H. and Simmonds, N.W. 1987. *Bananas,* 3rd edn. Longman, London.

Stow, J.R. 1989. Low ethylene, low O_2 CA storage of apples. *International Controlled Atmosphere Conference, Fifth, Proceedings, June 14–16, 1989, Wenatchee, Washington. Volume 1. Pome Fruits,* 325–332.

Stow, J.R. 1989a. Effects of O_2 concentration on ethylene synthesis and action in stored apple fruits. *Acta Horticulturae* **258**, 97–106.

Stow, J.R. 1996. Gala breaks through the storage barrier. *Grower* **126**, 26–27.

Stow, J.R. 1996a. The effects of storage atmosphere on the keeping quality of 'Idared' apples. *Journal of Horticultural Science* **70**, 587–595.

Stow, J.R. and Genge, P. 1993. The measurement of cell wall strength in stored apple fruits. *Sixth Annual Controlled Atmosphere Research Conference, Cornell University, Ithica, New York.*

Strehler, B.L. and Arnold, W.A. 1951. Light production by green plants. *Journal of General Physiology* **34**, 809.

Streif, J. 1989. Storage behaviour of plum fruits. *Acta Horticulturae* **258**, 177–184.

Strempfl, E., Mader, S. and Rumpolt, J. 1991. Trials of storage suitability of important apple cultivars in a controlled atmosphere. *Mitteilungen Klosterneuburg, Rebe und Wein, Obstbau und Fruchteverwertung* **41**, 20–26.

Strop, I. 1992. *Effects of plastic film wraps on the marketable life of asparagus and broccoli.* MSc thesis, Silsoe College, Cranfield Institute of Technology.

Sturgess, I.M. 1987. *The Wholesaling of Fruit and Vegetables.* Covent Garden Marketing Authority, London.

Suarez, M. and Duque, C. 1992. Change in volatile compounds during lulo Solanum vestissimum D. fruit maturation. *Journal of Agricultural and Food Chemistry* **40**, 647–649.

Subra, P. 1971. The cableway method of conveying bunches in a banana plantation. *Fruits Paris* **26**, 807–817.

Sugiyama, H.M. and Yang, K.H. 1975. Growth potential of *Clostridium botulinum* in fresh mushrooms packed in semi-permeable plastic film. *Applied Microbiology* **30**, 964–969.

Suhaila, M. 1990. Extending the shelf life of fresh durian *Durio zibethinus. Association of Southeast Asian Nations Food Journal* **2**, 78–80.

Suhaila, M., Khin, M.M.K., Idris, A.Z., Salmah, Y. Azizah, O. and Subhadrabandhu, S. 1992. Effects of various surface treatments (palm oil, liquid paraffin, Semperfresh or starch surface coatings and LDPE wrappings) on the storage life of guava (*Psidium guajava* L.) at 10°C. *Acta Horticulturae* **321**, 786–794.

Suhonen, I. 1969. On the storage life of white cabbage in refrigerated stores. *Acta Agricultural Scandanavica* **19**, 18.

Swarts, D.H. 1980. A method for determining the ripeness of avocados. *Information Bulletin, Citrus and Subtropical Fruit Research Institute* 90, 15–18.

Swarts, D.H. 1991. Effects of different shipping temperatures on cv. Queen pineapples. *Inligtingsbulletin Navorsingsinstituut vir Sitrus en Subtropiense Vrugte* **225**, 12–16.

Swinburne, T.R. 1976. Stimulants of germination and appressoria formation by *Colletotrichum musae* Berk. Curt. Arx. in bananas. *Phytopathologische Zeitschrift* Z. **87**, 74–90.

Tamas, S. 1992. Cold storage of watermelons in a controlled atmosphere. A gorogdinnye hutotarolasanak lehetosegei szabalyozott legterben. *Elelmezesi Ipar* **46**, 234–239, 242.

Tamim, M., Goldschmidt, E.E., Goren, R. and Shachnai, A. 2000. Potassium reduces the incidence of superficial rind pitting (nuxan) on 'Shamouti' orange. *Alon Hanotea* **54**, 152–157

Tan, S.C. and Mohamed, A.A. 1990. The effect of CO_2 on phenolic compounds during the storage of 'Mas' banana in polybag. *Acta Horticulturae* **269**, 389.

Tan, S.C., Lam, P.F. and Abdullah, H. 1986. Changes of the pectic substances in the ripening of bananas *Musa sapientum,* cultivar Emas after storage in polyethylene bags. *Association of Southeast Asian Nations Food Journal* **2**, 76–77.

Tan, X.P., Ueda, Y., Imahori, Y. and Chachin, K. 1999. Changes in pigments in pulp and aril of balsam pear *Momordica charantia* L. fruit during development and storage. *Journal of the Japanese Society for Horticultural Science* **68**, 683–688.

Tao, Y., Morrow, C.T. and Heinemann, P.H. 1990. Automated machine vision inspection of potatoes. *American Society of Agricultural Engineers Paper* 90–3531.

Tariq, M.A. 1999 *Effect of curing parameters on the quality and storage life of damaged citrus fruits.* PhD thesis, Cranfield University.

Tchango, J.T., Achard, R. and Ngalani, J.A. 1999. Study of harvesting stages for exportation to Europe, by ship, of three plantain cultivars grown in Cameroon. Etude des stades de recolte pour l'exportation par bateau, vers l'Europe, de trois cultivars de plantains produits au Cameroun. *Fruits Paris* **54**, 215–224.

Teixido, N., Usall, J. and Vinas, I. 1999. Efficacy of preharvest and postharvest *Candida sake* biocontrol treatments to prevent blue mould on apples during cold storage. *International Journal of Food Microbiology* **50**, 203–210.

Tellez, C.P., Fischer, G. and Quintero, C.O. 1999. Post harvest physiological, physical and chemical changes in the banana passion fruit *Passiflora mollissima* Bailey stored under refrigeration and ambient temperature. Comportamiento fisiologico y fisico quimico en la postcosecha de Curuba de Castilla *Passiflora mollissima* Bailey conservada en refrigeracion y temperatura ambiente. *Agronomia Colombiana* **16**, 13–18.

Terry, LA. and Joyce, D.C. 2000. Suppression of grey mould on strawberry fruit with the chemical plant activator acibenzolar. *Pest Management Science* **56**, 989–992.

Terwongworkule, A. 1995. *Grading apples using acoustic tests.* PhD thesis, Silsoe College, Cranfield University.

Testoni, A. and Eccher Zerbini, P. 1989. Picking time and quality in apple storage. *Acta Horticulturae* **258**, 445–454.

Tewfik, S. and Scott, E. 1954. Respiration of vegetables as affected by postharvest treatment. *Journal of Agricultural and Food Chemistry* **2**, 415–417.

Thomas, P. 1986. Radiation preservation of tropical fruits: bananas, mangoes and papayas. *Critical Reviews in Food Science and Nutrition* **23**, 147–205.

Thomas, T.H. 1981. Hormonal control of dormancy in relation to postharvest horticulture. *Annals of Applied Biology* **98**, 531.

Thompson, A.K. 1968. Stages of development of fruit of cashew *Anarcadium occidentale* Linn. *Proceeding of the Tropical Region of the American Society of Horticultural Science* **12**, 209–215.

Thompson, A.K. 1971. The storage of mango fruit. *Tropical Agriculture Trinidad* **48**, 63–70.

Thompson, A.K. 1971a. Transport of West Indian mango fruits. *Tropical Agriculture Trinidad* **48**, 71–77.

Thompson, A.K. 1971b. Report to the Government of Zambia on the postharvest handling of vegetables and fruits in Zambia. *Tropical Products Institute, London, Report* R100, 31 pp.

Thompson, A.K. 1972. Storage and transport of fruit and vegetables in the West Indies. *Proceedings of the Seminar/Workshop on Horticultural Development in the Caribbean, Maturin, Venezuela,* 170–176.

Thompson, A.K. 1972a. Report on an assignment on secondment to the Jamaican Government as food storage advisor October 1970–October 1972. *Tropical Products Institute Report* R278, 36 pp.

Thompson, A.K. 1978. Visit to Nepal. *Tropical Products Institute Report* R844, 23 pp.

Thompson, A.K. 1981. Reduction of losses during the marketing of arracacha. *Acta Horticulturae* **116**, 55–60.

Thompson, A.K. 1985. Postharvest losses of bananas, onions and potatoes in PDR Yemen. *Tropical Development and Research Institute, London, United Kingdom Contract Services Report* CO 485, 41 pp.

Thompson, A.K. 1987. The development and adaptation of methods for the control of anthracnose. In Prinsley, R.A. and Tucker, G. (Editors), Mangoes – *A Review.* Commonwealth Science Council, 29–38.

Thompson, A.K. 1996 *Postharvest Technology of Fruits and Vegetables.* Blackwell Science, Oxford, 410 pp.

Thompson, A.K. 1998. *Controlled Atmosphere Storage of Fruits and Vegetables.* CAB International, Oxford, 278 pp

Thompson, A.K. and Arango, L.M. 1977. Storage and marketing cassava in plastic films. *Proceedings of the Tropical Region of the American Horticultural Science* **21**, 30–33.

Thompson, A.K. and Burden, O.J. 1995. Harvesting and fruit care. In Gowen S. (Editor), *Bananas and Plantains.* Chapman and Hall, London, 403–433.

Thompson, A.K. and Lee, G.R. 1971. Factors affecting the storage behaviour of papaya fruits. *Journal of Horticultural Science* **46**, 511–516.

Thompson, A.K. and Seymour, G.B. 1982. Comparative effects of acetylene and ethylene gas on initiation of banana ripening. *Annals of Applied Biology* **101**, 407–410.

Thompson, A.K. and Seymour, G.B. 1984. Inborja CASCO banana factory at Machala in Ecuador. *Tropical Development and Research Institute Report,* 7 pp.

Thompson, A.K., Been, B.O. and Perkins, C. 1972a. Handling, storage and marketing of plantains. *Proceedings of the Tropical Region of the American Society of Horticultural Science* **16**, 205–212.

Thompson, A.K., Been, B.O. and Perkins, C. 1973. Reduction of wastage in stored yams. *Proceedings of the Third Symposium of the International Society for Tropical Root Crops, International Institute for Tropical Agriculture, Nigeria,* 443–449.

Thompson, A.K., Been, B.O. and Perkins, C. 1973a. Nematodes in stored yams. *Experimental Agriculture* **9**, 281–286.

Thompson, A.K., Been, B.O. and Perkins, C. 1974. Effects of humidity on ripening of plantain bananas. *Experientia* **30**, 35–36.

Thompson, A.K., Been, B.O. and Perkins, C. 1974a. Prolongation of the storage life of breadfruits. *Proceedings of the Caribbean Food Crops Society* **12**, 120–126.

Thompson, A.K., Been, B.O. and Perkins, C. 1974b. Effects of humidity on ripening of plantain bananas. *Experientia* **30**, 35–36.

Thompson, A.K., Been, B.O. and Perkins, C. 1974c. Storage of fresh breadfruit. *Tropical Agriculture Trinidad* **51**, 407–415.

Thompson, A.K., Been, B.O. and Perkins, C. 1977a. Fungicidal treatments of stored yams. *Tropical Agriculture Trinidad* **54**, 179–183.

Thompson, A.K., Booth, R.H. and Proctor, F.J. 1972. Onion storage in the tropics. *Tropical Science,* **XIV**, 19–34.

Thompson, A.K., El Berier, K. and Musa, S.K. 1973b. Quality and packing recommendations for air freight export of vegetables from the Sudan to Europe. *Proceedings of the First Seminar on the Export of Sudanese Fresh Fruits and Vegetables. Food Research Centre, Shambat, Sudan,* 34–38.

Thompson, A.K., Ferris, R.S.B. and Al Zaemey, A.B.S. 1992. Aspects of handling bananas and plantains. *Tropical Agriculture Association Newsletter* **12**, 15–17.

Thompson, A.K., King, P.I. and New, J.N. 1979. Report in fresh produce export marketing from Jamaica. *Tropical Products Institute, London, Report* R868, 79 pp.

Thompson, A.K., Lee, J.S., Shin, D.H. and Oh, S.R. 1977a. Storage of mushrooms in relation to subsequent losses during canning. *Proceedings of the Tropical Region of the American Society of Horticultural Science* **21**, 33–35.

Thompson, A.K., Magzoub, Y. and Silvis, H. 1974d. Preliminary investigations into desiccation and degreening of limes for export. *Sudan Journal Food Science Technology* **6**, 1–6.

Thompson, A.K., Mason, G.F. and Halkon, W.S. 1971. Storage of West Indian seedling avocado fruits. *Journal of Horticultural Science* **46**, 83–88.

Thompson, J.F., Rumsey, T.R. and Mitchell, F.G. 1998. Forced air cooling. *Tree Fruit Postharvest Journal* **9**, 3–15.

Thornton, N.C. 1930. The use of carbon dioxide for prolonging the life of cut flowers. *American Journal of Botany* **17**, 614–626.

Thorp, T.G. and Klein, J.D. 1987. Export feijoas: postharvest handling and storage techniques to maintain optimum fruit quality. *Orchardist of New Zealand* **60**, 164–166.

Tiangco, E.L., Agillon, A.B. and Lizada, M.C.C. 1987. Modified atmosphere storage of 'Saba' bananas. *Australian Centre for International Agricultural Research Banana Workshop, Kuala Lumpur, Malaysia, 17–19 February 1987.*

Tillet, R.D. and Batchelor, B.G. 1991. An algorithm for locating mushrooms in a growing bed. *Computers and Electronics in Agriculture* **6**, 191–200.

Tillet, R.D. 1991. Image analysis for agricultural processes: a review of potential opportunities. *Journal of Agricultural Engineering Research* **50**, 247–258.

Tillet, R.D. and Reed, J.N. 1990. Initial development of a mechatronic mushroom harvester. In *Mechatronics: Designing Intelligent Machines.* Mechanical Engineering Publications, Bury St Edmunds, 109–114.

Timbie, M. and Haard, N.F. 1977. Involvement of ethylene in the hardcore syndrome of sweet potato roots. *Journal of Food Science* **42**, 491.

Tindall, H.D. 1983. *Vegetables in the Tropics.* Macmillan Press, London.

Tirtosoekotjo, R.A. 1984. *Ripening behaviour and physicochemical characteristics of 'Carabao' mango Mangifera indica treated with acetylene from calcium carbide.* PhD thesis, University of the Philippines, Los Baños, Laguna.

Toet, A.J. 1982. Report on the Cameroons. *Food and Agriculture Organisation of the United Nations. Consultancy Report.*

Tollin, G., Fumimori, E. and Calvin, M. 1958. Action and emission spectra for the luminescence of green plant materials. *Nature* **181**, 1266.

Tomala, K., Andziak, J., Kobusinski, K., Dziuban, Z. and Sadowski, A. 1999. Influence of rootstocks on fruit maturity and quality of 'Jonagold' apples. *Apple Rootstocks for Intensive Orchards. Proceedings of the International Seminar, Warsaw Ursynow, Poland, 18–21 August, 1999,* 113–114

Tomkins, R.B. and Sutherland, J. 1989. Controlled atmospheres for seafreight of cauliflower. *Acta Horticulturae* **247**, 385–389.

Tomkins, R.G. 1957. Peas kept for 20 days in gas store. *Grower* 48, **5**, 226.

Tomkins, R.G. 1965. Deep scald in Ellison's Orange apples. *Annual Report of the Ditton Laboratory 1964–1965,* 19.

Tomkins, R.G. 1966. The storage of mushrooms. *Mushroom Growers Association Bulletin* **202**, 534, 537, 538, 541.

Tomkins, R.G. and Meigh, D.F. 1968. The concentration of ethylene found in controlled atmosphere stores. *Annual Report of the Ditton Laboratory 1967–68*, 33–36.

Tomlins, K.I., Ndunguru, G.T., Rwiza, E. and Westby, A. 2000. Postharvest handling, transport and quality of sweet potato in Tanzania. *Journal of Horticultural Science and Biotechnology* 75, 586–590.

Tongdee, S.C. 1988. Banana postharvest handling improvements. *Thailand Institute of Science and Technology Research Report.*

Tongdee, S.C. 1992. Postharvest handling and technology of tropical fruit. *Acta Horticulturae* **321**, 713–717.

Tongdee, S.C. 1993. Sulfur dioxide fumigation in postharvest handling of fresh longan and lychee for export. *Postharvest Handling of Tropical Fruit. Proceedings of an International Conference held in Chiang Mai, Thailand 19–23 July 1993. Australian Centre for International Agricultural Research Proceedings* **50**, 186–195.

Tongdee, S.C. 1997. Longan. In Mitra, S.K. (Editor), *Post Harvest Physiology and Storage of Tropical and Subtropical Fruits.* CAB International, Oxford, 335–346.

Tongdee, S.C., Suwanagul, A. and Neamprem, S. 1990. Durian fruit ripening and the effect of variety, maturity stage at harvest and atmospheric gases. *Acta Horticulturae* **269**, 323–334.

Tonini, G., Barberini, K., Bassi, F., Proni, R. and Retamales, J. 1999. Effects of new curing and controlled atmosphere storage technology on Botrytis rots and flesh firmness in kiwifruit. *Acta Horticulturae* **498**, 285–291.

Tonini, G., Brigati, S. and Caccioni, D. 1989. CA storage of kiwifruit: influence on rots and storability. *International Controlled Atmosphere Conference, Fifth, Proceedings, June 14–16, 1989, Wenatchee, Washington. Volume 2. Other Commodities and Storage Recommendations*, 69–76.

Townshend, G.K. Undated. *Packaging in Developing Countries.* World Packaging Organization, Geneva, 4 pp.

Tressl, R. and Drawert, F. 1973. Biogenesis of banana volatiles. *Journal of Agricultural and Food Chemistry* **21**, 560.

Tressl, R. and Jennings, W.G. 1972. Production of volatile compounds in the ripening banana. *Journal of Agricultural and Food Chemistry* **20**, 189–192.

Triglia, A., Malfa, G. la, Musumeci, F., Leonardi, C. and Scordino, A. 1998. Delayed luminescence as an indicator of tomato fruit quality. *Journal of Food Science* **63**, 512–515.

Tronsmo, A. 1989. Effect of weight loss on susceptibility to *Botrytis cinerea* in long termed stored carrots. *Norwegian Journal of Agricultural Sciences* **3**, 147–149.

Truter, A.B., and Combrink, J.C. 1992. Controlled atmosphere storage of peaches, nectarines and plums. *Journal of the Southern African Society for Horticultural Sciences* **2**, 10–13.

Truter, A.B., and Eksteen, G.J. 1986. Controlled amosphere storage of avocados and bananas in South Africa, *IFF/IIR FRIGAIR '86 Commissions C2*, E1, D1 Pretoria.

Tsay, L.M. and Wu, M.C. 1989. Studies on the postharvest physiology of sugar apple. *Acta Horticulturae* **258**, 287–294.

Tucker, G. and Grierson, D. 1987. Fruit ripening. In Davies, D.D. (Editor), *The Biochemistry of Plants. A Comprehensive Treatise*, Vol. 12. Academic Press, New York.

Tucker, W.G. 1977. The sprouting of bulb onions in storage. *Acta Horticulturae* **258**, 485–492.

Tucker, W.G. and Drew, R.L.K. 1982. Postharvest studies on autumm drilled bulb onions. *Journal of Horticultural Science* **57**, 339–348.

Tugwell, B. and Chvyl, L. 1995. Storage recommendations for new varieties. *Pome Fruit Australia* May, 4–5.

Turbin, V.A. and Volo. 1984. Storage of table grape varieties in a controlled gaseous environment. *Vinodelie i Vinogradarstvo SSSR* **8**, 31–32.

Turk, R. 1989. Effect of harvest time and precooling on fruit quality and cold storage of figs *F. carica* cv. 'Bursa Siyahi.' *Acta Horticulturae* **258**, 279–286.

Turner, D.W. 1997. Bananas and plantains. In Mitra, S.K. (Editor), *Post Harvest Physiology and Storage of Tropical and Subtropical Fruits.* CAB International, Oxford, 47–83.

Ukai, Y.N., Ishibashi, S., Tsutsumi, T. and Marakami, K. 1976. *Preservation of agricultural products.* US Patent 3 997 674.

Ulrich, R. 1970. Organic acids. In Hulme, A.C. (Editor), *The Biochemistry of Fruits and their Products*, Vol. 1. Academic Press, London, 89–118.

Umar, B. 1998. The use of solar cooling to minimize postharvest losses in the tropics. *Tropical Science* **38**, 74–77.

Umiecka, L 1973. Studies on the natural losses and marketable value of dill, parsley and chive tops in relation to storage conditions and type of packing. Badania nad ubytkami naturalnymi i wartoscia handlowa koperku, natki pietruszki i szczypiorku zaleznie od warunkow przechowywania i rodzaju opakowania. *Biuletyn Warzywniczy* **14**, 231–257.

Umiecka, L. 1989. The short term storage of vegetables. Krotkotrwale przechowywanie warzyw. *Biuletyn Warzywniczy, Supplement* **II**, 215–217.

Underhill, S.J.R., Bagshaw, J., Zauberman, G., Ronen, R., Prasad, A. and Fuchs, Y. 1992. The control of lychee *Litchi chinensis* Sonn. postharvest skin browning using sulphur dioxide and low pH. *Acta Horticulturae* **321**, 732–741.

Underhill, S.J.R., Critchley, C. and Simons, D.H. 1992a. Postharvest pericarp browning of lychee *Litchi chinensis* Sonn. *Acta Horticulturae* **321**, 718–725.

UNIFEM. 1988. *Fruit and Vegetable Processing.* United Nations Development Fund for Women, Food Technology Source Book 2, 67 pp.

Upadhyay, I.P., Noonhorm, A. and Ilangantileke S.G. 1993. Effects of gamma irradiation and hot water treatment on shelf life and quality of Thai mangoes. *Postharvest Handling of Tropical Fruit. Proceedings of an International Conference held in Chiang Mai, Thailand, 19–23 July 1993. Australian Centre for International Agricultural Research Proceedings* **50**, 348–351.

Urban, E. 1995. Postharvest storage of apples. Nachlagerungsverhalten von Apfelfruchten. *Erwerbsobstbau* **37**, 145–151.

Valdez, E.R.T. and Mendoza, D.B. Jr, 1988. Influence of temperature and gas composition on the development of senescent spotting. *Philippines Agriculture* **71**, 5–12.

Valero, D., Martinez Romero, D., Serrano, M. and Riquelme, F. 1998. Postharvest gibberellin and heat treatment effects on polyamines, abscisic acid and firmness in lemons. *Journal of Food Science* **63**, 611–615.

Van Den Berg J. and Lentz, C.P. 1972. Respiratory heat production of vegetables during refrigerated storage. *Journal of the American Society for Horticultural Science* **97**, 431–432.

Van der Merwe, J.A. 1996. Controlled and modified atmosphere storage. In Combink, J.G. (Editor), *Interated Management of Post Harvest Quality.* South Africa INFRUiTEC ARC/LNR, 104–112.

Van Doren, A. 1940. Physiological studies with McIntosh apples in modified atmosphere cold storage. *Proceedings of the American Society for Horticultural Science* **37**, 453–458.

Van Doren, A. 1952. The storage of golden delicious and red delicious apples in modified atmospheres. *Proceedings of the Forty-eighth Annual Meeting of the Washington State Horticultural Association*, 91–95.

Van Doren, A., Hoffman, M.B. and Smock, R.M. 1941. CO_2 treatment of strawberries and cherries in transit and storage. *Proceedings of the American Society for Horticultural Science* **38**, 231–238.

Van Eeden, S.J., Combrink, J.C., Vries, P.J. and Calitz, F.J. 1992. Effect of maturity, diphenylamine concentration and method of cold storage on the incidence of superficial scald in apples. *Deciduous Fruit Grower* **42**, 25–28.

Van Lelyveld, J., Visser, J.G., Swarts, D.H. and van Lelyveld, J. 1991. The effect of various storage temperatures on peroxidase activity and protein PAGE gel electrophoresis in Queen pineapple fruit. *Journal of Horticultural Science* **66**, 629–634.

Van Oirschot, Q.E.A., O'Brien, G.M., Dufour, D., El-Sharkawy, M.A. and Mesa, E. 2000. The effect of preharvest pruning of cassava upon root deterioration and quality characteristics. *Journal of the Science of Food and Agriculture* **80**, 1866–1873.

Van Oosten, H.J. 1986. A meadow orchard with *Carica pentagona* (Babaco). *Acta Horticulturae* **160**, 341.

Van Steekelenburg, N.A.M. 1982. Factors influencing external fruit rot of cucumbers caused by *Didymella bryoniae. Netherlands Journal of Plant Pathology* **88**, 47–52.

Varghese, Z., Morrow, C.T. and Heinemann, P.H. 1991. Automated inspection of golden delicious apples using digital image processing. *American Society of Agricultural Engineers Paper* 91–7002.

Variamov, G.P., Reinart, E.S., Sorokin, A.A. and Dolgosheev, A.M. 1999. Equipment for mechanized harvesting of the tubers of Jerusalem artichokes. *Traktory i El'skokhozyaistvennye Mashiny* **9**, 10–13.

Varoquaux, P., Gouble, B., Barron, C. and Yildiz, F. 1999. Respiratory parameters and sugar catabolism of mushroom *Agaricus bisporus* Lange. *Postharvest Biology and Technology* **16**, 51–61.

Vasquez Ochoa, R.I. and Colinas Leon, M.T. 1990. Changes in guavas at three maturity stages in response to temperature and relative humidity. *HortScience* **25**, 86–87.

Vega Pina, A., Nieto Angel, D. and Mena Nevarez, G. 2000. Effect of water stress and chemical spray treatments on postharvest quality in mango fruits cv. Haden, in Michoacan, Mexico. *Acta Horticulturae* **509**, 617–630.

Veierskov, B. and Hansen, M. 1992. Effects of O_2 and CO_2 partial pressure on senescence of oat leaves and broccoli miniflorets. *New Zealand Journal of Crop and Horticultural Science* **20**, 153–158.

Venkitanarayanan, K.S., Lin, C.M., Bailey, H. and Doyle, M.P. 2002. Inactivation of *Escherichia coli* O157:H7, *Salmonella enteritidis*, and *Listeria monocytogenes* on apples, oranges, and tomatoes by lactic acid with hydrogen peroxide. *Journal of Food Protection* **65**, 100–105.

Veras, M.C.M., Pinto, A.C. de Q. and de Meneses, J.B. 2000. Effect of growing season and maturation stage on sweet and acid passion fruit in cerrado conditions. Influencia da epoca de producao e dos estadios de maturacao nos maracujas doce e acido nas condicoes de cerrado. *Pesquisa Agropecuaria Brasileira* **35**, 959–966.

Vigneault, C. and Raghavan, G.S.V. 1991. High pressure water scrubber for rapid O_2 pull down in controlled atmosphere storage. *Canadian Agricultural Engineering* **33**, 287–294.

Vijaysegaran, S. 1993. Preharvest fruit fly control: strategies for the tropics. *Postharvest Handling of Tropical Fruit. Proceedings of an International Conference held in Chiang Mai, Thailand, 19–23 July 1993. Australian Centre for International Agricultural Research Proceedings* **50**, 288–303.

Vilcosqui, L. and Dury, S. 1997. Durian: a key fruit in Southeast Asia. Le durian: roi des fruits en Asie du Sud Est. *Fruits Paris* **52**, 47–57.

Villanueva Suarez, M.J., Redondo Cuenca, A., Rodriguez Sevilla, M.D. and Heredia Moreno, A. 1999. *Journal of Agricultural and Food Chemistry* **47**, 3832–3836.

Viraktamath, C.S. *et al.* 1963. Pre packaging studies on fresh produce. III. Brinjal eggplant *Solanum melongena. Food Science Mysore* **12**, 326.

Visintin, G., Falico, L., Garcia, B. and Garran, S. 2000. Biocontrol agents of *Penicillium* spp. in wax covers for citrus fruits. Agents biocontroladores de *Penicillium* spp. en la cera de cobertura para frutos citricos. *Fitopatologia* **35**, 163–168.

Voisine, R., Hombourger, C., Willemot, C., Castaigne, D. and Makhlouf, J. 1993. Effect of high CO_2 storage and gamma irradiation on membrane deterioration in cauliflower florets. *Postharvest Biology and Technology* **2**, 279–289.

Von Loesecke, H.W. 1949. *Bananas.* InterScience, London.

Wade, N.L. 1974. Effects of O_2 concentration and ethephon upon the respiration and ripening of banana fruits. *Journal of Experimental Botany* **25**, 955–964.

Waelti, H., Zhang, Q., Cavalieri, R.P. and Patterson, M.E.1992. Small scale CA storage for fruits and vegetables. *American Society of Agricultural Engineers Meeting Presentation Paper* 926568.

Wainwright, H. and Hughes, P.A. 1988. Objective measurement of banana pulp colour. *International Journal of Food Science and Technology* **24**, 553–558.

Wainwright, H. and Hughes. P.A. 1989. Characteristics of the banana fruit physiological disorder 'Yellow Pulp.' *Aspects of Applied Biology* **20**, 67–71.

Waller, J.M. 2001. Postharvest diseases. In *Plant Pathologist's Pocketbook.* CAB International, Wallingford.

Walsh, J.R., Lougheed, E.C., Valk, M. and Knibbe, E.A. 1985. A disorder of stored celery. *Canadian Journal of Plant Science* **65**, 465–469.

Wan, C.K. and Lam, P.F. 1984. Biochemical changes, use of polyethylene bags, and chilling injury of carambola *Averrhoa carambola* L. stored at various temperatures. *Pertanika* **7**, 39–46.

Wang, C.Y. 1980. Effect of CO_2 treatment on storage and shelf life of sweet pepper. *Journal of the American Society of Horticultural Science* **102**, 808.

Wang, C.Y. 1982. Physiological and biochemical responses of plants to chilling injury. *HortScience* **17**, 173–186.

Wang, C.Y. 1985. Effect of low O_2 atmospheres on postharvest quality of Chinese cabbage, cucumbers and eggplants. *Proceedings of the Fourth National Controlled Atmosphere Research Conference, North Carolina State University, Horticultural Report* **126**, 142.

Wang, C.Y. 1990. Physiological and biochemical effects of controlled atmosphere on fruit and vegetables. In Calderon, M. and Barkai Golan, R. (Editors), *Food Preservation by Modified Atmospheres.* CRC Press, Boca Raton, FL, 197–223.

Wang, C.Y. 2000. Effect of heat treatment on postharvest quality of kale, collard and Brussels sprouts. *Acta Horticulturae* **518**, 71–78.

Wang, C.Y. and Ji, Z.L. 1988. Abscisic acid and 1-aminocyclopropane 1-carboxylic acid content of Chinese cabbage during low O_2 storage. *Journal of the American Society for Horticultural Science* **113**, 881–883.

Wang, C.Y. and Ji, Z.L. 1989. Effect of low O_2 storage on chilling injury and polyamines in zucchini squash. *Scientia Horticulturae* **39**, 1–7.

Wang, C.Y. and Kramer, G.F. 1989. Effect of low O_2 storage on polyamine levels and senescence in Chinese cabbage, zucchini squash and McIntosh apples. *International Controlled Atmosphere Research Conference, Proceedings, Fifth, June 14–16, 1989, Wenatchee, Washington, Volume. 2. Other Commodities and Storage Recommendations*, 19–27.

Wang, G.X., Han, Y.S. and Yu, L. 1994. Studies on ethylene metabolism of kiwifruit after harvest. *Acta Agriculturae Universitatis Pekinensis* **20**, 408–412.

Wang, S.Y., Wang, P.C. and Faust, M. 1988. Nondestructive detection of watercore in apple with nuclear magnetic resonance imaging. *Scientific Horticulture* **35**, 227–234.

Wang, Z.Y. and Dilley, D.R. 2000. Initial low oxygen stress controls superficial scald of apples. *Postharvest Biology and Technology* **18**, 201–213.

Wang, Z.Y., Kosittrakun, M. and Dilley, D.R. 2000. Temperature and atmosphere regimens to control a CO_2-linked disorder of 'Empire' apples. *Postharvest Biology and Technology* **18**, 183–189.

Ward, C.M. and Tucker, W.G. 1976. Respiration of maleic hydrazide treated and untreated onion bulbs during storage. *Annals of Applied Biology* **82**, 135–141.

Wardlaw, C.W. 1937. Tropical fruits and vegetables: an account of their storage and transport. *Low Temperature Research Station, Trinidad Memoir 7. Reprinted from Tropical Agriculture Trinidad* **14**, 224 pp.

Wardlaw, C.W. and Leonard, E.R. 1936. *The Storage of West Indian Mangoes. Low Temperature Research Station Memoir* **3**, 47pp.

Wardlaw, C.W. and Leonard, E.R. 1940. The respiration of bananas during ripening at tropical temperatures. In *Studies in Tropical Fruits. Low Temperature Research Station, Memoir* 17

Waskar, D.P., Damame, S.V. and Gaikwad, R.S. 1998. Influence of packaging material and storage environments on shelf life of leafy vegetables. *Agricultural Science Digest Karnal* **18**, 264–266.

Waskar, D.P., Nikam, S.K. and Garande, V.K. 1999. Effect of different packaging materials on storage behaviour of sapota under room temperature and cool chamber. *Indian Journal of Agricultural Research* **33**, 240–244.

Waskar, D.P., Yadav, B.B. and Garande, V.K. 1999a. Influence of various packaging materials on storage behaviour of bottle gourd under different storage conditions. *Indian Journal of Agricultural Research* **33**, 287–292.

Watada, A., Anderson, R. and Aulenbach, B. 1979. Characteristiques sensorielles et composés volatils des peches stockées en atmosphère controlée. *Journal of the Society for Horticultural Science* **104**, 626–629.

Watada, A.E. 1986. Effects of ethylene on the quality of fruits and vegetables. *Food Technology* May, 82–85.

Watada, A.E. 1989. Non destructive methods of evaluating quality of fresh fruits and vegetables. *Acta Horticulturae* **258**, 321–330.

Watada, A.E. and Abbott, J.A. 1975. Objective method of estimating anthocyanin content for determing colour grade of grapes. *Journal of Food Science* **40**, 1278–1279.

Watada, A.E. and Morris, L. 1966. Effect of chilling and non chilling temperatures on snap bean fruits. *Proceedings of the American Society of Horticultural Science* **89**, 368–378.

Watada, A.E. and Worthington, J.T. 1974. Quality of tomatoes as related to appearence. *Proceedings of Tomato Quality Workshop, Florida*, 14–18.

Watada, A.E., Norris, K.H., Worthington, J.T. and Massie, D.R. 1976. Estimation of chlorophyll and carotenoid contents of whole tomato by light absorbtion technique. *Journal of Food Science* **41**, 329–332.

Watkins, C.B., Harman, J.E., Ferguson, I.B. and Reid, M.S. 1982. The action of lecithin and calcium dips in the control of bitter pit in apple fruit. *Journal of the American Society for Horticultural Science* **107**, 262–265.

Wei, Y. and Thompson A.K. 1993. Modified atmosphere packaging of diploid bananas Musa AA. *Post Harvest Treatment of Fruit and Vegetables. COST'94 Workshop, September 14–15, 1993, Leuven, Belgium*, 235–246.

Weichmann, J. 1973. The influence of different CO_2 partial pressures on respiration in carrots. Die Wirkung unterschiedlichen CO_2-Partialdruckes auf den Gasstoffwechsel von Mohren *Daucus carota* L. *Gartenbauwissenschaft* **38**, 243–252.

Weichmann, J. 1978. Lagerung von Gemuse V, Gemeinsame Lagerung verschiedener Arten. *Deutsher Gartenbau* **27**, 1124

Weichmann, J. 1981. CA storage of horseradish, *Armoracia rusticana* Ph. Gartn. B. Mey et Scherb. *Acta Horticulturae* **116**, 171–181.

Weichmann, J. 1987. *Postharvest Physiology of Vegetables.* Marcel Dekker, New York, 597 pp.

Wells, J.M. and Butterfield, J.E. 1999. Incidence of Salmonella on fresh fruits and vegetables affected by fungal rots or physical injury. *Plant Disease* **83**, 722–726.

Westercamp, P. 1995. Storage of President plums influence of the harvest date on storage of the fruits. Conservation de la prune President influence de la date de recolte sur la conservation des fruits. *Infos Paris* **113**, 34–37.

Whitney, B.D. 1993. Damage and disease detection in fruit and vegetables. *Postharvest Biology and Handling of Fruit, Vegetables and Flowers. Meeting of the Association of Applied Biologists, London, 8 December 1993.*

Wickham, L.D. 1988. Extension of dormancy in cush-cush yams (*Dioscorea trifida*) by treatment with gibberellic acid. *Tropical Science* **28**, 75–77.

Wickham, L.D. and Wilson, L.A. 1988. Quality changes during long-term storage of cassava roots in moist media. *Tropical Science* **28**, 79–86.

Wigginton, M.J. 1974. Effects of temperature, O_2 tension and relative humidity on the wound healing process in the potato tuber. *Potato Research* **17**, 200–214.

Wiid, M. 1994. Possible suitability of babaco for subtropical areas. Aanpasbaarheid van babako in subtropiese gebiede. *Inligtingsbulletin Instituut vir Tropiese en Subtropiese Gewasse* **268**, 17–19.

Wilcke, C. and Buwalda, J.G. 1992. Model for predicting storability and internal quality of apples. *Acta Horticulturae* **313**, 115–124.

Wild, B.L., McGlasson, W.B. and Lee, T.H. 1977. Long term storage of lemon fruit. *Food Technology in Australia* **29**, 351–357.

Wild, H. de and Roelofs, F. 1992. Plums can be stored for 3 weeks. Pruimen zijn drie weken te bewaren. *Fruitteelt Den Haag* **82**, 20–21.

Wilkinson, B.G. 1972. Fruit storage. *East Malling Reseach Station Annual Report for 1971*, 69–88.

Wilkinson, B.G. and Sharples, R.O. 1973. Recommended storage conditions for the storage of apples and pears. *East Malling Reseach Station Annual Report for 1972*, 212.

Willaert, G.A., Dirinck, P.J., Pooter, H.L. and Schamp, N.N. 1983. Objective measurement of aroma quality of Golden Delicious apples as a function of controlled atmosphere storage time. *Journal of Agricultural and Food Chemistry* **31**, 809–813.

Williams, B., Goodman, B.A. and Chudek, J.A. 1992. Nuclear magnetic resonance NMR micro imaging of ripening red raspberry fruits. *New Phytologist* **120**, 21–28.

Williams, J.A.H. 1975. *Mechanized gooseberry harvesting: economics, berry detatchment and design of a mobile harvester.* MSc thesis, Silsoe College, Cranfield Institute of Technology.

Williams, M.W. and Patterson, M.E. 1964. Nonvolatile organic acide and core breakdown of 'Bartlett' pears. *Journal of Agriculture and Food Chemistry* **12**, 89.

Williams, P.C. and Norris, K.H. (Editors). 1987. *Near Infra Red Technology in the Agriculture and Food Industries.* American Association of Cereal Chemists, St Paul, MN, 330 pp.

Williamson, B. 1993. NMR micro imaging: a non invasive approach to the study of soft fruit. *Postharvest Biology and Handling of Fruit, Vegetables and Flowers. Meeting of the Association of Applied Biologists, London, 8 December* 1993.

Willingham, S.L., Pegg, K.G., Cooke, A.W., Coates, L.M., Langdon, P.W.B. and Dean, J.R. 2001. Rootstock influences postharvest anthracnose development in 'Hass' avocado. *Australian Journal of Agricultural Research* **52**, 1017–1022.

Wills, R.B.H. 1990. Postharvest technology of banana and papaya in Association of Southeast Asian Nations: an overview. *Association of Southeast Asian Nations Food Journal* **5**, 47–50.

Wills, R.B.H., 1998. Enhancement of senescence in non climacteric fruit and vegetables by low ethylene levels. *Acta Horticulturae* **464**, 159–162.

Wills, R.B.H. and Tirmazi, S.I.H. 1979. Effects of calcium and other minerals on the ripening of tomatoes. *Australian Journal of Plant Pathology* **6**, 221–227.

Wills, R.B.H. and Tirmazi, S.I.H. 1981. Retardation of ripening of mangoes by postharvest application of calcium. *Tropical Agriculture Trinidad* **58**, 137–141.

Wills, R.B.H. and Tirmazi, S.I.H. 1982. Inhibition of ripening of avocados with calcium. *Scientia Horticulturae* **16**, 323–330.

Wills, R.B.H., Klieber, A., David, R., and Siridhata, M. 1990. Effect of brief pre marketing holding of bananas in nitrogen on time to ripen. *Australian Journal of Experimental Agriculture* **30**, 579–581.

Wills, R.B.H., McGlasson, W.B., Graham, D., Lee, T.H. and Hall, E.G. 1989. *Postharvest.* BSP Professional Books, Oxford, 174 pp.

Wills, R.B.H., Mulholland, .E., Brown, B.I. and Scott, K.J. 1983. Storage of two cultivars of guava fruit for processing. *Tropical Agriculture Trinidad* **60**, 175–178.

Wills, R.B.H., Poi, A., Greenfield, H. and Rigney, C.J. 1984. Postharvest changes in fruit composition of *Annona atemoya* during ripening and effects of storage temperature on ripening. *HortScience* **19**, 96–97.

Wills, R.B.H., Tirmazi, S.I.H. and Scott, K.J. 1982b. Effect of Postharvest application of calcium on ripening rates of pears and bananas. *Journal of Horticultural Science* **57**, 431–435

Wills, R.B.H., Warton, M.A., Mussa, D.M.D.N. and Chew, L.P. 2001. Ripening of climacteric fruits initiated at low ethylene levels. *Australian Journal of Experimental Agriculture* **41**, 89–92.

Wilson Wijeratnum, R.S., Jayatilake, S., Hewage, S.K., Perera, L.R., Paranerupasingham, S. and Peiris, C.N. 1993. Determination of maturity indices for Sri Lankan Embul bananas. *Postharvest Handling of Tropical Fruit. Proceedings of an International Conference held in Chiang Mai, Thailand, 19–23 July 1993. Australian Centre for International Agricultural Research Proceedings* **50**, 338–340.

Winston, J.R. 1955. The coloring or degreening of mature citrus fruits with ethylene. *US Department of Agriculture Circular* 961.

Witbooi, W.R., Taylor, M.A. and Fourie, J.F. 2000. *In vitro* studies on the effect of harvest maturity, relative humidity, wetness period and cold-storage temperature, on the infection potential of *Aspergillus niger* and *Rhizopus stolonifer* on Thompson Seedless table grapes. *Deciduous Fruit Grower* **50**, S1–S14.

Wojciechowski, J. 1989. Ethylene removal from gases by means of catalytic combustion. *Acta Horticulturae* **258**, 131–139.

Woltering, E.J. and Harkema, H. 1984. *Ethyleenschade bij snijbloemen en trekheesters. Onderzoekresultaten van Sierteelprodukten.* Sprenger Instituut, Wageningen.

Woltering, E.J., van Schaik, A.C.R. and Jongen, W.M.F. 1994. Physiology and biochemistry of controlled atmosphere storage: the role of ethylene. *COST 94. The Post Harvest Treatment of Fruit and Vegetables: Controlled Atmosphere Storage of Fruit and Vegetables. Proceedings of a Workshop, Milan, Italy, 22–23 April* 1993, 35–42.

Wood, G.A., Andersen, M.T., Forster, R.L.S., Braithwaite, M. and Hall, H.K. 1999. History of boysenberry and youngberry in New Zealand in relation to their problems with boysenberry decline, the association of a fungal pathogen, and possibly a phytoplasma, with this disease. *New Zealand Journal of Crop and Horticultural Science* **27**, 281–295.

Woodward, J.R. and Topping, A.J. 1972. The influence of controlled atmospheres on the respiration rates and storage behaviour of strawberry fruits. *Journal of Horticultural Science* **47**, 547–553.

Woolf, A.B., Wexler, A., Prusky, D., Kobiler, E. and Lurie, S. 2000. Direct sunlight influences postharvest temperature responses and ripening of five avocado cultivars. *Journal of the American Society for Horticultural Science* **125**, 370–376.

Workman, M. and Pratt, H.K. 1957. Studies on the physiology of tomato fruits. II. Ethylene production at 20°C as related to respiration, ripening and rate of harvest. *Plant Physiology* **32**, 330–334.

Workman, M.N. and Twomey, J. 1969. The influence of storage atmosphere and temperature on the physiology and performance of Russet Burbank seed potatoes. *Journal of the American Society for Horticultural Science* **94**, 260.

Workman, M.N. and Twomey, J. 1970.The influence of storage on the physiology and productivity of Kennebec seed potatoes. *American Potato Journal* **47**, 372–378.

Worrel, D.B. and Carrington, C.M.S. 1994. Post harvest storage of breadfruit. *Inter American Institute for Cooperation on Agriculture, Tropical Fruits Newsletter* **11**, 5.

Wright 1942, quoted by Burton, W.G. 1982. *Postharvest Physiology of Food Crops.* Longmans Ltd., London, 339 pp.

Wright, M.E., Hoehn, R.C., Coleman, J.R. and Brzozowski, J.K. 1979. A comparison of single use and recycled water leafy vegetable washing system. *Journal of Food Science* **44**, 381.

Wright, P.J., Grant, D.G. and Triggs, C.M. 2001. Effects of onion (*Allium cepa*) plant maturity at harvest and method of topping on bulb quality and incidence of rots in storage. *New Zealand Journal of Crop and Horticultural Science* **29**, 85–91.

Wright, R.C. 1942. The freezing temperatures of some fruit, vegetables and florists' stocks. *US Department of Agriculture Circular* 447, 12 pp.

Wu, C.E., Wang, W.S. and Kou, X.H. 2001. Effect of $CaCl_2$ and 6-BA on postharvest respiratory intensity and storage quality of Chinese jujube. *Scientia Agricultura Sinica* **34**, 66–71.

Wu, P. 1998. Studies on postharvest physiology of Chinese chive scapes. *Acta Horticulturae* **467**, 379–386.

Wu, P., Zheng, S.F., Huang, C. and Li, W. 1995. On postharvest physiology of Chinese kale *Brassica alboglabra. Acta Agriculturae Zhejiangensis* 7, 144–145.

Xu, L., Kusakari, S.I., Toyoda, H. and Ouchi, S. 1999. Role of fungi in the shattering of grape berries during storage. *Bulletin of the Institute for Comprehensive Agricultural Sciences, Kinki University* 7, 97–106.

Xuan, H. and Streif, J. 2000. Effect of pre- and postharvest applications of 'Biofresh' coating on keepability of apple fruits. *Acta Horticulturae* **513**, 483–492.

Xue, Y.B., Yu, L. and Chou, S.T. 1991. The effect of using a carbon molecular sieve nitrogen generator to control superficial scald in apples. *Acta Horticulturae Sinica* **18**, 217–220.

Yahia, E.M. 1989. CA storage effect on the volatile flavor components of apples. *International Controlled Atmosphere Research Conference, Fifth, Proceedings, June 14–16, 1989, Volume 1. Pome Fruits, Wenatchee, Washington.* 341–352.

Yahia, E.M. 1991. Production of some odor active volatiles by 'McIntosh' apples following low ethylene controlled atmosphere storage. *HortScience* **26**, 1183–1185.

Yahia, E.M. and Kushwaha, L. 1995. Insecticidal atmospheres for tropical fruits. Harvest and postharvest technologies for fresh fruits and vegetables. *Technologias de Cosecha y Postcosecha de Frutas y Hortalizas. Proceedings of a Conference held in Guanajuato, Mexico, 20–24 February 1995*, 282–286.

Yahia, E.M. and Tiznado Hernandez, M. 1993. Tolerance and responses of harvested mango to insecticidal low O_2 atmospheres. *HortScience* **28**, 1031–1033.

Yahia, E.M. and Vazquez Moreno, L. 1993. Responses of mango to insecticidal O_2 and CO_2 atmospheres. *Lebensmittel Wissenschaft und Technologie* **26**, 42–48.

Yahia, E.M., Medina, F. and Rivera, M. 1989. The tolerance of mango and papaya to atmospheres containing very high levels of CO_2 and/or very low levels of O_2 as a possible insect control treatment. *International Controlled Atmosphere Research Conference, Fifth, Proceedings, June 14–16, 1989, Wenatchee, Washington. Volume 2. Other Commodities and Storage Recommendations,* 77–89.

Yahia, E.M., Rivera, M. and Hernandez, O. 1992. Responses of papaya to short term insecticidal O_2 atmosphere. *Journal of the American Society for Horticultural Science* **117**, 96–99.

Yamashita, F. and Benassi, M. de T. 1998. Influence of different modified atmosphere packaging on overall acceptance of white guavas *Psidium guajava* L., var. Kumagai stored under refrigeration. Influencia de diferentes embalagens de atmosfera modificada sobre a aceitacao de goiabas brancas de mesa *Psidium guajava* L., var. Kumagai mantidas sob refrigeracao. *Alimentos e Nutricao* **9**, 9–16.

Yanez Lopez, L., Buescher, R.W. and Armella, M.A. 1998. Effect of temperature and 30% CO_2 on ethanol accumulation in cucumbers, lettuce and peaches. Efecto de la temperatura y 30% de CO_2 en la acumulacion de etanol en pepino, lechuga y durazno. *Agronomia Costarricense* **22**, 199–204.

Yang, C.X., Shewfelt, R.L. and Michalczuk, L. 1999. Effects of sealing of stem scar on ripening rate and internal ethylene, oxygen and carbon dioxide concentrations of tomato fruits. *Acta Horticulturae* **485**, 399–404.

Yang, Q. 1992. The potential for applying machine vision to defect detection in fruit and vegetable grading. *Agricultural Engineering* **47**, 74–79.

Yang, S.F. 1985. Biosynthesis and action of ethylene. *HortScience* **20**, 41–45.

Yang, Y.J. and Henze, J. 1987. Influence of CA storage on external and internal quality characteristics of broccoli *Brassica oleracea* var. *italica*. I. Changes in external and sensory quality characteristics. Einfluss der CA Lagerung auf aussere und innere Qualittsmerkmale von Broccoli *Brassica oleracea* var. *italica*. I. Veranderungen ausserer und degustativ erfassbarer Qualitätsmerkmale. *Gartenbauwissenschaft* **52**, 223–226.

Yang, Y.J. and Henze, J. 1988. Influence of controlled atmosphere storage on external and internal quality features of broccoli *Brassica oleracea* var. *italica*. II. Changes in chlorophyll and carotenoid contents. Einfluss der CA Lagerung auf aussere und innere Qualitätsmerkmale von Broccoli *Brassica oleracea* var. *italica*. II. Veranderungen der Chlorophyll und Carotinoidgehalte. *Gartenbauwissenschaft*, **53**, 41–43.

Yantarasri, T., Sornsrivichai, J. and Chen, P. 1998. X ray and NMR for nondestructive internal quality evaluation of durian and mangosteen fruits. *Acta Horticulturae* **464**, 97–101.

Yeatman, J.N. and Norris, K.H. 1965. Evaluating internal quality of apples with new automatic fruit sorter. *Food Technology* **19**, 123–125.

Young, N., de Buckle, T.S., Castel Blanco, H., Rocha, D. and Velez, G. 1971. Conservacion of yuca fresca. *Instituto Investigacion Tecnologia, Bogata, Colombia, Report.*

Ystaas, J. 1980. Effects of nitrogen fertilization on yield and quality of Moltke pear. In Atkinson, D., Jackson, J.E., Sharples R.O. and Waller, W.M. (Editors), *Mineral Nutrition of Fruit Trees.* Butterworths, London, 287–288.

Yu, K.S., Newman, M.C., Archbold, D.D. and Hamilton Kemp, T.R. 2001. Survival of *Escherichia coli* O157:H7 on strawberry fruit and reduction of the pathogen population by chemical agents. *Journal of Food Protection* **64**, 1334–1340.

Yuen, C.M.C. 1993. Calcium and postharvest storage potential. *Postharvest Handling of Tropical Fruit. Proceedings of an International Conference held in Chiang Mai, Thailand, 19–23 July 1993. Australian Centre for International Agricultural Research Proceedings* **50**, 218–227.

Yuen, C.M.C., Tan, S.C., Joyce, D. and Chettri, P. 1993. Effect of postharvest calcium and polymeric films on ripening and peel injury in Kensington Pride mango. *Association of Southeast Asian Nations Food Journal* **8**, 110–113.

Zagory, D. 1990. Application of computers in the design of modified atmosphere packaging to fresh produce. In *International Conference on Modified Atmosphere Packaging. Part 1.* Campden Food and Drinks Research Association, Chipping Campden, Gloucestershire.

Zagory, D. and Reid, M.S. 1989. Controlled atmosphere storage of ornamentals. *International Controlled Atmosphere Research Conference, Fifth, Proceedings, June 14–16, 1989, Wenatchee, Washington. Volume 2. Other Commodities and Storage Recommendations,* 353–358.

Zagory, D., Ke, D. and Kader, A.A. 1989. Long term storage of 'Early Gold' and 'Shinko' Asian pears in low oxygen atmospheres. *International Controlled Atmosphere Research Conference, Fifth, Proceedings, June 14–16, 1989,*

Wenatchee, Washington. Volume 1. Pome Fruits, 353–357.

Zamora-Magdaleno, T., Cardenas-Soriano, E., Cajuste-Bontemps, J.F. and Colinas-Leon, M.T. 2001. Anatomy of damage by friction and by *Colletotrichum gloeosporioides* Penz. in avocado fruit 'Hass.' *Agrociencia* **35**, 237–244.

Zanon, K. and Schragl, J. 1988. Storage experiments with white cabbage. Lagerungsversuche mit Weisskraut. *Gemuse* **24**, 14–17.

Zhang, D.L. and Quantick, P.C. 2000. Effect of low temperature hardening on postharvest storage of litchi fruit. *Acta Horticulturae* **518**, 175–182.

Zhao, H. and Murata, T. 1988. A study on the storage of muskmelon 'Earl's Favourite.' *Bulletin of the Faculty of Agriculture, Shizuoka University* **38**, 713.

Zheng, Y.H. and Xi, Y.F. 1994. Preliminary study on colour fixation and controlled atmosphere storage of fresh mushrooms. *Journal of Zhejiang Agricultural University* **20**, 165–168.

Zhou, H.W., Lurie, S., Lers, A., Khatchitski, A., Sonego, L. and Arie, R.B. 2000. Delayed storage and controlled atmosphere storage of nectarines: two strategies to prevent woolliness. *Postharvest Biology and Technology* **18**, 133–141.

Zhou, L.L., Yu, L. and Zhou, S.T. 1992b. The effect of garlic sprouts storage at different O_2 and CO_2 levels. *Acta Horticulturae Sinica* **19**, 57–60.

Zhou, L.L., Yu, L., Zhao, Y.M., Zhang, X. and Chen, Z.P. 1992a. The application of carbon molecular sieve generators in the storage of garlic sprouts. *Acta Agriculturae Universitatis Pekinensis* **18**, 47–51.

Zhou, T., Northover, J. and Schneider, K.E. 1999. Biological control of postharvest diseases of peach with phyllosphere isolates of *Pseudomonas syringae. Canadian Journal of Plant Pathology* **21**, 375–381.

Zlatev, S., Balinova, A. and Zlateva, M. 1976. Changes in the essential oil of dill plants during post harvest storage. *Rasteniev' dni Nauki* **13**, 51–57.

Zong, R.J. 1989. Physiological aspects of film wrapping of fruits and vegetables. *International Controlled Atmosphere Research Conference, Fifth, Proceedings, June 14–16, 1989, Wenatchee, Washington. Volume 2. Other Commodities and Storage Recommendations,* 29–39.

Index